Molecular Genetics in Diseases of Brain, Nerve, and Muscle

Molecular Genetics in Diseases of Brain, Nerve, and Muscle

Edited by

Lewis P. Rowland
Donald S. Wood
Eric A. Schon
Salvatore DiMauro

New York Oxford
OXFORD UNIVERSITY PRESS
1989

Oxford University Press

Oxford New York Toronto
Delhi Bombay Calcutta Madras Karachi
Petaling Jaya Singapore Hong Kong Tokyo
Nairobi Dar es Salaam Cape Town
Melbourne Auckland

and associated companies in
Berlin Ibadan

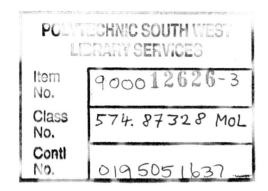

Published by Oxford University Press, Inc.,
200 Madison Avenue, New York, New York 10016

Oxford is a registered trademark of Oxford University Press

Library of Congress Cataloging-in-Publication Data

Molecular genetics in diseases of brain, nerve, and muscle.

Includes bibliographies and index.
1. Neuromuscular diseases—Genetic aspects.
2. Genetic disorders. 3. Molecular genetics.
I. Rowland, Lewis P. [DNLM: 1. Genetics, Biochemical.
2. Neuromuscular Diseases—familial & genetic.
WE 550 M7173]
RC925.5.M65 1989 616.7′4 88-12417
ISBN 0-19-505163-7 (alk. paper)

9 8 7 6 5 4 3 2

Printed in the United States of America on acid-free paper

To the wives, husbands, and children of the authors.
The support of families is essential for the progress
of science.

Preface

Molecular genetics is changing the theory and practice of clinical medicine. It affects all branches of medicine, but especially tumors, infections, and heritable diseases. Neurological diseases are especially common among the heritable diseases, and, among neurologic diseases, neuromuscular disorders have been at the forefront of progress in molecular genetics.

The papers come so fast, and from so many directions, that even investigators in the field find it difficult to keep up. And there are generations of physicians—those who graduated more than five years ago—who have not been trained in the language or concepts of molecular genetics. There are few readily available sources to guide them through the burgeoning literature.

It therefore seemed appropriate to develop a multiauthored book that would focus on neurologic and neuromuscular diseases, designed especially for clinicians and presuming no previous knowledge other than a medical education and some clinical experience.

With that in mind we resorted to a principle that had served well in the past—asking contributors to participate in a meeting and bring the manuscripts with them. The conference was designed to produce the book, and the papers were presented in the sequence they appear as chapters. Thus the book not merely is the proceedings of a conference, but was planned to provide fundamental information, followed by the examples of specific diseases.

We turned to friends and other experts in the field to provide the papers. The contributors were assembled from colleagues at the Columbia-Presbyterian Medical Center and other institutions, most of whom were supported in their research endeavors by the Muscular Dystrophy Association.

The first section of the book is intended as a primer, providing the basic information about molecular genetics and defining terms as they come. A comprehensive glossary is intended to assist the reader in learning this new language. Reading the introductory section should give the reader the necessary background to comprehend the later discussion of specific diseases.

The several neuromuscular diseases under the spotlight of molecular genetics include Duchenne muscular dystrophy, myotonic muscular dystrophy, familial amyloidotic polyneuropathy, Emery–Dreifuss muscular dystrophy, X-linked spinal muscular atrophy, and α-glucosidase deficiency (Pompe disease). We also chose the several diseases of the brain that have been subjected to DNA analysis—Huntington disease, Tay–Sachs disease, Alzheimer disease, phenylketonuria, ornithine transcarbamylase deficiency, Lesch–Nyhan syndrome, and mental retardation.

And there are some conditions that affect both central and peripheral structures. The mitochondrial "encephalomyopathies" incorporate that phenomenon in the name. But hexosaminidase deficiency, porphyria, and glycerol kinase deficiency may also affect brain or spinal cord as well as muscle or nerve.

Our goal was to make the fundamentals comprehensible to clinicians so that they could understand how genes for different diseases have been isolated and how that information can be used for antenatal diagnosis, for carrier detection, and for illuminating the pathogenesis of these diseases. A chapter on the use of DNA haplotypes for prenatal diagnosis and carrier detection has been provided for practical guidance. And we have included discussions of gene therapy and ethics, too.

The contributors met in February 1987, and the rest of the year saw more amazing progress—the power of reverse genetics in isolating the affected gene products in chronic granulomatous disease and Duchenne dystrophy, and evidence that the gene for familial Alzheimer disease is on chromosome 21. This has been incorporated into the book. We can be certain that there will be acceleration in the next few years.

This book is intended mainly for clinicians. We hope it will be of use not only to neurologists but also to psychiatrists, pediatricians, physiatrists, nurses, genetics counselors, and other health care workers who take care of patients with heritable diseases. Basic scientists are increasingly active in this field of clinical investigation, and we hope they will find parts of the book relevant.

We are grateful to the authors who contributed the "basic" chapters, even though they could just as well have written chapters on diseases in which they are also expert. And we are grateful to all the authors for their participation, and for their gracious patience with a heavy-handed clinical editor. Among them, we are especially grateful to C. Thomas Caskey for taking on so many chapters. Fortunately for us, he is among the world's leaders in molecular genetics in general and, in particular, the Lesch–Nyhan syndrome (lack of hypoxanthine phosphoribosyl transferase), the physical nature of changes in DNA that lead to clinical mutations, and the development of gene-replacement therapy.

We regret to report the untimely death of Dr. Shobhana Vora in an automobile accident that occurred shortly after she submitted her manuscript for this book. With few others, she had been responsible for elucidating the biochemical genetics of phosphofructokinase deficiency, and she was making excellent progress in defining the molecular biology of that disorder. She had started that work at the Columbia-Presbyterian Medical Center and was working at the University of California, San Diego, at the time of her death.

We are grateful to Mrs. Sheila Crescenzo for expert secretarial assistance in producing the manuscripts, to Ms. Esther Modell of the Continuing Medical Education office of the College of Physicians and Surgeons, to Mr. Jeffrey House and his colleagues at Oxford University Press, and to officials of the Muscular Dystrophy Association for supporting the project from the start.

Lewis P. Rowland, M.D.

Contents

DISEASES OF UNKNOWN GENE PRODUCT

IV. SOCIAL POLICY AND MOLECULAR GENETICS

Contributors

Frederick W. Alt, Ph.D.
Department of Biochemistry
Columbia-Presbyterian Medical Center
New York, New York 10032-3784

Richard J. Bartlett
Division of Neurology
Duke University Medical Center
Durham, North Carolina 27710

Thomas D. Bird
Division of Medical Genetics
University of Utah Medical Center
50 North Medical Drive
Salt Lake City, Utah 84132

Eduardo Bonilla
H. Houston Merritt Clinical Research Center
Columbia-Presbyterian Medical Center
New York, New York 10032-3784

Gail A. P. Bruns, M.D., Ph.D.
Genetics Division
The Children's Hospital
300 Longwood Avenue
Boston, Massachusetts 02115

Charles R. Cantor, M.D.
Department of Human Genetics and Development
Columbia-Presbyterian Medical Center
New York, New York 10032-3784

C. Thomas Caskey, M.D.
Departments of Medicine, and of Biochemistry and Cell Biology
Baylor College of Medicine
Texas Medical Center
Houston, Texas 77030

Phillip F. Chance, M.D.
Division of Medical Genetics
University of Utah Medical Center
50 North Medical Drive
Salt Lake City, Utah 84132

Stephen M. W. Chang
Institute for Molecular Genetics
Baylor College of Medicine
One Baylor Plaza
Houston, Texas 77030

P. Michael Conneally, Ph.D.
Department of Medical Genetics and Neurology
Indiana University School of Medicine
702 Barnhill Drive
Indianapolis, Indiana 46223

Kay E. Davies, Ph.D.
Neffield Department of Clinical Medicine
John Radcliffe Hospital
Headington
Oxford OX3 9DU
England

Darryl C. De Vivo
H. Houston Merritt Clinical Research Center
Columbia-Presbyterian Medical Center
New York, New York 10032-3784

Salvatore DiMauro, M.D.
H. Houston Merritt Clinical Research Center
Columbia-Presbyterian Medical Center
New York, New York 10032-3784

Wayne A. Fenton, Ph.D.
Department of Human Genetics
Yale University School of Medicine
333 Cedar Street
P.O. Box 3333
New Haven, Connecticut 06510

Kenneth H. Fischbeck, M.D.
Department of Neurology
University of Pennsylvania
3400 Spruce Street
Philadelphia, Pennsylvania 19104

R. A. Gibbs, M.D.
Institute for Molecular Genetics
Baylor College of Medicine
One Baylor Plaza
Houston, Texas 77030

T. Conrad Gilliam, Ph.D.
Departments of Psychiatry and Neurology
Columbia-Presbyterian Medical Center
New York, New York 10032-3784

Roy A. Gravel, Ph.D.
Department of Genetics
The Hospital for Sick Children
555 University Avenue
Toronto, Ontario M5G 1X8
Canada

Abraham Grossman
Department of Neurology
University of Texas Health Science Center
Southwestern Medical School
5323 Harry Hines Boulevard
Dallas, Texas 75235

James F. Gusella
Neurogenetics Laboratory
Massachusetts General Hospital
32 Fruit Street
Boston, Massachusetts 02114

Pamela Hawley
Genetics Division and Department of Pediatrics
The Children's Hospital
300 Longwood Avenue
Boston, Massachusetts 02115

J. Fielding Hejtmancik, M.D., Ph.D.
Institute for Molecular Genetics and Department of Medicine
Baylor College of Medicine
One Baylor Plaza
Houston, Texas 77030

Joseph Herbert, M.D.
H. Houston Merritt Clinical Research Center
Columbia-Presbyterian Medical Center
New York, New York 10032-3784

Rochelle Hirschhorn, M.D.
Department of Medicine
New York University
550 First Avenue
New York, New York 10016

R. Rodney Howell, M.D.
Department of Pediatrics
University of Texas Health Sciences Center at Houston
6431 Fannin
Houston, Texas 77030

William G. Johnson, M.D.
H. Houston Merritt Clinical Research Center
Columbia-Presbyterian Medical Center
New York, New York 10032-3784

Louis M. Kunkel, Ph.D.
Genetics Division
The Children's Hospital
300 Longwood Avenue
Boston, Massachusetts 02115

Samuel Latt
Genetics Division and Department of Pediatrics
The Children's Hospital
300 Longwood Avenue
Boston, Massachusetts 02115

James R. Lupski
Department of Biochemistry and Cell Biology
Baylor College of Medicine
Texas Medical Center
Houston, Texas 77030

Grant R. MacGregor, Ph.D.
Institute for Molecular Genetics
Baylor College of Medicine
One Baylor Plaza
Houston, Texas 77030

Don J. Mahuran
Department of Genetics
The Hospital for Sick Children
555 University Avenue
Toronto, Ontario M5G 1X8
Canada

Frank Martiniuk
Department of Medicine
New York University
550 First Avenue
New York, New York 10016

Edward R. B. McCabe
Department of Pediatrics
University of Colorado Health Sciences Center
Campus Box C 233
4200 East Ninth Avenue
Denver, Colorado 80262

Mark Mehler
Department of Medicine
New York University
550 First Avenue
New York, New York 10016

Armand F. Miranda
H. Houston Merritt Clinical Research Center
Columbia-Presbyterian Medical Center
New York, New York 10032-3784

Anthony P. Monaco, Ph.D.
Genetics Division
The Children's Hospital
300 Longwood Avenue
Boston, Massachusetts 02115

David L. Nelson
Institute for Molecular Genetics
Baylor College of Medicine
One Baylor Plaza
Houston, Texas 77030

P. I. Patel
Department of Biochemistry and Cell Biology
Baylor College of Medicine
Texas Medical Center
Houston, Texas 77030

Angel Pellicer
Department of Medicine
New York University
550 First Avenue
New York, New York 10016

Margaret A. Pericak-Vance, Ph.D.
Division of Neurology
Duke University Medical Center
Durham, North Carolina 27710

Peter N. Ray, M.D.
Department of Genetics
The Hospital for Sick Children
555 University Avenue
Toronto, Ontario M5G 1X8
Canada

Roger N. Rosenberg, M.D.
Department of Neurology
University of Texas Health Science Center
Southwestern Medical School
5323 Harry Hines Boulevard
Dallas, Texas 75235

Allen D. Roses, M.D.
Division of Neurology
Duke University Medical Center
Durham, North Carolina 27710

Lewis P. Rowland, M.D.
H. Houston Merritt Clinical Research Center
Columbia-Presbyterian Medical Center
New York, New York 10032-3784

Roy Schmickel, M.D.
Department of Human Genetics
University of Pennsylvania
196 Medical Labs Building
37th and Hamilton Walk
Philadelphia, Pennsylvania 19104

Eric A. Schon, Ph.D.
H. Houston Merritt Clinical Research Center
Columbia-Presbyterian Medical Center
New York, New York 10032-3784

William K. Seltzer, Ph.D.
Department of Pediatrics
University of Colorado Health Sciences Center
Campus Box C 233
4200 East Ninth Avenue
Denver, Colorado 80262

Serenell Servidei
H. Houston Merritt Clinical Research Center
Columbia-Presbyterian Medical Center
New York, New York 10032-3784

Cassandra L. Smith
Department of Human Genetics and Development
Columbia-Presbyterian Medical Center
New York, New York 10032-3784

James Speer
Genetics Division and Department of Pediatrics
The Children's Hospital
300 Longwood Avenue
Boston, Massachusetts 02115

Marcie Speer
Genetics Division and Department of Pediatrics
The Children's Hospital
300 Longwood Avenue
Boston, Massachusetts 02115

J. T. Stout
Department of Biochemistry and Cell Biology
Baylor College of Medicine
Texas Medical Center
Houston, Texas 77030

James E. Sylvester, Ph.D.
Department of Pathology
Hahnemann University
Mail Stop 435
Broad and Vine Streets
Philadelphia, Pennsylvania 19102

Umadevi Tantravahi
Genetics Division and Department of Pediatrics
The Children's Hospital
300 Longwood Avenue
Boston, Massachusetts 02115

Stephanie Tzall
Department of Medicine
New York University
550 First Avenue
New York, New York 10016

Shobhana Vora, M.D.*
Scripps Clinic and Research Foundation
Department of Basic and Clinical Research
1066 North Torrey Pines Road
La Jolla, California 92037

*Deceased

Pat Ward
Institute for Molecular Genetics
Baylor College of Medicine
One Baylor Plaza
Houston, Texas 77030

Nancy S. Wexler, Ph.D.
Columbia-Presbyterian Medical Center
324 Psychiatric Institute Annex
722 West 168th Street
New York, New York 10032

Savio L. C. Woo, Ph.D.
Department of Biochemistry and Cell Biology
Baylor College of Medicine
Texas Medical Center
Houston, Texas 77030

Donald S. Wood, Ph.D.
Director of Research
Muscular Dystrophy Association
810 Seventh Avenue
New York, New York 10019

Massimo Zeviani
H. Houston Merritt Clinical Research Center
Columbia-Presbyterian Medical Center
New York, New York 10032-3784

Neuro-Genetic Map 4/1987

Traits in Bold = Linkage Determined Since 4/1984

(Roswell Eldridge after Victor A. McKusick)

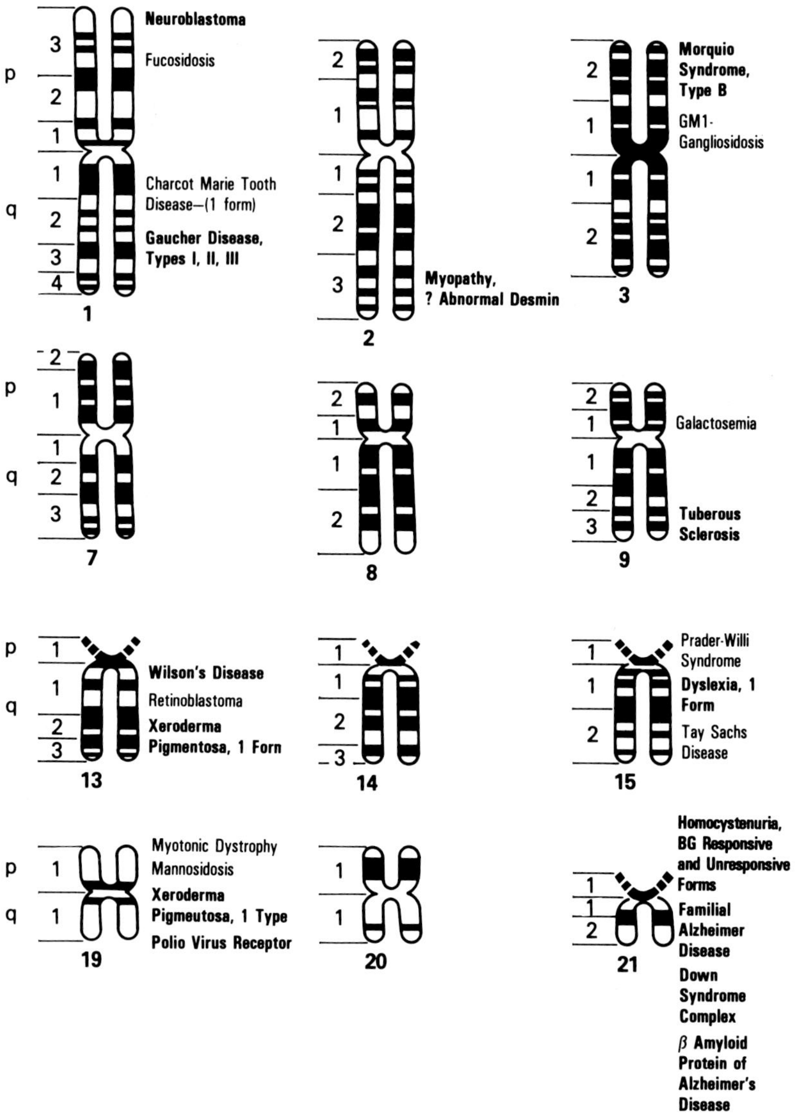

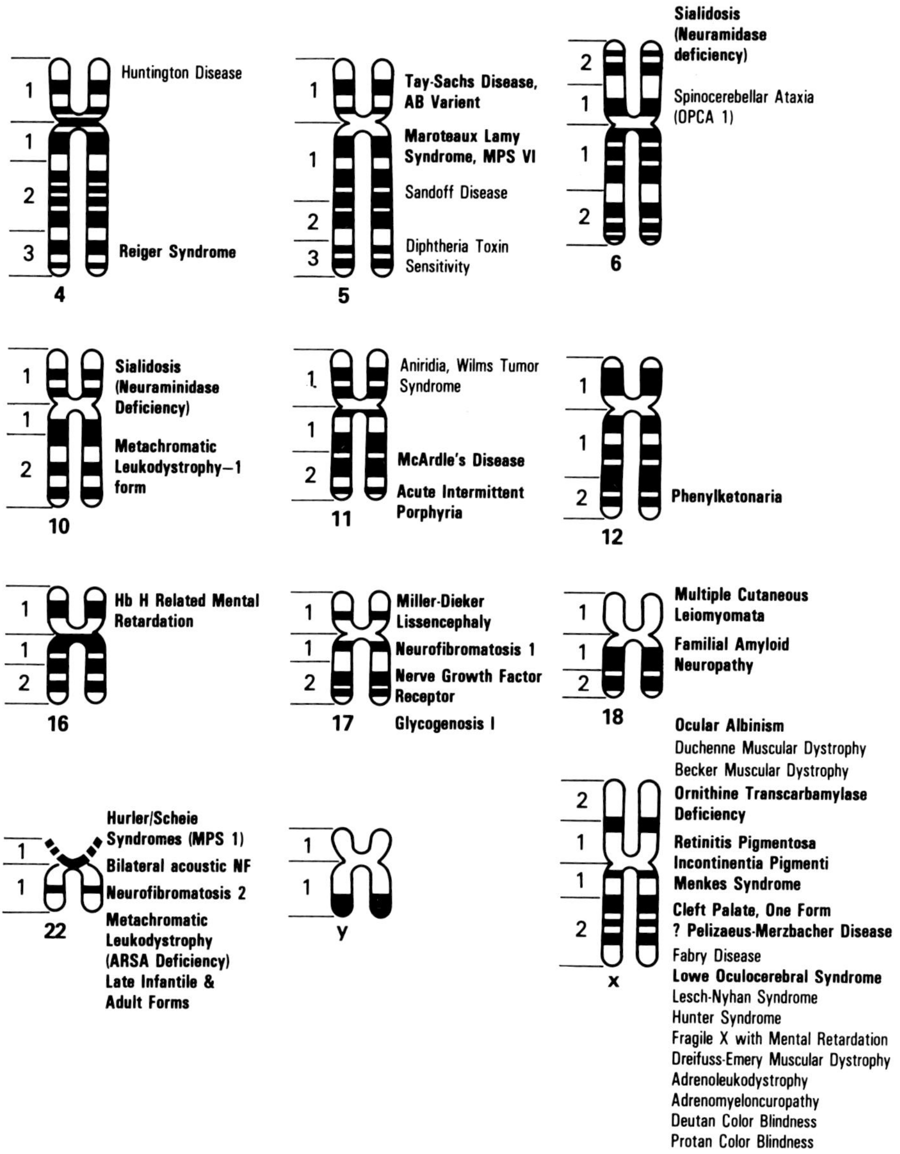

Huntington Disease
Reiger Syndrome
4
Tay-Sachs Disease, AB Varient
Maroteaux Lamy Syndrome, MPS VI
Sandoff Disease
Diphtheria Toxin Sensitivity
5
Sialidosis (Neuramidase deficiency)
Spinocerebellar Ataxia (OPCA 1)
6
Sialidosis (Neuraminidase Deficiency)
Metachromatic Leukodystrophy—1 form
10
Aniridia, Wilms Tumor Syndrome
McArdle's Disease
Acute Intermittent Porphyria
11
Phenylketonaria
12
Hb H Related Mental Retardation
16
Miller-Dieker Lissencephaly
Neurofibromatosis 1
Nerve Growth Factor Receptor
Glycogenosis I
17
Multiple Cutaneous Leiomyomata
Familial Amyloid Neuropathy
18
Hurler/Scheie Syndromes (MPS 1)
Bilateral acoustic NF
Neurofibromatosis 2
Metachromatic Leukodystrophy (ARSA Deficiency) Late Infantile & Adult Forms
22
y
Ocular Albinism
Duchenne Muscular Dystrophy
Becker Muscular Dystrophy
Ornithine Transcarbamylase Deficiency
Retinitis Pigmentosa
Incontinentia Pigmenti
Menkes Syndrome
Cleft Palate, One Form
? Pelizaeus-Merzbacher Disease
Fabry Disease
Lowe Oculocerebral Syndrome
Lesch-Nyhan Syndrome
Hunter Syndrome
Fragile X with Mental Retardation
Dreifuss-Emery Muscular Dystrophy
Adrenoleukodystrophy
Adrenomyeloncuropathy
Deutan Color Blindness
Protan Color Blindness
x

I

BACKGROUND

1

Molecular Genetics: Introduction

DONALD S. WOOD

A person has some physical characteristics that are unique, shared by no other human being, and others that are common, shared by every living and ancestral member of *Homo sapiens.* The determinant of both our unique individual traits and those we share as a species is the information encoded in DNA.

DNA (deoxyribonucleic acid) is the chemical substance from which genes—the units of heredity—are made. It was discovered by Friedrich Miescher in 1869, but its central role in genetics was not established until 1953 when the double helical structure was elucidated. The double helical structure of DNA gives it the capacity to perform the single most important function of a hereditary molecule: it can accurately copy, that is reproduce, itself to yield new DNA molecules identical in structure and information content to the parent molecule.

Mistakes in that copying process, mutations, are of fundamental interest because they can change the encoded information passed from one generation to the next. These mistakes can be beneficial and provide an important source of individual variation in plant and animal traits on which the evolution of new life forms is ultimately based. Mistakes can also be harmful; mutations in DNA are responsible for most types of human genetic disease. Developing the capability to recognize and locate DNA "variations" within and between species will greatly advance our ability to understand the basic processes of life. Developing the capability to correct DNA "defects" within individuals will probably be essential if cures for most genetic diseases are to be achieved.

A relatively short time ago the development of such capabilities appeared a long way off, if possible at all. The advent of recombinant DNA (rDNA) and related technologies, however, has greatly altered both the perception and the time frame; such capabilities are being developed and, in some cases, applied now. Recombinant DNA technology has unleashed a revolution in genetics by permitting direct manipulation of the genetic material itself. To appreciate the significance to science and medicine of being able to manipulate DNA directly requires some background.

DNA contains the information needed by all cells—the smallest irreducible units of life—to grow, divide, differentiate, mature, and even, probably, to die. The effort to determine how DNA "information" is translated into cellular "action" began in the 1940s with the development of radioisotope labeling techniques and other technologies that permitted separation, isolation, and study of large cellular molecules such as DNA and proteins. Using these methods, George Beadle and

Edward Tatum achieved one of the first important breakthroughs in understanding the nature of normal and abnormal gene function (1).

By their accounts, Beadle and Tatum's work was stimulated from observations of patients with inherited diseases reported by Archibald Garrod at the turn of the century. Garrod suggested that the symptoms of inherited disorders developed because the patient's tissue was biochemically different from that of unaffected people (2). He characterized these differences as "inborn errors of metabolism" and postulated that the cause was lack of a specific enzyme.

Beadle and Tatum analyzed the specific enzymic abnormalities underlying differences in the metabolic pathways between wild-type and mutant forms of the bread mold *Neurospora*. Their studies of inherited changes in *Neurospora* demonstrated that genes could regulate or control cell viability and nutritional requirements by determining the "specificities" of enzymes. For the first time, a biochemical effect, "protein production," of gene activity was established.

With Watson and Crick's identification of the double helical structure of DNA in 1953 (3), scientists knew what the genetic material was (DNA), how it replicated, and what it produced (protein). Throughout the next two decades investigators unraveled the details of how genes were expressed and how proteins were produced. It was during this period that the genetic code was "cracked."

The genetic code linking DNA "information" to protein "production" turned out to be simple and elegant. Both DNA and protein are assembled as linear molecules by connecting, chain-like, a series of molecular building blocks. The four building blocks of DNA are nucleotides: adenine (A), thymine (T), guanine (G), and cytosine (C); in the synthesis of a protein, the nucleotides encode 20 amino acids. A, T, G, and C serve as the alphabet of the genetic code. Certain three-letter sequences of these DNA nucleotides serve as the words of the genetic code. Each word specifies one amino acid or one instruction related to the making of a protein. For example, the DNA word CGA stands for the amino acid serine, whereas GCT stands for "end," a signal to stop synthesis of an amino acid chain. A gene is a DNA sentence that specifies the number and order of the amino acids in a protein segment known as a polypeptide chain. A protein may consist of one polypeptide chain, in which case it is specified by one gene. Some proteins, hemoglobin, for example, consist of more than one polypeptide chain and, therefore, are specified by more than one gene.

Elucidation of the genetic code and identification of the functional relationship between DNA and protein provided a rational framework on which to build an understanding of the etiology of genetic diseases. Throughout the late 1950s and 1960s some remarkable observations were made. For example, sickle cell anemia, a genetic disease affecting hemoglobin, was found to arise from a mutation in DNA that replaced one nucleotide with another, thereby changing the genetic code word from one amino acid to another. Thus, the change of only one amino acid in hemoglobin, a protein comprising almost 600 amino acids, accounted for the disease. Impressive though that knowledge is, there was little of medical value that could be done with it and, for the most part, human DNA was widely ignored as a subject for serious research investigation. On the one hand, it was too complex to handle routinely with existing technology and, on the other, there was apparently little genetic knowledge to be gained from its study that could not be obtained more effectively by analyzing the DNA of other organisms. The first noteworthy change in those attitudes began with the introduction of "somatic cell hybridization."

Somatic cell hybridization, combining the DNA of two somatic cells from different tissues or species, provided the means to analyze complex genetic organization and regulation in the DNA of higher organisms. Weiss and Green (4) were the first to recognize one of its most important applications: mapping human genes to specific chromosomes.

Chromosomes contain DNA and, as far as can be determined, only one molecule of DNA is contained in each chromosome. Human beings have 46 chromosomes and, therefore, 46 DNA molecules: 23 obtained from the father and 23 from the mother parent. At conception, the 23 chromosomes from each parent "pair up" to form 23 chromosomal pairs. With one exception, each chromosomal pair is a uniquely matched set of DNA, containing the same (homologous) number and type of genes. The one exception occurs in males whose sex/chromosome pair, X and Y, contains unmatched pieces of DNA with different, rather than homologous, genes.

The number of human genes within the chromosomal DNA is unknown. Estimates range from 50,000 to 100,000. It is known, however, that each gene has a fixed position on a specific chromosome making it possible to develop a map of the human genome. Human gene mapping has direct relevance to several clinical problems, including diagnosis and carrier detection. Until the advent of somatic cell hybridization, which provided a means to study gene expression on a single chromosome or part of a chromosome, only a handful of human genetic diseases had been "mapped." Somatic cell hybridization added dozens of human genes to that map. Still, it could not be used to locate, isolate, or analyze genes themselves.

It is striking to consider how much knowledge was obtained about genes, gene activity, and the genetic code without ever having the ability to study DNA directly. By the mid-1960s, however, the pace of DNA research had begun to slow considerably because the nucleotide sequences that made up both genes and their associated DNA regulatory elements could not be analyzed. The reason was technical; there was no systematic and reproducible way to isolate small DNA fragments of unique nucleotide sequence. The first enzymes, nucleases, found to break the connecting phosphodiester bonds between DNA nucleotides did so randomly, leaving behind a hopeless and nonreproducible "nucleotide jumble." The first real change came in 1970.

Smith and Wilcox (5) reported the discovery of a bacterial enzyme, *Hin*dII, that cleaved DNA only at sites having a specific nucleotide sequence. If that nucleotide sequence was not present, then *Hin*dII did not cut DNA. That discovery had two immediate consequences: it launched a search for other specific "restriction" nucleases, and it stimulated the development of DNA "restriction maps" in some bacterial and viral genomes to show the relative positions of nucleotide sites containing sequences recognized by restriction enzymes.

Today, dozens of restriction enzymes are known that together recognize almost 100 different nucleotide sequences as cleavage sites. Smith's discovery also spurred efforts to develop direct DNA-sequencing methods so that the nucleotides within, for example, a *Hin*dII fragment could be identified. Soon the gaps in restriction maps, which identified only the often widely spaced positions of restriction enzyme cutting sites, were filled with nucleotide sequences to become "nucleotide" maps.

Restriction enzymes, in addition to cutting DNA at specific nucleotide sequence sites, turned out to have another important feature. Some, like *Eco*RI, cut DNA in a staggered fashion leaving behind a "tail" of unconnected nucleotides. Mertz and Davis (6) in 1972 found that DNA fragments with unconnected nucleotide tails made by a restriction enzyme could be joined together, or recombined, using a different enzyme: DNA ligase.

Armed with restriction enzymes to cut DNA at known sites and ligases to recombine selected pieces of DNA, scientists quickly learned how to clone, that is, make exact copies of, DNA fragments. The method was relatively simple. An *Eco*RI DNA fragment from *any* DNA molecule, regardless of the cell, tissue, or species of origin, could be joined to an *Eco*RI fragment of a plasmid DNA (circular, self-replicating piece of nonchromosomal DNA) and inserted into bacteria. The "recombinant" plasmid DNA with the foreign DNA fragment would replicate inside the bacteria, producing an indefinite number of DNA copies. The era of recombinant DNA technology was born.

Recombinant DNA technology provides the means to locate, isolate, clone, and analyze virtually any gene or nongene DNA fragment from any organism, tissue, or cell. The technology has transformed human genetics, essentially a "nonexperimental" genetics, by making it possible to analyze the molecular nature of human DNA.

Exciting medical applications for this analysis abound. In October 1986, a worldwide research program spearheaded by the Muscular Dystrophy Association located the gene that, when defective, causes Duchenne muscular dystrophy (7), the first human gene identified solely by recombinant DNA technologies without reference to, or knowledge of, the gene's product. The approximate chromosomal locations of genes underlying dozens of other human diseases including, for example, cystic fibrosis, Huntington disease, and Charcot-Marie-Tooth disease, have also been identified through recombinant DNA technology, and the identification and isolation of these genes appears imminent. In 1987 gene products were identified from the DNA changes in Duchenne muscular dystrophy and chronic granulomatous disease.

Gene therapy for patients, correcting DNA defects that cause disease by repairing a gene or transplanting a normal functioning gene into cells containing one functioning improperly, is in the future. But experiments to achieve gene therapy are under way in the laboratory.

Concurrent with stimulating a new era in the study of human disease, rDNA technology has also launched renewed explorations of basic biological phenomena such as immune system regulation, tissue differentiation, membrane specialization, and hormonal control systems. Understanding these basic biological processes at the most fundamental molecular level, coupled with precise knowledge of the etiology of genetic diseases that affect them, will substantially improve virtually every facet of modern medicine, from diagnosis to treatment.

To speed identification of human genes underlying disease, to increase understanding of the genetic control mechanisms regulating human biology, and, ultimately, to effect cures for genetic diseases, it has been proposed that a nucleotide-by-nucleotide map be made of the entire human genome. This would be the largest and most complicated research effort in molecular biology ever undertaken. It is estimated that to complete the project would take several billion dollars and 30,000 scientist-years.

There are over three billion units of DNA in the human chromosome. If printed out in full the "book of humans" would stretch through 75 Manhattan phone directories. And what would that book reveal? No one knows. It would, after all, be an extraordinary type of knowledge about human beings for which there is no precedent. But one outcome of having that knowledge seems inevitable: DNA "defects" that give rise to genetic disease will eventually become treatable, thereby freeing future generations from the consequences of inherited "mistakes." For that accomplishment, the 1980s may well be remembered as the time when "genetic medicine" was born.

REFERENCES

1. Beadle, GW, Tatum EL. Genetic control of biochemical reactions in *Neurospora.* Proc Natl Acad Sci USA 1941; 27:499–506.

2. Garrod AE. The incidence of alkaptonuria: A study in chemical individuality. Lancet 1902; 2:1616–1620.

3. Watson J, Crick F. Molecular structure of nucleic acids—A structure for deoxyribose nucleic acid. Nature (London) 1953; 171:737–738.

4. Weiss MC, Green H. Human–mouse hybrid cell lines containing partial complements of human chromosomes and functioning human genes. Proc Natl Acad Sci USA 1967; 58:1104–1111.

5. Smith HO, Wilcox KW. A restriction enzyme from *Hemophilus influenzae,* I. Purification and general properties. J Mol Biol 1970; 51:379–391.

6. Mertz JE, Davis RW. Cleavage of DNA by RI restriction endonuclease generates cohesive ends. Proc Natl Acad Sci USA 1972; 69:3370–3374.

7. Monaco AP, Neve RI, Coletti-Feener CA, Bertelson CJ, Kurnit DM, Kunkel LM. Isolation of candidate cDNAs for portions of the Duchenne muscular dystrophy gene. Nature (London) 1986; 323:646–650.

2

The Transformation of Clinical Concepts and Clinical Practice by Molecular Genetics

LEWIS P. ROWLAND

In the past decade, the techniques of recombinant DNA have been applied to human disease, transforming clinical concepts and clinical practice. There have been few, if any, decades in the history of medicine in which so many radical changes with such far-reaching consequences have captured the attention of clinicians. And the reports now appear with accelerating flow.

The nervous system seems particularly vulnerable to genetic disorder, so it is not surprising that neurological diseases should be at the forefront of the advances of molecular genetics, or that neuromuscular diseases should be among those most thoroughly studied.

It is not possible to select a birth date for molecular genetics, but the first application of DNA probes to map a human gene was probably the 1978 report of Kan and Dozy (1) on sickle cell disease. Two years later, Botstein et al. (2) described the potential use of restriction fragment length polymorphisms (RFLPs) to find genes of unknown gene product. Their theory was greeted by some skepticism but, by 1983, Gusella and his colleagues (3) had found a probe for the Huntington disease gene.

By the end of 1986, genes had been localized for many diseases of known gene product, and haplotype DNA analysis was being used for antenatal diagnosis and carrier detection for families affected by Duchenne muscular dystrophy (DMD) (4,5) or myotonic muscular dystrophy (6,7). Candidate cDNAs had been obtained for chronic granulomatous disease (8) and for DMD (9).

One year later, by the end of 1987, the cDNA probes had been used to identify the gene products of the same two X-linked diseases: In chronic granulomatous disease (CGD) the affected gene product is a form of cytochrome *b*, a previously recognized protein that is part of a bacteriocidal oxidase system in polymorphonuclear leukocytes (10,11). In DMD, the gene product proved to be a protein that had not been previously identified and was therefore newly named as "dystrophin" (12). In 1983, we counted 16 neurological diseases that had been assigned to specific loci on human chromosomes (13). Four years later, there were 64 (Table 2-1), as listed by McKusick (14,15) and discussed by others (16–18). The precise number of loci may be slightly different, depending on the criteria used to choose diseases that qualify as "neurologic" or neuromuscular.

Table 2-1 Neurologic Diseases Assigned to Specific Loci on Specific Chromosomes, 1987

Chromosome	Disease
1	Charcot-Marie-Tooth-1 (11820)[a] Myoglobinuria-phosphofructokinase deficiency (23280)
2	Hypobetalipoproteinemia (14595) Abetalipoproteinemia (20010) Myopathy; Desmin deficiency (?) (12566) Carbamylphosphate synthetase deficiency (23730)
3	GM1 gangliosidosis (lack of β-galactosidase) (23050) Morquio syndrome (25301) Proprionic acidemia II (lack of proprionyl CoA-carboxylase) 23205) Postanesthetic apnea (succinylcholine sensitivity) (17740)
4	Huntington disease (14310) Aspartylgluosaminuria (20840) Phenylketonuria, atypical (lack of dihydropteridine reductase) (26163) Wolf–Hirschorn syndrome, with mental retardation
5	Sandhoff disease, hexosaminidase B deficiency (26880) Tay–Sachs disease, AB variant (lack of GM2 activator protein (27275)
6	Spinocerebellar ataxia (16440) Narcolepsy (16140)
7	Arginosuccinic aciduria (20790)
8	Renal tubular acidosis, with periodic paralysis, optic atrophy, or calcification of basal ganglia (11481)
9	Galactosemia (23640) Citullinemia (21570)
10	Ornithemia with gyrate atrophy of choroid and retina (and mental retardation) (25887) Metachromatic leukodystrophy (lack of sphingolipid activator protein) (24990) Myoglobinuria (lack of phosphoglycerate mutase) (26167, 17225)
11	Myoglobinuria, lactate dehydrogenase deficiency (15000) Myoglobinuria, phosphorylase deficiency (McArdle disease) (23260) Actue intermittent porphyria (17600) Pyruvate carboxylase deficiency (26615) WAGR syndrome (Wilms tumor, aniridia, genitalia abnormality, mental retardation) (19407)
12	Phenylketonuria (26160)
13	Wilson disease (27790)
14	Proprionacidemia (232000)
15	Prader–Willi syndrome (17627) Tay–Sachs disease, lack of hexosaminidase A (27280) Dyslexia-1 (12770)
16	Tyrosinemia (27660) Norum disease (24590) Hemoglobin H, with mental retardation (14175)
17	α-Glucosidase deficiency (23230) Galactosemia (23020)
18	Familial amyloidotic polyneuropathy (17630) Tourette's syndrome (13758)

Table 2-1 Neurologic Diseases Assigned to Specific Loci on Specific Chromosomes, 1987 (*Continued*)

Chromosome	Disease
19	Myotonic muscular dystrophy (16090)
	Mannisodosis (24850)
	Susceptibility to poliomyelitis (17385)
20	Galactosialidosis (25654)
21	Homocystinuria (23620)
	Familial Alzheimer disease (10476, 10430)
22	Familial acoustic neuroma (10100)
	Metachromatic leukodystrophy (25010)
	Familial meningioma (15610)
	Succinylpurinemic autism (10305)
	Hurler–Scheie syndrome (25280)
X	Glycerol kinase deficiency (30703)
	Ornithine transcarbamlyase deficiency (31225)
	Duchenne muscular dystrophy (31020)
	Becker muscular dystrophy (31010)
	Norrie syndrome (mental retardation) (31060)
	Menkes disease (30940)
	Charcot-Marie-Tooth, X linked (30280)
	Pelizaeus–Merzbacher disease (31208)
	Lowe disease (30900)
	Lesch–Nyhan disease (30800)
	Hunter disease (30990)
	Martin–Bell syndrome (fragile X mental retardation) (30955)
	Adrenoleukodystrophy (30010)
	Emery–Dreifuss muscular dystrophy (31030)
	Kennedy disease (spinal muscular atrophy) (30000)
	Sensorimotor peripheral neuropathy (30000)
	Spastic paraplegia (30000)

Source: From McKusick (14).

[a]Numbers in parentheses refer to the catalog numbers in McKusick (14).

Many of the chapters in this volume are devoted to descriptions of the techniques that have fueled this remarkable progress. It may be helpful to delineate some of the clinical changes that have resulted.

The most obvious change is the application of DNA techniques to prenatal diagnosis and carrier detection, not only for diseases of unknown gene product (such as DMD) but also diseases of known gene product, such as the Lesch–Nyhan syndrome, phenylketonuria, ornithine transcarbamylase deficiency, glycerol kinase deficiency (GKD), α-glucosidase deficiency, phosphorylase and phosphofructokinase deficiency, and hexosaminidase deficiency—all of which are described in detail in later chapters. Here, we will focus on a few other clinical transformations.

REVERSE GENETICS IS A REALITY: NO MORE FISHING EXPEDITIONS

A generation ago, my colleagues and I set out to find the abnormal gene product in DMD. We planned to evaluate each of the major muscle specific proteins—one by one. We started with the easiest protein to work with, myoglobin, and soon concluded that there was no abnormality of that pigmented protein in any muscular dystrophy, or even in any known human myopathy. We turned next to myosin, and, in what seem to have been the first electrophoretic studies of human myosin, we found no abnormality in DMD (19). Our attention was then diverted to studies of sarcolemmal proteins in DMD and to the enzymes of glycogen storage.

As late as 1984, some investigators still believed that DMD might be due to an abnormality of an identified muscle protein. At a meeting of developmental biologists who were studying these proteins, I was asked to review the clinical concepts of muscular dystrophy and to provide a summary of previous studies of possible gene products. The title I used was "Muscle-specific proteins, muscle development, and Duchenne muscular dystrophy." The original introduction read as follows:

> There will be a discontinuity between this paper and those that follow. The gap is a measure of the intensity of research achievements in elucidating the biology of muscle development, the genetic control and expression of genes in muscle, and the search for the genes involved in some forms of muscular dystrophy. In contrast, the search for the abnormal gene product in these diseases was never very vigorous and has now virtually ceased. That discontinuity may prove to be regretable because the search for the Duchenne gene, for instance, would be facilitated immeasurably if we knew the appropriate gene product.

The text then specified how virtually all of the known muscle-specific proteins had been studied and found to be normal in human dystrophies. But the editors of the proceedings of that conference changed both the title and the original text to more banal versions (20), presumably because there was still hope that some already-recognized protein might be important in a human disease.

My prescience was not precise. True, it would have helped the search for the gene for Duchenne dystrophy if the gene product had been known in advance—*forward genetics* defines the gene product (the biochemical abnormality) first, then the gene. But that meeting marked the end of an era, and there was little be deemed regrettable.

Within 3 years of that meeting, Orkin (10), Kunkel (12), and their colleagues proved resoundingly the practicality of *reverse genetics*, that it is possible to use restriction fragment length polymorphism (RFLP) analysis to identify the genes for diseases of unknown gene product and then to use one of two approaches to identify the gene product itself.

For chronic granulomatous disease, it was possible to discern the sequence of nucleotides in the cDNA. An attempt was then made to identify a known protein from the sequence of amino acids that would be encoded; that proved to be difficult because of problems identifying the reading frame. However, when the amino acid sequence of one candidate protein, a component of cytochrome *b,* was analyzed, it was found to match the nucleotide sequence.

For DMD, the sequence of amino acids deduced from the cDNA did not match any known muscle-specific protein, so fragments of cDNA were inserted into an expression vector and the fusion polypeptide was used to make antibodies that identified the product as a hitherto unknown protein, which they named "dystrophin" (12). Now, the task is to discover the normal function of that protein and how abnormalities of it might be related to the clinical manifestations of the disease.

CGD and DMD are the first successes of reverse genetics, and they show the power of molecular genetics in identifying the true gene products. Recombinant DNA methods have also been used to eliminate some candidate gene products.

For example, once the exons of the Duchenne gene had been identified and it was realized how large the protein might be, there were few known proteins of that size—more than 500,000 Da. But, two of the largest known human proteins were muscle specific: nebulin and titin.

Electrophoretic studies suggested that nebulin was absent or greatly reduced in DMD muscle (21). However, two groups of investigators then found that the gene for nebulin is on human chromosome 2 (22,23). Therefore, nebulin is not X-linked and it cannot be the affected gene product in DMD. [In the meantime, nebulin was found to be present, often in much reduced amount, in Duchenne muscle (12,24–26).]

The discovery of the DMD gene and dystrophin has also resolved another controversy about DMD, that it might be "neurogenic" and somehow related to defective innervation (27). Without question, the disease is a "myopathy," caused by an abnormality within muscle, and not secondary to a disorder in the nervous system, liver, or anywhere else.

Recombinant DNA methods have also been used to dispel erroneous concepts about other diseases. For example, nerve growth factor (NGF) and the NGF receptor were deemed possible gene products for neurofibromatosis, but RFLP studies showed that neither could be involved (28). The gene for neurofibromatosis is on chromosome 15, but the precise site is still uncertain and therefore the gene product is still not known.

Molecular genetics has narrowed the search for critical genes in other ways, too. For example, early evidence suggested that the gene for familial Alzheimer's disease might involve the amyloid protein that is deposited in the brain of patients with that disease. However, RFLP analysis indicated that the gene for amyloid protein is separated from the gene for the disease by 15 centimorgans (cM) (29,30).

Nevertheless, the research path for other diseases of unknown gene product is now illuminated. We know the principles now, how to find and how to test candidate gene products.

MOLECULAR GENETICS HAS CHANGED CLINICAL CONCEPTS: THE EXAMPLE OF DUCHENNE DYSTROPHY, AN Xp21 MYOPATHY

Until last year, most students of muscle disease viewed DMD as a homogeneous disorder, one that could be recognized clinically by eight characteristics: (1) X-linked inheritance; (2) manifestations in boys, not girls; (3) onset in early life, probably at birth, symptomatic as soon as the boys begin to walk; (4) first symptoms in legs, rather than arms; (5) calf hypertrophy; (6) high serum levels of creatine kinase (CK) and other sarcoplasmic enzymes; (7) relatively rapid progression to wheelchair by around age 12 years and death in the third or fourth decade; and (8) histological changes in muscle that included necrosis and evidence of regeneration but no storage of abnormal material or stucturally specfic abnormality.

Nevertheless, there was uncertainty about three problems: the relationship of DMD to a milder X-linked myopathy called Becker muscular dystrophy (BMD), the relationship of mental retardation to Duchenne dystrophy, and whether the same disease can appear in girls or women.

Molecular genetics has transformed the definition of DMD from a clinical–pathological description to the simple statement that "DMD is a myopathy due to a mutation at Xp21" (31). In changing the definition, molecular genetics has also illuminated each of these uncertainties about the disease.

The Problem of Mild Myopathy

As discussed in Chapter 27 by Kunkel, it is now evident that DMD and Becker dystrophy are allelic disorders. That conclusion contradicted earlier evidence that the two conditions might be encoded at different ends of the X-chromosome, and would therefore be attributable to disorders of two different gene products. Instead, the affected gene product must be identical in both conditions.

Additionally, DNA analysis has provided other examples of mild myopathies caused by mutations at Xp21 (Table 2-2). For example, mild myopathies are seen in obligatory carriers of the DMD gene, or in girls or women with translocations that involve Xp21. Mild myopathies are also seen in patients with glycerol kinase deficiency, as described in Chapter 28 by Seltzer and McCabe. The mildest clinical expression of the DMD gene may be calf hypertrophy without any detectable weakness, a syndrome seen in some carriers of the gene.

The slightest detectable expression of the DMD gene may be high serum values of CK, as seen in some, but not all, obligatory carriers of the gene. High serum levels of CK activity are also characteristic of another Xp21 mutation, the McLeod syndrome (32). Until now, that rare X-linked disorder has been diagnosed only in asymptomatic blood donors. Diagnosis has been restricted in that way because it has depended on blood-matching studies; McLeod subjects lack a precursor protein for the Kell antigen in red blood cells. Although it was originally defined serologically, the "McLeod phenotype" also includes acanthocytes (red blood cells with protrusions that create a thorny appearance under the microscope) and serum CK levels in the range of patients with DMD. All three components of the syndrome—absence of the Kell antigen, distorted red cells, and high CK levels—seem to be invariant, but that may change as more patients are studied. There is usually no clinical weakness (33), but the McLeod syndrome was one of the five X-linked disorders that were seen in BB (34), the fabled patient whose DNA was used by Kunkel et al. (35,36) to obtain the DNA probes for the DMD gene. The high serum CK values of the McLeod syndrome must mean that muscle membranes are affected by the condition, presumably because the Xp21 McLeod mutation overlaps the DMD gene.

Recognition of the relationship of the McLeod syndrome to Xp21 myopathies has affected another clinical problem, the occasional finding of very high serum CK values in an apparently healthy individual. We called that condition "idiopathic hyperCKemia" (37), but high serum levels of CK may be found in individuals who ultimately develop clinical manifestations of limb weakness (38). Even when the

Table 2-2 The Diverse Manifestations of Xp21 Myopathies

Either sex	
Male	DMD patients BMD patients
Female	Manifesting DMD carriers Manifesting BMD carriers Translocations involving Xp21 Myopathy with Turner XO syndrome Myopathy with GKD
Children or adults	
Children	DMD or BMD patients Some translocations Some GKD or CGD patients
Adults	Some BMD patients Some translocations Manifesting DMD carriers Manifesting BMD carriers
Mild or severe myopathy	
High serum CK alone	Some DMD carriers Some BMD carriers All McLeod subjects Presymptomatic BMD patients, possible
Calf hypertrophy alone	Some DMD carriers Some BMD carriers
Mild limb weakness	BMD patients Some DMD carriers Some BMD carriers Some translocations Some GKD patients Some CGD patients
Severe limb weakness	All DMD patients Some translocations Some GKD patients
Mental retardation	
Present or absent	DMD patients BMD patients Translocations GKD patients CGD patients

ABBREVIATIONS
BMD = Becker muscular dystrophy
DMD = Duchenne muscular dystrophy
GKD = glycerol kinase deficiency
CGD = chronic granulomatous disease

distribution of weakness differs from the typical pattern of DMD (38), the condition might prove to be due to an Xp21 mutation. If so, that would further distort the clinical definition of DMD.

The relationship of the McLeod syndrome to Xp21 has spurred interest in the disorder. The McLeod syndrome may also be linked to another neurological condition, chorea-acanthocytosis (39), which is usually thought to be autosomal recessive, but has not yet been studied by DNA probes. Sometimes, chorea-acanthocytosis is accompanied by high serum CK levels and those cases might be variants of the McLeod syndrome. In other cases, the serum levels of CK are normal. The chorea-acanthocytosis syndrome may therefore be heterogeneous, sometimes X-linked, and sometimes autosomal recessive, as a result of different mutations that affect different gene products.

On the other hand, molecular genetics has proven that not all X-linked myopathies are variants of DMD. For example, the demonstration that the gene for Emery–Dreifuss dystrophy is on the long arm of the X-chromosome specifies that it is an entirely different disease, not an allelic variant of DMD (40,41). Localization of the Emery–Dreifuss gene has resolved questions of the identity of cases that were originally thought to be different clinical syndromes (42,43).

In contrast, the possibility that some clinically identical DMD cases might be autosomal recessive (44) was negated by the development of additional DNA probes for the Xp21 area, and the resulting demonstration that more than 50% of DMD cases are due to deletions (45,46). There is no evidence that the typical Duchenne phenotype has been produced by mutation anywhere but at Xp21. But mutations at Xp21 do not always cause the typical DMD phenotype.

Duchenne Syndromes in Girls and Women

The second Duchenne problem is the appearance of a clinical myopathy in a girl or woman that, in strict theory, would not be anticipated in an X-linked disease. That puzzle has been explained by the syndromes that cause mild myopathy and also involve the X-chromosome at Xp21: translocations, obligatory carriers, and glycerol kinase deficiency. In all of these variants, the girl is left with one functioning X-chromosome, the one that bears the DMD mutation. In translocations, the break is at Xp21 and there is preferential inactivation of the normal X-chromosome, leaving the mutant gene. In obligatory carriers, there is presumably random inactivation of one X-chromosome, according to the principles of lyonization. Manifesting carriers have more of the mutant X than the normal gene in their multicellular muscle fibers. Also, for reasons that are not yet clear, mild myopathy may be seen discordantly in only one of identical twin girls or women who are obligatory carriers (47–49). There have been seven twin pairs of that kind, but only two pairs of Duchenne boys have been reported (47). Twinning seems somehow to enhance clinical manifestations in discordant fashion.

Mental Retardation and Duchenne Dystrophy

The third DMD problem is whether mental retardation is linked to the genetic fault that causes the muscular dystrophy. It has been thought that there is an undue frequency of retardation in Duchenne boys (50), and it has been suggested that mental retardation might signal genetic heterogeneity, dividing cases into those with and without mental retardation (51). DNA analysis of this problem is not yet complete, but in preliminary studies, Xp21 probes could not separate cases into those with or without retardation (52).

Information on the nature of the retardation has been provided by studies of glycerol kinase deficiency. Ordinarily, it is assumed that if mental retardation is seen in a child with a heritable biochemical abnormality, then the mental retardation must be caused by that biochemical abnormality. However, some individuals with virtually total lack of glycerol kinase have normal intelligence; the retardation cannot be due to the biochemical abnormality. More likely, there is a separate gene for mental retardation in the vicinity of Xp21 (31). As described in Chapter 29 by Davies, there are genes for at least 30 known mental retardation syndromes on the X-chromosome; there may be more, and one or more of these genes may be found in the region of Xp21. That would explain the inconstant appearance of mental retardation with either DMD or glycerol kinase deficiency.The ultimate clinical transformation of DMD is this: DNA analysis (or analysis of the gene product) is now needed for the diagnosis of an Xp21 myopathy. Diagnosis cannot be made clinically because so many clinical variations are seen with Xp21 mutations: mild or severe myopathy, asymptomatic rise in serum CK activity, onset in childhood, adolescence, or adult years, and myopathy in boys or men and girls or women (Table 2-2). Moreover, the cases in girls and women distort the usual patterns of X-linked recessive inheritance so that familial patterns might erroneously suggest autosomal recessive or dominant inheritance. Typical Duchenne boys can still be recognized, and unusual cases in their families can then be recognized, but DNA analysis or gene product identification is needed for other atypical cases. In fact, for precise diagnosis, no matter how much the clinical disorder differs from DMD, DNA analysis or evaluation of gene product will be needed in all cases of myopathy to be certain that there is no mutation at Xp21.

CLINICAL TRANSFORMATIONS IN OTHER DISEASES

Although DMD has been the center of much attention because the DNA progress has been so effective, DNA analysis has also affected concepts of other diseases. For example, research in Huntington disease has dispelled several earlier views. It had been thought that individuals who are homozygous for an autosomal dominant disease should have a much more severe clinical syndrome than the typical pattern in heterozygotes, but that was not substantiated by DNA studies of Huntington disease; the clinical picture was not different in homozygous children of two affected parents (53). It had also been thought that there might be two different syndromes in different families—some more rigid and others more choreic. But that was not substantiated, either (54). And there is still no evidence of new mutations of Huntington disease (53).

Clinical Molecular Genetics Has Transformed Research

The new genetics has spawned new patterns of research. Consider the following:

1. The achievements of molecular genetics could not have been attained by investigators working in isolation. Large teams of workers have been involved.

2. Never before have clinician and basic scientist cooperated so fully. Often, in the past, MD investigators have applied the results of basic science to clinical problems. Now, the basic scientists themselves have gone directly to the clinical problems.

   But they have not done that alone. Recognition of families, collection of samples from members of large families, and clinical analysis have been the province of clinical investigators. This has been seen most dramatically in the identification and study of patients with Huntington disease in Lake Maracaibo, Venezuela (3). That achievement of Dr. Nancy Wexler and her associates provided a key ingredient for the DNA studies of James Gusella and his associates.

   Basic science journals such as *Nature* and *Cell* have become required reading for clinicians. But reports are also likely to appear in general journals such as *Lancet* and basic scientists can be found reading journals intended for clinical neurologists or geneticists.
3. Cooperation has taken other forms. There were 70 authors of the paper on DNA deletions in Duchenne dystrophy (36). Exchange of samples from patients and exchange of DNA probes have become a new feature of the research trade, in spite of the competitive personality traits of biomedical investigators.
4. Commercialism has not yet had an impact on research in neuromuscular diseases, but the rise of biotechnology corporations in the United States has surely had an impact on research in molecular genetics. These corporations are usually directed by an advisory board of academic investigators, so new forms of cooperation and new conflicts of interest will have to be sorted out.
5. Rare diseases have always been important in the history of research on heritable diseases. It was alcaptonuria that captured the interest of Archibald Garrod, so starting biochemical genetics. Now, rare diseases have become important because the affected genes may be DNA neighbors of more common conditions, or because they provide examples for more common diseases. For example, the gene for glycerol kinase deficiency is an important neighbor of the DMD locus.

And DMD itself, although relatively rare, has surely been an important example for genetic research. Huntington disease is even rarer, but the experience with that disease proved the validity of the Botstein–Skolnick–White theory.

Another example. Familial amyloidotic polyneuropathy (FAP) is a rare condition that, as described in Chapter 22 by Herbert, appears in only a few specific countries—perhaps because of founder effects. There have been few recognized cases in the United States. Nevertheless, it is the first autosomal dominant disease for which presymptomatic diagnosis has been provided by DNA analysis. It is therefore the first autosomal dominant disease to generate the ethical problems of presymptomatic diagnosis that are described in Chapter 33 by Wexler.

FAP provides other important avenues of research opportunity. With probes now available for the affected gene, it may be possible to study the role of the gene product (transthyretin) in the pathogenesis of one form of inherited peripheral neuropathy (and we still do not understand the pathogenesis of even one of the numerous forms of heritable neuropathies). Workers in several laboratories have attempted to induce an experimental amyloidotic neuropathy by using transgenic mice, introducing the abnormal gene into the germ line cells of the animals.

SURPRISES

As more and more genes are cloned, there will be more information about the regulation and function of gene products. Some of this new information will be unexpected and there may be important surprises, some of clinical import.

For example, the gene for transthyretin (TTR) was cloned primarily because that serum protein is altered in familial amyloidotic polyneuropathy (55,56). When the gene became available, it was used for *in situ* experiments to determine where it is synthesized. There was earlier evidence that the protein was synthesized in the liver, and that was confirmed. However, it was also found that TTR is synthesized in the choroid plexus (57,58).

This observation clarified some earlier mysteries. An earlier name for TTR was "prealbumin," because it migrated more rapidly than albumin in studies of human blood serum (59). Prealbumin seemed to differ from almost all other serum proteins in two characteristics: (1) The concentrations of most proteins are higher in serum than in cerebrospinal fluid (CSF), but the CSF content of prealbumin is higher than in serum (60). (2) The CSF content of most proteins shows a gradient from lower values in the cerebral ventricles to higher values in the lumbar CSF; the content of TTR (prealbumin), however, shows a reverse pattern: higher in the ventricles than in lumbar CSF (60).

The combined evidence suggests that TTR is synthesized in the choroid plexus for export into the CSF. How synthesis is regulated and what TTR might be doing in CSF are now being investigated. How abnormal forms of TTR are deposited as amyloid and how that causes polyneuropathy are other important questions (61).

PROBLEMS YET TO BE RESOLVED

Molecular genetics has opened a new phase of clinical medicine. But there are still problems to be solved. Consider just a few:

1. Molecular genetics per se does not resolve questions of pathogenesis. Enzyme deficiencies have been recognized for 35 years. But we still do not know why lack of phenylhydroxylase causes mental retardation, or why lack of hypoxanthine phosphoribosyltransferase causes self-multilation and chorea.

For reverse genetics, there is the additional step: The gene product must first be isolated, then normal function of that protein has to be deciphered, and, finally, it must be determined why and how lack of that normal function causes the clinical syndrome. For instance, we shall have to learn how lack of dystrophin causes high serum levels of sarcoplasmic enzymes, progressive degeneration of skeletal muscle, and weakness that becomes progressively more severe for two or three decades.

2. Molecular genetics has not yet resolved the problem of clinical heterogeneity of allelic diseases. For instance, hexosaminidase deficiency may cause the devastating infantile brain disease that is manifest in Tay–Sachs or Sandhoff disease. But enzyme levels may be just as low in patients who pass infancy without symptoms and are then affected by an entirely different syndrome, one that affects only spinal motor neurons or both upper and lower motor neurons, the syndrome of amyotrophic lateral sclerosis.

Similar differences are seen in α-glucosidase deficiency, in which the infantile disorder is a fatal disease that affects both motor neurons and muscle, but adolescent and adult cases are restricted to muscle. And, as described in Chapter 22 by Herbert, similar amino acid abnormalities of transthyretin may give rise to quite different clinical disorders.

These variations are, in a sense, parallel to differences between Duchenne and Becker dystrophies. Monaco, Hoffman, Kunkel, and their colleagues in several institutions (62,63) have provided evidence that frameshift mutations may account for that difference and that DMD cases totally lack dystrophin, but there is residual protein in Becker cases. Similarly, residual enzyme is often found in the late-onset cases of α-glucosidase deficiency and hexosaminidase deficiency. In both groups of disorders, the syndromes of later onset are less severe than the infantile forms. However, as described in Chapters 18 and 19 by Gravel and Hirschorn and their colleagues, it has not been possible to define the DNA differences in early and late-onset cases. Moreover, there are families, presumably with the same mutation, in which there are mild and severe forms of the same disorder.

In Xp21 myopathies, the clinical course tends to be similar in each family (64), but there are families in which some members of the family have a Duchenne phenotype and other members of the family, siblings or cousins, have milder forms that would be compatible with Becker dystrophy (65–67). That conundrum is yet to be resolved.

3. Molecular genetics does not resolve questions of treatment. Whether and how gene therapy might be effective are discussed in detail by Caskey and his associates in Chapter 31. Other strategies may be useful if we can understand the pathogenesis of particular syndromes.

4. Health care systems have not yet adjusted to the availability of new diagnostic and new genetics techniques. Developing countries have other public health priorities. Technically advanced countries now seem to have difficulty supporting the costs of new technology in medical care, and medical research is also facing financial restriction. Diagnostic application of DNA analysis, as described by Hejtmancik and his associates in Chapter 14, is expensive and insurance systems have not yet approved this as a reimbursable test in the United States (68). The diagnostic technology must be improved to make it less expensive, safer for laboratory workers, and more rapidly applicable for antenatal diagnosis (17).

5. The ethical questions, discussed in detail in Chapters 32 and 33 by Wexler and Howell, are most dramatically illustrated by tests for presymptomatic diagnosis of autosomal dominant disease of late onset, particularly when there is no effective treatment. Huntington disease is one model; familial amyloidotic polyneuropathy is another, and it is the only one with precisely accurate DNA probes already available.

These challenges need not be daunting. There may always be biomedical problems to solve. But the rate of progress in the past decade gives promise that today's puzzles may not be so problematic tomorrow.

REFERENCES

1. Kan YW, Dozy AM. Antenantal diagnosis of sickle cell anemia by DNA analysis of amniotic fluid cells. Lancet 1978; 2:910–911.

2. Botstein D, White RL, Skolnick M, Davis R. Construction of a genetic map in man using restriction fragment length polymorphisms. Am J Hum Genet 1980; 32:314–321.

3. Gusella JF, Wexler NS, Conneally PM, et al. A polymorphic marker genetically linked to Huntington's disease. Nature (London) 1983; 306:234–238.

4. Hejtmancik JF, Harris SC, Tsao CC, Ward PA, Caskey CT. Carrier diagnosis of Duchenne muscular dystrophy using restriction fragment length polymorphisms. Neurology 1986; 36:1553–1562.

5. Darras BT, Harper JF, Francke U. Prenatal diagnosis and detection of carriers with NAD probes in Duchenne's muscular dystrophy. N Engl J Med 1987; 316:985–992.

6. Pericak-Vance MA, Yamaoka LH, Assinder BA, et al. Tight linkage of apolipporotein C2 to myotonic dystrophy on chromosome 19. Neurology 1986; 36:1418–1423.

7. Meredith AL, Huson SM, Lunt PW, Sarfarazi M, Harley HG, Brook JD, Shaw DJ, Harper PS. Application of a closely linked polymorphism of restriction fragment length to counselling and prenatal testing in families with myotonic dystrophy. Br Med J 1986; 293:1353–1356.

8. Royer-Pokora B, Kunkel IM, Monaco AP, et al. Cloning the gene for an inherited human disorder—chronic granulomatous disease—on the basis of its chromosomal location. Nature (London) 1986; 322:32–38.

9. Monaco AP, Neve R, Colletti-Feener C, Bertelson CJ, Kurnit DM, Kunkel LM. Isolation of candidate cDNAs for portions of the Duchenne muscular dystrophy gene. Nature (London) 1986; 323:646–650.

10. Dinauer MC, Orkin SH, Brown R, Jesaitis AJ, Parkos CA. The glycoprotein encoded by the X-linked chronic granulomatous disease locus is a component of the neutrophil cytochrome b complex. Nature (London) 1987; 327:717–720.

11. Teahan C, Rowe P, Parker P, Totty N, Segal AW. The X-linked chronic granulomatous disease gene codes for the B-chain of cytochrome b-245. Nature (London) 1987; 327:720–724.

12. Hoffman EP, Brown RH Jr, Kunkel LM. Dystrophin: The protein product of the Duchenne muscular dystrophy locus. Cell 1987; 51:919–928.

13. Rowland LP. Molecular genetics, pseudogenetics, and clinical neurology. Neurology 1983; 33:1179–1195.

14. McKusick VA. Mendelian Inheritance in Man, 7th ed. Baltimore: Johns Hopkins University Press, 1986.

15. McKusick VA. The morbid anatomy of the human genome. Medicine 1986; 65:1–33; 1987; 66:1–63, 237–295.

16. Cooper DN, Schmidtke J. Diagnosis of genetic disease using recombinant DNA. Supplement. Hum Genet 1987; 77:66–75.

17. Caskey CT. Disease diagnosis by recombinant DNA methods. Science 1987; 236:1223–1228.

18. Martin JB. Molecular genetics: Applications to the clinical neurosciences. Science 1987; 238:765–772.

19. Penn AS, Cloak R, Rowland LP. Myosin from normal and dystrophic human muscle: Immunochemical and electrophoretic study. Arch Neurol 1972; 27:159–173.

20. Rowland LP. Clinical perspective: Phenotypic expression in muscular dystrophy. In Strohman RC, Wolf S, eds. Gene Expression in Muscle. New York: Plenum Press, 1985, pp 3–14.

21. Wood DS, Zeviani M, Prelle A, Bonilla E, Salviati G, Miranda AT, DiMauro S, Rowland LP. Is nebulin the defective gene product in Duchenne muscular dystrophy? N Engl J Med 1987; 316:107–108.

22. Stedman H, Browning KM, Oliver N, Oronzi-Scott M, Fischbeck KH, Sarkar S, Sylvester J, Schmickel R, Wang K. Human nebulin cDNAs detect a 25-kilobase transcript distinct from that of the Duchenne muscular dystrophy gene. Genomics 1988; 22:249–256.

23. Zeviani M, Darris BT, Rizzuto R, et al. Cloning and expression of human nebulin cDNAs and assignment of the gene to chromosome 2q13-q33. Genomics, in press.

24. Sugita H, Nonaka I, Itoh Y, Asakura A, Hu DH, Kimura S, Maruyama K. Is nebulin the gene product of Duchenne muscular dystrophy? Proc Jpn Acad 1987; 63B:107–110.

25. Bonilla E, et al. Nebulin is present in Duchenne muscle. Neurology, in press.

26. Furst D, Nave R, Osborn M, et al. Nebulin and titin expression in Duchenne muscular dystrophy appears normal. FEB 1987; 224:49–53.

27. Rowland LP. Are the muscular dystrophies neurogenic? Ann New York Acad Sci 1974; 238:244–260.

28. Seizinger BR, Rouleau GA, Ozelius LJ, et al. Genetic linkage of von Recklinghausen neurofibromatosis to the nerve growth factor receptor gene. Cell 1987; 49:589–594.

29. Tanzi R, St. George-Hyslop PH, Haines JL, et al. The genetic defect in familial Alzheimer's disease is not tightly linked to the amyloid B-protein gene. Nature (London) 1987; 329:156–157.

30. Van Broecknoven C, Genthe A, Vanderberghe A, et al. Failure of familial Alzheimer's disease to segregate with the A4-amyloid gene in several European families. Nature (London) 1987; 329:153–155.

31. Rowland LP. The impact of molecular genetics on clinical concepts of Duchenne muscular dystrophy. Brain 1988; 111:479–495.

32. Marsh WL. Eievated serum creatine phosphokinase in subjects with McLeod syndrome. Vox Sang 1981; 40:403–411.

33. Swash, M. Scwaartz MS, Carter ND, Heath R, Leak M, Rogers KL. Benign X-linked myopathy with acanthocytes (McLeod syndrome). Its relationship to X-linked muscular dystrophy. Brain 1983; 106:717–733.

34. Francke U, Ochs, HD, de Martinville B, et al. Minor Xp21 deletion in a male associated with the expression of Duchenne muscular dystrophy, chronic granulomatous disease, retinitis pigmentosa and McLeod syndrome. Am J Hum Genet 1985; 37:250–267.

35. Kunkel LM, Monaco AP, Middlesworth W, Ochs SD, Latt SDA. Specific cloning of DNA fragments absent from the DNA of a male patient with an X-chromosome deletion. Proc Natl Acad Sci USA 1986; 82:4778–4782.

36. Kunkel LM, et al. Analysis of deletions in DNA from patients with Becker and Duchenne muscular dystrophy. Nature (London) 1986; 322:73–77.

37. Rowland LP, Willner J, Cerri C, DiMauro S, Miranda A. Approaches to the membrane theory of Duchenne muscular dystrophy. In Angelini C, Danielli GA, Fontanari D, eds. Muscular Dystrophy Research: Advances and New Trends, Amsterdam: Excerpta Medica, 1980, pp 3–13.

38. Galassi G, Rowland LP, Hays AP, Hopkins LC, DiMauro S. High serum levels of creatine kinase: Asymptomatic prelude to distal myopathy. Muscle Nerve 1987; 10:346–350.

39. Vance JM, Pericak-Vance MA, Bowman MH, Payne CS, Fredane L, Siddique T,

Roses AD, Massey EW. Chorea-acanthocytosis: A report of three new families and implications for genetic counselling. Am J Med Genet 1987; 28:403–410.

40. Boswinkel E, Walker A, Hodgson A, et al. Linkage analysis using 8 DNA polymorphisms along the length of the X-chromosome locates the gene for Emery–Dreifuss muscular dystrophy in distal Xq. Cytogenet Cell Genet 1985; 40:586.

41. Thomas NST, Ray PN, Belfal B, Duff C, Logan S, Oss I, Worton RG. Localization of the gene for Emery–Dreifuss muscular dystrophy to the distal long arm of the X chromosome. J Med Genet 1986; 23:596–598.

42. Rowland LP, Fetell MR, Olarte MR, Hays AP, Singh N, Wanat FE. Emery–Dreifuss muscular dystrophy. Ann Neurol 1979; 2:111–117.

43. Thomas PK, Calne DB, Elliott DF. X-linked scapulohumeral syndrome. J Neurol Neurosurg Psychiat 1972; 35:208–215.

44. Fischbeck KH, Ritter A, Kunkel LM, Bertelson C, Hejtmancik JF, Pericak-Vance MA, Boehm C, Ionasescu V. Genetic heterogeneity in Duchenne dystrophy. Neurology 1987; 37 supplement:116–117.

45. Koenig M, Hoffman EP, Bertelson CVJ, Monaco AP, Feener C, Kunkel LM. Complete cloning of the Duchenne muscular dystrophy (DMD) cDNA and preliminary genomic organization of the DMD gene in normal and affected individuals. Cell 1987; 50:509–517.

46. Den Dunnen JT, Bakker E, Breteler EGK, Pearson PL, van Ommen GJB. Direct detection of more than 50% of the Duchenne muscular dystrophy mutations by field inversion gels. Nature (London) 1987; 329:640–642.

47. Pena DJ, Karpati G, Carpenter S, Fraser FC. The clinical consequences of X-chromosome inactivation: Duchenne muscular dystrophy in one of monozygotic twins. J. Neurol Sci 1987; 79:337–344.

48. Younger DS, Warburton D, Tantravahi U, Hays AP, Lange DJ, Palllai M, Rowland LP. Monozygous twin carriers of the Duchenne gene; Discordant for clinical myopathy. Neurology 1987; 37 Suppl:222.

49. Burn J, Povey S, Boyd Y, Munro EA, West L, Harper K, Thomas D. Duchenne muscular dystrophy in one of monozygotic twins. J Med Genet 1986; 23:494–500.

50. Dubowitz V. Mental retardation in Duchenne muscular dystrophy. In Rowland LP, ed. Pathogenesis of Human Muscular Dystrophies. Amsterdam: Excerpta Medica, 1977, pp 685–698.

51. Emery AEH, Skinner R, Holloway S. A study of possible heterogeneity in Duchenne muscular dystrophy. Ann Hum Genet 1979; 15:444–449.

52. O'Brien T, Harper PS, Davies KEE, et al. Absence of genetic heterogeneity in Duchenne muscular dystrophy shown by a linkage study using two cloned DNA sequences. J Med Genet 1983; 20:249–251.

53. Young AB, Shoulson I, Penney JB, et al. Huntington's disease in Venezuela: Neurological features and functional decline. Neurology 1986; 36:244–249.

54. Folstein SE, Phillips JA, Meyers DA, et al. Huntington's disease. Two families with differing clinical features show linkage to the G8 probe. Science 1985; 229:776–779.

55. Sasaki H, Sasaki Y, Matsuo H, et al. Diagnosis of familial amyloidotic polyneuropathy by recombinant DNA techniques. Biochem Biophys Res Commun 1984; 125:636–642.

56. Mita S, Maeda S, Ide M. Tzuzuki T, Shinada K, Araki S. Familial amyloidotic polyneuropathy diagnosed by cloned human prealbumin cDNA. Neurology 1986; 36:298–301.

57. Soprano DR, Herbert J, Soprano KJ, Schon EA, Goodman DS. Demonstration of transthyretin mRNA in the brain and other extrahepatic tissues in the rat. J Biol Chem 1985; 260:11793–11798.

58. Herbert J, Wilcox JN, Pham KTC et al. Transthyretin: A choroid plexus-specific transport protein in human brain. Neurology 1986; 36:900–911.

59. Kabat EA, Moore DH, Landow H. An electrophoretic study of the protein components in cerebrospinal fluid and their relationship to the serum proteins. J Clin Invest 1942; 21:571–577.

60. Fishman RA, Ransohoff J, Osserman EF. Factors influencing the concentration gradient of protein in cerebrospinal fluid. J Clin Invest 1958; 37:1419–1424.

61. Herbert J. Chapter 22.

62. Monaco AT, Bertelson CJ, Liechti-Gallati S, Moser H, Kunkel LM. An explanation for the phenotypic differences between patients bearing partial deletions of the DMD locus. Genomics 1988; 2: 90–95.

63. Hoffman EP, Fischbeck KH, Brown RH, Johnson M, Medori R, Loike JD, Harris JB, Waterston R, Brooke M, Specht L, Lupsky W, Chamberlain J, Caskey CT, Shapiro F, Kunkel LM. Dystrophin characterization in muscle biopsies from Duchenne and Becker muscular dystrophy patients. N Engl J Med 1988; 318:1363–1367.

64. Hyser CL, Province M, Griggs RC, et al. Genetic heterogeneity in Duchenne dystrophy. Ann Neurol 1987; 22:553–555.

65. Furakawa R, Peter JB. X-linked muscular dystrophy. Ann Neurol 1977; 2:414–416.

66. Hausmanowa-Petrusewicz I, Borkawska J. Intrafamilial variability in X-linked muscular dystrophy. Mild and severe forms of X-linked muscular dystrophy in the same family. J Neurol Sci 1978; 28:43–50.

67. Forrest SM, Cross GS, Speer A, Gardner-Medwin D, Burn J, Davies KE. Preferential deletion of exons in Duchenne and Becker dystrophies. Nature (London) 1987; 329:638–640.

68. Pyeritz RE, Tumpson JE, Bernhardt BA. The economics of clinical genetics services. I. Preview. II. A time analysis of a medical genetics clinic. Am J Hum Genet 1987; 41:549–558; 559–565.

3

Mendelian and Non-Mendelian Inheritance

WILLIAM G. JOHNSON

An understanding of the principles of medical genetics helps both to diagnose neuromuscular diseases and to work with patients and families after the diagnosis has been made. In addition, the genetic basis of a disorder is of critical importance in understanding the cause and in developing specific treatment (1–4). This chapter focuses on the patterns of inheritance seen in human genetic disorders.

MENDELIAN INHERITANCE

At least 1906 human disorders have a Mendelian inheritance pattern (5). Among autosomal recessive disorders, about one-third have been "solved" biochemically; nearly all of them result from enzyme deficiencies (Table 3-1). Among autosomal dominant disorders, few have been "solved" biochemically; most of those solved so far are not due to lack of an enzyme (Table 3-1). The affected gene product is more often a structural protein or a protein of uncertain function. The picture for X-linked conditions is intermediate between that for dominant and recessive autosomal disorders. The inheritance pattern of a disorder, then, gives some clue to the cause, and the inheritance pattern is useful in addition for diagnosis and genetic counseling.

Table 3-1 Causes of Human Genetic Disorders

Disorders	Number
Dominant	Total 1172 (100.0%)
Disorders with enzyme deficiency	21 (1.8%)
Disorders with other biochemical marker	50 (4.3%)
Total number with biochemical marker	71 (6.1%)
Recessive	Total 610 (100.0%)
Disorders with enzyme deficiency	177 (29.0%)
Disorders with other biochemical marker	54 (8.8%)
Total number with biochemical marker	231 (37.9%)
X-Linked	Total 124 (100.0%)
Disorders with enzyme deficiency	19 (15.3%)
Disorders with other biochemical marker	9 (7.3%)
Total number with biochemical marker	28 (22.6%)

Autosomal Dominant Inheritance

Pattern of Transmission

Inheritance of a disorder transmitted in autosomal dominant fashion is vertical, that is, through successive generations, from parent to child to grandchild. Dominant inheritance, in the older terminology, is "hereditary" rather than "familial." Males and females are affected with equal frequency and with equal severity. There is no increased frequency of parental consanguinity as seen with autosomal recessive inheritance. Children of an affected parent have a 50% risk of receiving the harmful gene and therefore a 50% risk of being affected (see Penetrance). Half-sibs through the affected parent have the same risk of being affected as full sibs, in sharp contrast to autosomal recessive diseases in which there is little risk to half-sibs. Male-to-male transmission occurs in autosomal dominant inheritance and should always be considered in reviewing pedigrees. Male-to-male transmission cannot occur in X-linked dominant pedigrees and may not be seen in small autosomal dominant pedigrees simply by chance.

Definitions of Dominance: Genotypes versus Phenotypes

Critical to the understanding of the differences between autosomal dominant and autosomal recessive inheritance is the concept of genotype and phenotype.

*Genotype** is a shorthand statement of whether an individual's two gene copies are normal or abnormal. A gene is normally found at a particular spot or place *(locus)* on a particular chromosome. Genes have different forms *(alleles)* which may be normal (often called *wild type*) or abnormal; a gene may have several or even many abnormal forms or abnormal alleles. Since an individual has only two copies of a particular gene (two chromosomes with one copy each), there are only four possible genotypes:

1. The individual has two normal alleles ("normal homozygote").
2. The individual has one normal and one abnormal allele ("heterozygote").
3. The individual has two abnormal alleles ("abnormal homozygote").
4. The individual has one each of two different abnormal alleles (*"compound 1heterozygote"* or *"genetic compound"*).

Phenotype, on the other hand, is a property of the system in which the gene operates; usually the system is the patient or family member. The phenotype is either "normal or "abnormal." For example, the patient may have the abnormal phenotype because he has symptoms of the disorder, because he has the physical findings (signs) of that disorder, or because clinical tests (such as roentgenograms, electromyography, or serum phenylalanine level) are abnormal.

Now, the difference between autosomal dominant and autosomal recessive inheritance can be easily explained. In both autosomal dominant and autosomal recessive inheritance patterns, normal homozygotes have the normal phenotype and abnormal homozygotes have the abnormal phenotype. The difference between the dominant and recessive patterns results from the phenotype of the heterozy-

*Most italicized words in this and later chapters are defined in both the text and the Glossary.

gote; if the heterozygote has the abnormal phenotype, the disorder is dominant; if the heterozygote has the normal phenotype, the disorder is recessive.

Dominant Lethals

This group of dominant disorders is mentioned separately because the inheritance pattern is different. *Dominant lethal* disorders are those in which an affected individual does not reproduce. The disorder is not necessarily fatal for the patient. However, because the patients do not reproduce, the disease cannot be transmitted from parent to child. Therefore all cases (in fact, nearly all cases) are sporadic and result from a new mutation of a parental germ cell. When the literature is searched and all published cases of a disorder are found to be sporadic, it is not reasonable to conclude therefore that the disorder is nongenetic. The happy practical result of this situation, however, is that recurrence risk for the next pregnancy is (nearly) zero for parents who have had a child affected with a dominant lethal disorder.

Penetrance and Expressivity

These features are especially characteristic of dominant inheritance rather than recessive traits. *Penetrance* is the fraction of individuals with an abnormal allele that actually has the abnormal phenotype. In individuals who are known to carry the abnormal gene but who have the normal phenotype, the abnormal gene is said to be *nonpenetrant.* The vast majority of individuals affected with dominant disorders in human populations carry a single rather than a double dose of the abnormal gene; that is, they are heterozygotes rather than abnormal homozygotes. For this reason, other factors in addition to the abnormal gene can play a greater role in determining the phenotype. Examples of these other factors include a second, normal, copy of the gene, other genes at other loci, and environmental factors.

Obviously, whether or not an individual is "affected" can be defined or determined in different ways. Different ways of determining the abnormal phenotype will give different numerical values for penetrance. If, for example, in a group of individuals carrying the gene for neurofibromatosis, only individuals with overt neurofibromas are said to be affected, penetrance will be rather low. If a careful search for cafe-au-lait spots is made, additional affected individuals will be uncovered. If axillary freckles and Lisch nodules are carefully searched for, more affected individuals will be found. Finally, if procedures such as MRI of the spinal cord, examination for scoliosis or bony deformities, or complete autopsy are carried out, nearly all the individuals carrying the abnormal gene will be detected and the penetrance of the disorder will be high.

Expressivity is the degree of clinical involvement in an individual with the abnormal gene. It is particularly characteristic of dominant disorders that two individuals carrying the same abnormal gene, perhaps sibs in the same family, may vary greatly in the severity of the manifestations. Obviously, nonpenetrance is an extreme form of *variable expressivity.*

The Sporadic Case

The commonest form of genetic disorder in (small) human families is probably the sporadic case. If the clinician waited for a second case in a family before suspecting a genetic disorder, most genetic disorders would escape diagnosis. The following is a partial list of diagnostic possibilities for the sporadic case:

Autosomal dominant (reduced penetrance)
Autosomal dominant (new mutation)
Autosomal recessive
X-Linked recessive (male only)
X-Linked dominant (female or male)
Multifactorial-threshold disorder
Polygenic inheritance
Nonpaternity
Adopted child
Nongenetic (phenocopy)

Autosomal Recessive Diseases

Pattern of Transmission

Inheritance of a disorder transmitted in autosomal recessive fashion is horizontal rather than vertical. Affected individuals are usually seen only in a single sibship (or "family," accounting for the older name "familial"); the parents are unaffected but are both heterozygotes for the harmful gene. Cousins, uncles, or aunts in collateral sibships may occasionally be affected. Of course, disorders occurring in a sibship are not necessarily genetic; for example, infectious disorders may cluster in a sibship, or there may be exposure to an environmental toxin. In autosomal recessive conditions, males and females are affected with equal frequency and with equal severity.

In autosomal recessive disorders there is an increased incidence of parental consanguinity, unlike dominant disorders. In general, the rarer the disorder, the greater the fraction of families with parental consanguinity. A corollary of this is that rare recessive disorders are most likely to be found in inbred genetic isolates, such as the Amish people in Pennsylvania. Recessive disorders therefore show ethnic predilections, which is helpful in diagnosis. A corollary is that heterozygotes for some autosomal recessive disorders have increased frequency in specific ethnic groups, which has made possible carrier testing for disorders such as Tay–Sachs disease, thalassemia, and sickle cell disease.

Couples at risk (which means that both parents are heterozygotes) are usually ascertained only after the birth of the first affected child. Each additional child of a couple at risk has a 25% chance of being affected and a 50% chance of being a heterozygous carrier. Unaffected sibs of an affected individual have a 67% chance of being heterozygous carriers if they are old enough to be sure that they are not themselves affected. Couples at risk for having children affected with autosomal recessive disorders can be detected before the birth of the first child if a carrier test is available and if the heterozygous carriers are common or at least common in a defined ethnic group. Voluntary carrier testing for Tay–Sachs disease in individuals of Ashkenazi Jewish background has been successful; classical infantile Tay–Sachs disease is now rarely seen because prenatal diagnosis can be made by amniocentesis in families in which the parents have been identified as heterozygotes. Half-sibs of an affected individual with an autosomal recessive disease have a small chance of themselves being affected, in contrast to autosomal dominant diseases in which the risk to half-sibs through the affected parent is the same (50%) as that of full sibs.

Definitions of Recessive: Genotypes versus Phenotypes

Definitions of genotypes and phenotypes are the same as those previously discussed for autosomal dominant diseases. However, recessive diseases are those in which the heterozygote is unaffected; that is, the heterozygote has the normal phenotype.

In plant or animal genetics the definitions of dominant inheritance may be somewhat different. Not only must the heterozygote be affected with a trait or disease, but the phenotype of the affected heterozygote and the affected abnormal homozygote must be identical or indistinguishable. As already mentioned, in human populations, which are relatively small, abnormal homozygotes are rarely encountered and individuals affected with autosomal dominant diseases are nearly always heterozygotes.

Compound heterozygotes, that is individuals with two different abnormal alleles at a locus, are particularly important for autosomal recessive diseases. A genetic compound may have a phenotype quite different from that of either corresponding abnormal homozygote. Consequently, in taking the family history of an apparently autosomal disorder, care must be taken not to ignore a cousin or perhaps more distant collateral relative who is affected but appears to have a disorder different from that of the patient; both affected individuals may have different forms of the same disorder, perhaps the same enzyme deficiency, one individual being an abnormal homozygote, the other being a genetic compound. A genetic compound may also have a phenotype identical with that of either corresponding abnormal homozygote. In fact, many or even most individuals who are apparently abnormal homozygotes with autosomal recessive disorders may, in fact, be genetic compounds, except where there is parental consanguinity or where the parents are part of a genetic isolate or members of a defined population in which heterozygote frequency for that disorder is high. It is important to recognize genetic compounds, where possible, because carrier testing or prenatal diagnosis in that family may be difficult or impossible without carrying out special tests.

Other Features of Autosomal Recessive Disorders

Variable penetrance and variable expressivity are far less important for autosomal recessive disorders than for autosomal dominant disorders. Nonpenetrance is quite unusual and variation in expressivity tends to be much smaller. When major differences in expressivity are seen, there may be other explanations, such as the presence of a genetic compound.

X-Linked Diseases

Pattern of Inheritance

X-Linked Dominant Inheritance. X-linked dominant pedigrees show vertical transmission and, at first glance, look like autosomal dominant pedigrees. However X-linked dominant pedigrees can be identified by several characteristics:

1. Male-to-male transmission does not occur.
2. All daughters of an affected male are affected.
3. Females are more frequently affected than males.
4. Females are less severely affected than males.
5. Occasional female heterozygotes show nonpenetrance (probably as a result, by chance, of preponderant Lyon inactivation of the X-chromosome that carries the abnormal gene).

Because X-linked dominant pedigrees look so much like autosomal dominant pedigrees, X-linked dominant inheritance can easily be overlooked unless every apparently autosomal dominant pedigree is carefully examined for the features just mentioned.

X-Linked Recessive Inheritance. X-linked recessive inheritance somewhat resembles autosomal recessive inheritance, particularly if only a single sibship is considered. However, in larger pedigrees the pattern is different from either autosomal recessive or autosomal dominant inheritance. Transmission is "diagonal" rather than vertical or hozizontal; affected males are connected on the pedigree through unaffected females. Only males are affected, but there is no male-to-male transmission. Occasional female heterozygotes may be affected (probably as a result, by chance, of preponderant Lyon inactivation of the X-chromosome that carries the normal gene).

X-Linked Diseases: Genotypes versus Phenotypes

Definition of phenotypes is the same for X-linked disorders as for autosomal disorders; and the phenotype is either normal or abnormal (affected). For females, the possible genotypes are the same as for autosomal disorders: the normal homozygote, the heterozygote, the abnormal homozygote, and the genetic compound all as previously defined. However, definition of the genotypes is somewhat different for males. Because males have only one X-chromosome, they are hemizygotes; depending upon whether the X-chromosome carries the normal or abnormal allele, the male is a normal or abnormal hemizygote, for both genotype and phenotype.

The Lyon Hypothesis

A body of knowledge has accumulated about the peculiar behavior of the X-chromosome in mammalian females, a pattern explained by the hypothesis of Mary Lyon. X-inactivation in the mammalian female has the following features.

Early in the development of a female (about 15–20 days in humans), one of the two X-chromosomes is inactivated. The choice of which X-chromosome is inactivated is random. The X-inactivation is stable, and subsequent progeny of each X-inactivated cell have the same X-chromosome inactivated. Germ cell X-chromosomes are not inactivated. A small part of the inactivated X-chromosome remains active (or is not inactivated). This concept of X-inactivation has been well documented by a large body of work.

At least three aberrations of Lyonization could explain why some female heterozygotes are unaffected in X-linked dominant diseases, or why some female heterozygotes are affected in X-linked recessive diseases, as explained earlier.

First, Lyon inactivation is random, but this does not mean that exactly 50% of the cells in a female heterozygote have the normal allele active and 50% of the cells have the abnormal allele active. In fact, a small number of heterozygotes for X-linked recessive disorders have a preponderance of cells in which the abnormal allele is active, and the women may therefore be affected clinically. Also, a small number of heterozygotes for X-linked dominant disorders have a preponderance of cells in which the normal allele is active, and those women may therefore escape being affected clinically.

Second, some diseases affect the choice of which X-chromosome is inactivated. For example, in some types of X-autosome translocation, the translocated X-chromosome cannot be inactivated, and, therefore, the normal X-chromosome is always inactivated. This pattern explains why some girls or women are affected by X-linked recessive diseases—for example, Duchenne muscular dystrophy. If one X-chromosome has a damaged Duchenne gene because the gene is cut by a translocation, and if the normal X-chromosome therefore always undergoes Lyon inactivation, the girl has no functioning Duchenne gene and develops muscular dystrophy.

Third, although the initial X-inactivation is random, selection may act during development to affect cells that express different alleles of genes on the X-chromosome. For example, a cell expressing an abnormal X-linked allele may divide less rapidly than cells bearing the normal allele; as a result, relatively few of the abnormal cells are found in the adult heterozygous woman. Alternatively, cells expressing an abnormal X-linked allele may be selected *for*, so that the cells carrying the abnormal gene predominate in the adult woman. This may be the case in adrenoleukodystrophy, accounting for the fact that abnormal substrate accumulation is usually detected, even in the heterozygote.

NON-MENDELIAN INHERITANCE

A number of patterns of simple inheritance have been described that follow none of the classical Mendelian patterns discussed above. Three of these are discussed below: the metabolic interference pattern, the multifactorial-threshold inheritance pattern, and the mitochondrial inheritance pattern.

Metabolic Interference

Metabolic interference is a hypothetical mechanism in which two alleles at a locus, or two alleles of genes at different loci, cause a harmful effect only when they are present together in the same individual (6). That is, in the simple form of metabolic interference, only the heterozygote is affected; both of the corresponding homozygotes are unaffected. The pedigrees that result from this are in some instances rather conventional: for example, any apparently autosomal dominant pedigree could in fact result from metabolic interference. In some instances, however, these pedigrees would be strikingly unusual:

1. A disorder limited to females, apparently dominant or recessive, particularly a disorder passed to affected females through unaffected males (Table 3-2)

Table 3-2 Diseases Occurring Only or Largely in Females and Diseases Compatible with X-Linked Dominance But Not Worse in Males

Diseases occurring only or largely in females	
	Myopathy limited to females (16060, 30995)[a]
	Congenital cataract with microcornea or slight microphthalmia (30230)
	Agenesis of the corpus callosum with chorioretinal abnormality (30405)
	Focal dermal hypoplasia (Goltz syndrome) (30560)
	Incontinentia pigmenti (30830)
	Oral–facial–digital syndrome (31120)
	Sacral defect with anterior sacral meningocele (31280)
	Wildervanck syndrome (cervicooculoacoustic syndrome) (31460)
	Cunier color vision anomaly (9)
Diseases compatible with X-linked dominance	
	Albright hereditary osteodystrophy (30080)
	Hyperuricemia, ataxia, deafness (30720, 23995)
	Manic depressive psychosis (30920)
	Pterygium syndrome (X-linked form) (31215)

[a]Five-digit numbers refer to McKusick's catalogue (5).

(disorders limited to females may also result from X-linked dominant inheritance with male lethality, that is, the more severely affected males do not survive to term).

2. A disorder seen in all members of a large sibship with normal parents.
3. A disorder occurring in all members of a large sibship with one parent similarly affected.
4. An apparently dominant disorder with females more severely affected than males.
5. An apparently X-linked dominant disorder in which males are not more severely affected.

Several pedigrees have been reported that can be explained by the metabolic interference hypothesis but are not easily explained by any other known pattern (7,8).

Multifactorial-Threshold Diseases

Mechanism of Multifactorial-Threshold Inheritance

These diseases are rather common in the population, with frequencies of approximately one per thousand (0.1–5.0 per thousand births). However, they are often not recognized as being genetic because recurrence risks in a family are in the range of 2–20 per thousand, far lower than those for any of the Mendelian disorders discussed earlier.

The mechanism involves the interaction between a continuous variable (the genetic factor) and a discontinuous variable (an environmental factor). The genetic factor may be thought of as susceptibility to the disease; quantitatively, the degree of susceptibility is distributed in the population roughly according to a normal distribution curve. A threshold then operates on this susceptibility factor so that if the susceptibility is below the threshold, the phenotype is normal. However, if the susceptibility factor is above the threshold, the phenotype is abnormal. The presence

Table 3-3 Multifactorial Disorders

Neural tube defects	Peptic ulcer disease
Multiple sclerosis	Hirschprung disease
Mental retardation	Talipes equinovarus
The epilepsies	Cleft lip ± palate
Hydrocephalus	Cleft palate
Strabismus	Diabetes mellitus
Schizophrenia	Pyloric stenosis
Manic depressive (bipolar)	Rheumatic fever
Congenital heart malformations	Celiac disease
Congenital hip dislocation	Legg-Perthes
Scoliosis (idiopathic)	Exomphalos
HLA-related disorders	Psoriasis

of a critical threshold converts the continuous variation into a discontinuous variation. Examples of disorders that fit this model are neural tube defects, cleft lip, cleft palate, psoriasis, and pyloric stenosis among many others (Table 3-3).

Pattern of Inheritance

As mentioned, the pattern of multifactorial inheritance is quite different from the Mendelian patterns just discussed (Table 3-4). In Mendelian pedigrees, the risk to the next child is independent of the number of affected children. However, in a family with a multifactorial disease, the risk that the next child will be affected increases with the number of children in the family who are already affected. This is so because parents probably have a higher degree of genetic susceptibility if two children have already crossed the threshold and are affected than if only one child has crossed the threshold and is affected. Parents with two affected children are more likely to transmit greater genetic susceptibility to the next child than are parents with one affected child.

In addition, in Mendelian pedigrees, the risk to the next child is independent of the severity of the disease in a relative. However, in a family with a multifactorial disease, the risk that the next child will be affected increases if the already affected relative is severely affected rather than mildly affected. This is so because parents are likely to have higher genetic susceptibility if a child is severely affected than if a child is mildly affected. Parents of a severely affected child are more likely to transmit greater genetic susceptibility to the next child than are parents of a mildly affected child.

Table 3-4 Multifactorial-Threshold Pattern

Recurrence risk for next child is low compared with Mendelian traits, in the range of 0.5–3%
The more family members affected the higher the recurrence risk
The more severely affected the index case the higher the recurrence risk
Risk and frequency of disorder often vary with sex, geographic area, or population
Risk in relatives decreases faster than degree of relationship
Risk is higher if index case is of the less frequently affected sex

The frequency of multifactorial diseases characteristically varies between the sexes, between different ethnic groups, and between different geographical areas. For example, the incidence of pyloric stenosis is five times as great in males as in females. For anencephaly, the male-to-female incidence ratio is 0.6. Neural tube defects have increased incidence in Wales and Ireland.

A surprising result of these features is that risk for a multifactorial disease to relatives of a patient with that disease is higher if the patient is of the less frequently affected sex. For example, the risk of pyloric stenosis for the son of an affected father is 5.5 per thousand, whereas the risk to the son of an affected mother is 18.9 per thousand. This is so because females, less frequently affected, must carry a greater genetic susceptibility than males to cross the threshold and have the abnormal phenotype. Therefore, an affected female is like to carry greater genetic susceptibility than an affected male and is likely to transmit greater genetic susceptibility to her offspring than an affected male.

Mitochondrial Inheritance

Mitochondrial Inheritance Pattern

Although most mammalian DNA resides in the nucleus, mitochondria contain DNA that behaves, in some ways, differently from nuclear DNA (10,11). Although no human disease has been definitely shown to result from mutational damage to mitochondrial DNA, the pattern of segregation of mitochondrial markers in human pedigrees is known (12,13) from the study of normal RFLP variants. The pattern of mitochondrial inheritance is full maternal inheritance. All children, male and female, receive their mitochondria, and with it their mitochondrial DNA, from the mother. None of a father's children receives any of his mitochondrial DNA (to the limits of studies thus far). The inheritance pattern of mitochondrial diseases is expected to follow this pattern. That is, all children, male or female, of an affected mother should be affected. No children, male or female, of an affected father should be affected. In the case of inbred mouse strains, which differ with respect to mitochondrial markers, the cross A × B gives all progeny resembling A; the cross B × A gives all progeny resembling B.

Some human diseases (Table 3-5) show a maternal inheritance pattern that is compatible with that described for mitochondrial markers (14–18). Several other human diseases (Table 3-6) are suspected of having a mitochondrial component to the phenotype (see Chapter 21 by DiMauro).

Table 3-5 Possible Mitochondrial Diseases

Disease	Inheritance
Mitochondrial myopathies	Maternal
Leber optic atrophy	Maternal
Progressive myoclonus epilepsy	Maternal
Cardiomyopathy	Maternal
Chloramphenicol-induced blood dyscrasia	?Maternal

Source: Adapted from Merril and Harrington (14).

Table 3-6 Diseases with Possible Mitochondrial Influence

Disease		Inheritance
Huntington disease	AD	Late-onset form twice as commonly inherited from mothers
Myotonic dystrophy	AD	Congenital form inherited from mothers
Cerebellar ataxia	AD	Later age of onset and death if inherited from mother
Malignant melanoma	—	Increased severity in offspring of affected mothers
Neurofibromatosis	AD	Increased severity in offspring of affected mothers
Spina bifida and anencephaly	MF	Increased recurrence risk in maternal lineage
Abnormal systolic blood pressure	MF	Maternally inherited component

Note: AD, autosomal dominant; MF, multifactorial.
Source: Adapted from Merril and Harrington (14).

Disorders with Unexplained Inheritance Pattern

It is probably unwarranted to attempt to force all recalcitrant pedigrees into known inheritance patterns. Human inheritance is likely to contain further surprises and new mechanisms of inheritance may be found in the future. Some pedigrees are still unexplained (9).

REFERENCES

1. Johnson WG. Principles of genetics in neuromuscular disease. In Kelley VC, ed. Practice of Pediatrics, Vol 4. New York: Harper & Row, 1979.

2. Rosenberg RN. Neurogenetics: Principles and Practice. New York: Raven Press, 1986.

3. Vogel F, Motulsky AG. Human Genetics, Problems and Approaches. New York: Springer-Verlag, 1979.

4. Nora JJ, Fraser FC. Medical Genetics: Principles and Practice. Philadelphia: Lea & Febiger, 1981.

5. McKusick VA. Mendelian Inheritance in Man, 7th Ed. Baltimore: The Johns Hopkins University Press, 1986.

6. Johnson WG. Metabolic interference and the +/− heterozygote. A hypothetical form of simple inheritance which is neither dominant nor recessive. Am J Hum Genet 1980; 32:374–386.

7. Vaillaud JC, Martin J, Szepetowski G, et al. Le syndrome oro-facio-digital. Etude clinique et génétique à-propos de 10 cas observe dans une même famille. Rev Pediatr 1968; 4:303–312.

8. Rollnick B, Day D, Tissot R, Kaye C. A pedigree: Possible evidence for the metabolic interference hypothesis. Am J Human Genet 1981; 33:823–836.

9. Stern C, Walls GL. The Cunier pedigree of "color blindness." Am J Hum Genet 1957; 9:249–273.

10. Lane MD, Pedersen PL, Mildvan AS. The mitochondrion updated. Science 1986; 234:526–527.

11. Vawter L, Brown WM. Nuclear and mitochondrial DNA comparisons reveal extreme rate variation in the molecular clock. Science 1986; 234:194–196.

12. Giles RE, Blanc H, Cann HM, Wallace DC. Maternal inheritance of human mitochondrial DNA. Proc Natl Acad Sci USA 1980; 77:6715–6719.

13. Case JT, Wallace DC. Maternal inheritance of mitochondrial DNA polymorphisms in cultured human fibroblasts. Somatic Cell Genet 1981; 7:103–108.

14. Merril CR, Harrington MG. The search for mitochondrial inheritance of human diseases. Trends Genet 1985; 1:140–141.

15. Egger J, Wilson J. Mitochondrial inheritance in a mitochondrially mediated disease. New Engl J Med 1983; 309:142–146.

16. Holliday PL, Gilroy J. Mitochondrial inheritance. New Engl J Med 1983; 309:1583–1584.

17. Mechler F, Fawcett PRW, Mastaglia FL, Hudgson P. Mitochondrial myopathy—A study of clinically affected and asymptomatic members of a six-generation family. J Neurol Sci 1981; 50:191–200.

18. Rosing HS, Hopkins LC, Wallace DC, Epstein CM, Weidenheim K. Maternally inherited mitochondrial myopathy and myoclonic epilepsy. Ann Neurol 1985; 17:228–237.

II

THE LANGUAGE AND TOOLS OF MOLECULAR GENETICS

4

Prokaryotic and Eukaryotic Genomes

ERIC A. SCHON

The "central dogma" of molecular biology states that "DNA encodes RNA and RNA encodes protein." In other words, deoxyribonucleic acid, or DNA, which is the stable genetic material that is passed, via *replication,* from one generation to the next, is a template for the *transcription* of DNA into ribonucleic acid, or RNA. RNA is an essentially unstable compound that is rapidly degraded and cannot be inherited. Nevertheless, RNA literally "carries" the information contained in DNA throughout the cell. These "messenger" RNAs are templates for *translation* of the information into the synthesis of proteins, which are, directly or indirectly, the fundamental molecules of all life processes. All three molecules—DNA, RNA, and protein—are "isomorphs": they contain variations of the same information, as encrypted by variations in structure. In the same way, the notes of a musical composition (seen but not heard) and the music itself (heard but not seen) are isomorphic structures.

In this chapter I shall describe the structure and organization of DNA in both prokaryotes (organisms without a nucleus) and eukaryotes (organisms with a nucleus). The other elements of the central dogma—transcription and translation—are presented in the succeeding chapters.

THE STRUCTURE OF DNA

The fundamental unit of DNA is the nucleotide (Fig. 4-1). It consists of three elements: a cyclic base, either purine or pyrimidine, a pentose, or five-membered sugar ring, and a phosphate group.

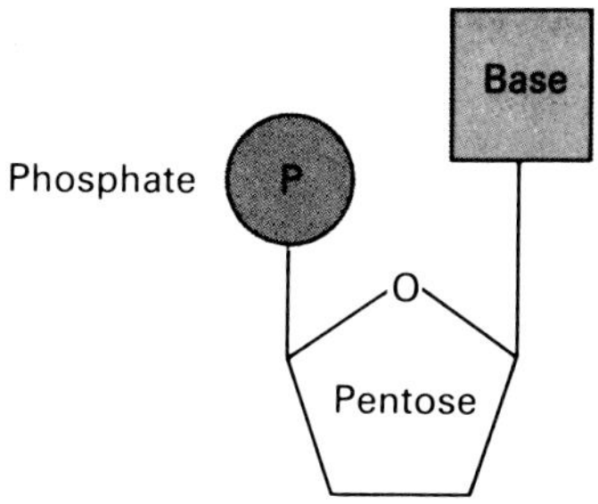

Fig. 4-1. The three elements comprising a nucleotide: a base and a phosphate group attached to a pentose sugar. (Adapted with permission from Molecular Cell Biology by J. E. Darnell et al., Copyright © 1986, Scientific American.)

There are five types of bases (Fig. 4-2). Two are purines: adenine (A) and guanine (G). The other three are pyrimidines: cytosine (C), thymine (T), and uracil (U). The carbon and nitrogen atoms of the ring structures are numbered from 1 through 9 in the bicyclic purines, and from 1 to 6 in the monocyclic pyrimidines. These bases are attached to the pentose of the nucleotide at the nitrogen in position #9 of the purine (N-9) or the #1 position of the pyrimidine (N-1).

To identify the carbon atoms of the pentose ring, a different set of numbers is used, the "prime" notation, from 1′ to 5′ (Fig. 4-3). The base is attached to the 1′-carbon of the pentose, while the phosphate moiety (Fig. 4-4) is attached to the 5′-carbon. The ionizable phosphate group makes both DNA and RNA *nucleic acids.*

In RNA, the pentose is *ribose* (Fig. 4-3); there are hydroxyl groups at both the 2′ and 3′ positions on the ring, as well as on the 5′-carbon. RNA is therefore called *ribonucleic acid.* However, in DNA, the OH group at the 2′ position is missing (it is replaced by a hydrogen atom); the pentose in DNA is thus *deoxyribose,* and the linear biopolymer is therefore called *deoxyribonucleic acid.*

The bases of the nucleotides of RNA are A, G, C, and U; those of DNA are A, G, C, and T. That is, uridine is found only in RNA, and thymidine only in DNA. In one of their major contributions, Watson and Crick saw the possible formation of hydrogen bonds between a purine and a pyrimidine (Fig. 4-5). In particular, the purine adenine can base pair with the pyrimidines thymine or uracil, and the purine guanine can base pair with the pyrimidine cytosine. Three pairs of atoms are involved in the hydrogen bonding of G-C, a stronger bond than in A-T or A-U, which have only two pairs of bonding atoms. The A-T, A-U, and G-C pairs are *complementary,* and are the most stable and thermodynamically favored; sometimes, however, there are other purine–pyrimidine pairs, particularly in RNA, but these are weaker than the "natural" Watson–Crick pairs.

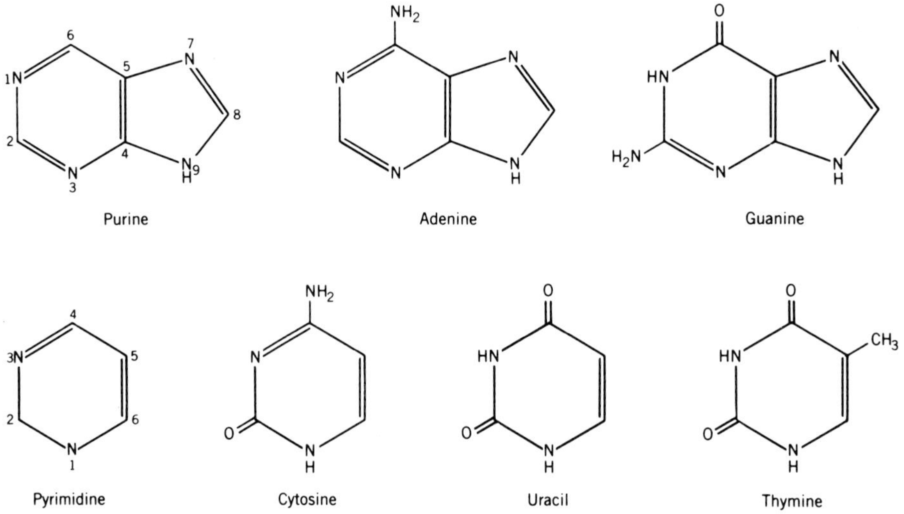

Fig. 4-2. The structure of the purine (adenine and guanine) and pyrimidine (cytosine, uracil, and thymine) bases in DNA and RNA. (Adapted with permission from Genes II by B. Lewin, Copyright © 1985, John Wiley.)

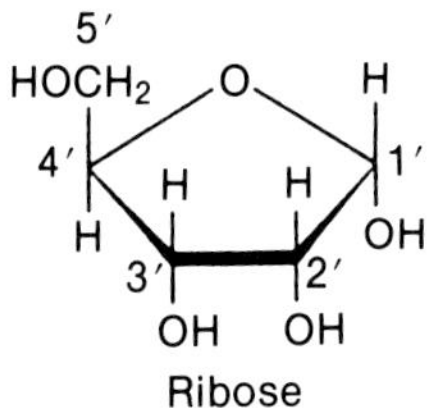

(a)

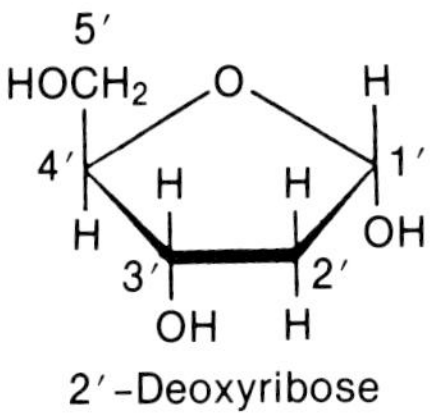

(b)

Fig. 4-3. The structure of (a) ribose and (b) 2′-deoxyribose.

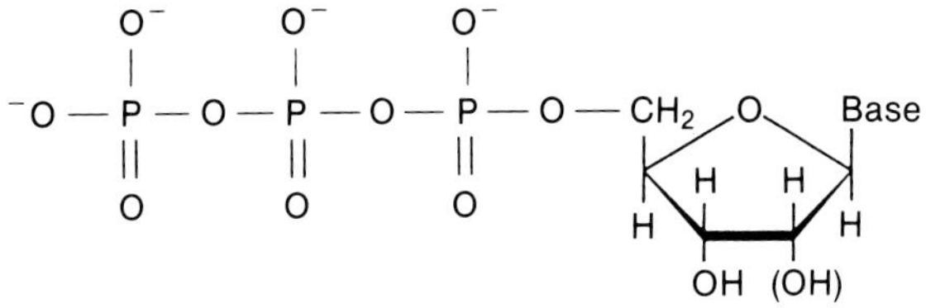

(a)

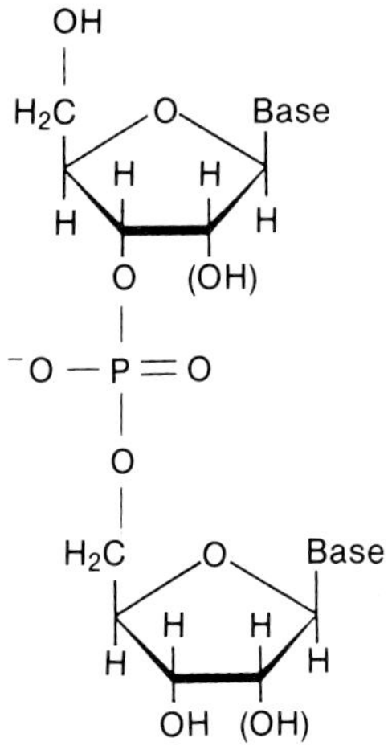

(b)

Fig. 4-4. The structure of the phosphate group. (a) As the triphosphate in free nucleotides, and (b) as the monophosphate in DNA/RNA polymers.

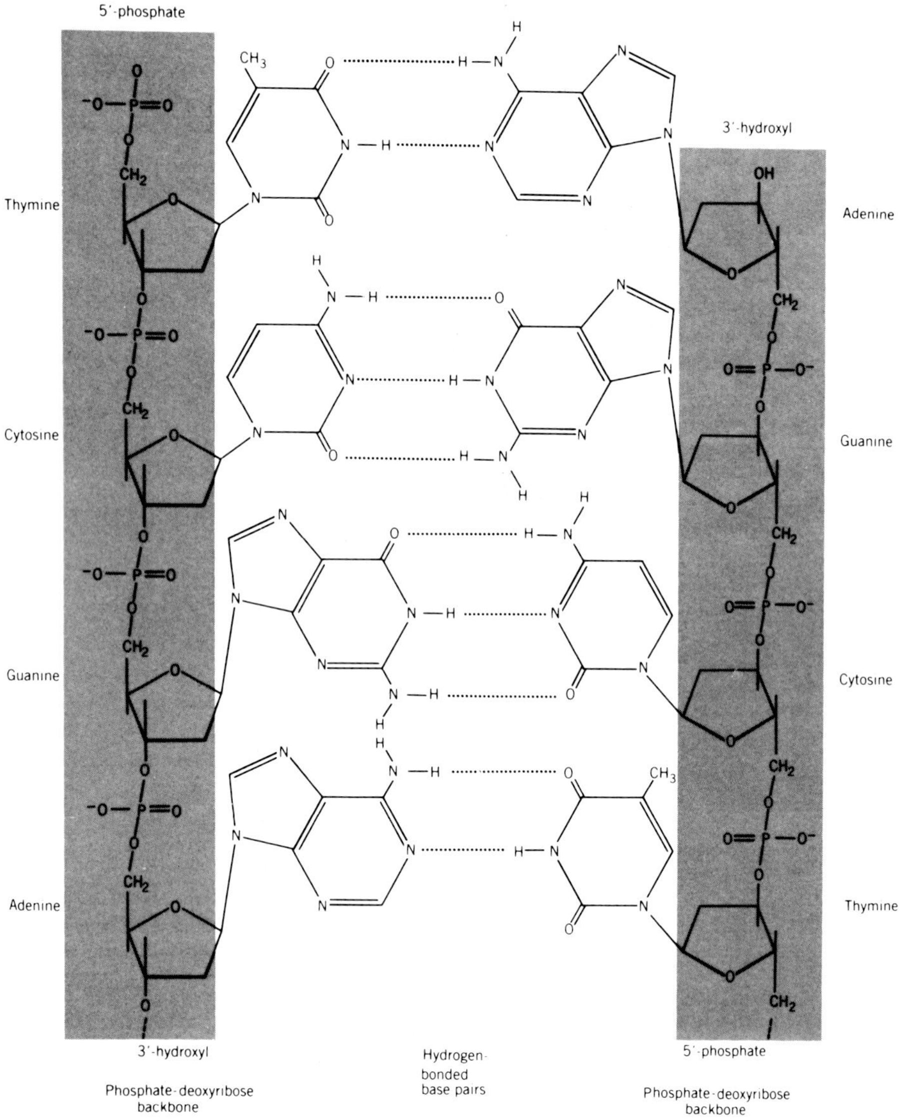

Fig. 4-5. Hydrogen bonding between A-T and G-C pairs (dotted lines). Note that there are two hydrogen bonds between A and T, but three between G and C. (Adapted with permission from Genes II by B. Lewin, Copyright © 1985, John Wiley.)

The individual nucleotides polymerize to form nucleic acids, by esterification between the phosphate group at the 5′ end of one nucleotide and the hydroxyl group at the 3′ end of a second nucleotide. These *phosphodiester bonds* link the nucleotides to form the linear polymer (Fig. 4-6). Because the phosphodiester bond between any two nucleotides involves a 5′–3′ link, the overall linear polymer has a *directionality;* the linkage is conventionally written in the direction 5′ to 3′. Two individual strands of nucleic acid with complementary bases can form a series of hydrogen-bonded bases that bring the two strands together to form a *duplex.* The

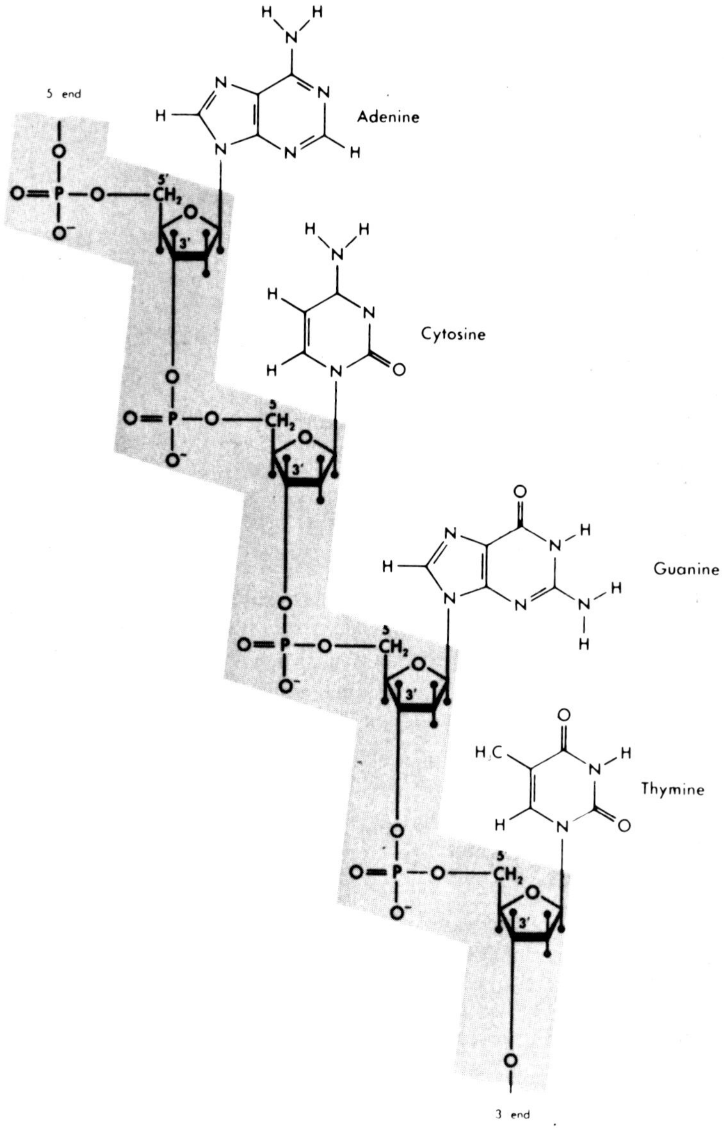

Fig. 4-6. A polymer of DNA consists of linked nucleotides running 5′ to 3′. (Adapted with permission from Molecular Biology of the Gene by J. D. Watson et al., Copyright © 1987, Benjamin/Cummings.)

two strands bond effectively only if the strand running 5′ to 3′ bonds the complementary strand running 3′ to 5′. In other words, the duplex consists of two *antiparallel* strands (Fig. 4-7).

In addition, the two strands do not line up side by side, but normally wind around each other in a right-handed spiral—the famous α *helix* (Fig. 4-8). The sugar phosphates lie on the exterior aspect of the helix; the bases are found in a planar orientation, perpendicular to the central helical axis, and face inward. There are just over 10 base pairs in each helical turn, with a "rise" of 0.34 nm between

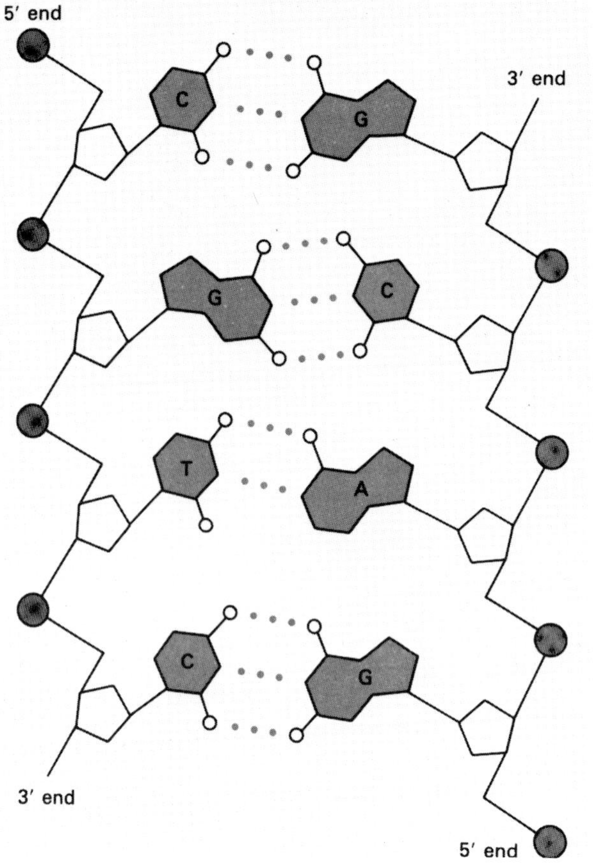

Fig. 4-7. A DNA duplex. Note that the two strands are antiparallel to each other. (Adapted with permission from Molecular Cell Biology by J. E. Darnell et al., Copyright © 1986, Scientific American.)

bases (about 3.4 nm/helical turn). Under some conditions the DNA is not in the usual orientation (known as B-DNA). Other, less common forms of right-handed DNA also exist, such as A-DNA, and there is a left-handed form called Z-DNA (Fig. 4-8), because of the zig-zag pattern of the sugar phosphate backbone.

The duplex nature of DNA with complementary strands is the basis of its genetic role. The two complementary strands are *isomorphs* and, if the sequence of bases on one strand is known, the base-pairing rules can be used to deduce automatically the [antiparallel] sequence of the other strand. Therefore, each strand can be replicated to make a new duplex; the information on one strand is imparted to make the complementary, isomorphic, partner.

DNA REPLICATION

How is replication accomplished? Most of what we know comes from studies of viral and bacterial replication. It is assumed that the process is virtually the same

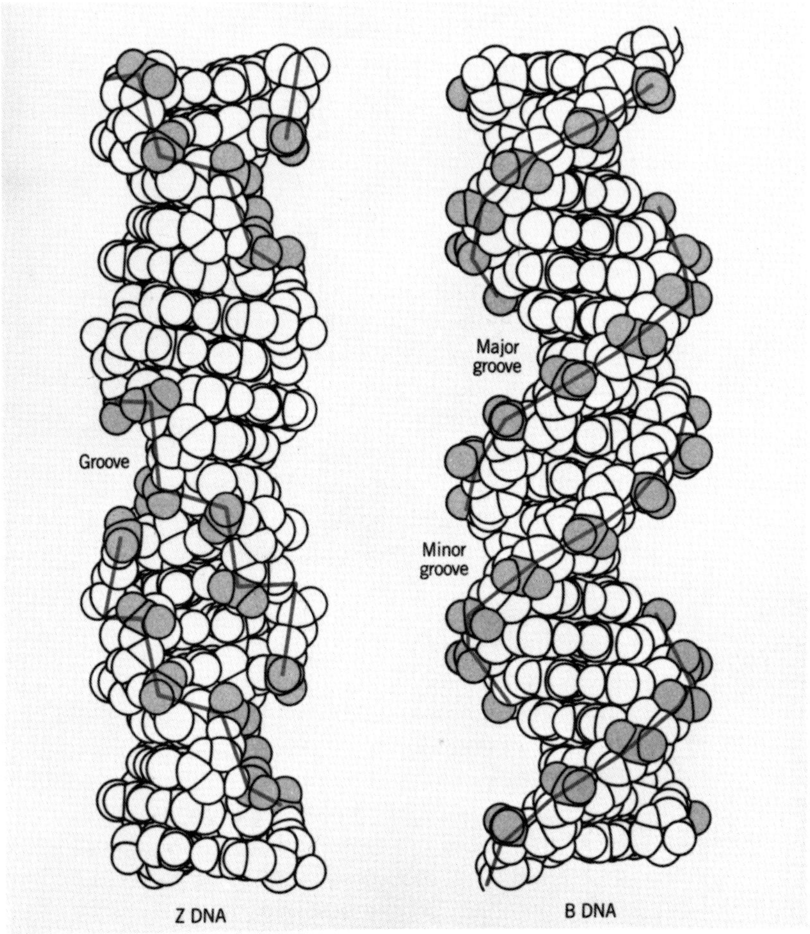

Fig. 4-8. A comparison of B- and Z-DNA. (Right) Duplex B-DNA shown as a right-handed α helix. Note the major (broad) and minor (narrow) grooves and the smooth pattern (dark lines) of the sugar phosphates (shaded atoms) on the exterior of the helix. (Left) Left-handed Z-DNA. Note the zig-zag pattern of the sugar phosphates. (Adapted with permission from Genes II by B. Lewin, Copyright © 1985, John Wiley.)

in eukaryotes in spite of the vastly greater amount of that is replicated in eukaryotic DNA, as discussed under Eukaryotic Genomes.

Although the enzymology is complex, the strategy of DNA replication is simple. The two strands of the duplex are unwound in a *replication bubble,* in which each exposed strand becomes a template for polymerization of new complementary DNA. As the unwinding proceeds, the bubble becomes ever larger, and a pair of new duplexes is formed behind each *replication fork.* This is a *semiconservative* mode of replication, that is, each new daughter duplex consists of one old parental strand and one newly synthesized strand (Fig. 4-9. The original strand is preserved, no matter how many times it serves as template for the synthesis of a new strand.

Replication from a replication bubble proceeds in both directions from the two

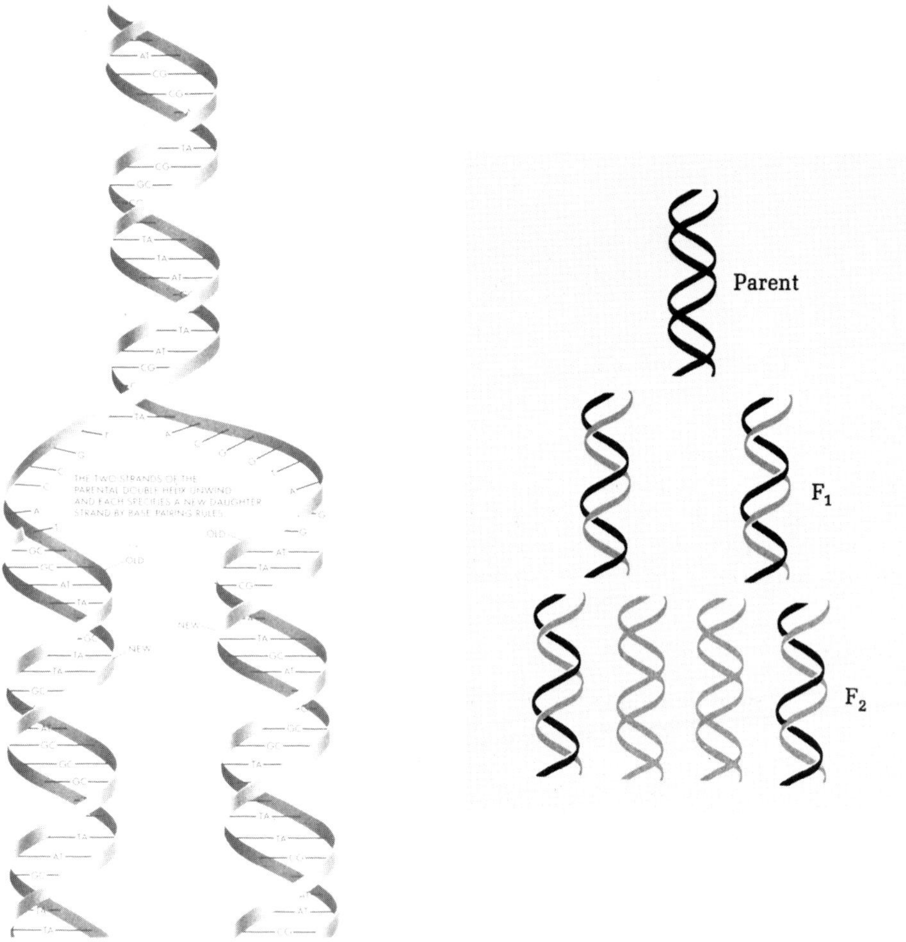

Fig. 4-9. Replication of DNA. (a) Each parental strand is the template for the synthesis of a new complementary strand. (Adapted with permission from Recombinant DNA: A Short Course by J. D. Watson et al., Copyright © 1983, Scientific American.) (b) Replication is semiconservative: each daughter strand consists of one old parental strand and one newly synthesized strand. F_1 and F_2 are successive generations. (Adapted with permission from Biochemistry by A. L. Lehninger, Copyright © 1975, Worth Publishers.)

forks growing to the left and right, called "Cairns" replication after the discoverer. If the DNA template is a circle, as in most viral and bacterial genomes, a single growing bubble forms a "theta structure" as the two forks advance around the circle; just before the two forks meet, there is a "figure-eight" structure (Fig. 4-10). In rapidly growing bacteria, new replication bubbles may appear on newly synthesized daughter duplexes even before the theta structure resolves into two new circular DNAs. Eukaryotes have much longer, linear DNAs; multiple replication forks are interspersed along the length of the template. As the bubbles grow, the forks advance to meet one another, all the while synthesizing new DNA behind them, until all the forks fuse to create two new daughter duplexes.

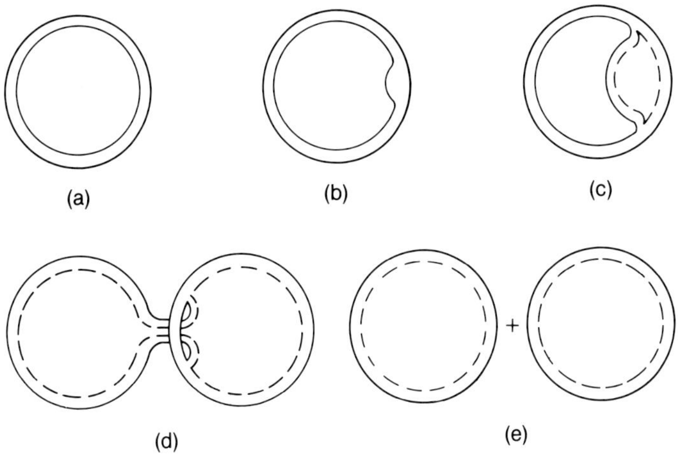

Fig. 4-10. "Cairns" replication. (a) The initial circular duplex; (b) a replication bubble opens up at a specific site on the circle—the "origin of replication"; (c) a "theta structure" early in replication showing parental (solid lines) and daughter (dotted lines) DNA; (d) a "figure-eight" structure late in replication; (e) resolution into two daughter circles.

GENOMES AND GENES

A *genome* can be defined as the total heritable informational content of an organism, as contained in the DNA. This definition is somewhat ambiguous in eukaryotes, which contain DNA not only in the nucleus, but also in other intracellular compartments, such as mitochondria and chloroplasts, as discussed under Eukaryotic Endosymbionts. It is therefore possible to consider the nuclear and mitochondrial genomes as separate entities, or to consider the entire genome as comprising all forms of DNA in the cell.

The definition of a *gene* is less precise, because the ever-increasing knowledge of genome organization makes it necessary to revise the concept. Perhaps the best definition is that a *gene* is an integral unit of "usable" genetic information. As such, the genome contains regions of DNA that are genes and regions that are not. In prokaryotes most of the genome is composed of genes, but in eukaryotes this is not so, because an enormous amount of intergenic (and even intragenic) DNA seems to serve no obvious role. The genes themselves consist mainly of DNA that encodes single polypeptide chains, but some genes encode RNA but not protein: for example, some genes encode only ribosomal and transfer RNAs that are used in protein translation.

It is convenient to represent the genome as a map, either linear or circular, with genes placed at appropriate intervals along its length. The earlier maps were *genetic maps,* in which the genes were ordered on the basis of *recombination frequency.* "Recombination" is a functional concept that will be discussed in Chapter 13 on *linkage.* Here it will suffice to say that the closer the interval between two genes on a stretch of DNA, the more likely they, and their associated phenotypic characteristics, will be transmitted together. Conversely, the farther apart the two genes, the more likely they will be transmitted separately, due to a "recombination event" that unlinks them. These distances are expressed in centimorgans (cM): 1 cM equals 1% of recombination. With the advent of recombinant DNA technology, the genetic maps were refined and became *physical maps,* in which the individual genes were ordered on the basis of known features or landmarks in the DNA, such as the location of restriction endonuclease sites, as described in Chapter 8. The ultimate physical map is the *DNA sequence,* in which every base of the DNA is known and ordered. Both the physical map and the total DNA sequence have now been established for several small genomes, including those of simian virus 40 (SV40) and bacteriophage λ.

PROKARYOTIC GENOMES

Genomic Organization

Prokaryotic genomes are small and compact. The genome of *Escherichia coli,* the most commonly used bacterium in molecular genetic studies, has a circular genome of 3–4 million base pairs (bp). There are about 3000 genes; half of them have been located on the circular map (most by classical genetic methods, but a growing number by recombinant DNA techniques). The length of an average gene is about 1500 bp, or 1.5 kilobase pairs (kb); therefore, almost the entire genomic complement of *E. coli* comprises structure genes, with little "nonessential" DNA.

The genomes of bacterial viruses, or *phages,* and of bacterial extrachromosomal, or *episomal,* elements, such as the circular *plasmids* that encode antibiotic resistance, are far smaller than the host bacterial genome, by a factor of 100–1000 (Table 4-1). Both phages and plasmids are used in recombinant DNA technology as *vectors* that can carry the recombinant, nonbacterial sequences of DNA (see Chapter 9).

Plasmids replicate autonomously within the bacterial host, but require some host protein functions for DNA replication, RNA transcription, and protein translation. The same may be said for the phages, except the relationship between phage and bacterium is far from benign; phages are true bacterial viruses and may kill the host bacteria.

Gene Structure

The organization of a prokaryotic gene is relatively simple (Fig. 4-11). Each protein-coding region of DNA is flanked at the 5′ and 3′ ends by regulatory regions of

Table 4-1 Some Phage and Episomal Genomes

Name	Genome type	Size (bp)	Comment
PiAN7	Circular plasmid	885	"Recombination" vector
pUC19	Circular plasmid	2,686	Cloning vector
pBR322	Circular plasmid	4,363	Cloning vector
φX174	Circular phage	5,386	
M13mp18	Circular phage	7,250	DNA sequencing vector
T7	Linear phage	39,936	
λ	Linear phage	48,502	Cloning vector
T4	Circular phage	165,000	

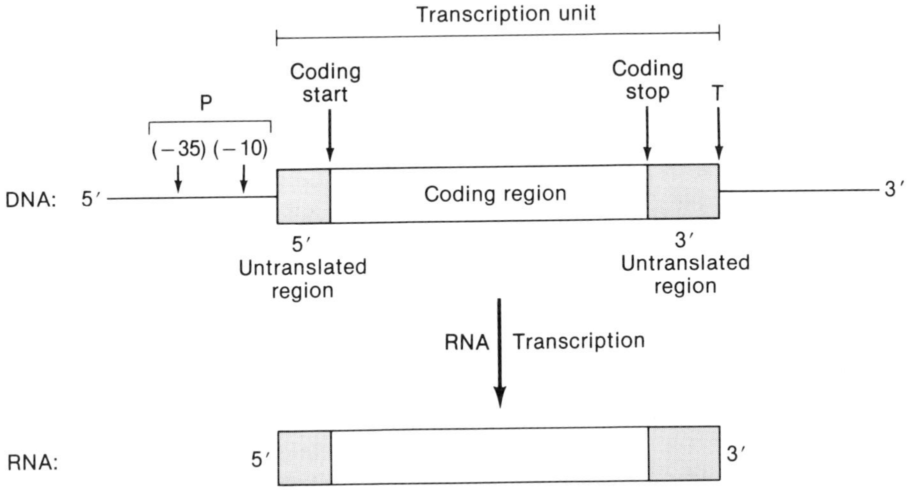

Fig. 4-11. Organization and transcription of a typical prokaryotic gene. The coding region (open box) is flanked by a transcribed but untranslated sequence (shaded boxes); it begins with the translational start codon (methionine) and ends with one of the three stop codons (UAA, UGA, or UAG). The nontranscribed region (solid lines) flanking the gene contains transcriptional promoter elements (P) 10 and 35 bp prior to the transcriptional start site at the 5′ end and termination signals (T) at the 3′ end.

DNA. The direction or polarity of genes is conventionally represented as reading 5′ to 3′, from left to right, in the "message-sense" of the transcribed RNA.

A *transcription unit* may be defined as the distance between sites of initiation and termination of a single RNA transcript; sometimes the transcript includes more than one gene. In structural protein-coding genes, the transcription unit contains signals that specify the beginning and the end of transcription; these are called the *promoter* and the *terminator.* The coding region has a "start" and "stop" signal for the final protein product. In the RNA, the *codon* triplet AUG encodes methionine (Met), which is the amino acid found at the amino terminus of most proteins. All proteins terminate at the carboxy end when the cell's translation system encounters one of the three "stop" codons, UAA, UAG, or UGA. Between the Met and the stop signals is the coding region itself, the RNA sequence that is read by the translation system in triplets; each triplet encodes a particular amino acid as defined by the genetic code (Fig. 4-12). The transcription unit, longer than the coding region, is flanked at both the 5′ and 3′ ends of the message by regions of 5′- and 3′-untranslated sequence that are specified by the promoter and terminator in the DNA.

EUKARYOTIC GENOMES

Three major features distinguish eukaryotic genomes from prokaryotic ones: differences in gene structure, genome size, and overall genomic organization.

FIRST BASE	SECOND BASE: U	SECOND BASE: C	SECOND BASE: A	SECOND BASE: G
U	UUU, UUC Phe UUA, UUG Leu	UCU, UCC, UCA, UCG Ser	UAU, UAC Tyr UAA, UAG TERM	UGU, UGC Cys UGA TERM UGG Trp
C	CUU, CUC, CUA, CUG Leu	CCU, CCC, CCA, CCG Pro	CAU, CAC His CAA, CAG Gln	CGU, CGC, CGA, CGG Arg
A	AUU, AUC, AUA Ile AUG Met	ACU, ACC, ACA, ACG Thr	AAU, AAC Asn AAA, AAG Lys	AGU, AGC Ser AGA, AGG Arg
G	GUU, GUC, GUA, GUG Val	GCU, GCC, GCA, GCG Ala	GAU, GAC Asp GAA, GAG Glu	GGU, GGC, GGA, GGG Gly

Fig. 4-12. The genetic code. Note that except for codons specifying methionine (AUG) and tryptophan (UGG), the code is degenerate—more than one codon can specify the same amino acid. The three termination codons are in upper case letters. (Adapted with permission from Genes II by B. Lewin, Copyright © 1985, John Wiley.)

Gene Structure

The first major difference between prokaryotic and eukaryotic DNA is the structure of most eukaryotic genes. Instead of a single contiguous piece of DNA that encodes a single polypeptide, almost all eukaryotic structural genes are "genes-in-pieces," with coding regions, or *exons,* separated by noncoding intervening sequences, or *introns* (Fig. 4-13). As described in Chapter 5, the RNA transcribed from a split gene also has this exon/intron organization. In the processing of RNA, the introns are spliced out; then the mature RNA transcript is exported from the nucleus into the cytoplasm, to be translated into protein. Some eukaryotic genes do not have introns; the genes encoding ribosomal RNA are examples. In most split genes the length of the introns is larger than the exons; sometimes there is more than 10 times as much intervening sequence as there is coding sequence. Most of the intron sequence seems to be "junk" DNA, although small regions of each intron contain signals that are required for correct splicing.

As with prokaryotic genes, eukaryotic genes have promoters at the 5′ ends and termination signals at the 3′ ends. However, different nucleotide sequences specify these signals, and the precise mechanisms of promotion and termination are still largely unknown. However, one hallmark of eukaryotic transcripts is worthy of note; they contain posttranscriptional modifications. A *cap* structure, consisting of an unusual guanine nucleotide, is attached to the 5′ end of the transcript RNA, and a long stretch of polyadenylic acid (anywhere from 50 to 200 Å) is attached to the

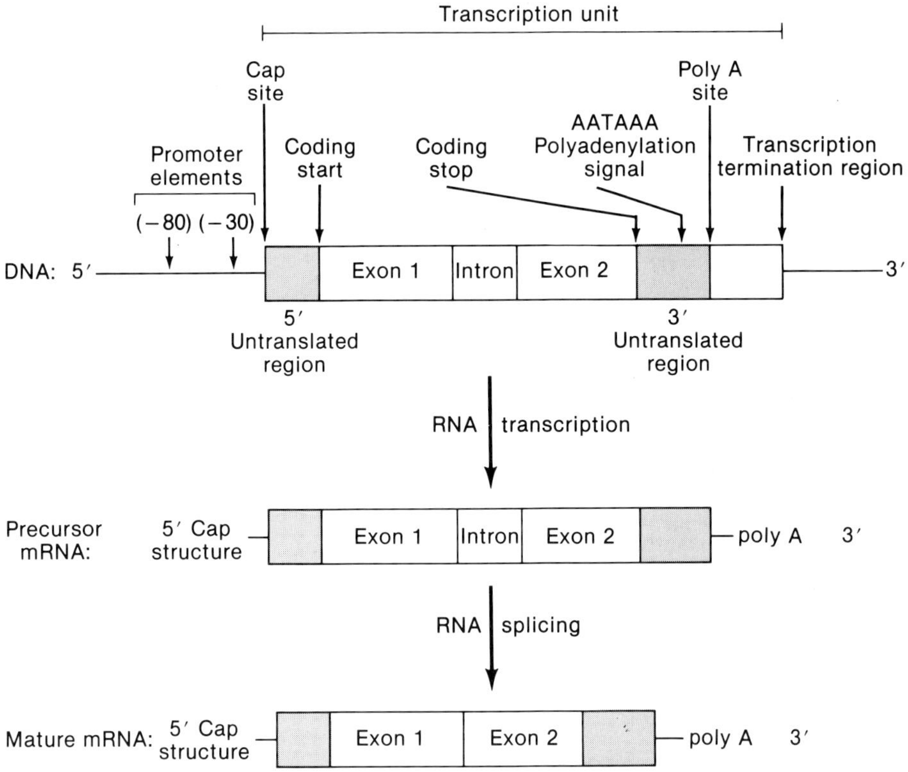

Fig. 4-13. Organization and transcription of a typical eukaryotic gene. The coding region, flanked by untranslated regions (shaded or open boxes), is in segments, or exons, separated by intervening sequences, or introns. The promoter elements 30 and 80 bp prior to the cap site and the cap and polyadenylation sites are indicated. The position of the −80 element is variable. AATAAA is the polyadenylation signal located near the end of the 3′ untranslated region. mRNA is messenger RNA. (Adapted with permission from Molecular Cell Biology by J. E. Darnell et al., Copyright © 1986, Scientific American.)

3′ end of the transcript. Transcription and transcriptional regulatory elements are discussed in Chapter 5.

In addition to structural genes, eukaryotic genomes contain *pseudogenes,* apparently inactive components of the genome derived by mutation of an ancestral functional gene. In most cases, both the active functional gene and the inactive pseudogene are found in the genome at the same time. Pseudogenes have no recognized role in gene function or regulation; they are regarded as evolutionary relics that litter the genome.

Pseudogenes come in two types (Fig. 4-14). For example, a functional gene may be duplicated by DNA-mediated recombination to create two functional genes that encode the same polypeptide. Since one of these genes is therefore superfluous, any deleterious mutation (see Chapter 7) that arises in one of the pair is tolerated by the organism, which can function with the still-effective second copy. The first gene slowly but surely accumulates mutations—it becomes a duplicated pseudogene—

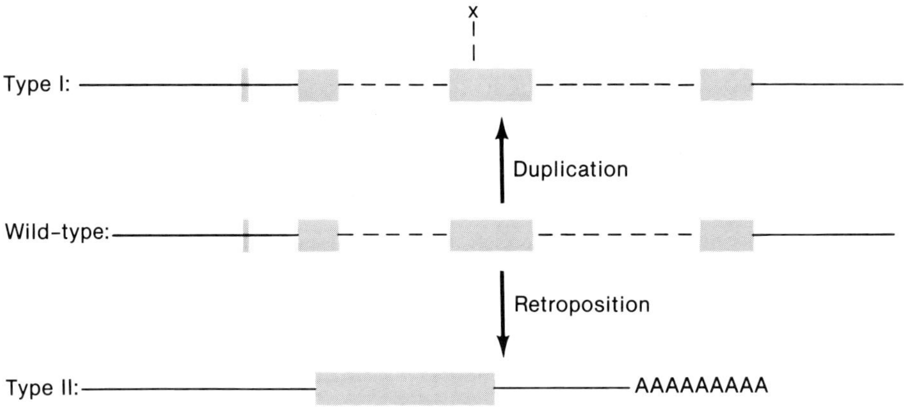

Fig. 4-14. The two types of pseudogenes. Type I is a duplicated functional gene that becomes disabled by mutation (x). Type II is a retroposed processed functional gene (see text). Shaded boxes are exons; dotted lines denote introns; the vertical bar denotes a promoter element.

whereas evolutionary pressure on the second gene maintains it in an essentially mutation-free functional state. Since it is a recombinatorial crossover event that gives rise to this type of pseudogene, both the active and inactive genes reside near each other at the same genetic locus. Examples of this type of pseudogene are found in the human α- and β-globin gene clusters (Fig. 4-15).

The second type of pseudogene arises through RNA-mediated *retroposition* (see Chapter 5). A mature RNA transcript derived from a functional gene is copied back into DNA, presumably by a cellular reverse transcriptase, and reinserted, or *retroposed,* back into the genome, usually at a site far removed from the original gene locus. Since it is derived from a mature RNA transcript, a retroposed pseudogene has several identifying characteristics: there are no introns, because reverse transcription takes place after introns have been excised, and there is a string of deoxyriboadenylic acid nucleotides at the 3′ end. It is not located near the cognate

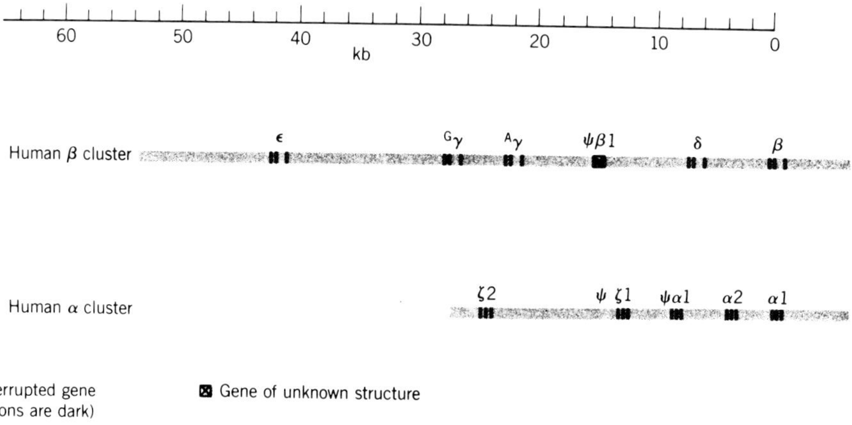

Fig. 4-15. Organization of the human α- and β-globin gene clusters. Functional genes and pseudogenes (denoted by a ψ) are shown. All the genes are transcribed 5′ to 3′, from left to right. (Adapted with permission from Genes II by B. Lewin, Copyright © 1985, John Wiley.)

functional gene. The locus of reinsertion contains a small duplicated DNA element that flanks the retroposed pseudogene, itself a hallmark of insertion. Finally, there is no DNA promoter element in the 5′ region immediately upstream of the site of reinsertion. Because these retroposed pseudogenes are ultimately derived from the RNA maturation process, they are also called *processed* pseudogenes.

Many, if not most, "single-copy" functional genes are members of gene families, including functional genes that encode proteins of related function and, also, pseudogenes. Duplicated psuedogenes and processed pseudogenes together may be found five times more often than functionally active genes.

The DNA Content of Eukaryotic Genomes

The second difference between prokaryotic and eukaryotic genomes is quantitative; prokaryotic genomes range in size from 10^3 to 10^7 bp, lower eukaryotes (algae, fungi, molds) contain up to 10^8 bp, and higher eukaryotes contain up to 10^{11} bp per haploid genome. Bacteria, and other organisms that reproduce asexually, are *haploid*—they contain genomes that consist of a single copy of DNA. *Diploid* organisms reproduce sexually and, except for the haploid sex cells, the cells carry two copies of the genome, one from each parent.

The DNA content of the haploid genome (Fig. 4-16) does not seem to be par-

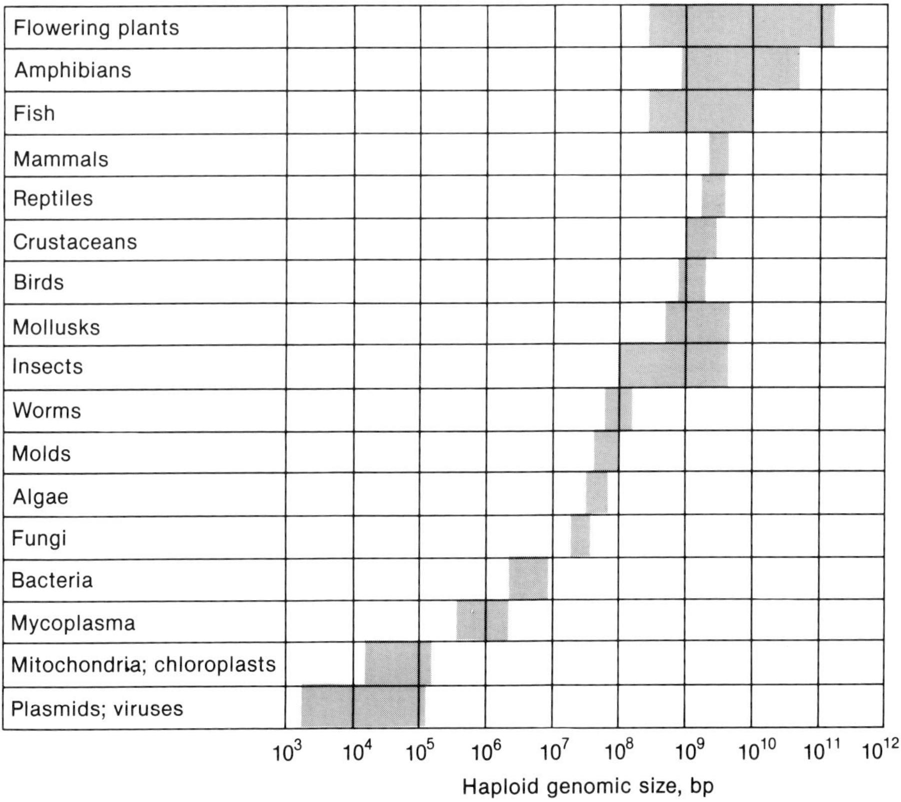

Fig. 4-16. The DNA content of the haploid genome in some phyla. (Adapted with permission from Genes II by B. Lewin, Copyright © 1985, John Wiley.)

ticularly related to anthropocentric concepts of organismic complexity or hierarchy in evolution. Most mammals, including humans, have a genomic content of about 3×10^9 bp, but some insects have twice as much, and many amphibians and plants have 50 times more DNA than humans.

It is estimated that the eukaryotic repertoire of genes is no larger than 50,000–100,000 genes. At an average length of 1500 bp/gene, the coding portion of the genome should contain, at most, 1.5×10^8 bp, or only 5% of the human genome, and far less than 1% of the genome in a salamander. What are the purpose and nature of the remaining 95–99% of the DNA? This question is the essence of the "C-value" paradox, in which C is genomic content or complexity.

The function of the excess DNA is not known, but there are at least three hypotheses. According to one view, the "excess" DNA has some as-yet-unknown function. Others believe that the excess is no more than "junk DNA" that has accumulated over the course of evolution. The third, most controversial, idea is that much of it is "selfish DNA" that propagates throughout the genome as mobile transposable elements, or *transposons* (see Chapter 5), with no intrinsic function.

We have more information about the two broad categories of sources of excess DNA: genomic increase from DNA-mediated recombinational events, and genomic increase from DNA- and RNA-mediated insertional events (Table 4-2). Both mechanisms ultimately result in formation of repetitive DNA elements. Some repetitive elements are few in number, 1–10 copies, but others are present in 50,000–1,000,000 copies.

Lower copy-number elements seem to result from a limited number of duplications or retropositions of what were originally single-copy elements. Examples include the following:

1. Some duplicated structural genes have essentially the same function, such as the two human adult α-globin genes.
2. Some duplicated structural genes have slightly different functions, such as

Table 4-2 Sources of "Excess" DNA in Eukaryotes

Duplication
Short repeated elements (e.g., satellites)
Duplicated genes with same function (e.g., ribosomal DNA)
Duplicated genes with similar function (e.g., α-globin)
Duplicated genes with modified function (e.g., fetal β-globin)
Duplicated genes with new function (e.g., LDL receptor/EGF)
Duplicated genes with no function (e.g., pseudogenes)
Retroposition
SINEs (e.g., Alu/B1 sequences)
LINEs (e.g., Kpn/L1 sequences)
Processed functional genes (e.g., rat preproinsulin)
Processed pseudogenes (e.g., almost any "single-copy" gene)
Retroviruses (e.g., cellular protooncogenes)
Transposition
Transposable elements (e.g., yeast Ty/Copia/P)
"Controlling" elements (e.g., maize Ds)

the human β-globin gene family, which includes one embryonic, two fetal, and two adult genes, all derived from a single ancestral β-globin gene.

3. Some duplicated structural *elements* of genes create new genes with new functions. For example, the gene for the human low-density lipoprotein (LDL) receptor and the gene for epidermal growth factor (EGF) both seem to contain exons derived from a single source (i.e., "exon shuffling").
4. Some duplicated structural genes have no function; these are the pseudogenes.
5. Some are processed functional genes, such as one of the two rat preproinsulin genes, which has lost one of the two introns and shows all the signs of retroposition.
6. Some are processed pseudogenes.
7. Some are functional but duplicated genes, such as the ribosomal RNA (rRNA) and transfer RNA (tRNA) genes; there are about 2300 rRNA genes and 1300 tRNA genes.

The high copy-number repeated sequences fall into three main classes: *satellite elements, short interspersed repeating elements* (SINEs), and *long interspersed repeating elements* (LINEs).

Satellite DNA is so named because it appears as a discrete band in buoyant density gradients of purified genomic DNA, of slightly different density than "main band" DNA (Fig. 4-17). Satellite DNA consists almost exclusively of multiple tandem repeats of the same sequence of bases, 7–10 bp in length, which evolved from multiple rounds of DNA duplication. The repeats are not always identical, but are related. There are at least 10 different types of satellite DNA, each with a different sequence, and each contains about a million tandem copies of the repeated unit. Most satellite DNA is concentrated near the centromere of a chromosome (see below).

SINEs are dispersed elements about 300 bp in length. The two best known SINE families are the *Alu elements* in human DNA and the *B1 family* of rodents. The

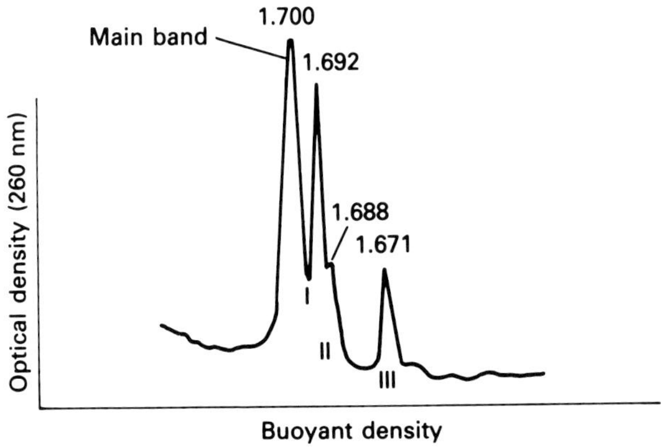

Fig. 4-17. The density of satellite DNA (I–III) is different from that of "main band" DNA, which comprises the bulk of the genome. (Adapted with permission from Molecular Cell Biology by J. E. Darnell et al., Copyright © 1986, Scientific American.)

human Alu family (so named because most elements contain an *Alu*I restriction endonuclease site) is large, with at least 300,000 individual members, and with any single member about 80–85% homologous to the *"consensus"* Alu element (the consensus was derived by comparing the DNA sequences of about 10–20 randomly isolated Alu elements). As opposed to satellite elements, SINEs probably arose from retroposition, which would explain why they are dispersed throughout the genome. For the Alu's, the primordial element was most likely a gene that encoded a small untranslated RNA, called 7SL RNA. Because they are randomly dispersed, individual Alu elements are distributed in the human genome at a frequency of about 1 per 10,000 bases (3×10^9 bp in the human genome $\div$ 3×10^5 Alu elements). Thus, Alu elements can be used to mark the presence of human genomic DNA in recombinant DNA experiments that involve transfer of genes from one species to another. Similarly, B1 elements can serve as markers of rodent DNA.

LINEs are also dispersed elements, longer and fewer in number than SINEs. In human DNA, the best known LINE is the Kpn family (again named after the restriction endonuclease, *Kpn*I, used to isolate the fragments of DNA). It is also known as L1 (for Line #1). There are about 50,000 Kpn members, of varying length. Some are as short as 500 bp and others as long as 7000 bp. Most of them have two features that imply that they are retroposed elements: a stretch of A's at the 3′ end and flanking DNA repeats. There are more short (2000 bp or less) L1 elements than long ones (2000–7000 bp), which may be related to the mechanism of retroposition. The consensus full-length L1 element in both human and rodent DNA contains a long *open reading frame;* that is, if this element were transcribed into RNA, a protein could be synthesized from it, because the translation machinery of the cell would "read" a long string of amino acid-encoding triplets before encountering a "stop" codon. The nature of the protein is not known, but one idea is that the encoded protein could be an L1 enzyme, responsible for its own retroposition!

Thus, there are many types and sources of "excess" DNA. Although estimates for each component (Table 4-3) could be in error by a factor of two, most of the excess DNA in the mammalian genome can be ascribed to duplicated and retroposed genes (active or inactive), introns, satellite DNA, SINEs, and LINEs.

Genomic Organization and Packaging

The third major difference between prokaryotic and eukaryotic DNA is related to size. Because there is so much DNA in the eukaryotic genome [six linear *feet* in

Table 4-3 An Estimate of the Total Amount of "Excess" DNA in the Human Genome

Functional genes: (100,000 genes) (1500 bp/gene)	150,000,000 bp
Pseudogenes: 5× functional genes	750,000,000 bp
Intervening sequences: (100,000 genes) (5000 bp/gene)	500,000,000 bp
SINEs: (300,000 Alu's) (300 bp/element)	100,000,000 bp
LINEs: (50,000 Kpn's) (5000 bp/element)	250,000,000 bp
Satellites: (10 types) (10 bp/type) (10^6 elements/type)	100,000,000 bp
Total	1,850,000,000 bp

one human cell!: (0.34 nm/bp) $\times$ (6 $\times$ 10^9 bp/diploid genome) = 2 $\times$ 10^9 nm = 2 m], the DNA must be packaged physically within the confines of the nucleus, a problem solved by the development of chromosomes. A *chromosome* may be defined as a discrete, large-scale unit of the genome, consisting of both genomic DNA and the associated proteins—*chromatin.* The chromosome is visible as a subcellular organelle, particularly during cell division. Organization of the genome into chromosomes confers two advantages. (1) The genome is broken down into manageable pieces (46 in the diploid human genome). (2) The chromosomal proteins provide a scaffold, around which DNA is wound into a compact structure of astonishingly small volume.

How is a chromosome constructed? It is much like fishing line on a reel, with the DNA as the line and the chromosomal proteins as the reel. This analogy, however, is too simplistic, because there is not just one folding of DNA around protein, but an intricate pattern of foldings upon foldings.

The lowest level of chromosomal organization is the *nucleosome* (Fig. 4-18). DNA (140 bp) is wrapped twice around a core of *histone* proteins (an octamer of four pairs of different histone types) and connected to the next nucleosome by 60 bp by DNA associated with a fifth histone. Histones are basic, positively charged proteins of high affinity for the acidic, negatively charged DNA. This "beads-on-a-string" chromatin structure is called the *10-nm fiber* (the diameter of naked DNA is 2 nm).

The 10-nm fiber is itself coiled into a second level of organization, the *30-nm fiber,* which takes the shape of a solenoid. The next level of chromatin folding goes from the 30-nm fiber to a *300-nm extended large-loop structure.* This, in turn, is then folded into a *700-nm condensed section* of the chromosome, which constitutes one of the two *sister chromatids* seen in the *1400 nm replicated metaphase chromosome.* The increase in the amount of DNA packaged with each increase in "diameter" is, of course, not arithmetic, but geometric (e.g., the 200 bp wrapped in a 10-nm fiber nucleosome repeat is not just tripled in one turn of the 30-nm fiber solenoid repeat, but goes up 6-fold). Thus, a mere four levels of chromatin folding can accommodate 100–300 million bp of DNA in a single metaphase chromosome. In other words, the 350-fold increase in packaging diameter from the 2-nm naked DNA to the 700-nm condensed chromatid results in a 10,000-fold increase in packaging efficiency; that is, a chromosome of average length is 5 μm long (about the size of a bacterium) and contains enough DNA in each chromatid to extend 5 cm in length if it were unraveled.

Chromosomes also contain nonhistone proteins; many of them are also basic. The function of these proteins is not clear, but some seem to be involved in gene regulation and transcription.

In the process of mitosis, the double helix of a chromosome is replicated to form two identical *chromatids,* each containing a double strand of DNA, and joined at a point called the *centromere;* the arms extend outward to form the four *telomeres.* When stained with appropriate dyes, the condensed sections of each chromosome display unique banding patterns, called G-bands when Giemsa stain is used or Q-bands with quinacrine dye. These bands are attributed to local differences in AT or GC richness. A single G-band contains 5–10 million bp of DNA.

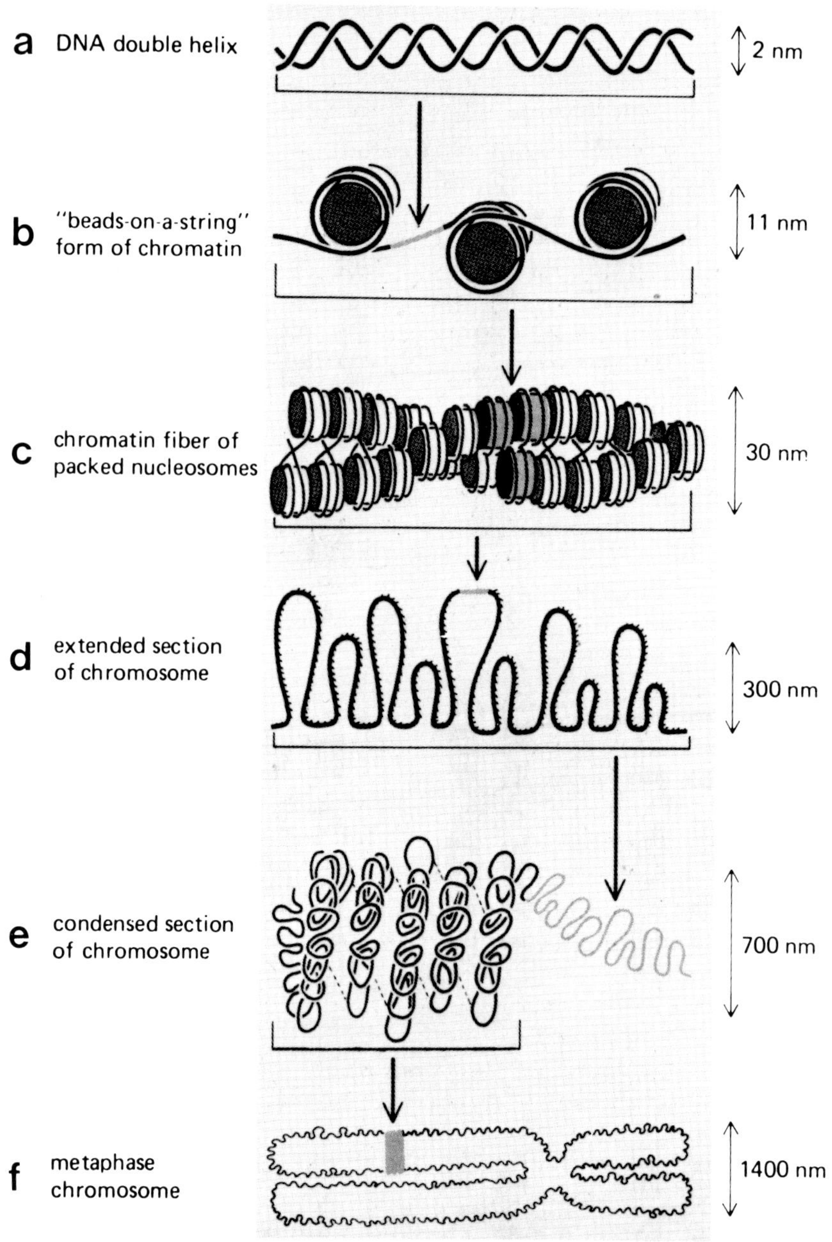

Fig. 4-18. Higher levels of chromatin organization. (a) Two-nanometer naked DNA; (b) the structure of a nucleosome—140 bp of DNA wrapped around a core octamer of histones—comprising the 10-nm fiber; (c) the 30-nm fiber as a solenoid of cylindrically folded nucleosomes; (d) the 300-nm extended section of the chromosome; (e) the 700-nm condensed section of one of the two sister chromatids; and (f) the 1400-nm metaphase chromosome consisting of two sister chromatids attached at the centromere. (Adapted with permission from Molecular Biology of the Cell by B. Alberts et al., Copyright © 1983, Garland Publishing.)

EUKARYOTIC ENDOSYMBIONTS

The unique identity of a eukaryotic organism resides in the information contained in the nuclear genomic DNA. However, all eukaryotic cells also contain extranuclear DNA in cytoplasmic organelles.

The cells of all eukaryotes, from molds and fungi to plants and animals, contain *mitochondria,* bacterium-sized organelles for the conversion of food to energy as ATP. Mitochondria contain a circular DNA genome (mtDNA) with relatively few genes. Each structural gene of mitochondria encodes one of three products: protein components of the respiratory chain (in which ATP is formed from metabolites of glycolysis and the citric acid cycle), mitochondrial ribosomal RNA (rRNA, transcribed from rDNA) used in translation on mitochondrial ribosomes, and a minimal set of tRNAs that is also used in mitochondrial protein translation.

The small size of the mitochondrion and the isolated pattern of independent replication, transcription, and translation are not just coincidentally reminiscent of bacteria. Mitochondria, in fact, are directly related to bacteria, somehow "captured" by the precursor of eukaryotes (perhaps by a mechanism no more complicated than ingestion). The parasitic "protomitochondrion" and the cell could have then developed a symbiotic relationship, the host providing nutrients and the bacterium providing energy. As long as the two organisms had independent functions to replicate, the symbiotic relationship could be interrupted at any time. However, over the eons of evolution, the bacterium lost most of its genome, retaining only some of the energy-transducing genes; in the process, the host cell became dependent on ATP-derived energy. The symbiosis became permanent, and the bacterium evolved into a new organelle—the mitochondrion.

The symbiotic relationship now means that more than 95% of the proteins required for mitochondrial function must be supplied by proteins derived from nucleus-encoded genes. Conversely, although mitochondria synthesize some respiratory chain proteins, they do not synthesize all of them, nor do they synthesize any of the enzymes of the citric acid cycle or lipid metabolism. Nevertheless, these activities are found only in mitochondria, and the cell would die without them.

The symbiotic relationship has been cemented even further by two other developments. First, all mitochondrial translation takes place on mitochondrial ribosomes, which are uniquely different from cytoplasmic ribosomes. Second, the genetic code of mitochondria (Table 4-4) differs from the "universal" genetic code that directs the translation of all nucleus-encoded eukaryotic genes and bacterial genes. The emergence of different genetic codes guaranteed the interdependence of the two DNA-containing organelles of the cell—the nucleus and the mitochondrion.

The number of mitochondria in each cell varies from tissue to tissue, but seems to be related to the energy requirements of each organ. There may be only 100 mitochondria in a skin fibroblast or quiescent tissue culture cell, 1000 in a liver cell, and up to 20,000 in a cardiac myocyte.

The mtDNAs of different eukaryotic phyla differ in size and gene organization. Human mtDNA is small, and has been completely sequenced. It is a circle about 16,500 bp long. In addition to the rRNA and tRNA genes, mtDNA encodes 13

Table 4-4 "Deviations" from the "Universal" Genetic Code (Capital Letters)

Codon[a]	Normal	Mitochondrial DNA				Ciliated protozoa
		Human	Fruit fly	Yeast	*N. crassa*	
UAR	Stop	Stop	Stop	Stop	Stop	Gln
UGA	Stop	Trp	Stop	Trp	Trp	Stop
CUX	Leu	Leu	Leu	Thr	Leu	Leu
AUA	Ile	Met	Ile	Met	Ile	Ile
AGA	Arg	Stop	Ser	Arg	Arg	Arg
AGG	Arg	Stop	Arg	Arg	Arg	Arg

[a]R, Purine; X, any base.

components of the electron transport system. All 13 have the structure of prokaryotic genes, with no introns. Furthermore, as with prokaryotes, the mitochondrial transcription units contain more than one gene (which are cleaved into individual RNAs posttranscriptionally).

Yeast, on the other hand, have mtDNAs that are more complex and have more of a eukaryotic appearance. The circular mtDNA is longer, about 84,000 bp, even though it encodes about an equivalent number of genes. In yeast, mtDNA genes have introns that encode portions of RNA to be spliced out, just as in nuclear genes. At least one gene, that encoding cytochrome *b,* encodes its own splicing enzyme, or "maturase," in an unusual intron that encodes a protein (no junk DNA here!).

As in prokaryotes, mitochondria divide by fission, but the replication of mtDNA is different. Each strand has a site at which replication commences and proceeds, not from a replication bubble but by a more complex "displacement" procedure. Moreover, there is more than one mtDNA genome in each mitochondrion. On average, each human mitochondrion contains four mtDNAs.

In plants, transfer of bacteria occurred at least one other time, resulting in the capture and evolution of a second DNA-containing organelle unique to photosynthetic plants—the *chloroplast.* As with mitochondria, chloroplasts have inner and outer membranes, contain multiple circular genomes, and play a role in energy transduction, in this case the conversion of light to reducing equivalents and ATP. Chloroplast DNA, on average, is about 10 times larger than mammalian mtDNA.

SUMMARY

We have seen how prokaryotic and eukaryotic genomes are organized, how the same underlying principles of DNA chemistry are used, and how the two groups have diverged in structure to accommodate their different needs.

To summarize these differences, we can say that

1. In viruses, plasmids, mitochondria, and chloroplasts
   There are 5–50 genes.
   The genes are "tightly spaced."
   The genome is 5,000–200,000 bp long.
   There is little or no "excess" DNA.

2. In bacteria
   There are a few thousand genes.
   The genes are "tightly spaced."
   The genome is 2–8 million bp long.
   There is little or no "excess" DNA.
3. In eukaryotes
   There are 50,000–100,000 genes.
   The genes are "widely spaced."
   The genome is 20 million–100 billion bp long.
   There are large amounts of "excess" DNA.
   The DNA is packaged as chromatin in chromosomes.

The knowledge of genomic structure can be used in the analysis of the molecular genetics of human disease, as is clearly demonstrated in the remainder of this book.

SUGGESTED READINGS

Because this chapter is in the nature of a review, I have not included specific references. Rather, for those interested in exploring the subject further, the following books are suggested:

1. Darnell J, Lodish H, Baltimore D. Molecular Cell Biology. New York: W. H. Freeman/Scientific American, 1986.
2. Lehninger A, Biochemistry, 2nd ed. New York: Worth, 1975.
3. Lewin B, Genes II. New York: John Wiley, 1985.
4. Watson JD, Hopkins NH, Roberts JW, Steitz JA, Weiner AM. Molecular Biology of the Gene, Vol. I. Menlo Park, CA: Benjamin/Cummings, 1987.
5. Watson JD, Tooze J, Kurtz DT. Recombinant DNA, A Short Course. New York: W. H. Freeman/Scientific American, 1983.

5

Transcription and Regulatory Elements

ERIC A. SCHON

Ribonucleic acid, or RNA, plays a central role in coordinating and transducing the flow of genetic information within the cell. In this role, RNA appears in many guises: It is the intermediary between the gene and its encoded protein product (messenger RNA, or mRNA). It is a structural gene product (e.g., ribosomal RNA, or rRNA). It mediates protein translation, by serving as an adaptor molecule for carrying amino acids to the growing polypeptide chain (transfer RNA, or tRNA). It can serve as an intermediate in the processing of other RNAs (e.g., small nuclear RNAs, or snRNA). It can even function as an enzyme ("self-splicing" RNAs).

Whatever the role, all RNAs must be synthesized as a ribonucleotide polymer, using a cognate DNA sequence as a template for synthesis; that process is known as *transcription,* and the fundamental enzyme responsible for transcription is *RNA polymerase.* Because RNA serves these many functions, transcription must be a regulated process.

This chapter will provide a brief overview of both the mechanics and the regulation of transcription. Although the bulk of the chapter stresses transcription in higher eukaryotes, other systems, such as bacteria, mitochondria, and yeast, have been studied more thoroughly and illuminate the problems involved in transcription.

NOMENCLATURE

In its simplest form, transcription begins when RNA polymerase binds to the "front," *"upstream,"* or 5′ end of a gene. As the enzyme travels forward on the DNA duplex, the polymerase "reads" one of the two strands of DNA as a template to produce a complementary single-stranded polyribonucleotide—the RNA. After the enzyme has traversed the gene, the RNA polymerase disengages from the DNA, and a full-length RNA is released. This *primary transcript,* also called *precursor RNA,* may be processed further, to produce a *processed* or *mature transcript.*

The sequence of bases in the DNA between the sites of initiation and termination of a single RNA primary transcript is called the *transcription unit;* sometimes the transcription unit includes more than one gene. The RNA is produced in the 5′ → 3′ direction: the numbering system of the nucleotides (nt) begins at +1. Numbering of RNA-coding sequences in the DNA also starts with +1 as the first nucleotide of the transcription unit; nucleotide positions upstream from that point are denoted by negative numbers. By convention, the upstream region on the DNA where RNA polymerase commences is called the "5′ end" and the downstream region, where it disengages from the DNA, is the "3′ end." The *coding strand* of the DNA duplex has the same 5′ → 3′ orientation as the encoded RNA, and is said to be in the *message sense.* In the convention of presenting DNA sequences, the message is usually read from left to right on the *top* strand. The *bottom* strand, therefore, is in the complementary *antimessage sense,* and reads 3′ → 5′ (from left to right). This antimessage sense strand is used as the template by RNA polymerase (see Fig. 5-1).

The RNA polymerase "knows" where to begin and where to end by recognizing transcription *start signals* and *stop signals* that are encoded by specific sequences of DNA at the 5′ and 3′ ends of the gene; these are called the *promoter* and *terminator,* respectively. In addition to these signals, other signals and factors regulate the time or place of transcription, particularly for tissue-specific or developmentally regulated genes. These elements act either *in cis* or *in trans.* A *cis-acting element* regulates transcription only when it is physically linked to the piece of DNA that is being transcribed; a *trans-acting element* regulates transcription of a gene without being linked to it, and is often a soluble factor that is the product of another gene remote from the target, sometimes on another chromosome.

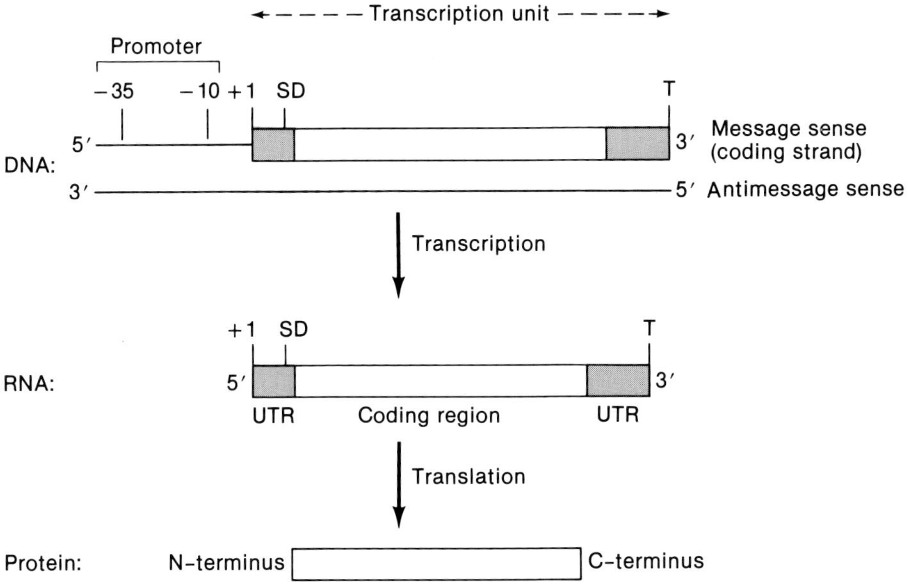

Fig. 5-1. Organization and transcription of a typical prokaryotic gene. The transcription unit is flanked by transcriptional promoter elements 10 and 35 bp prior to the transcriptional start site (nucleotide +1) at the 5′ end, and by termination signals (T) at the 3′ end. The coding region (open box), flanked by 5′ and 3′ untranslated regions (UTRs; shaded boxes), begins with the translational start codon (methionine) and ends with one of the three stop codons (UAA, UGA, or UAG). The Shine–Dalgarno sequence (SD) is a ribosome-binding site for initiation of translation.

There is only a single RNA polymerase in bacteria, but eukaryotic cells have three, each of whcih has a different role, and which can be differentiated from each other on the basis of sensitivity to a mushroom toxin called α-amanitin.

Note: all sequences in this chapter are written $5' \rightarrow 3'$, unless otherwise indicated; in presenting consensus sequences (a consensus or "ideal" sequence is derived by comparing nearly identical sequences from different sources), R denotes a purine (A or G), Y a pyrimidine (T, C, or U), and N any nucleotide.

TRANSCRIPTION IN PROKARYOTES

Prokaryotic transcription is relatively straightforward. RNA polymerase binds to the bacterial promoter, traverses the DNA, and is released at the terminator; there is no further modification of the RNA following transcription (Fig. 5-1). Because prokaryotes have no nuclei, all the factors required for both transcription and translation are present together in the cell. Thus, bacterial protein synthesis may commence even while an mRNA molecule is being transcribed. In fact, transcription and translation are often coupled.

The sequences responsible for promotion are well understood (1,2). The bacterial promoter comprises about 40 base pairs (bp) of upstream sequence, with two distinct elements (Fig. 5-1). First, there is a specific nucleotide sequence located about 10 bp upstream from the beginning of the transcription unit; this sequence has been called the *Pribnow box,* after its discoverer. It has a consensus sequence of TATAAT, and seems to be a recognition site for the binding of RNA polymerase. About 25 bp further upstream is a second DNA region, called the "−35 element," with a consensus sequence of TTGACA. The −10 and −35 elements, together, seem to be required for both accurate and efficient initiation of transcription. The RNA polymerase "core" protein can catalyze synthesis of an RNA chain at any point on the DNA; however, one particular subunit of the holoenzyme, called *sigma,* allows the RNA polymerase to recognize promoter sequences specifically and to initiate transcription at the correct site upstream of a gene (3). After transcription has started, sigma factor dissociates from the core enzyme and becomes free to attach to another core molecule (4).

Just before the end of the transcription unit, a termination sequence signals the RNA polymerase to disengage from the DNA template (5). Many prokaryotic transcripts have a GC-rich region of *dyad symmetry* (i.e., a self-complementary region in which the RNA can form a hairpin) in this region, followed by a series of U's (6). In terminators of this type, the hairpin probably causes the RNA polymerase to slow the rate of transcription, whereas the poly(U) sequence is most likely involved in release of the RNA polymerase from the DNA template (7,8). This type of termination does not seem to require any extra factors for release of the polymerase.

Other prokaryotic transcripts, however, do require *termination factors,* such as the protein called *rho* (9,10). Rho-dependent termination is less well understood than rho-independent termination, and has been studied mainly in bacteriophage λ. Rho-dependent terminators also exhibit dyad symmetry, but there is no poly(U) stretch present. The terminator region, by itself, cannot cause rho-dependent termination; a region of about 80 nt upstream of the 3′ end of the transcript encodes most of the signals required for termination, presumably for binding of rho factor.

Some hairpins are the sites for *antitermination,* in which termination is reduced or prevented by specific protein factors that either modify the RNA polymerase or "sequester" the termination signal; in an event called *readthrough,* RNA polymerase then continues past the termination point. Antitermination, an important regulatory device in bacteriophage λ (11), is one mechanism that allows transcription of a *polycistronic* message, in which the RNA encodes more than one discrete gene product. Antitermination is also used in a sophisticated control mechanism called *attenuation,* which is a mode of regulation that links transcription to translation (see below). The best characterized antiterminator factors are the λ N and Q proteins, which interact with RNA polymerase to render it terminator resistant (12, 13), so that the polymerase is unable to terminate when it transcribes the nut ("N utilization") or qut ("Q utilization") sites. Remarkably, the mechanisms of λ termination and antitermination are similar, as a single gene product, called Nun, both stimulates termination and interacts at the nut site to stimulate N-promoted antitermination (14).

RNAs destined to be translated into proteins are bound by ribosomes, and translation commences even before the RNA polymerase completes transcription. The binding of the ribosome to initiate translation on an mRNA probably involves base pairing between a short polypurine region at the 5′ end of the message just prior to the initiator Met AUG codon, called the *Shine–Dalgarno sequence* (5′-AGGAGG-3′), and a highly conserved region at the 3′ end of bacterial 16 S (S is Svedberg unit, a measure of molecular size in a centrifugal field) ribosomal RNA which is complementary to this sequence (3′-UCCUCC-5′) (15).

Regulation of prokaryotic transcription falls into two major types of control, positive and negative. *Positive-acting* proteins bind to DNA in the region of the promoter, thereby increasing the binding efficiency of the RNA polymerase to the promoter. *Negative-acting* proteins do the opposite; they bind to the promoter region and repress transcription, usually by physically blocking the promoter site and denying access to it by RNA polymerase. Such a negative-acting protein is called a *repressor,* and the site of binding of repressor to the DNA is called the *operator;* the operator and promoter sequences are adjacent. A third type of regulatory protein is an *effector,* which modulates the binding affinities of a repressor at the operator. Some effectors are called *inducers,* because they decrease binding affinity, so that polymerase can initiate. Alternatively, an effector may combine with a nonfunctional repressor, which then binds the operator. These effectors are called *corepressors.*

The classic example of negative regulation was demonstrated by Jacob and Monod with the lactose *(lac)* operon of *Escherichia coli* (16, 17). An *operon* is a complete unit of gene expression, including all the structural genes plus any regulator genes and control sequences. In the *lac* operon, this includes the promoter, the operator, three structural genes for the conversion of lactose to glucose and galactose, and the gene encoding the repressor. When lactose is absent, the *lac* repressor binds the operator and prevents transcription of *lac* mRNA. When lactose is present, however, the sugar acts as an inducer and binds the repressor; this complex cannot bind the operator, so transcription commences.

An example of positive control is the arabinose *(ara)* operon. The regulatory protein of the *ara* operon, *AraC,* is the product of a separate gene. When arabinose is present, it binds *AraC;* this complex then binds the promoter and stimulates transcription. *AraC* can also behave as a negative control element; when arabinose is absent, *AraC* binds simultaneously to an upstream region and to the promoter of the operon, thus blocking transcription (18, 19).

There are many variations on the theme of positive and negative control. Transcription of many operons may be under the control of one or a few trans-acting factors (for example, glucose and cyclic AMP control several sugar metabolism operons, by means of a "catabolite activator protein"). Alternately, a repressor may act at more than one (nonadjacent) site, so that widely scattered genes are regulated coordinately, as a *regulon.* An operon may have both a regulated promoter and a constitutive one, and some regulatory proteins seem to regulate their own transcription (20).

Transcriptional control may be intimately connected to translational control, as in a process known as *attenuation,* in which the progress of the ribosome is coordinated with that of the RNA polymerase (21). Attenuation is one of the modes of regulation of amino acid pathways genes, such as the phenylalanine (phe) and tryptophan (trp) operons (Fig. 5-2). After initiation of transcription in the phe operon,

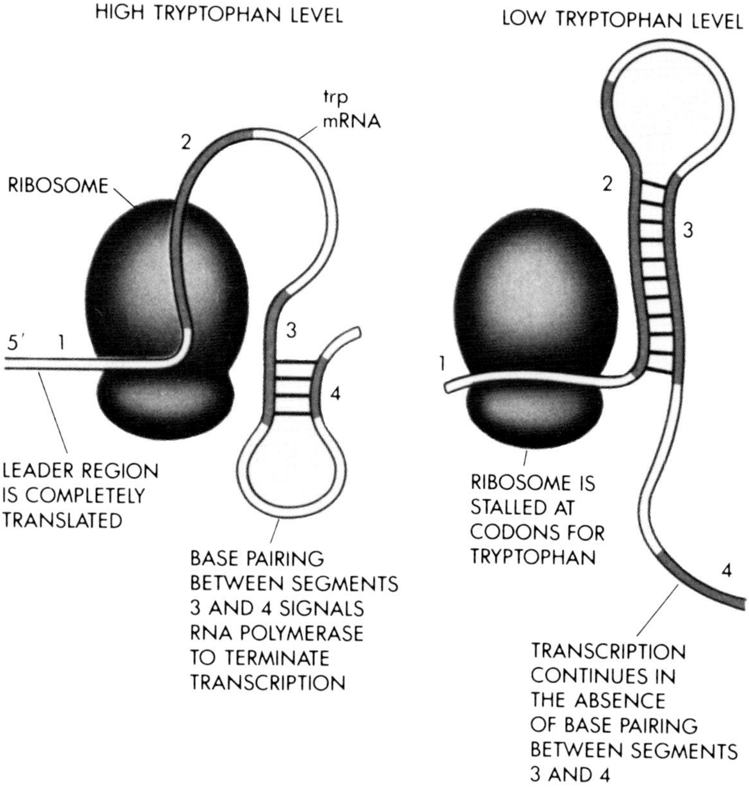

Fig. 5-2. Attenuation in the bacterial tryptophan operon. (Adapted with permission from Recombinant DNA: A Short Course by J. D. Watson et al., Copyright © 1983, Scientific American.)

RNA polymerase proceeds toward the attenuation region, which is downstream from a translatable sequence for a leader peptide; that leader sequence is rich in phe condons (7 of the 15 amino acids in the leader are phe). If phe is absent, the leader peptide cannot be translated, and the ribosome *stalls* in the leader-attenuator region, thus covering up a segment of RNA that is required for formation of the terminator hairpin; in the absense of the hairpin, the RNA polymerase continues into the structural region of the phe operon, and the cell can now synthesize phe. On the other hand, if phe is present, the leader peptide is easily translated by the ribosomes, which keep moving and no longer prevent hairpin formation; RNA polymerase now terminates transcription, and phe is no longer produced. Attenuation shows that termination is not just a mechanism to release RNA polymerase from the message, but is also a mechanism of metabolic regulation.

TRANSCRIPTION IN EUKARYOTES

In eukaryotes, the synthesis of the three major classes of transcripts—ribosomal, messenger, and transfer RNAs—is controlled by the three different polymerases, I, II, and III, respectively, which can be differentiated from each other by sensitivity to α-amanitin. Pol I is insensitive to α-amanitin, pol II is highly sensitive to α-amanitin, and poll III has intermediate sensitivity. Although the polymerases differ in specificity, most RNAs in all three share one feature in common; they are processed posttranscriptionally in some manner.

Of the three polymerases, pol I is the least well understood. Pol I operates only on ribosomal genes, which are tandem arrays of rDNA genes containing both internal transcribed and flanking nontranscribed spacer regions. Transcription of rDNA exhibits strong species selectivity. For example, pol I isolated from mouse will not transcribe human rDNA, and vice versa (22), although there are rare exceptions between mouse and frog (23). The minimal essential promoter region is in the area from -35 to $+5$ (24), and the only area of strong nucleotide sequence conservation in mammals is in the region $+2$ to $+18$ (25). The precursor message that pol I transcribes is a giant 45 S RNA, about 14 kb long. Each 45 S transcription unit is one member of the tandem array of ribosomal DNA genes, and each gene is separated by a nontranscribed *spacer* region (24). The 45 S pre-RNA consists of three smaller rRNAs (28 S, 18 S, and 5.8 S) that are separated by the transcribed spacer regions, and is terminated several hundred nucleotides (nt) downstream of the last rRNA gene, beyond the 28 S rRNA endpoint (26); as in prokaryotes, palindromic sequences seem to be important for termination (5). An endoribonuclease, or RNase, cleaves at the transcribed spacer-rRNA boundaries to produce the three mature rRNAs.

A fourth ribosomal RNA, called 5 S RNA (about 120 nt long), is transcribed not by pol I but by pol III, which also transcribes the tRNAs (about 80 nt long) and a number of short (300 nt or less) RNAs, including repetitive Alu elements in humans. Besides being small, pol III genes do not encode proteins, and are usually members of *repetitive gene families* that contain anywhere from a few dozen to many thousand individual, closely related, genes. The unique feature of pol III transcription is that the pol III promoter lies *inside* the structural portion of the gene (27,28). In the case of the 5 S rRNA, transcriptional factors bind to the DNA within the 5 S DNA gene region, and then "reach back" to interact with pol III at the 5′ end of the gene to initiate transcription (29). The 5 S rRNA thus formed is not processed further, and is ready for incorporation into the ribosome. In the case of the other main pol III gene product, tRNA, a small (10–15 nt) intervening sequence is spliced out of the pre-tRNA (see below), and a trinucleotide, CCA, is added to the 3′ end of the tRNA posttranscriptionally. As with the 5 S DNAs, the tDNAs also contain internal promoter regions for transcription. The simplest termination signals for pol III transcription are reminiscent of prokaryotic rho-independent termination; they comprise a short stretch of U's embedded in a GC-rich region (30).

Because it transcribes *heterogeneous nuclear RNA* (hnRNA), the precursor of mRNA, regulation of pol II transcription has been studied most intensively. Almost all pol II genes differ from prokaryotic protein-coding genes in one fundamental way: they contain *intervening sequences,* or *introns,* that vary in size from a few dozen bases to many kilobases. The *primary,* or *precursor transcript,* is a species of RNA that contains "coding regions" (encoded by the *exons* in the DNA) and interspersed "noncoding regions" (the *introns*). The intron-encoded regions in the RNA in the primary transcript are removed by *splicing* in the nucleus, leaving a processed message consisting only of RNA representing the exons, all in a single, contiguous, *reading frame* for translation of the message into a polypeptide (Fig. 5-3). Two other *posttranscriptional modifications* take place in conjunction with splicing. First, a *cap structure* is added at the 5′ end of the message, consisting of an unusual methylated G (7-methylguanosine triphosphate) that is covalently attached "backward" in a 5′–5′ triphosphate linkage to the first base of the transcript, which itself is methylated at the 2′ hydroxyl of the ribose (31). Second, a long stretch of *poly(A)* (50–200 A's) is attached to the last nucleotide at the 3′ end of the message. These modifications are believed to enhance stability of the message, transport of the spliced message from the nucleus to the cytoplasm, and initiation of translation by cytoplasmic ribosomes.

Two major classes of pol II transcripts are not polyadenylated: the histone genes, which encode the major structural proteins of chromatin, and the U-series of small nuclear RNAs *(snRNAs),* which are components of *ribonucleoprotein particles,* or *RNPs.* Termination of the nonpolyadenylated histone genes is associated with an RNA hairpin and a highly conserved purine-rich sequence (CAAGAAAGA) that are important for generating the 3′ end (32). However, since there are also histone transcripts that are polyadenylated and that lack these termination features, the hairpin/purine motif may not be important for termination per se, but for maturation of an even longer primary transcript. Termination signals in the DNA for the U-series snRNAs, such as U1 (33) and U2 (34), are also downstream from the 3′ end of the mature RNA.

There is only one known transcriptional signal within the transcription unit itself. This is the *polyadenylation signal* for addition of poly(A) to the transcript, by the enzyme *poly(A) polymerase.* It has a consensus sequence of AAUAAA (35) and is located about 25 nt upstream of the poly(A) addition point. This does *not* mean that the transcript ends at the point of poly(A) addition; the transcription unit actually extends for quite a long distance—up to 4 kb—beyond this point (36–

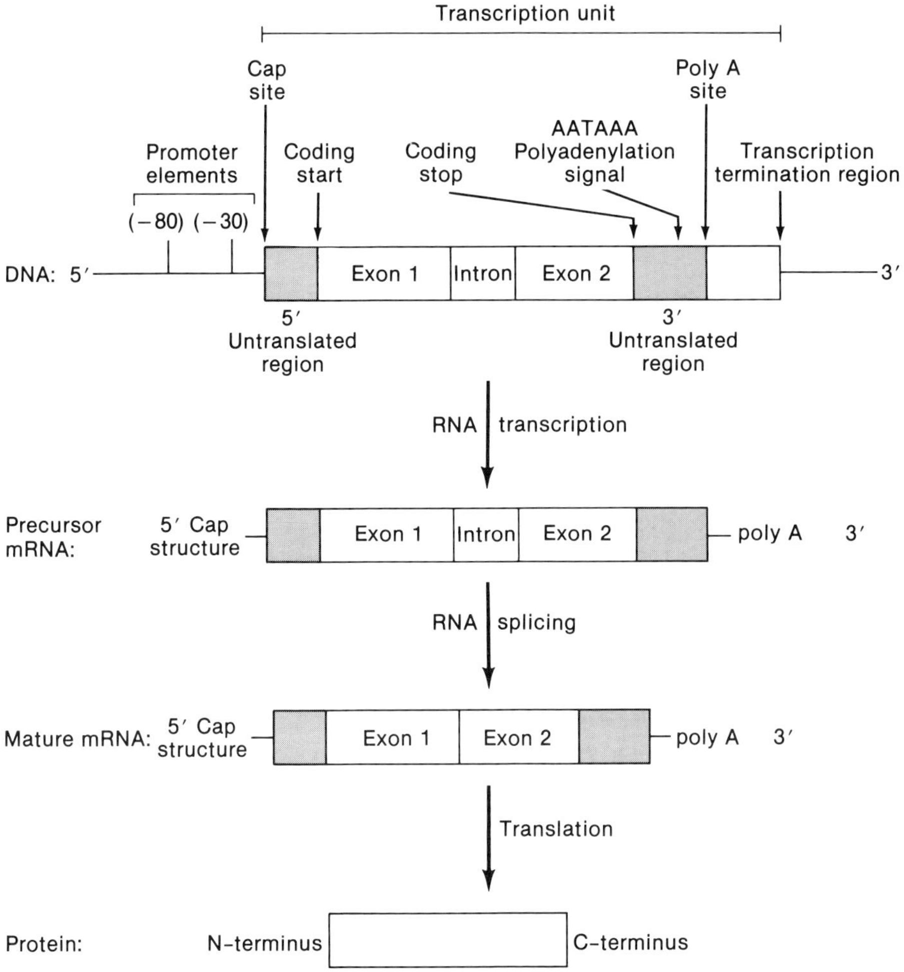

Fig. 5-3. Organization and transcription of a typical eukaryotic gene. The coding region, flanked by untranslated regions (shaded boxes), is in segments, or exons, separated by intervening sequences, or introns (open box). The promoter elements 30 and 80 bp prior to the cap site, and the cap and polyadenylation sites, are indicated. The position of the −80 element is variable. AATAAA is the polyadenylation signal located near the end of the 3′ untranslated region. (Adapted with permission from Molecular Cell Biology by J. E. Darnell et al., Copyright © 1986, Scientific American.)

38), with the AAUAAA sequence serving as a signal for endonucleolytic cleavage of the primary transcript as well as for polyadenylation. However, there is also an alternating GT sequence motif in the DNA, with a consensus YGTGTTYY immediately downstream from the polyadenylation site (39), which seems to be essential for the formation of the 3′ end; however, it is not clear whether this signal is required for polyadenylation, for cleavage of the precursor, or for termination. Some snRNAs may be required for termination and processing (40).

Regulation of transcription has focused mainly on the 5′ end of the transcription unit (41,42). The major cis-acting element at the 5′ end of pol II genes is the promoter, located immediately upstream of the cap site (i.e., nucleotide #1), in a region about 100 bp long (43,44). The promoter region contains modular elements that, taken together, appear to regulate both the efficiency and accuracy of transcription. The best-known promoter element is the so-called *TATA box,* also called the *Goldberg–Hogness box,* which is analogous to the prokaryotic Pribnow box. It is an AT-rich region (typical sequence ATATAAA) located around position −30, which is responsible for accurate initiation by pol II (45). Further upstream, around position −80, is the so called *CCAAT box* (46), which is one example of an *upstream promoter element,* or *UPE.* This UPE (typical sequence GGCCAATC) regulates the rate of transcription, and seems to operate independent of orientation. However, the distance between CCAAT and TATA is important, implying that the (protein) factors that bind at CCAAT interact physically with factors bound at TATA (47). Proteins that bind specifically to UPEs have been identified (48–50), and a number have been purified.

Other UPEs may lie beyond the CCAAT box. For example, sequences around −100 are responsible for glucocorticoid and metal-binding sites for the metallothionein gene, whose regulation is influenced by both hormonal and metal levels in the cell (51). These metallothionein UPEs may also be considered to be enhancers (see below), and, in fact, the distinction between promoters and enhancers has become blurred. Some UPEs are required for maximal *in vitro* transcription, including those encoding silkworm fibroin (52) and sea urchin histone (53). Yeast pol II genes also have UPEs, which are hundreds of base pairs upstream of the transcriptional start site. They fall into two classes: upstream activating sequences *(UASs)* and upstream repressing sequences *(URSs),* which, like bacteria, bind factors that turn transcription up or down (54).

The CCAAT/TATA motif for promoter elements seems to be prevalent in temporal- or tissue-specific genes. On the other hand, *housekeeping genes* (genes that are active in all cells because their products are necessary for normal, "nonspecial" functioning of the cell) have distinctly different promoter regions (55). They have no recognizable TATA or other AT-rich sequence; in fact, housekeeping gene promoters have a high GC content. For example, both dihydrofolate reductase (DHFR) (56) and hypoxanthine phosphoribosyltransferase (HPRT) (57) are involved in pathways of nucleotide metabolism, and both have GC-rich elements in the region around −40 to −50 that appear to behave analogously to CCAAT/TATA. In addition, the DHFR gene seems to have a GC-rich UPE around position −500. [As an exception to this rule, the human triosephosphate isomerase gene has a TATA box in addition to GC-rich elements (58).]

Besides the promoter elements, transcription of many eukaryotic genes is also regulated by another group of cis-acting elements, the *enhancers* (59–61). Enhancers are DNA sequences, 100–200 bp long, that activate transcription from a linked promoter in a manner independent of orientation. Although cis acting, enhancers can exert influence over a long distance—up to several kilobases away from the promoter—any may be located either 5′ or 3′ to the transcriptional start site (62). As with promoters, enhancers contain short distinct sequence-specific regions that bind or interact with trans-acting factors. The best-studied examples of enhancer action are two viruses, adenovirus-2 and simian virus 40 (SV40). The SV40 enhancer region contains both a "core" sequence (59) motif (AAG(C/A)ATGCA) and a repeated region with a GC-rich motif (GGGCGG).

Both temporal- and tissue-specific enhancers have been identified. The best characterized among these is the immunoglobulin (Ig) enhancer that is activated only after immunoglobulin gene DNA rearrangement in lymphoid cells (63) (see Chapter 6 by Alt). Some regions of the Ig enhancer are under negative control (64, 65); the enhancer includes five protein-binding sites, with a consensus sequence of CAGGTGGC (66, 67).

The heat shock enhancer is an example of an *inducible enhancer,* which responds to an activating factor ("heat-shock" genes are activated due to environmental stress, including, but not limited to, elevated temperature). Inducible enhancers activate promoters and, by definition, they act in a positive manner. However, that action can be achieved through either positive- and negative-acting mechanisms (e.g., negative control by inactivating a repressor, as in bacterial systems). Induction of the *Drosophila* heat shock response seems to involve positive regulation of an inducible enhancer (68); conversely, the inducible enhancer found in the human β-interferon gene (69) is under negative control, with the enhancer itself acting as a constitutive, or endogenous, positive control element (70).

The DNA sequences that are important in enhancer function bind to soluble factors. For example, the GC-rich element of the SV40 enhancer binds a soluble, trans-acting transcription factor called *Sp1* (43). Several genes contain enhancer regions that bind Sp1, and all of them contain a GC-rich sequence element, with an overall consensus sequence of (G/T)GGGCGGRRY. A factor similar to SP1 is involved in directing the heat-shock transcriptional response. This heat shock transcription factor (HSTF) has been isolated, and seems to interact with a UPE located about 40–95 bp upstream of TATA (71,72). The GC-rich regions of housekeeping gene promoters also contain sequence motifs for Sp1 binding and activation of transcription, particularly the sequence GGGGCGGAGC at position −33 of the mouse DHFR gene (73).

SPLICING

Splicing of precursor RNAs (74) occurs exclusively in the nucleus (75), and the process often seems to be required for export of the transcript to the cytoplasm (76). The 5′ and 3′ ends of the intron must be cleaved precisely, so that the flanking exons can be joined accurately.

The splicing machinery for transfer RNA introns is unique and differs from that for other RNAs. In yeast, a specific endonuclease recognizes and cleaves the intron boundaries, whereas a special ligase joins the 5′ and 3′ exons in an intramolecular reaction. A unique feature of the ligation reaction is that the phosphodiester bond is not in the "usual" 5′–3′ polarity, but 3′–5′, with an extra 2′ phosphate (derived from exogenous ATP) at the end of the 5′ exon (77). In vertebrates, the 3′–5′ linkage remains, but there is no 2′ phosphate group (78,79).

The specificity of the splicing reaction in messenger RNA precursors of higher eukaryotes is determined by consensus sequences at the two exon–intron boundaries, in what is called the "GT-AG" or "Chambon" rule (45): in the DNA, the 5′ end of the intron has a consensus of AG|GT(A)AGT and the 3′ end has a consensus of YYYYYYYYYYYNCAG|G. (The vertical line is the exon–intron boundary; the bracketed nucleotide may differ; the underlined nucleotides are mandatory for correct splicing.) In yeast the 5′ and 3′ splice junction consensus sequences are |GTATGT and YAG|, respectively (80). The GT dinucleotide is called the *donor* splice site and the AG is the *acceptor.* Not only are both important for determining the specificity of the splicing reaction, but they also participate in the mechanism of excision.

The excision reaction itself involves an unusual splicing intermediate—a *lariat*—formed at an internal *branch point* within the intron. The first modification of the pre-mRNA is cleavage at the 5′ splice site, releasing the 5′ exon (Fig. 5-4). The 5′ end of the intron is now joined via a 2′–5′ phosphodiester bond to an A residue located about 30–50 nt upstream of the 3′ splice junction, producing a tailed circular RNA, a "lariat," that includes the intron connected to the 3′ exon. This intermediate is then cleaved at the 3′ splice site, releasing both the intron (as a lariat) and the 3′ exon; concomitantly, the two exons are ligated. Thus, the final products of the reaction are the ligated exons and the excised intron lariat. In yeast, the internal branch point at which the 5′ end of the intron joins near the 3′ end to form the lariat has an obligate consensus sequence, UACUAAC, where the underlined A is the nucleotide to which the donor G is joined (81,82). In higher eukary-

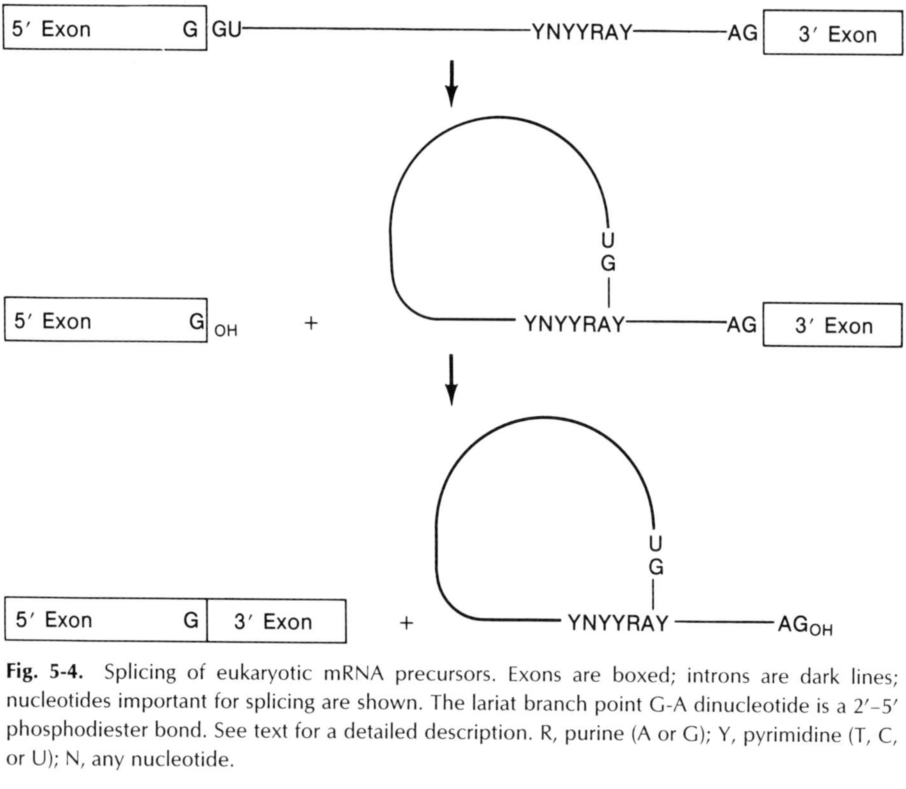

Fig. 5-4. Splicing of eukaryotic mRNA precursors. Exons are boxed; introns are dark lines; nucleotides important for splicing are shown. The lariat branch point G-A dinucleotide is a 2′–5′ phosphodiester bond. See text for a detailed description. R, purine (A or G); Y, pyrimidine (T, C, or U); N, any nucleotide.

otes, only the A residue is conserved, although a rough consensus sequence surrounding the A, YNYYRAY, can still be deduced (83).

The splicing mechanism described here physically separates the 5′ exon from the 3′ exon in an intermediate of the reaction. In genes with multiple exons and introns, the exons are never spliced out of order, and it is not clear how the splicing machinery ensures that a free 5′ exon is ligated to the 3′ exon that immediately follows. Most likely, factors involved in the splicing reaction hold the RNA substrates—the 5′ exon, the 3′ exon, and the intron—together in one complex, so that the flanking exons can be juxtaposed and then joined. This complex has been termed the *spliceosome* (84). It is large, sedimenting at 50–60 S, and contains both U1 and U2 snRNPs (85), and perhaps U5 as well (86). U1 RNA has strong sequence complementarity to the splice junctions (87,88), whereas U2 has sequence complementarity to the branch point (89), suggesting that snRNAs help to hold the intermediate together in the spliceosome. There is no strong evidence that the splicing reaction itself is a regulated process.

Exons are normally spliced in the correct order (i.e., no "exon skipping"), but orderly splicing of a multiexon transcript must be regulated by a mechanism other than simple 5′-to-3′ linear scanning, because partially spliced intermediate molecules with different combinations of persisting introns have been observed (90,91). Thus, the splicing mechanism may involve diffusion of a splicing complex along the RNA precursor.

A major piece of evidence in support of a trans-action model for splicing is the phenomenon of *alternative splicing* (92), the production of multiple mRNAs from a single transcriptional unit, and a mechanism of regulation at the posttranscriptional level. More than 50 genes have been identified in which different mature mRNAs arise from alternative splicing of the same precursor mRNA, usually in a developmental- or tissue-specific manner.

Alternative splicing can take many forms (Fig. 5-5). Let us examine a hypothetical gene transcript with a structure of XY-AaBbCcDdEeF-WZ, where X and Y are alternative promoters, A–E the exons, a–e the introns, and W and Z alternative polyadenylation signals. The normal precursor RNA would be YAaBbCcDdEeFZ, and the normal splicing pattern would result in a mature message of YABCDEFZ. However, usage of an alternative promotor would result in XABCDEFZ, whereas use of an alternative polyadenylation signal would result in YABCDEFW. A "cassette" pattern of alternative splicing would delete only one exon—a "cassette" of genetic information encoding a peptide subsegment of the total protein—to produce a mature message of the type YABDEFZ. A "mutually exclusive" pattern would delete either one exon or another, but never both: YABCEFZ or YABDEFZ, but never YABCDEFZ. A "retained intron" pattern would give YABCcDEFZ (the intron sequence, of course, would have to maintain the open reading frame). Other patterns are those in which a normally unused splicing donor or acceptor site within an exon is used, so that "half" of one exon is fused to the entire next exon, or vice versa.

One of the most complex families of alternatively spliced genes is that encoding troponin T in fast skeletal muscle (93). It contains 18 exons and 17 introns. Exons 1–3, 9–15, and 18 are always present, along with every possible combination of

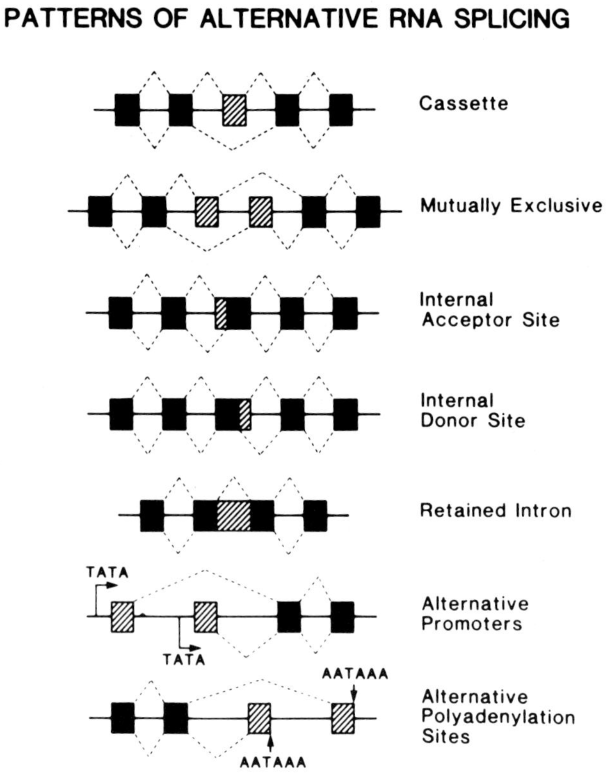

Fig. 5-5. Patterns of alternative RNA splicing (92). Constitutive exons (black boxes), alternative sequences (hatched areas), and introns (solid lines) are spliced by different pathways (dotted lines). Alternative promoters (TATA) and polyadenylation signals (AATAAA) are also shown. (Adapted with permission from Annual Reviews of Biochemistry, Copyright © 1987, Annual Reviews Inc.)

exons 4–8 and 16–17, resulting in 64 possible mutually exclusive mRNAs. Similarly, α-tropomyosin has seven alternative-splicing pathways, with each confined to a single cell type, such as smooth muscle, striated muscle, brain, myoblasts, and nonmuscle fibroblasts (92).

The mechanisms of alternative splice site selection are unknown, but probably involve both cis-acting information (e.g., secondary structure of the primary transcript) as well as trans-acting factors. Moreover, splices executed at one site in the RNA molecule may generate conformational changes that would affect the choice of a splice site elsewhere.

Although intramolecular splicing is the rule for all tRNAs and most mRNAs, an intermolecular mechanism has also been identified. *Trans splicing* of exons derived from two separate RNAs is important in the generation of the mature mRNAs of trypanosomes, including those encoding the variable surface glycoproteins (VSGs) expressed on the surface of the parasite (94). The 5′ ends of these mRNAs contain a common "leader" or "mini-exon-derived" transcript (medRNA), encoded by DNA sequences on a separate chromosome, juxtaposed to the major VSG mRNA (95). As an aside, one VSG transcription unit in trypanosomes provides the first evidence for the existence of a multicistronic transcription unit for cellular genes in eukaryotes; a single 60-kb RNA comprises at least one protein-coding message and yields seven other stable mRNAs (96).

The most unusual and, from an evolutionary standpoint, most provocative form of splicing is *self-splicing,* in which the RNA acts as a true enzyme, coined a *ribozyme*, to catalyze intron splicing (97–99). One such ribozyme is the self-splicing RNA of the intron in the 26 S rRNA precursor in *Tetrahymena,* a ciliated protozoan. In *Tetrahymena* self-splicing, the intron of the pre-rRNA is excised as a discrete linear molecule, which is subsequently converted to a circular form, requiring a G at the 5′ end of the intron (Fig. 5-6). Self-splicing introns of this type are termed Group I introns. Group I introns appear to be widespread, and have now been found in the tRNAs of plants (100). Another ribozyme, which cleaves but does not splice, is the RNA moiety of RNase P, the enzyme that produces the mature 5′ terminus of tRNA molecules.

Another class of ribozymes includes the so-called Group II introns, which are self-splicing mitochondrial introns of fungi. As with the classical mRNA splicing products (which are currently also classified as Group II introns; see Table 5-1), the group II splicing products are lariats (101).

The major category of RNA self-processing reactions [there are at least three (98)] is a two-step transesterification mechanism (Fig. 5-6). The 3′-OH of a guanosine molecule (G, GMP, or GTP) attacks the 5′ splice site, leaving a 3′-OH at the end of the 5′ exon. This OH group is then used in the subsequent ligation of the

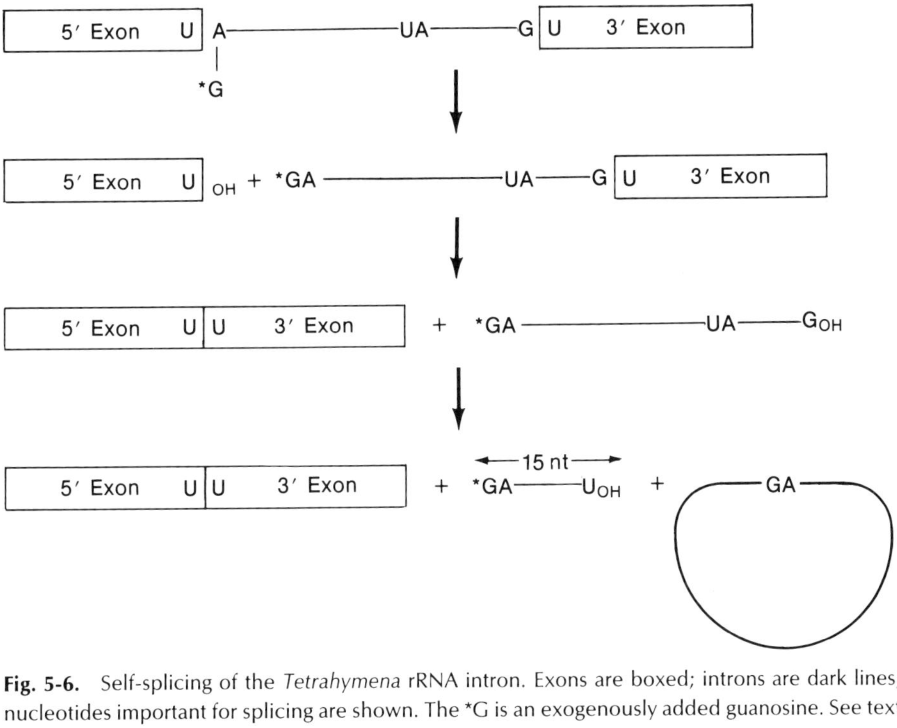

Fig. 5-6. Self-splicing of the *Tetrahymena* rRNA intron. Exons are boxed; introns are dark lines; nucleotides important for splicing are shown. The *G is an exogenously added guanosine. See text for a detailed description.

Table 5-1 Classification of Introns[a]

<table>
<tr><th rowspan="2">Group</th><th rowspan="2">Location</th><th rowspan="2">RNA type</th><th colspan="3">Splice sequence</th><th rowspan="2">Self-splicing?</th><th rowspan="2">Intron product</th><th rowspan="2">Comment</th></tr>
<tr><th>5′</th><th>. . .</th><th>3′</th></tr>
<tr><td>I</td><td>Nuclear</td><td>tRNA, rRNA, mRNA</td><td>U|</td><td>. . . . .</td><td>G|</td><td>Yes</td><td>Circle</td><td>Requires guanosine</td></tr>
<tr><td>II</td><td>Mitochondrial</td><td>mRNA</td><td>.|GUGCG</td><td>. . . . .</td><td>YAU|</td><td>Yes</td><td>Lariat</td><td></td></tr>
<tr><td>II</td><td>Nuclear</td><td>mRNA</td><td>.|GURAGU</td><td>. . . . .</td><td>YAG|</td><td>No</td><td>Lariat</td><td>Requires snRNPs</td></tr>
<tr><td>—</td><td>Nuclear</td><td>tRNA</td><td>Exon</td><td>. . . . .</td><td>Exon</td><td>No</td><td>Linear</td><td>Unique ligation</td></tr>
</table>

[a]Vertical lines denote exon–intron boundaries. Nucleotides obligately required for splicing are underlined. R, purine (A or G); Y, pyrimidine (T, C, or U).

two exons. The intron, still connected to the 3′ exon, is cyclized by ligation of the 3′-OH of the 3′ terminal G to a site 15 nt downstream from the 5′ end of the exon. The products of the reaction are thus the ligated exons, a circular intron, and a 15 nt linear intron oligonucleotide segment (102,103).

TRANSCRIPTION IN MITOCHONDRIA

Mitochondria are eukaryotic oraganelles that reside in the cytoplasm, and are almost certainly derived from a prokaryotic ancestor. Mitochondria contain their own DNA (mtDNA), which encodes a limited number of proteins. The human mtDNA genome is a small compact circle, about 16,500 bp long. Of the 37 genes (none of which has introns), 13 are structural components of the respiratory chain, 2 are rRNAs, and 22 are tRNAs for mitochondrial translation (104). Most of these genes are encoded on one DNA strand, the heavy (H) strand; only one structural gene and 8 tRNA genes are on the light (L) strand. The H-strand genes are "butt-jointed" one to the other, with little noncoding sequence between them. Another feature of the mtDNA genome is symmetry of transcription of the genes (105). Although the H-strand contains most of the coding information, both strands are transcribed over the entire length as two polycistronic messages of essentially equal size (106), using a single promoter region for bidirectional initiation of transcription (107,108). In addition, whereas all of the structural genes contain initiation codons (some contain ATA in the mtDNA instead of ATG because both AUA and AUG specify Met in mtRNA), most of the structural genes lack a termination codon; a T or TA follows the last sense codon in the DNA.

How are the 37 individual RNAs produced, and how are the messages that lack stop codons terminated? The tRNA sequences, almost regularly interspersed among the rRNA and mRNA genes, may function as signals for endonucleolytic cleavage (109,110). In the polycistronic message, these tRNAs would be excised by riboendonucleases (111), releasing both intact tRNAs and the structural messages between them. Individual message would then be cleaved by other specific RNases. As with eukaryotic pol II transcripts, the structural messages are then polyadenylated. Addition of a string of A's explains how the "stop codon-free" messages are terminated: The A's added to the U or UA in those mRNAs result in formation of an in-frame stop codon, UAA.

Mammalian mitochondrial genes do not contain introns, but those of many lower eukaryotes, such as yeast, do contain introns that display two remarkable features not seen in mitochondria of higher eukaryotes. First, several of these intron-containing mRNAs have conserved sequence elements that are also found in the *Tetrahymena* self-splicing RNA (112,113), and are therefore also termed Group I precursor RNAs. Many (but not all) yeast Group I introns are self-splicing, and the reaction product is a circular free intron. There are also mitochondrial Group II introns that self-splice, such as the last intron of the yeast *oxi3* gene, which encodes subunit I of cytochrome *c* oxidase; as with the non-self-splicing introns of nuclear mRNAs, the product of that splicing reaction is a lariat, not a circle (101).

Second, while most of the non-self-splicing introns are removed in the "usual" fashion, at least one actually encodes a protein that is required for the splicing process itself. The gene encoding cytochrome *b* has an intron with an open reading frame contiguous with the preceding exon (Fig. 5-7). This mRNA encodes a protein that is responsible for excision of the intron itself, and is therefore termed a *maturase* (114). This intron-encoded protein is responsible not only for forming the mature cytochrome *b* message, but also for the destruction of its own mRNA!

CONCLUDING REMARKS

We have seen how a few fundamental themes of transcription and its regulation are repeated in altered forms across widely divergent biological systems. These include the motifs of transcription by specific RNA polymerases, postive and negative regulation of promoters and terminators, action at a distance through both cis- and trans-acting elements, the use of simple sequences, dyad symmetry, and AT or GC richness as transcriptional signals, and apparently related modes of splicing. The prevalence of these mechanisms relates to their early origin in evolution.

Although the teleological basis for the existence of introns is still debated, the fact that most excised introns pass through a topologically closed intermediate suggests the possibility that the intron-splicing reaction may have had a common point

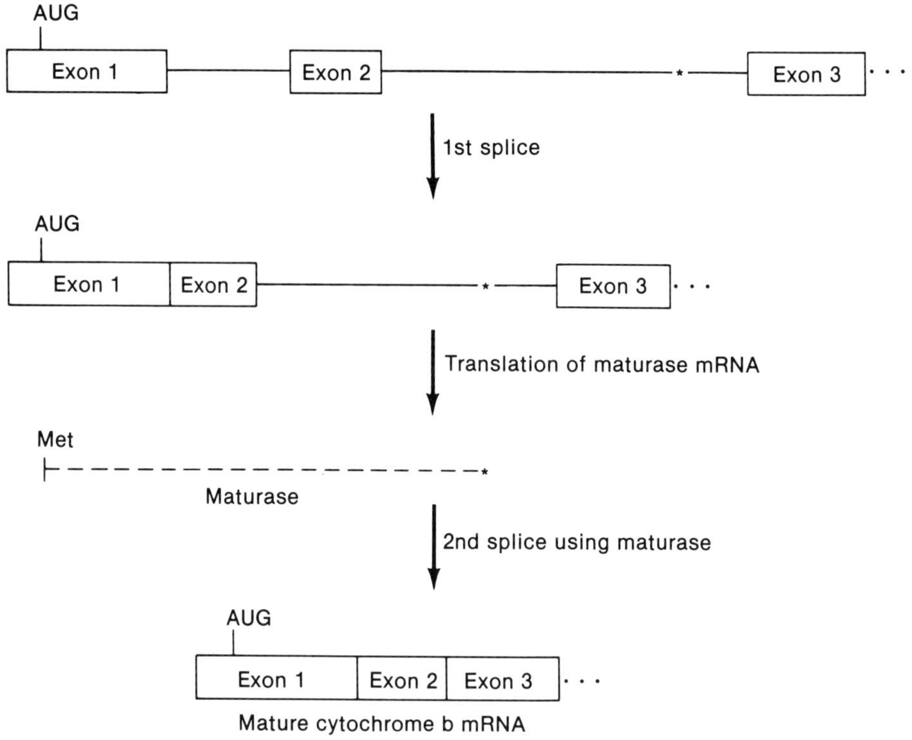

Fig. 5-7. Splicing of yeast cytochrome *b* mRNA using the intron-encoded maturase. Only the beginning of the cytochrome *b* gene is shown; the rest is indicated by the dotted line. The initiator Met (AUG) and maturase translation termination codon (*) are indicated. See text for a detailed description.

of origin early in evolution. Observations about the mechanisms of splicing have also reoriented our thinking about the origin of life, because the discovery of ribozymes provides evidence that the "DNA world" of today may have arisen three billion years ago in an "RNA world" of catalytic nucleic acids that did not require enzymatic proteins, which evolved later. Therefore, the search for meaning in transcription has enlarged our vision into the origin of life itself.

REFERENCES

1. Rodriguez R, Chamberlin M, eds. Promoters: Structure and Function. New York: Praeger, 1982.

2. Hawley DK, McClure WR. Compilation and analysis of *Escherichia coli* promoter DNA sequences. Nucleic Acids Res 1983: 11:2237–2255.

3. Johnson W, Moran C, Losick R. Two RNA polymerase sigma factors from *Bacillus subtilis* discriminate between overlapping promoters for a developmentally regulated gene. Nature (London) 1983; 302:800–804.

4. Travers AA, Burgess RR. Cyclic reuse of the RNA polymerase sigma factor. Nature (London) 1969; 222:537–540.

5. Platt T. Transcription termination and the regulation of gene expression. Annu Rev Biochem 1986; 55:39–72.

6. Platt T. Termination of transcription and its regulation in the tryptophan operon of *E. coli*. Cell 1981; 24:10–23.

7. Christie GE, Farnham PJ, Platt T. Synthetic sites for transcription termination and a functional comparison with tryptophan operon termination sites in vitro. Proc Natl Acad Sci USA 1981; 78:4180–4184.

8. Stroynowski I, Kuroda M, Yanofsky C. Transcription termination in vitro at the tryptophan operon attenuator is controlled by secondary structures in the leader transcript. Proc Natl Acad Sci USA 1983; 80:2206–2210.

9. Roberts JW. Termination factor for RNA synthesis. Nature (London) 1969; 224:1168-1174.

10. Galloway JL, Platt T. In Higgins CF, Booth IR, eds. Regulation of Gene Expression—25 Years On. New York: Cambridge University Press, 1986, pp 155–178.

11. Franklin NC. Conservation of genome form but not sequence in the transcription antitermination determinants of phages λ, Φ21 and P22. J Mol Biol 1985; 181:75–84.

12. Friedman DI, Gottesman ME. Lytic mode of lambda development. In Hendrix R, Roberts J, Stahl F, Weisberg R, eds. Lambda II. Cold Spring Harbor, NY: Cold Spring Harbor Laboratory, 1983, pp 21–52.

13. Barik S, Ghosh B, Whalen W, Lazinski D, Das A. An antitermination protein engages the elongating transcription apparatus at a promoter-proximal recognition site. Cell 1987; 50:885–899.

14. Robert J, Sloan SB, Weisberg RA, Gottesman ME, Robledo R, Harbrecht D. The remarkable specificity of a new transcription termination factor suggests that the mechanisms of termination and antitermination are similar. Cell 1987; 51:483–492.

15. Shine J, Dalgarno L. The 3′ -terminal sequence of *E. coli* 16S rRNA: Complementarity to nonsense triplets and ribosome binding sites. Proc Natl Acad Sci USA 1974; 71:1342–1346.

16. Jacob F, Monod J. Genetic regulatory mechanisms in the synthesis of proteins. J Mol Biol 1961; 3:318–356.

17. Miller J, Reznikoff W, eds. The Operon. Cold Spring Harbor, NY: Cold Spring Harbor Laboratories, 1980.

18. Lee N, Gielow W, Wallace R. Mechanism of AraC autoregulation and the domains of two overlapping promoters P_C and P_{BAD}, in the L-arabinose regulatory region of *Escherichia coli.* Proc Natl Acad Sci USA 1981; 78:752–756.

19. Dunn T, Hahn S, Ogden S, Schleif R. An operator at −280 base pairs that is required for repression of AraBAD operon promoter: Addition of DNA helical turns between the operator and promoter cyclically hinders repression. Proc Natl Acad Sci USA 1984; 81:5017–5020.

20. Ptashne M. Gene regulation by proteins acting nearby and at a distance. Nature (London) 322:697–701.

21. Yanofsky C, Kolter R. Attenuation in amino acid biosynthetic operons. Annu Rev Biochem 1982; 16:113–134.

22. Sollner-Webb B, McKnight S. Accurate transcription of cloned *Xenopus* rRNA genes by RNA polymerase I: Demonstration by S1 nuclease mapping. Nucleic Acids Res 1982; 10:3391–3405.

23. Culotta VC, Wilkinson JK, Sollner-Webb B. Mouse and frog violate the paradigm of species-specific transcription of ribosomal RNA genes. Proc Natl Acad Sci USA 1987; 84:7498–7502.

24. Sollner-Webb B, Tower J. Transcription of cloned eukaryotic ribosomal RNA genes. Annu Rev Biochem 1986; 55:801–830.

25. Financsek I, Mizumoto K, Mishima Y, Muramatsu M. Human ribosomal RNA gene: Nucleotide sequence of the transcription initiation region and comparison of three mammalian genes. Proc Natl Acad Sci USA 1982; 79:3092–3096.

26. Gurney T. Characterization of mouse 45S ribosomal RNA subspecies suggests that the first processsing cleavage occurs at 600 ± 100 nucleotides from the 5′ end and the second 500 ± 100 nucleotides from the 3′ end of a 13.9 kb precursor. Nucleic Acids Res 1985; 13:4905–4919.

27. Sakonju S, Bogenhagen DF, Brown DD. A control region in the center of the 5S RNA gene directs specific initiation of transcription. I: The 5′ border of the region. Cell 1980; 19:13–25.

28. Bogenhagen DF, Sakonju S, Brown DD. A control region in the center of the 5S RNA RNA gene directs specific initiation of transcription. II: The 3′ border of the region. Cell 1980; 19:27–35.

29. Biecker JJ, Martin PL, Roeder RG. Formation of a rate-limiting intermediate in 5S RNA gene transcription. Cell 1985; 40:119–127.

30. Bogenhagen DF, Brown DD. Nucleotide sequences in *Xenopus* 5S DNA required for transcription termination. Cell 1981; 24:261–270.

31. LeStourgeon WM, Lothstein L, Walker BW, Beyer AL. In Busch H, ed. The Cell Nucleus, Vol 9. New York: Academic Press, 1981, pp 49–87.

32. Birnstiel ML, Busslinger M, Strub K. Transcription termination and 3′ processing: The end is in site! Cell 1985; 41:349–359.

33. Hernandez N. Formation of the 3′ end of U1 snRNA is directed by a conserved sequence located downstream of the coding region. EMBO J 1985; 4:1827–1837.

34. Yuo C-Y, Ares M Jr, Kelly JD. Sequences required for 3′ end formation of human U2 small nuclear RNA. Cell 1985; 42:193–202.

35. Proudfoot NJ, Brownlee GG. 3′-Non-coding region sequences in eukaryotic messenger RNA. Nature (London) 1976; 263:211–214.

36. Hagenbuchle O, Wellauer PK, Cribbs DL, Schibler U. Termination of transcription in the mouse α-amylase gene Amy-2^a occurs at multiple sites downstream of the polyadenylation site. Cell 1984; 38:737–744.

37. Frayne EG, Leys EJ, Crouse GF, Hook AG, Kellems RE. Transcription of the mouse dihydrofolate reductase gene proceeds unabated through seven polyadenylation sites and terminates near a region of repeated DNA. Mol Cell Biol 1984; 4:2921–2924.

38. Citron B, Falck-Pederson E, Salditt-Georgieff M, Darnell JE Jr. Transcription termination occurs within a 1000 base pair region from the poly(A) site of the mouse β-globin (major) gene. Nucleic Acids Res 1984; 12:8723–8731.

39. McLauchlan J, Gaffney D, Whitton JL, Clements JB. The consensus sequence YGTGTTYY located downstream from the AATAAA signal is required for efficient formation of mRNA 3′ termini. Nucleic Acids Res 1985; 13:1347–1368.

40. Strub K, Galli G, Busslinger M, Birnstiel ML. The cDNA sequences of the sea urchin U7 small nuclear RNA suggests specific contacts between histone mRNA precursor and U7 RNA during RNA processing. EMBO J 1984; 3:2801–2807.

41. Sassone-Corsi P, Borrelli E. Transcriptional regulation by *trans*-acting factors. Trends Genet 1986; 2:215–219.

42. Maniatis T, Goodbourn S, Fischer JA. Regulation of inducible and tissue-specific gene expression. Science 1987; 236:1237–1245.

43. Dynan WS, Tjian R. Control of eukaryotic messenger RNA synthesis by sequence-specific DNA-binding proteins. Nature (London) 1985; 316:774–778.

44. McKnight S, Tjian R. Transcriptional selectivity of viral genes in mammalian cells. Cell 1986; 46:795–805.

45. Breathnach R, Chambon P. Organization and expression of eucaryotic split genes coding for proteins. Annu Rev Biochem 1981; 50:349–383.

46. Efstratiadis A, Posakony JW, Maniatis T, et al. The structure and evolution of the human β-globin gene family. Cell 1980; 21:653–668.

47. Takahashi K, Vigneron M, Matthes H, Wildeman A, Zenke M, Chambon P. Requirement of stereospecific alignments for initiation from the simian virus 40 early promoter. Nature (London) 1986; 319:121–126.

48. Galas DJ, Schmitz A. DNAase footprinting: A simple method for the detection of protein-DNA binding specificity. Nucleic Acids Res 1978; 5:3156–3170.

49. Fried M, Crother DM. Equilibria and kinetics of lac repressor-operator interactions by polyacrylamide gel eletrophoresis. Nucleic Acids Res 1981; 9:6505–6525.

50. Garner MM, Rezvin A. A gel electrophoresis method for quantifying the binding of proteins to specific DNA regions: Application to components of the *Escherichia coli* lactose operon regulatory system. Nucleic Acids Res 1981; 9:3047–3060.

51. Serfling E, Lubbe A, Dorsch-Hasler K, Schaffner W. Metal-dependent SV40 viruses containing inducible enhancers from the upstream region of metallothionein genes. EMBO J 1985; 4:3851–3859.

52. Tsuda M, Suzuki Y. Transcription modulation in vitro of the fibroin gene exerted by a 200-base-pair region upstream of the "TATA" box. Proc Natl Acad Sci USA 1983; 80:7442–7446.

53. Grosschedl R, Birnstiel ML. Delimitation of far upstream sequences required for maximal in vitro transcription. Proc Natl Acad Sci USA 1982; 79:297–301.

54. Guarente L. Yeast promoters: Positive and negative elements. Cell 1984; 36:799–800.

55. Dynan WS. Promoters for housekeeping genes. Trends Genet 1986; 2:196–199.

56. Dynan WS, Sazer S, Tjian R, Schimke RT. Transcription factor Sp1 recognizes a DNA sequence in the mouse dihydrofolate reductase promoter. Nature (London) 1986; 319:246–248.

57. Melton DW, McEwan C, McKie AB, Reid AM. Expression of the mouse HPRT gene: Deletional analysis of the promoter region of an X-chromosome linked housekeeping gene. Cell 1986; 44:319–328.

58. Brown JR, Daar IO, Krug JR, Maquat LE. Characterization of the functional gene

and several processed pseudogenes in the human triosephosphate isomerase gene family. Mol Cell Biol 1985; 5:1694–1706.

59. Khoury G, Gruss P. Enhancer elements. Cell 1983; 33:313–314.

60. Schaffner W, Serfling E, Jasin M. Enhancers and eukaryotic gene transcription. Trends Genet 1985; 1:224–230.

61. Serfling E, Jasin M, Schaffner W. Enhancers and eukaryotic gene transcription. Trends Genet 1985; 1:224–230.

62. Behringer RR, Hammer RE, Brinster RL, Palmiter RD, Townes TM. Two 3′ sequences direct adult erythroid-specific expression of human β-globin genes in transgenic mice. Proc Natl Acad Sci USA 1987; 84:7056–7060.

63. Gerster T, Matthias P, Thali M, Jiricny J, Schaffner W. Cell type-specificity elements of the immunoglobulin heavy chain gene enhancer. EMBO J 1987; 6:1323–1330.

64. Kadesch T, Zervos P, Ruezinsky D. Functional analysis of the murine IgH enhancer; evidence for negative control of cell-type specificity. Nucleic Acids Res 1986; 14:8209–8221.

65. Wasylyk C, Wasylyk B. The immunoglobulin heavy-chain B-lymphocyte enhancer efficiently stimulates transcription in non-lymphoid cells. EMBO J 1986; 5:553–560.

66. Ephrussi A, Church GM, Tonegawa S, Gilbert W. B lineage-specific interactions of an immunoglobulin enhancer with cellular factors in vivo. Science 1985; 227:134–140.

67. Church G, Ephrussi A, Gilbert W, Tonegawa S. Cell-type-specific contacts to immunoglobulin enhancers in nuclei. Nature (London) 1985; 313: 798–801.

68. Bienz M, Pelham HRB. Heat shock regulatory elements function as an inducible enhancer in the *Xenopus* hsp70 gene and when linked to a heterologous promoter. Cell 1986; 45:753–760.

69. Goodbourn S, Zinn K, Maniatis T. Human β-interferon gene expression is regulated by an inducible enhancer element. Cell 1985; 41:509–520.

70. Goodbourn S, Burstein H, Maniatis T. The human β-interferon gene enhancer is under negative control. Cell 1986; 45:601–610.

71. Parker CS, Topol J. A *Drosophila* RNA polymerase II transcription factor binds to the regulatory site of an hsp70 gene. Cell 1984; 37:273–328.

72. Pelham HRB. Activation of heat-shock genes in eukaryotes. Trends Genet 1985; 1:31–35.

73. McGrogan M, Simonsen CC, Smouse DT, Schimke RT. Heterogeneity at the 5′ termini of mouse dihydrofolate reductase. Evidence for multiple promoter regions. J Biol Chem 1985; 260:2307–2314.

74. Padgett RA, Grabowski PJ, Konarska MM, Seiler S, Sharp PA. Splicing of messenger RNA precursors. Annu Rev Biochem 1986; 55:1119–1150.

75. Nevins JR. Processing of late adenovirus nuclear RNA to mRNA. J Mol Biol 1979; 130:493–506.

76. Hamer DH, Leder P. Splicing and the formation of stable RNA. Cell 1979; 18: 1299–1302.

77. Greer CL, Peebles CL, Gegenheimer P, Abelson J. Mechanism of action of a yeast RNA ligase in tRNA splicing. Cell 1983; 32:537–546.

78. Laski FA, Fire AZ, RajBhandary UL, Sharp PA. Characterization of tRNA precursor splicing in mammalian extracts. J Biol Chem 1983; 258:11974–11980,.

79. Filipowicz W, Shatkin AJ. Origin of splice junction phosphate in tRNAs processed by HeLa cell extract. Cell 1983; 32:547–557.

80. Teem JL, Abovich N, Kauter NF, et al. A comparison of yeast ribosomal protein gene DNA sequences. Nucleic Acids Res 1984; 12:8295–8311.

81. Langford CJ, Gallwitz D. Evidence for an intron-contained sequence required for the splicing of yeast RNA polymerase II transcripts. Cell 1983; 33:519–527.

82. Pikielny CW, Teem JL, Rosbash M. Evidence for the biochemical role of an internal

sequence in yeast nuclear mRNA introns: Implications for U1 RNA and metazoan RNA splicing. Cell 1983; 34:395–403.

83. Zeitlin S, Efstratiadis A. In vivo splicing products of the rabbit β-globin pre-mRNA. Cell 1984; 39:589–602.

84. Grabowski PJ, Seiler SR, Sharp PA. A multicomponent complex is involved in the splicing of messenger RNA precursors. Cell 1985; 42:345–353.

85. Black DL, Chabot B, Steitz JA. U2 as well as U1 small nuclear ribonucleoproteins are involved in premessenger RNA splicing. Cell 1985; 42:737–750.

86. Chabot B, Black D, LeMaster DM, Steitz JA. The 3′ splice site of pre-messenger RNA is recognized by a small nuclear ribonucleoprotein. Science 1985; 230:1344–1349.

87. Lerner MR, Boyle JA, Mount SM, Wolin SL, Steitz JA. Are snRNPs involved in splicing? Nature (London) 1980; 283:220–224.

88. Rogers J, Wall R. A mechanism for RNA splicing. Proc Natl Acad Sci USA 1980; 77:1877–1879.

89. Keller EB, Noon WA. Intron splicing: A conserved internal signal in introns of animal pre-mRNAs. Proc Natl Acad Sci USA 1984; 81:7417–7420.

90. Berget SM, Sharp PA. Structure of late adenovirus 2 heterogeneous nuclear RNA. J Mol Biol 1979; 129:547–565.

91. Ryffel GU, Wyler T, Muellener DB, Weber R. Identification, organization and processing intermediates of the putative precursors of *Xenopus* vitellogenin messenger RNA. Cell 1980; 19:53–61.

92. Breitbart RE, Andreadis A, Nadal-Ginard B. Alternative splicing: a ubiquitous mechanism for the generation of multiple protein isoforms from single genes. Ann Rev Biochem 1987; 56:467–495.

93. Breitbart RE, Nguyen HT, Medford RM, Destree AT, Mahdavi V, Nadal-Ginard B. Intricate combinatorial patterns of exon splicing generate multiple regulated troponin T isoforms from a single gene. Cell 1985; 41:67–82.

94. Borst P. Discontinuous transcription and antigenic variation in trypanosomes. Annu Rev Biochem 1986; 55:701–732.

95. Van der Ploeg LHT. Discontinuous transcription and splicing in trypanosomes. Cell 1986; 47:479–480.

96. Johnson PJ, Kooter JM, Borst P. Inactivation of transcription by UV irradiation of *T. brucei* provides evidence for a multicistronic transcription unit including a VSG gene. Cell 1987; 51:273–281.

97. Cech TR, Bass BL. Biological catalysis by RNA. Annu Rev Biochem 1986; 55:599-629.

98. Cech TR. The generality of self-splicing RNA: Relationship to nuclear mRNA splicing. Cell 1986; 44:207–210.

99. Sharp PA, Eisenberg DE. The evolution of catalytic function. Science 1987; 238:729–730,807.

100. Bonnard G, Michel F, Weil JH, Steinmetz A. Nucleotide sequence of the split $tRNA^{Leu}$ gene from *Vicia faba* chloroplasts: Evidence for structural homologies of the chloroplast $tRNA^{Leu}$ intron with the intron from the autosplicable *Tetrahymena* ribosomal RNA precursor. Mol Gen Genet 1984; 194:330–336.

101. Peebles CL, Perlman PS, Mecklenburg KL, et al. A self-splicing RNA excises an intron lariat. Cell 1986; 44:213–223.

102. Zaug AJ, Cech TR. The intervening sequence excised from the ribosomal RNA precursor of Tetrahymena contains a 5′-terminal guanosine residue not encoded by the DNA. Nucleic Acids Res 1982; 10:2823–2838.

103. Zaug AJ, Grabowski PJ, Cech TR. Autocatalytic cyclization of an excised intervening sequence RNA is a cleavage-ligation reaction. Nature (London) 1983; 301:578–583.

104. Anderson S, Bankier AT, Barrell BG, et al. Sequence and organization of the human mitochondrial genome. Nature (London) 1981; 290:457–465.

105. Aloni Y, Attardi G. Symmetrical *in vivo* transcription of mitochondrial DNA in HeLa cells. Proc Natl Acad Sci USA 1971; 68:1757–1761.

106. Murphy WI, Attardi B, Tu C, Attardi G. Evidence for complete symmetrical transcription *in vivo* of mitochondrial DNA in HeLa cells. J Mol Biol 1975; 99:809–814.

107. Montoya J, Christianson T, Levens D, Rabinowitz M, Attardi G. Identification of initiation sites for heavy-strand and light-strand transcription in human mitochondrial DNA. Proc Natl Acad Sci USA 1982; 79:7195–7199.

108. Montoya J, Gaines G, Attardi G. The pattern of transcription of the human mitochondrial rRNA genes reveals two overlapping transcription units. Cell 1983; 34:151–159.

109. Ojala D, Merkel C, Gelfand R, Attardi G. The tRNA genes punctuate the reading of genetic information in human mitochondrial DNA. Cell 1980; 22:393–403.

110. Ojala D, Montoya J, Attardi G. The tRNA punctuation model of RNA processing in human mitochondira. Nature (London) 1981; 290:470–474.

111. Altman S, Bowman EJ, Garber RL, Kole R, Koski RA, Stark B. Aspects of RNase P structure and function. In Soll D, Abelson JN, Schimmel PR, eds. Transfer RNA: Biological Aspects. Cold Spring Harbor, NY: Cold Spring Harbor Laboratory, 1980, pp 71–82.

112. Davies RW, Waring RB, Ray JA, Brown TA, Scazzocchio C. Making ends meet: A model for RNA splicing in fungal mitochondria. Nature (London) 1982; 300:719–724.

113. Cech TR, Tanner NK, Tinoco I Jr, Weir BR, Zucker M, Perlman PS. Secondary structure of the *Tetrahymena* ribosomal RNA intervening sequence: structural homology with fungal mitochondrial intervening sequences. Proc Natl Acad Sci USA 1983; 80:3903–3907.

114. Lazowska J, Jacq C, Slonimski PP. Sequence of introns and flanking exons in wild-type and BOX-3 mutants of cytochrome b reveals an interlaced splicing protein coded by an intron. Cell 1980; 22:333–348.

6

Genetically Programmed and Adaptive Genome Plasticity in Mammalian Cells: Immunoglobulin Genes and Gene Amplification

Frederick W. Alt

Until the mid 1970s, the mammalian genome was thought to be inviolate throughout differentiation. Then observations in several different systems revealed remarkable plasticity. First, mammalian genes that encode the antigen receptors of lymphocytes were found in separate pieces in the germline; these are assembled, in an irreversible step of differentiation, to form different and complete antigen receptor genes in mature lymphoid cells. Then it was found that tumor cells, in response to environmental selection pressure, can produce and incorporate into their genomes many copies of specific genes; this process is called *gene amplification*. Other forms of genome plasticity include chromosomal transpositions and translocations and the movement of genetic material by retroviruses (RNA tumor viruses). Here we shall focus on the assembly of antigen receptor genes and gene amplification to illustrate some of the basic principles of the fluid genome; some of the other mechanisms will be mentioned in the context of these two systems or will be discussed in other chapters in this volume.

ASSEMBLY OF ANTIBODY GENES DURING B-LYMPHOCYTE DIFFERENTIATION

The Clonal Selection Mechanism

Vertebrates have two distinct systems of immunity, both of which are mediated by lymphocytes. Cell-mediated immunity involves T lymphocytes that react with foreign antigens on the surface of host cells in a process that eliminates fungi, parasites, intracellullar viral infections, foreign tissue or cell grafts, and even cancer cells. Humoral immunity is mediated by antibodies that are secreted by B lymphocytes to control the extracellular phases of bacterial or viral infection.

The immune system can respond specifically to a huge number of diverse antigens by a process called *clonal selection* (Fig. 6-1). Each lymphocyte is predetermined to carry a homogeneous set of antigen-binding receptors; the binding specificity of each clonally derived set of lymphocytes is unique. The fundamental

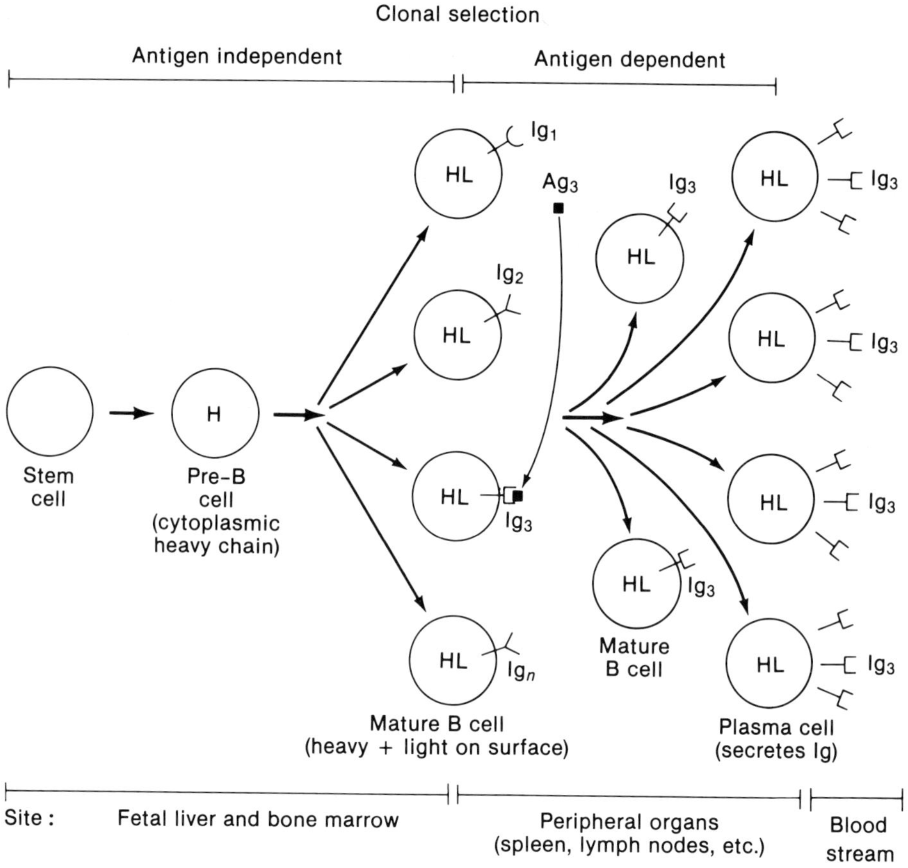

Fig. 6-1. A specific immune response: clonal selection. A clonally derived population of B lymphocytes is produced in primary differentiation organs in the absence of antigen. The total number of specificities encompassed by all B-lymphocyte clones represents the antibody repertoire. This repertoire is vast (10^9 or more unique specificities) and is schematized by Ig_1 through Ig_n. B lymphocytes migrate to peripheral lymphoid organs where they may come into contact with antigens. When antigen reacts with its cognate receptor on a B lymphocyte (illustrated for Ig_3), that cell matures into a cell that secretes into the blood stream antibody with the same specificity as the surface receptor on the progenitor cells. See text for other details.

aspect of the clonal selection mechanism is that all mature lymphocytes (i.e., those that carry an antigen receptor) arise in the absence of antigen. This "antigen-independent" phase of lymphocyte differentiation takes place in primary differentiation organs such as the thymus for T cells and the bone marrow for B cells. Once a lymphocyte matures (acquires its specific antigen receptor), it migrates from the primary differentiation organs to peripheral organs, such as spleen or lymph node, where it may come into contact with a cognate (specific) antigen. Binding of antigen to the specific receptor on the surface of a mature lymphocyte leads to the proliferation and maturation of that lymphocyte into a terminally differentiated effector cell. The receptor on a B lymphocyte is the antibody molecule itself (1); the receptor on a T lymphocyte is a different but related molecule (2). Both receptors are members of a greater immunoglobulin gene superfamily that includes diverse surface molecules found on other cells (2) as well as lymphocytes.

Antibody Structure (1)

Our primary focus shall be the B-lymphocyte lineage and the antibody. However, many of the molecular mechanisms described here also apply to the assembly and expression of T-cell receptor genes. The terminally differentiated cell of the B-cell lineage secretes into the blood large quantities of an antibody that has exactly the same specificity as the cell surface antibody of the parent B cell (Fig. 6-1), ensuring the specificity of the immune response. The molecular mechanism for this behavior will be considered below.

The antibody recognizes the cognate antigen by noncovalent reactions, much like those of enzyme–substrate interactions. The basic unit of an antibody (immunoglobulin) (Fig. 6-2) contains two identical heavy and two identical light polypeptide chains, held together by molecular complementarity and by disulfide bridges.

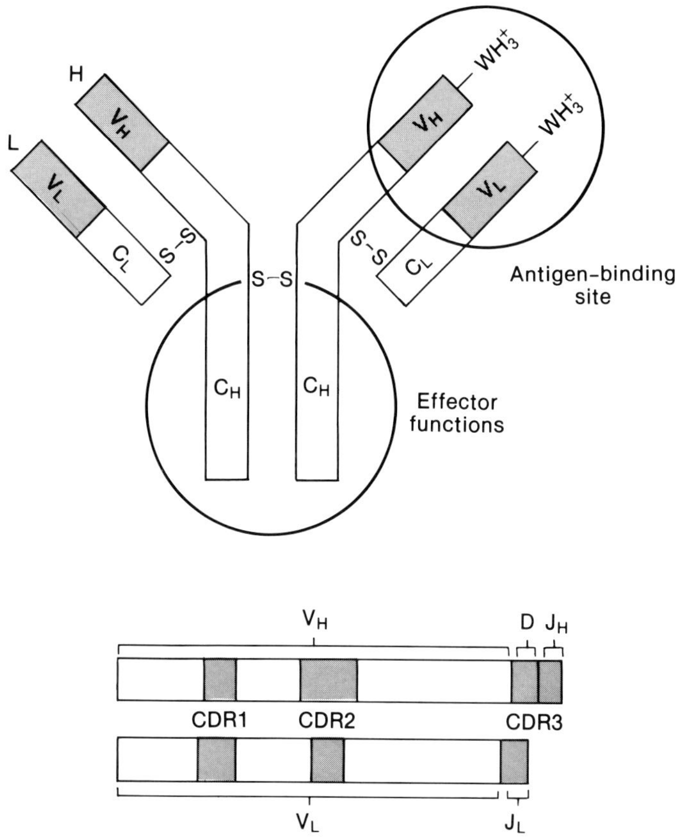

Fig. 6-2. Immunoglobulin (antibody) structure. Top: A schematic representation of a typical subunit of an immunoglobulin molecule. Bottom: A blow-up of the variable region of the heavy and light chain. Note that three hypervariable regions exist within the greater V region; these are referred to as CDRs. Also note that CDR3 is encoded by the V, (D), J junctional region.

Discrete domains of the antibody are involved with recognition and subsequent elimination of antigen. The N-terminus of each heavy or light chain contains a region of variable amino acid sequence *(variable region)* that interacts to form the *antigen-binding site.* Although most variable regions share a common framework of amino acids, there are three sets of *hypervariable* amino acids within the variable region (Fig. 6-2B). These hypervariable or *complementary determining regions* line the *antigen-binding pocket* in the three-dimensional antibody. Hypervariability of some regions is encoded in the germline, but others result from *diversification* mechanisms that act on the gene segments during lymphocyte differentiation. One form of diversification is the assembly of genes for a variable region from a set of germline variable gene segments during differentiation of B lymphocytes (see below).

The carboxyl terminus of the heavy or light chain is identical for each species of heavy or light chain of given class (see below) and determines the effector function of the antibody; this is the *constant region.* The distinct classes of heavy chain constant regions in mammals include μ, α, γ, δ, and ϵ. The type of constant region on the heavy chain influences the effector function of an antibody. That is, the constant region determines which types of cells the antibody will bind to and where it will go in the body. There are two types of light chain, κ and λ, but there is no known role for the different λ and κ constant regions.

The class of immunoglobulin can change during the course of a specific immune response. In other words, clonal lineages of cells can first make an antibody with one type of contant region (μ) and later "switch" to the production of the same specificity of antibody (variable region) attached to a different constant region (e.g., γ) (3). This phenomenon is the result of a programmed genetic rearrangement event that is separate from the one involved in the assembly of the genes for the variable region; as a result, the cell can make antigen-specific antibodies that carry out different effector functions.

Organization and Expression of Immunoglobulin Genes

Mammals can produce as many as 10^9 different antibody specificities, perhaps more. Generation of this diverse repertoire depends on several unusual genetic mechanisms. Comparative (cloning and sequencing) analysis of immunoglobulin structure in nonlymphoid cells and Ig-producing cells or their direct progenitors (usually in appropriate tumor cell analogs) has provided most of our current understanding of the molecular basis of specific Ig gene expression (1,4). The variable regions of heavy and light chains are somatically assembled from component gene segments. The genes that encode the immunoglobulin heavy chains and the genes for the two families of immunoglobulin light chains are each clustered on separate chromosomes. The heavy chain family is the most complex and well studied and will be described in greatest detail.

The variable region of an immunoglobulin heavy chain is encoded by three separate germline DNA gene segments (1). The bulk of the variable region is encoded by a *V_H segment* (96 amino acids), which is linked to a *D (diversity) segment* (1–15 amino acids), which is, in turn, linked to a *J_H (joining) segment* (10 amino acids) (Fig. 6-1). These three sets of segments are found in clusters along the chromosome

(Fig. 6-3). In the mouse, four different J_H segments lie about 7kb upstream from the exons that encode the μ constant region. About 10 diversity segments (each about 20 bp long) are scattered in 80 kb of DNA just upstream of the J_H segment cluster. This entire region of the chromosome has been cloned and mapped. It is estimated that there are as many as 1000 V_H segments in the mouse, spread over as much as 10 million base pairs or more; the average spacing between V gene segments is 10–15 kb. The heavy chain variable region locus in humans has been physically linked to the C_H locus by pulse-field gradient gel electrophoresis; the most proximal V_H segment lies within 80 kb of the 5 μ constant region gene (5).

The light chain variable region of either the κ or λ type is encoded by only two gene segments, the V_L and J_L (1) (Fig. 6-3). The general organization and composition of the two light chain families vary from species to species but, in general, they are similar to those of the heavy chain genes. For the κ locus in mouse, there are four functional J segments upstream from the gene segment that encodes the κ constant region. There are probably hundreds of $V\kappa$ gene segments, generally upstream from the J segments. Again, the physical linkage of V_K to J_K segments recently has been determined in humans; the nearest V_K segment is within 20 kb of the J_K segments (6).

The heavy and light chain variable region genes are assembled during the antigen-independent stages of B-lymphocyte differentiation (4,5) (Fig 6-3). The first known rearrangement involves the assembly of the heavy chain variable region. First a D segment is joined to a J_H segment, usually on both chromosomes. This leads to a precursor B lymphocyte with partially formed heavy chain genes; these are "DJ_H" intermediate rearrangements. The other gene segments have not been rearranged. The next step is the joining of a V_H segment to these preexisting DJ_H intermediates, leading to complete assembly of the heavy chain variable region gene (i.e., the V_HDJ_H). All of these events occur by a "deletion" mechanism; that is, intervening DNA between the joined segments is deleted from the chromosome and lost forever from the cell (and its progeny).

Once a complete gene for the heavy chain variable region has been assembled, it can be expressed by conventional mechanisms (4) (Fig. 6-3). Each germline V gene segment contains a transcriptional promoter. Control elements (enhancers) downstream of the J segments augment the activity of the promoter from the V segment employed in the V_HDJ_H rearrangement; thus, the assembled heavy chain variable region gene can be expressed at high levels in mature B cells whereas the promoters of unrearranged V_H segments are still silent. The primary RNA transcript runs downstream through the exons that encode the μ heavy chain constant region gene. Primary transcripts are "terminated" at one of two sites downstream of the constant region gene, leading to two different types of primary transcript (Fig. 6-3). Through RNA processing, intron sequences are removed to generate the final form of the heavy chain mRNA. Depending on which forms of the primary transcript are generated, the processed mRNA can have a final (3′) exon that encodes a membrane-bound or a secreted form of the protein (Fig. 6-3). At each stage of differentiation, the relative level of the two forms of the heavy chain mRNA depends on how often the primary transcript ends between the two alternative exons. Changes in the relative amounts of the two transcripts allow a B cell to produce primarily a membrane-bound form of the heavy chain and its clonally derived

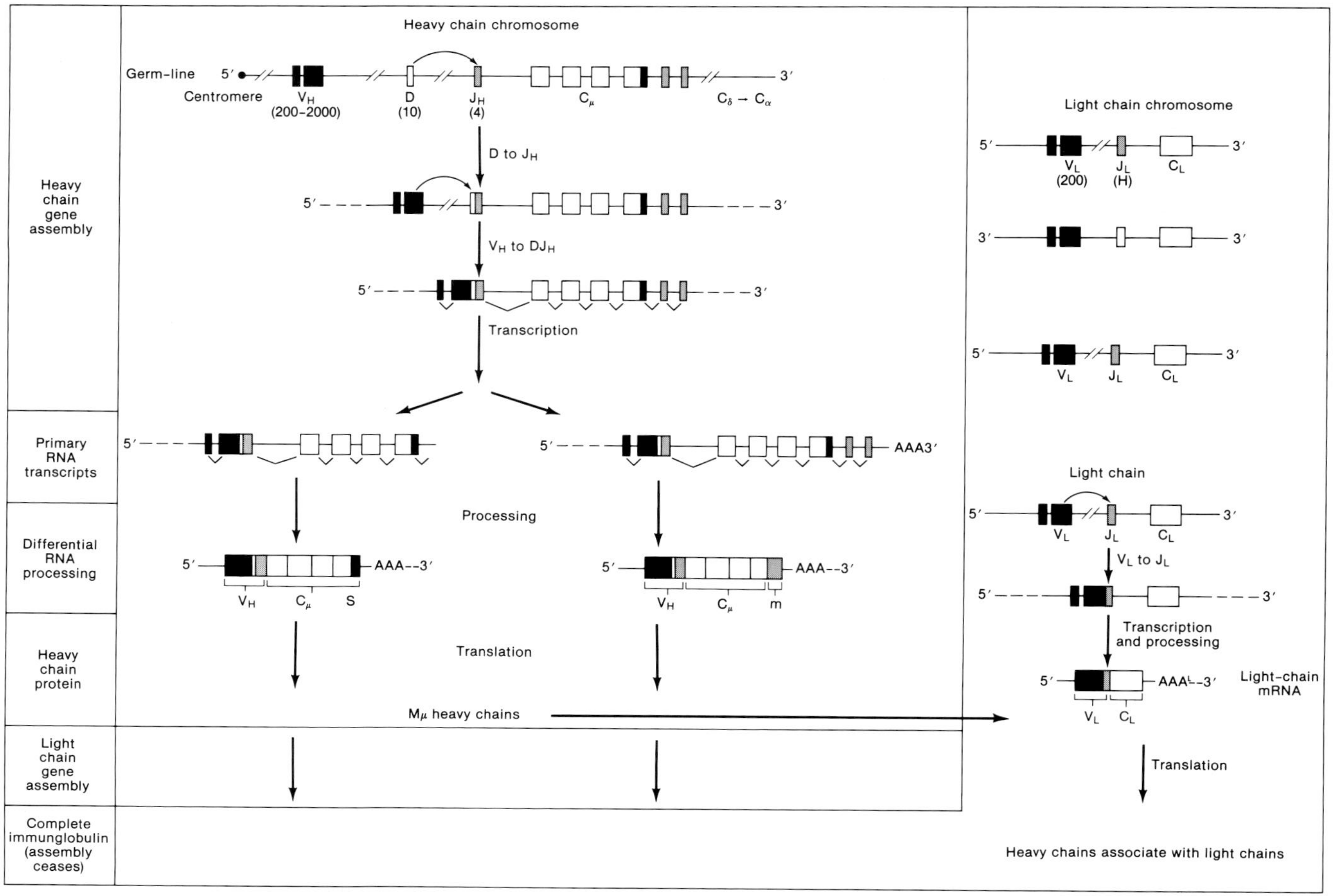
Heavy chain gene assembly
Primary RNA transcripts
Differential RNA processing
Heavy chain protein
Light chain gene assembly
Complete immunglobulin (assembly ceases)
Heavy chain chromosome
Germ-line
5′
3′
Centromere
V_H (200-2000)
D (10)
J_H (4)
C_μ
$C_\delta \rightarrow C_\alpha$
D to J_H
V_H to DJ_H
Transcription
AAA3′
Processing
AAA--3′
V_H
S
m
Translation
Mμ heavy chains
Light chain chromosome
V_L (200)
J_L (H)
C_L
Light chain
V_L to J_L
Transcription and processing
AAA--3′
Light-chain mRNA
Translation
Heavy chains associate with light chains

plasma cell progeny to produce primarily a secreted form of the heavy chain, with exactly the same variable region (specificity).

During B-lymphocyte differentiation, assembly of light chain variable region genes (V_L to J_L joining) generally follows assembly of the heavy chain variable region and expression of the complete heavy chain protein (4,5) (Fig. 6-3). Separating the heavy chain joining and expression phases from that of light chains may be necessary for variations in the details of the joining mechanisms and accurate control of the two processes (described later). Light chain genes are also expressed by conventional mechanisms, but only one form of the light chain constant region can be produced (Fig. 6-3).

Generation of Diversity

The immense diversity generated by the mammalian immune system relies in large part on the somatic process of gene assembly, generating a nearly limitless set of genes for different antigen receptors, assembled from a finite amount of DNA (1). Most mechanisms include (1) the diverse array of germline-encoded V, D, and J segments and (2) the apparent lack of restriction in the combinations of assembly of heavy and light chain V gene elements. For example, there are about 1000 V_H gene segments, 10 D segments, and 4 J_H segments; random assortment could lead to as many as 40,000 distinct combinations. Similar considerations lead to an estimate of nearly 1000 possible V_L to J_L combinations. If all of the different potential heavy and light chain genes are assembled in developing lymphocytes, and because the antigen-binding site is formed from the interaction of the variable regions of the heavy and light chains, this system alone could generate nearly 40,000,000 different antibody variable regions.

In addition to the combinational assortment of germline gene segments, the assembly of variable regions generates additional diversity, altering the nucleotide sequences of the interfaces of the junctional regions of the V_H, D, and J_H regions and of the V_L and J_L (1,4). This is achieved by removing bases and or adding bases *de novo* (without a template) to these junctions. The addition and loss of bases in a region of the gene that encodes as antigen-contact region (Fig. 6-1B) lead to changes in the structure of the protein in this region by changing the sequences of amino acids, or by loss or gain of amino acids. Thus, this process changes the

←

Fig. 6-3. Organization and expression of immunoglobulin genes. Some of the major phases or regulatory points in immunoglobulin gene rearrangement and expression during the antigen-independent phases of B-lymphocyte differentiation are indicated on the left. A schematic diagram of the general organization of heavy and light chain loci is indicated at the top; see text for details of this organization. The numbers in parentheses under the V_H, D, J_H, V_L, and J_L segments indicate the approximate numbers of those types of segments in the mouse genome; for V_L and J_L the numbers refer to the κ locus. Alternative RNA processing as a result of differential polyadenylation (termination) of primary transcripts leads to the generation of heavy chain mRNA sequences that encode the membrane-bound or secreted form of the heavy chain protein; these are indicated in schematic form where S stands for a secretion-specific mini-exon and M stands for a membrane-specific mini-exon. Note that light chain rearrangement generally occurs only after the assembly and expression of a heavy chain gene. Other details are contained within the text.

molecular nature of the antigen-binding pocket and, therefore, the specificity of the antibody.

Thus, somatic V gene assembly can generate billions of different antibodies from a few thousand gene segments. The small price that is paid for this remarkable diversity-generating system occurs because coding information is randomly added or lost at the interfaces of the segments. Given the triplet reading frame for translating the code into amino acids, approximately two-thirds of the assembled genes contain segments that are not aligned properly (4,5). These "aberrant" rearrangements do not yield RNA that can be translated into proteins. However, cells have two chromosomes for each immunoglobulin locus and there may be mechanisms to "salvage" aberrant rearrangements; it seems that much fewer than 50% of differentiating lymphocytes fail to assemble active immunoglobulin genes on one of their chromosomes. This is a small price to pay for a system in which the primary cells (mature B lymphocytes) actually have a short life span and are constantly regenerated throughout the life of an animal.

Allelic Exclusion (4,5)

The clonal selection mechanism of the specific immune response requires a given lymphocyte clone to express antigen receptors of unique specificity. This occurs by the process of *allelic exclusion.* That is, functional variable region genes are assembled at just one of the two heavy chain loci (chromosomes) and one of the several light chain loci (κ and λ, each on different chromosomes). Thus, even if the process for assembly of the variable region is completely random with respect to the V_H, D, and J_H segments that could be chosen on each chromosome, a given B-cell clone will make just one functional V_HDJ_H combination and one functional V_LJ_L combination. Therefore, each clone of B cells makes a single species of antibody. Although the mechanism that regulates this process has not been completely elucidated, it seems to work by a feedback mechanism. If a functional V_HDJ_H variable region is assembled on one chromosome, the heavy chain protein that derives from it will somehow give a signal that prevents further rearrangement on the heavy chain locus on the other chromosome. Multiple functional light chain rearrangements are similarly excluded by a feedback from complete immunoglobulins.

Heavy Chain Class Switching (3)

As mentioned earlier, the type of heavy chain constant region gene can be exchanged in a given clonal B-lymphocyte lineage without a change in the variable region specificity. This process is mediated by a recombination–deletion mechanism different from that involved with variable region gene assembly and that is referred to as *heavy chain class switching.*

The organization of the heavy chain constant region is outlined in Fig. 6-4. Just upstream of every constant region gene is a region of repetitive sequences (of up to several kilobases), a *switch region.* These switch region sequences apparently serve as targets for an entirely different recombination enzyme system: the *class switch recombinase.* Switching from the production of a μ heavy chain to a γ heavy chain involves the juxtaposition of μ and γ switch region sequences and the concomitant

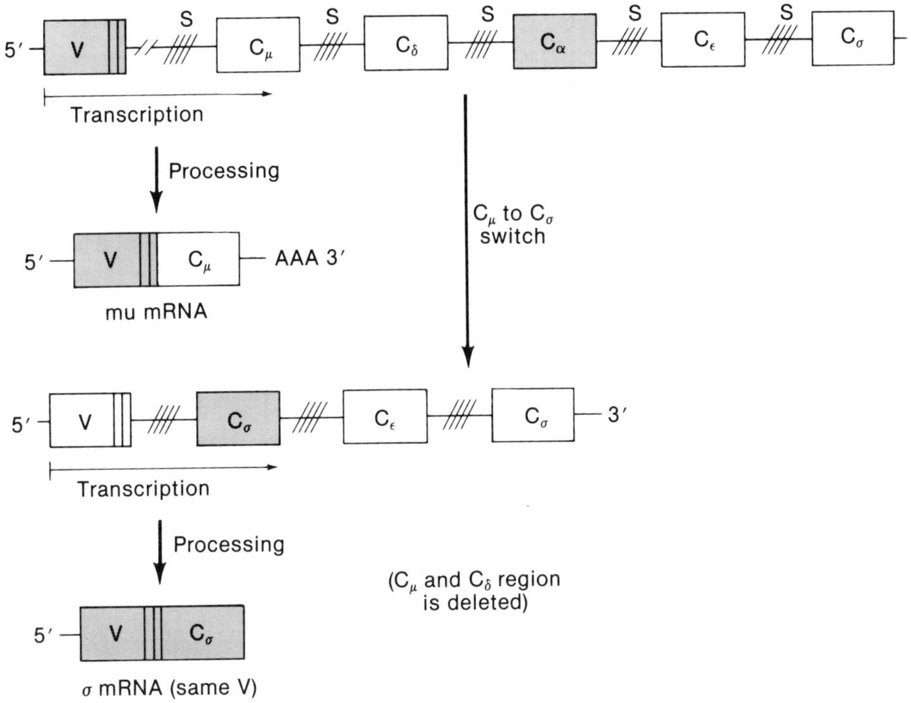

Fig. 6-4. Heavy chain class switching. A simplified diagram of the organization of the different gene segments that can encode the heavy chain constant region is shown at the top of the figure; the different constant region gene segments are spread out over more than 200 kb of DNA. After assembly of a V_HDJ_H variable region, the assembled variable region is expressed in association with the immediate downstream constant region (μ). At a later stage of maturation, a given B-cell clone can replace the μ constant region with a downstream constant region. This is illustrated for a μ to γ switch and involves a recombination event that juxtaposes the μ and γ switch regions (repetitive sequence regions indicated by S in the figure) and deletes the intervening DNA from the chromosome. After the switch, progeny cells can now express the γ constant region in association with the same V_HDJ_H variable region sequence that was originally expressed in association with a μ constant region.

deletion of all intervening genetic material, including the μ and δ constant region genes (Fig. 6-4). Now, the γ constant region can be expressed in association with the variable region that was first expressed with a μ constant region: maintaining specificity while changing effector function as outlined above. How the class switch recombination event occurs and how it is regulated are just beginning to be investigated in molecular detail.

Translocations into the Immunoglobulin Locus: Oncogene Activation (7–9)

Aberrant forms of variable region gene or class switch recombination have been implicated in some human diseases, particularly the chromosomal translocations seen with lymphoid malignancies. In many B lineage tumors, these translocations involve an immunoglobulin heavy or light chain locus on one chromosome. Several pre-B or B-cell leukemias (chronic lymphocytic leukemia, follicular lymphoma, acute lymphocytic leukemia, and endemic Burkitt's lymphoma) involve joining of the heavy J_H sequences to a restricted set of loci on other chromosomes that contain c-*myc* or other known or presumptive oncogenes (cellular cancer genes). Probably the best known and characterized translocations are those observed in murine plasmacytomas (myelomas) and sporadic human Burkitt's lymphoma. Most of these cases involve translocations of the oncogene locus on one chromosome into one of the heavy chain switch regions on a different chromosome. Molecular characteristics of the J_H or class switch associated translocation joints suggest a role for the respective recombinase systems in these aberrant translocation events (7). The occurrence of translocations into the J_H (V) locus in some tumors and into the switch regions in others seems to correlate with the differentiation stage of the transformed cell and implicates the prevailing recombinase activity at that stage (VDJ in earlier stages; class switch later) with at least a partial role in mediating the rearrangement event. It seems likely that aberrant translocation events are the result of the low frequency of breakdown in the normal recombination mechanism. However, if such a recombination event in some way activates a cellular growth-promoting locus (oncogene), it might give a powerful selective growth advantage to that cell, leading to the frequent observation of translocations in some types of malignant cells.

GENE AMPLIFICATION (10)

The discussion of oncogene activation by translocations in lymphoid tumors leads to the second topic: selective gene amplification. *Gene amplification* is a mechanism by which tumor cells achieve increased growth potential as well as a medium that may provide resistance to chemotherapeutic drugs. Thus, gene amplification is a fundamental problem in cancer.

In normal cells, the only documented examples of developmentally programmed gene amplification involve the generation of large numbers of extrachromosomal copies of the genes that encode rDNA in *Xenopos* oocytes and amplification of genes for chorion (an egg shell protein) in the follicular nurse cells of *Drosophila melanogaster*. Both of these states of amplification are transient; either the amplified copies or the cells are lost in development. Stable gene amplification in somatic mammalian cells has been observed only as an adaptive process, originally seen in studies of drug resistance in tumor cells (11).

Gene Amplification and Drug Resistance

The first demonstration of amplification in mammalian cells addressed the question of drug resistance. Mutants of cultured mouse cells become resistant to the growth inhibition by ever-increasing (stepwise) concentrations of methotrexate (mtx), a drug that blocks the activity of an essential cellular enzyme, dihydrofolate reductase (DHFR) (11). Methotrexate-resistant mutants produced increased levels of DHFR so that free enzyme was found in spite of high concentrations of inhibitor. This was the result of gene amplification; normal cells contain two copies of the DHFR gene, but the mtx-resistant cells contained hundreds to thousands of copies, with a corresponding increase in production of the DHFR enzyme and resistance to the drug (11). Amplification of many other genes has since been found in other types of drug-resistant cells selected by the stepwise procedure.

Increased production of a particular gene product can, in theory, occur by other mechanisms, including increased expression of a single gene copy as seen in some bacterial mutants. However, mutations that lead to gene amplification (perhaps just duplications or low-level multiplications) occur much more frequently than the mutations that lead to increased expression of a single gene (10). Thus, if a growing mammalian cell can grow better by producing increased amounts of a particular gene product, the most likely mechanism for achieving that overproduction in derivative mutant cells is by amplication of the gene, a fact of great significance in tumor progression, as described below.

Under appropriate selection pressures, it is possible to select cells that have extremely high copy numbers of specific genes. For example, some cells selected for growth in high levels of specific inhibitors can share 1000-fold or more amplification of genes that encode the target enzyme, producing nearly half of the total protein as the target enzyme. In addition, the length of the amplified unit of DNA can be very large. The target genes may ordinarily be only several kilobases long, but the amplification unit may contain up to 3000 kb of flanking DNA. These large units are attributed to the amplification mechanism and sometimes contain genes that are physically linked to the target gene but are not related to the selection process. Such genes are therefore called "passenger" genes (12). Amplification in cultured cells may arise by multiple rounds of DNA replication over a limited chromosomal region in a single cell cycle ("disproportionate replication") (10). This mechanism has been demonstrated in the amplification of the *Drosophila* gene for chorion. Agents that interfere with DNA replication such as viruses or drugs may augment the frequency of amplification. Adaptive gene amplification has been observed only in transformed cells; it is not known whether this occurs in normal cells.

Location of Amplified Genes

Amplified genes are usually found in one of two forms in the nucleus; they may be integrated into the chromosome as karyotypic abnormalities called *homogeneously staining regions (HSRs)* or they may reside in small extrachromosomal, but chromosome-like structures, that are called "double minutes" (10). HSRs are disruptions of the normal metaphase chromosome-banding patterns; such extended areas of the chromosome (Fig. 6-5A) contain an array of linked amplification units. HSRs are often found on chromosomes other than those that normally contain the amplified gene; thus, the amplification events frequently involve transpositions as well. The extrachromosomal double minutes (DMs) are small self-replicating bodies that are visible in the nucleus at metaphase (Fig. 6-5B). Whereas HSRs are stable structures, DMs can often be lost from cells. Thus, if cells containing DMs are removed from selection pressure, the DMs and their amplified genes may be rapidly lost. It is not known why some cells exhibit amplified genes in HSRs and others in DMs. However, given enough time in selection pressure, most cells are dominated by chromosomally integrated genes; the greater stability gives selective growth advantage.

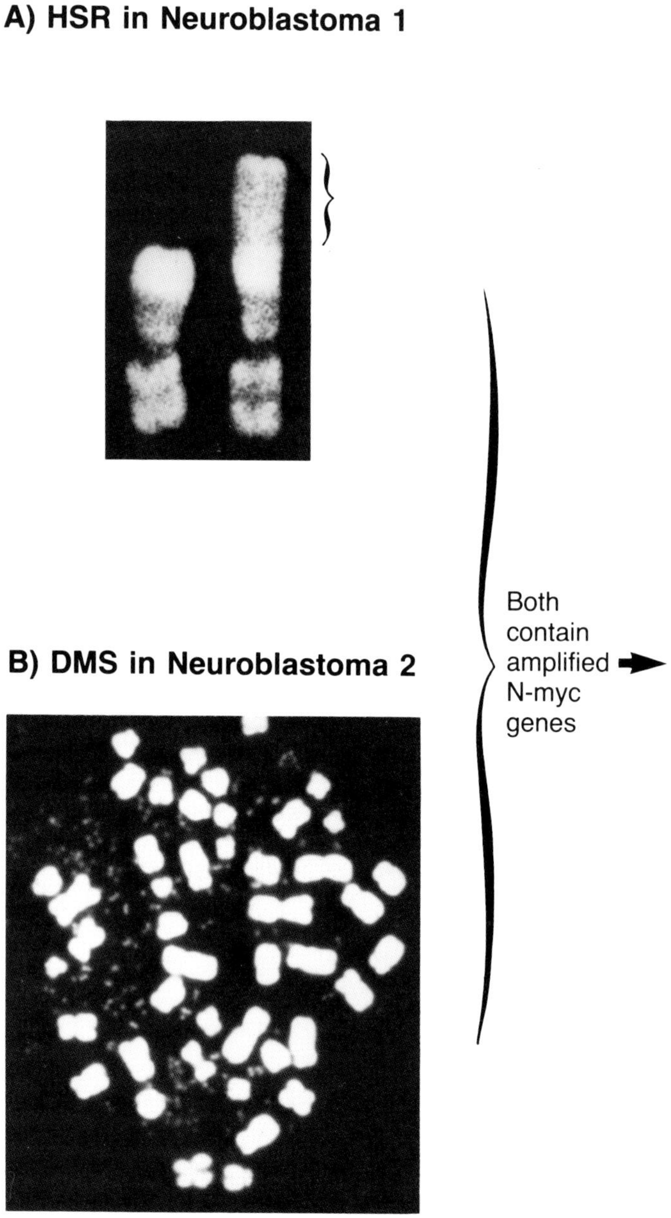

Fig. 6-5. Detection of amplified genes. (A) Banded human number 1 chromosomes from a neuroblastoma cell line. The chromosome at the right has the banding pattern of a normal number 1 chromosome. The chromosome at the left has an interruption of the bright band at the end of the long arm, which is visible as an extended area of lesser intensity staining (bracket). The extended area is a classical HSR; a chromosomal manifestation of amplified genes. (B) A metaphase spread of total chromosomes from a second human neuroblastoma cell line. No chromosomal anomalies are obvious, but the dozens of dot-like structures visible between and around the chromosomes

C) Southern blot assay

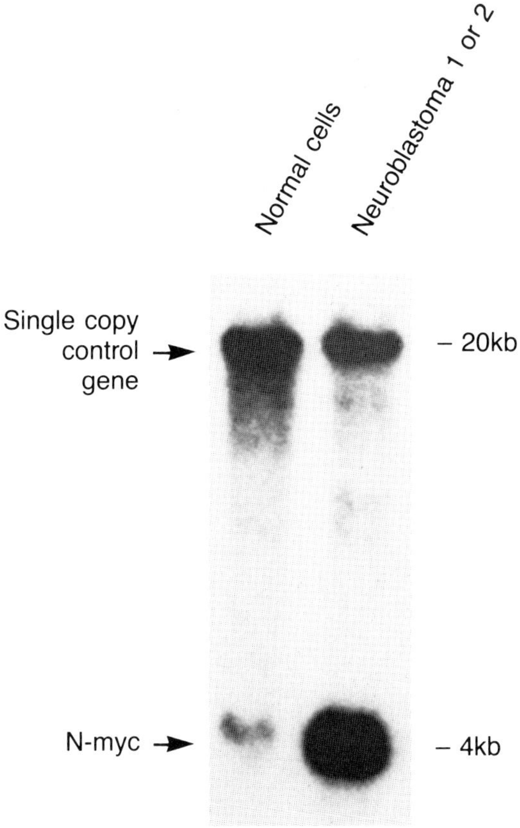

represent DMs; extrachromosomal manifestations of amplified genes. Both the HSRs and DMs in these lines were shown to contain amplified N-*myc* genes A (15). (C) A representative "Southern blotting" analysis of amplified N-*myc* oncogenes in human neuroblastoma DNA. Equal amounts of DNA (10 μg) from normal cells or neuroblastoma cells were cut with the restriction enzyme *Eco*RI and fractionated by electrophoresis through agarose gels. After electrophoresis, the DNA was transferred to nitrocellulose paper that was subsequently incubated with ^{32}P-labeled probes for the N-*myc* gene and for a cellular gene that is not amplified in either DNA source. After hybridization, the filter was washed to remove nonhybridized probes and the specifically hybridized probes visualized by autoradiography. As would be expected, the intensity of the band derived from hybridization of the single copy probe to its complementary restriction fragment (20 kb) was identical for the two samples and serves as an internal control. However, the intensity of the band that hybridized to the N-*myc* probe (4 kb) was more than 100-fold greater in the neuroblastoma DNA than in the normal cell DNA; this result indicates that the N-*myc* gene is amplified more than 100-fold in the neuroblastoma DNA.

Neither HSRs nor DMs have been observed in normal human cells. However, cytogeneticists have noted these chromosomal or extrachromosomal manifestations of amplified genes in tumor specimens from patients who had not had any drug therapy. The tumors may have had amplified genes for products involved in promoting growth of the tumors. Solid human tumors often display HSRs or DMs perhaps most striking in the rapidly progressive forms of neuroblastoma. These findings suggest that all progressive neuroblastomas may have amplified a common gene to account for the characteristics of that tumor.

Isolation of Amplified Genes

Different approaches have been used to isolate the amplified and overexpressed genes of drug-resistant cell lines or tumors. *Subtractive hybridization* (11) was used to identify and isolate the amplified dihydrofolate reductase genes from methotrexate-resistant cells. This method involves making cDNA copies of the RNA from cells that contain amplified genes, then hybridizing that cDNA with RNA from related cells that do not have amplified genes. If equivalent amounts of cDNA and RNA are used, all sequences expressed at similar levels in the two cell types should be driven into RNA/cDNA hybrids. On the other hand, most of the overexpressed sequences that arise from amplified genes will remain unhybridized because they are in excess. The unhybridized sequences can be separated from the hybridized sequences by chromatography and the purified single-stranded sequences can be used to make cDNA libraries. This procedure had been useful for cloning sequences from amplified genes for which no known probes were available (11–13).

For amplified genes of tumors cells that are not associated with drug resistance, the usual approach is to screen tumor cells by Southern blotting for amplification of sequences related to known viral or cellular oncogenes (Fig. 6-5C); cellular oncogenes are the normal counterparts to the transforming genes of RNA tumor viruses (14). This method is based on the fact that the DNA from related genes hybridizes with each other. This method has been widely used, but it is limited to the identification of known genes or their close relatives.

Oncogene Amplification in Tumor Cells

Amplified oncogenes in numerous tumors have been identified, including the amplification of genes for growth factor receptors in human adult brain tumors and genes that encode nuclear oncoproteins (i.e., *myc* genes) in other tumors (14,15). Some oncogenes frequently have amplification as a mechanism of activation, and some tissues frequently give rise to tumors that have amplified oncogenes, particularly tumors of "neural" origin.

Probably the best studied example of oncogene amplification in tumor cells is the neuroblastoma (15). Genomic blotting analysis of DNA from neuroblastomas that contained HSRs or DMs failed to identify amplification of any known cellular oncogenes, but did identify a sequence that was commonly amplified in neuroblastomas and that was related to the c-*myc* cellular oncogene (Fig. 6-5C). The c-*myc* gene is the mammalian counterpart of the transforming gene of an avian retrovirus; it encodes a nuclear protein that is thought to be involved in the regulation of cellular growth and proliferation (14). The best known form of deregulated c-*myc* expression involves the translocation of this gene to the immunoglobulin locus in Burkitt's lymphomas. However, deregulated expression of the c-*myc* gene by different mechanisms (including frequent amplification) has also been associated with several other tumors (14). The c-*myc*-related gene that was found to be amplified in human neuroblastomas was termed *N-myc,* based on the presence in neuroblastomas and homology to c-*myc* (16). The *N-myc* gene and gene product were found to be highly homologous to, but not identical with, c-*myc* (15). The *N-myc* gene was highly amplified (from 25 to 1000-fold) in all tested human neuroblastoma cell lines that had HSRs or DMs, but not in neuroblastomas that did not contain these chromosomal abnormalities (15,17).

Neuroblastomas can be classified into four clinical stages, based on patterns of tumor growth and metastasis. Stage 1 is a localized tumor whereas stage 4 has spread to remote sites, including the bone marrow. Amplification of the N-*myc* gene is correlated with the more advanced stages of the tumor; furthermore, at any stage of the tumor, N-*myc* amplification is associated with the worst prognosis (18). These correlations are probably the best example, to date, of the clinical usefullness of oncogene amplification as a diagnostic and prognostic indicator. Furthermore, these findings imply that N-*myc* is the "neuroblastoma-specific" oncogene.

A *myc* Oncogene Family

Additional support for the contention that N-*myc* amplification is important in oncogenesis came from the demonstration that the n- and c-*myc* genes are functionally related in oncogenic potential. For example, gene transfer of cloned c-*myc* genes (in appropriate expression vectors) alone into normal fibroblasts does not lead to oncogenic transformation. However, in the same cells, oncogenic transformation follows introduction of the c-*myc* expression vector in combination with a vector that expresses an activated oncogene derived from a member of the *ras* family of cellular oncogenes (19). An expression construct that contains the N-*myc* gene has transforming activity identical to c-*myc* in ability to "complement" the activated *ras* gene to cause oncogenic transformation of primary fibroblasts, indicating that the N-*myc* gene is a cellular oncogene even though it has not been found as a component of any tumor virus (20).

One other c-*myc*-related gene has been identified because it is frequently amplified in a form of small cell lung carcinoma (SCLC); correspondingly, the gene has been named L-*myc* (21). The sequence and organization of the l-*myc* gene are similar to the N- and c-*myc* genes, and L-*myc* shows a similar transforming activity in the complementation assay described above (15). Thus, there seems to be a family of *myc* genes that encodes nuclear oncoproteins; and others have been characterized at a preliminary level. Other than the nuclear localization of the proteins and the transforming activity in the complementation assay, little is known about the function of these proteins; controversial evidence suggests that they are involved in the regulation of specific gene expression. In spite of the apparent similarities in structure and function of *myc* family genes, high level expression of N- and L-*myc* genes is restricted with respect to tissue and developmental stage, whereas that of c-*myc* is more generalized (22). In addition, these genes are differentially expressed during the differentiation of specific cell lineages. These findings suggest that differential, or perhaps combinatorial, expression of *myc* family genes plays a fundamental role in normal cellular differentiation.

Differential Amplification of *myc* Family Genes

The N-*myc* gene is expressed, at least at low levels, in all tested neuroblastomas, including some tumors in which the gene is not amplified at any stage (12). By analogy to drug resistance gene amplification, N-*myc* amplification may not be a primary event in the generation of the tumor, but increased expression of the N-*myc* gene as a result of amplification may confer increased growth potential to derivative cells. This pattern would be consistent with the association between N-*myc* amplification, or oncogene amplification in general, and tumor progression. Despite extensive search, N-*myc* amplification has been found in few tumors other than neuroblastomas: only in a few retinoblastomas and the most severe form of SCLC. The L-*myc* gene has been found amplified only in SCLC.

Given the similar oncogenic characteristics of the three known *myc* family genes, it is not clear why activated N- and L-*myc* expression is limited to a few tumors and only one mechanism (amplification). The genes or the type of selective pressure that lead to amplification may play a special role in these tumors. Furthermore, amplification may be targeted by high level expression of the gene in normal precursor cells. Distinct expression patterns of these genes in normal development may denote the types of tumors in which they are expressed or activated (15). For example, the highest levels of N-*myc* and L-*myc* expression are found in developing neural tissues; neuroblastomas and retinoblastomas derive from the neuroectoderm. Also, L-*myc* expression is found in the developing lung and SCLC cells have some "neural" characteristics. Corresponding to the activation in different tumor types, the c-*myc* gene is expressed at high levels in many different developing tissues. Understanding the mechanisms that lead to amplification of this family of oncogens awaits a better understanding of the normal role of these genes.

ACKNOWLEDGEMENT

Supported by NIH grants CA42335, CA21112, CA40427, AI20047, CA23767, and ACS grant CD-269 and by the Mallinckrodt Foundation and the Howard Hughes Medical Organization.

REFERENCES

1. Tonegawa S. Somatic generation of antibody diversity. Nature (London) 1983; 302:575–581.

2. Hood L, Kronenberg M, Hunkapiller T. T cell antigen receptor and the immunoglobulin supergene family. Cell 1985; 40:225–229.

3. Shimizu A, Honjo T. Ig class switching. Cell 1984; 36:801–803.

4. Alt FW, Blackwell TK, Yancopoulos GD. Development of the primary antibody repertoire. Science 1987; 238:1079–1087.

5. Berman JE, Mellis SJ, Pollock R, Smith CL, Suh H, Heinke B, Kowal C, Surti U, Chess L, Cantor CR, Alt FW. Content and organization of the human Ig V_H locus: Definition of three new V_H families and linkage to the Ig C_H locus. EMBO 1988; 7:727–738.

6. Klobeck HG, Zimmer F-J, Combriato G, Zachau HG. Linking of the human immunoglobulin $V_K J_K C_K$ regions by chromosomal walking. Nucleic Acid Res 1987; 15: 9655–9665.

7. Haluska FG, Tsujimoto Y, Croce CM. Mechanisms of chromosome translocation in B and T-cell neoplasia. Trends Genet 1987; 3:11–15.

8. Rabbitts TH. The c-myc proto-oncogene: Involvement in chromosomal abnormalities. Trends Genet 1985; 1:327–331.

9. Klein G. The role of gene dosage and genetic transposition in carcinogenesis. Nature (London) 1981; 294:313–318.

10. Schimkel R. Gene amplification in cultured animal cells. Cell 1984; 7:705.

11. Alt FW, et al. Selective multiplication of dihydrofolate reductase genes in methotrexate-resistant variants of cultured murine cells. J Biol Chem 1978; 253:1357–1371.

12. Kohl N, Gee C, Alt F. Activated expression of the N-myc gene in human neuroblastomas and related tumors. Science 1984; 226:1335.

13. Michitsch RW, Montgomery KT, Melera P. Expression of the amplified domain in human neuroblastoma cells. Mol Cell Biol 1984; 4:2370.

14. Varmus H. The molecular genetics of cellular oncogenes. Annu Rev Genet 1984; 18:553.

15. Alt FW, DePinho RA, Zimmerman K, Legouy E, Hatton K, Ferrier P, Tesfaye A, Yancopoulos GD, Nisen P. The human myc-gene family. Cold Spring Harbor Quant Biol, in press.

16. Kohl NE, Kanda N, Schreck RR, Bruns G, Latt SA, Gilbert F, Alt FW. Transposition and amplification of oncogene-related sequences in human neuroblastomas. Cell 1983; 35:359–367.

17. Brodeur GM, Seeger RC, Schwab M, Varmus HE, Bishop JM. Amplification of N-myc in untreated human neuroblastomas correlated with advanced disease stage. Science 1984; 224:1121–1124.

18. Seeger R, Brodeur G, Sather H, Dalton A, Siegel S, Wong K, Hammond O. Association of multiple copies of the N-myc oncogene with rapid progression of neuroblastomas. New Engl J Med 1985; 313:1111–1119.

19. Land H, Parada LF, Weinberg RA. Tumorigenic conversion of primary embryo fibroblasts requires at least two co-operating oncogenes. Nature (London) 1983; 304:596–601.

20. Yancopoulos GD, Nisen PD, Tesfaye A, Kohl NE, Goldfarb MP, Alt FW. N-myc can cooperate with ras to transform normal cells in culture. Proc Natl Acad Sci USA 1985; 82:5455–5459.

21. Nau M, Brooks B, Battey J, Sausville E, Gasdar A, Kirsh I, McBride O, Bertness V, Hollis G, Minna J. L-myc, a new myc-related gene amplified and expressed in human small cell lung cancer. Nature (London) 1985; 318:69–73.

22. Zimmerman KA, Yancopoulos GD, Collum RG, et al. Differential expression of myc family genes during murine development. Nature (London) 1986; 319:780–783.

7

Molecular Mechanisms and Detection of Mutations Leading to Human Disease Phenotypes

JAMES R. LUPSKI AND C. THOMAS CASKEY

Molecular medicine began with Pauling's discovery that hemoglobin from normal people and patients with sickle cell disease migrated to different positions when subjected to gel electrophoresis (1). Ingram (2) found that this was caused by a single amino acid change (glutamic acid to valine) in the β chain of the hemoglobin molecule. With the discovery of DNA, a physical basis was given to the hereditary unit—the gene. The double helical structure of the Watson–Crick model also postulated that spontaneous mutations could result from tautomeric shifts (enol to keto tautomerization) that allowed a different base pairing to occur (3,4). After the genetic code was deciphered in the 1960s, it was found that base pair changes could lead to amino acid substitutions or mutations in bacterial proteins. The development of recombinant DNA technology in the 1970s, first in prokaryotic systems, was rapidly extended to hitherto intractable eukaryotic problems, many of medical significance.

As of the 1986 Cold Spring Harbor Symposium of Quantitative Biology on the Molecular Biology of *Homo sapiens* (5), about 1000 human genes had been isolated and mapped to different chromosomes. Human genes and highly polymorphic unique anonymous DNA segments from humans are being isolated at a logarithmic rate (6,7), and most defective human genes are associated with a disease (8). All human disease may be genetic in the sense that heritable diseases, cancer, and autoimmune and infectious diseases (considering genetically determined immune responsiveness and HLA) are influenced by an individual's DNA makeup (9).

In this chapter the physical basis of mutation will be explored and the mechanisms for generating mutations will be described. Examples will be given of human diseases that result from mutations, emphasizing genes that have emerged as model systems: the β-globin gene in hemoglobinopathies, the Factor VIII gene in hemophilia, and the hypoxanthine phosphoribosyltransferase (HPRT) gene in Lesch–Nyhan disease. Studies on the Duchenne muscular dystrophy (DMD) locus will be described to demonstrate the molecular basis of a "new mutation" disease. Finally, methods to detect these mutations will be described since mutations leading to the disease state appear to occur at all levels ranging from a single base to entire chromosomes.

MUTATIONAL TYPES

Point Mutations

Point mutations are single base pair changes in the DNA double helix. In the Watson–Crick model, the base guanine (G) always forms a base pair by a triple hydrogen bond with cytosine (C), and adenine (A) always pairs by a double hydrogen bond to thymine (T). When a tautomeric shift occurs, a base pair other than the normal Watson–Crick pair may form. This will result in a mutation phenotype only if it disrupts the natural flow of genetic information in the cell. Most point mutations do not yield detectable phenotypic changes. However, point mutations become important in generating restriction fragment length polymorphisms which will be discussed later.

To appreciate how point mutations can lead to human disease phenotypes, it is necessary to trace the flow of genetic information in a cell. The fundamental theorem of molecular biology explains how information is stored, passed on, and retrieved in living systems (10). The information content in DNA is passed on to successive generations via DNA replication wherein a faithful copy of the DNA molecule is made. Within a given generation, information is transduced from the stored form (DNA) into protein by first being transcribed into RNA, which, in turn, is translated into protein. Once translation has occurred and the information is passed into protein, it cannot return to DNA. This is the central dogma of molecular biology.

Point mutations at several places in the DNA primary sequence may disrupt the natural flow of information (11–13) (Fig. 7-1). An individual gene contains signals in the primary sequence of DNA that allow expression of the gene. Transcription of a gene is initiated when RNA polymerase binds to the promoter at the 5′ end of a structural gene (14). The primary sequence motifs, TATA and CAAT, play an important role in this protein–nucleic acid interaction. A primary transcript of ribonucleotides is polymerized in the 5′ to 3′ direction. Termination of transcription occurs by a mechanism that is better understood in prokaryotic systems than in eukaryotic systems (15,16).

The primary transcript is processed by the recognition of a short *consensus sequence* (5′ donor GT and 3′ acceptor AG) that is conserved in evolution at exon–intron boundaries. The intervening sequences or introns are spliced out and a cap (7-methylguanosine triphosphate) is added to the 5′ end; a poly(A) sequence of up to 200 A residues is added to the 3′ end (17). The mature messenger RNA is transported to the cytoplasm to be translated by ribosomes.

The ribosomes must recognize and bind to the mRNA again by a protein–nucleic acid interaction that involves the primary nucleotide sequence information. Ribosome binding is still ill defined in eukaryotes but has been demonstrated to involve a specific nucleotide sequence in prokaryotes (18), called the ribosome-binding site (RBS). The mRNA RBS interacts with the 3′ terminal sequence of 16 S mRNA, the Shine–Dalgarno sequence (19). Translation begins with the initiation

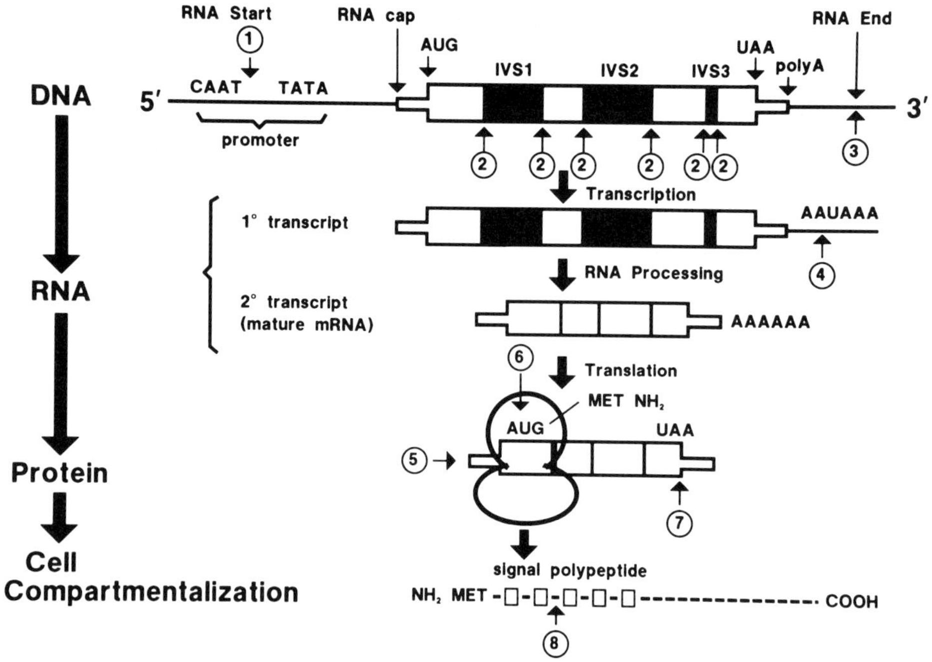

Fig. 7-1. The flow of genetic information within a cell. Above is shown a typical eukaryotic gene. To the left or 5′ side of the gene transcription is initiated by binding of RNA polymerase to the promoter. The DNA sequence motifs TATA and CAAT are important elements to most eukaryotic promoters. The coding sequences are depicted by wide boxes; intervening sequences (IVS1, IVS2) or introns are denoted by shaded boxes. Also shown are the translational initiation codon AUG and termination codon UAA of this idealized gene. Transcription of this DNA leads to a primary transcript that is processed to a mature mRNA. The sequence motif AAUAAA is the recognition site for poly(A) addition. Below is shown a ribosome translating the mRNA into a protein. At the bottom of the figure the synthesized protein is shown with its amino(NH_2) terminal hydrophobic signal polypeptide, which is important to directing the protein to its proper subcellular compartment. The circled numbers refer to points in the information flow diagram where point mutations (base pair change) may have profound effects on gene expression. ① Promoter mutations alter levels of transcription (11). ② Point mutations affecting splice donor acceptor site will alter processing (11). ③ Aberrant transcription termination will change the structure of the mRNA. ④ Altered poly(A) addition will affect transport of the mRNA to the cytoplasm. ⑤ If the ribosomes do not bind to the mature mRNA translation may not occur. ⑥ A base pair change would not allow translation to begin from its proper place (12). ⑦ Improper translation termination will yield an extended protein. ⑧ An altered signal sequence may place the protein in its wrong subcellular compartment (16). Effects of point mutations within coding sequences are described in Fig. 7-2.

codon AUG that encodes the amino terminal initial amino acid (Met) of the protein. Translation is terminated by one of three translation termination or *stop codons* (UAG, amber; UAA, ochre; UGA, opal). Many proteins are then directed to specific cell compartments by a hydrophobic signal polypeptide consisting of several amino acids at the amino terminal end of the protein.

In the globin gene system, point mutations in different regions of the gene have been found to affect each of the steps outlined in Fig. 7-1. Several mutations due to base pair changes have been found to inactivate a promoter and therefore mRNA is not made (11). The thalassemias are primarily due to aberrant splicing events; 18 of the 37 sequenced β-globin mutations are point mutations that disrupt splicing (20–22). These may occur within the splice junction, which tends to result in a more severe disease phenotype, or there may be mutations in both exons and introns that activate cryptic splice sites.

Point mutations within the coding region (exon) of a structural gene may have different effects (Fig. 7-2A). For instance, a base pair change in the third position of a triplet codon may still encode the original amino acid because of "redundancy" of the genetic code and because several "degenerate" codons may specify the same amino acid. This is called a *silent mutation* (Fig. 7-2A).

Base pair changes that lead to substitution of a different amino acid are called *missense mutations* (23). If the base pair change leads to a triplet that encodes a similar amino acid (Fig. 7-2B), there may be no functional change in the protein. However, if that amino acid plays a crucial role in the active site of the protein, the amino acid substitution may have profound consequences. Base pair changes in the first and second position of a triplet codon may lead to substitution of a dissimilar amino acid (Fig. 7-2C) that might affect enzymatic function if the amino acid has an important structural or functional role.

A base pair change that generates a *termination codon* will also have profound consequences (Fig. 7-2D). These *nonsense mutations* cause premature translational chain termination and therefore greatly disrupt the flow of genetic information (11,13).

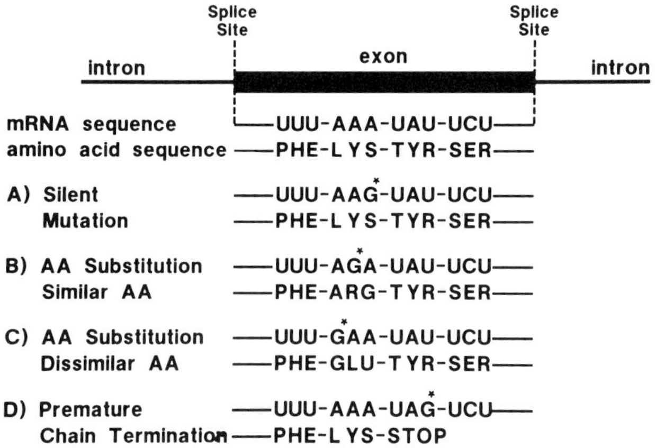

Fig. 7-2. Point mutations in expressed DNA sequences. Above is shown a typical eukaryotic gene with an expessed DNA sequence, exon, surrounded by two intervening sequences or introns. Below that is shown a hypothetical mRNA encoded by the DNA and the amino acid as translated via the genetic code. (A) A silent mutation is a base pair change that does not lead to an amino acid change and therefore the primary structure of the protein is not altered. (B) A base pair change that leads to the substitution of a similar amino acid. (C) A → G transition substitutes an acidic amino acid for a basic amino acid; both (B) and (C) are missense mutations (20). (D) A U → G transversion generates a nonsense codon and leads to premature termination of transcription.

Many known mutations have affected proteins with more of an enzymatic than a structural role. In general, point mutations that lead to amino acid substitutions may be less likely to lead to a disease phenotype in structural proteins than proteins involved in enzyme functions. Nevertheless, sickle cell anemia is a flagrant exception to that rule. Point mutations have been implicated in the hemoglobinopathies (11,21), hemophilia (21–25), osteogenesis imperfecta (26), familial amyloidotic polyneuropathy (27), ornithine transcarbamylase deficiency (28), the Lesch–Nyhan syndrome (29), and in the activation of some oncogenes (30).

Point mutations seem to occur with increased frequency at CpG dinucleotides (30,31). Many DNA polymorphisms are seen with the restriction endonucleases, *Taq*I and *Msp*I, and both of them have CpG as part of the recognition sequence (*Msp*I: CCCG; *Taq*I: TCGA). The CpG dinucleotide is not found as frequently as other dinucleotides in the mammalian genome, and the C is usually methylated (32). In bacteria, 5meC residues are *hot spots* for *transition mutations* (33). The 5meC sequence in a CpG dinucleotide may also be a mutational hot spot in mammalian DNA, with an increased frequency of transition mutations at this nucleotide: CpG → TpG or CpA. The CpG dinucleotide is thought to be about 10 times more likely to be polymorphic than other dinucleotides so *Msp*I and *Taq*I restriction sites are lost more often than other restriction sites. This has been confirmed in several human gene systems (11,25,30,31).

Frameshift Mutations

These mutations occur within a coding sequence and are the result of a 1- or 2-base pair insertion, or the deletion of 1 or 2 bp (Fig. 7-3). This shifts the *reading frame* to result in a different nucleotide triplet codon being read, and usually a different

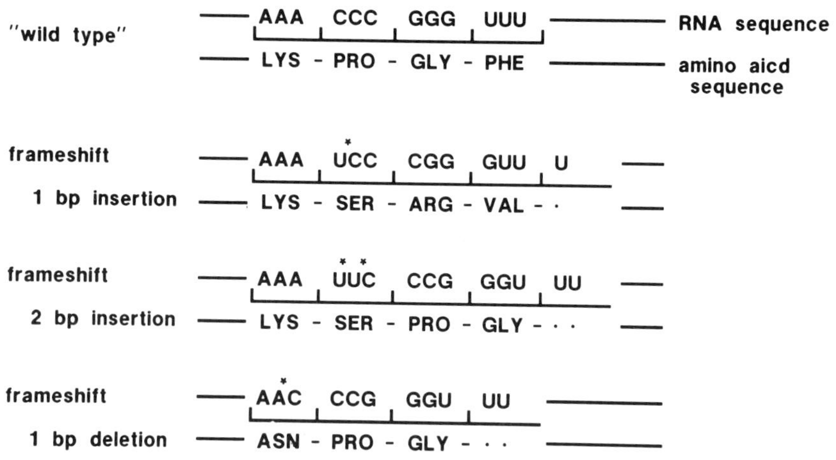

Fig. 7-3. Frameshift mutations. A frameshift mutation shifts the reading frame of the mRNA. This leads to the translation of different triplet codons and therefore different amino acids are placed in the primary sequence of the protein. As shown in this figure insertion of 1 or 2 bp or deletion of 1 or 2 bp will lead to a shift in the reading frame. If insertion or deletion of base pairs occurs in multiples of three, no frameshift will occur since the genetic code is a triplet code. Frameshift mutations generally involve just a few base pairs.

amino acid sequence being specified. Frameshift mutations dramatically alter the information content of a protein because the entire phase is changed distal to the site of mutation. The earlier or the more toward the 5′ end a frameshift mutation occurs in the gene, the more severe the change in information content of the proteins. Frameshift mutations have been detected in many isolates of β-thalassemia (11).

DNA Rearrangements

Gross loss or rearrangement of DNA sequences can generate human disease phenotypes. The plasticity of a given genome was not generally appreciated until the discovery of transposable genetic elements and the study of other *illegitimate (nonhomologous) recombination* events (34). DNA rearrangements generally involve a number of base pairs or long stretches of DNA (Fig. 7-4). *Insertions* may comprise repetitive DNA [e.g., Alu sequence (35)] or exogenous retroviral DNA. *Deletions* may span only a portion of a gene or the entire gene. Partial or total gene deletions are the mechanisms responsible for mutations in several genes causing disease, including the HPRT gene (36–39) in the Lesch–Nyhan syndrome, the genes for Factor VIII (11,40) in hemophilia A, and the low-density lipoprotein (LDL) receptor in familial hypercholesterolemia, respectively. *Endoduplications* may involve specific exons of a gene (36). *Amplification* of an individual gene leads to the gen-

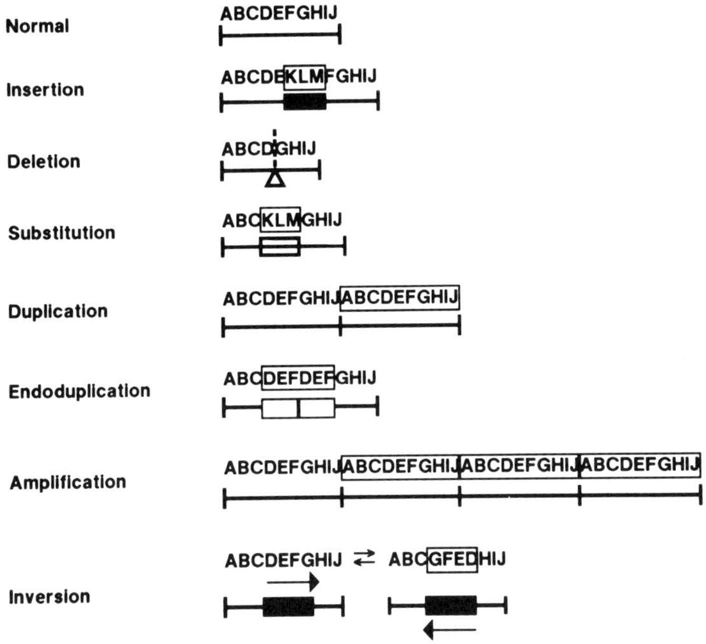

Fig. 7-4. DNA rearrangements. Mutations resulting from DNA rearrangements usually involve gross loss or rearrangement of many base pairs. The letters are used to depict several DNA bases. Below each set of letters are lined drawings of the DNA sequences. Each alteration in DNA sequence resulting from a particular rearrangement is surrounded by dashed lines. The triangle denotes a deletion.

eration of many copies of the gene. Amplification of some oncogenes has been found to be a mechanism for generating malignancy (41–43).

DNA rearrangements seem to occur with increased frequency at specific areas of the genome that are hot spots for recombination. In lower life forms this has been associated with the presence of a transposable genetic element (34,44). Mutational hot spots for DNA rearrangements in the human genome may someday be found to be associated with an endogenous transposable element (45–47) or a cryptic integrated retrovirus (48). Alu sequences (49,50) and a repetitive CACA sequence may be hot spots for DNA rearrangements.

Chromosomal Aberrations

The term *chromosomal aberrations* will be used for those that can be seen by light microscopy in karyotype analysis. After the proper chromosome number [46:XY, (51)] was assigned for humans, one of the first germline chromosomal aberrations to be identified was trisomy 21 or Down syndrome (52). As a result of nondisjunction in meiosis, all cells of the offspring have an extra copy of all the genes on chromosome 21.

A gene dosage imbalance leads to many anomalies. An interesting example is the *ets* oncogene on chromosome 21. Down syndrome patients have a 20- to 50-fold increased chance of developing leukemia. In several cases of Down syndrome with leukemia cytogenetic analysis of leukemic cells revealed tetrasomy 21, but other somatic cells were trisomy 21 (53,54). We have followed two patients with Down syndrome and a blood dysplasia that progressed to full-blown leukemia, accompanied by a karyotypic change to tetrasomy 21 in the leukemia cells (J. Lupski and S. Welty unpublished observations).

Some individuals with normal chromosome complements have leukemia, in which malignant cells are either trisomy 21 (55) or contain a translocation of 21 to another autosome (56). Trisomies other than trisomy 21 are found at much lower frequencies because they are likely to be embryonic lethals and would go undetected. One possible reason why trisomy 21 is the commonest human trisomy seen in live births is that chromosome 21 is the smallest human chromosome; it contains only 1.5% of the total genetic material (57). The frequency of trisomy 21 is 1.5 per 1000 births, whereas that for trisomy 18 is 0.3 and 13 is 0.2 per 1000 births (58). A general rule may be that the smaller the DNA information content, the more frequent the finding of trisomy of that specific chromosome.

One unusual chromosomal mechanism for generating human disease is *uniparental disomy*. An individual inherits two copies of the same parental chromosome and therefore expresses the disease phenotype of any recessive gene on that chromosome. This was found in an individual with cystic fibrosis and short stature (59).

Another chromosomal mechanism for generating mutations involves *Lyonization*. The Lyon hypothesis states that in order to maintain proper gene dosage in females, one X-chromosome undergoes heterochromatinization to form an inactive Barr body (60). In some X-linked recessive diseases girls or women have the disease because of nonrandom inactivation of X with expression of the disease phenotype from the mutant gene on the active (nonheterochromatinized) X chromosome. Examples include hemophilia in females who are heterozygous at the factor

VIII locus, where deletion in the X-chromosome with a normal factor VIII locus leads to nonrandom heterochromatinization of that chromosome and expression of the disease allele from the other X-chromosome (61).

X-linked recessive diseases that are expressed in females are usually associated with X-autosome translocation. Two mechanisms may be involved and both involve heterochromatinization of the nontranslocated X-chromosome. Heterochromatinization of the translocated X would lead to inactivation of the autosome and, presumably, cell lineages descending from this event would be lost. In the first mechanism, the X-linked recessive gene is disrupted by the X-autosome translocation. The second mechanism involves expression of the mutant sex-linked recessive gene located at a point other than the translocation breakpoint, with heterochromatinization of the normal X.

Turner syndrome (XO) may also lead to expression of an X-linked recessive disease in a female (62). Other germline chromosomal rearrangements have been associated with specific syndromes, e.g., deletion within the short arm of chromosome 5 in Cri-du-Chat (63) and deletion 15q11.2 in Prader–Willi syndrome (64). Physical characteristics of a chromosome may increase the relative frequencies of rearrangements. Large *inverted duplications* have been associated with gene amplification (65,66) and also with recombination events that result in chromosomes which contain both duplicated and deleted regions.

In addition to germline chromosomal aberrations, mutations in individual somatic cells can lead to human disease. *Somatic chromosomal mutations* are probably responsible for most human cancers (67). For some time it had been argued that malignancy was more likely to be generated by recessive mutations in somatic cells than by dominant ones, and that the recessive mutations might be unmasked by genetic events that resulted in loss of the unaffected gene on the homologous chromosome (68). The hemizygous state could arise from loss of the homologous chromosome itself.

For instance, retinoblastoma occurs in two forms, a sporadic form and an autosomal dominant form. It was postulated that the genetic loss was the same in both forms of the tumor, but that, in the heritable form, one of the alleles was already mutated or otherwise rendered nonfunctional in the germline (69). Thus, a single event that inactivates the homologous locus in the unaffected chromosome would generate the tumor in predisposed individuals whereas in sporadic cases both alleles would have to be inactivated by two separate genetic events. Twelve years later, experimental confirmation of this hypothesis was obtained when it was found that deletion of a region on chromosome 13 (13q14) leads to predisposition to retinoblastoma, whereas homozygosity of deletion of that locus is found in retinoblastoma tumor cells (70). The gene that predisposes to retinoblastoma and osteosarcoma has been isolated (71).

Other somatic cell chromosomal aberrations have been associated with specific malignancies. Loss of genes on chromosome 22 has been associated with familial acoustic neuroma (72), and loss of genes on the short arm of chromosome 11 has been associated with bladder cancer (73). Familial osteosarcoma has been associated with 13;14 chromosomal rearrangement (74). Amplification of chromosomal sequences that include the N-*myc* oncogene is responsible for development of neuroblastoma (75). Many leukemias seem to be associated with chromosomal anomalies in the leukemic cell (67), the classic example being chronic myelogenous leukemia, in which the Philadelphia chromosome (22 → 18) translocation leads to activation of the c-*myc* oncogene.

PHENOTYPIC VARIATION

Although individuals in a single family may have a genetic predisposition to a specific inherited disease, affected members may have widely different phenotypes. This is particularly true of autosomal dominant diseases. Geneticists have classically used the global terms *variable expressivity* and *decreased penetrance* to describe this diversity. A heuristic example to explain this terminology is polydactyly. If an individual inherits the gene for autosomal dominant polydactyly but has only five fingers, there is said to be decreased penetrance for the disorder in that individual. If another individual in that family has a remnant extra digit, the term *variable expressivity* applies. Charcot-Marie-Tooth disease and myotonic dystrophy are clinical examples in which individuals from a single family may display a heterogeneous spectrum of symptoms. Both penetrance and variable expressivity are poorly understood at the molecular level.

Mutations in single alleles can also result in different phenotypes in different families. For example, DMD has been differentiated from the less severe Becker muscular dystrophy. Yet, the same DNA fragments that map to the DMD locus are also deleted in about 10% of the cases of Becker muscular dystrophy. In one patient with Becker muscular dystrophy, a deletion removed a small exon without affecting splice donor–acceptor sets, or affecting the reading frame. In contrast two DMD deletions examined in detail were found to produce frame shifts (76).

Lack of the enzyme hypoxanthine phosphoribosyltransferase (HPRT gene) gives a broad spectrum of disease. Complete deficiency is associated with the Lesch–Nyhan syndrome whereas partial deficiency is associated with gout (38). Individuals with total HPRT deficiency display a wide spectrum of symptoms. Detailed molecular analysis of different mutations is likely to increase understanding of phenotypic variation.

MUTATIONS LEADING TO DUCHENNE MUSCULAR DYSTROPHY

Studies of DMD have emerged as a model system for molecular approaches to the dissection of a human genetic disease. One of the first chromosomal localizations of the DMD locus came by studying females with DMD in which a translocation of the X-chromosome to an autosome was found, with the translocation breakpoint at Xp21 (77). Serendipitously a patient (BB) was found who had four X-linked inherited diseases: DMD, chronic granulomatous disease (CGD), McLeod syndrome, and an inherited form of retinitis pigmentosa (RP) (78). Cytogenetic studies revealed a deletion of band Xp21 (78). A DNA subtraction library was constructed to isolate DNA sequences (pERT clones—phenol enhanced reassociation technique) specific to Xp21 using DNA isolated from patient BB (79). These pERT clones were demonstrated to define specific mutations in DMD and Becker muscular dystrophy (80) and to isolate a candidate cDNA for the DMD locus (81). Other markers around the DMD locus have been isolated by X-chromosome-specific libraries (82–84) or by the isolation of translocation breakpoints in DMD females (85,86). These probes have all been useful in the prenatal diagnosis of DMD (76,87,89).

The high frequency (1 in 4000 males) of DMD, the observation of independent translocation events causing the disease, the fact that more than 50% of the mutations involve gene deletions (88), and demonstration of recombination within a region of closely linked molecular markers (76,87,89) suggest that this locus is a hotspot for recombination. The DMD locus is enormous, almost 1,000,000 base pairs (86). For comparison, the entire *Escherichia coli* chromosome is 4,000,000 base pairs and contains over 2000 genes. Because of the enormous size of the DMD locus, the simple target theory of induced mutations may partially explain the high frequency of mutation.

X-linked recessive diseases, such as DMD, in which males often do not live to reproductive age, are new-mutation diseases. This is true of several other X-linked disorders for which molecular probes are now available, including Lesch-Nyhan syndrome and Factor VIII and IX deficiency. New mutations, which are of gametic origin, are manifested as sporadic cases in pedigrees with no known history of the disease. This is in contrast to inherited disorders such as sickle cell disease in which a single mutation has been carried through many generations. The first new mutation to be described by molecular techniques was an HPRT gene alteration of grandparental origin in a Lesch–Nyhan family (36). Maternal gametic mosaicism can also be responsible for new-mutation diseases (76).

METHODS OF DETECTING MUTATIONS

Most methods of DNA analysis require the separation of DNA fragments by electrophoresis through a gel matrix. To separate different sized molecules, the matrix can be varied (agarose or acrylamide of differing percentages) or the potential difference across the gel can be varied because DNA is a charged molecule. DNA molecules migrate through a gel as function of relative molecular weight. An approximate range of DNA sizes and the method used to separate molecules in a given range are given in Fig. 7-5. The ultimate method to detect point mutations, albeit very labor intensive, is DNA sequence analysis (90,91). On the other end of the scale, karyotype analysis may reveal chromosomal aberrations. Many forms of mutation are amenable to molecular analysis (92–94), several prenatally (95,96). A brief description of the presently utilized methods is given below.

Restriction Fragment Length Polymorphisms (RFLPs)

RFLPs are differences in the size of fragments that result from the digestion of the corresponding region of DNA from homologous chromosomes (97). RFLPs are generally detected directly in genomic DNA by restriction enzyme digestion, followed by gel fractionation of resulting fragments, transfer of the DNA to a solid support by Southern blotting (98), and hybridization to a specific cloned probe for

DNA Size	Enzyme Used	Gel Separation Method
1 bp	DNA pol I	12% PAGE
10 bp-100 bp	Restriction endonuclease with 4 bp recognition sites (4 bp cutters)	8% PAGE
100-10,000 bp	6 bp cutters	1% agarose gel electrophoresis
1,000-100,000 bp	6 bp cutters	0.35% agarose gel electrophoresis
10,000-10,000,000 bp	8 bp cutters	PFGE
10,000-10,000,000 bp	8 bp cutters	Field inversion gels
1,000,000-100,000,000 bp	--	High resolution chromosome banding
10,000,000-100,000,000 bp	--	In situ hybridization
100,000,000 bp	--	Karyotyping
>1,000,000,000 bp	--	Human genome

Fig. 7-5. Size ranges of DNA separation methods. On the left is shown the approximate size range of the DNA molecules in base pairs (bp). The middle column shows the enzyme used to generate fragments within the given size range on the left. The last column lists the method used to perform the separation.

the locus in question (99). The RFLPs are visualized as different fingerprint patterns on a Southern blot (9) and were first used in the physical mapping of temperature-sensitive mutation of adenovirus (100). In humans, RFLPs are a measure of polymorphisms of normal DNA and are inherited according to Mendelian principles (97). The level of human DNA polymorphism is remarkably high; one base in 250–500 differs in any pair of chromosomes chosen randomly (101). When RFLPs are found in close association with cloned genes or genomic segments, they can be used for prenatal diagnosis of genetic disease and in the detection of carriers of the disease gene. Many human polymorphic gene loci have been described (102) and the list is expanding at a rapid rate.

The physical basis for an RFLP is a difference in the DNA sequence of different individuals. It may be as little as a single base pair change or a point mutation that destroys or creates a new restriction endonuclease recognition site, or a DNA rearrangement may have occurred around the probe hybridization site. These mutations lead to a mobility change in the piece of DNA that is detected by the specific probe on an agarose gel because there is a change in DNA cutting (or size), giving distinct patterns in the Southern blot. In some regions of the human genome, DNA sequences with tandem repeat structures of a core repeat sequence may lead to unequal exchanges in recombination, generating altered restriction fragments and a high degree of polymorphism at these locations. Cloned DNA probes that detect these regions [*variable number tandem repeat probes;* VNTRs (103)] reveal high degrees of polymorphism and may be more informative than a two-allele system, such as detection of loss or gain of a restriction site (99).

RFLPs were first used in DNA diagnosis to detect sickle cell disease (104,105) with the cloned β-globin gene as a probe; this was followed by other applications

(106–109). *Anonymous DNA segments* have also been used to detect RFLPs associated with disease loci (110,111). The use of the actual cloned gene locus as a probe to detect an RFLP is better than a linked anonymous DNA segment because the possibility of recombination occurring in the region between the linked RFLP and the disease locus may decrease diagnostic accuracy (87,89).

The ultimate test, in terms of predictive value, occurs when a given mutation responsible for a genetic disease is due to a point mutation within the sequence of a specific restriction site. This occurs in sickle cell anemia in which an *Mst*II restriction site has been lost as a result of an A→T transversion that is responsible for the substitution of valine in the place of glutamic acid in the β chain of the hemoglobin molecule (112–114 and see Fig. 7-6).

Synthetic DNA Probes

If the precise point mutation is known and it is responsible for the phenotype in most cases, synthetic oligonucleotide probes can be used to detect the mutation. Separate allele-specific oligodeoxynucleotides (ASO) complementary to the disease or normal DNA sequence are synthesized with a sequence that is an oligonucleotide of 19 bp in which the point mutation is flanked by 9 bases on either side. Hybridization conditions are chosen so that ASO probes complementary to the normal or to the mutant allele in the region of the point mutation can distinguish between the two alleles. DNA from an individual homozygous for the normal gene will hybridize to the gene probe with the normal complementary DNA sequence but not to the probe that contains the point mutation complement, and DNA from those homozygous for the mutation will hybridize to the mutation probe but not to the normal probe. Only DNA from heterozygotes will hybridize to both probes. This method has been effective in detecting sickle cell anemia (115) α_1-antitrypsin

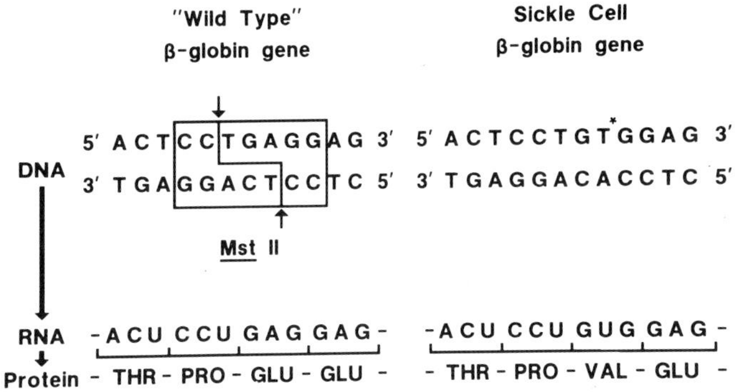

Fig. 7-6. The mutation that leads to sickle cell anemia. One of the most prevalent hemoglobinopathies, sickle cell anemia, is due to a base pair change A → T transversion that leads to substitution of valine for glutamic acid in the sixth position of the B chain of the hemoglobin molecule (114). On the left is shown a wild-type B globin gene DNA sequence around the sickle mutation. The amino acid sequence it encodes is shown below. Note the recognition site surrounded by lines and the cleavage site demonstrated by arrows for restriction endonuclease *Mst*II. On the right of the figure is shown the sickle cell mutation. It alters the DNA sequence in such a way that it can no longer be cut by *Mst*II.

deficiency (116), phenylketonuria (PKU, 117) and several variants of β-thalassemia (11).

Polymerase Chain Reaction (PCR)

This method enables selective amplification of specific gene sequence *in vitro*. As with synthetic oligonucleotide probes, the DNA sequence information of the gene has to be known. The method uses synthetic oligonucleotide probes as primer for DNA polymerase (118). Primer probes are hybridized to different regions of a gene, oriented so that the polymerase extension products overlap. A series of repetitive cycles of denaturation, hybridization, and DNA polymerization is carried out; in these reactions, the primer probes act as primers for the initiation of DNA replication from that site. The result is amplification of the DNA sequences between the original oligonucleotide primers, sometimes up to 2,000,000-fold (118). Any mutation within that amplified stretch of DNA is faithfully copied and more readily analyzed after amplification. The method was first applied to the diagnosis of sickle-cell anemia and will probably be useful in conjunction with allele-specific oligodeoxynucleotides (92).

Ribonuclease Cleavage

When the nucleotide sequence of a disease gene is known, the ribonuclease A (RNase A) cleavage method can detect single-base mismatches between the radioactive RNA probe and the patient's (17) genomic or transcribed sequences (92). Thus, it can detect single base substitutions (point mutations) in cloned and genomic DNAs (119–121). The method relies on the enzymatic cleavage of RNA at a single-base mismatch in an RNA–RNA or RNA–DNA hybrid. The method was made possible by the development of methods for synthesizing highly radioactive RNA probes and on the observation that RNase A cleaves at certain types of single-base mismatch.

A ^{32}P-labeled RNA probe is synthesized from a wild-type DNA template. This RNA probe is hybridized to denatured DNA or to mRNA which has a point mutation, and the hybrids are treated with RNase. The RNA products are then subjected to denaturing gel electrophoresis. The RNase will cleave the RNA–RNA or RNA–DNA hybrids at the position of the point mutation. The sizes of the bands in the gel indicate where the initial mutation occurred. The method has been applied to the identification and localization of mutations at the Lesch-Nyhan locus (122).

Denaturing Gradient Gel Electrophoresis

Double helical DNA molecules can be separated according to the nature of the sequence of bases (123). This procedure differs from conventional gel methods, which separate DNA on the basis of relative molecular weight. In model experiments, denaturing gradient gel electrophoresis could detect point mutations, single base deletions, or insertions and could also separate DNA fragments with hundreds of base pair differences. The theoretical basis of the method is that the melting properties of a double helix differ as a result of mutations (123). The relevant gene must be available as a probe, but the DNA sequence does not have to be known.

Pulse Field Gel Electrophoresis (PFGE)

PFGE (124,125) and the similar field inversion gel electrophoresis (FIGE) (126) separate DNAs of very large molecular weight and can span great distances of the human genome. Development of this method was a tremendous advance because it enabled the analysis of DNA fragments of a size too large for conventional agarose gel electrophoresis but too small to be detected by cytogenetic techniques. PFGE has been used to construct a restriction map of the entire 4 million base pairs of the *E. coli* chromosome (127). It is likely to be even more useful in conjunction with a new strategy for rapid analysis and sorting of large genomic libraries that has been used to determine a complete restriction map of *E. coli* (128). The gel matrix is agarose and the potential difference across the gel is alternately pulsed (PFGE) to invert the electric field. By altering the pulsing characteristics, it is possible to change the size range of DNA molecules that can be separated. This method may be useful in looking for mutations distant from the probe hybridization site on the genome, because the region of analysis will be extended.

Chromosome-Specific Libraries

Chromosome-specific libraries have been used to map specific genes or probes for RFLPs to particular chromosomes and also to give a regional map of small chromosomal deletions. These libraries are constructed by first purifying specific metaphase chromosomes by fluorescence-activated flow cytometry and then making phage recombinant libraries of the individual purified chromosomes. The recombinant clones from these libraries can be used to analyze specific regions of individual chromosomes (129). Recombinant clones from a chromosome 13-specific library were useful in delineating deletions in regions 13q12–13q22 involved in retinoblastoma. A chromosome 15 library was important to the study of the Prader–Willi syndrome (deletion 15q11.2), previously shown to be associated with chromosome deletions (43).

High-Resolution Chromosome Banding

If cell division is arrested in early prophase rather than metaphase, chromosomes are in a more extended state. Chromosome banding can obtain about 10 times the number of bands to increase the resolution by almost 10-fold (130) so that small deletions can be detected in individual chromosomes (112). A remarkable example of the power of this technique was demonstrated with studies of the DMD locus (8).

DISCUSSION

Mutations that cause human disease phenotypes can be generated by many mechanisms, which may involve only a single base pair or an entire chromosome. Single base pair changes in the Watson–Crick double helix lead to point mutations that can have different effects, depending on their location within a transcriptional unit or gene. Point mutations that are not in coding sequences disrupt the flow of genetic information at the level of transcription or processing of RNA. Point mutations within coding sequences usually affect the genetic information flow at the level of translation in the case of nonsense mutations, but missense mutations disrupt only the genetic information inherent in the protein structure. Frameshift mutations also alter the flow of genetic information at the level of translation. The CpG dinucleotide seems to be a hot spot for point mutations.

Many mutational events involve more than a single base or a few base pairs. Mechanisms for understanding these DNA rearrangements in the human genome are still poorly understood. In prokaryotes and lower eukaryotes, illegitimate recombination via movable genetic elements is responsible for a wide variety of biological phenomena by allowing for the integration of nonhomologous segments of genetic information into the genome. What role the human genome structure plays, with repetitive sequences (35,49,50) and variable number of tandem repeat sequences (VNTRs) (103), in generating DNA rearrangements has not been elucidated. Perhaps these sequences act as a nidus for recombination events. Gene families and pseudogenes as well as other divergent DNA sequence structures can lead to increased legitimate (homologous) as well as illegitimate (nonhomologous) recombination events (132,133). Chromosomal structures, such as fragile sites, may affect mutation frequency on a DNA structural level (134) as well as on a chromosomal level.

Regardless of the mechanisms responsible for generating mutations, techniques to detect them are becoming available at an ever-increasing rate. A combination of approaches involving biochemical studies, DNA molecular analysis, and cytogenetic studies of chromosomes from patients with a given genetic disorder seems to be a fruitful approach as exemplified by the studies of Wilms tumor-aniridia syndrome and the mapping of chromosome 11. The future promises even greater understanding of human genetic disease.

ACKNOWLEDGMENTS

We are very grateful to Drs. Pragna I. Patel and Gordon Schutze for critically reviewing the manuscript and to Lillian Tanagho and Elsa Perez for typing it. This work has been supported by the Muscular Dystrophy Association Task Force on Genetics, NIH grant DK31428, and the Howard Hughes Medical Institute.

REFERENCES

1. Pauling, L, Itan HA, Singer SJ, and Wells IC. Sickle cell anemia, a molecular diease. Science 1949; 110:543–548.

2. Ingram, VM. A specific chemical difference between the globins of normal human and sickle cell anemia hemoglobin. Nature (London) 1956; 178:792–794.

3. Watson JD, Crick FHC. A structure for deoxyribose nucleic acid. Nature (London) 1953; 171:737–738.

4. Watson JD, Crick FHC. The structure of DNA. Cold Spring Harbor Symp Quant Biol 1953; 18:123–132.

5. Watson JD, Siniscalco M. Molecular biology of *Homo sapiens.* Cold Spring Harbor Symp Quant Biol 1986; 51:703–1208.

6. Schmidtke J, Cooper DN. A list of cloned human DNA sequences. Human Genet 1983; 65:19–26.

7. Beaudet AL. Bibliography of cloned human and other selected DNAs. Am J Hum Genet 1985; 37:386–406.

8. McKusick, VA. Mendelian Inheritance in Man, 7th ed. Baltimore: The Johns Hopkins University Press, 1986.

9. Lupski JR. Genetic engineering in medicine. In Chopra VL, ed. Genetic Engineering and Biotechnology: Concepts Methods and Applications. New Delhi, India: Oxford and IBH Publishing Co., in press.

10. Hunt T. The general idea. In Hunt T, Prentis S, Tooze J, eds. DNA Make RNA Makes Protein. Amsterdam: Elsevier Biomedical Press, 1983.

11. Kazazian HH, Antonarakis SE, Youssoufian H, et al. Comparison of deficiency alleles of the β-globin gene and factor VIII: C genes: New lessons from a giant gene. Cold Spring Harbor Symp Quant Biol 1986; 51:371–379.

12. Morle F, Lopez B, Henni T, Godet J. Alpha-thalassemia associated with the deletion of two nucleotides at position -2 and -3 preceding the AUG codon. EMBO J 1985; 4:1245–1250.

13. Russel DW, Lehrman MA, Sudhof TC, et al. The LDL receptor in familial hypercholesterolemia; use of human mutations to dissect a membrane protein. Cold Spring Harbor Symp Quant Biol 1986; 51:811–819.

14. Ziff EB. Transcription and RNA processing by the DNA tumor viruses. Nature (London) 1980; 287:491–499.

15. Brosius J. Toxicity of an overproduced foreign gene product in *Escherichia coli* and its use in plasmid vectors for the selection of transcription terminators. Gene 1984; 26:57–68.

16. Lupski JR, Ruiz AA, Godson GN. Promotion, termination, and antitermination in the *rpsU-dnaG-rpoD* macromolecular synthesis operon of *E. coli* K-12. Mol Gen Genet 1984; 195: 391–401.

17. Darnell JE. RNA Sci Am 1985; 253:68–78.

18. Steitz JA. Nucleotide sequences of the ribosomal binding sites of bacteriophage R17 RNA. Cold Spring Harbor Symp Quant Biol 1969; 34:621–630.

19. Shine J, Dalgarno L. The 3′-terminal sequence of *Escherichia coli* 16S ribosomal RNA: Complementarity to nonsense triplets and ribosome binding sites. Proc Natl Acad Sci USA 1974; 71:1342–1346.

20. Gitschier J, Wood WI, Shuman MA, Lawn RM. Identification of a missense mutation in the factor VIII gene of a mild hemophiliac. Science 1986; 232:1415–1416.

21. Antonarakis SE, Kazazian HH, Orkin SH. DNA polymorphism and molecular pathology of the human globin gene clusters. Hum Genet 1985; 69:1–14.

22. Antonarakis SE, Waber PG, Kittur SD, et al. Hemophilia A: Detection of molecular defects and of carriers by DNA analysis. N Engl J Med 1985; 313:842–848.

23. Gitschier J, Wood WI, Tuddenham EGD, et al. Detection and sequence of mutations in the factor VIII gene of haemophiliacs. Nature (London) 1985; 315:427–430.

24. Lawn RM. The molecular genetics of hemophilia: Blood clotting factors VIII and IX. Cell 1985; 42:405–406.

25. Lawn RM, Wood WI, Gitschier J, et al. Cloned factor VIII gene and the molecular genetics of hemophilia. Cold Spring Harbor Symp Quant Biol 1986; 51:365–369.

26. Cohn DH, Byers PH, Steinmann B, Gelinas RE. Lethal osteogenesis imperfecta resulting from a single nucleotide change in one human pro alpha 1 (I) collagen allele. Proc Natl Acad Sci USA 1986; 83:6045–6047.

27. Maeda S, Mita S, Araki S, Shimada K. Structure and expression of the mutant preal-

bumin gene associated with familial amyloidotic polyneuropathy. Mol Biol Med 1986; 3:329–338.

28. Nussbaum RL, Boggs BA, Beaudet AL, Doyle S, Potter JL, O'Brien WE. New mutation and prenatal diagnosis in ornithine transcarbamylase deficiency. Am J Hum Genet 1986; 38:149–158.

29. Reddy EP, Reynolds RK, Santos E, Barbacid M. A point mutation is responsible for the acquisition of transforming properties by the T24 human bladder carcinoma oncogene. Nature (London) 1982; 300:149–152.

30. Youssoufian H, Kazazian HH, Phillips DG, et al. Recurrent mutations in hemophilia A give evidence for CpG mutation hotspots. Nature (London) 1986; 27:380–382.

31. Barker D, Schafer M, and White R. Restriction sites containing CpG show a higher frequency of polymorphism in human DNA. Cell 1984; 36:131–138.

32. Wallace RB, Petz LD, Yam PY. Application of synthetic DNA probes to the analysis of DNA sequence variants in man. Cold Spring Harbor Symp Quant Biol 1986; 51:257–261.

33. Miller JH. Mutational specificity in bacteria. Annu Rev Genet 1983; 17:213–238.

34. Campbell A. Some general questions about movable genetic elements and their implications. Cold Spring Harbor Symp Quant Biol 1981; 45:1–19.

35. Jelinek WR, Toomey TP, Leinwand L, et al. Ubiquitous interspersed repeated sequences in mammalian genomes. Proc Natl Acad Sci USA 1980; 77:1398–1402.

36. Yang TP, Patel PI, Chinault AC, et al. Molecular evidence for new mutation at the hprt locus in Lesch-Nyhan patients. Nature (London) 1984; 310:412–413.

37. Patel PI, Caskey CT. HPRT and the Lesch-Nyhan syndrome. BioEssays 1986; 2:4–8.

38. Stout JT, Caskey CT. HPRT: Gene structure, expression and mutation. Annu Rev Genet 1985; 19:127–148.

39. Wilson JM, Stout JT, Palella TD, Davidson BL, Kelley WN, Caskey CT. A molecular survey of hypoxanthine-guanine phosphoribosyltransferase deficiency in man. J Clin Invest 1986; 77:188–195.

40. Youssoufian H, Antonarakis SE, Aronis S, Tsiftis G, Phillips DG, Kazazian HH. Characterization of five partial deletions of the factor VIII gene. Proc Natl Acad Sci USA 1987; 84:3772–3776.

41. Ibson JM, Waters JJ, Truentyman PR, Bleehen NM, Rabbits PH. Oncogene amplification and chromosomal abnormalities in small cell lung cancer. J Cell Biochem 1987; 33:267–288.

42. Alitalo K. Amplification of cellular oncogenes in cancer cells. Med Biol 1984; 62:304–317.

43. Latt SA, Lalande M, Donlon T, et al. DNA-based detection of chromosome deletion and amplification: Diagnostic and mechanistic significance. Cold Spring Harbor Symp Quant Biol 1986; 51:299–305.

44. Lupski JR. Molecular mechanisms for transposition of drug-resistance genes and other movable genetic elements. Rev Infect Dis 1987; 9:357–368.

45. Sankaranarayanan K. Transposable genetic elements, spontaneous mutations and the doubling-dose method of radiation genetic risk evaluation in man. Mutat Res 1986; 160:73–86.

46. Hoegerman SF, Rary JM. Speculation of the role of transposable elements in human genetic disease with particular attention to achondroplasia and the fragile X syndrome. Am J Med Genet 1986; 23:685–699.

47. Kole LB, Haynes SR, Jelinek WR. Discrete and heterogeneous high molecular weight RNAs complementary to a long dispersed repeat family (a possible transposon) of human DNA. J Mol Biol 1983; 165:257–286.

48. Clements JE. Hypothesis on the molecular basis of nononcogenic retroviral diseases. Rev Infect Dis 1985; 7:68–74.

49. Lehrman MA, Russell DW, Goldstein JL, Brown MS. Alu-Alu recombination deletes splice acceptor sites and produces secreted low density lipoprotein receptor in a subject with familial hypercholesterolemia. J Biol Chem 1987; 262:3354–3361.

50. Lehrman MA, Goldstein JL, Russell DW, Brown MS. Duplication of seven exons in the LDL receptor gene caused by Alu-Alu recombination in a subject with familial hypercholesterolemia. Cell 1987; 48:827–835.

51. Tjio HJ, Levan A. The chromosome numbers of man. Hereditas 1956; 42:1–6.

52. Lejeune J, Gautier M, Turpin MR. Etude des chromosomes somatiques de neut enfants mongoliens. CR Acad Sci 1959; 248: 1721–1722.

53. Jabs EW, Stamberg J, Leonard CO. Tetrasomy 21 in an infant with Down syndrome and congenital leukemia, Am J Genet 1982; 12:91–95.

54. Hecht F, Hecht BK, Morgan R, Sandberg AA, Link MP. Chromosome clues to acute leukemia in Down syndrome. Cancer Genet Cytogenet 1986; 21:93–98.

55. Ferster A, Verhest A, Varmos E, DeMaertelaere E, Otten J. Leukemia in a trisomy 21 mosaic: Specific involvement of the trisomic cells. Cancer Genet Cytogenet 1986; 20:109–113.

56. Sacchi N, Watson DK, Guerts-van-Kessel AH, et al. Hu-*ets*-1 and Hu-*ets*-2 genes are transposed in acute leukemias with (4;11) and (8;21) translocations. Science 1986; 231:379–382.

57. Patterson D. The causes of Down syndrome. Sci Am 1987; 257:52–60.

58. Smith DW. Recognizable Patterns of Human Malformation. Genetic Embryologic and Clinical Aspects, 3rd ed. Philadelphia: W. B. Saunders, 1982.

59. Spence JE, Perciaccante RG, Greig GM, et al. Uniparental disomy as a mechanism for human disease. Am J Human Genet 1988; 42:217–226.

60. Ferrier P, Bamatter F, and Klein D. Muscular dystrophy (Duchenne) in a girl with Turner syndrome. J Med Genet 1965; 2:38–46.

61. Lyon MF. Chromosomal and subchromosomal inactivation. Annu Rev Genet 1968; 2:31–52.

62. Nisen P, Stamberg J, Ehrenpreis R, et al. The molecular basis of severe hemophilia B in a girl. N Engl J Med 1986; 315:1139–1142.

63. Lejeune J, Lafourcade J, Berger R, et al. Trois cas de deletion partielle du bras court d'un chromosome 5. CR Acad Sci 1963; 257:3098–3101.

64. Ledbetter DH, Riccardi VM, Airhart SD, Strobel RJ, Keenan BS, Crawford JD. Deletions of chromosome 15 as a cause of Prader-Willi syndrome. N Engl J Med 1981; 304:325–329.

65. Ford M, Fried M. Large inverted duplications are associated with gene amplification. Cell 1986; 45:425–430.

66. Herrmann BG, Barlow DP, Lehrach H. A large inverted duplication allows homologous recombination between chromosomes heterozygous for the proximal t complex inversion. Cell 1987; 48:813–825.

67. Yunis JJ. Genes and chromosomes in human cancer. Prog Med Viral 1985; 32:58–71.

68. Ohno S. Genetic implications of karyological instabilities of malignant somatic cells. Physiolog Rev 1971; 51:496–526.

69. Knudson AG. Mutation and cancer: Statistical study of retinoblastoma. Proc Natl Acad Sci USA 1971; 68:820–823.

70. Cavenee WK, Dryja TP, Phillips RA, et al. Expression of recessive alleles by chromosomal mechanisms in retinoblastoma. Nature (London) 1983; 305:779–784.

71. Friend SH, Bernards R, Rogelj S, et al. A human DNA segment with properties of the gene that predisposes to retinoblastoma and osteosarcoma. Nature (London) 1986; 323:643–646.

72. Seizinger BR, Martuza RL, Gusella JF. Loss of genes on chromosome 22 in tumorigenesis of human acoustic neuroma. Nature (London) 1986; 322:644–647.

73. Fearon ER, Feinberg AP, Hamilton SH, Vogelstein B. Loss of genes on the short arm of chromosome 11 in bladder cancer. Nature (London) 1985; 318:377–380.

74. Gilman PA, Wang N, Fan SF, Reede J, Khan A, Leventhal BG. Familial osteosarcoma associated with 13;14 chromosomal rearrangement. Cancer Genet Cytogenet 1985; 17:123–132.

75. Brodeur GM, Seeger RC, Schwab M, et al. Amplification of N-myc in untreated neuroblastomas correlates with advanced disease stage. Science 1984; 224:1121–1124.

76. Witkowski JA, Caskey CT. Duchenne muscular dystrophy: DNA diagnosis in practice. In Appel S, ed. Current Neurology 1988; 9:1–36.

77. Jacobs PA, Hunt PA, Mayer M, Bart RD. Duchenne muscular dystrophy (DMD) in a female with an X/autosome translocation: Further evidence that the DMD locus is at Xp21. Am J Human Genet 1981; 33:513–518.

78. Franke U, Ochs HD, deMartinville B, et al. Minor Xp21 chromosome deletion in a male associated with expression of Duchenne muscular dystrophy, chronic granulomatous disease, retinitis pigmentosa, and McLeod syndrome. Am J Hum Genet 1985; 37:250–267.

79. Monaco AP, Bertelson CJ, Middlesworth W, et al. Detection of deletions spanning the Duchenne muscular dystrophy locus using a tightly linked DNA segment. Nature (London) 1985; 316:842–845.

80. Kunkel LN, Hejtmancik JF, Caskey CT et al. Analysis of deletion in DNA from patients with Becker and Duchenne muscular dystrophy. Nature (London) 1986; 322:73–77.

81. Monaco AP, Neve RL, Colletti-Feener C, Bertelson CJ, Kurnit DM, Kunkel LM. Isolation of candidate cDNAs for portions of the Duchenne muscular dystrophy gene. Nature (London) 1986; 323:646–650.

82. Davies KE, Young BD, Elles RG, Hill ME, Williamson R. Cloning of a representative genomic library of the human X chromosome after sorting by flow cytometry. Nature (London) 1981; 293:374–376.

83. Kunkel LM, Tantravahi V, Eisenhard M, and Latt SA. Regional localization on the human X of DNA segments cloned from flow sorted chromosomes. Nucleic Acids Res 1982; 10:1557–1578.

84. Hofker MH, Wapenaar MC, Goor N, et al. Isolation of probes detecting restriction fragment length polymorphisms from X-chromosome specific libraries; potential use for diagnosis of Duchenne muscular dystrophy. Human Genet 1985; 70:148–156.

85. Worton RG, Duff C, Sylvester JE. Duchenne muscular dystrophy involving translocation of the dmd gene next to ribosomal RNA genes. Science 1984; 224:1447.

86. Ray PN, Belfall B, Duff C. Cloning of the breakpoint of an X;21 translocation associated with Duchenne muscular dystrophy. Nature (London) 1985; 318:672.

87. Darras BT, Harper JF, Franke U. Prenatal diagnosis and detection of carriers with DNA probes in Duchenne's muscular dystrophy. N Eng J Med 1987; 316:985–992.

88. Koenig M, Hoffman EP, Bertelson CJ, et al. Complete cloning of the Duchenne muscular dystrophy (DMD) cDNA and preliminary genomic organization of the DMD gene in normal and affected individuals. Cell 1987, 50:509–517.

89. Hejtmancik JF, et al. Prenatal diagnosis of Duchenne muscular dystrophy. Submitted.

90. Sanger F, Nicklen S, Coulson AR. DNA sequencing with chain-terminating inhibitors. Proc Natl Acad Sci USA 1977; 74:5463–5467.

91. Maxim AM, Gilbert W. Sequencing end-labeled DNA with base-specific chemical cleavages. In Grossman L, Moldane K, eds. Methods in Enzymology, Vol. 65. New York: Academic Press, 1980, p 499.

92. Caskey CT. Disease diagnosis by recombinant DNA methods. Science 1987; 236:1223–1228.

93. Cooper DM, Schmidtke J. Diagnosis of genetic disease using recombinant DNA. Human Genet 1986; 73:1–11.

94. Davies KE, Robson KHJ. Molecular analysis of human monogenic diseases. BioEssays 1987; 6:247–253.

95. Caskey CT. Antenatal diagnosis of neurologic disorders. Pediat Neurol 1986; 13–27.

96. Caskey CT. Recombinant DNA methods for prenatal diagnosis. Ann Int Med 1983; 99:718–719.

97. Botstein D, White RL, Skolnick M, Davis RW. Construction of a genetic linkage map in man using restriction fragment length polymorphisms. Am J Human Genet 1980; 32:314–331.

98. Southern, EM. Detection of specific sequences among fragments separated by gel electrophoresis. J Mol Biol 1975; 98:503–517.

99. Gusella JF. DNA polymorphisms and human disease. Annu Rev Biochem 1986; 55:831–854.

100. Grodzicker T, Williams J, Sharp P, Sambrook J. Physical mapping of temperature-sensitive mutations of adenoviruses. Cold Spring Harbor Symp Quant Biol 1974; 39:439–446.

101. Cooper DN, Schmidtke J. DNA restriction length polymorphisms and heterozygosity in the human genome. Human Genet 1984; 66:1–16.

102. Willard HF, Scolnick MH, Pearson PL, Mandel JL. Report of committee on human gene mapping by recombinant DNA techniques. Cytogenet Cell Genet 1985; 40:360–489.

103. Jeffreys AJ, Wilson V, Thein SL. Hypervariable 'minisatellite' regions in human DNA. Nature (London) 1985; 314:67–73.

104. Kan YW, Dozy AM. Polymorphism of DNA sequence adjacent to human β-globin structural gene: Relationship to sickle mutation. Proc Natl Acad Sci USA 1978; 75:5631–5635.

105. Kan YW, Dozy AM. Antenatal diagnosis of sickle-cell anemia by DNA analysis of amniotic fluid cells. Lancet 1978; 2:910–911.

106. Hejtmancik JF, Harris SG, Tsao CC, Ward PA, Caskey CT. Carrier diagnosis of Duchenne muscular dystrophy using restriction fragment length polymorphisms. Neurology 1986; 36:1553–1562.

107. Kidd VJ, Woo SL. Recombinant DNA probes used to detect genetic disorders of the liver. Hepatology 1984; 4:736–756.

108. Kidd VJ, Woo SL. Molecular diagnosis of human genetic disease. Modern Cell Biol 1984; 3:113–129.

109. Beaudet AL. Molecular genetics and medicine. In Braunwald E, Isselbacher K, Petersdorf R, Wilson E, Martin J, Fauci A, eds. Harrison's Principles of Internal Medicine 11th Ed. New York: McGraw-Hill, 1987.

110. Donis-Keller H, Barker DF, Knowlton RG, Schumm JW, Braman JC, Green P. Highly polymorphic RFLP probes as diagnostic tools. Cold Spring Harbor Symp Quant Biol 1986; 51:317–324.

111. Gusella JF, Wexler NS, Conneally PM, et al. A polymorphic DNA marker genetically linked to Huntington's disease. Nature (London) 1983; 306:234–238.

112. Nienhus AW. Mapping the human genome. N Engl J Med 1978; 299:195–196.

113. Orkin SH, Little PFR, Kazazian HH, Boehm CD. Improved detection of the sickle mutation by DNA analysis: Application to prenatal diagnosis. N Eng J Med 1982; 307:32–36.

114. Marota CA, Wilson JT, Forget BG, Weisman SM. Human β-globin messenger RNA. J Biol Chem 1977; 252:5040–5053.

115. Conner BJ, Reyes AA, Morin C, et al. Detection of sickle cell β^S-globin allele by hybridization with synthetic oligonucleotides. Proc Natl Acad Sci USA 1983; 80:278–282.

116. Kidd VJ, Wallace RB, Itakura K, Woo SCL. Alpha 1-antitrypsin deficiency detection by direct analysis of the mutation in the gene. Nature (London) 1983; 304:230–234.

117. DiLella AG, Marrot J, Lidsky AS, Guttler F, Woo SLC. Tight linkage between a splicing mutation and a specific DNA haplotype in phenylketonuria. Nature (London) 1986; 322:799–803.

118. Mullis K, Faloona F, Scharf S, Saiki R, Horn G, Ehrlich H. Specific enzymatic amplification of DNA in vitro: The polymerase chain reaction. Cold Spring Harbor Symp Quant Biol 1986; 51:263–273.

119. Myers RM, Larin Z, Maniatis T. Detection of single base substitutions by ribonuclease cleavage at mismatches in RNA:DNA duplexes. Science 1985; 230:1242–1246.

120. Myers RM, Lumelsky N, Lerman LS, Maniatis T. Detection of single base substitutions in total genomic DNA. Nature (London) 1985; 313:495–498.

121. Myers RM, Maniatis T. Recent advances in the development of methods for detecting single-base substitutions associated with human genetic diseases. Cold Spring Harbor Symp Quant Biol 1986; 51:275–284.

122. Gibbs RA, Caskey CT. Identification and localization of mutations at Lesch-Nyhan locus by ribonuclease A cleavage. Science 1987; 236:303–305.

123. Lerman LS, Silverstein K, Grinfeld E. Searching for gene defects by denaturating gradient gel electrophoresis. Cold Spring Harbor Symp Quant Biol 1986; 51:285–297.

124. Schwartz DC, Cantor CR. Separation of yeast chromosome-sized DNA by pulsed field gradient gel electrophoresis. Cell 1984; 37:67–75.

125. Carle GF, Olson MV. Separation of chromosomal DNA molecules from yeast by orthogonal-field-alteration gel electrophoresis. Nucleic Acids Res 1984; 12:5647–5664.

126. Carle GF, Frank M, Olson MV. Electrophoretic separation of large DNA molecules by periodic inversion of the electric field. Science 1986; 232:65–68.

127. Smith CL, Econome JG, Schutt A, Klco S, Cantor CR. A physical map of the *Escherichia coli* K12 genome. Science 1987; 236:1448–1453.

128. Kohara Y, Akiyama K, Katsumi I. The physical map of the whole *E. coli* chromosome: Application of a new strategy for rapid analysis and sorting of a large genomic library. Cell 1987: 50:495–508.

129. Deaven LL, van Dilla MA, Bartholdi MF, et al. Construction of human chromosome specific libraries from flow sorted chromosomes. Cold Spring Harbor Symp Quant Biol 1986; 51:159–167.

130. Yunis JJ. High resolution of human chromosomes. Science 1976; 191:1268–1270.

131. Lupski JR. Transposon Tn*5* mutagenesis. In Chopra VL, ed. Genetric Engineering and Biotechnology: Concepts Methods and Applications. New Delhi, India: Oxford and IBH Publishing Co., in press.

132. Nathans J, Piantanida TP, Eddy RL, Shows TB, Hogness DS. Molecular genetics of inherited variation in human color vision. Science 1986; 232:203–210.

133. Nathans J, Thomas D, Hogness DS. Molecular genetics of human color vision: The genes encoding blue, green, and red pigments. Science 1986; 232:193–202.

134. Purrello M, Alhadeff B, Esposito D, et al. The human genes for hemophilia A and hemophilia B flank the X-chromosome fragile site at Xq27.3. EMBO-J 1985; 4:725–729.

135. Housman DE, Glaser T, Gerhard DS, et al. Mapping of human chromosome 11: Organization of genes within the Wilms tumor region of the chromosome. Cold Spring Harb Symp Quant Biol 1986; 51:837–841.

8

The Tools of Molecular Biology: Restriction Enzymes, Blots, and RFLPs

PETER N. RAY

The 1950s and 1960s witnessed the emergence of molecular genetics and its rapid growth to become a major area of biological research. In this period great advances were made in deducing the structure of genes and chromosomes and how they were regulated. Most of the activity was centered on prokaryotic organisms and their viruses, particularly *Escherichia coli* and bacteriophage λ. However, by the early 1970s, molecular genetics was in decline and many biologists left the field. The available technology could not cope with the immense size of the eukaryotic genome (1000 times larger than *E. coli* and 100,000 times larger than λ). Then, there were two major technological breakthroughs. First, the discovery of restriction endonucleases made it possible to cut DNA into a reproducible set of small manageable molecules. Second, methods were devised to visualize specific DNA fragments in a genomic digest that contained several million unique fragments. These discoveries gave rise to the explosion in recombinant DNA technology that has made molecular biology the most exciting area of science today.

RESTRICTION ENZYMES

Restriction endonucleases are a group of enzymes, isolated from prokaryotes, that cut duplex DNA at defined base sequences. The *restriction sites* or *recognition sites* are usually four or six bases long, although some five-base and eight-base sites are known, and are usually palindromic, with the same sequence of bases in either direction. Some enzymes allow degeneracy in the recognition sequence. For example, the enzyme *Hin*dII (Fig. 8-1) recognizes the sequence GTPyPuAC, in which the internal two bases can be either pyrimidine next to either purine. Most enzymes do not permit that kind of redundancy, and, for the most of them, any base change inactivates the restriction site.

Restriction enxymes introduce double-stranded breaks in the DNA, usually within the recognition site. The cleavage site may be in the center of the recognition sequence as shown for *Hin*dII (Fig. 8-1), generating fragments with "flush" or "blunt" ends. Other enzymes create breaks that are staggered around the axis of

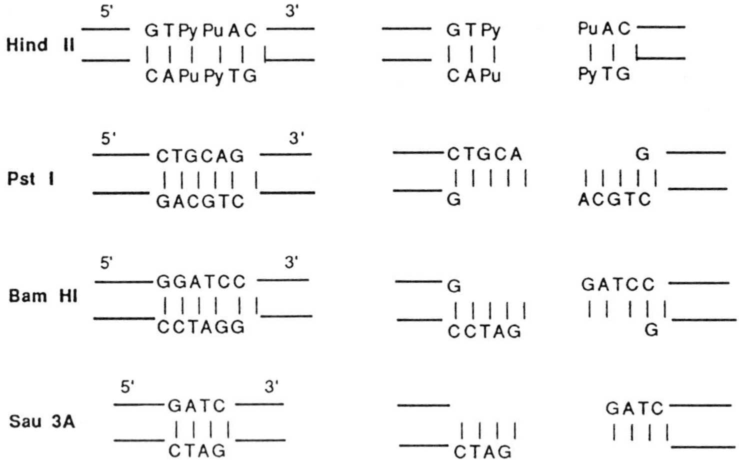

Fig. 8-1. Recognition sites and cleavage pattern for some typical restriction endonucleases. The fragments generated by restriction enzymes have either blunt ends (*Hin*dII), 3′ overhangs (*Pst*I), or 5′ overhangs (*Bam*HI). Some enzymes that recognize four base sequences (*Sau*3A) generate the same "sticky end" as some enzymes with six base recognition sequences (*Bam*HI). These fragments can be efficiently joined to each other to produce recombinant DNA molecules.

symmetry resulting in fragments with protruding single-stranded ends. The ends may then have a 3′ overhang (such as *Pst*I) or a 5′ overhang (such as *Bam*HI). The enzymes that create staggered cuts are particularly useful for DNA cloning experiments because DNA fragments with protruding ends or "sticky" ends can anneal through specific base pairing of the four-base overhang to other fragments cut with the same enzyme. Once annealed, these fragments can be covalently linked with the enzyme *DNA ligase.* DNA fragments can be joined in a specific way to generate new *recombinant* DNA molecules.

Sometimes, the tetranucleotide recognition sequence of one enzyme occurs within the hexanucleotide recognition sequence of another enzyme. For example, *Sau*3A recognizes the sequence GATC whereas *Bam*HI recognizes the sequence GGATCC (Fig. 8-1). Both enzymes produce fragments with the same sticky end, so fragments from either digestion can be ligated to the other. Since the recognition site of *Sau*3A is only four bases long, it should occur frequently in the genome, on average once in 4^4 (1:256) base pairs, whereas the *Bam*HI site would be encountered much less frequently, once in 4^6 (1:4096). Thus, relatively small DNA fragments can be generated with *Sau*3A and cloned into unique *Bam*HI sites in plasmid or bacteriophage vectors. This is the principle that underlies the construction of most genomic libraries, as described in Chapter 9. Note, however, that since there is no specificity in the base that precedes the *Sau*3A recognition site, most *Sau*3A/*Bam*HI ligations do not reconstitute the *Bam*HI site, making it difficult to remove a *Sau*3A fragment once it is inserted into a vector.

The first restriction endonuclease, *Hin*dII, was isolated in 1970 (1). To date, several hundred such enzymes have been isolated, recognizing well over 100 different sites (2). In practice, however, fewer than 50 are commonly used.

The role of these enzymes in prokaryotes is thought to be the equivalent of the mammalian immune system; the enzymes degrade the DNA of any invading virus that enters the bacterial host. Obviously then, the prokaryote must have some way to keep its own DNA from being degraded by its own restriction enzymes. This is accomplished by modifying the restriction recognition sites. Site-specific DNA methylases recognize the same sequences as the corresponding restriction enzyme and add a methyl group to one of the bases within the sequence. Once methylated, these sites cannot be recognized by the restriction enzyme. For example (Fig. 8-2), *Eco*RI recognizes the sequences GAATTC. *Eco*RI methylase recognizes the same sequence and methylates the internal adenine residues, so blocking the action of *Eco*RI. Several methylases have now been isolated and are used extensively in the laboratory to protect recombinant DNA molecules during subsequent cloning operations.

Some restriction enzymes are not sensitive to methylation of a cytosine residue in the recognition site. For example, the enzyme *Msp*I cuts the sequence CCGG whether or not the internal cytosine is methylated (Fig. 8-3). In contrast, *Hpa*II, an isoschizomer of *Msp*I (*isoschizomers* are enzymes isolated from different sources that have the same recognition sequence), cuts the sequence only if it is unmethylated. Pairs of enzymes such as these have been used to study methylation patterns in eukaryotic genomes (3).

Some CpG doublets in higher eukaryotes contain 5-methylcytosine. The pattern of methylation is both tissue and developmentally specific and is thought to be involved in the regulation of gene activity (3). By comparing the digestion pattern produced by *Msp*I and *Hpa*II, it is possible to determine which CpG doublets are methylated.

Another major use of restriction enzymes is in genomic mapping. Because of the highly specific nature of the recognition sites, restriction enzymes cut DNA molecules into a reproducible set of discrete fragments. The fragments can be separated on the basis of size by electrophoresis through an agarose gel, then visualized by staining with the fluorescent dye ethidium bromide. If a DNA molecule is digested and analyzed with a number of restriction enzymes, using one enzyme at

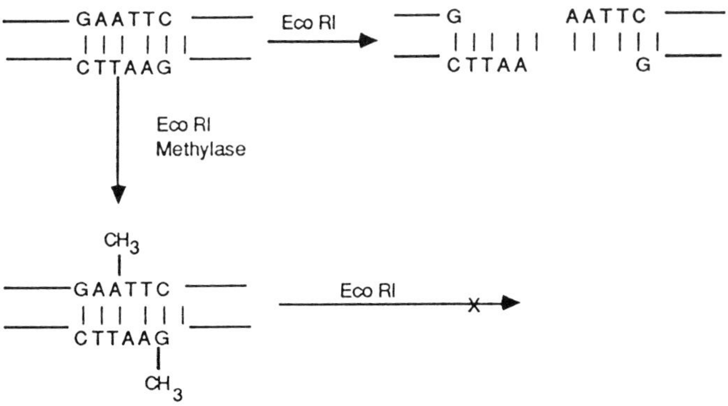

Fig. 8-2. The *Eco*RI restriction–modification system. The sequence GAATTC can be recognized by both the *Eco*RI endonuclease and by the *Eco*RI methylase. In the former case the sequence will be cut; in the latter case the internal adenosine residue will be methylated. When methylated, the sequence cannot be cut by the restriction enzyme.

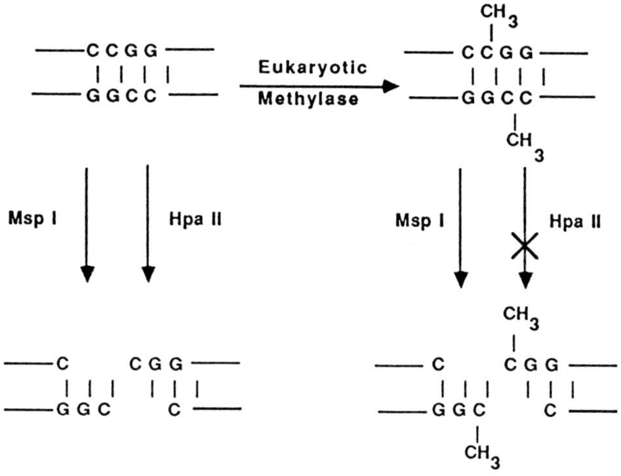

Fig. 8-3. Methylated bases in the eukaryotic genome can often be detected by their resistance to restriction enzyme cutting. The cytosine residue of CpG dinucleotides is often methylated in higher organisms. If the internal cytosine of a CCGG sequence is methylated, the sequence can be cut with the enzyme *MspI*, but not *HpaII*.

a time, or two or three enzymes together (double and triple digests), it is possible to construct a map of the restriction sites. With this type of analysis many viral and some bacterial genomes have been completely mapped (4).

SOUTHERN BLOTS

In studying complex genomes, it is not possible to visualize individual restriction fragments on agarose gels by staining with ethidium bromide. When the human genome, consisting of 3×10^9 base pairs, is digested with a restriction enzyme such as *Eco*RI, that recognizes a hexanucleotide, approximately 7×10^5 fragments of average size, $4–10 \times 10^3$ base pairs are generated. No agarose gel can resolve so many fragments; the digestion products appear as a smear that runs from the top of the gel to the bottom. The technology required to visualize a unique fragment in such a smear was developed by E. M. Southern and is called the *Southern blot* (5).

Basically, the DNA is transferred from an agarose gel to a solid support (nitrocellulose), and specific bands are then detected by hybridization with complementary radioactive probes (Fig. 8-4). This is achieved as follows: the DNA fragments are separated by electrophoresis on an agarose gel, then denatured by soaking the gel in alkali and neutralizing *in situ.* The gel is transferred to a blotting platform, which consists of a pad of absorbent paper that is soaked in high salt buffer and covered with a sheet of nitrocellulose filter membrane. A stack of paper towels is placed above the nitrocellulose and a light weight is placed on top to be certain that all layers maintain contact. As the solution is drawn up into the pad of paper towels, the DNA fragments are carried out of the gel into the nitrocellulose membrane to which they bind under the high salt conditions. After blotting for a suitable time,

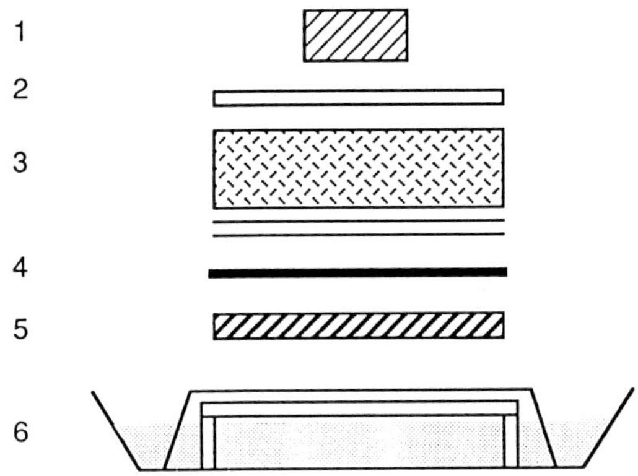

Fig. 8-4. Schematic representation of a Southern blotting apparatus. The agarose gel (5) is placed on an absorbent pad on a blotting platform (6). Buffer is wicked up into the pad from a reservoir around the platform. A nitrocellulose or nylon membrane (4) is placed on the gel and is overlaid with paper towels (3). A glass plate (2) and a weight (1) keep all layers in contact.

usually overnight, during which most of the DNA has been transferred to the nitrocellulose, the apparatus is disassembled and the DNA is fixed to the filter by baking *in vacuo*. The result is a stable replica of the gel on a solid support with the DNA in a single-stranded form.

To visualize the fragments of interest, specific DNA probes of 100–2000 base pairs are made radioactive by the incorporation of ^{32}P-labeled nucleotides and hybridized to the filter under conditions of salt and temperature that permit annealing of the probe to the complementary sequences found on the filter.

Hybridization is usually carried out by sealing the filter and probe in a plastic bag and submerging it in a water bath. After hybridization, the filter is washed to remove unbound radioactive probe and exposed to X-ray film to reveal where the probe has bound.

Nowadays, nitrocellulose is being replaced by nylon membranes that are more durable and can be reprobed more often. The procedures used are similar to those with nitrocellulose.

NORTHERN BLOTS

A variant of Southern blotting has been developed to examine messenger RNA fragments (6). The method can be used to determine whether the tissue being examined contains the mRNA of interest and, if so, in what quantities, and of what size the molecules are. mRNA is electrophoresed on denaturing agarose gels, usually containing formaldehyde or methyl mercury hydroxide, which melt out any secondary structure of the RNA so that the molecules are separated on the basis of length. The RNA is transferred to a nitrocellulose or nylon membrane and hybridized to appropriate probes, essentially as described for DNA.

The method is sensitive, can detect low levels of transcription, and can reveal variations in transcription and mRNA processing within the cell.

The combination of restriction enzymes and Southern blotting has made it possible to develop physical maps of many human genes and even gene clusters. However, it is still difficult to extend these techniques to map larger regions, such as those defined by genetic recombination. New technology, such as pulse field gel electrophoresis (7), makes it possible to separate and visualize DNA fragments in the range of 10^6 base pairs and it may soon be possible to make physical maps of whole chromosomes.

RFLPs

The most profound clinical applications of the technology described above is the prenatal detection of genetic disease. For several diseases, the gene involved has been cloned and the mutation causing the disease has been identified by DNA sequencing. These mutations can often be detected by direct analysis of the gene of the individual at risk. The tests may involve using a short oligonucleotide spanning the mutation site as a probe under conditions in which it will hybridize to the "normal" gene in which the DNA sequence homology is 100%, but not to the mutant gene in which the homology is less than 100%. Some mutations causing phenylketonuria (8) or α_1-antitrypsin deficiency (9) can be detected this way. Other diseases, such as sickle cell anemia, can be detected directly because the mutation has inactivated a restriction enzyme recognition site within the gene. For example, when digested with the enzyme *Mst*II, the normal globin gene gives a 1.1 kb band on a Southern blot whereas the sickle cell mutant gene gives a 1.3 kb band (10). However, in most genetic diseases, the affected gene is either unknown (as in cystic fibrosis or Huntington disease) or the mutations are too numerous (e.g., Duchenne muscular dystrophy) to make direct detection feasible. In these cases, there is an indirect method that is based on genetic linkage of RFLPs that can often be used for prenatal diagnosis.

There are naturally occurring variations within the DNA sequence of the human genome that can be detected by restriction endonuclease analysis as polymorphic restriction fragment band lengths. Probes that reveal these polymorphisms can be used as genetic markers to follow the segregation of individual chromosomes in families. The polymorphisms sometimes consist of variable numbers of tandemly repeated short DNA sequences *(hypervariable regions).* In others, single base variations alter the recognition site of a restriction endonuclease. Hypervariable tandem repeats are particularly useful because the high degree of fragment length variability at the site makes it likely that the two chromosomes of an individual will show different fragment lengths. In fact, using a combination of hypervariable region probes, A. J. Jeffries calculated that no two individuals would have the same banding pattern (11). Unfortunately, such highly polymorphic regions have not been found near the genes for most heritable disorders and the second type of polymorphism, resulting from single base changes, is used most often.

This type has only two forms: the restriction site is present and the band is short, or the restriction site is absent and the band is large. In either case, the polymorphisms are inherited in a simple Mendelian fashion and can be used as genetic markers for prenatal diagnosis (Fig. 8-5). In this type of analysis, we are not actually looking at the mutant gene but at a polymorphic probe site that is closely linked to

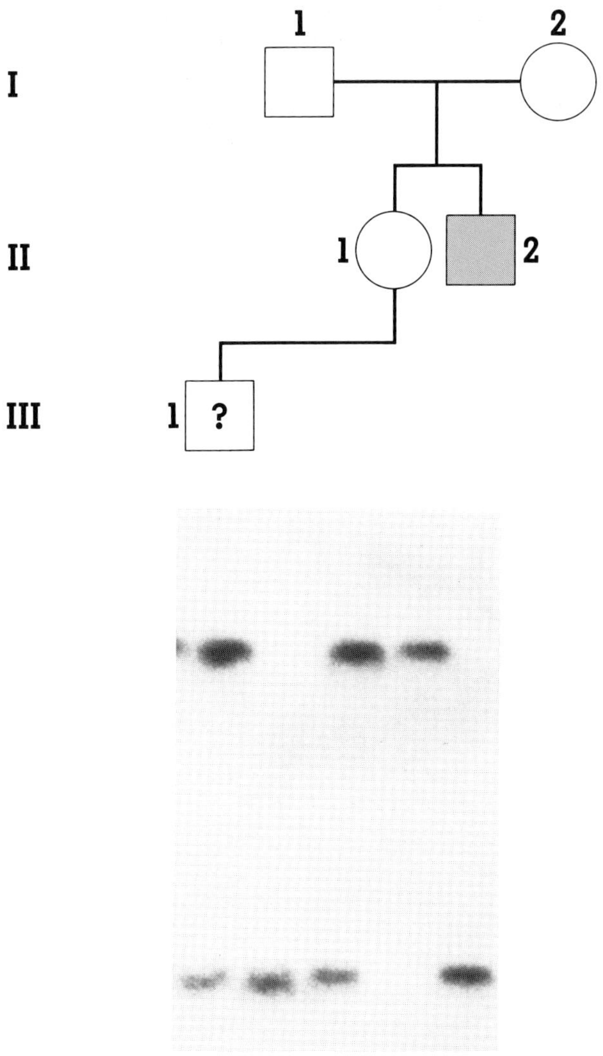

Fig. 8-5. Prenatal diagnosis of Duchenne muscular dystrophy by linkage analysis to a restriction fragment length polymorphism detected by the probe DXS164. The autoradiogram shows the distribution of the two polymorphic fragments in the family. The grandmother, I-2, is informative (i.e., heterozygous) for this RFLP and has passed the X-chromosome carrying the large fragment and the DMD mutation to her affected son, II-2. The daughter, II-1, has received an X-chromosome bearing the short fragment from her father, I-1, and the chromosome with the large fragment from her mother. She is therefore a carrier of the disease. The fetus, III-1, has inherited the grandmaternal X-chromosome and is thus predicted to have Duchenne muscular dystrophy.

the gene. By tracking the transmission of the RFLP through a family that includes identified clinically affected and heterozygote individuals, it is possible to determine which chromosome carries the defective gene and thus predict whether a fetus in the family will be affected or normal.

Several difficulties arise in this procedure. First, the appropriate individuals in the family must be *informative,* or heterozygous for the RFLP marker, to distinguish between the two chromosomes. For hypervariable probes, this is not generally a problem but with the restriction site RFLPs, the percentage of informative individuals is not more than 50%, and for most probes, 25–35%. Therefore, several probes must be linked to a disease to provide diagnosis for most families.

The second problem is the possibility of recombination between the probe site and the disease locus. The best probes are those closely linked to the disease, but in many cases the only available probes or the only informative probes show significant recombination frequencies with the disease. The recombination rates can be as high as 5–15%, yielding risk estimates that are of limited value. This problem can often be overcome by using two probes that flank the disease locus; the resulting risk estimate is the product of the recombinational frequencies of the two probes with the disease.

The third problem arises from the need to determine which RFLP allele is associated with the disease mutation, the *phase* of the RFLP. To do this, it is necessary to analyze the DNA of several family members, which is often difficult, particularly if affected individuals do not live long and are no longer available for DNA analysis.

In spite of these difficulties, the RFLP linkage method of prenatal diagnosis can be applied to many common genetic disorders to give highly accurate risk estimates. At the moment, RFLPs are used to diagnose cystic fibrosis, Duchenne muscular dystrophy, Huntington disease, phenylketonuria, polycystic kidney disease, α_1-antitrypsin deficiency, ornithine transcarbamylase deficiency, and several hemoglobinopathies. This list is expanding rapidly as the technology improves and more probes are developed.

REFERENCES

1. Smith HO, Wilcox KW. A restriction enzyme from *Hemophilus influenzae,* I. Purification and general properties. J Mol Biol 1970; 51:379–391.
2. Roberts RJ. Restriction and modification enzymes and their recognition sequences. Nucleic Acids Res 1985; 13:r165–200.
3. Razin A, Riggs AD. DNA methylation and gene function. Science 1980; 210:604–609.
4. Genetic Maps. O'Brien SJ, ed. CSH Laboratory NY, 1986.
5. Southern EM. Detection of specific sequences among DNA fragments separated by gel electrophoresis. J Mol Biol 1975; 98:503–517.
6. Alwine, JC, Kemp DJ, Stark GR. Method for detection of specific RNAs in agarose gels by transfer to diazobenzyloxymethyl—paper and hybridization with DNA probes. Proc Natl Acad Sci USA 1977; 74:5350–5354.
7. van Ommen GJB, Verkerk JMH. Restriction analysis of chromosomal DNA in a size range up to two million base pairs by pulse field gradient electrophoresis, chap. 8. In Davies KE, ed. Human Genetic Diseases, a Practical Approach. Oxford: IRL Press, 1986, pp 113–135.
8. DiLella AG, Marvit J, Lidshy AS, Guttler F, Woo SLC. Tight linkage between a splicing mutation and a specific DNA haplotype in phenylketonuria. Nature (London) 1986; 322:799–803.

9. Kidd VJ, Golbus MS, Wallace RB, Itakura K, Woo SLC. Prenatal diagnosis of α_1-antitrypsin deficiency by direct analysis of the mutation site in the gene. New Eng J Med 1984; 310:639–642.

10. Orkin SH, Little PFR, Kazazian HH, Boehm CD. Improved detection of the sickle mutation by DNA analysis: Application to prenatal diagnosis. New Eng J Med 1982; 307:32–36.

11. Jeffreys AJ. Hypervariable DNA and genetic fingerprints. In Lerman LS, ed. DNA Probes: Application in Genetic and Infectious Disease and Cancer. Cold Spring Harbor, NY: Cold Spring Harbor Laboratory, pp. 57–63.

9

Making Libraries: Using Vectors with Genomic DNA or cDNA

JAMES SYLVESTER AND ROY SCHMICKEL

A library is a collection of recombinant DNA molecules that can serve as a source of specific clones. The term "library" is not quite accurate because the clones are not catalogued. The term is descriptive in the sense that there is no guarantee that the item sought is in the library. It is also appropriate in that the library contains items for borrowing and community use. Here we shall describe the contents of a DNA library and how libraries are constructed.

A library is a collection of recombinant molecules that can serve as a source of clones; it is therefore necessary to describe recombinants and clones. Although DNA can recombine in many ways, the recombinants used in making a library are specific; they are composed of gene sequences that can self-propagate in bacteria plus a segment of DNA from another species of interest. The result is that the DNA of interest and the self-propagating DNA replicate as a single covalent segment of DNA. In library terminology, the recombinant is like a book. It has a cover to help carry it around and it is filled with the pages that a reader will be interested in reading.

The *self-replicating DNA* is derived from a plasmid or virus. The most commonly used self-replicating DNAs are the plasmid pBR (1), the viral phage λ (2), and the single-strand DNA MS2. Since MS2 is used mainly for sequencing not for libraries, we will not speak of it further. The size of plasmid DNA is 4.5 kilobase pairs (kbp); phage is 48 kpb. These self-replicating structures are called *vectors,* and the DNA of interest is the *insert.* The recombinant molecule replicates in *Escherichia coli* or other bacteria. The *E. coli* is referred to as the host.

Another vector, the cosmid, has been created by genetic engineering (3). It is a combination of the ends of a λ phage and the middle of a plasmid, created to accommodate larger inserts. Plasmids can comfortably accommodate up to 7 kb of DNA and phage up to 20 kb. Cosmids can accommodate 35–45 kb. In addition to the size of inserts that can be accommodated, each vector has other important properties that will be discussed.

The recombinants are made by using the restriction enzymes discussed in Chapter 8. Both inserts and vector are cut or "restricted" with compatible enzymes and then ligated. Of course all combinations of ligation do not result in self-replicating recombinants, but the ones that are viable can be selected by transfecting or infecting them into a host. In *transfection,* pure DNA is taken up by an *E. coli* that has been "shocked." Shocking is accomplished by washing *E. coli* cells in a calcium buffer, mixing the washed cells with the desired recombinant DNA, and heating the two briefly at 42°C. Under these conditions about 1 particle of DNA in 100,000 finds its way into a bacterium and replicates. The DNA of phage or cosmids can be *infected* into hosts by a clever technique called *in vitro packaging* (4). The recombinant phage or cosmids are mixed with separate protein building blocks of the phage. Under appropriate conditions *in vitro* the phage will self-assemble with the recombinant DNA and become an infectious particle that can infect at much higher efficiences: 1 particle in 20 infects the host organism compared to an efficiency of 1 in 2000 for transfection. In practical terms, transfection yields about 10^5 plasmid or phage clones/μg of DNA and *in vitro* packaging yields about $10^8/\mu$g.

The hosts are types of *E. coli* that permit phage recombinants to grow but do not allow nonrecombinants to grow. Conditions are selected so that recombinant and host can grow together but the host cannot grow alone. Typically the plasmid has a gene for antibiotic resistance, and the host organism is sensitive to that antibiotic. Only hosts that bear an antibiotic-resistant plasmid will grow when the antibiotic is present. In other systems, the plasmid or phage may carry a lethal termination mutation that can be suppressed by the host. This prevents the plasmid from spreading to nonengineered, natural hosts.

CLONES

Once the DNA recombinant has been formed and the host has been transfected, individual molecules can be isolated. For the most part, only one molecule transfects or infects each individual *E. coli.* Collectively, the *E. coli* organisms form colonies or plaques on an agar plate; each colony or plaque contains several thousand identical copies of the recombinant DNA. The plaques or colonies can be isolated and grown in bulk amounts to yield $10^9 - 10^{12}$ identical recombinants/ml. After the organisms grow for a few hours, the hosts or lysates of the organism may provide 200 μg of DNA. The yield may be milligrams of a homogeneous DNA segment that originally existed in a heterogeneous cell population of over 100,000 different segments. If that DNA segment is a gene we would say "we have cloned a gene." Cloning is therefore the process of obtaining replicas of the DNA from a single cell or organism. In the example given, each bacterial colony originates from a single bacterium that may contain a particular fragment of human DNA.

MAKING A LIBRARY

To make a library the most appropriate vector is chosen and combined with the insert, which may be either genomic DNA or cDNA. The essential difference is that genomic DNA includes the total genomic DNA of a species, including introns as well as exons: cDNA, on the other hand, includes only the expressed DNA exons that have been transcribed as RNA, which in turn served as the template to synthesize the cDNA. Of course the genomic DNA differs in different species. In general, human and other mammalian genomes contain 3×10^9 bp of DNA. This means 3×10^6 clones of 10^3 bp in length, 10^5 clones of 10^4 bp in length, and so forth. cDNA is a complementary DNA copy of mRNA. In general, the cytoplasm contains 1,000 to 10,000 different mRNAs in each cell.

GENOMIC LIBRARIES

An instructive example is the construction of a complete human genomic library in a λ phage vector (5) (Fig. 9-1). Early work in recombinant DNA research was halted until "safe" vector–host systems could be developed (1,2). To that end, an essential gene of the viral vector was mutated, and a specific bacterial host was required for survival of the virus. Typically the sequence of a codon was altered from sense to nonsense by changing the position of a termination codon; that change rendered the virus noninfectious in wild-type bacteria. Some hosts could suppress the mutation by supplying suppressor tRNAs to allow the infection. The bacterial hosts were "auxotrophic"; they also required the addition of specific nutrients and were sensitive to antibiotics. The development of new λ vectors has improved the ease of growth and versatility. In λ vectors the unique restriction enzyme sites must be arranged so that exogenous DNA can either be inserted into the phage DNA or replace segments of nonessential phage DNA. These are called insertional or replacement vectors (5). The end product must be between 42,000 and 52,000 bp to produce viable phage particles.

Replacement vectors are generally more useful; Charon 30 is an example. The size of Charon 30 is 46.76 kpb and it can be digested with *Bam*HI, *Eco*RI, *Hin*dIII, *Sal*I, *Xho*I, or combinations of these restriction enzymes; as a result, two phage arms of 21 and 12–15 kpb are produced. Therefore between 10 and 13 kbp of internal DNA is removed to be replaced by exogenous DNA. Since the virus can hold a total of 52 kbp, 16–19 kbp can replace the removed DNA.

The replacement or insert DNA is DNA obtained from the species to be cloned. Highly purified human genomic DNA must be isolated from cultured white blood cells, or fibroblasts, or from a surgical specimen (6). DNA is removed by lysing cells in a detergent–EDTA–proteinase K buffer and extracting the proteins and lipids with phenol and chloroform. RNA is then removed by digestion with ribonuclease. After more phenol extractions, the DNA is dialyzed against a storage buffer and is ready for restriction. As discussed in the previous chapter, restriction enzyme recognition sites are spread randomly throughout the genome. On the average, 6-base recognition enzymes cleave every 4000 bp; 4-base enzymes cleave every 250 bp. Each enzyme yields a Poisson distribution of fragments including some fragments that are too large or too small to fit into a given phage vector.

Complete digestion of genomic DNA does not give a complete library because some fragments do not fit into the vector. The most common means to circumvent that problem is to do a partial digest at a frequent sequence (7). For instance, *Sau*3A recognizes the sequence GATC and leaves a single strand with that sequence. This enzyme is used and a partial reaction is established so that the average length of the fragments is 17 kbp. The average restriction fragment contains

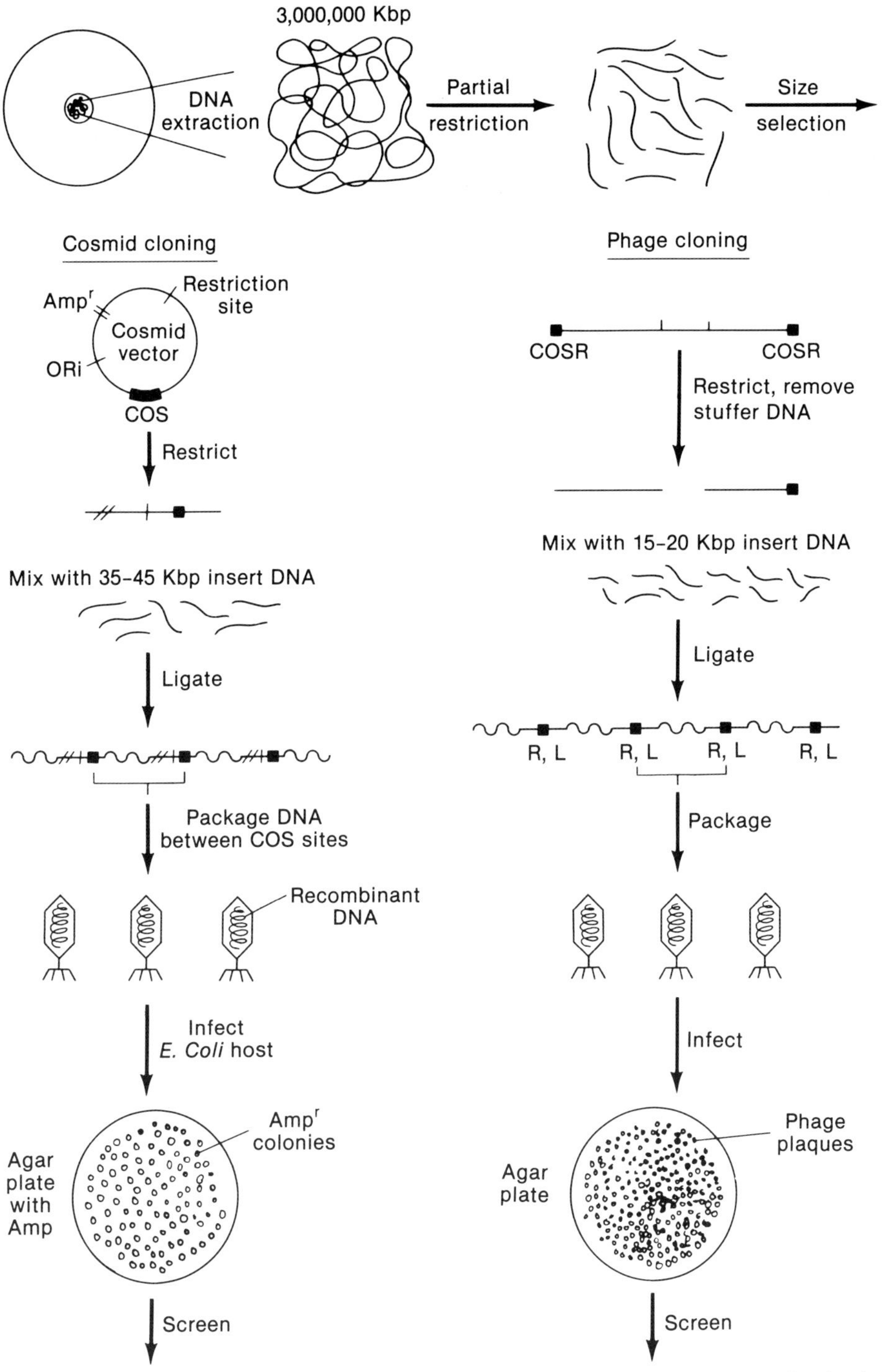

Fig. 9-1. Genomic DNA library: a generalized scheme for making recombinant molecules for their cloning. After DNA extraction and partial restriction, the DNA fragments (inserts) are size selected by either preparative gel electrophoresis or sucrose gradient ultracentrifugation. Only viable recombinant molecules are shown, i.e., arm-to-arm or insert-to-insert ligation products are not shown. See Chapter 10 for screening the library.

17,000/250 or 68 *Sau*3A sites. However the cleavage sites are random; a sequence of interest may be found in different, overlapping restriction fragments. The *Sau*3A partial digests can be inserted into a vector that has been restricted with *Bam*HI because the sticky ends generated by *Bam*HI and *Sau*3A are identical.

To summarize the important features of genomic library construction, one must match the vector and insert by size and restriction enzyme.

Once the phage vector and enzyme have been chosen, experimental conditions must be considered to produce the recombinants and introduce them into bacteria (5). The actual formation of recombinant DNA or covalent joining of viral and insert DNA is done in a *ligation reaction.* Purified λ phage arms (right and left each with complementary cohesive or "COS" ends) are mixed with the appropriately restricted and size-selected insert DNA. Size-selection is achieved by partially restricting small amounts of genomic DNA with decreasing amounts of enzyme (*Sau*3A) or with a fixed amount of enzyme for different times. The reaction is terminated and analyzed on agarose gels. Reaction conditions that produce the correct insert size are scaled to use 200–500 μg of DNA. The 15- to 20-kb fragments can be concentrated by use of sucrose gradients or agarose gel electrophoresis. The purified arms and insert DNA are mixed at optimal relative amounts to produce the most recombinants. This results when the reactants are mixed such that one right arm, one left arm, and one insert are ligated by T_4 DNA ligase to form one molecule; the molar ratio of phage arms to inserts is near 2:1. This ratio makes it more likely that recombinants form, instead of joining inserts-to-inserts or arms-to-arms. A high concentration of DNA in the reaction promotes the annealing of the cohesive ends of individual phage for the packaging reaction. However, theory does not correspond exactly to experimental design; pilot ligations and packaging at substrate ratios bracketing 2:1 are performed to determine the best conditions for a large scale experiment.

The most effective way of introducing foreign DNA into cells is by viral infection of bacterial cells. If recombinant molecules are packaged within the phage head *in vitro* then the recombinants can be introduced into the *E. coli* by infection. λ virus is packaged by proteins that scaffold the DNA, enzymes that insert the λ DNA into the head, and then cleave the concatenated DNA near the COS ends.

To prepare protein extracts that can package λ DNA, it is necessary to separate these proteins from endogenous viral DNA. This is accomplished by using mixtures of extracts from phage mutants that complement each other for packaging. If extracts are capable of forming 10^9 virus/μg of DNA, then ligation must produce only 0.001 μg of recombinant DNA to obtain a library of 10^6 clones from the human genome.

Special attention is given in choosing the phage vector or enzyme digest. For most purposes, a general partial *Sau*3A library could be cloned in the most versatile *Bam*HI phage vector. However, it might be necessary to clone DNA from a single person to study a particular genetic trait. To that end, it would be necessary to make a specific library, not necessarily a complete one. Preliminary characterization of the genome by Southern blotting could determine what enzyme digest would permit cloning of the genetic locus in question. Preparative gel electrophoresis and DNA fractionation would enrich the concentration of genomic DNA for that particular size of restriction fragment. If a phage vector is available to accommodate fragments generated by that enzyme, a much smaller library of fewer recombinants can be made and screened. Therefore phage vectors are constantly being engineered to be as flexible as possible for fragment insertion and genomic walking.

Plasmids or cosmids can also be used to make genomic DNA libraries. Plasmids are usually easier to work with than λ phage but they do not accept large fragments, and the recombinants are more difficult to screen than phage. But under special conditions, particularly when highly enriched DNA is used, plasmid cloning is practical. Methods that introduce plasmid DNA into bacterial cells are being improved and this approach is becoming more popular.

Cosmid vectors enjoy the theoretical advantages of both phage and plasmid systems but are difficult to use experimentally (3). Cosmids are plasmids that include a unique cloning site, antibiotic resistance genes, an origin of replication, and phage COS (cohesive ends) sequences. Because of the COS site, the recombinant cosmid can be packaged *in vitro* in a phage head. Once inside the cell by infection, the cosmid acts as a plasmid. Antibiotic-resistant *E. coli* colonies grow instead of phage plaques. Since cosmids are small vectors (4–7 kb) and packaging requires 40–50 kb molecules, more than 40 kb can be inserted into these vectors (twice that of phage vectors). The more DNA per recombinant the less screening is required. However, there is more DNA and therefore more risk of DNA rearrangement, deletion, and/or recombination in the host. Hosts that are recombination negative (REC^-) may only partially alleviate this problem.

OTHER LIBRARIES

Construction of genomic libraries has become commonplace in molecular genetic laboratories. However, the exercise is not yet trivial and there are many libraries, so it is often best to clone by mail. Most libraries have been made available (with few imposed conditions) through the generosity of the originators. Someone else's library can be used to characterize a particular genome; if data from the cloned DNA do not fit observations made on intact genomic DNA the differences must be reconciled. It may be necessary to use libraries constructed in different laboratories and in different host–vector systems to elucidate the genetic locus in question.

The U.S. Department of Energy has contracted National Laboratories at Los Alamos and Lawrence-Livermore to construct human chromosome-specific libraries. Chromosomes from cultured cells were sorted and isolated by an automated fluorescence method. The DNA from each chromosome was isolated to make *Hin*dIII and *Eco*RI libraries in phage vectors. These libraries are now available to researchers at nominal cost and can be used to isolate and localize human genes as discussed in Chapter 10.

cDNA LIBRARIES

We have discussed only a library of the entire genetic information of the nucleus. Biological questions of temporal, developmental, and tissue-specific expression can best be answered by construction and analysis of cDNA libraries. In this approach, complementary DNA (5) is cloned to evaluate the RNA population of a specific cell at a specific time in development or in response to environmental stimuli. Coupled with data from Northern blots and translation experiments, cDNA cloning can be used to characterize the RNA content of cells. These patterns can explain the transcriptional and posttranscriptional events that determine how the individual cell behaves. Characterization of housekeeping genes and developmental genes can provide the molecular details of regulation and coordinated expression. cDNA cloning includes only genetic regions that are expressed as RNA; walking in libraries of genomic DNA must be used to isolate regions that control the expression of RNA. Because of the intron–exon organization of genes, cDNA clones (exons) are spatially separated fragments in the actual genome; these cDNA clones can be used as probes to detect the bordering DNA (introns) in genomic DNA. The ability to clone RNA as cDNA has vastly extended the capabilities of molecular biologists.

RNA is much more difficult to use experimentally than DNA for two reasons. First, most RNA is itself short lived; often the half-life is measured in minutes. Second, there are many stable ribonucleases that degrade RNA. Care must be taken to work quickly and as free of RNase as possibly by using RNase inhibitors. DNA sequences of interest may be present in low abundance in the genome, but this corresponding mRNA may account for 1–2 to 50% or more of the total RNA in a cell. It is not possible to be certain that all the RNA of a given cell is completely represented in the cDNA library. Therefore the RNA population of cells must be evaluated by other techniques.

To begin cDNA cloning (Fig. 9-2) RNA is isolated by solubilizing the homegenized tissue or cells of interest in a 4 *M* solution of guanidinium isothiocyanate, then centrifuging the RNA as a pellet through a gradient of cesium chloride. Protease digestion, phenol extraction, and ethanol precipitation are often used instead of centrifugation through cesium. Separating nuclei from cytoplasm can also help to differentiate the species of RNA being isolated. Of cellular RNA 95% consists of ribosomal and low-molecular-weight [transfer (t) and small nuclear (sn)] RNA, but messenger RNA (mRNA) is the target of a cDNA preparation.

mRNA is heterogeneous in size, and has a long polyadenylate [poly(A)] tract at the 3′ end. It can be enriched by repeated passage through oligo(dT) cellulose affinity columns. The poly(A) RNA is bound to the column, and the abundant tRNA, snRNA, and rRNA pass through the column. The eluting buffer is then changed to release the mRNA for collection and concentration.

For cloning of RNA it is necessary to synthesize a complement to all of the RNA molecules in the preparation. This is accomplished by annealing oligo(dT) to the poly(A) stretches of the mRNA. The oligo(dT) serves as a primer for the action of reverse transcriptase to make a complementary strand. After the reverse transcriptase reaction, the RNA is hydrolyzed to leave single-stranded DNA. The 3′ end of this DNA folds back upon itself to form a short double-stranded region at that end of the molecule. This doubled end can then be extended by either reverse transcriptase or a DNA polymerase called the Klenow fragment (a portion of DNA polymerase from which exonuclease activity has been removed); the result is an intrastrand base paired molecule with a single-stranded bubble at one end. The single strand is susceptible to cleavage by S1 nuclease to yield a double-stranded

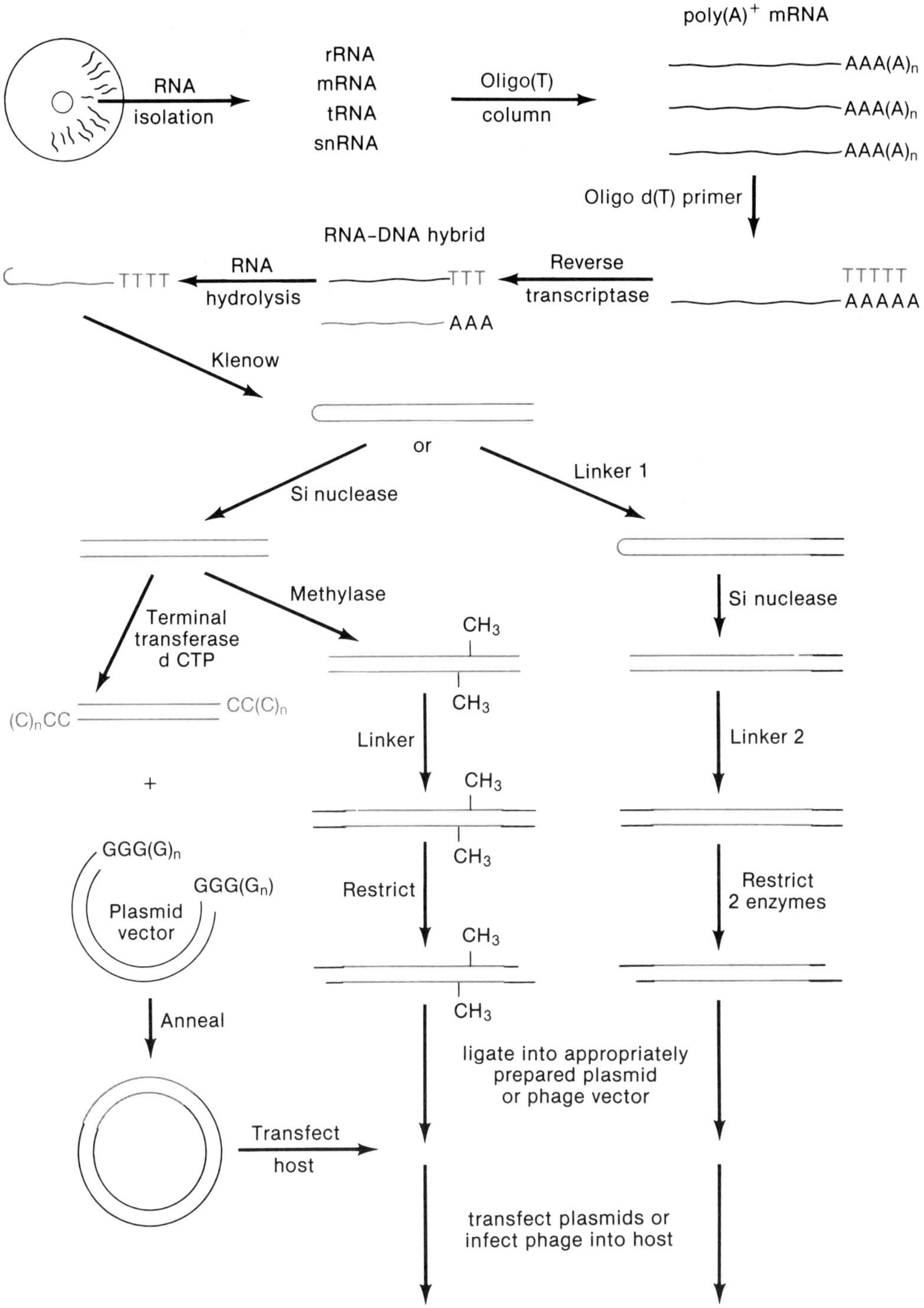

Fig. 9-2. cDNA library: alternate schemes for constructing cDNA libraries are shown. Methylation of internal restriction sites is necessary when the linker contains the same restriction site. After preparation of the insert molecules and the formation of recombinants (with vector), the introduction into the host is similar to that of genomic libraries. See Chapter 10 for the use of specialized expression vectors for cDNAs.

molecule. In a good library all of the RNAs are represented and the copies of each are close to full length (8).

The pool of these cDNA molecules is now prepared for insertion into a particular vector. If a plasmid vector is used, a string of approximately 20 dC residues may be added to each 3′ end of the insert DNA. A *Pst*I-digested plasmid vector then has a string of approximately 20 dG residues added to each 3′ end of the linearized plasmid. These reactions are done with an enzyme called terminal deoxynucleotidyltransferase that adds the appropriate deoxynucleotide to the 3′ ends of DNA strands. The tailed plasmid and insert are annealed to each other to produce a closed circular recombinant DNA molecule. These molecules can be used to transform competent *E. coli* bacteria. The inserted DNA can be removed from the isolated plasmid preparations by digestion with *Pst*I restriction endonuclease.

A second method of preparing DNA for insertion into plasmid or phage vectors is by attaching synthetic *linkers* or adaptors to each end of the insert DNA. Linkers are short 8- to 12-bp oligonucleotides that contain one or two restriction sites. The choice of restriction site depends on availability of the vector and the lack of such sites in the insert DNA. Alternatively one can add linkers to the vector or use the appropriate modification enzyme (*Eco*RI methylase) to methylate internal restriction sites, making them refractory to digestion by that enzyme. The linkers are attached by DNA ligase and must be digested with the enzyme to produce suitable sticky ends for cloning.

Adaptors are short double-stranded DNA oligomers that are blunt ended on one side and sticky on the other. Using one linker (or adapter) before and another one after S1 digestion, cloning can be made directional; ligation to the vector can then orient the DNA to promote transcription from bacterial sequences through the inserted sequences. If fusion peptides are produced by the growing *E. coli* and these peptides can be detected by antibodies, the vectors are termed *expression vectors.*

Expression vectors contain *E. coli* RNA polymerase start sites (promoters), 5′ untranslated ribosome-binding sites, and translation start sites *in frame with* a series of restriction sites in which insert cDNA can be ligated. The cDNA must be ligated in the correct translational reading frame with the vector sequences to ensure production of the desired protein as discussed in Chapter 10.

Although some techniques increase the likelihood of cloning mRNA of low abundance, the cDNA library may not represent all possible full-length, intact mRNA in the original cell. It may be necessary to make a second library from the same source, using primers that anneal randomly throughout the sequence to generate a "randomly primed" [rather than an oligo(dT)-primed] cDNA library. This is analogous to a partial restriction genomic library and can also be used to walk in cDNA libraries. Screening both cDNA and genomic DNA libraries with appropriate probes is used to detect both coding and noncoding regions, and also to provide additional probes for cloning an entire genetic area in steps.

PLUS–MINUS SCREENING

All the expressed sequences of DNA at a specific point in the life cycle of a cell population may be cloned. Tissue-specific and developmental questions can be evaluated by comparing the contents of two different libraries. Clones common to both libraries can be recognized, as well as those that are not represented in the alternate library. The simplest way is to use probes against mRNA isolated from two sources; cDNA may be represented in one cell but not another.

For example, if a muscle-specific mRNA is sought, it might be possible to select probes that hybridize with muscle mRNA but not fibroblast mRNA. This pattern would eliminate all housekeeping genes from analysis; housekeeping genes encode enzymes for ubiquitous pathways of oxidative phosphorylation or glycolysis that are needed by all mammalian cell types.

A second technique can detect a specific mRNA that is induced by a specific stimulus such as the action of a hormone or heat shock, or detect the mRNA that is otherwise at higher concentration in one RNA pool than in another (9). ^{32}P-labeled cDNA from each of the two RNA populations is used separately to screen replica platings of the induced cDNA library (Fig. 9-3). The induced RNA population should generate probes that are not found in the noninduced population. The induced mRNA may be represented by clones that are detected by one set of probes and not the other.

An alternative approach is to enrich the single-stranded cDNA population from one source for a specific message found only in that source (10). The cDNA population is hybridized to an excess of mRNA from the undifferentiated cell (Fig. 9-4). The cDNA–mRNA hybrids are removed by equilibrium centrifugation or by

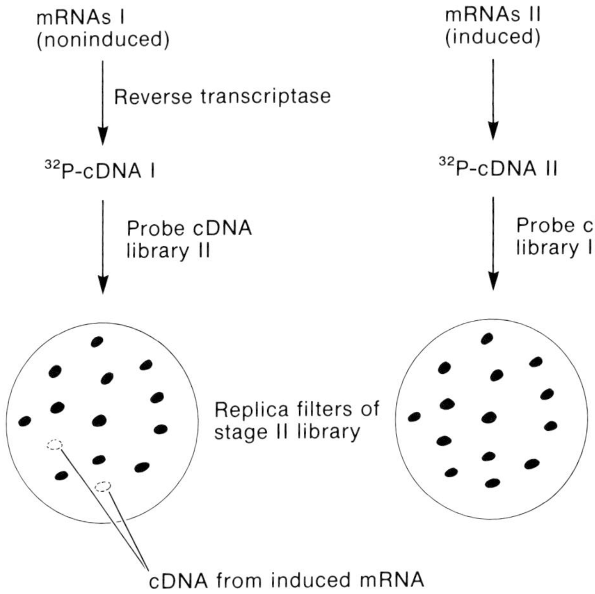

Fig. 9-3. Inducible RNA. cDNAs are made from both the noninduced and induced RNA pools. Only the induced cDNA as probe can detect its own recombinant when replica filters of the induced library are screened.

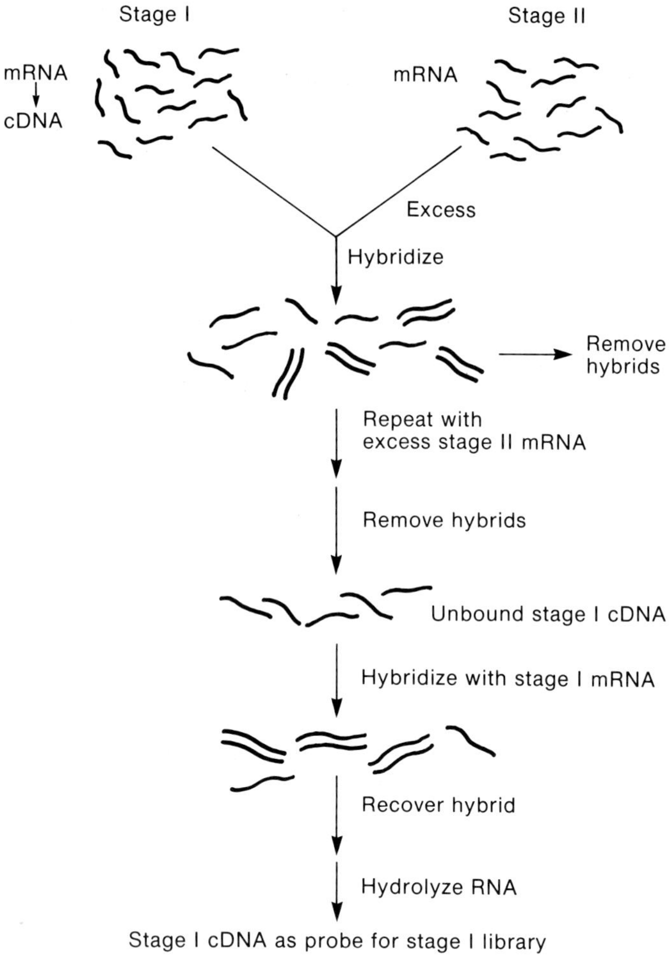

Fig. 9-4. Differential hybridization. This method removes nonspecific (housekeeping and/or common RNAs) and concentrates those mRNAs (as cDNA) that are present or specific to one RNA pool but not another.

binding to hydroxyapatite columns. Successive hybridizations to an excess of mRNA from the undifferentiated source removes most of the RNA that is common to both differentiated and immature cells. This leaves behind a cDNA population that is highly enriched for the differentiated source. This residual cDNA population is then hybridized to an mRNA source from differentiated cells and the hybrids are recovered and used to make cDNA clones or to make probes to screen a library of cDNA from differentiated cells. Developmental changes in RNA populations can be detected and specific mRNA can be represented in cDNA clones. The genes can then be identified by sequencing DNA and protein; regulating features may be discovered once genomic clones are analyzed.

PITFALLS

There are several pitfalls associated with a library. The first is underrepresentation. The probability of a clone being present is related to the number of separate clones involved and the size of the fragment being cloned. The number of separate clones can be respresented as (5)

$$N = \ln(1 - p)/\ln(1 - f)$$

where N is the number of recombinants needed to be 99% certain (p) of including a fragment by random distribution. For a library of 17-kbp fragments from the human genome of 3×10^9 bp, the fraction of the genome in a single recombinant, f, is $1.7 \times 10^4/3 \times 10^9$ or 5.6×10^{-6}.

Thus

$$N = \ln(1 - 0.99)/\ln(1-5.6 \times 10^{-6}) = 810{,}000$$

To be 99% certain of having the human genome completely represented one would need a library of 810,000 different clones.

If all clones were randomly represented, the mathematical approach would suffice. However there may still be some clones that are not represented in a partial *Sau*3A library. There may be some sequences in which *Sau*3A sites are very rare or very frequent. For this reason, it may be necessary to pool the products from a complete digestion and a very brief digestion to provide representation from those categories.

Some sequences may be poisonous to the host cell by producing expressed products or by binding important host proteins. However the sequences may propagate when grown in a nonsense direction or if fragmented and cloned as small pieces.

Another problem is rearrangment. Tandemly repeated sequences are stable in the genome but may be unstable in a prokaryotic host. These clones may need to be grown in a recombination-deficient host.

A third problem is contamination. It is not easy to eliminate prokaryotic or plasmid DNA from the library that is not intended to carry these. This can occur when antibiotics fail, special vectors revert, or glassware is contaminated by extraneous DNA. Since the DNA sought is represented in picomolar amounts, a small contamination can be disastrous.

If a library is made from a single transfection or infection, each recombinant molecule will be represented by the number of phage in a plaque or plasmids in a colony. For phage, the number of phage may be from 10^4 to 10^6 per plaque and similar numbers for plasmids per colony. The number of phage or plasmids can vary at least 100-fold from plaque to plaque or colony to colony. If the original infection is amplified by a second round of growth the variation will be 10,000 fold and after a few rounds the differences in representation will be enormous. For this reason, amplification of libraries should not be done for more than a single round and only from the original stock.

SUMMARY

The object of a library is to collect a complete set of genes or expressed genes in a recombinant form. These recombinants are the source of specific packages of information that can be withdrawn and examined one at a time. The knowledge of the size of the complete set and a knowledge of the average size of each recombinant will permit a reasonable evaluation of the completeness. The library can be assayed for size and diversity after the library is complete. The correct composition of the library depends on a random distribution of restriction sites. The equal representation of all fragments assumes that each fragment can be grown in a host without poisoning the host or without deleterious rearrangement. These conditions are not always met. For cDNA clones, there is no a priori guarantee that a given mRNA is present or if present is in sufficient quantity to be found. The presence of the mRNA of interest must be ascertained prior to the construction of the library. Northern analysis and *in vitro* translation are the best ways to assure yourself that a particular message is present in sufficient quantities. Utmost care is required for library construction and controls should be built into every step.

If you are successful in constructing a library you will surely have a valuable community resource. There are always more borrowers than there are libraries. If you have a reliable library you are now ready to explore the genome, walk along chromosomes, and locate genes.

REFERENCES

1. Armstrong KA, Hershfield V, Helinski DR. Gene cloning and containment properties of plasmid col E1 and its derivatives. Science 1977; 196:172–174.
2. Leder P, Tiemeier D, Enquist L. EK2 derivatives of bacteriophage lambda useful in cloning DNA from higher organisms: The λgt WES system. Science 1977; 196:175–177.
3. Collins J, Hohn B. Cosmids: A type of plasmid gene-cloning vector that is packagable *in vitro* into bacteriophage λ heads. Proc Natl Acad Sci USA 1978; 75:4242–4246.
4. Hohn B, Murray K. Packaging recombinant DNA molecules into bacteriophage particles *in vitro*. Proc Natl Acad Sci USA 1977; 74:3259–3263.
5. Maniatis T, Fritsch EF, Sambrook J. Molecular Cloning: A Laboratory Manual. Cold Spring Harbour, NY: Cold Spring Harbor Laboratory, 1982.
6. Blin N, Stafford DW. Isolation of high-molecular-weight DNA. Nucleic Acids Res 1976; 3:2303–2308.
7. Seed B, Parker RC, Davidson N. Representation of DNA sequences in recombinant DNA libraries prepared by restriction enzyme partial digestion. Gene 1982; 19:201–209.
8. Lund H, Grez M. Hauser H, Lindenmaier W, Schutz G. 5′ Terminal sequences of eucaryotic mRNA can be cloned with high efficiency. Nucleic Acids Res 1981; 9:2251–2266.
9. Taniguchi T, Fujii-Kuriyama Y, Muramatsu M. Molecular cloning of human interferon cDNA. Proc Natl Acad Sci USA 1980; 77:4003–4006.
10. Zimmerman CR, Orr WC, Leclerc RF, Barnard EC, Timberlake WE. Molecular cloning and selection of genes regulated in *aspergillus* development. Cell 1980; 21:709–715.

10

The Tools of Molecular Genetics: DNA Probes, Library Screening, DNA Sequencing, and Expression Vectors

ANTHONY P. MONACO

It is now considered conventional technology in human genetics to acquire chromosome-specific DNA segments (probes) and analyze them for subchromosomal location and DNA polymorphisms for genetic linkage to human disease loci. A major emphasis of this chapter is the strategy used to isolate DNA probes and expand them through chromosome walking (1) in genomic DNA libraries. The molecular analysis of a genomic region obtained through chromosome walking can uncover a human gene locus (2–5). A gene locus contains both introns and exons of a gene that, after mutation, gives rise to a human disease. To isolate a copy of the gene without the noncoding intron sequences, it is necessary to have a cDNA library constructed from mRNA via the action of reverse transcriptase. The cDNA clones isolated from these libraries are sequenced to read the genetic code that predicts the amino acid sequence of the encoded protein. The cDNA clones are ligated into expression vectors for transcription and translation by bacterial cells. The synthesized protein is produced in sufficient quantities to study its structure and function or to obtain antibodies against the protein.

DNA PROBES: ACQUISITION AND ANALYSIS

Widely used sources of chromosome-specific DNA probes are genomic libraries constructed in small-insert phage vectors from flow-sorted metaphase chromosome DNA (6,7). A chromosome-specific phage library is constructed from a complete restriction enzyme digest of flow-sorted chromosome DNA that is ligated into phage vectors that accept inserts of 1–9 kilobases (kb). The libraries are screened by absorbing the recombinant phages on a bacterial host and plating them on agar plates that contain a nutrient medium. Clear plaques appear on a lawn of bacteria after overnight growth and nitrocellulose filters are then placed on top of the plate to make a replica of the phage plaques (8). The nitrocellulose filters are removed from the plate and denatured in an alkaline-salt solution to completely lyse the phage heads and make the DNA single stranded. The filters are brought to neutral pH and baked in a vacuum oven for several hours to ensure that the phage DNA adheres to the nitrocellulose. The filters can be screened with radioactively labeled totally cleaved human DNA so that repeated DNA sequences in the phage inserts are hybridized and identified by autoradiography. Phage clones that do not hybridize with highly repeated sequences in total human DNA probably contain DNA inserts that are unique in the human genome. These phage clones can be plaque purified; the DNA is then isolated (9) and analyzed for single-copy DNA fragments.

The small DNA inserts must be tested for chromosome location. This is accomplished by hybridization to Southern blots (10) which contain DNA that has been totally digested by a restriction enzyme; the DNA should come from the appropriate human chromosome segregated by human–rodent somatic cell hybrids (11). The DNA probes can be made by directly labeling the phage insert DNA, or the DNA can first be cloned in plasmid vectors such as pBR322 (12) or pUC 18 (13) for amplification, digestion by a restriction enzyme, and radioactive labeling. Different DNA probes give different patterns of hybridization to Southern blots.

In one study, we used four different probes (Fig. 10-1). *Hin*dIII-digested human and somatic cell hybrid DNA samples were used to localize DNA probes to the

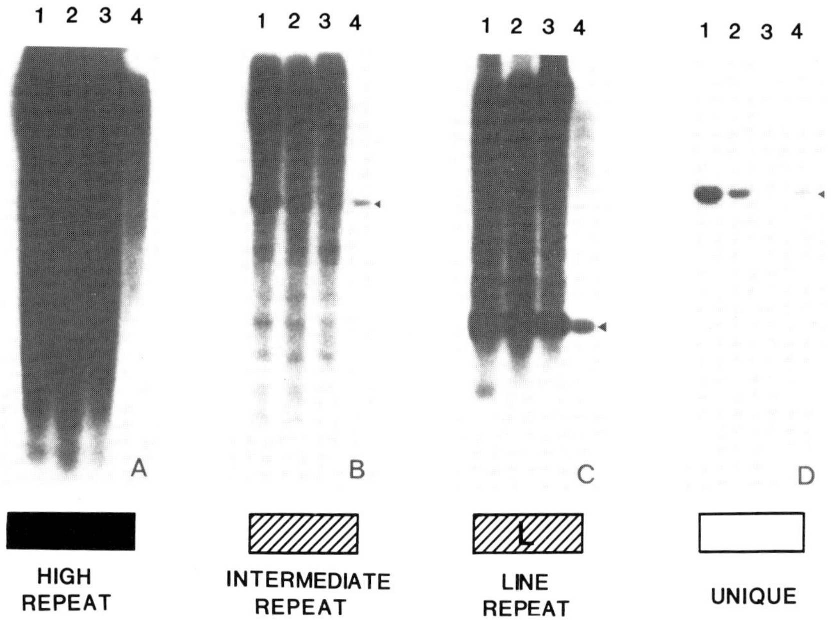

Fig. 10-1. Patterns of DNA probe hybridization to genomic Southern blots for localization of DNA fragments to the X-chromosome short arm. (A) Probe A, a highly repeated DNA fragment. (B) Probe B, an intermediate repeat with an X-specific band marked with an arrowhead. (C) Probe C, the LINE sequence repeat (14) with the 1.9-kb *Hin*dIII repeat indicated with an arrowhead. (D) Probe D, a single-copy DNA fragment (arrowhead) that localizes to band Xp21. *Hin*dIII digested genomic samples: lane (1) 49, XXXXY DNA; lane (2) normal 46, XY male DNA; lane (3) 46, XY (del X) DNA isolated from patient BB, who has a visible Xp21 deletion and three X-linked diseases (15); lane (4) G89, a somatic cell hybrid with a human X-chromosome segregated in a hamster background.

short arm of the X-chromosome. The highly repeated sequences in probe A produced a darkened autoradiograph of the Southern blot because they hybridized with a repeated sequence, with about $1-5 \times 10^5$ copies in the human genome. That repeat was probably a short interspersed repeat of the *Alu*I family (16). Prescreening the small-insert phage library with radioactively labeled human DNA would limit the number of such highly repeated DNA probes that enter the final analysis. An intermediate repeat was demonstrated by probe B because it hybridized with an X-chromosome-specific fragment as well with many other nonspecific DNA fragments. A single-copy subfragment of that DNA probe could usually be identified by eliminating the nonspecific repeated sequences with further digestion of the DNA probe with other restriction enzymes. A specific intermediate repeat was identified by probe C, which is called the LINE sequence (long interspersed nucleotide element) (14) that has a common *Hin*dIII fragment size of 1.9 kb and a full repeat length of almost 6.0 kb. That repeat is found about $1-5 \times 10^4$ times in the human genome and similar family members are seen in other mammalian genomes. An example of a single-copy DNA probe was given by probe D that hybridized to only one DNA fragment in a *Hin*dIII digest of genomic DNA and was localized to the short arm of the X-chromosome band Xp21 (Fig. 10-1). Single-copy DNA probes such as probe D can be used to search for DNA polymorphisms in genetic linkage studies or they can be used as hybridization probes in large-insert genomic libraries to start chromosome walking.

CHROMOSOME WALKING

Large-insert genomic libraries are screened to expand small DNA probes when the surrounding region has to be analyzed for DNA polymorphisms or genes involved in disease loci. Large-insert genomic libraries can also be screened to obtain the genomic locus (introns and exons) for a known gene, using previously isolated cDNA clones as hybridization probes. To construct large-insert libraries, total or chromosome-specific human DNA is partially digested (usually with *Mbo*I), size selected, and ligated into phage vectors that accept up to 22 kb of DNA [Charon series (17) or EMBL series (18)] or cosmid vectors that accept inserts of 40 kb or more (19). The large-insert phage libraries are prepared for screening by a DNA probe by pulling nitrocellulose filter replicas from plates of phage plaques on bacterial lawns, as described for small-insert phage libraries. Cosmid libraries do not form plaques on bacterial lawns and are amplified as circular plasmids inside the bacterial host. This can be achieved by growing the bacteria on top of nitrocellulose filters placed over agar plates, with the correct nutrient medium and antibiotic. Once the bacteria have grown, the filters are removed from the plates and replicas are made for storage. One set of filters is floated on an alkaline-Triton detergent solution to lyse the bacteria and denature the DNA. The filters are brought to neutral pH, treated with proteinase K to degrade bacterial proteins, and baked in a vacuum oven to fix the cosmid DNA on the filters. Both phage and cosmid library filters are prehybridized with the appropriate buffer to minimize nonspecific probe hybridization.

The basic strategy for screening both large-insert phage and cosmid libraries is outlined in Fig. 10-2. The overall goal of chromosome walking (1) is to acquire the most new DNA, in both directions, without walking backward through DNA that had been isolated previously. A small chromosome-specific DNA probe is radioactively labeled, denatured, and hybridized to the primary filter replicas of the library. The reaction is performed in a hybridization buffer at a temperature about 25°C below the melting temperature of the expected duplex of probe and phage insert DNA. After hybridization is complete (usually overnight), the filters are washed at high stringency to remove any background nonspecific hybridization between the probe and the filters, air dried, and exposed for autoradiography. The area on the plate or filter beneath a positive autoradiograph signal is picked off; the positive clones are purified through two to three rounds of dilution platings, nitocellulose filter replicas, and DNA probe hybridization. The resulting pure clones are amplified and DNA is isolated (9).

The DNA is digested with at least three different restriction enzymes in all combinations of one- and two-enzyme digests. The enzyme digests are size separated by electrophoresis through agarose gels and the restriction fragment sizes are estimated by comparison with *Hin*dIII-digested λ DNA as a size marker. A linear map of the DNA clone is constructed from the pattern of the single- and double-enzyme digests. The restriction pattern for individual positive clones that hybridized with the original DNA probe can be compared to overlap the clones. The restriction enzyme digests can then be used to subclone the smaller fragments into plasmid

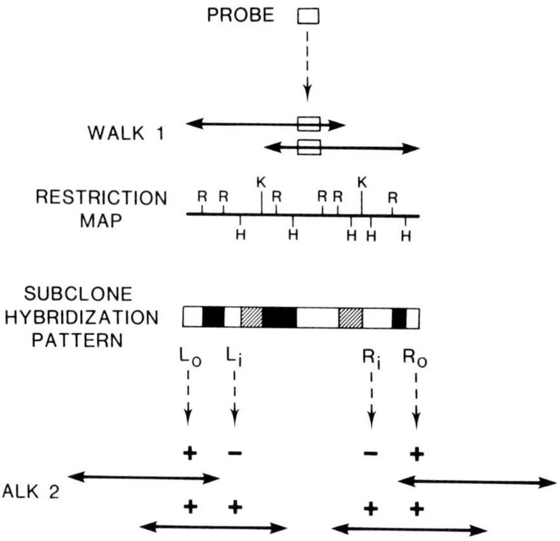

Fig. 10-2. Chromosome walking strategy. A single-copy DNA probe is used as a hybridization probe in a large-insert genomic library. Positive clones are purified and DNA is isolated and digested with restriction enzymes to create an overlapping map. The small digested DNA fragments are subcloned into plasmid vectors and tested for hybridization patterns on genomic Southern blots as shown in Fig. 10-1. Single-copy DNA fragments at the extremes of the walk 1 clones (L_o and R_o) are used as pooled hybridization probes for the next walking step. DNA probes L_i and R_i are used on duplicate filters of second round purification platings to differentiate those clones that extend the farthest in the desired direction.

vectors for further mapping with other restriction enzymes that cleave more frequently in the genome. The smaller DNA fragments (subclones) can be radioactively labeled to reveal their hybridization pattern in Southern blots of genomic DNA (Fig. 10-1). The Southern blots should contain the appropriate genomic samples to map the DNA fragments to the chromosome of interest.

To make the next walking step, single-copy DNA fragments from the extreme outside of both the left and right overlapping clones (probes L_o and R_o in Fig. 10-2) are used as pooled hybridization probes on the same primary filter replicas of the large-insert phage or cosmid library. New hybridizing clones are picked and plated for a secondary screening and purification step. On this plating, however, duplicate filter replicas are made so that the clones positive for probe L_o or probe R_o can be differentiated. The secondary filters that are negative for one side of the walk can then be rehybridized with single-copy DNA fragments a few kilobases inside the extreme outside probes. Those clones that are positive for the outermost probes (L_o or R_o) but are negative for the inner probes (L_i or R_i) break between the probes and extend farthest in the desired direction.

This *plus–minus screening* can be done in both directions on the same set of duplicate secondary filters. This eliminates most of the clones that extend backward over DNA already isolated in the previous walk. The tertiary screening and purification step is therefore less cumbersome and more directly isolates clones that give the largest walking step into new DNA. The clones are amplified and DNA is isolated and analyzed as described above for the third bidirectional walk.

SCREENING cDNA LIBRARIES

To find the exons of a gene in a large genomic region acquired by chromosome walking, the individual DNA clones are either analyzed for expression in RNA from different tissues or they are used in searches for conservation of DNA sequences in different mammalian species. The conservation strategy was conceived on the assumption that DNA sequences that encode amino acids (exons) have been conserved in the evolution of mammalian species whereas noncoding sequences (introns) have diverged with time. This approach was successful in finding exons of large genomic regions that contain the genes for Duchenne muscular dystrophy (3) or retinoblastoma (4). Once an exon has been identified in a DNA clone, an appropriate tissue-specific cDNA library is screened. The same overall strategies and methods for screening genomic DNA libraries apply to the screening of cDNA libraries. Most cDNA libraries are currently constructed in the λ phage vectors gt10 and gt11 (20) because of the high efficiency of *in vitro* packaging of λ DNA. The resulting libraries are screened by DNA probes by plaque-filter hybridization (8), as described for genomic phage libraries. Some cDNA libraries are constructed in plasmid vectors and are screened by colony *in situ* hybridization, as described for genomic cosmid libraries. Positive cDNA clones are purified by successive rounds of dilution platings, filter replicas, and DNA probe hybridizations. The purified cDNA clones are amplified and DNA is isolated and tested by hybridization to Southern blots for chromosomal location. The cDNA clones are also hybridized to Northern blots of RNA (21) to ensure that the cDNA clones recognize the correct size transcript with appropriate tissue specificity. If the primary cDNA clones do not cover the entire length of the transcript, then the same cDNA library can be rescreened with these clones to acquire more cDNA probes that extend in both directions toward the 5′ and 3′ ends of the gene. The strategies employed for genomic chromosome walking can also be applied for walking the entire length of the cDNA representation of the transcript.

DNA SEQUENCING

Both genomic DNA fragments and cDNA clones can be used for DNA sequence analysis. Genomic DNA fragments that contain intron:exon boundaries are usually sequenced to identify the 5′ and 3′ splice consensus sequences (22). *Consensus splice sequences* are found in the genomic introns that surround exons that encode open reading frames for uninterrupted stretches of amino acids. The nucleotides of cDNA clones are sequenced to use the genetic code to predict the amino acid sequence without the interruptions of noncoding introns. The sequencing strategies for both genomic DNA fragments and cDNA clones are essentially the same. First, a detailed restriction enzyme map of the fragment of interest is needed to help in isolating and purifying small subfragments.

Two sequencing methods are generally used: (1) Maxam–Gilbert (23,24) base modification and (2) Sanger (25,26) chain termination. Maxam–Gilbert sequencing uses a chemical modification of a purified double-stranded DNA or cDNA fragment, followed by degradation of that fragment at the modified bases. One strand at either the 5′ or 3′ end of the DNA fragment is labeled and the labeled ends are separated by digesting with a restriction enzyme that cleaves between the labeled ends or by physically separating the fragments by electrophoresis on a gel. The specific base modifications are performed in separate tubes for G, G + A, C, and C + T in a partial chemical reaction so that not every base is modified on each strand of DNA. Strand scissions are performed that cleave the DNA at the modified bases and the resulting DNA fragments are fractionated on a polyacrylamide gel and visualized by autoradiography (Fig. 10-3A).

Sanger chain termination is a synthetic approach to DNA sequencing that depends on the incorporation of 2′,3′-dideoxynucleoside triphosphates (ddNTPs) into a newly synthesized strand of DNA so that molecules of different lengths are produced. When the ddNTP is added to a molecule of DNA by the large (Klenow) fragment of DNA polymerase I, another deoxynucleoside triphosphate (dNTP) cannot be added because there is no 3′-hydroxyl in the incorporated ddNTP necessary for chain extension and elongation. A purified DNA or cDNA fragment of interest is subcloned by ligation into an appropriate vector for sequencing (Fig. 10-3B and C).

Double-stranded plasmid vectors, such as Bluescribe and Bluescript (Stratagene), contain polylinkers with many common restriction enzyme sites for ligation of DNA inserts. The polylinker is bracketed by a T3 and T7 RNA polymerase promoter on opposite sides that are used to make strand-specific radioactively labeled RNA probes from the cloned insert. Synthetic oligonucleotides (17 nucleotides

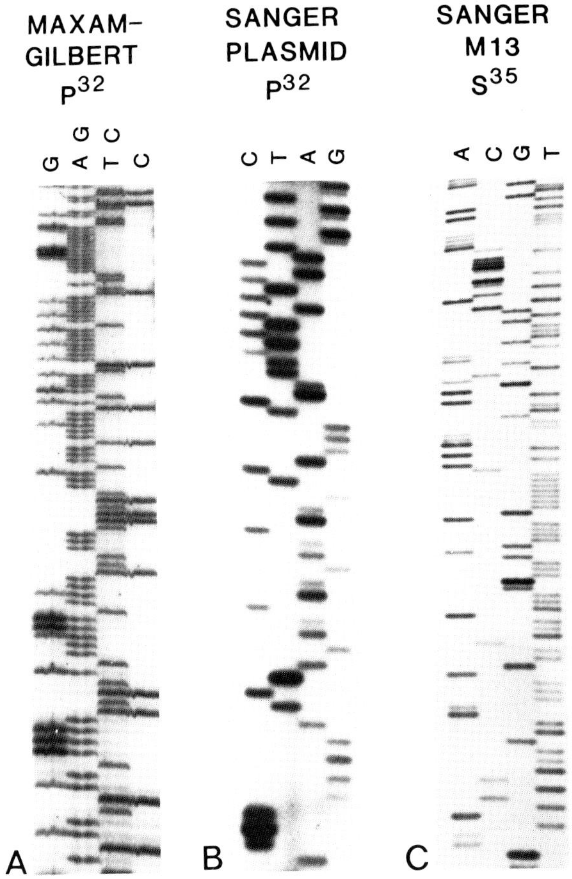

Fig. 10-3. DNA sequencing. (A) Maxam-Gilbert sequence of a mouse cDNA fragment, 3′ end labeled with [α-^{32}P]dCTP by filling in a 5′ overhand of a restriction site with T4 DNA polymerase[9] (provided by Eric Hoffman). (The gel is read 5′ to 3′ from top to bottom because a 3′ end labeled larger fragment was partially degraded to smaller fragements.) (B) Sanger dideoxy sequencing of a human cDNA fragment subcloned into the double-stranded Bluescribe vector (Stratagene) using [α-^{32}P]dATP as the labeled nucleotide. (C) Sanger dideoxy sequencing of a genomic DNA fragment subcloned into the single-stranded M13 vector and using [α-^{35}S]dATP as the labeled nucleotide (provided by Rob Fuller).

long) that are homologous to each of these RNA polymerase promoters are used as primers for DNA synthesis by the large fragment of DNA polymerase I. The double-stranded DNA plasmid that contains the fragment of interest is denatured in NaOH, and the appropriate primer (T3 or T7) is reannealed to the now single-stranded DNA. The dideoxy sequencing reactions are performed in separate tubes for A, T, C, and G with the appropriate ddNTP and a single radioactive dNTP (either ^{32}P or ^{35}S). The resulting DNA molecules synthesized by the large fragment of DNA polymerase I are fractionated on a polyacrylamide gel and visualized by autoradiography. A dideoxy-generated sequence from the double-stranded plasmid vector Bluescribe is shown in Fig. 10-3B using [α-^{32}P]dATP as the labeled nucleotide. Note that the nucleotide sequence is read 5′ to 3′ from bottom to top of the gel since the DNA fragments are synthesized in the 5′ to 3′ direction from a small 17 nucleotide primer.

A second vector used for Sanger dideoxy sequencing is M13 (27,28), a filamentous phage of male *Escherichia coli* that has a single-stranded circular form of packaged viral DNA. The M13 virus is nonlytic to *E. coli* so that infected bacteria grow, divide, and secrete virus particles into the medium for easy purification. In its replicative form inside the bacteria, M13 viral DNA is a double-stranded circle. Isolation of the double-stranded form of the virus from the bacteria allows DNA fragments of interest to be ligated into polylinkers that have been artificially introduced into the M13 viral DNA. To sequence a DNA fragment ligated into M13, single-stranded viral DNA is purified from secreted viruses in the liquid culture medium. Dideoxy sequencing reactions are performed using a universal M13 primer that is located 5′ to the polylinker region (Fig. 10-3C). An example of dideoxy sequencing using the phage M13 as a vector is shown in Fig. 10-3C. In this reaction the radioactive nucleotide used was [α-^{35}S]dATP. Since the energy of decay of ^{35}S-labeled nucleotides is weaker than that of ^{32}P, the use of ^{35}S results in sharper, thinner bands. As with plasmid dideoxy sequencing, single-stranded M13 sequencing is read 5′ to 3′ from the bottom to the top of the gel.

Different strategies are used to obtain sequences beyond a normal gel reading (about 250–300 nucleotides). Maxam–Gilbert sequencing employs frequently cutting restriction enzymes to generate DNA fragments that are small enough to sequence different regions and to give complete overlapping data. Any gaps in the sequence can be filled by selecting restriction enzymes after examining the already determined sequence at the borders of these gaps. Sanger dideoxy sequencing allows for many approaches to obtain sequence data for regions far removed from the restriction enzyme sites used for ligation into the polylinker sequences. A set of nested and overlapping deletions generated by Bal31 exonuclease or exonuclease III are the most popular enzymatic methods of sequencing large DNA fragments. The new availability of automatic DNA synthesizers, however, allows the synthesis of oligonucleotide primers from existing sequence data to prime DNA synthesis into areas of unknown sequence. In the same manner as the T3/T7 and M13 primers, the synthetic primers are reannealed to the fragment of interest and used to prime DNA synthesis by the large fragment of DNA polymerase I. Combining both Maxam–Gilbert and Sanger sequencing methods is one of the best and most rapid ways to obtain complete sequence information from both strands of the fragment of interest in large sequencing projects.

The complexity of data generated by a large sequencing project requires computer programs to store and analyze the data. These programs merge individual gel readings into a complete overlapping sequence and analyze assembled sequences for restriction enzyme sites and amino acid coding regions. The hydrophobicity plot of the predicted amino acid sequence is calculated, as well as the most probable structure of the polypeptide domains (α-helix, β-sheet, turn); both of these characteristics give insight into the possible function of the protein. The DNA nucleotide sequence and predicted amino acid sequence can be compared to databases of other known DNA and protein sequences to look for other genes that might be related in certain domains of the protein. These databases are operated by NIH and are available through many distributors for use in a microcomputer.

EXPRESSION VECTORS

There are two general uses for expression vectors for cDNA and genomic DNA fragments. The first use is a convenient vector for the construction of cDNA libraries that can be screened with antibodies to antigens of interest. The second use is to provide a vehicle for the overproduction of polypeptides that are encoded by the foreign DNA or cDNA inserts. Expression vectors produce sufficient quantities of the protein for analysis and for production of antibodies. The λ phage vector gt11 (20) may fulfill both expression goals. A unique *Eco*RI restriction site located near the C-terminal end of the *lacZ* gene in λgt11 is used to insert foreign DNA or cDNA fragments. Foreign DNA is inserted into the structural *lacZ* gene for β-galactosidase and fusion proteins are produced when the insert DNA is introduced in the correct open reading frame with the *lacZ* gene codons. The fusion of the polypeptide encoded by foreign DNA sequences with the C-terminal end of the β-galactosidase protein ensures an adequate level of transcription and translation of the DNA insert in *E. coli* and increases the stability of the fusion protein, a combination of the "foreign" polypeptide with *E. coli* galactosidase. cDNA libraries constructed in λgt11 can be screened with antibodies to proteins of interest. This is done directly on nitrocellulose filter replicas of the phage plaques and lysed bacteria that have released the recombinant fusion protein. After the specific antibody is bound to the protein antigen of interest, a second radioactively labeled antibody or ^{125}I-labeled *Staphylococcus aureas* protein A is used to visualize the site of primary antibody binding. The phage plaque beneath the positive autoradiograph signal is purified by subsequent rounds of dilution platings, filter replicas, and antibody bindings.

λgt11 is also used to produce recombinant polypeptides from the cloned DNA or cDNA insert as a fusion protein with β-galactosidase. The fusion protein is produced by a λgt11 recombinant clone after it has integrated into the *E. coli* genome (i.e., a lysogen). The *E. coli* lysogen containing the recombinant λgt11 clone is grown and induced to express the β-galactosidase fusion protein. The bacterial cells are lysed to recover the recombinant polypeptide. The recombinant fusion protein is isolated on SDS–polyacrylamide gels or by column chromatogrpahy and then used to characterize the protein and produce serum antibodies in rabbits.

Several plasmid vectors are used to produce fusion proteins between β-galactosidase and the polypeptide encoded by the DNA or cDNA fragment of interest (29). These plasmid vectors are constructed for the cloning of foreign DNA inserts into the 5′ or 3′ end of the *lacZ* gene and expressed as fusion proteins at the N-terminal or C-terminal portion of β-galactosidase, respectively. Multiple restriction enzyme cloning sites are inserted in the plasmid in all three reading frames. This allows a specific plasmid construction to be chosen for ligation depending on the amino acid coding frame determined by the DNA sequence of the insert. The recombinant fusion protein is directly isolated from the bacterial host cells and purified as described above for serum antibody production in rabbits.

SUMMARY

This chapter reviews the isolation and analysis of chromosome-specific DNA probes and their expansion by chromosome walking in large-insert genomic phage and cosmid libraries. The procedures and strategies of cloning cDNAs for exons represented in large genomic regions are quite similar to the screening of genomic libraries. The DNA sequencing methods for both Maxam–Gilbert base modification and Sanger chain termination are described for both genomic DNA and cDNA sequencing projects. The expression of DNA and cDNA inserts in both phage and plasmid expression vectors is described for producing recombinant polypeptides for antibody production. These methods are conventional technologies in molecular human genetics and are applicable to the cloning and analysis of genes involved in human disease loci.

ACKNOWLEDGMENTS

Thanks to Eric Hoffman and Robert Fuller for providing examples of Maxam–Gilbert and M13 sequences and to Louis M. Kunkel, Corlee J. Bertelson, and Chris Colletti-Feener for a critical reading of the manuscript. A.P.M. is supported by PHS NRSA (2T 32 GM07753-07) from the National Institute for General Medical Sciences and The Muscular Dystrophy Association.

REFERENCES

1. Bender W, Arkham M, Karch F, et al. Molecular genetics of the bithorax complex in *Drosophila melanogaster*. Science 1983; 221:23–29.
2. Royer-Pokora B, Kunkel LM, Monaco AP, et al. Cloning the gene for an inherited human disorder (chronic granulomatous disease) on the basis of its chromosomal location. Nature (London) 1986; 322:32–38.
3. Monaco AP, Neve RL, Colletti-Feener C., Bertelson CL, Kurnit DM, Kunkel LM. Isolation of candidate cDNAs for portions of the Duchenne muscular dystrophy gene. Nature (London) 1986; 323:646–650.
4. Friend SH, Bernards R, Rogelj S, et al. A human DNA segment with properties of the gene that predisposes to retinoblastoma and osteosarcoma. Nature (London) 1986; 323:643–646.
5. Orkin S. Reverse genetics and human disease. Cell 1986; 47:845–850.
6. Young BD Chromosome analysis by flow cytometry: A review. Bas Appl Histochem 1984; 28:9–19.
7. Latt SA, LaLande M, Flint A, et al. Metaphase chromosome flow sorting and cloning; rationale, approaches, and applications. In Gray JW, ed. Flow Cytogenetics. New York: Academic Press, in press.
8. Benton WD, Davis RW. Screening lambda-gt recombinant clones by hybridization to single plaques in situ. Science 1977; 196:180–182.
9. Maniatis T, Fritsch EF, Sambrook J Molecular Cloning: A Laboratory Manual. New York: Cold Spring Harbor Laboratory, 1982.

10. Southern EM. Detection of specific sequences among DNA fragments separated by gel electrophoresis. J Mol Biol 1975; 98:503–517.

11. McKusick VA, Ruddle FH. The status of the gene map of the human chromosomes. Science 1977; 196:390–405.

12. Bolivar F, Rodriguez PJ, Greene MC, et al. Construction and characteristics of new cloning vehicles. II. A multipurpose cloning system. Gene 1977; 2:95–113.

13. Vierra J, Messing J. The pUC plasmids, an M13 mp7-derived system for insertion mutagenesis and sequencing with synthetic universal primers. Gene 1982; 19:259–268.

14. Singer MF. SINEs and LINEs: Highly repeated short and long interspersed sequences in mammalian genomes. Cell 1982; 28:433–434.

15. Francke U, Ochs HD, deMartinville B, et al. Minor Xp21 chromosome deletion in a male patient associated with expression of Duchenne muscular dystrophy, chronic granulomatous disease, retinitis pigmentosa, and McLeod syndrome. Am J Hum Genet 1985; 37:250–267.

16. Schmid DW, Jelnick WR. The Alu family of dispersed repetitive sequences. Science 1982; 216:1065–1070.

17. Blattner FR, Williams AE, Blechl K, et al. Charon phages: Safer derivatives of bacteriophage lambda for DNA cloning. Science 1977; 196:161–169.

18. Frischauf AM, Lehrach H, Poutska A, Murray N. Lambda replacement vectors carrying polylinker sequences. J Mol Biol 1983; 170:827–842.

19. Collins J, Hohn B. Cosmids: A type of plasmid gene-cloning vector that is packageable in vitro in bacteriophage lambda heads. Proc Natl Acad Sci USA 1978; 75:4242–4246.

20. Huynh TV, Young RA, Davis RW. Constructing and screening cDNA libraries in lambda gt10 and lambda gt11, chap 2. In Glover DM, ed. DNA Cloning: A Practical Approach, Vol 1. Oxford, England: IRL Press, 1985, pp 49–78.

21. Thomas PS. Hybridization of denatured RNA and small DNA fragments transferred to nitrocellulose. Proc Natl Acad Sci USA 1980; 77:5201–5205.

22. Mount SM. A catalogue of splice junction sequences. Nucleic Acids Res 1982; 10:459–472.

23. Maxam AM, Gilbert W. A new method for sequencing DNA. Proc Natl Acad Sci USA 1977; 74:560–564.

24. Maxam AM, Gilbert W. Sequencing end-labeled DNA with base specific chemical modifications. Methods Enzymol 1980; 65:499–559.

25. Sanger F, Nicklen S, Coulson AR. DNA sequencing with chain terminating inhibitors. Proc Natl Acad Sci USA 1977; 74:5463–5467.

26. Sanger F. Determination of nucleotide sequences in DNA. Science 1981; 214:1205–1210.

27. Messing J, Gronenborn B, Muller-Hill B, Hofschneider PW. Filamentous coliphage M13 as a cloning vehicle: Insertion of a Hind III fragment of the lac regulatory region in the M13 replicative form in vitro. Proc Natl Acad Sci USA 1977; 74:3642–3646.

28. Messing J. New M13 vectors for cloning. Methods Enzymol 1983; 101:20–78.

29. Koenen M, Griesser HW, Muller-Hill B. Immunological detection of chimaric β-galactosidases expressed by plasmid vectors, chap 4. In Glover DM, ed. DNA Cloning: A Practical Approach, Vol 1. Oxford, England: IRL Press, 1985, pp 89–100.

11

Assigning Genes to Chromosomes: Family Studies, Somatic Cell Hybridization, Chromosome Sorting, *in situ* Hybridization, Translocations

GAIL A. P. BRUNS

The human gene map has been developed by combined application of genetic linkage analysis in kindreds, somatic cell hybrids, cytogenetic analysis of chromosomal variants and rearrangements, and, most recently, recombinant DNA technology. The current map includes more than 3400 loci, an important adjunct for the study and diagnosis of inherited human disorders.

EARLY LINKAGE ANALYSIS IN KINDREDS

In 1911, E. B. Wilson (1) linked color blindness to the X-chromosome. It was the first human gene assigned to a specific chromosome (Table 11-1). Because the single X-chromosome in males produces a characteristic inheritance pattern for X-linked disorders, at least 60 X-linked genetic loci were known (2) before the first autosomal locus was assigned to a specific chromosome.

In the early 1950s, family studies used blood group antigens as genetic markers to establish linkage of the locus for the Luthern blood group (LU) and the ABO secretor locus (SE) (3), of the Rh blood group locus (RH) to dominantly inherited elliptocytosis (EL1) (4), and of the ABO blood group locus (ABO) to the Nail–

Table 11-1 Landmarks in Human Gene Mapping

Date	Event
1911	First gene assigned to a human chromosome (E. B. Wilson)—color blindness to the X
1951	First autosomal linkage group (J. Mohr), Luthern blood group, and ABO secretor loci
1960	First somatic cell hybrids (G. Barski) Spontaneous fusion of two different mouse cells
1967	First human–rodent somatic cell hybrids segregating human chromosomes (Weiss and Green)
1968	First autosomal gene assigned to a human chromosome (Donahue et al.)—Duffy blood group linked to a chromosome 1 centromeric polymorphism
1968	First gene assigned to a chromosome with somatic cell hybrids (Migeon and Miller)—human thymidine kinase on an E group metacentric (#17 or #18)

Patella syndrome (5). However, these pairs of linked autosomal loci could not be assigned to any particular chromosome by kindred analysis.

The first chromosomal assignment of an autosomal gene was accomplished in 1968 (6) by linkage of the Duffy blood group locus (FY) to a cytogenetically identified polymorphism of the centromeric heterochromatic region of chromosome 1. A locus for congenital zonular pulverulent cataract (CAE) had also been linked to FY (7). Hence a group of three loci was assigned to chromosome 1. That linkage group was expanded to include the related salivary and pancreatic amylase loci (AMY1, AMY2) when FY was linked to AMY (8). The α-haptoglobin locus was similarly assigned to a chromosome by linkage to translocations and cytogenetic polymorphisms of chromosome 16 (9,10).

For linkage studies in kindreds, genetic variation in the population is essential for distinguishing the two chromosome homologues of an individual. By the early 1970s about 60 polymorphic marker loci had been identified (1,11) for this purpose; the loci included cell surface antigens, serum proteins, constitutive metabolic enzymes, and more than 10 chromosomal variants such as hereditary fragile sites and polymorphisms of centromeric heterochromatic regions. Ten additional autosomal linkage groups had also been defined (12). Despite these successes, many extensive kindred analyses served only to exclude disease genes of interest from the vicinity of these marker loci, rather than identifying a linked gene or the chromosome of origin of the gene.

DEVELOPMENT OF SOMATIC CELL HYBRIDS

Beginning in 1960, the nearly simultaneous development of methods to make human–rodent somatic cell hybrids (13) and cytogenetic staining techniques for identification of each human chromosome (14) ushered in a period of rapid progress in the development of the human gene map. Barski et al. (15) first observed the rare spontaneous formation of hybrid cells between two related mouse tumor cell lines that were growing in a mixed culture. The hybrids, identified and isolated because they grew more rapidly than the parental cells, contained a total number of chromosomes that was nearly equal to the sum of the chromosomes of each of the parental cell lines and included essentially all the marker chromosomes of both cell lines.

Two modifications were needed for the subsequent widespread use of somatic cell hybrids: the hypoxanthine–aminopterin–thymidine (HAT) selection system first used by Littlefield (16) for selective isolation of hybrid cells, and enhancement of the frequency of hybrid formation by inactivated Sendai virus in 1965 (17) and later by polyethylene glycol (18).

Littlefield grew two sublines of mouse L cells in mixed culture. One lacked thymidine kinase (TK); the other lacked hypoxanthine phosphoribosyltransferase (HPRT). Single clones of hybrid cells were isolated with a frequency of 10^{-6} (16) in medium supplemented with hypoxanthine, aminopterin, thymidine, and glycine (HAT medium). In the presence of the folate antimetabolite, aminopterin, a blocker of *de novo* purine and pyrimidine synthesis, neither parental cell line could grow because each lacked one enzyme needed for utilization of thymidine or hypoxanthine by an alternative pathway of nucleotide synthesis (the salvage pathway). Only hybrids that had both TK and HPRT, reconstituting the salvage pathway by genetic complementation, could survive. That selection system, modified (13) to generate hybrids from an established rodent cell line that lacked either TK or HPRT and a diploid cell incapable of continuous growth, such as a fibroblast or lymphocyte (Fig. 11-1), has been used to generate most of the human–rodent somatic cell hybrids used for gene mapping studies.

The first viable human–rodent hybrids were isolated in 1967 by Weiss and Green (19) with the HAT selection system from mixed cultures of TK-deficient mouse clone 1-D cells and human embryonic lung diploid fibroblasts. Surprisingly, when the number of chromosomes was tabulated shortly after the clones were isolated, there were only 2–15 human chromosomes per clone but nearly all the mouse chromosomes persisted. With prolonged growth, the number of human chromosomes was further reduced to 1–3 per clone. Although some chromosome loss (10–20%) had been observed among interspecies rodent hybrids, this was the first time that extensive preferential loss of chromosomes of one parental species was

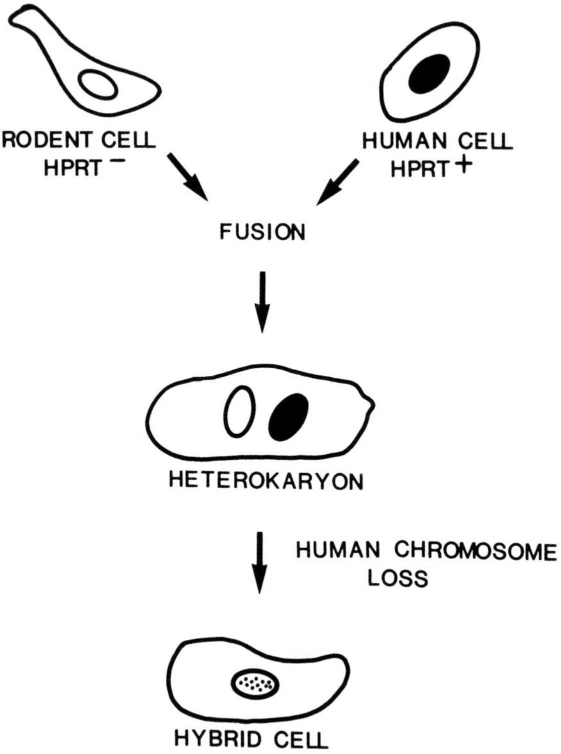

Fig. 11-1. Half selection system for human–rodent hybrid cells. Cells from an established mouse or hamster line deficient in HPRT, TK, or another selectable marker are mixed with diploid human cells of limited growth potential (fibroblasts, leukocytes) and briefly exposed to polyethylene glycol to promote cell fusion. A fraction of the cells form heterokaryons with one rodent and one human nucleus. With a frequency of 10^{-3} to 10^{-6}, viable hybrids are obtained following nuclear fusion and subsequent segregation of human chromosomes.

observed in hybrids. A small human metacentric chromosome was found in many clones that grew in HAT media but was not seen when the cells were back-selected for loss of human TK. That chromosome was found to be chromosome 17 (20,21), the first assignment of a gene to a human chromosome by somatic cell hybrids.

Subsequent progress was rapid. The first autosomal *synteny* (a term for genes on the same chromosome) defined by somatic cell genetics was that for lactate dehydrogenase B and peptidase B, identified in 1970 (22,23) by cosegregation of expression of the two human isozymes in a series of independent hybrid clones. Nonselectable genes such as those for lactate dehydrogenase A and other constitutive metabolic enzymes were directly assigned to chromosomes beginning in 1972 (24) by correlation of human enzyme expression with the human chromosome complements of hybrid clones. By 1977 at least one isozyme locus had been assigned to each of the human autosomes.

Hybrid cell analysis had other effects, including the assignment to particular chromosomes of autosomal linkage groups that had been previously defined by family studies. In kindred studies, linkage can be demonstrated only over distances less than 50 cM (roughly equal to 5×10^7 base pairs) (1). In hybrids, on the other hand, genes on the same chromosome cosegregate even if they are separated by as much as 250 cM (2.5×10^8 base pairs). For example, the Rh blood group (RH) had been linked to elliptocytosis-1 (EL1) (4) and the locus for the enzyme phosphogluconate dehydrogenase (PGD) (25) by family studies. In hybrids, PGD was found to be syntenic with the enzymes phosphoglucomutase-1 (PGM1) and peptidase C (PEPC) (26,27). When PEPC was assigned to chromosome 1 by cytogenetic analysis of a series of hybrid clones (28), the PGD, RH, EL1, PGM1 linkage group could also be assigned to chromosome 1.

Similarly, the HLA, PGM3 linkage group was assigned to chromosome 6 and the ABO, Nail Patella, AK1 linkage group to chromosome 9 when the loci for the enzymes PGM3 and AK1 were assigned to chromosomes with somatic cell hybrids (29,30). Regional localization of these and other genes soon followed as hybrids were made with human cells from donors with inherited chromosomal translocations. Alternatively, the occasional spontaneous chromosome rearrangements that occur in hybrid cells were used to include or exclude a gene from a chromosome region. The first regional assignment with somatic cell hybrids was that of glucose-6-phosphate dehydrogenase (G6PD), phosphoglycerate kinase (PGK), and HPRT to the long arm of the human X-chromosome with a translocation involving chromosomes X and 14 in 1972 (31).

GENE MAPPING WITH HYBRIDS

In practice, human–rodent hybrids are prepared by mixing human diploid fibroblasts or lymphocytes with mouse or hamster cells from an established cell line that lacks either TK or HPRT and briefly exposing the mixture of cells to 50% polyethylene glycol (PEG) (32), which, by an unknown mechanism, promotes fusion of cell membranes (Fig. 11-1). The PEG is quickly removed and the cells are allowed to grow for a day or two before the culture is exposed to the HAT selection system. The rodent parental cells begin to die within 3–7 days and small colonies of hybrid cells become visible at 7–14 days. To be certain that each hybrid clone is an independent cell line, only one colony is isolated from each dish. The clones are grown in mass culture for a few generations to stabilize the human chromosome complement and then frozen in aliquots at −80°C or in liquid nitrogen for future regrowth.

The human chromosome complement of each clone is determined by analyzing 10–20 metaphases per clone after Giemsa banding of the chromosomes (33) or differential staining of rodent and human chromosomes (34). Hybrid clones with extensive rearrangements or fragmentation of human chromosomes are discarded. Before DNA probes for each human chromosome were available, the human chromosome complements of newly isolated hybrid clones were also monitored by electrophoresis of cell extracts from each clone for human isozymes that had been assigned to each autosome and the X-chromosome.

An example of this method is provided by isozyme analysis of human–mouse hybrids for cytoplasmic isocitric dehydrogenase (IDH1) and the mitochondrial form (IDH2) (Fig. 11-2). The IDH1 locus had been assigned to the long arm of chromosome 2 and the IDH2 locus to chromosome 15. The human and rodent isozymes were clearly separable for both enzymes. Hybrids that lacked human chromosome 2 gave an IDH1 pattern comparable to the mouse parental cell line. In contrast, hybrids with human chromosome 2 showed a component with a mobility intermediate to the mouse and human isozymes and often showed the human isozyme as well. A similar phenomenon was observed for the IDH2 isozymes. Both IDH1 and IDH2 are homodimers, so the additional hybrid components were heteropolymers composed of one human and one mouse polypeptide chain. Heteropolymeric enzymes of this kind are not seen when the two parental cell lines are

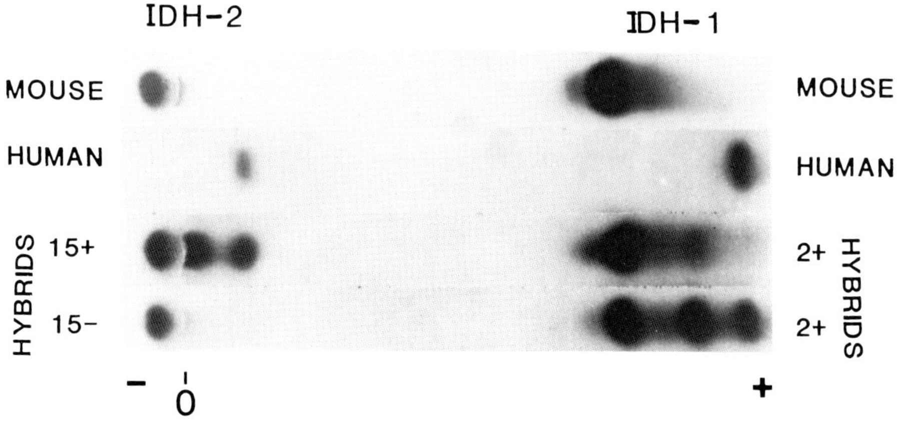

Fig. 11-2. Rodent and human isocitrate dehydrogenase isozymes in hybrid cells. Cytoplasmic (IDH1) and mitochondrial (IDH2) isozymes of isocitrate dehydrogenase were identified by enzymatic staining following starch gel electrophoresis. Both hybrids with human chromosome 2 exhibit a human–rodent heteropolymeric IDH1 isozyme in addition to the mouse isozyme and one hybrid also has the human IDH1 homopolymer. Similarly the hybrid with human chromosome 15 has the IDH2 heteropolymeric and human isozymes, whereas only the mouse isozyme is observed in the hybrid lacking this chromosome.

mixed; their presence is direct evidence of the activity of both rodent and human genomes in hybrid cells.

During the 1970s more than 60 different starch gel, agarose, and cellulose acetate membrane electrophoresis systems were developed for separation of rodent and human isozymes to map genes with hybrid cells. About 40 electrophoresis systems were used frequently to monitor the human chromosome complements of hybrids. Currently, isozyme analysis is rarely used to determine the human chromosome complement of hybrid cells. Rather, DNA from each cell line is analyzed by hybridization techniques with DNA probes for several different regions of each chromosome (Fig. 11-3).

For chromosome assignment, the segregation of the human gene of interest in a series of primary hybrid clones can be directly correlated with the human chromosome complements of the hybrids (a primary assignment) or, alternatively, with the segregation of an isozyme or DNA probe previously assigned to a chromosome (a secondary assignment). The human gene can be monitored by a gene product (an isozyme or specific antigen) or by the hybridization signal of a cDNA or genomic clone for the gene with DNAs from each hybrid clone. The observed segregation pattern of the human gene of interest in a hybrid panel is compared with the segregation of the human chromosomes of the hybrids. A gene on chromosome 10, for example, should cosegregate with that chromosome (Table 11-2) in all but the occasional cell line in which a spontaneous rearrangement of the chromosome had occurred. Although only five hybrids with the proper complements of human chromosomes could suffice for a unique binary signature for 32 chromosomes (2^5) (1), in practice a larger panel of hybrids (10–15 cell lines) is used to provide enough redundancy to compensate for the subtle chromosome rearrangements that occur in hybrids.

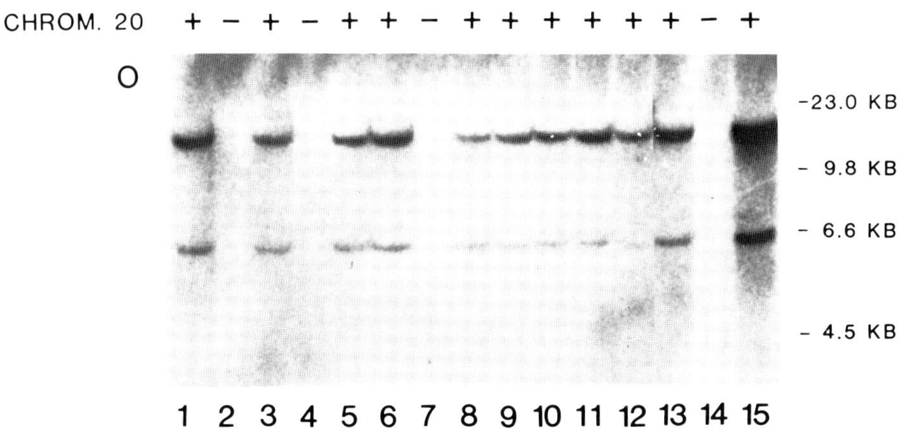

Fig. 11-3. Use of a cloned DNA probe from human chromosome 20 to monitor the chromosome complement of hybrid cells. *Eco*RI-digested DNAs from 13 human–hamster hybrids (lanes 1–13), a hamster cell line (lane 14), and a human cell line (lane 15) were separated by agarose electrophoresis, transferred to nitrocellulose, and hybridized with a DNA probe previously assigned to human chromosome 20. Hybrids with this chromosome exhibit 2 bands at approximately 6.5 and 14 kb, whereas no hybridization is observed with DNA from hybrids lacking chromosome 20.

Table 11-2 Somatic Cell Hybrid Panel for Chromosome Mapping

Hybrid clone	Human chromosomes											
	1	2	3	4	5	6	7	8	9	10	11	12
G1	+	+	+	−	−	+	+	−	−	+	−	−
G2	−	+	+	−	−	−	+	+	−	+	+	+
G3	+	−	+	+	−	+	−	+	+	−	−	−
G4	+	−	−	−	+	+	−	−	+	−	+	+
G5	−	−	−	−	+	+	−	+	−	−	+	+
G6	−	−	−	+	−	+	+	−	−	+	+	−
G7	−	−	−	−	−	−	+	−	−	−	+	+
G8	−	−	−	+	+	−	−	−	−	+	−	+
G9	+	−	−	−	−	−	−	−	+	−	−	−

Table 11-3 shows one way to present data from a mapping experiment. First, for each human chromosome, we calculate the number of hybrid clones in which the gene of interest and the chromosome are both present together or absent together (concordant clones). Then, we calculate the number of clones in which the gene is present and the chromosome is absent or, conversely, where the gene is missing but the chromosome is present (discordant clones). In most cases, genes on the same chromosome show concordant segregation in at least 90% of hybrid clones. The exceptional clones are not disregarded because they imply that chromosome rearrangements have occurred, events that may localize the gene to a subchromosomal region.

The loss of human chromosomes from hybrids is not completely random. It is therefore necessary to establish that the gene of interest segregates with only one human chromosome. Specifically, the gene should show concordant segregation with one chromosome in nearly every hybrid, and discordant segregation with each of the other chromosomes in at least 20–25% of the clones.

MAPPING WITH FLOW-SORTED CHROMOSOMES

With the development of single (Fig. 11-4) and dual laser methods (36), chromosome sorting has become an alternative to somatic cell hybrids for chromosome mapping of gene probes. With combinations of two dyes such as chromomycin A3 and DIPI (37) or chromomycin A3 and Hoechst 33258 (38), all the human chromosomes except chromosomes 9–12 can be resolved by dual laser flow cytometry of metaphase chromosomes isolated from lymphocyte cell lines. Chromosomes are isolated in a polyamine-containing buffer and stained with one of the two dye pairs. Thirty thousand chromosomes of each type are directly sorted onto small nitrocellulose or nylon filters (Fig. 11-5). The chromosomal DNA is denatured on the filter

Table 11-3 Segregation of a DNA Probe with Human Chromosome 10

	Ch 10+	Ch 10−
Probe signal +	14	0
Probe signal −	0	9

Human chromosomes												Human gene or DNA probe
13	14	15	16	17	18	19	20	21	22	X	Y	
+	+	+	+	+	+	+	+	−	+	+	−	+
−	−	−	−	−	+	+	+	−	−	−	+	+
−	+	−	+	−	−	+	+	+	−	+	−	−
−	−	+	+	−	+	+	+	−	−	+	−	−
+	+	−	+	−	+	+	+	+	+	−	+	−
+	+	+	−	−	+	+	−	−	+	+	−	+
+	−	−	+	+	−	+	−	−	−	+	−	−
−	−	−	−	−	+	+	−	+	+	−	−	+
−	−	−	−	+	−	−	−	−	−	+	−	−

with a NaOH–NaCl mixture and the filter is then neutralized for hybridization with a radiolabeled DNA probe. This method, called *spot blotting,* was first used to assign the gene defective in McArdle syndrome, skeletal muscle glycogen phosphorylase, to chromosome 11 (37). It is particularly useful for mapping gene probes that are homologous to only one chromosomal locus.

GENE ASSIGNMENT BY *IN SITU* HYBRIDIZATION

Starting in the early 1970s, direct visualization of the chromosomal location of gene families with multiple identical or closely related sequences at the same site was

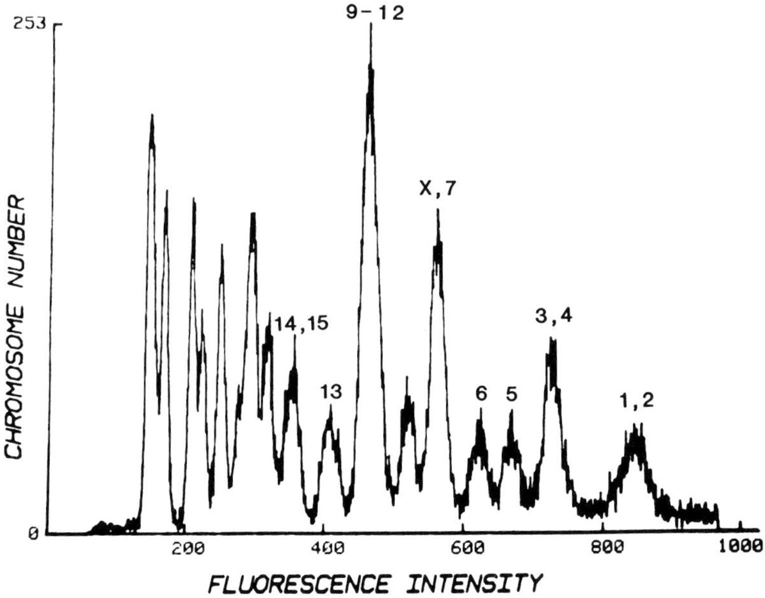

Fig. 11-4. One-dimensional flow histogram of human metaphase chromosomes from a 49, XXXXY lymphoblastoid cell line stained with the dye 33258 Hoechst (35). The chromosome composition of the peaks of fluorescence intensity are indicated by the letter or number designations.

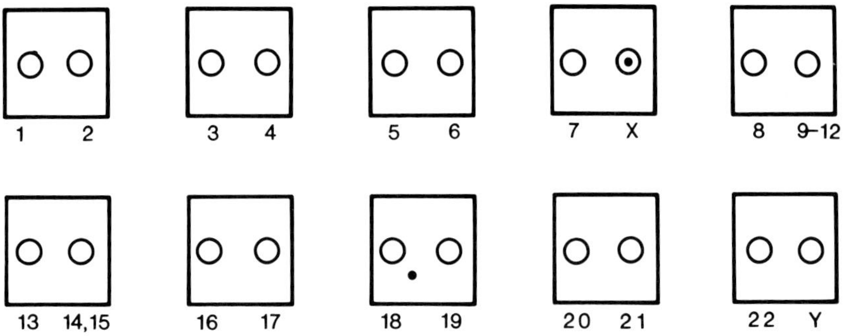

Fig. 11-5. Diagramatic representation of hybridization of an X-chromosome DNA probe to spot blots of flow sorted human chromosomes. Thirty thousand of each chromosome type are flow sorted directly onto nitrocellulose or Nylon filters (37,38) which are hybridized with a radiolabeled DNA probe after denaturation and neutralization of the chromosomal DNA. The circles on each filter denote the position of the sorted chromsomes, identified by the numbers or letters under each circle. Specific hybridization of a probe to its chromosome of orgin is indicated by a hybridization signal within a circle, whereas nonspecific or background hybridization is indicated by the dot near, but outside the chromosome 18 circle.

achieved by autoradiography. Human satellite DNAs (39) and the ribosomal genes (40) were localized that way. However, mapping of single copy genes on metaphase chromosomes by *in situ* hybridization of a labeled gene probe could not be consistently carried out until technical advances were made by Harper and Saunders (41) in 1981.

In current practice (42), human metaphase chromosomes on microscope slides are treated with ribonuclease A, gently denatured and then hybridized at 42°C for a number of hours with a highly radiolabeled DNA or RNA probe for the gene to be mapped in the presence of formamide, dextran sulfate, and salts that promote annealing of the probe to homologous sequences of chromosomal DNA. The slides are washed to remove excess probe, dipped in an autoradiographic emulsion, and exposed for 7–14 days at 4°C. After the emulsion is developed, the chromosomes are stained by one of several methods that provide the banding patterns needed to identify each human chromosome. The locations of silver grains from 50–100 metaphases are depicted on an ideogram of the human karyotype (Fig. 11-6) and the gene is mapped to the region or regions of maximum grain density.

This method, although technically demanding, is particularly useful for localization of a gene to a small region of a chromosome. *In situ* hybridization is often used to confirm the chromosome assignment of a gene first made with somatic cell hybrids or flow-sorted chromosomes and to localize the gene to a chromosome band or region.

CHROMOSOMAL TRANSLOCATIONS AND GENE MAPPING

Reciprocal chromosomal translocations change the genetic segregation relationships of the chromosome segments involved in the translocation. In one example (Fig. 11-7), nearly the entire long arm of the X-chromosome was joined to the short

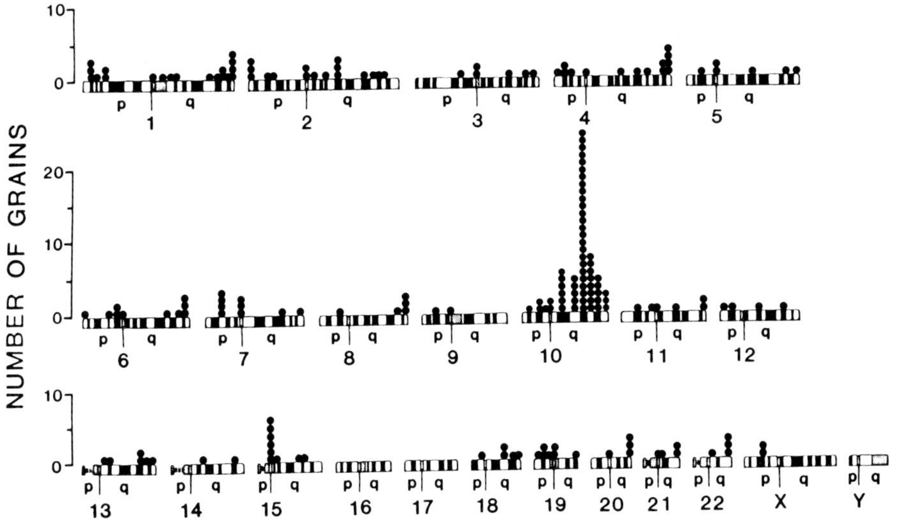

Fig. 11-6. Histogram of the *in situ* hybridization pattern of a chromosome 10 probe. For chromosomal assignment, the location of each silver grain in 50–100 metaphases is depicted on a ideogram of the human karyotype and the position(s) of maximal hybridization identified. For this chromosome 10 probe, hybridization was maximal at 10q21–24 (42).

arm of chromosome 19 (the der19 chromosome) and a segment of the chromosome 19 short arm was joined to the remainder of the X-chromosome (the derX chromosome). In a kindred or in hybrids, an allele on the X-chromosome segment of the der19 chromosome would segregate with genes on all but the distal short arm of chromosome 19. Conversely, alleles on the tip of the short arm of the chromosome 19 involved in the translocation would segregate with the short arm of the X-chromosome.

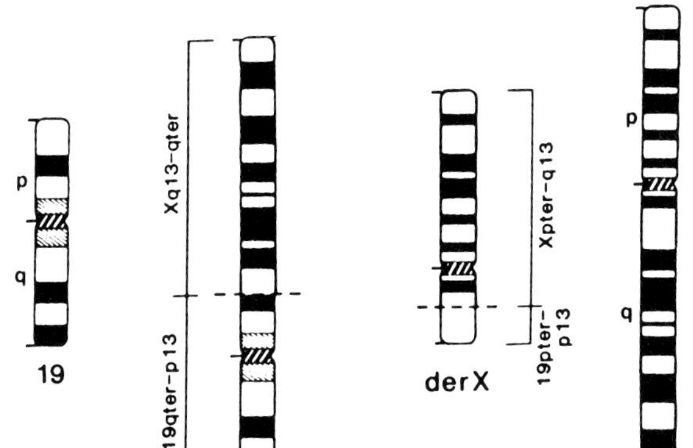

Fig. 11-7. Schematic representation of an X/19 translocation. In this balanced translocation, the X-chromosome long arm (Xq13-qter) and the distal short arm of chromosome 19 (19p13-pter) have been exchanged to form the derX and der19 chromosomes. The breakpoints on the X and on chromosome 19 are indicated by the dotted lines.

The changes in the physical and segregation relationships of genes on chromosomal segments involved in translocations can be used for assignment of genes to translocation boundary regions by linkage studies in kindreds, or for regional localization of genes with somatic cell hybrids, spot blots, and *in situ* hybridization. As previously noted, the second gene assigned to a specific human chromosome, the α-haptoglobin locus, was assigned to chromosome 16 by cosegregation with a chromosome 16 translocation in a kindred (9). Finding X-linked disorders in females with X/autosome translocations led to localization of several loci on the human X-chromosome, including the gene for Duchenne muscular dystrophy (43).

Translocations involving any chromosome that can be selectively retained in somatic cell hybrids by complementation of a metabolic or growth requirement of the rodent parental cell line can be used to generate a panel of hybrid cells for localization of genes to multiple regions of that chromosome (Fig. 11-8). Conversely, a panel with different regions of an otherwise nonselectable chromosome can be constructed with translocations that join parts of the chromosome of interest, chromosome 7, for example, to a selectable region of another chromosome. The human genes necessary to complement the metabolic deficiency (auxotrophy) produced by aminopterin in the HAT system, the most common selective system used for hybrids, are on the X-chromosome (the HPRT gene) and chromosome 17 (the TK gene). Therefore, translocations that involve these two chromosomes are used most often to create somatic cell hybrid regionalization panels. Translocations that shift the size of one of the derivative chromosomes can also be used for regionalization of genes with spot blots of flow-sorted chromosomes (44). Similarly, the rearrangement of gene relationships in translocation chromosomes makes it possible to confirm gene assignments made by *in situ* hybridization techniques with chromosome from normal individuals. For example, hybridization of a probe for a gene near the centromere of chromosome 19 would be expected on both the normal 19 and the der19 chromosome of the translocation shown in Fig. 11-7. In contrast, a probe for a gene on the tip of the short arm of chromosome 19 should hybridize both to the derX chromosome and the normal 19.

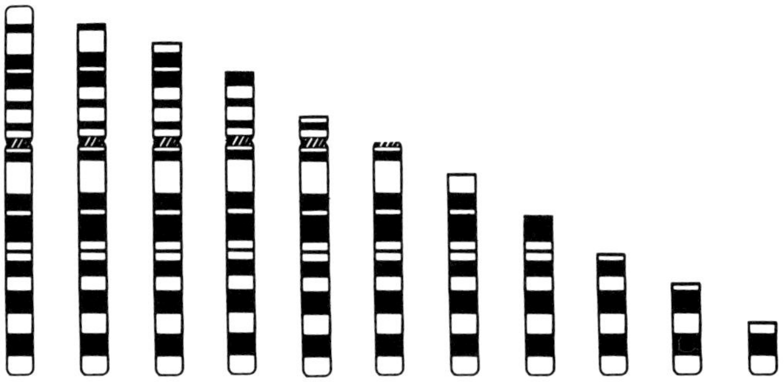

Fig. 11-8. A panel for localizaiton of genes on the X-chromosome constructed with hybrids carrying different X/autosome translocations. The regions of the X-chromosome retained in the human–rodent hybrids of this panel range from the entire X (extreme left) to Xq26-qter (extreme right). X-Chromosome DNA probes can be localized to 11 regions of the chromosome with this hybrid panel.

Table 11-4 Chromosome Assignment of Genes for Neurological Syndromes

Chromosome	Neurological disorder
1p	Fucosidosis
1p	Neuroblastoma oncogene (N-*ras*)
1q	Charcot-Marie Tooth disease-1
1q	Gaucher disease
1q	Muscle phosphofructokinase deficiency
2p	N-*myc* (neuroblastoma)
3p	GM1 gangliosidosis
4p	Huntington disease
5	Maroteaux–Lamy disease
5q	Sandhoff disease
7q	Mucopolysaccharidosis VII
11p	Manic-depressive disorder
11q	Acute intermittent porphyria
11q	McArdle disease
12q	Phenylketonuria-1
15q	Prader–Willi syndrome
15q	Tay–Sachs disease
17	Von Recklinghausen neurofibromatosis
17q	Pompe disease
18	Familial amyloid polyneuropathy
19	Mannosidosis
19	Myotonic dystrophy
19	Polio virus sensitivity
21	Down syndrome
21	Familial Alzheimer disease
22	Hurler disease
22	Bilateral acoustic neurofibromatosis
22q	Metachromatic leukodystrophy
Xp21	Duchenne muscular dystrophy
Xp21	Becker muscular dystrophy
Xp21	Glycerol kinase deficiency
Xp21	OTC deficiency
X	Menkes disease
X	Kennedy disease
Xq	Pelizaeus–Merzbacher disease
Xq	Emery–Dreifuss muscular dystrophy
Xq26–q27	Lesch–Nyhan disease
Xq27–q28	Fragile X syndrome (mental retardation)

THE HUMAN GENE MAP

Rapid progress in developing the human gene map followed the assignment of the Duffy blood group to chromosome 1 in 1968 and the TK locus to chromosome 17 in 1971. At the first international workshop on human gene mapping in 1973, 56 loci had been assigned to the human autosomes, the great majority with somatic cell hybrids (12). The tabulation in 1987 lists 3400 loci (45). Many of these are cloned genes or DNA segments that identify restriction fragment length polymorphisms (RFLPs) of importance for diagnosis of human genetic disorders and mapping of disease genes by linkage studies in kindreds. The assignment of a locus for von Recklinghausen neurofibromatosis to chromosome 17 (46,47) and a locus for familial Alzheimer disease to chromosome 21 (48) illustrates the power of kindred analysis using DNA probes previously assigned to particular chromosomes with somatic cell hybrids and *in situ* hybridization.

Although less than 5% of the 3×10^9 DNA base pairs of the human genome has been cloned, more than 1100 chromosome-specific DNA probes that identify RFLPs and hence polymorphic loci within the human genome are now available for linkage studies (45). Many RFLPs are informative in only a fraction of individuals and not every autosome has a complement of polymorphic loci evenly distributed along the length of the chromosome. However, the goal of mapping the locus for any inherited disease within 10 cM of a polymorphic locus, first proposed by Botstein et al. (49), is within sight (50). Development of the human gene map has been an essential step toward this goal.

The chromosome assignments of loci implicated in human neurological disorders are shown in Table 11-4. DNA probes for these loci, many already isolated, will be important tools for carrier detection and genetic counseling in families with these disorders. As many of the genes important for neuronal structure and function are only now being defined, rapid growth of the map of nerve and muscle specific loci can be expected.

REFERENCES

1. McKusick VA, Ruddle FH. The status of the gene map of the human chromosomes. Science 1977; 196:390–405.
2. McKusick VA. Mendelian Inheritance in Man. Baltimore: The Johns Hopkins University Press, 1986, pp 1313–1471.
3. Mohr J.A search for linkage between the Luthern blood group and other hereditary characters. Acta Path Microbiol Scand 1951; 28:207–210.
4. Lawler SD, Sandler M. Data on linkage in man: Elliptocytosis and blood groups. 4. Families 5, 6 and 7. Ann Eugen (London) 1954; 18:328–334.
5. Renwick JH, Lawler SD. Genetical Linkage between the ABO and Nail-Patella loci. Ann Hum Genet 1955; 19:312–331.
6. Donahue RP, Bias WB, Renwick JH, McKusick VA. Probable assignment of the Duffy blood group locus to chromosome 1 in man. Proc Natl Acad Sci USA 1968; 61:950–955.
7. Renwick JH, Lawler SD. Probable linkage between a congenital cataract locus and the Duffy blood group locus. Ann Hum Genet (London). 1963; 27:67–84.
8. Merritt AD, Lovrien EW, Rivas ML, Conneally PM. Human anylase loci: Genetic linkage with the Duffy blood group locus and assignment to linkage group I. Am J Hum Genet 1973; 25:523–538.
9. Robson EB, Polani PE, Dart SJ, Jacobs PS, Renwick JH. Probable assignment of the alpha locus of haptoglobin to chromosome 16 in man. Nature (London) 1969; 223:1163–1165.
10. Magenis RE, Hecht F, and Lovrien EW. Heritable fragile site on chromosome 16: Probable localization of haptoglobin locus in man. Science 1970; 170:85–87.
11. Francke U. Gene mapping. In Emery AEH, Rimoin, DL, eds. Principles and Practice of Medical Genetics. Edinburgh: Churchill Livingstone, 1983, pp 91–110.

12. New Haven Conference (1973). First International Workshop on Human Gene Mapping. Cytogenet Cell Genet 1974; 13:9–216.

13. Ephrussi B. Hybridization of Somatic Cells. Princeton: Princeton University Press, 1972, pp 5–56.

14. Caspersson T, Zech L, Johansson C. Analysis of human metaphase chromosome set by aid of DNA-binding fluorescent agents. Exp Cell Res 1970; 62:490–492.

15. Barski G, Sorieul S, Cornefert F. Production dans des cultures in vitro de deux suches cellulaires en association, de cellules de caractere "hybride." CR Acad Sci (Paris) 1960; 251:1825–1827.

16. Littlefield J W. Matings of fibroblasts in vitro and their presumed recombinants. Science 1964; 145:709–710.

17. Harris H, Watkins JF. Hybrid cells derived from mouse and man: Artificial heterokaryons of mammalian cells from different species. Nature (London) 1965; 205:640–646.

18. Pontecorvo G. Production of mammalian somatic cell hybrids by means of polyethylene glycol treatment. Somat Cell Genet 1975; 1:397–400.

19. Weiss MC, Green H. Human-mouse hybrid cell lines containing partial complements of human chromosomes and functioning human genes. Proc Natl Acad Sci USA 1967; 58:1104–1111.

20. Migeon BR, Miller CS. Human-mouse somatic cell hybrids with a single human chromosome (group E): Link with thymidine kinase activity. Science 1968; 162:1005–1006.

21. Miller OJ, Allderdice PW, Miller DA, Breg WR, Migeon BR. Human thymidine kinase gene locus: Assignment to chromosome 17 in a hybrid of man and mouse cells. Science 1971; 173:244–245.

22. Santachiara AS, Nabholz M, Miggiano V, Darlington AJ, Bodmer W. Genetic analysis of man-mouse somatic cell hybrids. Nature (London) 1970; 227:248–251.

23. Ruddle FH, Chapman VM, Chen TR, and Klebe RJ. Linkage between human lactate dehydrogenase A and B and peptidase B. Nature (London) 1970; 227:251–257.

24. Boone C, Chen TR, Ruddle FH. Assignment of three human genes to chromosomes (LDH-A to 11, TK to 17, and IDH to 20) and evidence for translocation between human and mouse chromosomes in somatic cell hybrids. Proc Natl Acad Sci USA 1972; 69:510–514.

25. Weitkamp LR, Guttormsen SA, Greendyke RM. Genetic linkage between a locus for 6-PGD and the Rh locus: Evaluation of possible hetergeneity in the recombination fraction between sexes and among families. J Hum Genet 1971; 23:462–470.

26. Westerveld A, Visser RPLS, Meera Khan P, Bootsma D. Loss of human genetic markers in man–Chinese hamster somatic cell hybrids. Nature (London) New Biol. 1971; 234:20–24.

27. Van Cong N, Billardon C, Picard JY, Feingold J, Frezal J. Liaison probable (linkage) entre les locus PGM1 et Peptidase C chez l'gomme. C. R. Acad Sci (Paris) 1971; 272:485–487.

28. Ruddle F, Riciuti F, McMorris FA, Tischfield J, Creagan R, Darlington G, Chen T. Somatic cell genetic assignment of peptidase C and the Rh linkage group to chromosome A-1 in man. Science 1972; 176:1429–1431.

29. van Someren H, Westerveld A, Hagemeijer A, Mees JR, Meera Khan P, Zaalberg OB. Human antigen and enzyme markers in man/Chinese hamster somatic cell hybrids. Evidence for synteny between the HL-A, PGM3, ME1 and IPO-B loci. Proc Natl Acad Sci USA 1974; 71:962–965.

30. Westerveld A, Jongsma APM, Meera-Khan P, van Someren H, Bootsma D. Assignment of the AK1:Np:ABO linkage group to human chromosome 9. Proc Natl Acad Sci USA 1976; 73:895–899.

31. Grzeschik KH, Allderdice PW, Grzeschik A, Opitz JM, Miller OJ, Siniscalco M.

Cytological mapping of human X-linked genes by use of somatic cell hybrids involving an X-autosome translocation. Proc Natl Acad Sci USA 1972; 69:69–73.

32. Davidson RL, Gerald PS.: Improved techniques for the induction of mammalian cell hybridization by polyethylene glycol. Somat Cell Genet 1976; 2:165–176.

33. Summner AT, Evans HJ, Buckland RA. New technique for distinguishing human chromosomes. Nature (London), New Biol 1971; 232:31–32.

34. Friend KK, Dorman BP, Kucherlapati RS, Ruddle JH. Detection of interspecific translocations in mouse-human hybrids by alkaline giemsa staining. Exp Cell Res 1976; 99:31–36.

35. Latt SA, Lalande M, Kunkel LM, Schreck R, Tantravahi U. Applications of fluorescence spectroscopy to molecular cytogenetics. Biopolymers 1985; 14:77–95.

36. Gray JW, Langlois RG, Carrano AV, Burkhart-Schultz K, Van Dilla MA. High resolution chromosome analysis: One and two parameter flow cytometry. Chromosoma 1979; 73:9–27.

37. Lebo RV, Gorin F, Fletterick RJ, Kao FT, Cheung MC, Bruce BD, Kan YW. High resolution chromosome sorting and DNA spot-blot analysis assign McArdle's syndrome to chromosome 11. Science 1984; 225:57–59.

38. Neve RL, Harris P, Kosik KS., Kurnit DM, Donlon TA. Identification of cDNA clones for the human microtubule associated protein tau and chromosomal localization of the genes for tau and microtubule-associated protein 2. Mol Brain Res 1986; 1:271–280.

39. Godson JR, Mitchell AR, Buckland RA, Clayton RP, Evans HJ. The location of four human satellite DNAs on human chromosomes. Cytogenet Cell Genet 1975; 14:338–339.

40. Evans JH, Buckland RA, Pardue ML. Location of the genes coding for 18S and 28S ribosomal RNA in the human genome. Chromosoma 1974; 48:405–426.

41. Harper ME, Saunders GF. Localization of single copy DNA sequences on G-banded human chromosomes by in situ hybridization. Chromosoma 1981; 83:431–439.

42. Bruns G, Stroh H, Velman G, Latt SA, Floros J. The 35kDa pulmonary surfactant associated protein is encoded on chromosome 10. Hum Genet 1987; 76:58–62.

43. Jacobs PA, Hunt PA, Mayer M, Bart RD. Duchenne muscular dystrophy (DMD) in a female with an X/autosome translocation: Further evidence that the DMD locus is at Xp21. Am J Hum Genet 1981; 33:513–518.

44. Lebo RV, Cheung M-C, Bruce BD, Riccardi VM, Kao F-T, Kan YW. Mapping parathyroid hormone, β-globin, insulin, and LDHA genes within the human chromosome 11 short arm by spot blotting sorted chromosomes. Hum Genet 1985; 69:315–320.

45. Human Gene Mapping 9. Paris Conference (1987). Ninth International Workshop on Human Gene Mapping. Cytogenet Cell Genet 1987; 46:1–762.

46. Barker D, Wright E, Nguyen K, Cannon L, Fain P, Goldgar D, Bishop T, Carey J, Baty B, Kivlin J, Willard H, Waye JS, Greig G, Leinwand L, Nakamura Y, O'Connell P, Leppert M, Lalouel J-M, White R, Skolnick M. Gene for von Recklinghausen neurofibromatosis is in the pericentromeric region of chromosome 17. Science 1987; 236:1100–1102.

47. Seizinger BR, Rouleau GA, Ozelius LJ, Lane AH, Faryniarz AG, Chao MV, Huson S, Korf PR, Parry DM, Pericak-Vance MA, Collins FS, Hobbs WJ, Falcone BG, Iannazzi JA, Roy JC, St. George-Hyslop PH, Tanzi RE, Bothwell MA, Upadhayaya M, Harper P, Goldstein AE, Hoover DL, Bader JL, Spence MA, Mulvihill JJ, Aylsworth AS, Vance JM, Rossenwasser GOD, Gaskell PC, Roses AD, Martuza RL, Breakefield X. O, Gusella JF. Genetic linkage of von Recklinghausen neurofibromatosis to the nerve growth factor receptor gene. Cell 1987; 49:589–594.

48. St. George-Hyslop P, Tanzi RE, Polinsky RJ, Haines JL, Nee L, Watkins PC, Myers RH, Feldman RG, Pollen D, Drachman D, Growdon J, Bruni A, Foncin JF, Salmon D, Frommelt P, Amaducci L, Sorbi S, Placentini S, Stewart G, Hobbs WJ, Conneally PM,

Gusella JF. The genetic defect causing familial Alzheimer's disease maps on chromosome 21. Science 1987; 235:885–890.

49. Botstein D, White RL, Davis R. Construction of a genetic linkage map in man using restriction fragment length polymorphisms. Am J Hum Genet 1980; 32:314–331.

50. Donis-Keller H, Green P, Helms C, Cartinhour S, Weiffenbach B, Stephens K, Keith TP, Bowden DW, Smith DR, Lander ES, Botstein D, Akots G, Rediker KS, Gravius T, Brown VA, Rising MB, Parker C, Powers JA, Watt DE, Kauffman ER, Bricker A, Phipps P, Muller-Kahle H, Fulton TR, Ng S, Schumm JW, Braman JC, Knowlton RG, Barker DF, Crooks SM, Lincoln SE, Daly MJ, Abrahamson J. A genetic linkage map of the human genome. Cell 1987; 51:319–337.

12

Construction of Physical Maps and Their Application in Finding Genes

CASSANDRA L. SMITH AND CHARLES R. CANTOR

A *physical map* of a genome is the structure of its DNA. There are various types of physical maps that differ in their resolution and in the way they are actually displayed or stored. The DNA sequence is the ultimate physical map: it has a resolution of a single base pair. At the other extreme, the least informative physical map would just describe the size of each chromosome. Between these extremes lie two types of maps that are usually referred to when the term physical map is used. These are restriction and linked cosmid maps (Fig. 12-1). In this chapter we shall explain how these maps can be made for small regions of the human genome. We will then illustrate how new methods promise to accelerate greatly the process of making the maps; it is reasonable to anticipate that the entire human genome will be mapped within a decade. Finally, we shall describe how maps expedite the identification of previously unknown genes, including those responsible for inherited neurological or psychiatric disorders.

A *restriction map* is an ordered set of DNA fragments. In ordinary DNA analysis these fragments are generated by cleaving DNA with enzymes that have *recognition sequences* of four or six bases. These cuts occur, on average, every 256 or 4096 base pairs for random sequence DNA. The resulting fragments fall in this general range of size, and the resolution of the map is accordingly a few hundred to a few thousand base pairs.

The map is constructed by separating the pieces according to size and using different strategies to identify neighboring fragments. It is now a routine process to make restriction maps of DNAs as large as 100,000 base pairs (100 kb) but there are few maps for much larger stretches of DNA. When a restriction map is made for a simple viral DNA or plasmid, the DNA fragments are usually analyzed directly. The size of each fragment is known, and each fragment is usually available as a purified individual species that can be used directly for further study. However, restriction maps of segments of more complex genomes are constructed by indirect methods that will be described later. These indirect methods never actually produce purified individual DNA pieces; the resulting map is a representation of the fragments by size and linear order, but it is not actually a collection of the pure fragments.

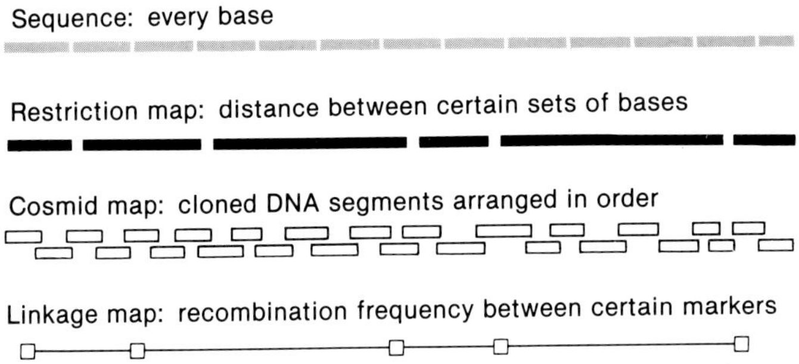

Fig. 12-1. Four types of genome maps illustrated schematically.

A *linked clone map* is a set of isolated and characterized cloned DNA fragments that together span the entire DNA region of interest. In this type of map, each segment of DNA is actually isolated. The earliest linked clone maps were made by a process called *chromosome walking;* that is, cloned DNA fragments are used to screen a library of clones in order to find adjacent fragments. This works because two clones that share part of their sequences will cross-hybridize. In this way it is possible to move outward from a given clone to identify DNA segments that are progressively more and more distant. Chromosome walking is tedious and the presence of highly repeated DNA sequences in typical mammalian genomes creates serious obstacles, because the repeated sequences lead to multiple cross-hybridizing clones, but only one is the true neighbor. However, in spite of these difficulties it is usually possible to walk hundreds of kilobases from the starting reference point.

Human chromosomes range in size from about 50 million base pair (Mb) to 250 Mb. The scale of ordinary physical maps is minute by comparison. However, techniques developed in the last few years have made it possible to construct physical maps of regions of DNA millions of base pairs long. Restriction maps have been extended to much larger regions by increasing the size of the individual DNA fragments analyzed. For example, restriction nucleases with eight-base recognition sequences would be expected, by chance, to yield DNA fragments averaging 64 kb in size. In practice, however, the two known enzymes with eight-base recognition often yield much larger fragments. From human DNA, for example, the enzyme *Not* I produces DNA fragments that average more than 1 Mb; the other enzyme, *Sfi* I, yields fragments of about 250 kb (1). Also, several of the six-base-specific restriction enzymes have fewer recognition sites than expected. The most extreme of these is *Mlu* I, which yields fragments that are as large as those generated by the 8-base *Not* I.

The new techniques make it possible to work with these large fragments without breaking them, and to fractionate the fragments by size with high resolution (2). Large DNA pieces are prepared by first suspending live cells in liquid agarose. After the agarose gels become solid, enzymes and detergents are allowed to diffuse into the agarose, lysing the cells and removing all cellular constituents except the DNA. These procedures work because the DNA is much larger than the agarose pores and cannot diffuse; in contrast all other reagents are much smaller than the pores.

Still in agarose, the pure DNA is cut into discrete fragments by diffusing restriction enzymes into the agarose and allowing them to act there. The enzymes are then removed and the mixture of fragments is separated by a special type of electrophoresis (3), pulsed field gel electrophoresis (PFG). In ordinary electrophoresis, all the DNA fragments are made to move in a single direction by the application of an electrical field. All DNAs larger than about 50 kb move at the same velocity, so no separations can occur. In PFG, the electrical field direction is switched periodically. The molecules must change direction to continue to move in response to the changing electric field. Larger molecules change direction more slowly than smaller ones. The overall motion of the large fragments is therefore retarded and, as a result, it is possible to separate DNAs ranging in size from 50 kb up to 10 Mb.

There have been many different experimental variations on the original apparatus for PFG (4). However all have one characteristic, a periodic change in the direction of the electrical field with pulse times varying from a few seconds (for highest resolution in the 50–100 kb size range) to an hour, for DNAs in the 1–10 Mb range (5) (Fig. 12-2). For general analysis of large restriction fragments of human DNA, pulse times of 25 seconds to 100 seconds are optimal.

Macrorestriction maps can be constructed by using available probes to identify the location of particular large fragments. Each probe is hybridized to a blot of a PFG-fractionated restriction digest in a manner essentially the same as that used for ordinary Southern blotting. If a sufficient number of DNA probes are available for the region of interest, it is possible to identify all of the large fragments. If the order of DNA probes is known, the hybridization results indicate the order of the large restriction fragments. In other words, if a genetic map is available with a sufficient number of cloned DNA markers, it is possible to use that map to construct the corresponding physical map. This approach was first developed to map most of the *E. coli* genome (6); it was then used to determine maps of several regions of the human genome (7) including the MHC on chromosome 6 (8,9) and the DMD gene on the X-chromosome (10–12).

There are two major limitations to this approach. It can be used only where there is already a sufficiently detailed genetic map, and it is difficult to prove that the map is complete. Fragments will be missed if they happen to lack sequences corresponding to the available cloned probes. It is possible to circumvent this problem, in part, by using more than one restriction nuclease. Then different enzymes are used to generate DNA fragments that overlap, by confirming these parts of the map and filling in the missing sectors. However, the procedure is still not formally sufficient to guarantee that a complete map will be produced.

Two advances in the basic strategy for constructing macrorestriction maps sidestep or eliminate all of the limitations described above. These are illustrated in Fig. 12-3. *Linking probes* are cloned small pieces of DNA that happen to contain the recognition sequence for enzymes that cut the genome only rarely (2). When each linking probe is hybridized to a blot of DNA fragments generated by the same enzyme, it identifies two different fragments (Fig. 12-3). These fragments must be adjacent. The complete set of linking probes therefore contains enough information to place all of the fragments of the genome in order. Methods for efficient production of libraries of linking probes are being developed. For example, it is possible to start with an existing ordinary DNA library in a circular vector and cut those fragments with the enzyme of interest. Only clones that contain the rare recognition site will be linearized, and only those will then accept an insert with the appropriate

Fig. 12-2. An example of pulsed field gel separation of large DNAs. The center lane is tandem oligomers of 48.5 kb λ DNA monomer. The left lane contains chromosomal DNAs of the yeast *S. cerevisiae,* which range in size from 200 kb to more than 1.3 Mbp. The right lane is a *NotI* digest of total human DNA.

end and containing a selectable marker. In that way, it is possible to select only the linking clones from a total genomic library.

The second restriction mapping strategy shown in Fig. 12-3 is *partial digestion* (1,2,13). Here, a single probe can be used to identify a whole set of neighboring fragments. The critical information about these fragments is contained in the precise size of the DNA bands seen in the partial digest. It is therefore important to have separations that extend to very large sizes, and also to have accurate length standards throughout the size range of interest. In ordinary DNA electrophoresis, resolution decreases sharply with increasing size of the DNA fragments, a charac-

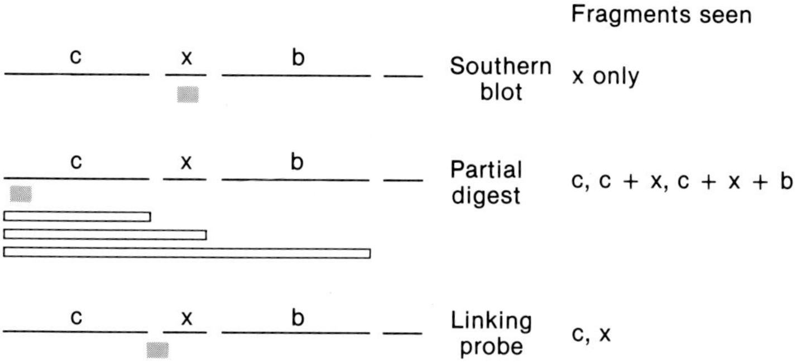

Fig. 12-3. Three of the strategies used to make macrorestriction maps of large regions of DNAs. The hatched bar indicates the location of particular cloned DNA probes used to analyze the region. At right the sizes of the DNA bands visualized by these probes are listed.

teristic that has discouraged use of partial digest methods. However, with PFG, resolution is essentially constant across a wide range of sizes, and annealed tandem λ DNA oligomers provide a reliable size standard throughout the range (Fig. 12-2). Partial digestion has proven to be a powerful and efficient method for the construction of macrorestriction maps; each probe can identify a set of fragments and prove that they are neighbors. Fewer probes are needed to complete a map. The most effective strategies combine linking probes and partial digests, because the halves of each linking probe (generated by cleaving the probe at the rare cutting site it must contain) are particularly effective probes for the analysis of partial digests (1).

The second general genome mapping method in current use is to assemble a *linked clone library* by identifying clones that are neighbors. If two clones contain partially overlapping segments of the genome, they will have sequences and restriction enzyme recognition sites in common. At least two types of methods can be used to screen for these common sequence elements (Fig. 12-4). In principle, these methods can be used on all types of clones. In practice, with all other things equal, it is more efficient to use the largest possible cloned DNA fragments to minimize the total number of clones that have to be linked up. For that reason, cosmid clones are preferred, and the resulting library is called *a linked (or overlapping) cosmid library.*

The *fingerprinting method* has been used in all large scale attempts to construct linked cosmid clone libraries (14,15). In the fingerprinting method, each clone is digested with a mixture of several restriction enzymes that are frequently cutting and the resulting mixture is separated by length by ordinary gel electrophoresis at the highest possible resolution. The lengths of the fragments generated are a signature of the sequences contained in the cosmid. If there are too many fragments, selective radiolabeling methods can be used to view just a discrete set of them. Clones that overlap will have some fragment lengths in common.

The *oligonucleotide hybridization method* has been proposed as a more efficient strategy than fingerprinting (16). Here a set of specific DNA probes is hybridized, one at a time, to the entire cosmid library. The pattern of probes that hybridize to each clone consists of a list of those probes that do hybridize and those that do not. This pattern will be different for every clone, if enough probes are used. However,

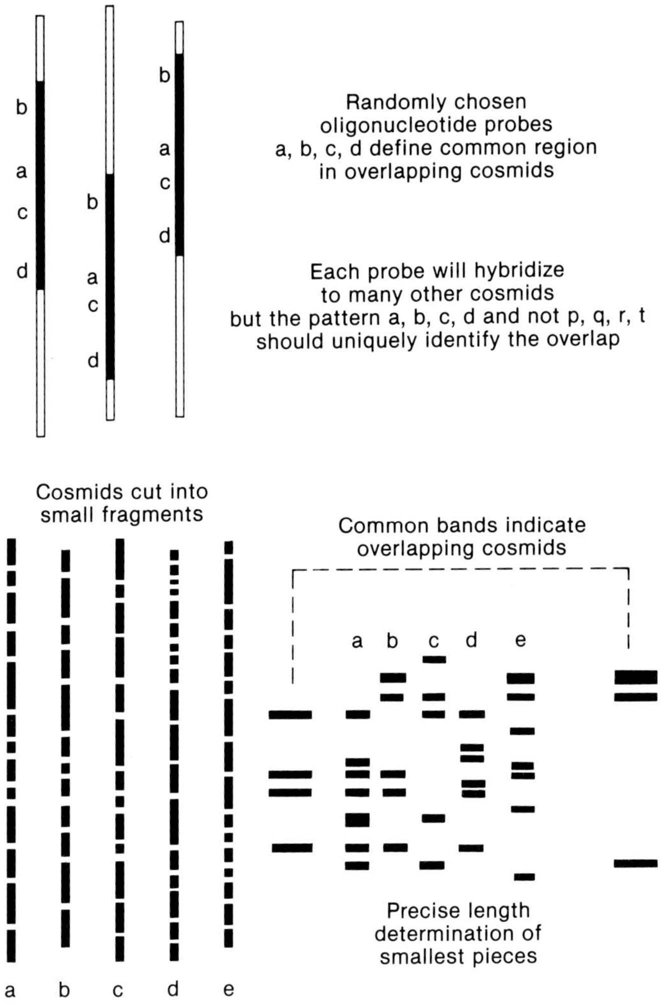

Fig. 12-4. Two strategies that can be used to construct linked cosmid libraries are illustrated schematically.

overlapping clones will have similar patterns and, in principle, there is enough information in the patterns to determine all of the overlaps. The oligonucleotide method is attractive, but it has not yet been subjected to a large-scale experimental test.

Whichever method is used to link cosmids, the result is increasing information on contiguous pairs of clones, then larger contiguous sets of clones, until, finally, in principle, an entire linked clone map is reached. In practice, the early stages of cosmid link up proceeds rapidly and efficiently. Later stages require increasingly more effort, and it has not yet been possible to complete most maps. This occurs for two reasons. In the fingerprinting approach, new cosmids are examined a few at a time, selected at random. As the map nears completion, most or all of the cosmids that are tested in each round have already been seen before, so no new information is gained. This is analogous to the slow ending of a race epitomized by Zeno's paradox. A more serious problem is bias in the original cosmid library. If some regions cannot be cloned into cosmids, the map will never be completed. Even if all regions are clonable, but some clones are seriously underrepresented in the library, it will become virtually impossible to locate them.

With current cosmid-mapping methods, libraries that are 70–80% complete can be constructed readily. New methods are needed to devise efficient ways to move closer to 100%. The simplest approach seems to involve use of the large DNA fragment methods (described above) to identify contiguous blocks of cosmids that lie on the same large DNA fragment, even though they are not sufficiently complete to span that entire fragment.

Cosmid mapping is a bottom-up or constructionist strategy. It can be used without any prior knowledge about a genome, and it can be used, in principle, on an entire genome at once. In contrast, most strategies for making entire physical maps more rapidly and efficiently are those that involve reductionist strategies (17). That is, the genome is divided into pieces that are successively subdivided until there are segments small enough to be completely mapped directly, either by macrorestriction mapping or cosmid mapping. Genomes are naturally divided into chromosomes, so reductionist strategies inevitably start by dividing the genome into separate chromosomes. There are two basic ways to do this. The first is direct physical separation of chromosomes. With simple organisms this can be accomplished directly by PFG (Fig. 12-2). With mammalian genomes the chromosomes are too large to resolve directly by current PFG techniques. However, metaphase human chromosomes can be separated effectively by *fluorescence-activated flow sorting* (18). Fluorescently stained chromosomes are passed through a laser beam in a liquid stream, one at a time. When a signal corresponding to a particular chromosome type is detected, the stream is deflected to capture that individual chromosome. Almost all of the human chromosomes can be so purified.

The amounts of material provided by flow sorting are small, typically a microgram of DNA or less per chromosome. However, the DNA can be amplified by cloning fragments of the DNA in a suitable vector. Single-chromosome libraries are now available routinely and are valuable for both physical and genetic mapping of individual chromosomes. In principle, it is possible to combine flow sorting of chromosomes with large DNA techniques. For example, flow-sorted material could be used to prepare a chromosome-specific linking library, or flow-sorted DNA could be cut with restriction enzymes such as *Not* I, then separating the resulting chromosome-specific large fragments of DNA. These combined approaches are limited by the amounts of DNA that are made available by current flow-sorting methods; advances, either in sorting yield or the efficiency of DNA fragment detection would be helpful in making these approaches more generally useful.

The alternative approach to chromosome separation is the use of *human–rodent cell lines* (usually mouse or hamster). Human and rodent cells are fused. Genetic markers are then used to select hybrid cells that have retained, ideally, only a single intact human chromosome. There are many pitfalls. Frequently more than one human chromosome remains, and there is always the risk that the human chromosomes may have been altered by rearrangements. However, once appropriate hybrid cells are available, they provide a long-term source of ample amounts of DNA. Hybrid cells can be used, in flow sorting, to purify human chromosomes that are difficult to separate from other human chromosomes. Alternatively, the hybrids can be used directly for a cloning and mapping by analysis with human-specific DNA probes. In that way, the rodent background disappears and the only DNA fragments detected on a gel, or as a clone, are those that contain human DNA. An efficient way to accomplish this is by using human repeated sequences. The *Alu repeat,* for example, is so frequent that it occurs on most segments of human DNA, but it is absent in rodent DNA.

Enzymes such as *Not* I cut the human genome into about 3000 fragments, too many to resolve most individual pieces by PFG. Thus, when a macrorestriction digest of total human DNA is probed with Alu, which detects most of the fragments, what is seen on the gel is a smear, not too different from a direct ethidium-stained pattern of the DNA itself (Figs. 12-2 and 12-5).

In contrast, when the same experiment is performed with a hybrid cell line that contains only one human chromosome, ethidium still shows a smear because it detects both human and rodent DNA, but the Alu hybridization now gives a discrete set of bands. In this manner, it is possible to identify, in a single gel lane, most of the large DNA fragments that arise from an individual human chromosome.

This is the first stage of the reductionist approach, which can be carried further by subdividing each individual chromosome into regions. For example, when cell lines are available with specific deletions of parts of a chromosome, they can be

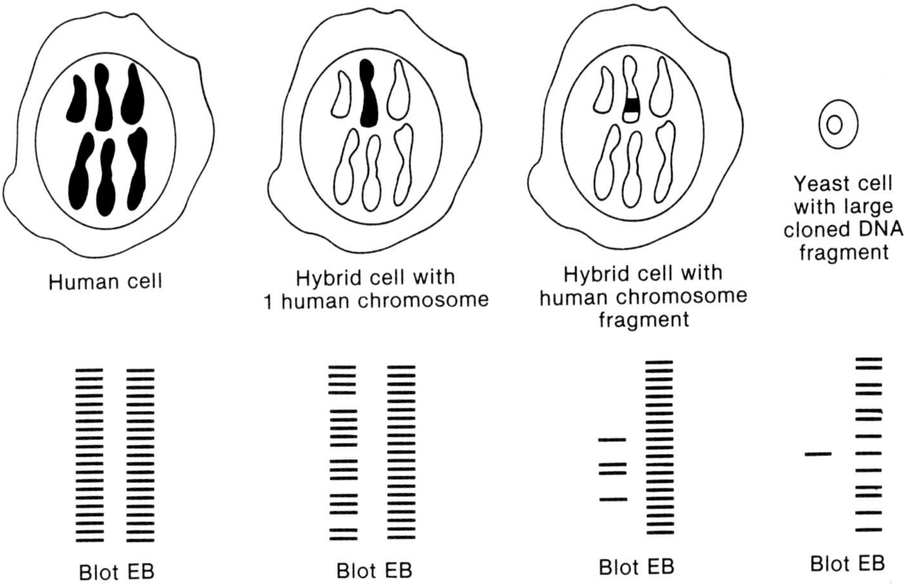

Fig. 12-5. Four different stages in subdividing the human genome illustrated along with the expected ethidium-stained macrorestriction digest for each one. Also shown is the expected pattern of hybridization of the digest with a highly repeated human DNA probe such as an Alu sequence.

used to map either the large DNA fragments or particular cloned DNA pieces to regions of the chromosome. Potentially more powerful is the use of hybrid cell lines that are designed to contain only small fragments of individual human chromosomes. These are prepared by starting with cell lines that contain only human chromosomes, breaking up the chromosomes into fragments by X-ray irradiation, for example, and then fusing the cells with unirradiated rodent cells. Again, a selectable marker is helpful to ensure retention of a particular human DNA segment, while others are lost by segregation. Cell lines of this kind provide an excellent source of material for PFG analysis because they contain just a few large human DNA fragments (Fig. 12-5) and a physical map of the human component of each line can be constructed. However, the techniques used to generate the cell lines often generates chromosome rearrangements. It will probably be necessary to examine several different cell lines for each region in order to be certain that the normal, unrearranged map has been revealed.

Development of cell lines with DNA fragments from only one human chromosome is really a method of cloning megabase segments of DNA. However, cell lines have several disadvantages compared to ordinary cloning procedures. The human DNA is still in a rodent background, risking cross-hybridization with many probes, and also making it more difficult to purify the human DNA. The cloned piece is present only in a single genome equivalent and it is diluted by the whole rodent genome. Clearly it would be better to be able to clone large fragments of DNA in simple organisms; this has been done directly in yeast and indirectly in *Escherichia coli* (19, 20).

Development of these methods promises to revolutionize techniques for physical mapping (Fig. 12-5). Each macro-DNA clone can be restriction mapped easily, because it consists of just a few large restriction fragments. Most or all of these will be visible by direct ethidium staining, so the mapping process will not be heavily dependent on the availability of cloned probes for the region of interest. Adjacent macroclones can be placed in order in the genome by fingerprinting or by reference to a coarser physical map. Whichever approach is used, the key ingredient is the small number of macroclones needed to span the genome correspondingly reducing the time and effort needed to handle the clones. Large DNA cloning will effectively span the gap between constructionist cosmid mapping and reductionist restriction mapping.

The methods described here are, today, allowing construction of the first physical maps of major regions of the human genome. They are already sufficient to provide maps of entire human chromosomes and advances in technology are already beginning to appear, promising to make the mapping task progressively even easier. Physical maps of a few hundred kilobases resolution should be available for most, possibly all, of the human genome within 5 years. The question we shall explore in the remainder of this chapter is the implications these maps will have for locating human genes and for diagnosing human disease (Fig. 12-6).

Linkage-mapping, using restriction fragment length polymorphisms (RFLPs) or other genetic markers, can reveal the approximate chromosomal location of the gene responsible for a particular phenotype, such as a genetic disease. The precision of linkage mapping will vary, depending on the size and informativeness of avail-

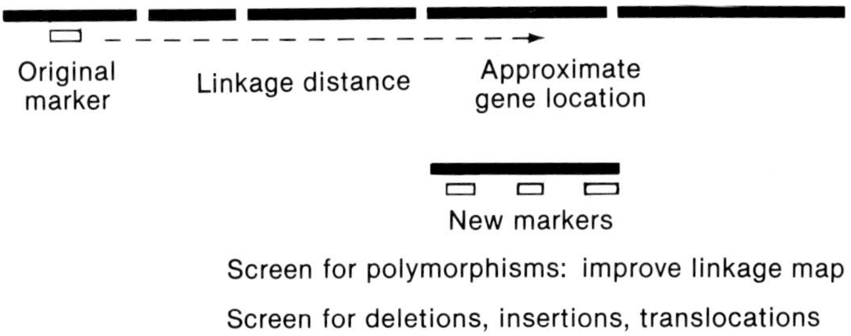

Fig. 12-6. A schematic illustration of some of the ways in which a physical map will help determine the location of a gene responsible for a particular disease.

able pedigrees and the fortuitous occurrence of recombination events in the region of interest. At present, genes can be located within a few centimorgans of genetic distance from the nearest available marker. This genetic distance should correspond, on average, to a few million base pairs of physical distance. Unfortunately, the uncertainty in this estimate is considerable and, in extreme cases, a centimorgan might include as little as 100 kb or as much as 10 Mb.

The challenge, once a linked marker has been found, is to move rapidly toward the gene itself, ultimately identifying that gene. A physical map assists this process because the map can be used to select, immediately, pieces of DNA that should be closer to the gene than the original marker. These pieces can be screened for informative polymorphisms and, if any are found, the resolution of the linkage map can be improved. If two linked markers flank the gene, knowledge of the physical and genetic distance between them can be used to estimate the average recombination frequency in the region. This permits a more accurate estimate of the physical location of the gene. An average mammalian gene is about 30 kb, and the combination of physical and linkage mapping data will probably narrow the location of the gene of interest to a region of 100 kb, which will only contain a few genes. In many cases the location of these genes will be signaled by easily detected sequences such as *HpaII tiny fragment (htf) islands* (7). Thus, the amount of candidate DNA that will have to be studied in detail to identify the actual gene will be limited in most cases.

Where the deleterious allele or alleles responsible for disease are RFLPs, the physical map will probably pinpoint the actual gene. In the 100 kb size range, the resolution of PFG is just a few kilobases. Thus, any significant insertion, deletion, or more complex rearrangement should be apparent even in a heterozygote. It is premature to guess what fraction of human disease phenotypes involves size polymorphisms rather than simple point mutations, but there is precedent already in DMD and familial hypercholesterolemia (21).

Physical maps will be useful in searching for disease genes even in the absence of linkage data. Genes are not arranged randomly; many are clustered in families that are related by evolution or by function. The pattern of organization, in local regions, is similar in different mammals. The physical map of the human genome will provide clues to guide linkage studies in the future or more direct searches for specific genes, just as a geological map can provide different insights about the likely characteristics of particular regions.

REFERENCES

1. Smith CL, Lawrance SK, Gillespie GA, Cantor CR, Weissman SM, Collins FS. Strategies for mapping and cloning macro-regions of mammalian genomes. In Gottesman M., ed. Methods in Enzymology, Vol. 151. San Diego: Academic Press, 1987, pp 461–489.

2. Smith CL, Warburton PE, Gaal A, Cantor CR. Analysis of genome organization and rearrangements by pulsed field gradient gel electrophoresis. Genet Eng 1986; 8:45–70.

3. Schwartz DC, Cantor CR. Separation of yeast chromosome-sized DNAs by pulsed field gradient gel electrophoresis. Cell 1984; 37:67–75.

4. Anand R. Pulsed field gel electrophoresis: A technique for fractionating large DNA molecules. TIG November 1986; 278–283.

5. Smith CL, Matsumoto T, Niwa O, Klco S, Fan J-B, Yanagida M, Cantor CR. A molecular karyotype for *Schizosaccharomyces pombe* by pulsed field gel electrophoresis. Nucleic Acids Res 1987; 15:4481–4489.

6. Smith CL, Econome J, Schutt A, Klco S, Cantor CR. A physical map of the *Escherichia coli* genome. Science 1987; 236:1448–1453.

7. Brown WR, Bird AP. Long-range restriction site mapping of mammalian genomic DNA. Nature (London) 1986; 322:477–481.

8. Lawrance SK, Smith CL, Weissman SM, Cantor CR. Macro-organizational analysis of the HLA complex. Science 1987; 235:1387.

9. Hardy DA, Bell JI, Long EO, Lindsten T, McDevitt HO. Mapping of the class II region of the human major histocompatibility complex by pulsed-field gel electrophoresis. Nature (London) 1986; 323:453–455.

10. van Ommen GJB, Verkerk JMH, Hofker MH, et al. A physical map of 4 million bp around the Duchenne muscular dystrophy gene on the human X-chromosome. Cell 1986; 47:499–504.

11. Burmeister M, Lehrach H. Long-range restriction map around the Duchenne muscular dystrophy gene. Nature (London) 1986; 324:582–585.

12. Kenwrick S, Patterson M, Speer A, Fischbeck K, Davies K. Molecular analysis of the Duchenne muscular dystrophy region using pulsed field gel electrophoresis. Cell 1987; 48:351–357.

13. Cantor CR, Smith CL, Argarana C. Strategies for finishing physical maps of macro-DNA regions. In Symposium on Integration and Control of Metabolic Processes: Pure and Applied. Cambridge, England: ICSU Press, 1987, pp 427–438.

14. Olson MV, Dutchik JE, Graham MY, et al. Random-clone strategy for genomic restriction mapping in yeast. Proc Natl Acad Sci USA 1986; 83:7826–7830.

15. Coulson A, Sulston J, Brenner S, Karn J. Toward a physical map of the genome of the nematode *Caenorhabditis elegans.* Proc Natl Acad Sci USA 1986; 83:7821–7825.

16. Poustka A, Pohl T, Barlow DP, et al. Molecular approaches to mammalian genetics. Cold Spring Harbor Quant Biol 1986; 50, in press.

17. Smith CL, Yu MT, Cantor CR. Phased approaches to mapping the physical structure of entire chromosomes. In Kahl G, ed. Architecture of Eukoryotic Genes. Weinheim FRG: VCH Publisher, 1987, pp 489–496.

18. Gray JW, Langlois RG. Chromosome classification and purification using flow cytometry and sorting. In Engelman D, Cantor CR, Pollard TD, eds. Annual Review of Biophysics and Biophysical Chemistry, Vol. 15. Palo Alto, CA: Annual Reviews, Inc., 1986, pp 195–235.

19. Gaitanaris GA, McCormick M, Howard BH, Gottesman ME. Reconstitution of an operon from overlapping fragments: Use of the λSV2 integrative cloning system. Gene 1986; 46:1–11.

20. Burke DT, Carle GF, Olson MV. Cloning of large segments of exogenous DNA into yeast by means of artificial chromosome vectors. Science 1987; 236:806–812.

21. Lehrman MA, Russell DW, Goldstein, JL, Brown MS. Exon-Alu recombination deletes 5 kilobases from the low density lipoprotein receptor gene, producing a null phenotype in familial hypercholesterolemia. Proc Natl Acad Sci USA 1986; 83:3679–3683.

13

Assigning Genes: Linkage, Crossing Over, Lod Scores

P. MICHAEL CONNEALLY

Assignment of genes to specific chromosomes using family studies has been the classical approach to human gene mapping. This method usually requires a few large families or numerous small families to be successful in finding the chromosomal location of the gene in question. Before, however, describing the methods involved in the analysis of family data for linkage, we need to define some terms.

Linkage may be defined as the occurrence of two loci that are sufficiently close together on a chromosome such that their assortment is recognized as being non-independent; that is, they are passed together in meiotic replication of the chromosome. Loci on the same chromosome are said to be *syntenic* (1). Linked loci are always syntenic, but syntenic loci are not necessarily linked. For example, the loci for Duchenne muscular dystrophy (DMD) and chronic granulomatous disease are both syntenic and linked. However, the loci for DMD and hemophilia A are syntenic but not linked.

Crossing over is the process by which homologous segments of paired maternal and paternal chromosomes are exchanged. Crossing-over occurs at the "four-strand" stage of meiosis. The *frequency of recombination* is defined as the proportion of strands that experience an odd number of crossover events between the loci; recombination frequency is the unit of measurement of genetic map distance. The unit of map distance, a *Morgan,* is defined as the length of chromosomal segment that, on the average, experiences one exchange per strand. For short chromosome segments, the frequency of recombination is directly proportional to the map distance [expressed in *centiMorgans (cM)*]. The physical map (number of base pairs) and genetic map (cM) are not parallel because, for longer intervals, due to the occurrence of multiple crossovers and interference, the relationship is nonlinear and a mapping function is necessary to relate the frequency of recombination to map distance. *Interference* (positive or negative) is the tendency for an already established crossover to suppress or enhance the formation of other crossovers in nearby regions.

In a double heterozygote (AaBb) there are two *linkage phases,* one with alleles A and B on one chromosome and a and b on the other. This is arbitrarily termed the *coupling phase.* The other, termed the *repulsion phase,* has A and b on one chromosome and a and B on the other.

Two terms, *linkage* and *association,* are sometimes confused. A number of diseases have an association with specific antigens at the ABO or HLA loci. This is simply a statistical association in human populations. The diseases that are associated with ABO or HLA antigens usually do not have a simple mode of inheritance; indeed, most have only a small genetic component. In the case of *linked loci,* the disease is not associated with a specific allele at the other locus (except, of course, within sibships) unless there is linkage disequilibrium, which is defined below. Linkage disequilibrium is seen commonly with closely linked loci and may mimic association in the population.

Linkage disequilibrium is the term used for the condition in which the frequency of a specific haplotype or chromosome is not equal to the product of the individual gene frequencies. For example, the frequency of chromosome Ab is equal to the frequency of allele A times the frequency of allele b if the loci are in equilibrium. If two (or more) loci are closely linked, the loci are usually not in equilibrium; in other words, there is linkage disequilibrium and the two characteristics tend to be inherited together. If, on the other hand, the loci are in equilibrium, the frequency of coupling and repulsion genotypes will be equal. In multipoint mapping of several loci, we cannot assume that there is genetic equilibrium (unless that has actually been demonstrated) because that assumption could lead to erroneous linkage results.

A *polymorphic locus* is one in which there are at least two alleles or haplotypes where the most frequent one is encountered in not more than 99% of a population. The more polymorphic a marker locus, the more useful that marker is for linkage analysis. In many cases a restriction endonuclease cuts at several but varying sites in the area of a probe. This pattern gives rise to multiple haplotypes and increases the frequency of heterozygosity.

The LOD score method of linkage analysis was introduced by Morton (2) in 1955. He combined the probability ratio test of Haldane and Smith (3) with the method of sequential analysis to give the first practical approach to human linkage analysis. The *probability ratio* (Pr) is the probability (or likelihood) of the data, given a specified value of the recombination frequency (θ_i), divided by the probability of the data if there were independent assortment ($\theta = \frac{1}{2}$). To add scores among families, the log of the probability ratio is used, resulting in the term *lod* (log odds). Thus a lod score (Z) is defined as

$$Z = \log[\Pr(\text{data} \mid \theta = \theta_i)/\Pr(\text{data} \mid \theta = \tfrac{1}{2})]$$

Morton suggested the following criteria to evaluate the significance of a lod score: (1) If, after the lod scores (log of the probability ratio) are accumulated for several families the total is 3 or more, we can conclude that the frequency of recombination is significantly less than 0.5 and that there is linkage. (2) If the total is -2 or less, we conclude that the frequency of recombination is significantly more than the value of θ for which the lods were calculated; that is, linkage can be ruled at that value of θ. (3) If the total lod score lies between -2 and $+3$, we suspend judgment about linkage until further data lead to a decision. In other words, the statistical test for linkage is conservative; a lod score of 3 (or odds of 1000:1) is required to show linkage. This rule avoids, as much as possible, conclusions that loci are linked when, in fact, they are not.

Morton provided tables of lod scores that were available for sibships but not for large families because it would be impractical to include lod scores for all possible combinations in large families. Morton's method also did not allow for situations such as reduced penetrance or late-onset disorders. These situations are commonplace in human disease studies.

Elston and Stewart in 1971 (4) described a general approach to obtain the likelihood of a pedigree, i.e., the probability of individuals having given phenotypes, affected or unaffected, at the disease locus and also the types AA, AB, or BB at the marker locus. They devised an efficient "peeling" algorithm. Likelihood is computed recursively, beginning with the most recent generation and working backward to the most remote. The advantage of this method is that the likelihood for an individual can be calculated first, and the result can then be attached as a factor to the appropriate term for his or her parent. The individual is then no longer needed in further computation and may be "peeled off," reducing the computational complexity.

In theory, the likelihood can be obtained for a pedigree of any complexity for both qualitative and quantitative traits. Unknown phenotypes, late onset, and variable modes of inheritance can be easily dealt with in using this algorithm, but these considerations do increase the complexity of calculation.

A computer program called LIPED (5) is based on the Elston–Stewart algorithm and has been widely used (with modifications) for practically all two-point linkage analyses. Figure 13-1 is an example of a pedigree with a dominant disorder and a marker with two alleles. The frequencies of alleles at the disease locus are p_1 and q_1 and, for the marker locus, p_2 and q_2.

The probability of the four children's status, at the disease and marker loci (three noncrossovers and one crossover) given these parents is $[(1-\theta)/2]^3 \times \theta/2$. The probability that their father is affected is $p_2^2q_1^2$. These terms $[1-\theta/2]^3 \times \theta/2$ and $p_2^2q_1^2$ would normally be attached to the mother. The probability for the mother is ½; her affected father was homozygous at the marker locus, so that there cannot be any recombination and θ is not involved. Finally, the probability that the grandmother is type AA and unaffected and the grandfather is type BB and affected is $p_1^2q_2^2 \times q_1^2(1 - q_2^2)$. That term is then combined with the others to give the final probability for the complete family of $p_1^2q_2^2 \times q_1^2(1 - q_2^2) \times ½ \times p_2^2q_1^2 \times [(1 - \theta)/2]^3 \times \theta/2$. To obtain lod scores for this family for a specific value of the recombination fraction θ, say 0.1, the numerator is the above with $\theta = 0.1$ whereas the denominator is the above with $\theta = ½$. In this case the population gene frequencies

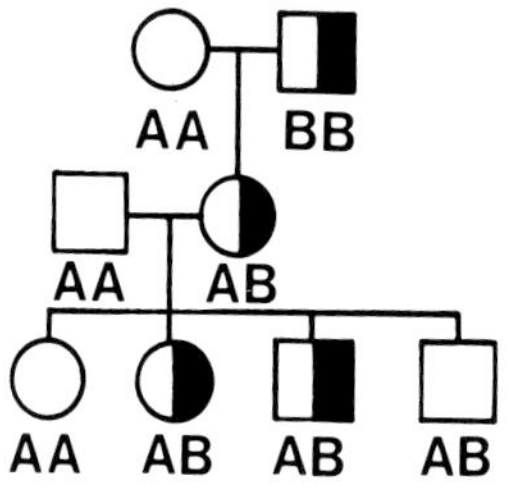

Fig. 13-1. Pedigree showing dominant inheritance of a disorder together with a marker locus. Half-dark individuals are affected. Genotypes for marker locus are given under each individual.

are not required because they cancel out in the numerator and denominator. The likelihood ratio and lod scores for four values of θ for this family are given below.

	θ			
	0.1	0.2	0.3	0.4
Likelihood ratio	1.166	1.638	1.646	1.382
Lod score	0.067	0.214	0.217	0.141

As the family size increases, it quickly becomes extremely tedious if not impossible to compute the lod scores by hand, so computer programs such as LIPED become important.

The development of the molecular technology to generate large numbers of anonymous DNA markers (*restriction fragment length polymorphisms,* RFLPs) throughout the genome means that once a gene is linked to an RFLP and then assigned to a chromosome, other linked RFLPs can be used to provide more precise location of the disease gene. From an applied point of view, the method can identify flanking markers that are important for gene walking or hopping and for predictive purposes.

Renwick and Bolling (6) were the first to produce an algorithm for *multipoint mapping,* i.e., linkage analysis with more than two loci. Their method relied on the lod scores for each pair of linked loci and thus did not make full use of the data. For example, they could not use information on triple heterozygotes. Meyers et al. (7) used a parametric non-Bayesian approach to obtain a multipoint map of chromosome 1, including data from triply and quadruply heterozygous crosses, together with independent two-point data from nuclear families. These approaches could not analyze data from large families such as those allowed in the LIPED program.

This deficiency was overcome by Lathrop et al. (8), who developed the program package, LINKAGE, that can simultaneously obtain likelihoods of pedigrees for an arbitrary number of loci. However, if more than three loci are included, the analysis requires large amounts of computer time, and the practical limit is generally five loci. Multipoint analysis allows full use of triply and quadruply informative matings to determine order and map distance. This method is useful in determining the location of one gene among a previously determined "map" of linked marker loci.

In multipoint mapping a number of hypotheses may be tested. For example, we might consider two marker loci A and B and the disease locus D. One test might be $\theta_{AB} = \theta_1$, $\theta_{AD} = \theta_2$, and $\theta_{BD} = \theta_3$ versus the possibility that each of the loci is independent of the other. This is the usual example given for multipoint mapping. In practice, however, A and B are usually known to be linked and D will have been linked to one or both by two-point analysis. Then, it is necessary only to compare the order ABD versus DAB versus ADB. The statistical methods involved in that calculation are beyond the scope of this chapter but are described in detail by Ott and Lathrop (9).

One problem in human linkage analysis is the possibility of heterogeneity if the same phenotype can result from genes at two or more loci. Recognition of genetic heterogeneity is important in finding the gene and, particularly, for presymptomatic or prenatal diagnosis. The first test of heterogeneity used "large sample theory," as follows:

$$X^2 = 2 \log_e 10(\Sigma\, \hat{z}_i - \hat{Z})$$

with $n - 1$ degrees of freedom where n is the number of families, $\hat{z}_i$ is the maximum lod score for the ith family, and $\hat{Z}$ is the maximum lod score for all families combined. A more appropriate test, better for the small sample sizes of human families, was described by Smith (10), then refined by Ott with a computer program (HOMOG) to implement the algorithm. Assume a proportion (α) of families belong to the linked group; then the posterior probability that the ith family belongs to the linked type, given the data, is expressed as

$$W_i(\alpha,\theta) = \alpha L_i(\theta)/[\alpha L_i(\theta) + 1 - \alpha]$$

where $L_i(\theta)$ is the likelihood of the ith family for a given value of θ. This would appear to be the best choice. HOMOG estimates the proportion of "linked" families jointly with the recombination fraction (θ) in these families.

The confidence intervals for the frequency of recombination (θ) are important and should always be reported. A relatively simple method to obtain these is to construct the lod score curve and then draw a horizontal line at a distance of one lod score below the peak lod score (Fig. 13-2). The two points of intersection between the straight line and the lod score curve mark the two end-points of the

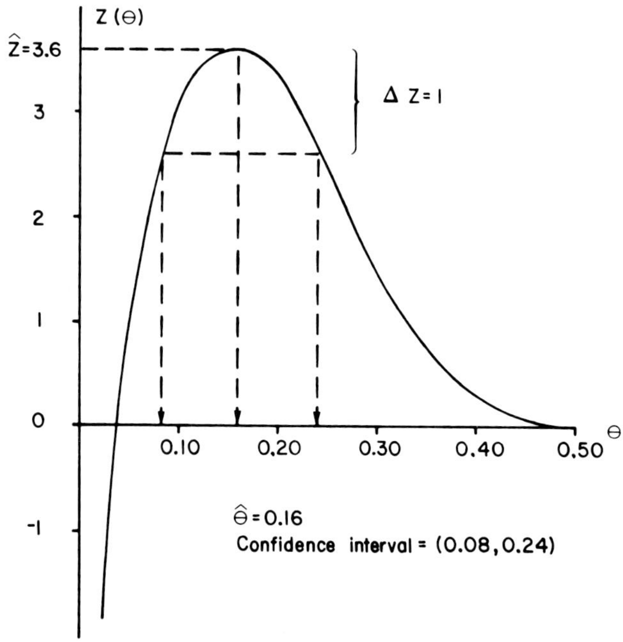

Fig. 13-2. Illustration of method used to construct a confidence interval for the recombination fraction, θ.

confidence interval. The associated confidence coefficient is at least 90% and, in large data sets, approaches a value better than 95%. This problem is discussed elsewhere in more detail (11).

What are the criteria for an optimal linkage study? Although each disorder has its own specific issues, there are general criteria for all cases.

The first step in a linkage study is the selection of appropriate families. For autosomal dominant disorders, large mutligeneration families with many affected individuals are ideal. Much more information on linkage is obtained from a large pedigree than from the same number of individuals in separate pedigrees (12). This makes the study more economical because fewer individuals have to be sampled and phenotyped. A second reason to examine large families is the possibility of heterogeneity. If more than one locus is involved in the disease, a significant deviation from linkage will usually be detected only if the family is large.

For autosomal recessive and some X-linked disorders, it is difficult to find multigeneration-affected families, and attempts are therefore made to find large multiply-affected sibships. For creating RFLP maps, pedigrees with many phase-known meioses are needed.

Proper clinical diagnosis is also critical in a linkage study. At least one affected member of the family must have been unequivocally shown to be affected with the disease if the family is to be used in linkage analysis. Some misclassification is acceptable for a beginning study. This will have the effect of giving a larger estimate of the recombination fraction (since misclassification is much more likely to result in a recombinant) and will decrease the lod score. Once a linkage is established and attempts are made to "walk" to the disease gene, even one misdiagnosed individual can have serious consequences. Because a small linkage distance (1 cM) is the equivalent of a large physical distance (one million base pairs), the importance of precision in diagnosis must not be underestimated but it is often overlooked by geneticists.

For late-onset disorders, early diagnosis or techniques that detect asymptomatic gene carriers greatly enhance the power of the linkage analysis by decreasing the number of uncertain genotypes in the sample.

The choice of markers to be used in the linkage analysis should not be arbitrary. Standard markers, such as red cell antigens and serum isozymes, are less expensive than DNA markers and serve a second purpose, paternity testing. If a linkage is found with a standard marker, this will usually allow chromosomal localization, which means that only a limited number of probes will be needed to find a more closely linked DNA marker. The choice of DNA marker should, at first, be a cDNA that may have relevance to the disease in question, i.e., a candidate gene. Even if the marker has no relevance to the disease it is just as likely to be linked to the disease as an anonymous probe, and thus is no worse a choice.

Finally, proper analysis of the data is important. Fortunately, there are several excellent programs available. The best known are LIPED for two-point analysis and LINKAGE for multi-point analysis. However, proper incorporation of age-at-onset curves, linkage disequilibrium, gene frequencies, recombinational sex differences, and mutation makes these final analyses more complex than they might otherwise seem.

The human gene map is extending rapidly because many DNA markers are now available with advances in molecular laboratory techniques, and also because of the development of more efficient and complete mathematical techniques for linkage analysis.

REFERENCES

1. Renwick JH. Progress in mapping human autosomes. Br Med Bull 1969; 26:65–73.

2. Morton NE. Sequential tests for the detection of linkage. Am J Hum Genet 1955; 7:277–318.

3. Haldane JBS, Smith CAB. A new estimate of linkage between the genes for colourblindness and hemophilia in man. Ann Eugen 1947; 14:10–31.

4. Elston RC, Stewart J. A general model for the analysis of pedigree data. Hum Hered 1971; 21:523–542.

5. Ott J. Estimation of the recombination fraction in human pedigrees: Efficient computation of the likelihood for human linkage studies. Am J Hum Genet 1974; 26:588–597.

6. Renwick JH, Bolling DR. An analysis procedure illustrated on a triple linkage of use for prenatal diagnosis of myotonic dystrophy. J Med Genet 1971; 4:399–406.

7. Myers DA, Conneally PM, Lovrien EW, et al. Linkage group I. The simultaneous estimation of recombination and interference. In Baltimore Conference (1975): Third International Workshop on Human Gene Mapping. Birth Defects: Original Article Series, Vol. 11. New York: The National Foundation, 1976, pp 335–339.

8. Lathrop GM, Lalouel JM, Julier C, Ott J. Strategies for multilocus linkage analysis in humans. Proc Natl Acad Sci USA 1984; 80:4808–4812.

9. Ott J, Lathrop GM. Goodness-of-fit tests for locus order in three-point mapping. Genet Epidemiol 1987; 4:51–57.

10. Smith CAB. Testing for heterogeneity of recombination fraction values in human genetics. Ann Hum Genet 1963; 27:175–182.

11. Conneally PM, Edwards JH, Kidd KK, et al. Human Gene Mapping 8: Report of the committee on methods of linkage analysis and reporting. Cytogenet Cell Genet 1985; 40:356–359.

12. Conneally PM, Rivas ML. Linkage analysis in man. In Harris H, Hirschhorn K, eds. Advances in Human Genetics, Vol. 10. New York: Plenum Publishing Corp, 1980, pp 209–266.

14

Genetic Analysis: A Practical Approach to Linkage, Pedigrees, and Bayesian Risk

J. FIELDING HEJTMANCIK, PAT WARD, UMADEVI TANTRAVAHI, PAMELA HAWLEY, JAMES SPEER, SAMUEL LATT, MARCIE SPEER, AND MARGARET A. PERICAK-VANCE

DNA-based diagnosis of Duchenne muscular dystrophy (DMD) is now possible for many families. If a woman is interested in her DMD heterozygote status or if she is at risk for giving birth to a boy with DMD, she should contact a genetics clinic. There, a genetic counselor determines whether the family structure is appropriate for DNA diagnosis. The DNA diagnostic laboratory or medical director should assist in that evaluation.

For example, a reasonable family for analysis of an X-linked recessive trait would be an obligate carrier in a multi-generation pedigree with most members available, including at least one affected first degree male relative and the maternal grandfather.

DNA analysis can also be used for carrier detection in autosomal dominant disorders, such as myotonic dystrophy (DM). An optimum family with DM would include three-generation data, including clinical evaluations.

However, there are no strict rules for suitability, and each family is judged separately. Both parents are not required for analysis, and successful diagnosis (e.g., of a fetus at risk for DMD) has been carried out without the previously affected relative being available. As the pedigree becomes less complete, the analysis becomes more complex and diagnostic accuracy may decrease. Creatine kinase (CK) data on relevant females may be valuable in families at risk for DMD, particularly if blood has been obtained before the pregnancy.

Once a case has been accepted for DNA analysis, we generate an intake form that contains all information necessary to receive and process samples from members of the family. We send the referring center an information packet that contains shipping instructions (often with packing material and sample containers) and consent forms. Optimally, DNA-based data on family members are obtained before the woman at risk becomes pregnant.

Amniotic fluid and chorionic villus samples are shipped in sterile containers because they are usually cultured, even if direct analysis of the sample is also carried out. Blood samples from family members can be shipped in tubes that have been anticoagulated with citrate, but heparinized blood is preferred if the cells are to be transformed by Epstein-Barr virus.

DNA is prepared from samples by repeated proteinase digestions and organic solvent extractions to separate nucleic acids from other cellular constituents. DNA is then digested with restriction endonucleases. Then, the DNA fragments are size fractionated by agarose gel electrophoresis and transferred to a nylon or nitrocellulose membrane. Fragments of interest are identified by hybridization with cloned DNA, typically radioactively labeled, for analysis by autoradiography.

The results are then interpreted by the laboratory director and reviewed by the medical director and genetics counselor. For prenatal diagnosis, the counselor relays the results by telephone. The laboratory director generates a formal report that is then reviewed by the counselor and medical director before it is sent to the referring physician. These diagnostic tests are complex and correct interpretation requires sophistication. Best results are obtained when cases are referred from, and patient counseling is also carried out by, teams at genetics centers where the members are conversant with recombinant DNA technology.

GENETIC LINKAGE ANALYSIS

Genetic linkage is observed when two genes are so close on the same chromosome that they do not segregate independently. Linked loci are *syntenic* (on the same chromosome), but syntenic loci are not always linked. Linked loci are transmitted as a unit to the same gamete during meiosis, except when there has been an odd number of crossover events between them. Crossing over between chromatids of homologous chromosomes at meiosis results in the exchange of chromatin between the two chromatids.

The average number of interchanges between two loci on a given chromatid (the formal *genetic map distance* in Morgans) is low for small interlocus separations. The number increases as the distance between the two loci increases. A *recombinant (R) gamete* (defined for a given pair of loci) is one that contains one allele from each parental homologue. A *nonrecombinant (NR) gamete* contains a parental allele configuration. Linkage is estimated from measurement of the fraction of recombinant gametes (θ). For small values of θ, θ approximates the genetic map distance in Morgans. Hence $\theta = 0.1 = 10\%$ is approximately equal to a map separation of 0.1 M or 10 cM. Conversely, a value of 0.5 for θ implies free recombination, resulting from an interlocus separation too great to detect linkage (1).

Family (pedigree) linkage studies have been important in human gene mapping. Family studies are particularly useful for mapping genetic disorders that are not expressed biochemically in cultured cells. The disease is analyzed for linkage with polymorphic marker loci. A DNA sequence variation of restriction fragment length polymorphism *(RFLP)* is recognized in a population if there are two or more genotypes at a single locus, if neither genotype is rare. If the rarer allele is found in at least 1% of a population, the frequency of that allele is 0.01 and the locus is said to be *polymorphic*. In addition to RFLPs, marker loci may include phenotypic measurements, such as blood group types or serum isoenzymes (2).

The *most probable value* of θ (called $\hat{\theta}$) can be estimated from pedigree analysis by using *likelihood* calculations. We calculate the likelihood of observing the particular pedigree by comparing values for free recombination (no linkage) and the particular family over a selected range of θ (0.0–0.5). The logarithm of the ratio of the likelihoods is the *lod score (log of the odds of linkage), z:*

$$z = \log_{10} \frac{L\,(\text{given } \theta = 0.0.\text{–}0.5)}{L\,(\theta = 0.5)}$$

If the sum of lod scores for a marker in different families exceeds +3.00, the odds are 1000:1 that there is linkage. Lod scores greater than +3.0 are therefore considered indicative of linkage. Similarly, a score of −2.00 implies 100:1 odds against linkage and is taken as indicating no linkage (3). Confidence intervals can be determined for $\hat{\theta}$, the *maximum likelihood estimate of the recombination fraction* (4), and are used in the clinical application of linkage data.

The amount of information obtained from any particular family depends on the number of individuals tested, the number of meiotic events, and the extent of polymorphism at the marker loci. Pedigree studies evaluate the number of recombinant and nonrecombinant offspring of informative matings. A *mating is informative* only if one of the parents is a double heterozygote; that is, the person is heterozygous at both the disease and the marker locus. RFLPs are ideal marker loci when they are highly polymorphic. In addition, several RFLPs at the same locus can be combined into a *haplotype,* increasing the heterozygosity [i.e., for X-linked conditions, the *polymorphism information content (PIC)* (5)] of the marker locus.

Calculation of lod scores and estimation of θ are facilitated by computer analysis. One of the most commonly used programs is the LIPED program written by Ott (6), based on the algorithm of Elston and Stewart, a general method for obtaining the likelihood of the whole pedigree. Another linkage program, developed by Lathrop and Lalouel (7), can be generalized to include data from multiple loci. Both programs include all family members in the linkage analysis. Large families provide more information and are important in discerning genetic heterogeneity if more than one gene results in the same clinical phenotype (5,8).

Once linkage has been established between a disease and a marker locus, the information can be used for carrier detection and prenatal diagnosis. Use of linkage in genetic counseling is illustrated by considering a hypothetical family with Huntington disease (HD), an autosomal dominant condition. A 20-year-old asymptomatic son of a woman with HD wants to know his risk of having inherited the HD gene. His *prior possibility* of having inherited the HD gene from his mother is 0.50. If the HD locus (Hh alleles) were known to be linked to some marker locus (alleles AA′), it is possible to refine his risk estimate. The closer the linkage, the more accurate the prediction. This is facilitated if members of several generations of the family can be analyzed, so that the individual's maternal and paternal alleles can be identified. In other words, the *"phase is known."*

For example, if θ (the recombinant fraction) = 0.01, if the mating of his parents was AH/A′h × A′h/A′h (phase known), and if he had alleles A′A′ at the marker locus, his risk of having inherited the HD gene (H) would be 0.01. One percent is substantially less than the prior risk of 50%. Conversely, if he had been marker type AA′ his risk would have been 99%. In either case, the estimate would be more helpful.

The usefulness of linkage in genetic counseling depends on the mating types and the frequency of the alleles for the polymorphism in question. Once a tightly linked RFLP-disease combination is found, it is desirable to expand the RFLP into a highly polymorphic haplotype system and, also, to generate still more linked RFLPs, ideally including markers for loci that flank the gene. This increases the number of informative matings and improves the accuracy of genotypic determination. If flanking markers are available, false-negative predictions for the disorder could occur only if there were double crossovers between the two markers. If the markers are close to the disease gene, there is little chance of a double crossover. Conversely, however, single crossovers between flanking marker loci greatly reduce the usefulness of the loci.

We shall now discuss specific aspects of the use of DNA analysis for diagnosis and carrier detection of DMD. DNA technology can also be applied in a similar fashion to myotonic dystrophy with more than 98% accuracy and it can be used for other diseases, too. DMD exemplifies the process.

PROBLEMS IN PEDIGREE AND HAPLOTYPE ANALYSIS IN DMD

Application of linkage analysis to DMD diagnosis continues to improve. Still, there are some problems, some of a general nature and others specific for the DMD locus.

Problems in pedigree and haplotype analysis in DMD arise from several sources. One is lack or misidentification (e.g., nonpaternity) of one or more key members of a pedigree. Alternatively, DNA markers or CK data may be lacking, indeterminate, or conflicting. Finally, erroneous assumptions might be used in interpreting the data.

DNA marker information can be subdivided into RFLP and deletion data. Because the DMD gene is so large (9,10), informative markers flanking the DMD gene may be lacking, or there may be uncertainty about the location of "intragenic" marker loci in relation to the particular DMD gene alteration in that pedigree. Deletion data should suffice to detect affected males. Deletions involving the loci DXS 164, defined primarily by the pERT 87 series of probes (9,11), and DXS 206, defined by the XJ (12) series of probes, can at present be detected in 7–8% of DMD boys tested. However, a much higher percentage of DMD mutations may result from deletions yet to be detected; the number of deletions detected increases as new probes are developed. Deletion analysis is particularly useful in families with an isolated case of DMD, or if the pedigree is not suitable for linkage analysis. However, in the absence of suitable RFLPs for a deleted allele, *dosage blotting* (in which the number of gene copies present is estimated from the density of bands on an autoradiogram) is required to identify DMD heterozygotes (13). Also, deletions may be small, detected by only a subset of available markers.

Accessory data, such as CK measurement, may give false-positive or, more often, false-negative impressions. This must be considered in decision making, which may be difficult if CK and DNA data conflict.

DNA haplotype analysis may be confused by recombination between flanking markers. Alternatively, assumptions used to interpret haplotype data may be incorrect. A trivial example is the use of an incorrect map distance or lack of a reasonable estimate of the uncertainty of map distance. Another possibility is *recombinational interference* (which results if there are multiple recombination events between closely linked markers more often predicted by the rate of single recombination events); interference therefore distorts estimates of the likelihood of multiple recombinational events. Finally, mutations may occur in some parental germ cells but not leukocytes or other somatic tissues. This is termed *germline mosaicism,* which, whatever the cause, may distort risk estimates for either the proband or future family members.

In an ideal situation, CK data are accurate and several probes that flank the DMD locus are informative in all important family members. For most diseases, linkage analysis is dominated by the closest informative marker or, at least, the closest informative markers on either side of the gene. In DMD, intralocus data are also important. For example, flanking markers outside the gene might sense a region of unusual genetic instability, a *recombinational "hot spot"* near the DMD gene, but the evidence has been conflicting (14). The DMD gene may extend for at least 2 megabases or 2–4 cM. The "sidelines" of an informative marker within the gene and the side or extent of the genetic defect in a pedigree may be unknown; that circumstance would invalidate the *"point source" marker location* assumption, which implies that recombination does not occur within a genetic locus; no formal method has yet been developed for combining intralocus and extralocus data.

To minimize problems, it is advisable to use as many DNA probes as possible, including all the polymorphic pERT 84 and 87 probes (9,11), XJ probes (12), and at least two or three outside markers on either side of the DMD locus. More often than not, however, only a few DNA probes are informative. DNA information is combined with CK data for carrier detection and prenatal diagnosis.

HAPLOTYPE ANALYSIS BASED ON RFLPs

In nondeletion cases, RFLPs can still be useful in prenatal diagnosis of DMD. For example (Fig. 14-1), an analysis was performed when only pERT 87.8, 87.15, and

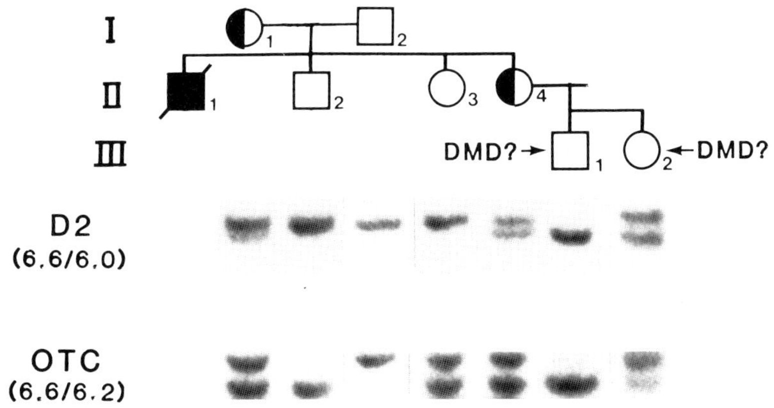

Fig. 14-1. Prenatal diagnosis of DMD in a fetus (III1) based on inheritance of DNA markers distal (D2) and proximal (OTC) to the DMD locus. The DNA blots are positioned in this and subsequent figures under the corresponding family member symbol.

87.1 were available for the DXS 164 locus. None of those probes was informative in that family.

I1 was almost certainly a carrier because she had high CK values and an affected son. In that family only two probes, D2 (distal to DMD locus, θ = 15 cM) and OTC (proximal to DMD θ = 10 cM) were informative. I1 was heterozygous for both markers. II4, the consultant, inherited the paternal X-chromosome that carried the 6.6-kb allele of D2 and the 6.6-kb allele of OTC. She inherited the maternal X-chromosome that carried the 6.0-kb allele of D2 and the 6.2-kb allele of OTC, both of which were coupled to DMD. III1 inherited the pair of alleles coupled with DMD, and had a 97% chance of inheriting the disease.

In a later pregnancy of II4, the fetus was a girl who inherited alleles that were identical to those in the maternal X-chromosome that was not coupled with DMD. She was therefore almost certainly ($p > 0.97$) not a DMD carrier.

If RFLPs are informative on only one side of the DMD locus, partial information can be obtained about the family (Fig. 14-2). In this second family only OTC, proximal to the DMD locus, was informative. The female fetus inherited the maternal X-chromosome that carried the OTC alleles that were not coupled with DMD; she was probably not a DMD carrier.

Problems arise in haplotype analysis for carrier detection and prenatal diagnosis if there is recombination between flanking markers. Even then, the outcome can be estimated, particularly if there is no recombination between the markers within the DMD locus. A more serious problem, although rare, is recombination between markers that are tightly linked to the DMD locus (14).

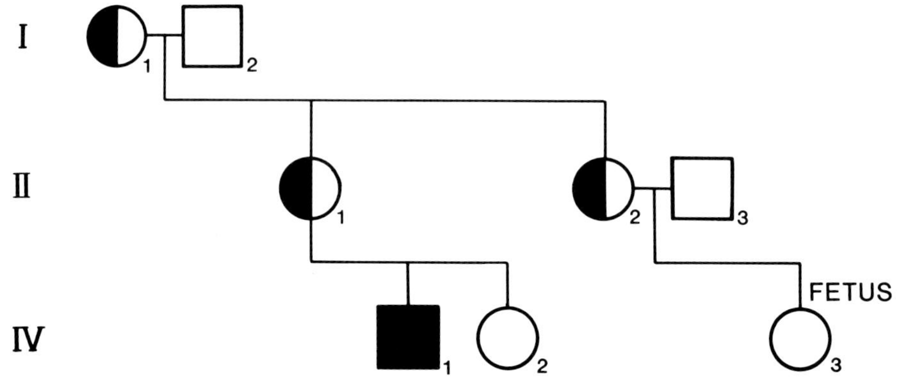

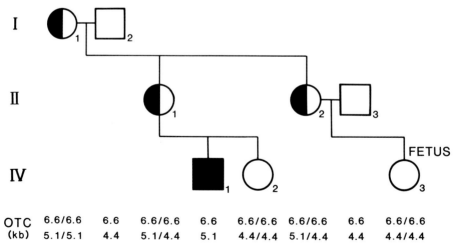

OTC	6.6/6.6	6.6	6.6/6.6	6.6	6.6/6.6	6.6/6.6	6.6	6.6/6.6
(kb)	5.1/5.1	4.4	5.1/4.4	5.1	4.4/4.4	5.1/4.4	4.4	4.4/4.4

Fig. 14-2. An example of RFLP analysis in which only one DNA marker (OTC) is informative (proximal to DMD locus) in a family with DMD. The X-chromosome carrying the alleles for OTC, 6.6 and 5.1, in affected boy IV1 is in coupling with DMD in I1, II1, and II2. The fetus IV3 inherits the X-chromosome carrying the alleles 6.6 and 4.4, which is not in coupling with DMD in the mother. She is homozygous for both alleles since her other X-chromosome from the father also carries the same alleles.

HAPLOTYPE ANALYSIS BASED ON DELETIONS; WITH OR WITHOUT RFLPs

Deletions that cause DMD vary in size and location. Therefore, all available pERT, XJ, and other DXS 164 and DXS 206 probes must be used in attempts to detect deletions (9,11). It is difficult to depend on deletions for carrier detection because the analysis depends on dosage analysis of hybridization intensity. For DNA dosage analysis, equal amounts of DNA must be applied to each lane on a Southern blot. Moreover, the same blot must be tested with a probe that is not on the X-chromosome to control for the amount of DNA loaded onto the gel (13). It is better to couple deletion analysis for carrier detection with RFLPs that lie within the deletion.

In contrast, deletion analysis is simple in affected boys, needing only the presence or absence of hybridization (12). In an example (Fig. 14-3), probe 87.15 was deleted in an affected boy (II2, lane 7). The fetus, III1, also lacked 87.15 sequences (lane 1). The extent of deletion was not known because the study did not include all pERT and XJ probes.

Some deletions exceed 200 kb. In one family with a DMD deletion (Fig. 14-4), it was simple to determine the carrier status of the mother because the probes that detected deletions in the affected boy were also associated with RFLPs in the mother. The affected boy, IV1, carried deletions from 87.JBir to 87.1 (lane 6) or about 200 kb. His mother, III4, had both alleles for probes 87.30, 87.15 (*Taq*I), 87.8 (*Bst*XI), and 87.1 (*Bst*NI) (lane 5). Therefore, the mother was probably not a DMD carrier, but she was advised to have prenatal diagnosis in future pregnancies because she could have carried the deletion in a mosaic pattern in her germ cells (15). That is, her somatic cells (lymphocytes and fibroblasts) did not show the deletion, but some of her oocytes might carry the deletion and she might be at risk for passing that deletion to a child in a future pregnancy.

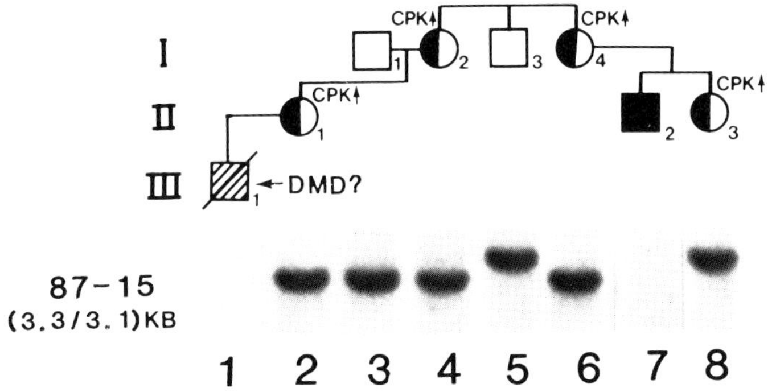

Fig. 14-3. Prenatal diagnosis of a DMD fetus based on deletion analysis. Sequences that are detected by probe 87.15 which is tightly linked to the DMD locus are absent in II2 (lane 7), the affected boy, and III1 (lane 1), the fetus, by lack of hybridization. Note the intensity of hybridization in males and females in this family for this probe is the same, suggesting that the females also lack the sequences on one of their X-chromosomes. Coupled with the elevated CK values, the females studied in this family also appear to have a deletion for this probe.

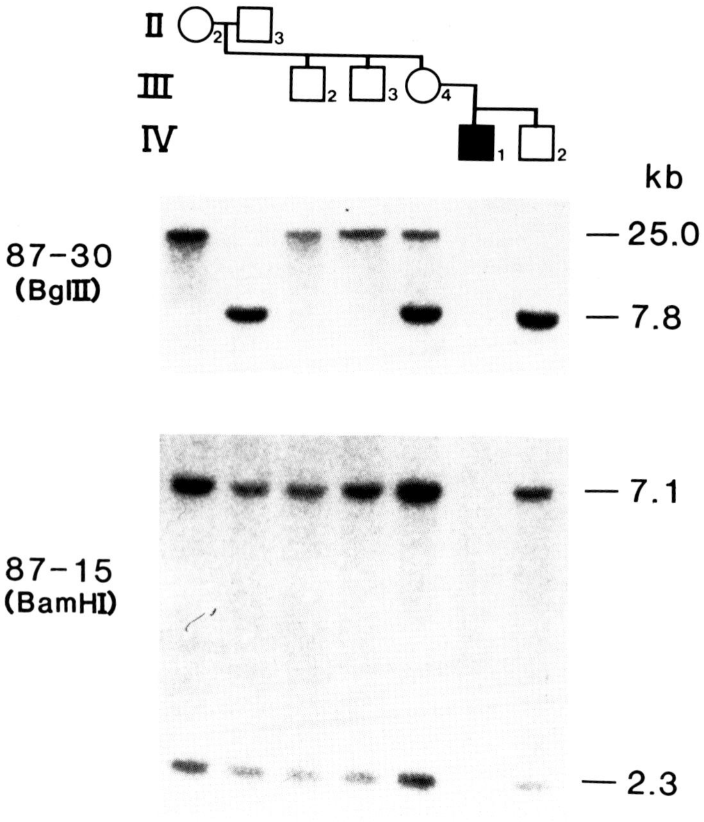

Fig. 14-4. Detection of carrier status involving a large deletion of sequences detected by several probes, some of which are associated with RFLPs (lanes from left to right correspond to 1 to 7). Note the absence of hybridization of probes 87.30 and 87.15 (A) and 87.1 and 87.JBir (B) in IV1 (lane 7). The mother III4 has both alleles for 87.30 and 87.15 by polymorphisms and for 87.1 and 87.JBir by dosage. Probe 87.42 (B), which is distal to 87.1, is not deleted in any of the family members.

In an example of a small deletion (Fig. 14-5), IV1, the affected boy (lane 5) showed a deletion for probes 87.8, 87.1, and 87.42 (spanning about 50 kb). If fewer pERT probes had been used to study that family, the small deletion might not have been recognized.

Haplotype analysis will improve in the future because more polymorphic probes are continually being isolated. If several probes are used to detect RFLPs in DNA cut by the same restriction enzyme, fewer blots are needed. Reproducible, sensitive nonisotopic labeling becomes increasingly important as the number of informative probes increases. The use of site-specific oligomers or DNA probes that detect a single base pair mismatch can also reduce uncertainty, particularly when a mutation affects several family members in more than one generation and, neglecting the remote possibilty of multiple new mutations, it is desired to examine DNA from other members of this pedigree. Ultimately, haplotype analysis will be replaced by a more direct analysis of the gene product. However, if the gene prod-

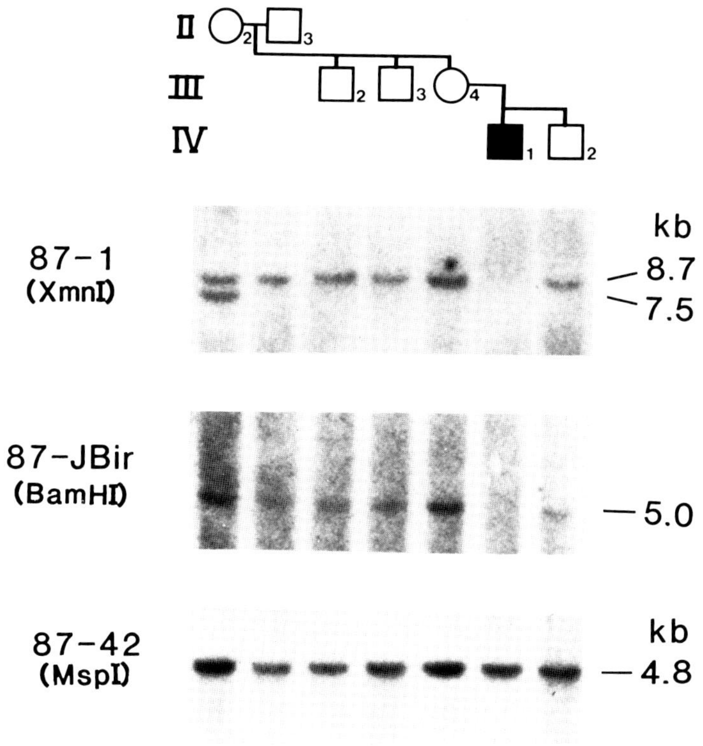

Fig. 14-4. *(Continued)*

uct is a muscle protein, DNA analysis may still be needed for prenatal diagnosis, because fetal muscle cannot be biopsied. Some means of "tricking" the DNA sequence into directing the synthesis of a muscle-specific protein (e.g., a tissue-specific factor acting on a gene enhancer) coupled with a functional or immunological assay for the gene product might become feasible.

BAYESIAN ANALYSIS AS APPLIED TO DNA DIAGNOSIS

If DNA probes always recognized the mutation causing a disease, or if the probes were extremely close to that locus, diagnostic accuracy would be 100%. However, for most diseases, we have to use probes that, although linked to the gene of interest, may recombine to cause an error in diagnosis. The impact of recombination on diagnostic accuracy varies with pedigree, genetic distance of the probes from the disease gene, or even with the population frequencies of alleles for the RFLPs. Diagnostic accuracy must therefore be calculated in each case.

Information from other sources may help diagnosis. For example, pedigree information can help determine carrier risks. Coagulation factor activity or CK activity may be needed. Linked probes can facilitate diagnosis only by demonstrating inheritance of the same allele carried by an affected relative diagnosed by tra-

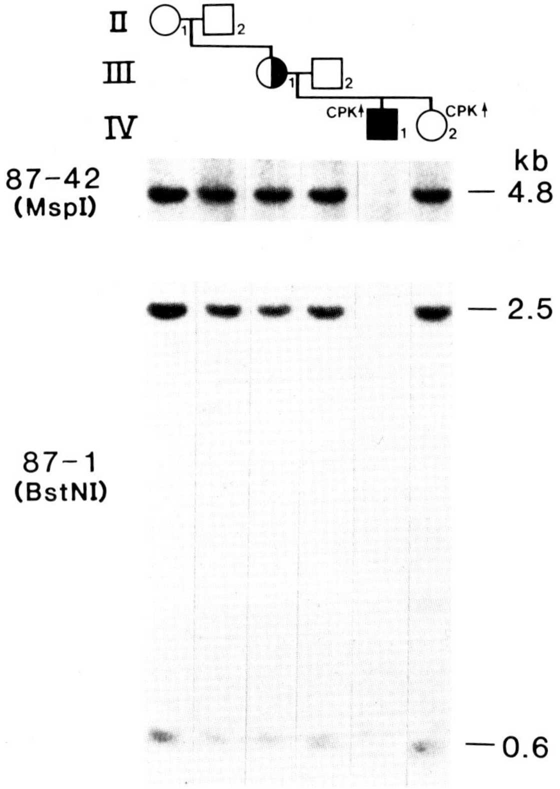

Fig. 14-5. An example of a relatively small deletion in a family with DMD. The deletion includes 87.8, 87.1, and 87.42. Note the absence of hybridization of probes 87.42 and 87.1 in the affected boy IV1 (lane 5). Lanes from left to right correspond to 1 to 5.

ditional criteria. Estimation of a new mutation rate may be needed to provide risk assessment in a lethal X-linked disease (17). These diverse types of information must be combined into a single risk estimate to provide the most accurate diagnosis.

One way to do this is by Bayesian analysis (18,19). The Reverend Thomas Bayes, an English clergyman, was a fellow of the Royal Society in the first half of the eighteenth century. He derived an algorithm for scientific inference, essentially a way to answer this question: What is the probability of an event A (e.g., the presence of a disease gene) or B (e.g., the absence of a disease gene) given the knowledge of an event C (e.g., an observable CK level) or event D (e.g., an RFLP result).

The probability of event A given C [$P(A|C)$], or the probability of event B given C [$P(B|C)$] is called a *conditional probability*. In this case, the set (A,B) is all inclusive. However, the mathematical treatment can be expanded to include a larger set of mutually exclusive events whose total probability sums to unity.

Bayes formalized an algorithm that can be used by physicians to make diagnostic decisions based on observables, e.g., clinical features and laboratory test results. That is, given a set of clinical and laboratory findings, what is the probability that they are due to one particular disease versus another (or a "normal" state)? That analysis can be used in a wide range of diagnostic and therapeutic decisions, and it is also excellent for analyzing data generated by linked probe studies.

Bayesian analysis uses two additional types of probability expressions. First is the *marginal probability*, P(A), P(B), P(C) etc. that a given event or state of affairs occurs at all. Second is the *compound* or *joint probability*, e.g., P(A and C), that two events occur together. This is equal to the product of the individual probability of one and the conditional probability of the second, given the first. The order in which these events occur is irrelevant. Hence;

$$P(\mathrm{A}) \text{ and } \mathrm{C} = P(\mathrm{A}) \times P(\mathrm{C} \mid \mathrm{A}) = P(\mathrm{C}) \times P(\mathrm{A} \mid \mathrm{C}) \tag{1}$$

also

$$P(\mathrm{B} \text{ and } \mathrm{C}) = P(\mathrm{B}) \times P(\mathrm{C} \mid \mathrm{B}) = P(\mathrm{C}) \times P(\mathrm{B} \mid \mathrm{C}) \tag{2}$$

By rearranging one obtains

$$P(\mathrm{A} \mid \mathrm{C}) = \frac{P(\mathrm{A})\, P(\mathrm{C} \mid \mathrm{A})}{P(\mathrm{C})}$$

the usual form of Bayes theorem, where P(C) equals the sum of the conditional probabilities considered. Here, to reiterate:

1. P(A) is referred to as the *prior probability* of hypothesis A. This is the probability without knowledge of event A. For DNA diagnosis, it could be taken from population figures such as carrier rates, mutation rates, or principles of Mendelian inheritance. In diagnostic analysis, the two hypotheses usually considered are the presence or absence of a mutant allele for a particular gene.
2. $P(\mathrm{C} \mid \mathrm{A})$ is the *conditional probability* of event C if A is true. In DNA diagnosis it might be the probability of a normal (or abnormal) laboratory value or an asymptomatic clinical state in individuals who carry the disease gene.
3. $P(\mathrm{A})\, P(\mathrm{C} \mid \mathrm{A})$ is the *joint probability* of A and C, whereas P(A) and P(C) are called the *marginal probabilities* that the individual events will occur.
4. $P(\mathrm{A} \mid \mathrm{C})$ is the *posterior probability* of A. It is adjusted so the sum of posterior probabilities of all hypotheses equals one, because one hypothesis is assumed to be correct.

If A and C are independent these simplify to

$$P(\mathrm{A} \text{ and } \mathrm{C}) = P(\mathrm{A}) \times P(\mathrm{C}) \text{ etc.}$$

and Bayesian analysis is not required.

Dividing the elements of Eq. (1) by those of Eq. (2), one can write

$$\frac{P(\mathrm{A})\, P(\mathrm{C} \mid \mathrm{A})}{P(\mathrm{B})\, P(\mathrm{C} \mid \mathrm{B})} = \frac{P(\mathrm{C})\ P(\mathrm{A} \mid \mathrm{C})}{P(\mathrm{C})\, P(\mathrm{B} \mid \mathrm{C})}$$

$$\frac{P(\mathrm{A} \mid \mathrm{C})}{P(\mathrm{B} \mid \mathrm{C})} = \frac{P(\mathrm{A})\, P(\mathrm{C} \mid \mathrm{A})}{P(\mathrm{B})\, P(\mathrm{C} \mid \mathrm{B})}$$

where $P(\mathrm{A} \mid \mathrm{C}) + P(\mathrm{B} \mid \mathrm{C}) = 1$. The ratio $P(\mathrm{A})/P(\mathrm{B})$ reflects the ratio of a priori probabilities. The ratio $P(\mathrm{A} \mid \mathrm{C}/P(\mathrm{B} \mid \mathrm{C})$ contains the ratio of a posteriori probabilities. $P(\mathrm{C} \mid \mathrm{A})/P(\mathrm{C} \mid \mathrm{B})$ is a measure of the information contained in the test, "C."

A few additional points deserve mention before considering examples. First, computer programs can assist in risk calculations for X-linked diseases (6,7,20–22), especially LIPED (6), and LINKAGE (7). Second, our pedigree examples are relatively simple and designed to illustrate the Bayesian method. A more comprehensive treatment of complex pedigrees for X-linked diseases is available (23).

If D is a second, independent observable, e.g., from linkage, the calculation is as follows:

$$\frac{P(\mathrm{A}|\mathrm{C,D})}{P(\mathrm{B}|\mathrm{C,D})} = \frac{P(\mathrm{A})}{P(\mathrm{B})}\,\frac{P(\mathrm{C}|\mathrm{A})}{P(\mathrm{C}|\mathrm{B})}\,\frac{P(\mathrm{D}|\mathrm{A})}{P(\mathrm{D}|\mathrm{B})}$$

where

$$P(\mathrm{A}|\mathrm{C,D}) + P(\mathrm{B}|\mathrm{C,D}) = 1$$

Example 1

This is one of the simplest types of cases to analyze (Fig. 14-6A). The mother, individual 2, is an obligate carrier because she has an affected brother and son. Her daughter, individual 5, is initially at 50% risk for inheriting the DMD allele, i.e.,

$$\frac{P(\mathrm{A})}{P(\mathrm{B})}\,\frac{0.5}{0.5} = 1$$

She has inherited the same pERT 87 allele as the affected son (event D), a situation with a 95% probability if the daughter is a carrier [$P(\mathrm{D}|\mathrm{A}) = 0.95$] because the pERT 87 probe location does not recombine with an arbitrary DMD mutation approximately 95% of the time (24). Similarly, inheritance of the risk pERT 87 allele would be expected 5% of the time if the daughter is not a carrier or $P(\mathrm{D}|\mathrm{B}) = 0.05$.

However, this woman (#5) has normal CK values (event C), which is found in about one-third of obligatory DMD carriers $P(\mathrm{C}|\mathrm{A}) = 1/3$. Normal CK values also occur in 95% of noncarriers so that $P(\mathrm{C}|\mathrm{B}) = 0.95$. To take both factors into account, we modify initial risk [$P(\mathrm{A})$] by the CK information $P(\mathrm{C}|\mathrm{A})$ and the link-

A.

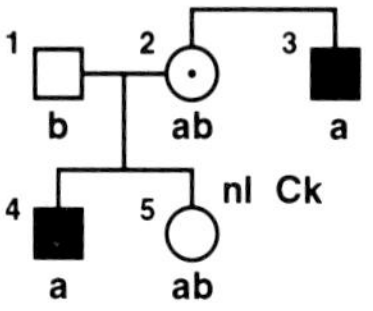

	Carrier		Noncarrier	
Prior Probability	1/2		1/2	
CK	1/3		.95	
DNA	.95		.05	
Compound Probability	.158	+	.024	= .182
Posterior Probability	.158/.182	= .87	.024/.182	= .13

Fig. 14-6. Hypothetical pedigrees (A–D) for Bayesian analysis. The compound probability is often referred to as the joint probability. See text for details.

B.

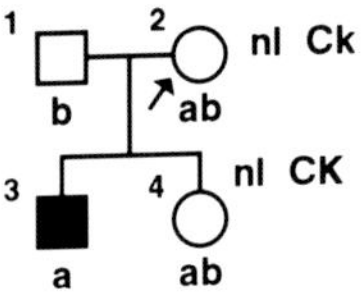

Mother	Carrier		Noncarrier	
Prior Probability	2/3		1/3	
Normal CK	1/3		.95	
Compound Probability	.22	+	.32	= .54
Posterior Probability	$\frac{.22}{.54}$	= .41	$\frac{.32}{.54}$	= .59

Note: Daughters normal CK will be used in the next calculation.

C.

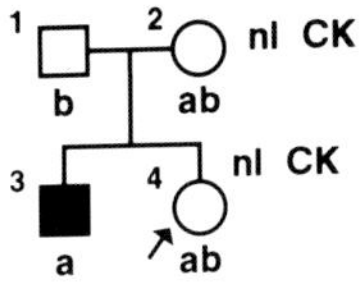

Daughter	Carrier		Noncarrier	
DNA	.905(.41)		.59 + (.41).095	
Normal CK	1/3		.95	
Compound Probability	.12	+	.60	= .72
Posterior Probability	$\frac{.12}{.72}$	= .17	$\frac{.60}{.72}$	= .83

D.

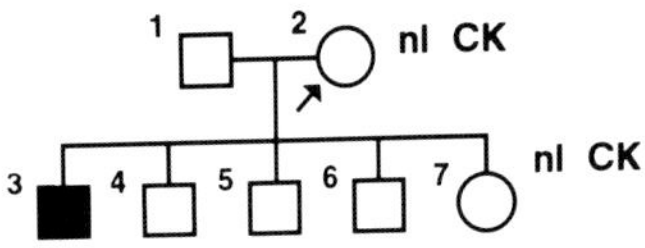

	Carrier		Noncarrier	
Prior Probability	2/3		1/3	
Normal CK	1/3		.95	
3 Unaffected Sons	$(1/2)^3$		1	
Daughter with nl CK	.65		.95	
Compound Probability	.018	+	.30	= .318
Posterior Probability	$\frac{.018}{.318}$	= .06	$\frac{.30}{.318}$	= .94

age information $P(D|A)$ by taking the product of their probabilities [$P(C|A)$ and $P(D|A)$]. This results in the compound conditional probability, $P(A|C,D)$, divided by $P(C)P(D)$ (which cancel out later). Similarly

$$\frac{P(B|C,D)}{P(C)P(D)} = P(B)\ P(C|B)\ P(D|B)$$

Since $P(A|C,D) + P(B|C,D) = 1$, and $P(C)\ P(D)$ is a common factor, the probability products can then be adjusted so that their sum is one. In doing so we derive the posterior probability that individual #5 is a carrier (87%) $P(A|C,D)$ or is not a carrier (13%) $P(B|C,D)$.

Example 2

This situation (Fig. 14-6B) is slightly more complex, and demonstrates two new points. Individual 3 is an isolated case of DMD, and we must determine how this fact affects the carrier risk for individual 4. In addition, the carrier risk for the mother, individual 2, must be calculated to use the information in calculating her daughter's carrier risk. The a priori risk that individual 2 is a carrier, assuming that DMD is a genetic lethal and that the male and female mutation frequencies are approximately equal, is 67%. This must be modified by the effect of a normal CK value on her risk. Her carrier risk could also be decreased because she has a daughter with normal CKk values (see below).

However, we are interested in deriving the daughter's risk, and the daughter's CK value will be used in the second step of the calculation. Therefore, it will not be included here. This is an important general point: Information is used only one in a risk calculation—it should not be given a "double impact." When these factors are considered in calculating the mother's risk, as we did in Example 1, we obtain the a posteriori probability of 41% that she is a carrier. This is then used to calculate the daughter's risk (Fig. 14-6C).

The daughter, individual 4, inherited the same pERT 87 allele as her affected brother, individual 3. She would be a carrier if her mother is a carrier and if there has been no recombination event between the pERT 87 locus and DMD. The individual probability that the mother is a carrier is 0.41 and the individual probability that there has not been a single recombination is 0.905. The joint probability is 0.371. The value 0.905 is used here rather than 0.95, the recombination frequency for pERT 87 and DMD, because a recombination event could occur in either individual 3, 4, or both. Thus, 0.905 ($= 0.95 \times 0.95, + 0.05 \times 0.05$), which estimates the probability that a single recombination event has not occurred, is the appropriate value. This is always true if the phase relationship between DMD and a pERT 87 allele is set only by one child. If individual 2 is not a carrier or if there has been a recombination event, individual 4 would probably not be a carrier. In addition, individual 4 has normal CK values and these modify her risk to the final estimated value of 0.17.

Example 3

Normal sons and daughters affect the carrier risks for DMD. In this family (Fig. 14-6D) the a priori probability that the mother of a boy with DMD is a carrier is

0.67. However, this mother also has three unaffected sons and a daughter with normal CK values. If she were a carrier, the probability that each son is not affected is 0.5. The probability that her daughter is a noncarrier would also be 0.5, but the probability that a carrier would have a normal CKk value would be roughly one-third. The overall probability would be about 0.65 that her daughter would have normal CK values. If the mother were not a carrier, the probability that her other sons would be affected would be close to zero, and the probability that her daughter would have normal CK values would be about 95%. If these factors are combined in the usual fashion, there is an overall probability of 6% that the mother is a carrier.

These risks would change if DNA analysis were carried out. If the mother were a carrier and if all four siblings had inherited the normal maternal pERT allele (not the one with the DMD mutation), then the probability that each would be unaffected or a noncarrier would be 95% even if the mother were a carrier. The carrier risk of the mother would be close to that calculated in example 2. In contrast, if two unaffected sons inherited the same maternal pERT 87 allele as the affected boy, and if the third unaffected son inherited the alternative maternal pERT 87 allele, the carrier risk for individual 2 would be less than 1%. Thus, DNA analysis of normal siblings can raise or lower the carrier risk from original estimates.

Comments

There are limits in applying Bayesian analysis to prenatal diagnosis. The calculations include data from different sources and of varying reliability. A 5% risk of DMD might be considered insignificant in statistical terms, but 1 of 20 patients will give birth to an affected boy. For them, it is as though the risk had been 100% during the pregnancy. Instead of quoting statistical risks, counselors might advise that risks are low, moderate, and high. Similarly, these risks are subject to compound uncertainty limits (difficult to calculate) because there is some uncertainty in the linkage and other types of data.

If possible, we should use flanking probes that increase the accuracy of diagnosis by pointing out recombination events in the area. When specific genes causing a disease can be used in diagnosis the accuracy increases dramatically, although DMD, which can undergo recombination within the gene itself, may prove to be more difficult than most other diseases in this respect. It is probably best to consider risk calculation by Bayesian analysis an interim solution, to be used with caution, and only until a means of arriving at a more accurate diagnosis can be achieved.

REFERENCES

1. Conneally PM, Rivas ML. Linkage analysis in man. In Harris H., Hirschhorn K., eds. Advances in Human Genetics, Vol 10. New York: Plenum, 1982, pp 209–266.

2. McKusick VA. Mendelian Inheritance in Man. Baltimore: The Johns Hopkins University Press, 1986.

3. Morton NE. Sequential tests for the detection of linkage. Am J Hum Genet 1955; 7:277–318.

4. Conneally PM, Edwards HJ, Kidd KK, Lalouel JM, Morton NE, Ott J., White R. Report of the committee on methods of linkage analysis and reporting. Cytogenet Cell Genet (Helsinki Conf) 1985; 40:356–360.

5. Chakravarti A, Buetow KH. A strategy for using multiple linked markers for genetic counseling. Am J Hum Genet 1985; 37:984–997.

6. Ott J. Estimation of the recombination fraction in human pedigrees: Efficient computation of the likelihood for human linkage studies. Am J Hum Genet 1974; 26:588–597.

7. Lathrop GM, Lalouel JM. Easy calculations of lod scores and genetic risks on small computers. Am J Hum Genet 1984; 36:460–465.

8. Ott J. Analysis of Human Genetic Linkage. Baltimore: The Johns Hopkins University Press, 1985.

9. Monaco AP, Bertelson CJ, Middlesworth W, Colletti CA, Aldridge J, Fischbeck KH, Bartlett R, Pericak-Vance MA, Roses AD, Kunkel LM. Detection of deletions spanning the Duchenne muscular dystrophy locus using a tightly linked DNA segment. Nature (London) 1985; 316:842–845.

10. Monaco AP, Neve RL, Colletti-Feener C, Bertelson CJ, Kurnit DM, Kunkel LM. Isolation of candidate cDNA clones for portions of the Duchenne muscular dystrophy gene. Nature (London) 1986; 323:646–650.

11. Kunkel LM. Analysis of deletions in the DNA of patients with Becker and Duchenne muscular dystrophy. Nature (London) 1986; 322:73–77.

12. Ray PN, Belfall B, Dukff C, Logan C, Kean V, Thompson MW, Sylvester JE, Borski JL, Schmickel RB, Worton RG. Cloning of the breakpoint of an X:21 translocation associated with Duchenne muscular dystrophy. Nature (London) 1985; 318:672–675.

13. Tantravahi U, Kirschner DA, Beauregard L, Page L, Kunkel LM, Latt SA. Cytological and molecular analysis of 46, XXq-cells to identify a DNA segment that might serve as a probe for a putative human X chromosome inactivation center. Hum Genet 1983; 64:33–38.

14. Bertelson CJ, Bartley JA, Monaco AP, Colletti-Feener C, Fischbeck K, Kunkel LM. Localization of Sp21 meiotic exchange points in Duchenne muscular dystrophy families. J Med Genet 1986; 23:531–537.

15. Lanman JT, Pericak-Vance MA, Bartlett RJ, Chen JC, Yamaoka L, Koh J, Speer MC, Hung WY, Roses AD. Familial inheritance of a DXS164 deletion mutation from a heterozygous female. Am J Hum Genet 1987; 41:138–144.

16. Wood DS, Zeviani M, Prelle A, Bonilla E, Salviati G, Miranda AF, DiMauro S, Rowland LP. Is nebulin the defective gene product in Duchenne muscular dystrophy? New Engl J Med 1987; 316:107–108.

17. Caskey CT, Nussbaum RL, Cohan LC, Pollack L. Sporadic occurrence of Duchenne muscular dystrophy: Evidence for new mutation. Clin Genet 1980; 18:329–341.

18. Murphy EA, Chase GA. Principles of Genetic Counselling. Chicago: Year Book Medical Publishers, 1975, pp 70–74.

19. Thompson JS, Thompson MW. Genetics in Medicine, 3rd ed. Philadelphia: WB. Saunders, 1980, pp 340–343.

20. Sarfarazi M, Williams H. A computer program for estimation of genetic risk in X linked disorders, combining pedigree and DNA probe data with other conditional information. J Med Genet 1986; 23:40–45.

21. Clayton JF. A computer program to calculate risk in X linked disorders using multiple marker loci. J Med Genet 1986; 23:35–39.

22. Andrews DF, Brasher PMA, Manchester KE, Percy ME, Rusk ACM, Soltan HC, Trueman DW. DUCHEN: An interactive computer program for calculating heterozygosity

(carrier) risks in X-linked recessive lethal diseases, and its application in Duchenne muscular dystrophy. Am J Med Genet 1986; 25:211–218.

23. Clark AG. The use of multiple restriction fragment length polymorphisms in prenatal risk estimation. I. X-linked diseases. Am J Hum Genet 1985; 37:60–72.

24. Fischbeck KH, Ritter AW, Tirschwell DL, et al. Recombination with pERT 87 (DXS164) in families with X-linked muscular dystrophy. Lancet 1986; 2:104.

III

CLONED GENES FOR HUMAN NEUROLOGIC DISEASES

15

Lesch–Nyhan Syndrome: Human Hypoxanthine Phosphoribosyltransferase Deficiency

R. A. GIBBS, P. I. PATEL, J. T. STOUT, AND C. THOMAS CASKEY

Hypoxanthine phosphoribosyltransferase (HPRT) is an important enzyme in purine metabolism, where it participates in the salvage of hypoxanthine and guanine for the synthesis of DNA and RNA. Deficiency of HPRT is therefore associated with abnormalities of purine metabolism, including excess uric acid production and hyperuricemia. In patients with partial HPRT activity, this leads to the painful condition of gouty arthritis (1). Total HPRT deficiency results in the Lesch–Nyhan (LN) syndrome, a severe neurological disorder that is characterized by self-mutilation, choreoathetosis, mental retardation, and spasticity (2,3). Little is known about the pathophysiology of the LN syndrome, and how the metabolic abnormality causes neurological disorder. However, HPRT deficiency stands out from other heritable neurological disorders because so much is known about the molecular basis of HPRT gene expression in both normal and HPRT mutant cells. The LN syndrome is therefore a useful model to examine the neurological effects of a defect in a single gene.

A key development in the expansion of information about HPRT expression was molecular cloning of the HPRT gene (4). Cloned HPRT DNA has been used in studies of induced and spontaneous HPRT mutations in cultured cells (5,6) to assess the nature of somatic mutations leading to the HPRT$^-$ phenotype in circulating T lymphocytes (7,8), to examine the role of DNA methylation in X-chromosome activation (9,10), and, most recently, to attempt to modify HPRT gene expression by the use of antisense HPRT RNA (11) and the introduction of HPRT genes in retroviral vectors for gene therapy (see Chapter 31 by MacGregor et al.).

This chapter briefly reviews the clinical and neurological aspects of LN, describes the molecular structure and pattern of expression of the human HPRT gene, and provides an up-to-date survey of the diagnostic potential of new molecular methods for the detection of HPRT deficiency.

GENETICS AND CLINICAL PATHOLOGY OF THE LESCH–NYHAN SYNDROME

The HPRT gene locus maps to the distal end of the long arm of the human X-chromosome (12). LN is a consequence of HPRT deficiency and, hence, the disease has an X-linked, recessive mode of inheritance. Female carriers are unaffected, and, with a single exception, all affected individuals have been males (13). The disease occurs with a frequency of about 1 in 100,000 live births (14). This low frequency is, at least in part, a consequence of the genetically lethal nature of the disease (15). Some affected individuals have had parents who are not carriers of the disease, indicating that LN is a good model to study new mutations in the human germline. The disease occurs with equal frequency in a wide range of racial groups, and there is no greater frequency in inbred populations (14). This pattern differs from some other human X-linked diseases that are the result of persistent mutant alleles. Cystic fibrosis, for example, occurs at a frequency of 1 in 2000 and shows a marked preference for white populations (16).

The biochemical abnormalities associated with LN are usually evident at birth and lead directly to hyperuricacidity, uric acid crystals, and stone formation. Developmental delay is evident by age 3–6 months. Signs of corticospinal and extrapyramidal motor syndromes are apparent before age 1 year. The first signs are usually fine athetoid movements of the hands and feet. Later, these movements have a choreic character, similar to those seen in patients with caudate and putamen lesions. Death usually occurs in the second or third decade from infection or renal failure. There have been no reports of gross anatomic abnormalities at autopsy of patients with the LN syndrome.

However, three lines of evidence implicate cells of the basal ganglia in pathogenesis. First, dissection and biochemical analysis of normal postmortem human brain tissue reveal high levels of HPRT in the area of the basal ganglia (17–20). Conversely, normal levels of amidophosphoribosyltranferase (AMPRT, the rate-limiting enzyme of *de novo* purine synthesis) are lowest in this area, suggesting that this region may be particularly dependent on purine salvage (17).

Second, the involuntary movements in other diseases of the extrapyramidal motor system are essentially the same as those in the LN syndrome: postural tremor, chorea, athetosis, and dystonia. Normally, there is a balance in the activities of dopaminergic suppressor and facilitatory mechanisms of the extrapyramidal motor systems. Destruction of one extrapyramidal nucleus could cause an imbalance of these mechanisms, leading clinically to a specific form of involuntary movement. Chorea and athetosis are seen with lesions of the caudate and putamen (21,22).

Third, postmortem analysis of brain tissue from LN patients showed that indices of dopamine function were reduced by 70–90% in the basal ganglia (23).

Although these lines of evidence suggest involvement of the basal ganglia in the LN syndrome, it is not known whether glial cells or neurons are primarily affected. In the brain of one LN patient, there was demyelination in the area of the basal ganglia and cerebellum (24). Degeneration of the cerebella granule and Purkinge cells has also been reported (25). It is not known whether these changes are caused directly by the biochemical disorder or whether they are secondary effects of the hyperuricemia or other biochemical abnormalities.

HYPOXANTHINE PHOSPHORIBOSYLTRANSFERASE

In normal individuals, HPRT is a soluble cytoplasmic enzyme with a subunit size of about 24,500 daltons. The active protein is a tetramer and the normal HPRT amino acid sequence has been determined (26,27). Most LN patients totally lack or have negligible levels of HPRT protein in all tissues. A few patients have had HPRT antibody cross-reacting material (CRM) but no HPRT enzyme activity. Single amino substitutions have been identified by peptide sequencing in least four of the CRM-negative cases (28).

Molecular cloning of the HPRT gene has improved prospects for the characterization of LN cases in which no HPRT protein is present. Clones of cDNA for HPRT were first obtained from a mouse neuroblastoma cell line (NBR4) that overproduced mRNA for HPRT 20- to 50-fold (29). First, a cDNA library was constructed using mRNA from the NBR4 cells. Next, probes were prepared from normal and NBR4 mRNA and used for differential hybridization to the NBR4 cDNA recombinant library (4). The cDNA clones that were so isolated were then used to obtain cDNAs corresponding to sequences that encoded Chinese hamster and human HPRT (30,31). A human HPRT cDNA clone was also obtained by a method of serial gene transfer (32). Each of the cDNAs has 651 nucleotides that can code for a protein that contains 217 amino acid residues, and the predicted human cDNA encoded peptide conforms precisely to the amino acid sequence described by Wilson et al. (26).

The HPRT cDNA clones have been used to characterize the genomic organization of the murine and human HPRT coding sequences (33–36). The human gene is distributed in nine exons, and each of the exon/intron boundaries has been sequenced. The total size of the human HPRT gene is about 44 kb (Fig. 15-1). The promoter region of the human gene has been examined extensively and start sites for HPRT transcription have been identified. The HPRT promoter is unusual in that it lacks some commonly found sequences (CAAT box, TAATA box), but it does contain another, less frequently found sequence (CCGCCC, GC box) that may substitute for the function of governing the efficiency of the initiation of transcription to mRNA.

Functional analysis of the DNA sequences flanking the 5′ end of the human HPRT gene has been achieved by construction of expressing HPRT minigenes that

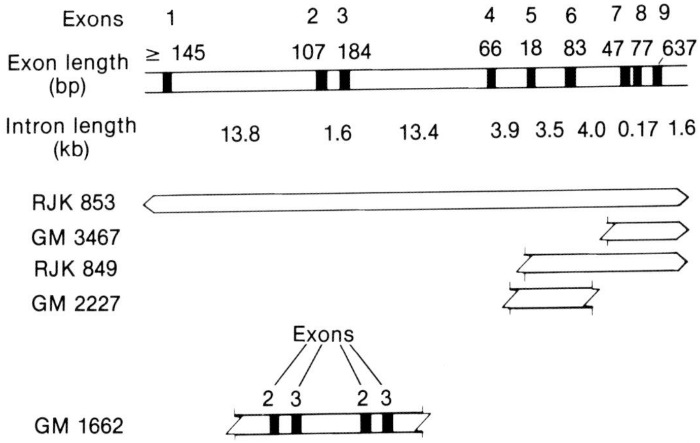

Fig. 15-1. HPRT gene structures in normal humans (top) and in Lesch–Nyhan patients. The bars indicate the regions of gene deletion in RJK 853, GM 3467, and RJK 849 and the regions of alteration or duplication in GM 2227 and GM 1662, respectively (39).

contain varying lengths of 5′ sequence and are linked to the HPRT coding DNA. Introduction of these minigenes into $HPRT^-$ cells by calcium-phosphate-mediated gene transfer or microinjection localized the promoter element to within 200 bp of the initiation codon, and also identified a putative negative regulatory element immediately upstream of the promoter region. The tissue-differential expression of HPRT with different levels of enzyme activity in different organs may be regulated by the interaction of trans-acting factors with these regulatory sequences (37).

MOLECULAR DIAGNOSIS OF LESCH–NYHAN

Several mutant HPRT genes have now been examined at the nucleic acid level by Southern blotting (38). This procedure can identify any DNA alterations that result in a size change of greater than 50–200 bp in the DNA fragments generated by restriction endonuclease digestion of normal DNA. Large gene deletions are readily detected, but single base changes are found only if they fortuitously occur in known restriction enzyme recognition sites. In analysis of LN patients (more than 85%), there were no detectable Southern blotting changes in more than 85% of the cases (39). In 15% of cases, there were molecular lesions ranging from small to complete deletions or duplications of DNA sequences (Fig. 15-1). Most of the LN cases that showed no molecular alteration by Southern blotting also show a normal mRNA by Northern blotting, a technique that can resolve mRNA changes of more than 50–200 bases. Together, these observations suggest that point mutations may be the predominant lesions leading to LN.

Currently, the method of choice for diagnosis of the LN syndrome is an assay for HPRT activity using radioactive hypoxanthine as a substrate (40,41). The assay is sensitive and requires only a few milligrams of tissue, including peripheral white blood cells, amniocytes, or chorionic villi. The enzyme assay is less useful for the detection of gene carriers. The female carriers are genetically mosaic; after inactivation of one X-chromosome early in development, the active X-chromosome in some cells is the one that bears the mutant HPRT allele, but, in other cells, the normal X-chromosome is active. Therefore, several different tissue sites must be sampled to obtain an indication of the HPRT activity status of clonal cell populations. Perhaps, surprisingly, lymphocyte clones from suspected carrier blood samples cannot be used as a representative source of test material, because there seems to be an *in vivo* selection of lymphocytes that have normal HPRT activity (42). The most successful approach has been hair follicle assay; several individual hair follicles are assayed to record the distribution of HPRT activity. Bimodal activity distributions indicate that some cells have low HPRT levels, indicating that individual is a carrier (43).

The diagnosis of LN by recombinant DNA methods has several advantages over enzyme assays. DNA techniques give qualitative information about the genetic status of the patient or carrier while the enzyme assays must be carefully interpreted. DNA diagnosis is now widely used, and can be compared to protein activity measurements. Unfortunately, the frequency, predominance, and the heterogeneous nature of new HPRT mutations have hampered the adaptation of recombinant DNA techniques for the routine diagnosis of LN and identification of LN carriers.

The most frequently used approach for DNA diagnosis is association of the mutant allele with a restriction fragment length polymorphism (RFLP) (44). In spite of extensive search, few RFLPs have been identified in tight linkage with the human HPRT gene locus, and the ones that are available are rarely informative. The alternative approach, direct examination of HPRT alleles is useful only in the small fraction of cases with an HPRT mutation that can be detected by Southern or Northern blotting. Accordingly, we have been exploring methods to facilitate identification of "small" genetic lesions, that are below the resolution of Southern or Northern blotting.

Three recently described techniques may improve DNA diagnoses of HPRT deficiency. Two procedures, ribonuclease A (RNase A) cleavage (45,46) and denaturing gradient gel electrophoresis (47), can detect point mutations in mammalian cell DNA or RNA without prior knowledge of the precise location of the substitutions. The third procedure, the polymerase chain reaction (PCR) for the amplification of DNA (48), can be used to enrich total genomic DNA samples for HPRT gene sequences, improving the sensitivity of subsequent assays for HPRT mutations. We have focused attention on the RNase A cleavage procedure, which has allowed a considerable improvement in the fraction of LN cases that can be diagnosed at the molecular level.

In RNase A cleavage analysis, radiolabeled RNA is used to probe complementary RNA or DNA. The hybrid molecules are then treated with RNase A, which digests single-stranded regions of RNA, but does not degrade double-stranded hybrids. Single base mismatches, within the double-strand regions, are cleaved in 30–50% of cases. Deletions of one or more bases cause the probe to loop out, in a single strand and therefore also result in cleavage. The RNase A cleavage assay provides a test for exact complementarity of a particular nucleotide sequence and a radiolabeled probe.

We examined RNA from 14 LN cases, using radiolabeled antisense RNA probes that had been generated from HPRT cDNA and cloned in transcription vectors. Five samples revealed distinctive RNase A cleavage patterns (Fig. 15-2), and two of the five were found to have deletions in the HPRT message; the other three were either HPRT point mutations or deletions too small to be recognized (Fig. 15-3). These 14 cases were chosen because they showed no alterations in the Southern or Northern blotting patterns. The RNase A cleavage assay, in combination with the previous Southern and Northern analysis, therefore brought the total fraction of LN cases directly diagnosed by DNA analysis to 50% (49).

These developments have improved the predictive assessments for LN families. Direct application of RNase A cleavage analysis of HPRT RNA to identify LN carriers is not yet a practical alternative to the HPRT enzyme assays, but it does make it possible to localize the genetic lesion to a small part of the mutant HPRT gene in families if samples are available from a proband. This information can then be used to expedite later prenatal diagnosis or carrier assessment with DNA techniques. If a single base substitution in the HPRT message is identified in an LN patient, for example, it is possible to synthesize an oligonucleotide that spans the mutation, but is complementary to the normal sequence. The oligonucleotide can

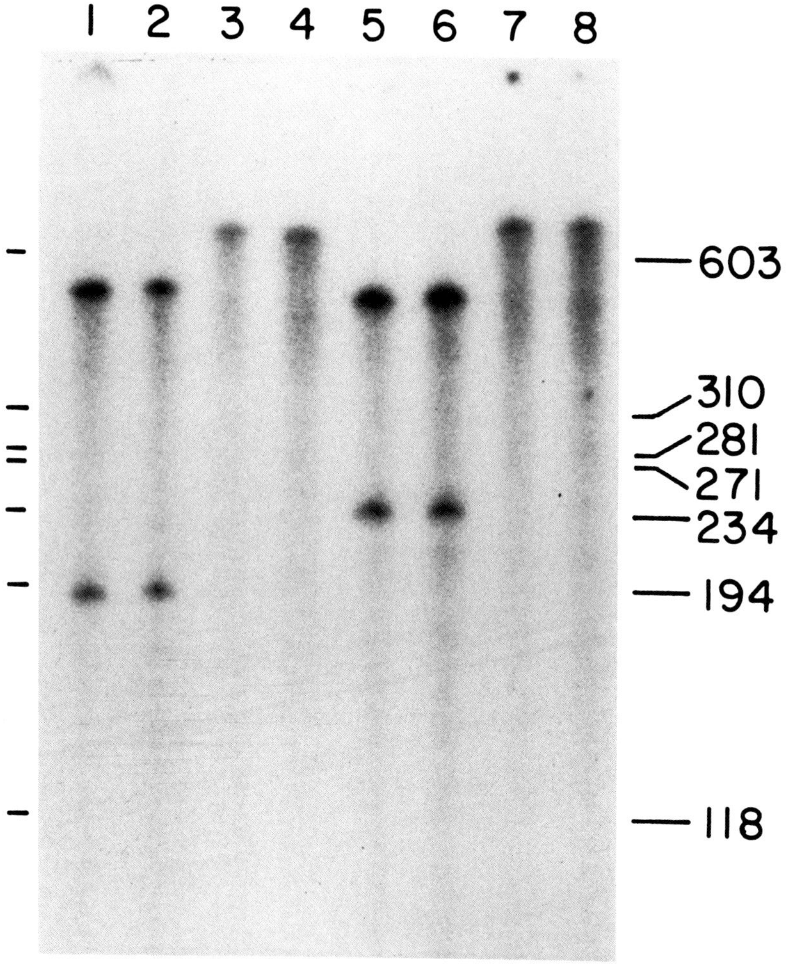

Fig. 15-2. RNase A cleavage analysis of RNA from four HPRT-deficient individuals. Of total RNA isolated from lymphoblastoid cells 100 μg was hybridized to an antisense riboprobe representing 685 bp of HPRT coding sequence as previously described (49). The hybridization mixture was passed through messenger-affinity paper (Amersham), the hybrids eluted with water, and digested with either 2 μg/ml (lanes 1, 3, 5, 7) or 10 μg/ml (lanes 2, 4, 6, 8) of RNase A. The protected fragements were analyzed on a 5% acrylamide:7 *M* urea gel. Lanes 1, 2; RJK 894; lanes 3, 4; RJK 896; lanes 5, 6; RJK 906; lanes 7, 8; RJK 950. Complete cleavage at the site of mutation occurred with RNA from RJK 894 and 906 whereas no cleavage was observed with RNA from RJK 896 and 950.

then be used to discriminate mutant and normal sequences by differential hybridization under stringent washing conditions (50).

Other developments in techniques for identifying nucleic acid changes are likely to improve the ease and certainty of genetic assessment for LN. Our long-term goal is to provide accurate and rapid diagnosis of LN patients and carriers, without necessarily involving any family members other than the individual subject. To this end we are now exploring applications of RNase A cleavage analysis and denatur-

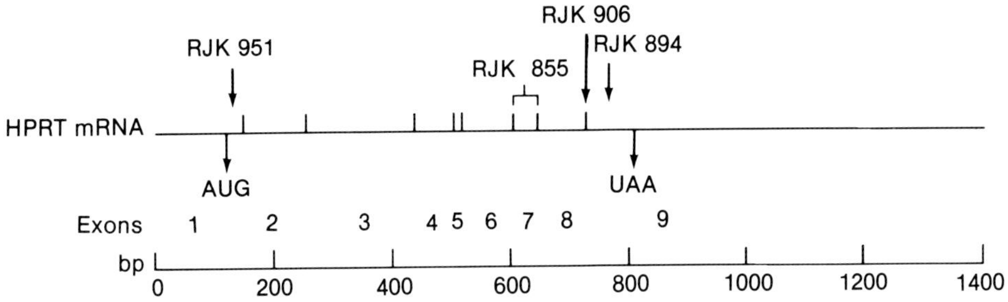

Fig. 15-3. Approximate location of mutations within the HPRT coding sequence of four HPRT-deficient humans identified by RNase A cleavage analysis conducted as described in the legend to Fig. 15-2.

ing gradient gel electrophoresis directly to DNA, so that HPRT mutations can be easily identified.

SUMMARY

The LN syndrome is a rare but severe genetic disease and no cure is yet available. The defective gene has been thoroughly characterized in some LN patients, using nucleic acid probes for the HPRT gene locus. Diagnosis of LN and assessment of carrier status are now achieved by HPRT enzyme assays. Efforts to improve diagnosis by recombinant DNA methods have been hampered by several factors, including the heterogeneity of mutations and the high incidence of HPRT mutations that are not detected by Southern blotting. The development of new techniques to detect *de novo* point mutations, such as RNase A cleavage, is the basis for optimism about improving diagnosis of carriers and affected individuals.

REFERENCES

1. Kelley WN, Rosenbloom FM, Henderson JF, Seegmiller JE. A specific enzyme defect in gout associated with overproduction of uric acid. Proc Natl Acad Sci USA 1967; 57:1735–1739.

2. Lesch M, Nyhan WL. A familiar disorder of uric acid metabolism and central nervous system function. Am J Med 1964; 36:561–570.

3. Seegmiller JE, Rosenbloom FM, Kelley WN. Enzyme defect associated with a sex-linked human neurological disorder and excessive purine synthesis. Science 1967; 155:1682–1684.

4. Brennand J, Chinault AC, Konecki DS, Melton DW, and Caskey CT. Cloned cDNA sequences of the hypoxanthine guanine phosphoribosyltransferase gene from a mouse neuroblastoma cell line found to have amplified genomic sequences. Proc Natl Acad Sci USA 1982; 79:1950–1954.

5. Fuscoe JC, Fenwick RL, Ledbetter DM, Caskey CT. Deletion and amplification of the HGPRT locus in Chinese hamster cells. Mol Cell Biol 1983; 3:1086–1096.

6. Gibbs RA, Camakaris J, Hodgson GS, Martin RF. Molecular characterization of ^{125}I decay and X-ray induced HPRT mutants in CHO cells. Int J Radiat Biol 1987; in press.

7. Turner DR, Morley AA, Haliandros M, Kutlaca R, Sanderson BJ. *In vivo* somatic mutations in human lymphocytes frequently result from major gene alterations. Nature (London) 1985; 315:343–345.

8. Albertini RJ, O'Neill JP, Heintz NH, Kelleher PC. Alterations of the HPRT gene in human in vivo-derived 6-thioguanine resistant T lymphocytes. Nature (London) 1985; 316:369–371.

9. Yen PH, Patel P, Chinault AC, Mohandas T, Shapiro LJ. Altered methylation of hypoxanthine phosphoribosyltransferase genes on active and inactive human X chromosomes. Proc Natl Acad Sci USA 1984; 81:1759–1763.

10. Lock LF, Melton DW, Caskey CT, Martin GR. Methylation of the mouse hprt gene differs on the active and inactive X chromosomes. Mol Cell Biol 1986; 6:914–924.

11. Stout JT, Caskey CT. Anti-sense RNA inhibition of endogenous genes. Methods Enzymol 1987; in press.

12. Becker MA, Yen RCK, Itkin P, et al. Regional localization of the gene for human phosphoribosylpyrophosphate synthetase on the X chromosome. Science 1979; 203:1016–1019.

13. Ogasawara N, Kashiwamata S, Oishi H, et al. Hypoxanthine-guanine phosphoribosyltransferase (HGPRT) deficiency in a girl. Adv Exp Med Biol 1984; 165:13–18.

14. Kelley WN, Wyngaarden JB. Clinical syndromes associated with hypoxanthine-guanine phosphoribosyltransferse deficiency. In Stanbury JB, Wyngaarden JB, Frederickson DS, Goldstein JL, Brown MS, eds. The Metabolic Basis of Inherited Disease, 5th ed. New York: McGraw Hill, pp 1115–1143.

15. Haldane JBS. The rate of spontaneous mutation of a human gene. J Genet 1935; 31:317–326.

16. Talamo RC, Rosenstein BJ, Berninger RW. Cystic fibrosis. In Stanbury JB, Wyngaarden JB, Frederickson DS, Goldstein JL, Brown, MS, eds. The Metabolic Basis of Inherited Disease, 5th ed. New York: McGraw Hill, pp 1889–1917.

17. Watts RW, Spellacy E, Gibbs DA, et al. Clinical, postmortem, biochemical and therapeutic observations on the Lesch-Nyhan syndrome with particular reference to the neurological manifestations. Q J Med 1982; 51:43–78.

18. Howard WJ, Kerson LA, Appel SH. Synthesis de novo of purines in slices of rat brain and liver. J Neurochem 1970; 17:121–123.

19. Kelley WN, Greene ML, Rosenbloom FM, et al. Hypoxanthine-guanine phosphoribosyltransferase deficiency in gout. Ann Intern Med 1969; 70:155–206.

20. Krenitsky TA. Tissue distribution of purine ribosyl- and phosphoribosyl transferases in the Rhesus monkey. Biochim Biophys Acta 1969; 179:506–509.

21. Vinken PJ, Bruyn GW. Handbook of Clinical Neurology, Vol 6. Amsterdam: North Holland Publishing, 1968, p 889.

22. Denny-Brown D. The Basal Ganglia, 1982. London: Oxford, p. 144.

23. Lloyd KG, Hornykiewicz O, Davidson L, et al. Biochemical evidence of dysfunction of brain neurotransmitters in the Lesch-Nyhan syndrome. N Engl J Med 1981; 305:1106–1111.

24. Sass JK, Itabashi HH, Dexter RA. Juvenile gout with brain involvement. Arch Neurol 1965; 13:639–655.

25. Partington MW, Hennen BK. The Lesch-Nyhan syndrome: Self destructive biting, mental retardation, neurological disorder and hyperuricemia. Dev Med Child Neurol 1967; 9:563–572.

26. Wilson JM, Tarr GE, Mahoney WC, Kelley WN. Human hypoxanthine-guanine phosphoribosyltransferase: Complete amino acid sequence of the erythrocyte enzyme. J Biol Chem 1982; 257:10978–10985.

27. Johnson GG, Eisenberg LR, Migeon BR. Human and mouse hypoxanthine-guanine phosphoribosyltransferase: Dimers and tetramers. Science 1979; 203:174–176.

28. Wilson JM, Young AB, Kelley WN. Hypoxanthine-guanine phosphoribosyltransferase deficiency. N Engl J Med 1983; 309:900–910.

29. Melton DW, Konecki DS, Ledbetter DH, Hejtmancik JF, Caskey CT. *In vitro* translation of hypoxanthine-guanine phosphoribosyltransferase mRNA: Characterization of a mouse neuroblastoma cell line that has elevated levels of hypoxanthine-guanine phosphoribosyltransferase protein. Proc Natl Acad Sci USA 1981; 78:6977–6980.

30. Konecki DS, Brennand J, Fuscoe JC, Caskey CT, Chinault AC. Hypoxanthine-guanine phosphoribosyltransferase genes of mouse and Chinese hamster: Construction and sequence analysis of cDNA recombinants. Nucleic Acids Res 1982; 10:6763–6775.

31. Brennand JB, Konecki DS, Caskey CT. Expression of human and Chinese hamster hypoxanthine-guanine phosphoribosyltransferase cDNA recombinants in cultured Lesch-Nyhan and Chinese hamster fibroblasts. J Biol Chem 1983; 258:9593–9596.

32. Jolly DJ, Okayama H, Berg P, et al. Isolation and characterization of a full-length expressible cDNA for human hypoxanthine phosphoribosyltransferase. Proc Natl Acad Sci USA 1983; 80:477–481.

33. Melton DW, Konecki DS, Brennand J, Caskey CT. Structure, expression and mutation of the hypoxanthine phosphoribosyltransferase gene. Proc Natl Acad Sci USA 1984; 81:2147–2151.

34. Patel PI, Nussbaum RL, Framson PE, et al. Organization of the HPRT gene and related sequences in the human genome. Somat Cell Mol Genet 1984; 10:483–493.

35. Patel PI, Framson PE, Caskey CT, Chinault AC. Fine structure of the human hypoxanthine phosphoribosyltransferase gene. Mol Cell Biol 1986; 6:393–403.

36. Kim SH, Moores JC, David D, et al. The organization of the human HPRT gene. Nucleic Acids Res 1986; 14:3103–3108.

37. Patel PI, Tsao TY, Caskey CT, Chinault AC. 5′ regulatory elements of the human HPRT gene. In Granner D, Rosenfeld MG, Chang S, eds. Transcriptional Control Mechanisms, 1987. New York: Alan R. Liss, pp 45–55.

38. Southern EM. Detection of specific sequences among DNA fragments separated by gel electrophoresis. J Mol Biol 1977; 98:503–507.

39. Yang TP, Patel PI, Chinault AC, et al. Molecular evidence for new mutation at the HPRT locus in Lesch-Nyhan patients. Nature (London) 1984; 310:412–414.

40. Gillin FD, Roufa DJ, Beaudet AL, Caskey CT. 8-Azaguanine resistance in mammalian cells. I. Hypoxanthine-guanine phosphoribosyltransferase. Genetics 1972; 72:239–252.

41. Stout JT, Jackson LL, Caskey CT. First trimester diagnosis of Lesch-Nyhan syndrome: Applications to other disorders of purine metabolism. Prenatal Diagnos 1985; 5:183–189.

42. McKeran RO, Howell A, Andrews TM, Watts RWE, Arlett CF. Observations on the growth *in vitro* of myeloid progenitor cells and fibroblasts from hemizygotes and heterozygotes for "complete" and "partial" hypoxanthine-guanine phosphoribosyltransferase (HGPRT) deficiency, and their relevance to the pathogenesis of brain damage in the Lesch-Nyhan syndrome. J Neurol Sci 1974; 22:183–195.

43. Gartler SM, Scott RC, Goldstein JL, Campbell B. Lesch-Nyhan syndrome: Rapid detection of heterozygotes by use of hair follicles. Science 1971; 172:572–574.

44. Botstein D, White RL, Skolnick M, Davis RW. Construction of a genetic linkage map in man using restriction fragment length polymorphisms. Am J Hum Genet 1980; 32:314–331.

45. Myers RM, Larin Z, Maniatis T. Detection of single base substitutions by ribonuclease cleavage at mismatches in RNA:DNA duplexes. Science 1985; 230:1242–1246.

46. Winter E, Fumiichiro Y, Almoguera C, Perucho M. A method to detect and characterize point mutations in transcribed genes: Amplification and overexpression of the mutant c-Ki-*ras* allele in human tumor cells. Proc Natl Acad Sci USA 1985; 82:7575–7579.

47. Myers RM, Lumelsky N, Lerman LS, Maniatis T. Detection of single base substitutions in total genomic DNA. Nature (London) 1985; 313:495–498.

48. Mullis KB, Faloona FA. Specific synthesis of DNA *in vitro* via a polymerase catalyzed chain reaction. Methods Enzymol 1987; in press.

49. Gibbs RA, Caskey CT. Identification and localization of mutations at the Lesch-Nyhan locus by ribonuclease A cleavage. Science 1987; in press.

50. Kidd VJ, Wallace RB, Itakura K, Woo SL. Alpha 1-antitrypsin deficiency detection by direct analysis of the mutation in the gene. Nature (London) 1983; 304:230–234.

16

Phenylketonuria: Molecular Basis and Clinical Applications

SAVIO L. C. WOO

Phenylketonuria (PKU) is a metabolic disorder due to an inborn error in amino acid metabolism. Pathologically, cerebral myelination is inadequate, leading to severe mental retardation in untreated children. The biochemical lesion is a deficiency of hepatic phenylalanine hydroxylase (PAH). Genetically, PKU is transmitted as an autosomal recessive trait. The prevalence is about 1 in 10,000 among Caucasians, with a carrier frequency of about 1 in 50 (1,2).

There have been several landmarks in PKU research (Fig. 16-1). First, Fölling discovered the disease in 1934. Three years later, Jervis identified the enzyme deficiency. In 1963, Guthrie (3) developed the simple test that is still used for PKU screening of all newborn infants in Western countries. In 1954, Bickel successfully implemented the first dietary treatment of PKU. In the 1970s, Seymour Kaufman clarified the aromatic amino acid hydroxylation system, which is relatively complex. For full activity, the enzyme requires a cofactor, tetrahydrobiopterin, this is recycled by a second enzyme, dihydropteridine reductase (Fig. 16-2).

This chapter will focus on the molecular genetics of PKU: (1) gene isolation and characterization, (2) prenatal diagnosis by restriction fragment length polymorphism(RFLP) analysis, (3) mutation identification and population genetics, (4) carrier detection without a proband, and (5) possible somatic gene therapy.

ISOLATION OF THE PAH GENE

Several years ago, Dr. Katherine Robson constructed phenylalanine hydroxylase cDNA clones from rat and human liver (4). The full-length human PAH cDNA clone is about 2400 base pairs in length and contains an open reading frame that encodes 451 amino acid residues (Fig. 16-3). From the DNA sequence, the primary structure of the enzyme was deduced (5).

The human phenylalanine hydroxylase gene has been detected in genomic DNA using the cDNA as a hybridization probe (6,7). That analysis uncovered RFLPs within the PAH gene. For example (Fig. 16-4), DNA from three normal individuals

1. Discovered, Fölling (1934)
2. Biochemical defect identified, Jervis (1937)
3. Neonatal screening test developed, Guthrie (1963)
4. Dietary treatment successful, Bickel (1964)
5. Hydroxylation system clarified, Kaufman (1970s)

Fig. 16-1. Landmarks in the history of PKU research.

showed three different patterns. All three contained an 11-kb band, but individual 1 had a single additional band at 7.0 kb, individual 2 had a 9.7-kb band, and individual 3 had both the 9.7- and 7.0-kb bands. That pattern implied that two PAH alleles were responsible for the polymorphic restriction patterns (Fig. 16-4). In addition to *Sph*I, several other restriction enzymes show similar polymorphisms (3).

These patterns were seen in normal individuals, so that this kind of polymorphism has nothing to do with the mutation that causes PKU. Nevertheless, the RFLPs can be used to trace the transmission of mutations in PKU families. In one family, for example (Fig. 16-5), the enzyme *Hin*dIII detected two different genetic alleles: individual I/I was homozygous for a 4.2 band; and individual II/II was homozygous for a 4.0 band. This analysis has been applied for prenatal diagnosis of PKU for a pregnancy at risk. In this particular family, both parents were heterozygotes for the 4.2 and 4.0 band. The affected child (proband) in this family was homozygous for the 4.0 band. Therefore, we could conclude that the two 4.0 bands in both parents must have been derived from the mutated genes and the 4.2 bands from the normal genes. This was inferred because PKU is an autosomal recessive disorder and both parents must be obligate carriers.

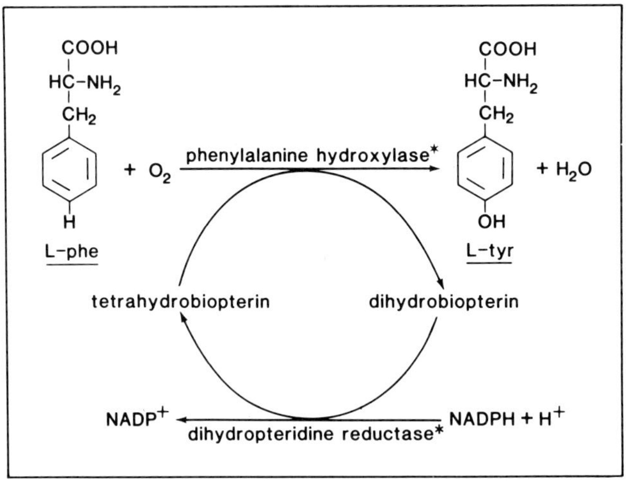

* Classical Phenylketonuria : deficiency of phenylalanine hydroxylase
* Hyperphenylalaninemia : reduced levels of phenylalanine hydroxylase
* Atypical Phenylketonuria : deficiency of dihydropteridine reductase

Fig. 16-2. The hydroxylation system for phenylalanine in humans.

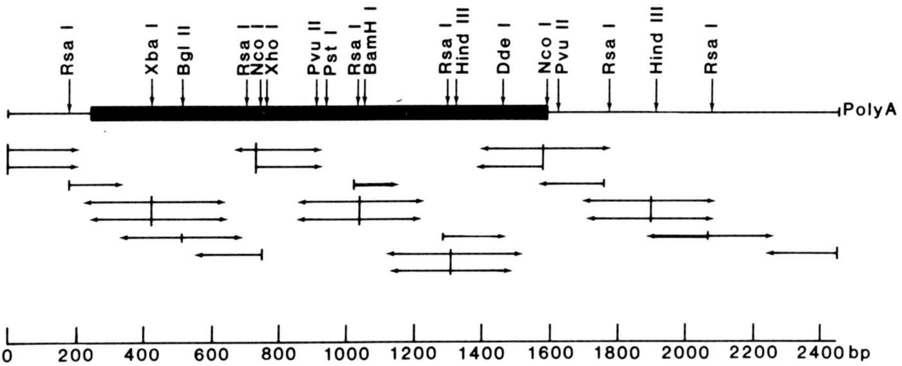

Fig. 16-3. A schematic representation of a full-length human phenylalanine hydroxylase cDNA clone. The black box represents an open reading frame that encodes 451 amino acid residues. The symbols above the box represent different restriction enzyme recognition sequences. The arrows under the box represent the strategy of determining the nucleotide sequence of the entire cDNA clone.

PRENATAL DIAGNOSIS

We performed prenatal diagnosis for the same family; DNA analysis of amniotic cells indicated that the fetus was also homozygous for the 4.0 band (Fig. 16-5). Therefore, the fetus must have inherited both sets of the mutant alleles from the two parents and was presumed to be affected. The pregnancy was not interrupted and, after a full-term pregnancy, the prenatal diagnosis was confirmed by biochem-

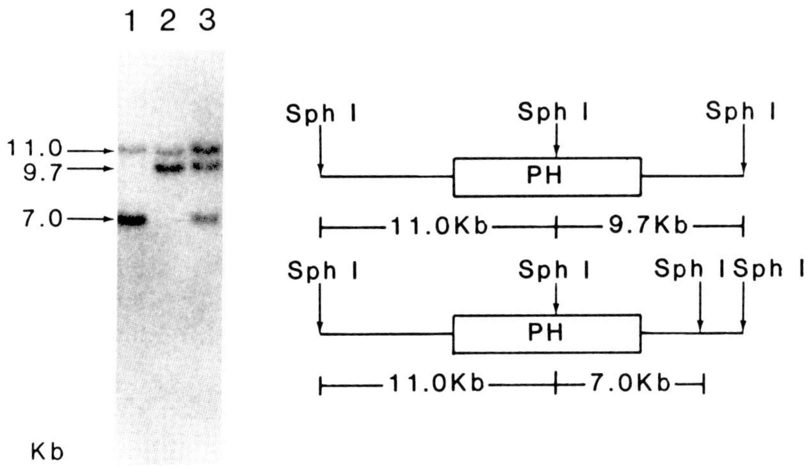

Fig. 16-4. RFLP in the human PAH gene as revealed by the enzyme *SphI*. Left: lanes 1–3 represent genomic DNAs from three normal Caucasians. Right: schematic representations of the two polymorphic alleles.

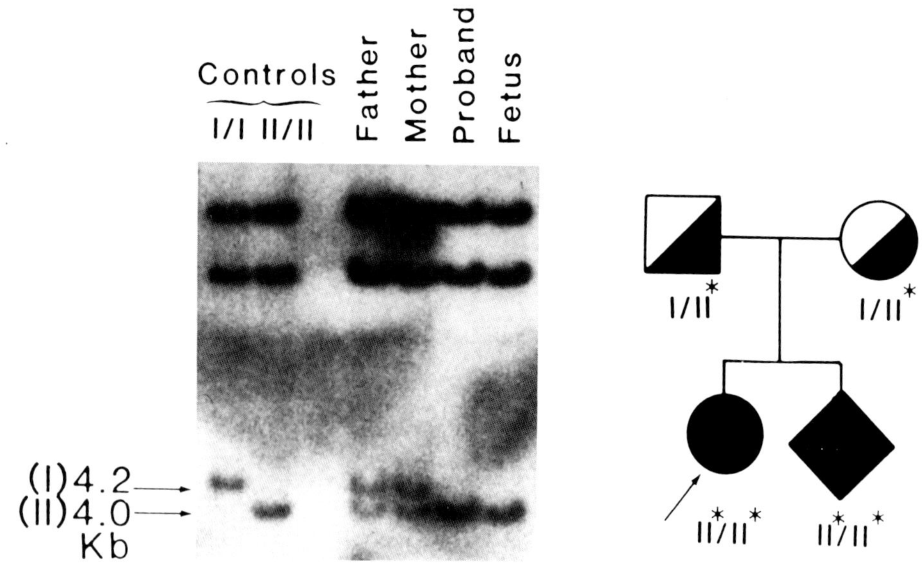

Fig. 16-5. Prenatal diagnosis of PKU by *Hin*dIII polymorphism in a PKU family. This is an autoradiogram of a hybridization experiment using the cloned phenylalanine hydroxylase cDNA to hybridize with genomic DNA isolated from two normal individuals shown on the two left-hand lanes, as well as those isolated from a PKU family with a pregnancy at risk shown on the four right-hand lanes. The schematic panel on the right represents the transmission of two mutant alleles to the fetus as determined by the RFLP analysis.

ical evidence of hyperphenylalaninemia in the neonate (8). This was the first case of prenatal diagnosis of PKU.

This kind of analysis, however, has severe limitations. There must already be a proband in the family before the analysis can be considered. Without a proband, it is not known whether a fetus with a homozygous 4.0 DNA pattern has inherited any normal or mutant genes at all.

POPULATION GENETICS

Something more must be done to extend this type of analysis to families without previous cases of PKU. To do that, we embarked on "RFLP haplotyping" analysis. For example, the phenylalanine hydroxylase gene must have two *Hin*dIII recognition sites that give a 4.2 or a 4.0 band. On the same gene, there must also be another pair of PAH enzyme sites for *Sph*I that are 9.7 kb apart or 7.0 kb apart. The two *Hin*dIII alleles times two *Sph*I alleles give four possible combinations, or four RFLP haplotypes (Fig. 16-6). Basically, an RFLP haplotype is just a combination of individual RFLP patterns, each created by a single enzyme.

Alan Lidsky carried out this kind of haplotyping analysis in Denmark, a relatively isolated genetic area. Dr. Flemming Güttler collected about 90% of all PKU family samples there and only 12 haplotypes were found by RFLP analysis (Table 16-1).

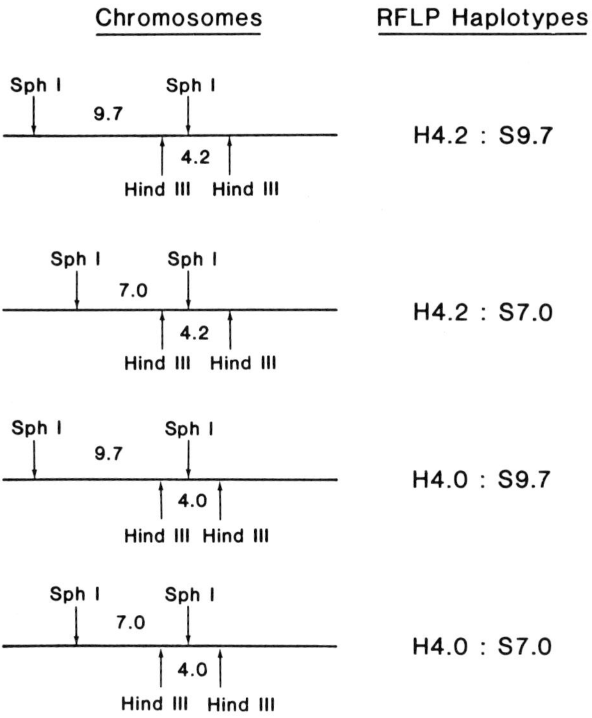

Fig. 16-6. RFLP haplotypes in the human phenylalanine hydroxylase gene as defined by the composite profiles of two independent restriction fragment-length polymorphisms.

In a large population haplotypes may increase by exponential powers (2), and we could have expected theoretically over a thousand different haplotypes. Because so few haplotypes were detected, there must have been linkage disequilibrium between some of these polymorphic sites; that is, some of them segregated together as units (9).

When we counted the haplotypes, haplotypes 1 and 4 were most prevalent among the normal chromosomes; haplotypes 2 and 3 were relatively rare (Table 16-1). Among the PKU chromosomes, in contrast, haplotypes 1, 2, 3, and 4 were the most common ones. Together, they made up more than 90% of all of the PKU chromosomes in that population.

In β-thalassemia, Orkin and Kazazian (10) found that specific mutations were associated with different RFLP haplotypes of the human β-globin gene (10). It is not yet known whether the four RFLP haplotypes we found are associated with four different mutations in the PAH gene. If so, it would be possible to devise specific tests for these four mutation alleles, so detecting 90% of the PKU chromosomes in the population. Carrier detection and prenatal diagnosis would now be possible without an earlier case in the family.

To test this hypothesis, the first mutation chromosome we analyzed obviously was haplotype 3, which constitutes about 40% of all the mutation chromosomes in the Danish population. Tony DiLella and Joshua Marvit isolated this mutation allele from a patient who was homozygous for mutant haplotype 3 to establish the

Table 16-1 RFLP Haplotypes of the Phenylalanine Hydroxylase Genes in PKU Families from Denmark

Haplo-									Number of Genes	
types	*Pvu*II$_a$	*Bgl*II	*Pvu*II$_b$	*Eco*RI	*Xmn*I	*Msp*I	*Hind*III	*Eco*RV	Normal	PKU
1	+	−	−	−	−	+	−	−	23	12
2	+	−	−	−	−	+	+	+	3	13
3	+	−	−	+	+	−	−	−	2	25
4	+	−	−	+	+	−	+	+	21	9
5	−	+	+	+	−	+	−	+	7	0
6	−	+	+	+	−	+	−	−	0	2
7	−	+	−	+	+	−	−	−	7	1
8	+	−	−	+	−	+	−	+	1	0
9	+	+	−	+	−	+	−	+	0	1
10	+	−	−	+	−	+	−	−	1	0
11	−	+	−	+	−	+	−	+	1	1
12	+	−	−	−	−	+	−	+	0	2
									66	66

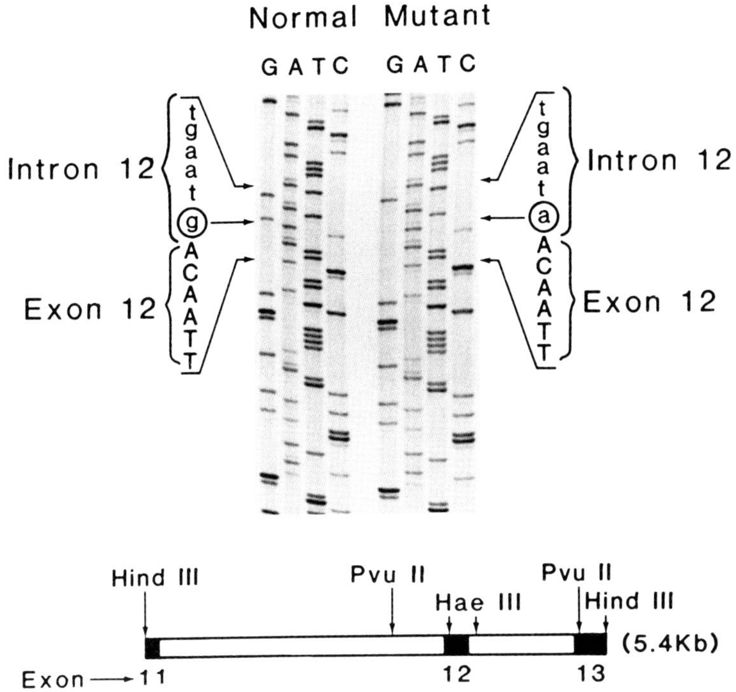

Fig. 16-7. Mutation sequence in a mutant haplotype 3 allele of the human phenylalanine hydroxylase gene. The single nucleotide substitution from G in the normal sequence to A in the mutant sequence is shown by circles.

structure of the gene (11). The regions encoding mRNA were found in 13 exons separated by 12 introns. The entire gene is contained in a total of about 90,000 bp of DNA.

The sequences of all of the RNA-encoding regions in the mutant gene were identical to the normal gene (Fig. 16-7) except for the sequence following exon number 12. At the beginning of intron number 12 in the normal gene, the obligate GT was the donor for RNA splicing. In the mutation chromosome, however, the G had been changed to A giving AT in the donor-splicing site of intron 12 (12).

We carried out many biochemical characterizations to establish the effect of this mutation. In RNA splicing the preceding exon is skipped, resulting in an unstable PAH protein in the cell that gives the PKU phenotype. This phenotype is therefore designated "$mRNA^+$, CRM^-, and PAH $activity^-$" (13).

Having characterized this mutation, we tried to perform some genetic analysis. Because the mutation does not change a restriction site, we synthesized two specific oligonucleotide probes, one for the normal and one for the mutant sequence. The two probes are identical except in the middle, where there is a difference of a single nucleotide. The probes are 21 nucleotides long and the cloned normal and mutant genes showed identical patterns after cutting with *Pvu*II (Fig. 16-8). We know the 2-kb band is the one that contains the part of the gene that contains the mutation.

With the normal probe, hybridization is specific for the normal gene and there

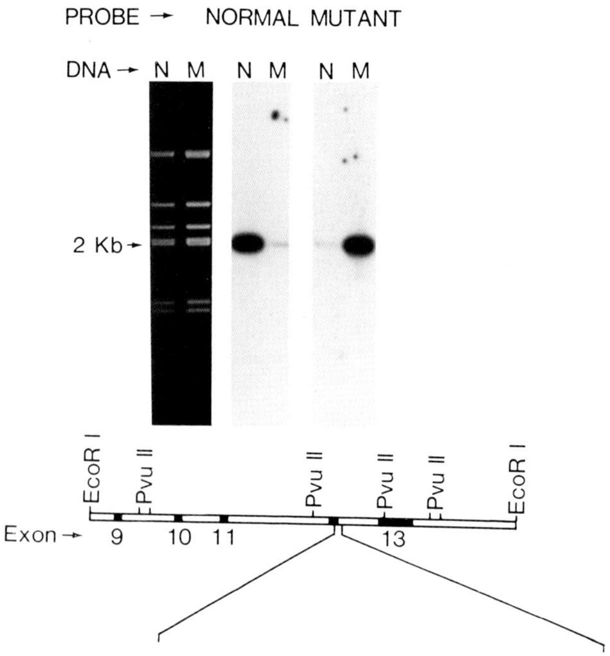

Normal Sequence	5′...TCC ATT AAC AGT AAG TAA TTT...3′
Normal Probe	3′...AGG TAA TTG TCA TTC ATT AAA...5′
Mutant Sequence	5′...TCC ATT AAC A*AT AAG TAA TTT...3′
Mutant Sequence	3′...AGG TAA TTG TTA TTC ATT AAA...5′
Mutant Probe	5′...TCC ATT AAC AAT AAG TAA TTT...3′
Normal Sequence	3′...AGG TAA TTG T*CA TTC ATT AAA...5′

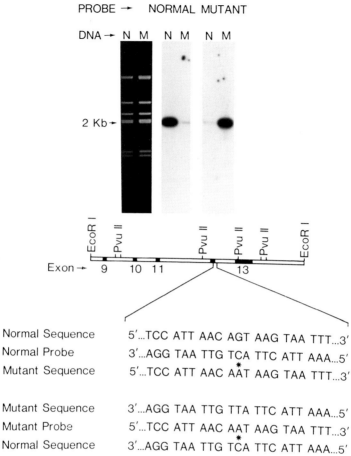

Fig. 16-8. Demonstration of hybridization specificity using a normal and a mutant oligonucleotide as probes to hybridize with cloned normal and mutant human phenylalanine hydroxylase genes.

is little reaction with the mutant gene. On the contrary, the mutant probe shows specific hybridization only with the mutant gene and not with the normal gene. These autoradiograms were purposefully overexposed to show a slight hybridization signal with the wrong allele to indicate the level of specificity in terms of differential signals, which is in excess of 100-fold.

Using this technology, we analyzed all of the affected individuals in the Danish population, including patients who had haplotype 3 and those who did not (Fig. 16-9). All affected individuals with the haplotype 3 mutant allele showed hybridization specifically with the mutant probe. All of those who did not contain the mutant haplotype 3 allele did not show hybridization (12).

That experiment led to two fundamental conclusions. (1) All of the mutation PAH alleles associated with haplotype 3 seem to be caused by the same mutation. (2) PKU must be caused by more than one mutation and, therefore, it must be a heterogeneous disease.

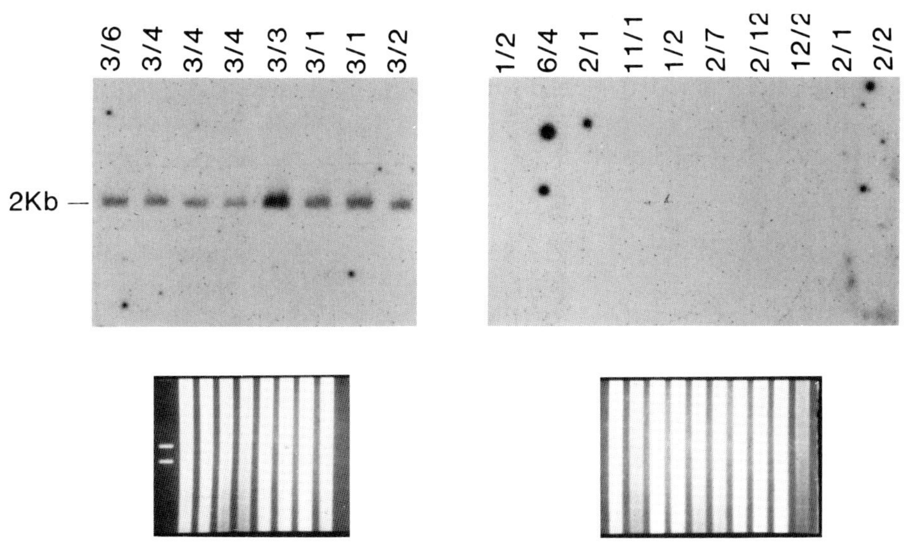

Fig. 16-9. Detection of the splicing mutation in PKU patients bearing a mutant haplotype 3 allele (left) and the absence of this particular mutation in patients who do not bear the mutant haplotype 3 allele (right).

At this point we wanted to know whether the mutation chromosome had spread throughout the white race. Therefore, we analyzed patients from different countries: Denmark, Scotland, Germany, Switzerland, Italy, and a few others. In Denmark, the mutant haplotype 3 allele accounted for 38% of the mutation chromosomes; in Scotland, 33%; Germany, 20%; and Italy, 40%. The only country in which it was infrequent was Switzerland, about 5%.

We analyzed all the probands in these different countries, with exactly the same results. Therefore, the same mutation is associated with the haplotype allele throughout the white race. What does that mean?

1. There is a high level of association between specific RFLP haplotypes and specific mutations in the PAH gene.
2. The association, due to linkage disequilibrium is both inclusive and exclusive.
3. The distribution of specific mutant genes in different populations could be the result of "founder effects." That is, there must have been a single mutational event on a normal chromosome of haplotype 3 background, and that mutant allele then spread throughout the white population.

CARRIER DETECTION WITHOUT A PROBAND

These results led to the final conclusion: Mutant genes can be detected in individuals without a family history of PKU. Therefore, it will be possible to detect carriers in a population. That is, if we continue to characterize the other major mutant alleles in the population and can detect 90% of the mutants, we would have 90% accuracy in carrier detection.

We have also characterized the mutation allele associated with haplotype 2. It is a point mutation that causes a single amino acid substitution in the enzyme, resulting in an unstable protein and the PKU phenotype. Genetic analysis is essentially identical to that observed for mutant haplotype 3 in that it is also a single mutation allele that has been spread throughout the white race (14).

We can now detect about 60% of the mutant alleles in the population; we are now characterizing haplotypes 1 and 4 so that we can achieve the goal of carrier detection without prior PKU history.

CLINICAL APPLICATION

Güttler (15) has defined three different classes of hyperphenylalanemia. "Classical PKU" includes individuals with a blood phenylalanine exceeding 20 mg/dl. They need rigorous dietary correction. "Hyperphenylalanemia" is the term for those with phenylalanine levels less than 10 mg/dl; these patients usually do not require treatment. There is also an intermediate class of patients, with a phenylalanine level of 10–20 mg/dl; the treatment regimen for these patients is less stringent.

We would like to know whether these different phenotypes of hyperphenylalanemia are associated with different genotypes (Table 16-2). We found that if an

Table 16-2 The Parental Haplotypes Related to the Phenotype of Their PKU or HPA Child

Haplotypes	Haplotype 1	Haplotype 2	Haplotype 3	Haplotype 4
1		m[a]	m m m m h	m m h h
2		c c	c c	
3			c c	c
4			h	h

[a] c, classical PKU; m, mild PKU; h, HPA.

individual is homozygous for haplotypes 2 or 3, there is always classical PKU. Individuals who are compound heterozygotes for haplotypes 2 and 3 also have classical PKU (Table 16-2). Knowing the mutations and the biochemical CRM^- phenotypes this pattern was expected because we already know that haplotype 2 and 3 mutations gave no CRM and biochemical PAH activity.

All patients with a haplotype 1 or 4 mutant allele, however, have had either the mild PKU type or the hyperphenylalanemia type, suggesting that the mutations associated with the haplotypes 1 and 4 alleles are of the less severe type. If that can be verified by expanding the correlation study of clinical phenotypes with genotypes, it may be of clinical value in terms of prescribing different dietary regimens (16).

The following can therefore be concluded: (1) Several different mutations in the PAH gene can cause PKU. Some mutations result in zero enzymatic activity; others can give rise to some residual activity, with less severe clinical disorder. (2) Combinations of mutation genotypes can lead to different clinical phenotypes. The major question is whether the identification of genotypes in patients can predict the clinical cause of the metabolic disorder. That question cannot be answered yet, but we are working with the "PKU Collaborative Study Group" headquartered in Los Angeles to determine the genotype of all patients they have treated in the past 20 years, correlating the known clinical course with the genotype.

SOMATIC GENE THERAPY

The disease is caused by lack of the enzyme that converts phenylalanine to tyrosine in the liver. If we could transduce the normal gene into the liver or any other tissue we could restore some of the enzymatic activity, perhaps enough to convert the PKU phenotype to the asymptomatic hyperphenylalanemia phenotype. In other words, there would be a clinical cure. That is our goal, but we are not near that goal yet. However, we are making progress.

Retroviral vectors are excellent, efficient vehicles for gene transfer into mammalian cells. Dr. Richard Mulligan at MIT (17) put a mouse leukemia virus genome into a mouse NIH 3T3 cell so that the cell made all of the viral proteins: core protein, envelope protein, reverse transcriptase, and everything else (Fig. 16-10). The only thing that this viral genome lacked was a signal for packaging the viral RNA transcript by viral proteins to yield an infectious virus.

We could theoretically make a recombinant viral construct that contained the cDNA for human PAH, with the long terminal repeats on both ends and the packaging signal. If both cDNA and packaging signal were inserted together into a cell, the recombinants would produce viral proteins and packageable RNA transcripts resulting in a cryptic virus that could infect mammalian cells in culture. Yet it could not replicate because it would lack all the other viral functions.

This kind of "cryptic" virus could be used to infect fresh mammalian cells in culture. The mammalian cells we have used are from a mouse hepatoma cell line that is deficient in PAH, a cell line we obtained from Dr. Gretchen Darlington at Baylor College of Medicine.

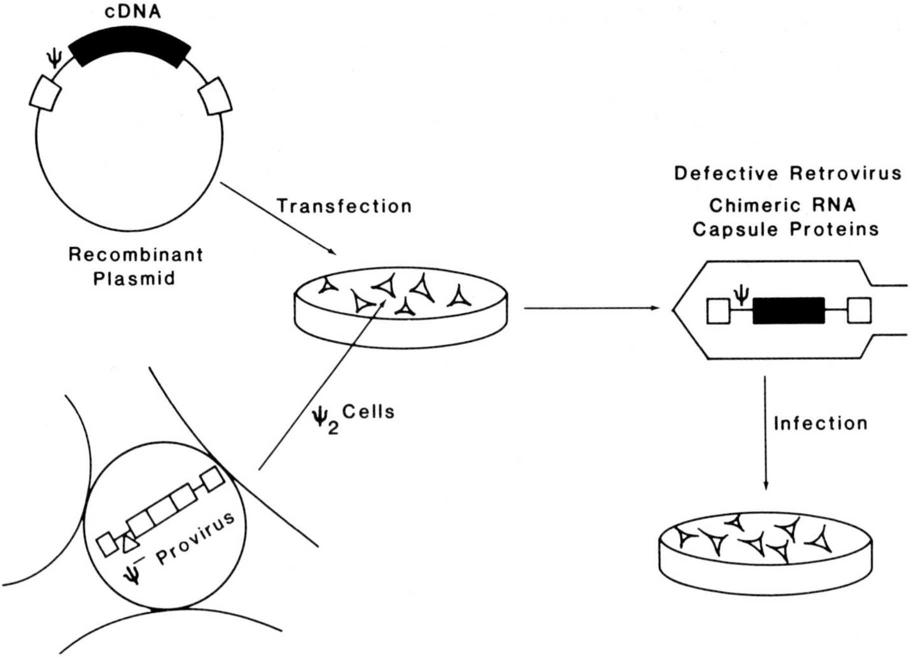

Fig. 16-10. Schematic representation of the construction of cryptic retroviruses for transduction of human phenylalanine hydroxylase cDNA into cultured mammalian cells via viral infection.

We obtained hepatoma cells that were infected by the cryptic virus and they produced active PAH enzyme in the cytoplasm (17). But does that mean the cell could convert phenylalanine to tyrosine? Fred Ledley performed most of these gene transfer analyses. He used a medium deficient in tyrosine; the cells would survive only if they could hydroxylate phenylalanine. When he cultured the parental hepatoma cells in this medium, the cells all died after a few passages. After infection with a retrovirus that contained the human PAH gene, however, the cells grew (Fig. 16-11). Therefore, the infected cells not only made active enzyme in the cytoplasm, but they could also survive in a tyrosine-free medium and must therefore have been able to convert phenylalanine to tyrosine *de novo.*

In the future, we shall attempt to transduce the active human phenylalanine hydroxylase gene into an experimental animal and to see if we could get active enzyme in the liver or other organs to pursue our fantasy of somatic gene therapy for PKU. This could be a model for other human metabolic disorders.

REFERENCES

1. Scriver CR, Clow CL. Phenylketonuria and other phenylalanine hydroxylation mutants in man. Annu Rev Genet 1980; 4:179–202.

2. Scriver CR, Clow CL. Phenylketonuria: Epitome of human biochemical genetics. N Engl J Med 1980; 303:1336–1343.

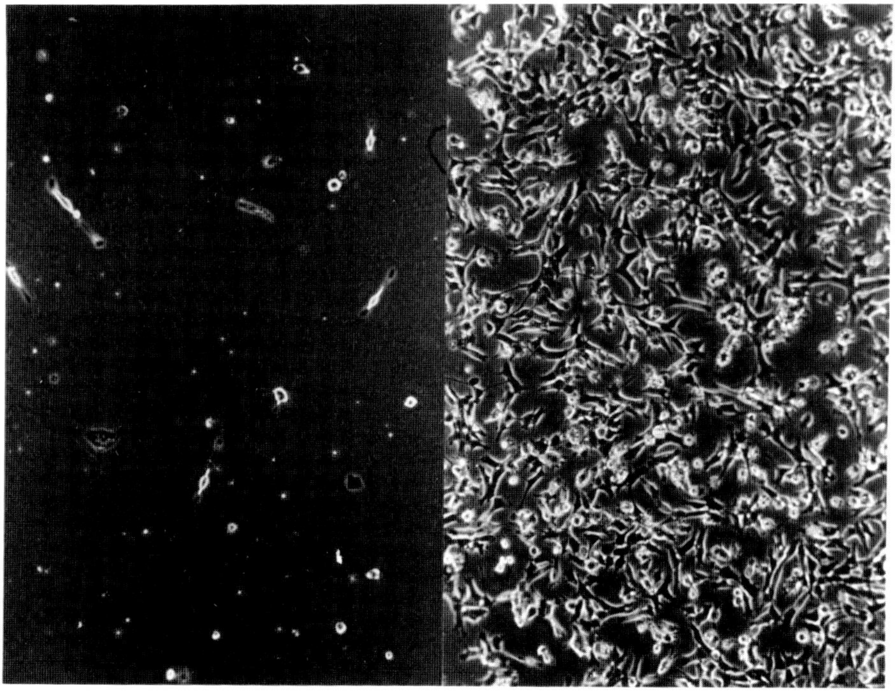

Fig. 16-11. Culture of a phenylalanine hydroxylase-deficient mouse hepatoma cell line in tyrosine-free medium before (left) and after (right) infection with the cryptic retrovirus containing the full-length human phenylalanine hydroxylase cDNA.

3. Guthrie R, Susi A. A single phenylalanine method for detecting phenylketonuria in large populations of new born infants. Pediatrics 1963; 32:338–343.

4. Robson K, Chandra T, MacGillivray R, Woo, SLC. Polysome immunoprecipitation of phenylalanine hydroxylase mRNA from rat liver and cloning of its cDNA. Proc Natl Acad Sci USA 1982; 79:4701–4705.

5. Kowk SCM, Ledley FD, DiLella AG, Robson KJH, Woo SLC. Nucleotide sequence of a full-length cDNA clone and amino acid sequence of human phenylalanine hydroxylase. Biochemistry 1985; 24:556–561.

6. Woo SLC, Lidsky AS, Güttler F, Robson K. Cloned human phenylalanine hydroxylase gene allows prenatal diagnosis and carrier detection of classical phenylketonuria. Nature (London) 1983; 306:151–155.

7. Lidsky A, Ledley FD, DiLella AG, Kwok SCM, Daiger SP, Robson KJH, Woo SLC. Molecular genetics of phenylketonuria: Extensive restriction site polymorphisms in the human phenylalanine hydroxylase locus. Am J Hum Genet 1985; 37:619–634.

8. Lidsky AS, Güttler F, Woo SLC. Prenatal diagnosis of classical phenylketonuria by DNA analysis. Lancet 1985; 1:549–551.

9. Chakraborty R, Lidsky AS, Daiger SP, Güttler F, Sullivan S, DiLella AG, Woo SLC. Polymorphic DNA haplotypes at the human phenylalanine hydroxylase locus and their relationship with phenylketonuria. Human Genet, in press.

10. Orkin SH, Kazazian HH. The mutation and polymorphism of the human β-globin gene and its surrounding DNA. Annu Rev Genet 1984; 18:131–172.

11. DiLella AG, Kwok SCM, Ledley FD, Marvit J, Woo SLC. Molecular structure and polymorphic map of the human phenylalanine hydroxylase gene. Biochemistry 1986; 25:743–749.

12. DiLella AG, Marvit J, Lidsky AS, Güttler F, Woo SLC. Tight linkage between a splicing mutation and a DNA haplotype in phenylketonuria. Nature (London) 1986; 322:799–803.

13. Marvit J, DiLella AG, Brayton K, Ledley FD, Robson KJH, Woo SLC. A mutant human phenylalanine hydroxylase cDNA results from an error in messenger RNA processing. Mol Cell Biol, submitted.

14. DiLella AG, Marvit J, Brayton K, Woo SLC. An amino acid substitution in phenylketonuria is in linkage disequilibrium with DNA haplotype-2. Nature (London), submitted.

15. Güttler F. Hyperphenylalaninemia: Diagnosis and classification of the various types of phenylalanine-hydroxylase deficiency in childhood. Acta Pediat Scand Suppl. 1980; 280:1–80.

16. Cepko CL, Roberts BE, Mulligan RC. Construction and applications of a highly transmissible murine retrovirus shuttle vector. Cell 1984; 37:1053–1062.

17. Ledley FD, Grenett HE, McGinnis-Shelnutt M, Woo SLC. Retroviral mediated gene transfer of human phenylalanine hydroxylase into NIH3T3 and hepatoma cells. Proc Natl Acad Sci USA 1986; 83:409–413.

17

Muscle Phosphofructokinase and Phosphorylase Deficiency: Biochemical and Molecular Genetics

SHOBHANA VORA*

This chapter will review the biochemical and molecular genetics of two enzymes of carbohydrate metabolism, myophosphorylase (EC 2.4.1.1; PPL) and muscle-type phosphofructokinase (EC 2.7.1.11; PFK). Lack of either enzyme causes an almost identical clinical syndrome of exertional muscle pain and myoglobinuria. PPL deficiency results in the more common McArdle disease (glycogenosis type V). PFK deficiency causes the less common Tarui disease (glycogenosis type VII) (1).

PHOSPHOFRUCTOKINASE DEFICIENCY

PFK catalyzes one of the rate-limiting reactions of glycolysis, conversion of fructose 6-phosphate (Fru-6-P) to fructose 1,6-bisphosphate (Fru-1,6-P_2) in the presence of ATP and Mg^{2+}; the reaction is essentially irreversible *in vivo.* The enzyme is a tetrameric protein with a subunit M_r of ~85,000. In several mammals, including humans, there are three unique subunits: M (muscle), L (liver), and P (platelet or brain) types, each under separate genetic control (2–4). Each organ may express one, two, or all three genes; random tetramerization of the resultant subunits produces 1, 5, and 15 homo- or heterotetrameric isozymes. The isozymes have been distinguished by a combination of ion-exchange chromatography and subunit-specific monoclonal and polyclonal antibodies (3,5).

Human organs exhibit tissue-specific isozyme distribution patterns. For example, skeletal muscle exclusively expresses M_4; hepatocytes express only the L_4 isozyme. Red cells exhibit a five-membered set that is composed of M and L subunits in the following ratios: M_4, M_3L, M_2L_2, ML_3, and L_4 (6). In contrast, cultured fibroblasts and lymphoblastoid cell lines express all three subunits in 15 species (7).

Muscle PFK deficiency exhibits an autosomal recessive mode of inheritance; most patients in the United States have been of Ashkenazi Jewish ancestry (2,8). It is characterized by childhood onset of exercise-induced myalgia and myoglobinuria. For reasons unknown, they have hyperuricemia and gout (8). Hemolysis is always present but compensated. Enzyme activity in muscle is almost absent, but

* Deceased.

red cell activity is about 50% of normal. Metabolic studies of muscle reveal an increase of glycogen content and a block in glycolysis at the level of PFK.

Among the known 43 patients in 32 families, the myopathic symptoms have varied in severity, age at onset, and rate of progression (2,8). Although episodic weakness is typical, a few patients show permanent and progressive weakness. Muscle PFK deficiency may also result in a fatal infantile myopathy (9,10) or progressive weakness beginning in the seventh or eighth decade of life, the late-onset variant (11). In addition to these clinical variations there is biochemical evidence of heterogeneity, with differences in stability or kinetics, subunit sizes, or aggregation properties of the mutant M subunits (2), implying that the genetic lesions must be heterogeneous at the molecular level.

In contrast to the normal red cell PFK, which consists of a five-membered set of M and L subunits, the residual red cell PFK from patients consists exclusively of an L_4-type isozyme (Fig. 17-1), indicating total lack of the M subunit. However, one late-onset patient showed 40% residual activity of the M subunit-containing isozymes (11). Although catalytic activity is absent, antibodies detect muscle PFK-related antigen (8,12) (Fig. 17-2). Therefore, the structural gene that encodes the M subunit is intact and the mutations must reside in functionally critical parts of the PFK molecule, such as catalytic sites or subunit–subunit interaction sites, to render the protein dysfunctional.

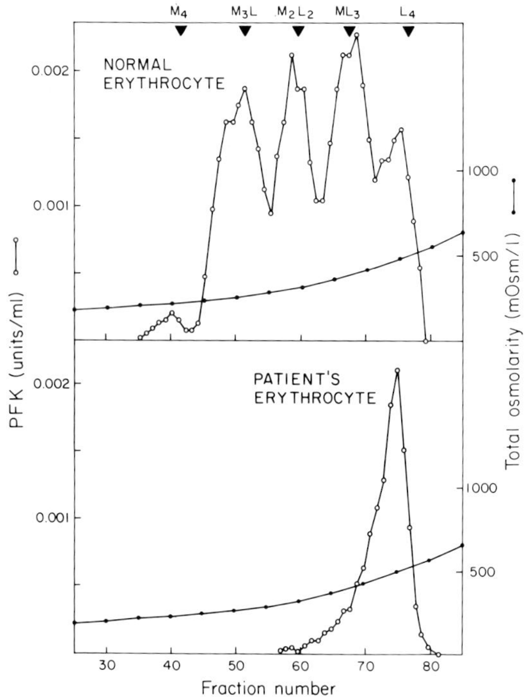

Fig. 17-1. Top: The chromatographic resolution of normal red cell PFK into five isozymes. Bottom: Chromatogram of residual red cell PFK from patients showing a single peak in the position of liver-type isozyme. [Reprinted from Vora et al. (6), with permission.]

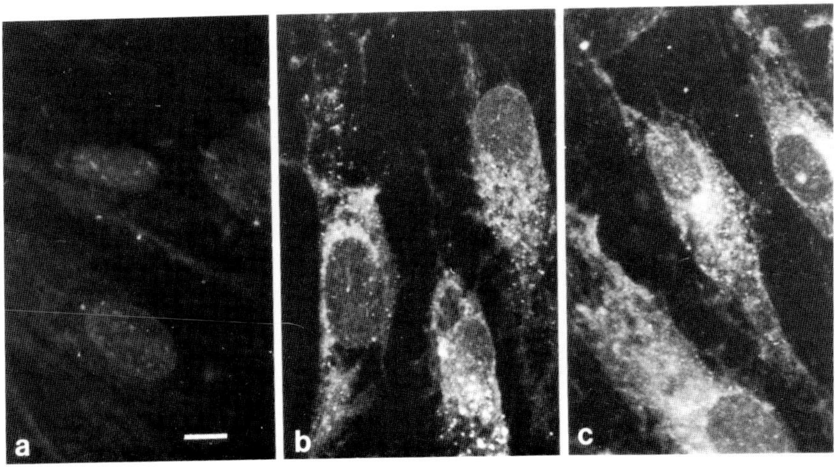

Fig. 17-2. Indirect immunofluorescence staining of cultured human skin fibroblasts using anti-M monoclonal antibody. (a) Normal cells stained with nonimmune mouse ascites show minimal nonspecific staining. (b) Normal cells stained with anti-M show distinct cytoplasmic fluorescence. (c) Muscle PFK-deficient cells show staining similar to (b), indicating the presence of immunoreactive M-type subunits. Baris 10 μm. [Reprinted from Vora et al. (6), with permission]

Inherited muscle PFK deficiency has been discovered among English Springer Spaniel dogs (13). Although the genetic lesion is identical to the human and muscle-type subunits are absent, these animals differ from human patients in that the animals have no apparent muscle disorder, only chronic hemolytic anemia with acute hemolytic crises and hemoglobinuria. We (4) found that the canine pattern can be explained as follows: the M subunit makes a greater contribution to canine red cell PFK, accounting for more severe PFK deficiency in red cells and more severe hemolytic syndrome. The absence of muscle symptoms is attributed to the higher oxidative potential of canine muscle and the presence of liver-type PFK in the PFK-deficient muscle.

My colleagues and I have isolated a cDNA for the M subunit to investigate (1) the nature of human mutations, (2) organ-specific distribution patterns of isozymes, (3) regulation by hormones, nutritional states, and development, (4) altered gene expression during malignant transformation, and (5) concordant or discordant gene expression under regulatory conditions (2). Therefore, we also cloned genes for normal nonmuscle PFKs.

Since no information was then available on the amino acid sequence of PFK from any species (except for the enzyme from *Bacillus stearothermophilus* (14), we used antibodies as probes to isolate cDNAs for muscle PFK from expression vector libraries. We first screened a HeLa cell cDNA library made in the plasmid pUC-9 and a human fibroblast cDNA library in the bacteriophage λgt11 using a rabbit anti-M antibody. The antibody was raised against affinity-purified human muscle PFK and was absorbed extensively using the host bacteria. We obtained 10 putative positive clones from the pUC-9 library that gradually lost immunoreactivity for reasons not clear; there was either a down-regulation of the expression of PFK or there was degradation of the foreign protein by the bacterial host.

However, we obtained three consistently cross-hybridizing clones from the λgt11 library. These clones did not represent the human muscle PFK message because there was no hybridization with the rabbit muscle cDNA (S. Vora, Unpublished data). The cDNA for rabbit muscle PFK was found indirectly. In 1984, the amino acid sequence of the rabbit enzyme was determined by the diligent efforts of two groups of investigators (15). Meanwhile, cDNAs for several muscle-specific proteins were obtained by a "shot-gun approach" (16), randomly sequencing ~180 cDNAs from a rabbit muscle library and then matching the nucleotide sequences with known amino acid sequences of muscle proteins. This approach was successful; Kolb et al. (14) had cloned several muscle-specific proteins, including myophosphorylase and PFK. Since the primary structure of muscle PFK had just been elucidated, the corresponding cDNA was identified—the perfect case of cloning by computer.

We therefore used the rabbit cDNA probe to isolate the human gene (17). We had previously found strong immunochemical homologies among 18 vertebrate muscle PFKs from five different classes (5) implying structural conservation of vertebrate muscle enzymes and also the feasibility of using the rabbit probe to isolate the human gene. As a prelude to screening a human cDNA library, we first surveyed human genomic and cDNA libraries of cross-hybridizing sequences. The rabbit probe hybridized with both rabbit and human genomic DNAs (Fig. 17-3) (17). As expected, the rabbit genomic DNA yielded more intense signals than the human DNA. Only a single band was visible in either DNA after the action of several restriction enzymes. The bands were not seen in the lanes loaded with human and rabbit DNAs that had been restricted by *Hin*dIII or *Pst*I, indicating that there were multiple internal restriction sites. The presence of a single major genomic DNA fragment hybridizing with the rabbit probe suggested that there was probably a single muscle PFK gene per haploid genome in human and rabbit.

Using this probe, we screened a human fibroblast cDNA library constructed in the Okayama and Berg vector (18). That vector was designed to enhance transcription of full-length cDNAs. Knowing that PFK mRNA would be at least 3 kb long, we used this library to obtain full-length or nearly full-length cDNAs. Human fibroblasts express all three PFK genes and if there are sequence homologies among them, a single screening might detect cDNAs for all three isozymes, because all three mammalian PFK isozymes show immunochemical homologies that indicate structural homologies and, therefore, a shared ancestry (3,5). In other words, they arose by gene duplication and divergence, so we expected nucleotide homologies. In fact, the library gave two intensely hybridizing recombinants, pO4 and pO6 (18). The inserts of these clones were ~2 kb in length; they were identical in preliminary restriction mapping and nucleotide sequencing. For the latter, three nonoverlapping *Pst*I fragments were subcloned into bacteriophage M13mp8 and multiple sequencing runs were conducted in either direction, using a dideoxy chain termination technique and synthetic oligonucleotides as primers. The pO4 cDNA included about ~1400 bp of the coding sequence of the protein, ~400 bp of the 3′ untranslated region, and ~150 bp of the vector sequences (17).

Two lines of evidence indicated that cDNA pO4 encoded human muscle PFK mRNA (17). First, comparison of the deduced 458 amino acids showed ~95% perfect match with the published protein sequence of rabbit muscle PFK. This high

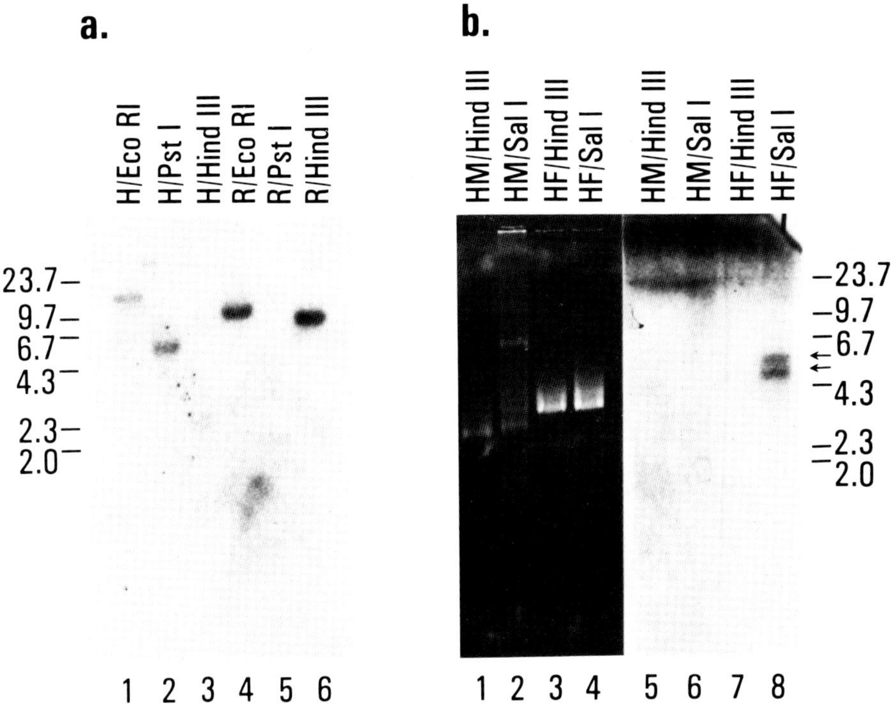

Fig. 17-3. Electrophoresis and Southern blots of genomic and linearized plasmid DNAs. (a) Ten-microgram aliquots of human (H) (lanes 1–3) and rabbit (R) (lanes 4–6) genomic DNAs was digested with *Eco*RI, *Pst*I, and *Hin*dIII. (b) One-microgram aliquots of total plasmid DNA from human muscle (HM) (lanes 1–2) and human fibroblasts (HF) (lanes 3–4) cDNA libraries was digested with *Hin*dIII and *Sal*I. The sizes of the cross-hybridizing DNA fragments were deduced from concurrently electrophoresed DNA markers. [Reprinted from Vora et al. (17), with permission.]

level of homology seems to be the rule among glycolytic enzymes implying evolutionary constraint on structure divergence of the PFK protein molecule. The second line of evidence was that the pO4 cDNA was assigned to human chromosome 1 by Southern analysis of DNAs from somatic cell hybrids that carried chromosomes 1, 10, and 21. Using somatic cell hybrids and subunit-specific antibodies, we had previously assigned the gene loci encoding muscle, brain, and liver isozymes to chromosomes 1, 10, and 21, respectively (19–21). pO4 cDNA hybridized to the genomic DNA from the hybrid that contained human chromosome 1 and 21 (Fig. 17-4), but not to DNAs from Chinese hamster fibroblasts or somatic cell hybrids that contained chromosomes 10 or 21 (17).

Based on internal homologies of the C-terminus half and the N-terminus half of the muscle PFK protein, the gene for mammalian muscle PFK may have originated by tandem duplication of the primordial prokaryotic PFK gene (Fig. 17-5) (15). Bacterial PFK subunits are ~34,000 or about half the size of the mammalian enzyme. The two halves of the rabbit PFK molecule also share 35–45% amino acid sequence homology with PFK from *B. stearothermophilus,* supporting this hypothesis. The amino acid sequence of the bacillus PFK had been determined (14). *Esch-*

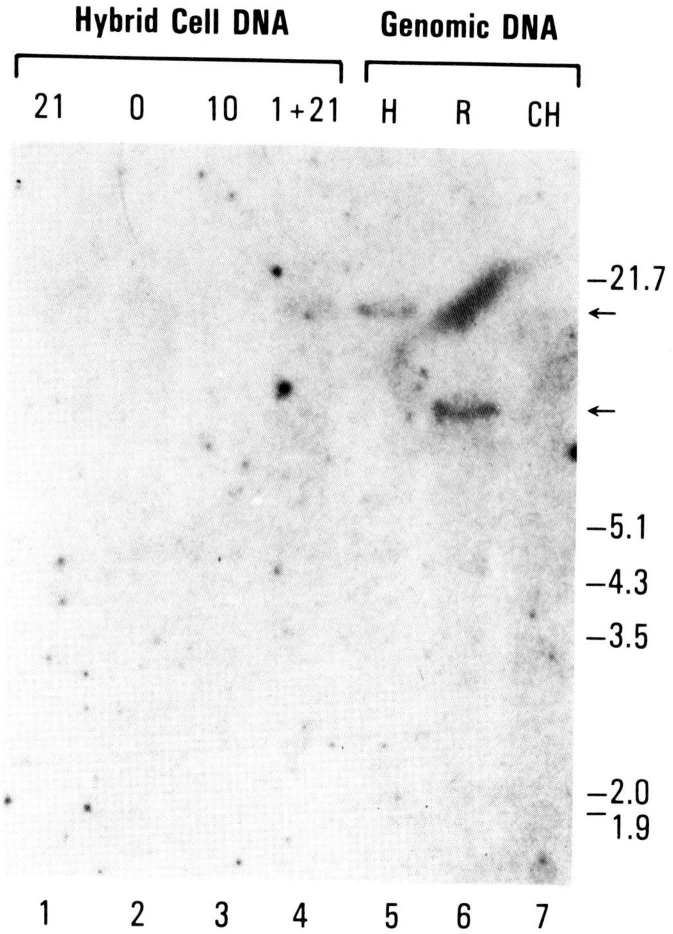

Fig. 17-4. Assignment of p04 cDNA to human chromosome 1 by Southern analysis of DNAs from somatic cell hybrids. Lanes 1–4 were loaded with 10 μg of DNAs from hybrids bearing at least one copy of human chromosome 1, or 10, or 21. The relevant chromosomal complement is indicated at the top of the lanes. Lanes 5–7 were loaded with 10 μg of human (H), rabbit (R), and Chinese hamster (CH) genomic DNAs, respectively. Note that the p04 cDNA hybridizes to the genomic DNAs from humans and the rabbit, which yielded the fragments of ~16.5 and 9.7 kb in length, respectively (lanes 5 and 6). Only the hybrid bearing chromosomes 1 and 21 yielded an identical ~16.5-kb fragment (lane 4). Neither Chinese hamster DNA (lane 7) nor hybrids bearing chromosome 21 or 10 (lanes 1 and 3) showed the presence of cross-hybridizing fragments. [Reprinted from Vora et al. (17), with permission.]

erichia coli has two tetrameric isozymes, A and B, composed of two unique subunits (M_r 34,000 and 36,000, respectively) that are under separate genetic control (22). The isozymes A and B contribute ~90% and ~10% of the total PFK activity. The isozymes differ in physicochemical, immunochemical, and kinetic-regulatory properties indicating that they are not ancestrally related proteins. Moreover, the two subunits do not form heterotetramers implying differences in secondary and tertiary structure. Both *E. coli* PFKs have been cloned (23,24). The deduced amino

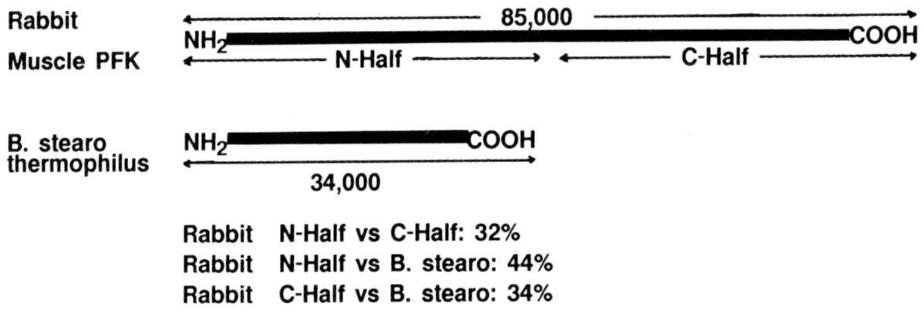

Fig. 17-5. Evolution of mammalian muscle PFK gene. Based on the internal homologies at the amino acid level of the N-terminus and C-terminus halves of rabbit muscle PFK and those between the bacterial PFKs and the rabbit PFK, it has been proposed that the mammalian muscle PFK arose by gene duplication and divergence of the primordial prokaryotic enzyme. The fact that the prokaryotic enzymes are approximately half the size of the mammalian enzymes lends further support to this hypothesis.

acid sequence of the A isozyme shows extensive homology with the enzyme from *B. stearothermophilus,* both halves of the rabbit enzyme (24), and also the C-terminus half of human muscle PFK (S. Vora, unpublished). However, the B isozyme of *E. coli* shares only minor or no homologies, indicating that it probably evolved separately (24). These results also imply that vertebrate PFK arose from the bacterial enzyme by tandem gene duplication and divergence.

The ligand-binding sites of the various PFKs are highly conserved (24). The Fru-6-P, ATP and ADP binding sites of the four mammalian and bacterial PFKs show essentially the same amino acids. The critical residues are shared by the C-termini of the rabbit and human enzymes, which differ here from the prokaryotic enzymes (S. Vora, unpublished).

To obtain the missing 5′ sequence of muscle cDNA and cDNAs of liver and brain isozymes, we screened additional cDNA libraries, a randomly primed fibroblast library and a hepatoma library; we isolated 10 and 2 cross-hybridizing cDNAs, respectively. On nucleotide sequencing, the 3 of 10 cDNAs from the randomly primed cDNA library hybridizing to the 5′-most fragment of muscle cDNA (Fig. 17-6; clones A, B, and C) were found to be 80–90% homologous with the muscle cDNA and mapped within the same. The two hepatoma cDNAs were found to be equally homologous to muscle cDNA and mapped at the 3′ end of the muscle cDNA (Fig. 17-6; clones D and E). These results were strongly suggestive of the fact that these cDNAs represent mRNAs of related PFK isozymes, brain and liver types.

The 5′ end of all five cDNAs abruptly ended at about the same point on the PFK mRNA, suggesting that reverse transcription was hindered at that point because of the formation of stem-loop or hairpin structures. Other factors, listed below, add to the difficulty of isolating a full-length cDNA.

First, muscle PFK is a large protein with an M_r of ~85,000, that translates into a message of ~3 kb, including ~2.5 kb of coding region and 0.5 kb of the 5′ and 3′ untranslated regions. Since reverse transcription is generally not efficient, there are many partial cDNAs for large proteins. Second, PFK is not a highly expressed enzyme compared to other enzymes. Third, the paucity of the message is made worse by the limited tissue distribution and expression of the muscle isozyme,

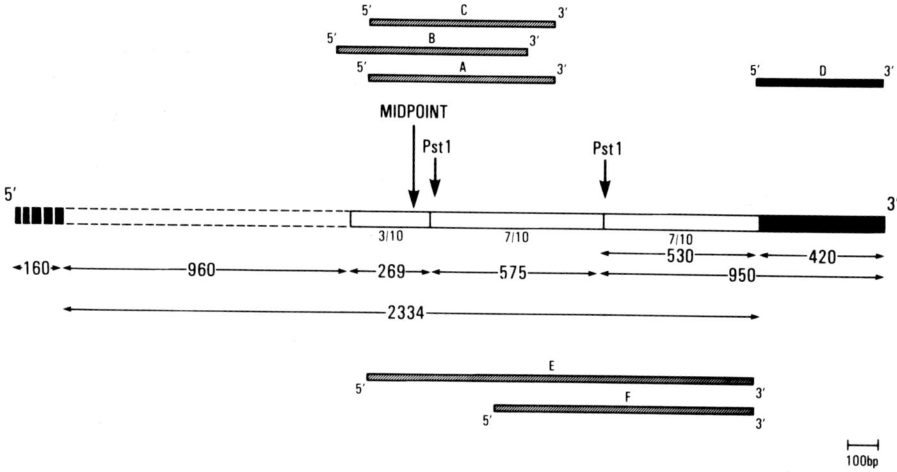

Fig. 17-6. Map of the cDNA for human muscle PFK. The missing 5′ sequences and 5′-untranslated region are shown by the dashed lines. The cDNAs, A–D, were isolated from a randomly primed human fibroblast cDNA library whereas E and F were isolated from an oligo(dT)-primed hepatoma cDNA library. The cDNA, D, was isolated from the fibroblast library by using the 5′-most *PstI* fragment of the cDNA. Note that all of these cDNAs that are derived from three independent libraries and most likely representing all three PFK isozymes end abruptly at the approximate midpoint of the molecule indicating an impediment to reverse transcription.

which is primarily expressed by skeletal muscle, brain, heart, and red cells. Lastly, PFK is largely inactivated or minimally expressed in cultured cells.

The by-product of this endeavor has been the isolation of partial cDNAs, for liver- and brain-type PFKs. These cDNAs show 70–80% homology with muscle PFK in coding and 3′-untranslated regions and efforts are currently under way to obtain a 5′ sequence of the cDNAs of all three isozymes.

PHOSPHORYLASE DEFICIENCY

Despite the clinical similarities of PPL and muscle PFK deficiency, there are some differences (1). Patients with PPL deficiency show the "second wind" phenomenon; those with PFK deficiency do not. PPL deficiency is either mild or asymptomatic in the first two decades of life, is similar to PFK deficiency in the third and fourth decades of life, and may culminate in fixed weakness in the fifth and sixth decades. In contrast, PFK deficiency is invariably symptomatic from early childhood. Moreover, unlike PFK deficiency, there is no hemolysis in PPL deficiency.

Phosphorylase catalyzes the key rate-limiting reaction of glycogenolysis, cleavage of glucosyl units from the nonreducing ends of the α-1,4-glucosyl chains of glycogen, liberating glucose 1-phosphate; the reaction is irreversible *in vivo.* The enzyme is a dimeric protein with a subunit M_r of ~97,000. In several mammals, including humans, there are three unique muscle, liver, and brain-types subunits under separate genetic control (25–27). As with PFK, a given organ may express one, two, or all three phosphorylase genes: random dimerization of the resultant subunits produces one, three, and six homo- or heterodimeric isozymes. As the names imply, these genes are predominantly or exclusively expressed in the respective organs, as in skeletal muscle and liver. The isozyme systems and genetics of PFK and PPL are therefore similar.

In resting muscle, PPL exists in the inactive or *b* form. Conversion of *b* to the active form, or *a,* follows phosphorylation by phosphorylase *b* kinase (25–27). PPL kinase, as with PPL, exists in two interconvertible forms. The inactive form is converted to the active form by cAMP-dependent PPL-*b*-kinase-kinase. Activation of muscle PPL by epinephrine is mediated by cAMP. Liver PPL is affected not only by epinephrine but also by glucagon, a hormone that mobilizes hepatic glycogen as blood sugar.

Both muscle and liver-type PPLs have been molecularly cloned (28–31). The cloning strategies were essentially identical to those used to clone the PFKs and results have been similar, including the problems. PPL seems to be "better expressed" than PFK. The estimated frequency of the mRNA for PPL is between 0.1 and 0.2% of total mRNA (28), about 10–20 times more than mRNA for PFK. Moreover, the inability to find 5′ sequences of PFK cDNAs beyond the midpoint of the mRNA from three separate fibroblast cDNA libraries suggested problems in reverse transcription of the PFK mRNAs, but those problems were not encountered in work with the PPL genes.

The complete amino acid sequence of rabbit muscle PPL was elucidated in 1977 (32), information used in isolating the cDNA. Both immunological (31) and non-immunological (28–30) approaches have been used successfully to isolate cDNAs for muscle and liver PPLs. Robert Fletterick's group (28–30) at the University of California, San Francisco, isolated the cDNAs as follows. First, they size-fractionated total mRNA from rabbit skeletal muscle into a high M_r fraction (enriched for PPL messages) and a low M_r fraction that had little or no mRNA for PPL. The two mRNA fractions were used to prepare two types of radiolabeled probes that were used to screen ~2000 unique members of a rabbit muscle cDNA library. Twenty cDNAs selectively hybridized to the PPL-enriched cDNAs. On sequencing, 3 of 20 cDNAs encoded PPL by comparison with the known amino acid sequence. The largest cDNA was then used to isolate the full-length cDNA of ~2.9 kb in length from ~20,000 recombinants of another rabbit muscle cDNA library. The full-length muscle PPL cDNA was used to isolate partial cDNAs for the muscle isozymes from human and rat, and also a full-length cDNA for human liver isozyme by screening appropriate libraries.

The following account recapitulates the experiences of investigators who successfully used the immunological approach to clone PPL cDNAs (31). They used λgt11 expression vector that contained rat muscle and liver cDNA libraries. They also used antibodies that were isozyme specific and purified on PPL affinity columns. They obtained four liver-specific cDNAs from 3.2×10^5 plaques and two muscle-specific cDNAs from 1.6×10^5 plaques. On sequencing, all were homologous to rabbit muscle PPL, proving that they were PPL cDNAs. All of these cDNAs were partial with maximal lengths of about 1 kb; the cDNAs represent 3′-termini of muscle and liver cDNAs. The rat cDNAs were essentially the same as the full-length cDNAs of rabbit and human PPLs.

Rabbit muscle PPL cDNA encodes 843 amino acids (29). In the nascent polypeptide the first residue is methionine, which is removed posttranslationally so that serine forms the NH_2 terminus. As with most other muscle-specific proteins, including PFK, the amino-terminus residue is acetylated. Only minor discrepancies were found between the primary structures determined by protein sequences and DNA sequencing.

In analysis of cDNA sequences for 24 liver and 13 muscle proteins in different species (30), liver sequences contained about 50% G + C overall and 60% G + C in the third codon positions, whereas muscle sequences contained 60% G + C overall and 80% in the third position (Fig. 17-7). Muscle- and liver-specific isozymes of two other enzymes, i.e., creatine kinase and aldolase, also showed these patterns (30).

The nature of this tissue-specific codon usage is unclear. Organisms such as thermophilic bacteria and the protozoan *Leischmania,* like skeletal muscle, are subject to environmental stresses of high temperature and low pH and have a high G + C content in their coding sequences. Newgard et al. (30) therefore postulated that the muscle genes selectively maintain a high G + C content in response to environmental stress. G · C base pairs are more stable than A · T base pairs, which might aid in gene replication and transcription. Alternatively, the strong G + C bias in the noncoding regions could play a role in tissue-specific gene expression by generating consensus sequences that are recognized by regulatory proteins in a tissue-specific manner (30).

Nucleotide sequences of muscle PPLs from humans, rabbit, and rat, and liver PPLs from humans and rat indicate that these protein structures have been highly conserved (28–31). The muscle PPLs show 96% amino acid homology and 90% nucleotide homology. Most of the amino acid substitutions are conservative; they do not alter the tertiary structure of the enzyme. As a general rule, homologous proteins show more sequence divergence on the surface of the molecule and less internally, where the catalytic sites, allosteric sites, and subunit–subunit interaction

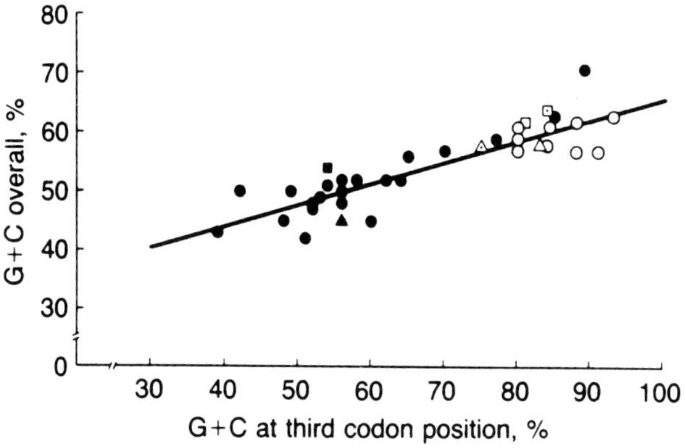

Fig. 17-7. G + C percentage at the third codon position vs the overall G + C percentage in liver and muscle cDNA sequences. Closed and open symbols represent liver and muscle proteins, respectively, which inlcude PPLs from humans and the rabbit (for details, see ref. 30). The slope of the line is 0.37; the correlation coefficient is 0.88. [Reprinted from Newgard et al. (30), with permission.]

sites are located. This pattern seems to be unusual, with a high degree of conservation among homologous PPLs in regions of the molecule that can tolerate evolutionary drift, i.e., the surface.

The primary structures of two nonvertebrate PPLs, potato tuber (33,34) and *E. coli* (35) show considerable homology (34). The three were about 40–50% homologous; 31% of the residues were shared by all three enzymes. The potato enzyme was more closely related to the rabbit enzyme. *E. coli* was more distantly related to the other two. The regions involved in the catalytic reaction were well conserved but not those involved in regulation. The potato enzyme was larger than the mammalian enzyme, primarily because there was a 78-residue insertion in the middle of the polypeptide chain. The large insertion was accommodated on the surface without altering the tertiary structure (34). The insertion was close to the glycogen storage site of the rabbit muscle enzyme, and could interfere sterically with the anchoring of glycogen, impeding the catalytic activity. Interference in the formation of the glycogen-storage site in the potato enzyme (36) could also explain the divergent kinetics of that enzyme (33,34).

Plants and vertebrate PPLs are homologous to those of *E. coli* and *Klebsiella pneumoniae;* prokaryotic enzymes are smaller in size (M_r 81,000). It has therefore been proposed that present-day PPL arose by tandem gene duplication and divergence (34) as did present-day PFK (16). However, there are no clear candidates for the ancestral short PPL gene and there is no internal sequence repetition in the present-day PPL, casting doubt on that hypothesis.

In summary, there are many parallels between PFK and PPL. Both are oligomeric proteins and are encoded by three distinct gene loci. The organ distributions of muscle, liver, and brain-type isozymes are essentially identical. Heritable deficiency of muscle isozymes causes a clinically identical myopathy. In spite of the evolutionary distance, the enzymes from bacteria, plants, and mammals show structural homologies, indicating ancestral relatedness and also an evolutionary constraint on the divergence of protein structure of the enzyme.

Genes for the muscle and liver isozymes of these enzymes from different species have been cloned. Strong evidence exists for the evolutionary origin of mammalian PFK, i.e., gene duplication and divergence of the ancestral prokaryotic enzyme. Similar evidence is lacking for mammalian PPL. Finally, availability of the cDNAs for the isozymes of these enzymes should help to unravel problems of regulation and disease related to both enzymes.

REFERENCES

1. Howell RR, Williams JC. The glycogen storage diseases. In Stanbury JB, Wyngaarden JB, Fredrickson DS, Goldstein JL, Bowen MS, eds. The Metabolic Basis of Inherited Disease. New York: McGraw-Hill, 1983.
2. Vora S. Isozymes of phosphofructokinase. Isozymes: Curr. Top. Biol. Med. Res. 1982; 6:119–167.
3. Vora S, Oskam R, Staal GEJ. Isozymes of phosphofructokinase in the rat: Demonstration of the three non-identical subunits by biochemical, immunochemical and kinetic studies. Biochem J 1985; 229:333–341.
4. Vora S, Giger U, Turchen S, Harvey J. Characterization of the enzymatic lesion in

inherited phosphofructokinase deficiency in the dog: An animal analogue of human glycogen storage disease type VII. Proc Natl Acad Sci USA 1985; 82:8109–8113.

5. Vora S, Wims LA, Durham S, Morrison SL. Production of monoclonal antibodies to the subunits of human phosphofructokinase: New tools for the immunochemical and genetic analyses of isozymes. Blood 1981; 58:823–829.

6. Vora S, Seaman C, Durham S, Piomelli S. Isozymes of human phosphofructokinase: Identification and subunit structural characterization of a new system. Proc Natl Acad Sci USA 1980; 77:62–66.

7. Vora S. Isozymes of human phosphofructokinase in blood cells and cultured cell-lines: Molecular and genetic evidence for a trigenic system. Blood 1981; 57:724–732.

8. Vora S, Davidson M, Seaman C, Miranda AF, Noble NA, Tanaka KR, Frenkel EP, DiMauro S. Heterogeneity of the molecular lesions in inherited phosphofructokase deficiency. J Clin Invest 1983; 72:1995–2006.

9. Danon MJ, Schlisefeld L. A fatal infantile glycogen storage disease with deficiencies of phosphofructokinase and phosphorylase b kinase. Neurology 1979; 29:564 (abstr.).

10. Servidei S, Bonilla E, Diedrich RG, Kornfeld M, Oates JD, Davidson M, Vora S, DiMauro S. Fatal infantile form of muscle phosphofructokinase deficiency. Neurology 1986; 36:1465–1470.

11. Vora S, DiMauro S, Spear D, Harker D, Danon MJ. Characterization of the enzymatic defect in late-onset muscle phosphofructokinase deficiency: A new sub-type of glycogen storage disease type VII. J Clin Invest 1987; 80:1479–1485.

12. Davidson M, Miranda AF, Bender A, DiMauro S, Vora S. Muscle phosphofructokinase deficiency: Biochemical and immunological studies of phosphofructokinase isozymes in muscle culture. J Clin Invest 1983; 72:45–55.

13. Giger U, Harvey JW, Yamaguchi RA, McNulty PK, Chiapella A, Beutler E. Inherited phosphofructokinase deficiency in dogs with hyperventilation-induced hemolysis: Increased in vitro and in vivo alkaline fragility of erythrocytes. Blood 1985; 65:345–351.

14. Kolb E, Hudson PJ, Harris JI. Phosphofructokinase: Complete amino-acid sequence of the enzyme from *Bacillus stearothermophilus.* Eur J Biochem 1980; 108:587–597.

15. Poorman RA, Randolph A, Kemp RG, Heinrikson RL. Evolution of phosphofructokinase-gene duplication and creation of new effect sites. Nature (London) 1984; 309:467–469.

16. Putney SD, Herlihy WC, Schimmel P. A new troponin T and cDNA clones for 13 different muscle proteins, found by shotgun sequencing. Nature (London) 1983; 302:718–721.

17. Vora S, Hong F, Olender E. Isolation of a cDNA for human muscle phosphofructokinase. Biochem Biophys Res Commun 1986; 135:615–621.

18. Okayama H, Berg P. A cDNA cloning vector that permits expression of cDNA inserts in mammalian cells. Mol Cell Biol 1983; 3:280–289.

19. Vora S, Durham S, de Martinville B, George DL, Francke U. Assignment of the human gene for muscle-type phosphofructokinase (PFKM) to chromosome 1 (cen → q32) using somatic cell hybrids and monoclonal anti-M antibody. Somatic Cell Genet 1982; 8:95–104.

20. Vora S, Miranda A, Hernandez E, Francke U. Assignment of the human gene for platelet-type phosphofructokinase (PFKP) to chromosome 10p: Novel use of polyspecific rodent antisera in mapping human enzyme genes. Hum Genet 1983; 63:374–379.

21. Vora S, Francke U. Assignment of the human gene for liver-type phosphofructokinase isozyme (PFKL) to chromosome 21 by using somatic cell hybrids and monoclonal anti-L antibody. Proc Natl Acad Sci USA 1981; 78:373–428.

22. Fraenkel DG. The biochemical genetics of glycolysis in microbes. In Hollaender A,

ed. Trends in the Biology of Fermentation for Fuel and Chemicals. New York: Plenum Press, 1981, pp. 201–215.

23. Daldal F. Nucleotide sequence of gene pfkB encoding the minor phosphofructokinase of *Escherichia coli* K-12. Gene 1984; 28:337–342.

24. Hellinga HW, Evans PR. Nucleotide sequence and high-level expression of the major *Escherichia coli* phosphofructokinase. Eur J Biochem 1985; 149:363–378.

25. Hers HP. The control of glycogen metabolism in the liver. Annu Rev Biochem 1976; 45:167–189.

26. Fletterick RJ, Madsen NB. The structures and related functions of phosphorylase α. Annu Rev Biochem 1980; 49:31–61.

27. Dombradi V. Structural aspects of the catalytic and regulatory function of glycogen phosphorylase. Int J Biochem 1981; 13:125–139.

28. Hwang PK, See YP, Vincentini AM, Powers MA, Fletterick RJ, Crerar MM. Comparative sequence analysis of rat, rabbit, and human muscle glycogen phosphorylase cDNAs. Eur J Biochem 1985; 152:267–274.

29. Nakano K, Hwang PK, Fletterick RJ. Complete cDNA sequence for rabbit muscle glycogen phosphorylase. FEBS Lett 1986; 204:283–287.

30. Newgard CB, Nakano K, Hwang PK, Fletterick RJ. Sequence analysis of the cDNA encoding human liver glycogen phosphorylase reveals tissue-specific codon usage. Proc Natl Acad Sci USA. 1986; 83:8132–8136.

31. Osawa S, Chiu RH, McDonough A, Miller TB, Jr, Johnson GL. Isolation of partial cDNAs for rat liver and muscle glycogen phosphorylase isozymes. FEBS Lett 1986; 202:282–288.

32. Titani K, Koide A, Hermann J, Ericsson LH, Kumar S, Wade RD, Walsh KA, Neurath H, Fischer EH. Complete amino acid sequence of rabbit muscle glycogen phosphorylase. Proc Natl Acad Sci USA. 1977; 74:4762–4766.

33. Nakano K, Tashiro Y, Kikumoto Y, Tagaya M, Fukui T. Amino acid sequence of cyanogen bromide fragments of potato phosphorylase. J Biol Chem 1986; 261:8224–8229.

34. Nakano K, Fukui T. The complete amino acid sequence of potato α-glucan phosphorylase. J Biol Chem 1986; 261:8230–8235.

35. Palm D, Goerl R, Burger KJ, Buhner M, Schwartz M. Chemical and Biological Aspects of Vitamin B6 Catalysis, Part A. New York: Alan R. Liss, 1984, pp 209–221.

36. Shimomura S, Fukui T. A comparative study on α-glucan phosphorylases from plant and animal: Interrelationship between the polysaccharide and pyridoxal phosphate binding sites by affinity electrophoresis. Biochemistry 1980; 19:2287–2294.

18

Acid α-Glucosidase Deficiency: Molecular Basis and Studies of Patients

ROCHELLE HIRSCHHORN, FRANK MARTINIUK, STEPHANIE TZALL, MARK MEHLER, AND ANGEL PELLICER

Inherited deficiency of lysosomal acid α-glucosidase (GAA) results in glycogen storage disease type II, encompassing a spectrum of myopathies of varying severity, ranging from a rapidly fatal infantile disorder (Pompe disease), to a slowly progressive adult onset myopathy (1–4 reviewed in 5).

The infantile syndrome is characterized by massive accumulation of glycogen in cardiac and skeletal muscle, as well as in other tissues. In the adult-onset myopathy, involvement is usually limited to skeletal muscle. Childhood cases (juvenile form) are a clinically and biochemically heterogeneous group. Both the adult and the infantile forms are also biochemically heterogeneous. For example, patients with the adult-onset form often have higher residual acid α-glucosidase activity than patients with the infantile form. However, some adult-onset patients have essentially no detectable enzyme activity and are indistinguishable by enzyme assay from the infantile patients. Patients with the infantile form are heterogeneous with respect to presence or absence of "CRM" or protein that cross-reacts with antibody to GAA (6).

Acid α-glucosidase or acid maltase (EC 3.2.1.3) is a lysosomal enzyme that primarily hydrolyzes 1,4-α-glucose polymers ranging in size from maltose to glycogen. The enzyme is translated, processed, and sorted along a pathway that is shared by other lysosomal enzymes. The enzyme is first translated as a glycosylated precursor protein, estimated by SDS–PAGE as 100–110 kDa in size (7–10 reviewed in 11). The gene for human acid α-glucosidase has been regionally localized to q21–q23 of chromosome 17 (12–14), designated as GAA on the human gene map (15).

GAA is also genetically polymorphic in the normal population (16,17). One of the three normal alleles encodes a protein with normal affinity for low-molecular-weight substrates but with markedly diminished affinity for high-molecular-weight substrates including starch and glycogen (16,18).

To examine the molecular basis for the heterogeneity of normal and mutant GAA, we sought to obtain the gene for acid α-glucosidase.

APPROACHES TO ISOLATION OF GENES

Approaches to the isolation of genes have been discussed in Chapters 8, 9, and 10. Nonetheless, we would like to discuss methods briefly, in order to provide a background for the two approaches we have used.

There are several ways to obtain either the expressed, exonic segments of a gene (cDNA) or the total structural gene (including introns as well as exons). To a large extent, the approach chosen to isolate a specific gene depends on the information available. Thus, genes can be broadly divided into two classes: the smaller class where the gene product is known and the more numerous conditions where the gene product is unknown. In the class of unknown gene product, there is a subgroup in which the chromosomal location of the gene is also unknown but the gene confers a specific and recognizable phenotype. In a second subgroup, the chromosomal location is known from linkage studies but there is no recognizable biochemical phenotype. Many diseases do not fall into any of these categories.

If the protein gene product is known, the most straightforward approach is to purify the protein and obtain the amino acid sequence. A synthetic oligonucleotide (in reality, probably a mixture of oligonucleotides, because of the degeneracy of the genetic code) can then be synthesized, radiolabeled, and used to screen a recombinant cDNA library. If the protein cannot be sequenced even partially, but the protein can be purified to homogeneity and a highly specific polyclonal antibody of high titer can be generated, the antibody can be used to screen a recombinant library. The recombinant library that expresses the protein product is screened; for example, a λgt11 expression library can be used (19). Ideally, the library should be constructed by using the mRNA from a tissue that expresses large amounts of the protein. If the protein can only be partially purified but can be used to generate several different monoclonal antibodies that recognize different epitopes, a mixture of monoclonal antibodies can be used instead of the polyclonal antibody to "screen" an expression library.

A totally different approach has been used successfully when the protein cannot be purified but confers a known phenotype that can be "selected for." This approach, termed *DNA-mediated transformation,* is based on the observation that when total human cellular (genomic) DNA is added to mouse cells that do not express the protein or phenotype, under special conditions (e.g., calcium phosphate), a small proportion of the exogenous DNA is taken up by a subpopulation of the cells (20). The mRNA and protein encoded by the foreign DNA are transiently expressed by the recipient cells or *(transient DNA-mediated transformation)* (21–24). A few cells continue to express the specific protein for a longer time (21). If the cells that express the specific protein can be selected physically or biochemically, then the DNA for the gene of interest can eventually be isolated. That approach was used to isolate the gene for thymidine kinase (25–27), which confers a biochemically selectable phenotype. It has also been used to isolate genes for cell surface markers that can be selected physically using fluorescence-activated cell sorters (23–28).

Lastly, some genes whose protein products confer a biochemically selectable phenotype have been isolated as a result of amplification of the gene that can occur following prolonged and increasingly stringent biochemical selection. This increase in the DNA and RNA of the gene allows for eventual isolation using several strategies (29).

Genes can also be isolated even if the product of the gene is not known. The gene may confer a phenotype that is selectable but by an unknown mechanism; for example, the gene may confer a growth advantage. If so, DNA-mediated transformation can be used. That approach was used most strikingly to isolate a series of oncogenes (30). If the gene confers a recognizable but not a selectable phenotype, DNA-mediated transformation can, in theory, still be used if purified DNA encoding a dominant selectable gene is added to mouse cells together with the total human genomic DNA. The two types of DNA enter a cell to form a large concatemer. Selecting for the simultaneously added selectable gene will yield mouse cell clones that contain both types of foreign DNA and that can then be tested for the recognizable phenotype. That method has been called *DNA-mediated cotransformation* (31–35).

The last group of genes we will discuss are those for which neither the protein product nor an *in vitro* phenotype is known, but where defects in the gene result in disease. If the chromosomal location of the gene is known, several approaches are possible. If the critical site is included in a chromosomal deletion, as in Duchenne dystrophy, the gene can be isolated. This is "reverse genetics" in action, not just in theory (36,37). If the gene is involved in a translocation, genomic libraries can be screened for the "breakpoint," i.e., the site at which DNA from the two different chromosomes is in close proximity. Similarly, if linkage studies have identified DNA segments that cosegregate with the disease, those DNA segments can be used to walk or hop along the chromosome until the gene is found.

An even larger group of genes does not fall into any of these groups.

ISOLATION OF SEQUENCES ENCODING FOR ACID α-GLUCOSIDASE

We have used two different approaches to isolate the gene for acid α-glucosidase. (1) Because we had purified the protein and raised an antibody, we used the straightforward and obvious approach of screening a cDNA expression library with antibody to the enzyme (10,14,38). (2) We used DNA-mediated cotransformation of mouse cells selecting transformants that expressed the human enzyme, aiming to obtain the structural gene (39).

We were successful in isolating a cDNA for human acid α-glucosidase with the antibody and an expression library (38) and we will devote most of this chapter to a description of those results. However, in retrospect we also seem to have isolated a partial genomic fragment by DNA-mediated cotransformation. For the purposes of completeness, we will also briefly describe these experiments.

ISOLATION OF A cDNA FOR ACID α-GLUCOSIDASE

Acid α-glucosidase is readily purified to apparent homogeneity by affinity chromatography (7,10,16,40). We had previously used the purified enzyme to raise high-titer specific polyclonal antibody. We further purified that antibody by binding it to and eluting it from an enzyme-affinity column (19), then using it, screened a λgt11 expression library.

Our first strategy was to identify recombinant phage that produced a *fusion protein* reacting with the affinity-purified polyclonal antibody to GAA, and then to retest the phage fusion protein for reactivity with two different monoclonal antibodies, each recognizing a different epitope (41). We screened about one million recombinant phage from a human liver cDNA λgt11 expression library with the affinity-purified antibody to human placental GAA (14). Thirty phage that reacted strongly with the polyclonal antisera were plaque purified. The recombinant phage were then analyzed for the size of the cDNA insert and tested for reactivity with two monoclonal antibodies. The cDNA inserts ranged in size from 0.2 to 2.0 kb. The phage clone (GAA-67) that contained the largest insert (2.0 kb) produced a fusion protein that was recognized by one of the two monoclonal antibodies. (None of the other 29 recombinant phage cross-hybridized with the 2.0-kb insert.)

To provide additional evidence that the 2-kb cDNA encoded human GAA, we tested the insert for ability to hybridize specifically to DNA from a panel of mouse × human somatic cell hybrids that retained different portions of human chromosome 17, including the locus for GAA. We also tested hybrids selected for loss of chromosome 17. We had previously regionally localized the gene for GAA to 17q21–q23 by determining expression of human GAA in somatic cell hybrids that contained informative translocations of chromosome 17 (13,14). We hybridized the 2.0-kb radiolabeled insert to Southern blots of *Eco*RI-digested DNA from those hybrids and from human and mouse cells. The cDNA hybridized to a 20-kb *Eco*RI restriction fragment in human DNA and to a 10-kb fragment in mouse DNA, allowing for easy identification of the human fragment. The human 20-kb fragment was found in DNA from a thymidine kinase-deficient mouse × human somatic cell hybrid that contained a 17pter-q25 segment of human chromosome 17 (14). A bromodeoxyuridine (BUdR)-resistant subclone of that hybrid was isolated. This BUdR-resistant hybrid also had selectively lost chromosome 17 and, as a result, lost expression of human GAA. DNA from the hybrid also lost the human 20-kb fragment but retained the mouse 10-kb fragment. DNA from a somatic cell hybrid containing a 17pter-q23 segment of chromosome 17 and retaining expression human GAA still showed the human band.

Finally, DNA from the most informative hybrid that expressed human GAA but retained only a segment (17q21–q23) of chromosome 17 that encoded GAA translocated to a mouse chromosome (and no other human chromosomes), still contained the 20-kb *Eco*RI human fragment. We estimate that the hybrid contained between 0.1 and 0.3% of the human genome. Thus, the 2.0-kb cDNA specifically recognized a human sequence that was present on the small portion of human chromosome 17 containing the locus for GAA.

Our results were partially consistent with others (42,43) who reportedly cloned a partial cDNA for monkey GAA. Their cDNA hybridized to a 20-kb *Eco*RI fragment in human DNA, as did the cDNA we isolated. However, using the monkey cDNA as a probe for *in situ* hybridization, they regionally localized the human gene to 17q25-qter. We and others had localized the gene to 17q21–q25 (14,44–46) and we have further localized the gene to 17q21–23 (13). The discrepancy could be due to the limits of the techniques used, such as estimates of gene localization in somatic cell hybrids that included translocation chromosomes as compared to *in situ* hybridization.

We next sought to determine whether the cDNA hybridized to an mRNA of appropriate size to code for GAA and, most importantly, whether the cDNA would detect any abnormality of this mRNA in patients with GAA deficiency. RNA from normal and three GAA-deficient human cell lines was electrophoresed in agarose-formaldehyde gels, transferred to nitrocellulose filters, and hybridized to the labeled 2.0-kb cDNA insert. The cDNA hybridized to an mRNA from normal cells of approximately 3.4 kb, consistent with the size expected for a protein of approximately 105 kDa. RNA from cells of one patient with infantile-onset GAA deficiency (GM 244) showed a normal size and amount of the 3.4-kb mRNA. In contrast, RNA from a second infantile-onset GAA-deficient cell line (GM 4912) did not contain any detectable mRNA for GAA. RNA from an adult-onset GAA-deficient cell line (GM 1464) showed reduced amounts and smaller size (approximately 2.9 kb) of GAA mRNA. Essentially equivalent amounts of undegraded mRNA were found in all of the samples, as demonstrated by hybridization to a probe for adenosine deaminase (47).

Thus, we had isolated a partial cDNA clone that encoded GAA based on the following four lines of evidence: (1) The protein encoded by the cDNA was recognized by a monoclonal antibody to the enzyme. (2) The cDNA specifically hybridized to DNA from a small portion of chromosome 17 that contains the locus for GAA. (3) The cDNA recognized an mRNA (3.4 kb) of size sufficient to encode a 105-kDa protein (4). Most significantly, the cDNA detected abnormalities in the mRNA of patients with GAA deficiency.

We have started to use the partial cDNA for GAA to study additional patients, to isolate a full length cDNA, and to isolate the structural gene (including introns and exons).

FURTHER STUDIES OF GAA-DEFICIENT PATIENTS

Preliminarily, we have used the partial cDNA to evaluate heterogeneity of the disease at the level of gross aberrations of RNA and DNA in cells from 11 additional patients. We can summarize these findings as follows: 14 patients have been examined with respect to mRNA for GAA by Northern analysis. Of the 10 infantile-onset patients, half had grossly normal mRNA and half had no detectable mRNA. Adult-onset patients showed either grossly normal mRNA (one patient) or reduced amount and smaller size of mRNA (3 of 4). In all cases, mRNA for GAA was quantitated relative to mRNA for adenosine deaminase in the same blot.

DNA from 14 GAA deficients was digested with *Eco*RI, *Sac*I, or *Pst*I and hybridized to the 2-kb cDNA for GAA. DNA from 12 patients showed patterns of hybridization that were identical to DNA from normals. Two infantile-onset GAA deficients showed an extra *Sac*I band that was not seen in 50 normal individuals examined. Thus, the extra band is not a common restriction fragment length polymorphism (RFLP) but could be a disease-specific mutation. Further studies are needed to define a possible insertion or deletion or disease-specific alteration in a *Sac*I site.

In preparation for screening DNA from GAA-deficient patients for gross deletions or rearrangements that may not be due to normal genetic polymorphisms (RFLPs), we screened DNA from five independent individuals for high-frequency RFLPs after digestion with seven enzymes. Preliminarily, we have found RFLPs for *Pst*I, *Sac*I, *Taq*I, and *Msp*I. The delineation of high-frequency RFLPs will be useful for linkage analysis to the neurofibromatosis gene (48).

ISOLATION OF A 20-kb GENOMIC FRAGMENT

To isolate the 20-kb genomic fragment containing a portion of the structural gene, we constructed a human genomic library using size-fractionated genomic DNA that had been digested to completion with *Eco*RI and ligated into λEMBL4. We isolated a phage that contained a 20-kb *Eco*RI fragment. The pattern of hybridization observed following digestion with several restriction enzymes was consistent with that seen with human genomic DNA. Additionally, hybridization with a probe for human Alu sequences indicated that two of four *Sac*I fragments contained human repetitive Alu sequences (49). These results were consistent with the results of the cotransformation studies that are described in more detail below.

DNA-MEDIATED COTRANSFORMATION WITH A DOMINANT SELECTABLE GENE: AN APPROACH TO ISOLATION OF THE STRUCTURAL GENE

In experiments conducted simultaneously with the attempts to obtain the cDNA (described above), we tried an alternative approach to isolate the structural gene directly. Although we are no longer using this latter approach, we believe it is informative to describe the results of those experiments. We cotransformed mouse 3T3 cells with calcium phosphate-precipitated human genomic DNA or DNA derived from a mouse × human hybrid that contained human chromosome 17 together with the dominant selectable bacterial gene *Neo* (35) (contained in a pBR322 plasmid). The presence of the *Neo* gene conferred resistance to the antibiotic G418. We tested G418-resistant clones for expression of human GAA (39) as first detected by heat inactivation since the human enzyme is more resistant to heat than the mouse enzyme. We then confirmed the presence of the human enzyme by RIE (rocket immunoelectrophoresis) using a human-specific antibody (14). We isolated several *Neo*-resistant clones that expressed human GAA.

To evaluate the presence and amount of human DNA in the mouse cell clones expressing the human enzyme, we tested DNA from these clones for repetitive DNA sequences (Alu) (49) that are specific for human DNA. We analyzed DNA from one of those clones that expressed human GAA and was derived from cotransformation with hybrid cell DNA. The isolated DNA contained only three to five human repetitive sequences (Alu) and, therefore, very little human DNA.

That DNA contained an Alu band that by its presence or absence appeared to correlate with expression of human GAA (39).

We next constructed a recombinant phage library using genomic DNA from a GAA-positive subclone that contained few human sequences. We screened that genomic library with pBR322-Alu (to detect pBR-*Neo* sequences and human Alu repetitive sequences) and we isolated 20 positive recombinant phage. At this point, the next step would have required a strategy to determine which of these 20 recombinant phage actually contained portions of the structural gene for GAA. Although we had simultaneously succeeded in isolating a cDNA for GAA and, therefore, departed from this approach, we performed some preliminary experiments to determine whether we had isolated a portion of the structural gene of GAA. We tested DNA from cotransformants and from the positive recombinant phage for hybridization with the cDNA that encodes human GAA. The cDNA for GAA recognized fragments generated by digestion of DNA from the cotransformants with the restriction endonuclease *Pst*I. Several of the generated fragments comigrated with fragments generated by digestion of genomic human DNA with *Pst*I and also differed in size from fragments generated by digestion of mouse DNA. Therefore, the DNA from the cotransformed cells, which contained less than 1% of the human genome, did contain the gene for GAA. Additionally, one of the recombinant phage contained a 6-kb *Eco*RI fragment that was recognized by the cDNA for GAA and by a probe for Alu. That 6-kb genomic DNA insert was probably a fragment of the structural gene for GAA.

SUMMARY

The isolation of a cDNA for human GAA provides a powerful tool for analysis at the molecular level of genetic heterogeneity in the human GAA gene in normals and in patients with acid α-glucosidase deficiency. Examination of mRNA in patients has already revealed such heterogeneity; several infantile-onset patients have shown no detectable mRNA. The absence of detectable mRNA without a gross deletion of the gene in some patients could imply an unstable mRNA or, less likely, a regulatory mutation. The presence of mRNA in other infantile-onset patients, some with no discernible CRM (5), could result from synthesis of an incomplete protein (nonsense mutation) or an unstable protein (missense mutation) or the antibody used could not recognize the mutant protein. The presence of mRNA in the adult-onset patients was expected because of the delayed onset of disease and, in most patients, the presence of detectable enzyme activity. The reduced amounts of mRNA in one patient correlated with the markedly reduced enzyme activity we found in that cell line. However, the smaller size of the mRNA in two patients was unexpected and provided a third type of molecular heterogeneity detected by gross studies of the mRNA. One explanation for the smaller mRNA could be an RNA-processing defect due to point mutations that give rise to an mRNA lacking an exon or prematurely terminated in a cryptic polyadenylation site. Alternatively, the shorter mRNA could result from an as yet undetected deletion in the DNA. No major deletion or rearrangement of the GAA gene was seen in the first three GAA-deficient cell lines we analyzed, but gross DNA alterations have already been seen in infantile-onset patients. Other deletions or rearrangements may be detected by studies with full-length cDNA probes.

The isolation of the cDNA for GAA will enable comparison with similar studies of other lysosomal genes (50–52) and could clarify the basic mechanisms of lysosomal enzyme processing and trafficking, besides providing a method to dissect the molecular basis for clinical and biochemical heterogeneity in the acid α-glucosidase deficiency syndromes.

REFERENCES

1. Pompe JC. Over Idiopatische Hypertrophie van het hart. Ned Tijdschr Generskd 1932; 76:304–311.
2. Courtecuissf V, Royer F, Habib R, Monnifer C, Denvos J. Glycogenose musculaire par deficit d'alpha 1-4-glucosidase simulant une dystrophie musculaire progressive. Arch Franc Pediat 1965; 22:1153–1164.
3. Engel AG, Gomez MR, Seybold ME, Lambert EH. The spectrum and diagnosis of acid maltase deficiency. Neurology 1973; 23:95–106.
4. Hers, HG. Alpha-Glucosidase deficiency in generalized glycogen-storage disease (Pompe's disease). Biochem J 1963; 86:11–16.
5. Howell RR, William JC. In Stanbury JO, Wyngaarden JB, Fredrickson DS, eds. The Metabolic Basis of Inherited Disease. New York: McGraw-Hill, 1983; pp 142–166.
6. Beratis N, La Badie GU, Hirschhorn K. Characterization of the molecular defect in infantile and adult acid alpha-glucosidase deficiency fibroblasts. Am J Hum Genet 1978; 62:1264–1274.
7. Bruni CB, Auricchio F, Covelli I. Alpha-Glucosidase glucohydrolase from cattle liver. J Biol Chem 1969; 244:4735–4742.
8. Brown BI, Brown DH. The subcellular distribution of enzymes in type II glycogenosis and the occurrence of oligo alpha-I,4-glucan glucohydrolase in human tissues. Biochim Biophys Acta 1981; 110:124–133.
9. Hasilik A, Neufeld EF. Biosynthesis of lysosomal enzymes in fibroblasts. J Biol Chem 1980; 255:4937–4945.
10. Martiniuk F, Hirschhorn R. Characterization of neutral isozymes of human alpha-glucosidase. Biochim Biophys Acta 1981; 658:248–261.
11. Kornfeld S. Trafficking of lysosomal enzymes in normal and disease states. J Clin Invest 1986; 77:1–6.
12. Solomon E, Swallow D, Burgess S, Evans L. Assignment of the human acid alpha-glucosidase gene (alpha-GLU) to chromosome 17 using somatic cell hybrids. Ann Hum Genet 1979; 42:273–281.
13. Martiniuk F, Ellenbogen A, Hirschhorn K, Hirschhorn R. Further regional localization of the genes for human acid alpha glucosidase (GAA), peptidase D (PEPD), and alpha mannosidase B (MANB) by somatic cell hybridization. Hum Genet 1985; 69:109–111.
14. Honig J, Martiniuk F, D'Eustachio P, Zamfirescu C, Hirschhorn K, Hirschhorn LR, Hirschhorn R. Confirmation of the regional localization of the genes for human acid alpha-glucosidase (GAA) and adenosine deaminase (ADA) by somatic cell hybridization. Ann Hum Genet 1984; 48:49–56.
15. Fifth International Workshop on Human Gene Mapping, 1979.
16. Swallow DM, Corney G, Harris H, Hirschhorn R. Acid alpha-glucosidase: A new polymorphism in man demonstrable by "affinity" electrophoresis. Ann Hum Genet 1975; 38:391–406.

17. Nickel BE, McAlpine PJ. Extension of human acid alpha-glucosidase polymorphism by isoelectric focusing in polyacrylamide gel. Ann Hum Genet 1982; 46:97–103.

18. Beratis NG, La Badie GU, Hirschhorn K. Genetic heterogeneity in acid alpha-glucosidase deficiency. Am J Hum Genet 1980; 35:21–33.

19. de Wet JR, Fukushima HJ, Dewji MN, Wilcox E, O'Brien JS, Helinski DR. Chromogenic immunodetection of human serum albumin and alpha-L-fucosidase clones in a human hepatoma cDNA expression library. DNA 1984; 6:437–447.

20. Wigler M, Pellicer A, Silverstein S, Axel R, Urlaub G, Chasin L. DNA-mediated transfer of the adenine phosphoribosyltransferase locus into mammalian cells. Proc Natl Acad Sci USA 1979; 76:1373–1376.

21. Pellicer A, Robins SD, Wold B, Sweet R, Jackson J, Lowy I, Roberts JM, Sim GK, Silverstein, S, Axel R. Altering genotype and phenotype by DNA-mediated gene transfer. Science 1980; 209:1414–1422.

22. Chang LJA, Gamble CL, Izaguirre CA, Minden MD, Wak TW, McCulloch EA. Detection of genes coding for human differentiation markers by their transient expression after DNA transfer. Proc Natl Acad Sci USA 1982; 79:146–150.

23. Berman JW, Basch RS, Pellicer A. Gene transfer in lymphoid cells: Expression of the Thy-1.2 antigen by Thy-1.1 BW5147 lymphoma cells transfected with unfractionated cellular DNA. Proc Natl Acad Sci USA 1984; 81:7176–7179.

24. Milman G, Herzberg M. Efficient DNA transfection and rapid assay for thymidine kinase activity and viral antigenic determinants. Somatic Cell Genet 1981; 7:161–170.

25. Perucho M, Hanahan D, Lipsich L, Wigler M. Isolation of the chicken thymidine kinase gene by plasmid rescue. Nature (London) 1980; 285:207–210.

26. Wigler M, Silverstein S, Lee LS, Pellicer A, Cheng Y, Axel R. Transfer of purified herpes virus thymidine kinase gene to cultured mouse cells. Cell 1979; 11:223–232.

27. Lin P-F, Zhao S-Y, Ruddle FH. Genomic cloning and preliminary characterization of the human thymidine kinase gene. Proc Natl Acad Sci USA 1983; 80:6528–6532.

28. Kavathas P, Herzenberg LA. Stable transformation of mouse L cells for human membrane T-cell differentiation antigens, HLA and beta-2-microglobulin: Selection by fluorescence-activated cell sorting. Proc Natl Acad Sci USA 1983; 80:524–528.

29. Yeung C-Y, Frayne EG, Al-Ubadi MR, et al. Amplification and molecular cloning of murine adenosine deaminase gene sequences. J Biol Chem 1983; 258:15179–15185.

30. Pulciani S, Santos E, Lauve AV, Long LK, Robbins KC, Barbacid M. Oncogenes in human tumor cell lines: Molecular cloning of a transforming gene from human bladder carcinoma cells. Proc Natl Acad Sci USA 1982; 79:2845–2849.

31. Murray MJ, Kaufman RJ, Latt SA, Weinberg RA. Construction and use of a dominant, selectable marker: A Harvey sarcoma virus-dihydrofolate reductase chimera. Mol Cell Biol 1983; 3:32–43.

32. Robins DM, Ripley S, Henderson AS, Axel R. Transforming DNA integrates into the host chromosome. Cell 1981; 23:29–39.

33. Perucho M, Hanahan D, Wigler M. Genetic and physical linkage of exogenous sequences in transformed cells. Cell 1980; 22:309–317.

34. Wigler M, Sweet R, Sim GK, Wold B, Pellicer A, Lacy E, Maniatis T, Silverstein S, Axel R. Transformation of mammalian cells with genes from procaryotes and eucaryotes. Cell 1979; 16:777–785.

35. Colbere-Garapin F, Horodniceanu F, Kourilsky P, Garapin AC. A new dominant hybrid selective marker for higher eukaryotic cells. J Mol Biol 1981; 150:1–14.

36. Monaco AP, Neve RL, Colletti-Feener C, Bertelson CJ, Kurnit DM, Kunkel LM. Isolation of candidate cDNAs for portions of the Duchenne muscular dystrophy gene. Nature (London) 1986; 323:646–650.

37. Royer-Pokora B, Kunkel LM, Monaco AP, Goff SC, Newburger PE, Baehner RL,

Cole FS, Curnutte JT, Orkin SH. Cloning the gene for an inherited human disorder—chronic granulomatous disease—on the basis of its chromosomal location. Nature (London) 1986; 322:32–38.

38. Martiniuk F, Mehler M, Pellicer A, Tzall S, La Badie G, Hobart C, Ellenbogen A, Hirschhorn R. Isolation of a cDNA for human acid alpha-glucosidase and detection of genetic heterogeneity for mRNA in three alpha-glucosidase deficient patients. Proc Natl Acad Sci USA 1986: 83:9641–9644.

39. Martiniuk F, Pellicer A, Mehler M, Hirschhorn R. Detection, frequency, and stability of cotransformants expressing nonselectable human enzymes. Somatic Cell Mol Genet 1986; 12:1–12.

40. Martiniuk F, Honig J, Hirschhorn R. Further studies of the structure of human placental acid alpha-glucosidase. Arch Biochem Biophys 1984; 231:454–460.

41. La Badie GU, Harris H, Beratis NG, Hirschhorn K. Monoclonal antibodies to acid alpha-glucosidase: Further evidence for genetic heterogeneity in Pompe disease. Am J Hum Genet 1985; 37:A12 (Abstr.).

42. Konings A, Hupkes P, Grosveld G, Reuser A, Galjaard H. Cloning a cDNA for the lysosomal alpha glucosidase. Biochem Biophys Res Commun 1984; 119:252–258.

43. Halley DJJ, Konings A, Hupkes P, Galjaard H. Regional mapping of the human gene for lysosomal alpha-glucosidase by in situ hybridization. Hum Genet 1984; 67:326–328.

44. Weil D, Van Cong N, Gross MS, Frezal J. Localisation du gene de l'alpha-glucosidase acide (alpha-GLUa) sur le segment q21 qter du chromosome 17 par l'hybridation cellulaire interspecifique. Hum Genet 1979; 52:249–257.

45. Sandison A, Broadhead DM, Bain AD. Elucidation of an unbalanced chromosome translocation by gene dosage studies. Clin Genet 1982; 22:30–36.

46. Nickel BE, Chudley AE, Pabello PD, McAlpine PJ. Exclusion mapping of the GAA locus to chromosome 17q21-q25. Cytogenet Cell Genet 1982; 32:303–305.

47. Orkin SH, Daddona PE, Shewach DS, Markham AF, Bruns GA, Goff SC, Kelley WN. Molecular cloning of human adenosine deaminase gene sequences. J Biol Chem 1983; 258:12753–12756.

48. Seizinger BR, Rouleau GA, Ozelius LJ, et al. Genetic linkage of von Recklinghausen neurofibromatosis to the nerve growth factor receptor gene. Cell 1987; 49:589–594.

49. Schmid CW, Jelinek WR. The Alu family of dispersed repetitive sequences. Science 1982; 216:1065–1070.

50. Ginns EI, Choudary PV, Martin BM, Winfield S, Stubblefield B, Mayor J, Merkle-Lehman D, Murray GJ, Bowers LA, Barranger JA. Isolation of cDNA clones for human beta-glucocerebrosidase using the lambda-gt11 expression system. Biochem Biophys Res Commun 1984; 123:574–580.

51. Fukushima H, de Wet JR, O'Brien JS. Molecular cloning of a cDNA for human alpha-L-fucosidase. Proc Natl Acad Sci USA 1985; 82:1262–1265.

52. Calhoun DH, Bishop DF, Bernstein HS, Quinn M, Hantzopoulos P, Desnick R. Fabry disease: Isolation of a cDNA clone encoding human alpha-galactosidase A. Proc Natl Acad Sci USA 1985; 82:7364–7368.

19

β-Hexosaminidase Deficiency: Molecular Genetics and Enzyme Processing

ROY A. GRAVEL AND DON J. MAHURAN

The lysosomal storage diseases are inherited metabolic disorders with a special characteristic: accumulated metabolites cannot leave the cells in which they are produced. The resulting lysosomal swelling and cellular destruction produce a constellation of storage-related organ damage and morphological change. Many of these disorders are fatal, but less severe forms impair growth and development, often with neurological damage. The prototype lysosomal storage disease is Tay–Sachs disease, resulting from a defect involving β-hexosaminidase [*N*-acetylhexosaminidase (EC 3.2.1.52)].

Infantile amaurotic idiocy was independently described by Warren Tay (1) and Bernard Sachs (2) about 100 years ago. However, it was not until 1968 that the biochemical basis of the disorder was established when Robertson and Stirling (3) found two forms of hexosaminidase on starch gel electrophoresis and O'Brien et al. (4) found that one of these isozymes was missing in Tay–Sachs disease. We now know that there are two major isozyme forms of hexosaminidase: hexosaminidase A (Hex A) made up of α- and β-subunits and Hex B containing only β-subunits (Fig. 19-1).

Both enzymes hydrolyze terminal *N*-acetylhexosamines in oligosaccharides, glycolipids, glycoproteins, and glycosaminoglycans. However, only Hex A, in combination with a specific activator protein, can hydrolyze GM_2 ganglioside, a glycosphingolipid that is generated primarily in brain and peripheral neurons. The inability to hydrolyze GM_2 ganglioside is the major cause of storage, and it is also responsible for the neural degeneration in hexosaminidase deficiency.

Several distinct inborn errors have been identified with hexosaminidase deficiency. As a group, they are known as the *GM_2 gangliosidoses.* The most common is Tay–Sachs disease, a debilitating, neurodegenerative disease that is invariably fatal by age 3–4 years in its severe form. It is caused by a mutation of the α chain.

Patients with Sandhoff disease have a similar clinical phenotype, but the mutation affects the common β-subunit and, as such, both Hex A and Hex B are affected. In a third form, the AB variant of GM_2 gangliosidosis, an activator protein, is deficient (5). These patients make both Hex A and Hex B but cannot form the activator–GM_2 complex that is necessary for hydrolysis by Hex A.

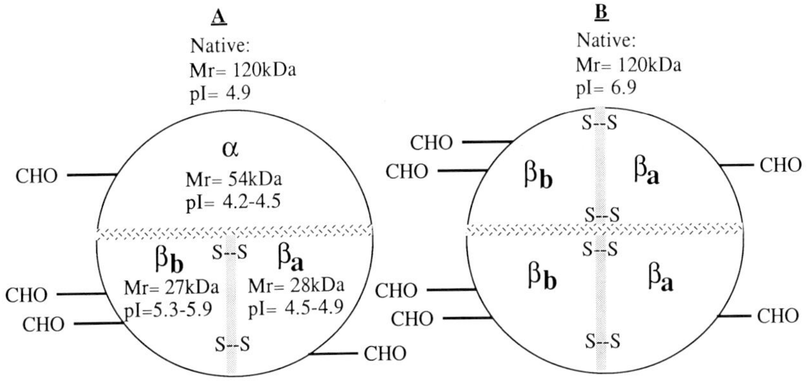

Fig. 19-1. Diagrammatic structure of Hex A and B, showing that Hex A is a heteropolymer composed of an α- and β-subunit and that Hex B is a homopolymer composed of two β-subunits. The β-subunit is made up of two polypeptides, linked by disulfide bridges, derived from the proteolytic cleavage of a single pro-β polypeptide in the lysosome. The defect in Tay–Sachs is in the α-subunit; the defect in Sandhoff is in the β-subunit.

One additional genetic disease affecting hexosaminidase is I-cell disease, with a less severe variant, pseudo-Hurler polydystrophy. In these disorders, there is a defect of the phosphorylation of lysosomal hydrolases (6). In the absence of phosphorylation, the enzymes cannot bind to a specific receptor that is required for delivery of the hydrolases to the lysosome. The affected lysosomes are devoid of most hydrolases, but there are abnormally high levels of the enzymes in serum and urine. The widespread inability to degrade oligosaccharide-containing molecules leads to the Hurler-like clinical features that are characteristic of mucopolysaccharide storage diseases.

We have investigated the structure and function of the hexosaminidase isozymes and the nature of the mutations resulting in GM_2 gangliosidosis. We will now review the early structural findings and the isolation of cDNA clones that encode the polypeptides of hexosaminidase. We have provided a nearly complete description of the biosynthesis and processing of the enzyme and we have used cDNA clones as probes to examine the mutations.

Early investigations of the structure of hexosaminidase were hampered by the need to do many protein fractionation steps that led to partial degradation of the enzyme. Those studies led to several models for the structure of the hexosaminidases that were not fully compatible with biochemical or genetic data. With the development of an affinity ligand by Geiger et al. (7), it was finally possible to recover enough unaltered enzyme for direct examination of structure. However, the subunit structure of the mature hexosaminidase isozymes remained controversial until we discovered that the diffuse "single" band on SDS gels corresponding to the reduced β-subunit comprised two distinct polypeptides of similar mobility (8). Using isoelectric focusing and ion-exchange high-performance liquid chromatography (HPLC) to resolve the denatured β polypeptides, we found that the β-subunit is composed of two polypeptides, β_a and β_b, held together by disulfide bonds. Thus, Hex A has the structure $\alpha[\beta_a\beta_b]$ and Hex B has the structure $2[\beta_a\beta_b]$

(Fig. 19-1). Pulse chase and peptide mapping experiments with human fibroblasts (9,10) suggested that the two β polypeptides arise from a common pro-β precursor. A contrasting view was proposed by Hasilik and Neufeld (9), who suggested that the β-subunit was composed of a major β polypeptide and smaller fragments. Resolution of the precise structure of the β-subunit and the mechanism of its maturation would require knowledge of the complete amino acid sequence.

ISOLATION OF cDNA CLONES USING OLIGONUCLEOTIDE PROBES

Several methods have proved effective for the isolation of clones that encode mammalian enzymes. The most commonly used procedures depend on the availability of either a specific antibody or partial amino acid sequence of the enzyme. Both approaches have been used successfully to isolate cDNA clones coding for hexosaminidase. We favored the use of oligonucleotide probes, determined from the partial sequence of peptides generated from purified enzyme, to screen cDNA libraries directly. In that procedure, it is possible to confirm the nature of a cDNA isolate by sequencing a probe-binding restriction fragment of the cDNA insert, then comparing the deduced amino acid sequence to the peptide sequence data already generated.

The key to any cDNA isolation is the availability of a good library. We selected the simian virus 40 (SV40)-transformed fibroblast cDNA library of Okayama and Berg (11). It contains over one million independent clones and even "rare" mRNAs are represented. In addition, the library was constructed by vector priming, in which the oligo(dT) primer normally used to prime the synthesis of the cDNA copy of the mRNA is made part of the cloning vector. That feature assures that all cDNA inserts contain a poly(A) tail at their 3′ end and that the inserts are always in the same orientation in the vector. The pcD library provides for a rapid evaluation of cDNA clones, sometimes taking as little as 6 weeks to isolate and prove a clone by DNA sequencing and comparison with peptide data.

We obtained amino acid sequence data from peptides generated by CNBr or tryptic cleavage of each subunit of hexosaminidase after isolation by reverse-phase HPLC. DNA sequences were deduced from the peptide sequences and the best ones were used to specify oligonucleotide probes (Fig. 19-2). Two characteristics were important in selecting probes. The first involves the redundancy of the genetic code. Since most amino acids are encoded by more than one triplet codon, we preferred to use peptide sequences that contain only one or two codon alternatives in order to minimize the number of oligonucleotides that would have to be combined to account for all the possible coding sequences. Highly favored are Met or Trp, which are encoded by single codons, or nine other amino acids that are encoded by two codons, with differences in the third or "wobble" position of the triplet.

The second characteristic we used to select oligonucleotide probes is length. The longer the oligonucleotide, the more the sequence is likely to be unique. Of course, the longer the probe the larger the number of oligonucleotides that would have to be specified in a mixture to account for the wobble positions in the sequence. In our experience, the number of combinations of sequences in the oligonucleotide probe mixture need not be a serious hindrance to screening as long as there is suf-

β SUBUNIT

```
Amino acid
sequence        Met-Val-Ile-Glu-Tyr-Ala

mRNA
sequence     5'-AUG-GUN-AUA-GAA-UAC-GCN-3'
                          C   G   U
                          U

Probe
sequence     3'-TAC-CAN-TAN-CTT-ATG-CG-5'        17MER/64
                              C   A
```

α SUBUNIT

```
Amino acid
sequence        Ser-Tyr-Gly-Pro-Asp-Trp-Lys

mRNA
sequence     5'-TCN-TAT-GGN-CCN-GAT-TGG-AAA-3'
                AGT   C           C       G
                  C

Probe 1      3'-AGN-ATA-CCN-GGN-CTA-ACC-TT-5'   20mer/256
                      G           G

Probe 2      3'-TCA-ATA-CCN-GGN-CTA-ACC-TT-5'   20mer/128
                  G   G           G
```

Fig. 19-2. Oligonucleotide probes used for the isolation of cDNA clones coding for the β- (top) and α- (bottom) subunits of hexosaminidase. The amino acid sequences from peptide sequence data, the corresponding mRNA sequences, and selected oligonucleotide sequences are shown. The β-subunit probes were divided into two sets to reduce the number of alternatives used in screening cDNA libraries.

ficient length. We have successfully used oligonucleotides with as few as 12 bases, but we have found that at least 17 and preferably 20 or 23 bases are optimal. A 17-mer might generate 5–10 classes of cDNA clones, but a 23-mer might generate only one or two classes, considerably simplifying the analysis. Figure 19-3 compares the clones detected with a 12-mer containing 8 combinations of oligonucleotides versus a 17-mer with 256 combinations (12). In spite of the much smaller number of oligonucleotides making up the 12-mer probe, the sequence is so short that all the clones on the filter show up with the probe, with the correct clones giving a relatively stronger signal. In contrast, the use of 256 alternative oligonucleotides did not obscure the resolution of the 17-mer on the same filter. Only the correct clones were detected in this case (the probes correspond to different sequences within the same cDNA).

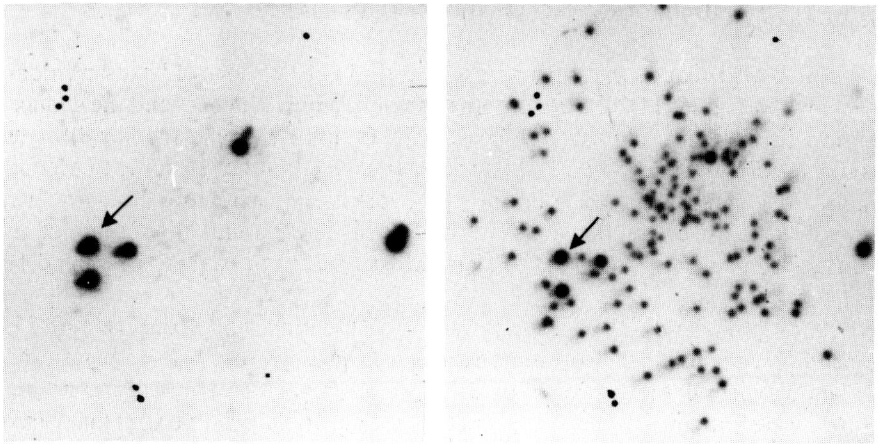

Fig. 19-3. Illustration of the effect of length and complexity of mixtures of probes on colony hybridization. The panels are photographs of the exposed X-ray films from the same blot after probing with the indicated oligonucleotides. The 12-mer (8 combinations) and 20-mer (256 combinations) hybridize to the same cDNA at different sites. The blots are from a secondary round of purification of a positive colony.

Proving the identity of a cDNA clone can often be the most difficult part of the process. For the α- and β-subunits of hexosaminidase, the deduced amino acid sequences of the peptides used to specify the oligonucleotides were identified in the cDNA clones, as were the sequences of several additional peptides in each case (13, 14). In the absence of peptide sequence data, a number of approaches can be used to either prove the clone or at least eliminate false clones. If the enzyme has already been mapped to a human chromosome, Southern blotting with DNA from somatic cell hybrids can be used to ascertain that sequences homologous to the cDNA are found on the proper chromosome. Such an experiment can reduce the several independent classes or isolates of cDNAs to a single choice for sequencing. Of the 6 candidate cDNA clones for the α-subunit, only one mapped to human chromosome 15 when we examined DNA of hybrids with or without that chromosome (14).

Another way to validate the cDNA clones is to examine Northern blots of RNA from fibroblast cultures derived from patients who lack the enzyme. In almost all inborn errors, some patients show reduced amounts or absence of mRNA. Less frequently, the DNA is also altered. This can be detected by aberrant patterns on Southern blots, but they must be distinguished from "normal" polymorphisms. Although demonstrating these alterations would seem to be conclusive proof of the nature of a clone, it must be remembered that the clone will in turn be used to evaluate the RNA or DNA of patients. This "proof" is therefore circular, but it is difficult to imagine that a patient with a seemingly absent RNA could "confirm" the wrong clone. Myerowitz and Proia (15) used this type of study on Tay–Sachs patients to evaluate a cDNA clone that encoded the α chain of hexosaminidase and which had been isolated by an antibody procedure.

STRUCTURE AND PROCESSING OF HEXOSAMINIDASE

The amino acid sequences of the prepro-α and -β polypeptides of hexosaminidase are shown in Fig. 19-4 (14). The combination of peptide sequences and the deduced amino acid sequence identifies the beginning (amino) and end (carboxyl) of the proteins, as well as the internal *clip site* that is severed in the lysosome to produce the final enzyme protein. Isolation of glycopeptides, determination of amino acid sequence, and use of NMR to characterize the oligosaccharides have identified the positions and final structures of the oligosaccharide side chains (Figs. 19-4 and 19-5).

The prepro-α and -β polypeptides show 57% amino acid sequence and 55% nucleotide sequence homology. Such extensive homology was not expected because the two polypeptides are encoded on different chromosomes. However, some similarities were anticipated because both polypeptides contain active sites that cleave artificial substrates and the terminal sugar of asialo-GM_2 (GA_2) (16). They differ only because Hex A is active against GM_2 (16) and Hex B is not. The actual homologies suggest that the two polypeptides have a common evolutionary origin, presumably by gene duplication and transposition to separate chromosomes.

The primary sequence of the prepro-α and -β polypeptides can be deduced from the cDNA clones by searching for an appropriately long open reading frame flanked on either side by stop codons. The presence of a complete reading frame is what usually defines a full-length clone. The first in-frame Met residue after the stop codon at the 5′ end almost always corresponds to the initiator amino acid. In the α chain, the only Met residue after stop codons (17) and before the N-terminus of the mature protein is indicated as the expected start (Fig. 19-4). In the β chain, our longest cDNA contains an open reading frame to the 5′ end of the clone, although we were able to find a stop codon in the same reading frame nearby in a genomic clone. The first Met residue in the cDNA is shown as the putative start, although there are additional Met residues before the mature N-terminus of the β-subunit (Fig. 19-4).

Glycoproteins require an N-terminal signal peptide that directs synthesis of the nascent peptide into the lumen of the endoplasmic reticulum (ER). Although no common sequence is found among the signal peptides of glycoproteins, they range from 20 to 30 amino acids in length and Von Heijne (18) has identified rules, from the examination of about 75 signal peptide sequences, that define permissible sequences. Both the predicted α and β signal peptide sequences, indicated by the filled boxes in Fig. 19-4, are acceptable within the von Heijne constraints. These rules predict signal peptide cleavage sites after residue 22 for the α and 24 for the β chain. To evaluate these sites, it will be necessary to determine the amino acid sequence of the translated product of natural mRNA.

The aligned prepropolypeptides show C-termini in approximately the same position (Fig. 19-4). In the β-subunit, there seems to be no C-terminal processing because, in the mature enzyme, we recovered a peptide that contained the C-terminal sequence deduced from the cDNA. However, the most C-terminal peptide identified thus far for the α chain (LSHFI in Fig. 19-4) still leaves some 25 amino acids unaccounted for. Thus, there could be C-terminal processing in the α chain.

Polypeptide Chain | Amino Acid #

```
α                     MTSSRLWFSLLLAAAFAGRATA-------------------    22
                       *    *             **
β                     MELCGLGLPRPPMLLALLLATLLAAMLALLTQVALVVQVAEA   42

α  -------------LWPWP-QNFQTSDQRYLYPNNFQFQYDVSSAAQPGCSVLDEAFQRYR   68
                ***  *      *     *  *  **       *   *   *   *   *  ***  **
β  ARAPSVSAKPGPALWPLPLSVKMTPNLLHLAPENFYISHSPNSTAGPSCTLLEEAFRRYH   102
                                              @
                                                      Gn-Gn-(Man)3

α  DLLFGSGSWPRPYLTGKRHTLEKNVLVVSVVTPGCNQLPTLESVENYTLTINDDQCLLLS  128
      **   *         *   **      *   *   * * ***        *
β  GYIFGFYKWHHEPAEFQAKTQVQQLLVSITLQSECDAFPNISSDESYTILVKEPVAVLKA  162
                                                Gn-Gn-(Man)3
                            @
α  ETVWGALRGLETFSQLVWKSAEGTFFINKTEIEDFPRFPHRGLLLDTSRHYLPLSSILDT  188
    ***************      *** **   *  * *** *** * ********   ** *
β  NRVWGALRGLETFSQLVYQDSYGTFTINESTIIDSPRFSHRGILIDTSRHYLPVKIILKT  222

                            Gn-Gn-(Man)5-7

α  LDVMAYNKLNVFHWHLVDDPSFPYESFTFPELMRKGSYNPVTHIYTAQDVKEVIEYARLR  248
   ** ** ** ** *** *** **** * *****  ****     * **   **  ********
β  LDAMAFNKFNVLHWHIVDDQSFPYQSITFPELSNKGSYSLS-HVYTPNDVRMVIEYARLR  281

                                                  @
α  GIRVLAEFDTPGHTLSWGPGIPGLLTPCYSGSEPSGTFGPVNPSLNNTYEFMSTFFLEVS  308
   ***** ************ *   *******         *** ** ** ** *  *** * *
β  GIRVLPEFDTPGHTLSWGKGQKDLLTPCYSRQNKLDSFGPINPTLNTTYSFLTTFFKEIS  341
                     <----- βb |   | βa -----> Gn

α  SVFPDFYLHLGGDEVDFTCWKSNPEIQDFMRKKGFGEDFKQLESFYIQTLLDIVSSYGKG  368
    ****   ******* * ** *** ****** **** *** *******  ***     **
β  EVFPDQFIHLGGDEVEFKCWESNPKIQDFMRQKGFGTDFKKLESFYIQKVLDIIATINKG  401

α  YVVWQEVFDNKVKIQPDTIIQVWREDIPVNYMKELELVTKAGFRALLSAPWYLNRISYGP  428
    ******* ***  * **  **  *     *  **  **  **    *******  ****
β  SIVWQEVFDDKVKLAPGTIVEVWK-DSA--YPEELSRVTASGFPVILSAPWYLDLISYGQ  458

α  DWKDFYVVEPLAFEGTPEQKALVIGGEACMWGEYVDNTNLVPRLWPRAGAVAERLWSNKL  488
   **    * **** * **   ** * ****** ****** *** ******* ** ***** *
β  DWRKYYKVEPLDFGGTQKQKQLFIGGEACLWGEYVDATNLTPRLWPRASAVGERLWSSKD  518
                                         @ 1

α  TSDLTFAYERLSHFRCELLRRGVQAQPLNVGFCEQEFEQT                      528
    *   ** **   **    **  ****   *  *   *
β  VRDMDDAYDRLTRHRCRMVERGIAAQPLYAGYCNHENM                        556
```

Putative Signal Peptides (shaded box)

Partial Amino Terminal Sequences of the Mature Polypeptide Chains (rectangular box)

Carboxy-Most Peptides Isolated (rounded box)

@ Putative Glycosylation Sites for which no corresponding glycopeptide was isolated

— — Amino Acid Sequence Obtained from Isolated Peptides, Gaps Represent "Blanks" from the Sequenator Caused by Attached Oligosaccharides.

1. No "Blank" was found at the 'N' of this peptide confirming that it is not glycosylated.

For example, an RR (Arg-Arg) sequence near the C-terminus could be a candidate for a cathepsin B or a more general, tryptic-like cleavage. Such dipeptides of basic amino acids are prevalent in the cathepsin B-like processing of neuropeptides.

N-terminal sequencing of the mature protein revealed the sites of proteolytic cleavage of the propolypeptides to generate the final enzyme structure (19). Three N-termini were identified in Hex A and two in Hex B (Fig. 19-4, open boxes). The N-termini of the mature α and β_b polypeptides were in identical positions in the aligned sequences, indicating that, in the lysosome, 65 amino acids had been removed from the pro-α and 97 amino acids from the pro-β sequences. The N-terminus of the α chain was "ragged" [about half starting at Thr (T) and half at the next amino acid, Leu (L)] implying that the initial cleavage site was more N-terminal with subsequent exopeptidase digestion. There are several candidate cathepsin B or tryptic-like sites within the removed sequence for both propolypeptides that could be the initial site of cleavage.

Next, using the N-terminal sequences as indicators, the β_b peptide was placed at the N-terminal half and the β_a peptide at the C-terminal half of the pro-β sequence. Both peptides were found in the β-subunit of Hex A and Hex B, but they contained one striking difference. The N-terminus of β_a was displaced by one amino acid residue depending on its isozyme of origin. Thus, in Hex B, the N-terminus occurred uniformly at Lys (K), whereas in Hex A, the N-terminal amino acid was at the next residue, Leu (L) (Fig. 19-4, β_a open box). This difference suggests a similar but not identical conformational presentation of the site of cleavage in Hex A and Hex B.

The mechanism of the internal cleavage to generate the β polypeptides can be inferred by inspection of the sequence of the C-terminal peptide of the β_b (N-terminal half of the pro-β chain) polypeptide. The most C-terminal peptide sequenced from β_b (open oval in Fig. 19-4) should have terminated with an arginine residue (R) but stopped abruptly with the identified serine (residues 304–311, Fig. 19-4). The sequence unaccounted for between the two peptides is Arg-Gln-Asn-Lys (RQNK) for Hex A and Arg-Gln-Asn (RQN) for Hex B. Each one of these amino acids would carry a strong positive charge in the acidic milieu of the lysosome. We surmise that in the β-subunit of the proisozymes, this sequence must be extended at the surface of the protein and subject to proteolytic attack. In contrast, the corresponding sequence in the pro-α polypeptide is probably buried and not subject to cleavage. On either side of the cleavage site in the β-subunit (asterisks in Fig. 19-4) there was extensive homology between the two subunits. However, there was no homology at all between the deduced sequences of the pro-α and -β polypeptides at the cleavage site itself.

The characteristics of these cleavage sites led us to conclude that they are clipped solely because they present vulnerable faces to the lysosomal milieu; that

←

Fig. 19-4. Alignment of the deduced amino acid sequences of prepro-α and -β polypeptides of hexosaminidase. The amino termini of the mature α, β_a, and β_b polypeptides as well as the postulated signal peptides and the internal peptide removed during β_a and β_b formation are indicated. Also, the carboxyl most tryptic or CNBr peptides that we have isolated thus far are shown.

is, hexosaminidase is subjected to proteolytic attack in the lysosome. This is suggested by the lack of absolute uniformity at two of the N-termini and also by the highly hydrophilic nature of the internal cleavage site in the pro-β polypeptide. Furthermore, the secreted form of the enzyme, which does not pass through the lysosome, remains in the pro form, yet is enzymatically active and kinetically similar to the lysosomal forms (20). We therefore believe that the mature lysosomal forms of hexosaminidase and, perhaps, lysosomal hydrolases in general represent the residual, proteolytically resistant structures of the proenzymes.

OLIGOSACCHARIDES IN HEXOSAMINIDASE

The oligosaccharides in hexosaminidase are ultimately responsible for directing the enzyme to the lysosome. That process is initiated by transfer of a nine-mannose and two-glucose moiety to specific asparagine residues [sequence, Asn-X-Ser(Thr) in the ER] (21). All glycoproteins undergo this process. The two glucoses and one or two mannoses are removed before the protein leaves the ER. In the Golgi, the high-mannose structure is progressively trimmed and processed to form either a complex oligosaccharide that also contains other sugars, or to a partially trimmed high mannose structure that contains phosphate in the 6-position of some mannose residues (Fig. 19-5). Phosphorylation is catalyzed by the phosphotransferase, men-

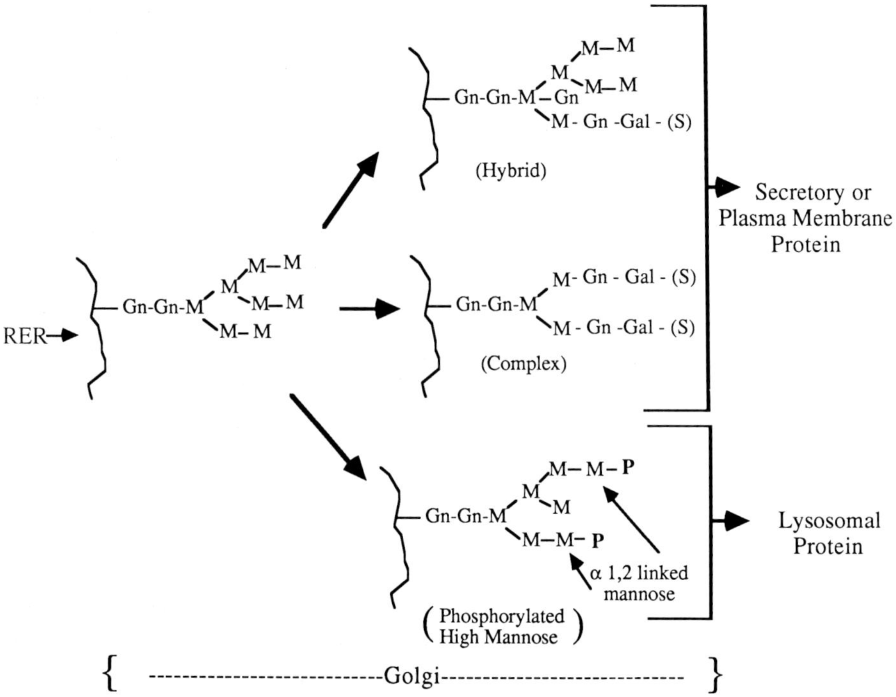

Fig. 19-5. Processing of oligosaccharides in glycoproteins. The diagram shows the status of the high-mannose structure postulated on leaving the rough endoplasmic reticulum (RER) and the resultant processing to complex, phosphorylated high-mannose, or hybrid structures occurring in the Golgi. M, mannose, Gn, *N*-acetylglucosamine, Ga, galactose.

tioned above, adding a phospho-*N*-acetylglucosamine group, followed by cleavage of the terminal sugar to leave an exposed mannose 6-phosphate (22). The phosphorylated oligosaccharide is the structure through which protein binds to the mannose 6-phosphate receptor in the Golgi. The enzyme–receptor complex is then vesicularly transported to the lysosome. These vesicles appear to be subsequently acidified, or they fuse with acidic prelysosomes (endosomes) that in turn fuse with, or become, through further acidification, the mature lysosome.

The major enigma is what determines which glycoproteins are phosphorylated and sent to the lysosome? Not all glycoproteins enter the lysosome, but all begin with the same high-mannose structure. Binding the phosphotransferase must, in part, depend on a specificity toward the peptide component of the glycoprotein (23). However, no universal amino acid sequence has been found in lysosomally destined glycoproteins.

The evidence that most lysosomal hydrolases are subject to this type of phosphorylation and receptor binding is demonstrated by I-cell disease (6). Because the phosphotransferase activity is lacking, 15–20 lysosomal proteins are not incorporated into lysosomes and, instead, are found in serum and urine. Some enzymes, notably β-glucocerebrosidase, a membrane-associated hydrolase, are not secreted, implying that there is more than one lysosomal pathway. In addition, there are at least two distinct mannose 6-phosphate receptors (24) and there may also be phosphate-independent pathways (25). The mechanism of lysosomal sequestration is therefore complex, but it is clear that hexosaminidase is processed via the mannose 6-phosphate receptor and that this pathway is disrupted in I-cell disease (6).

We used two methods to isolate the glycopeptides of hexosaminidase. In the simpler approach, we used the affinity of mannose-containing glycoproteins to bind to the plant lectin, concanavalin A (Con A) (26,27). From a Con A-Sepharose column, we isolated three distinct glycopeptides and the oligosaccharides were cleaved with glycosidases. The resulting peptides were sequenced to localize them within the deduced amino acid sequence, and the oligosaccharide structures were determined by NMR. Complex oligosaccharides were detected by separating tryptic peptides by HPLC and identifying glycopeptides by glucosamine determinations. Four glycopeptides were identified, one on the α chain, two on the β_b chain, and the last, representing a single attached GlcNAc (Gn) residue, on the β_a chain. All are located at asparagine residues and follow the Asn-X-Ser(Thr) rule (Fig. 19-6) (27).

Although all three oligosaccharides with sufficient structure show the core $GlcNAc_2$-Man_3 typical of Asn-linked oligosaccharides, none has the end point structure of compounds synthesized in the Golgi (26,27) (compare Figs. 19-5 and 19-6). Instead, they seem to be degradation products formed by glycosidic attack within the lysosome. The size of the oligosaccharide may give an indication of the extent to which it is exposed on the surface of the protein. For example, the small Man_3 and GlnNAc structures imply that those oligosaccharides are extensively degraded. In contrast, the $Man_{5,\ 6,\ \text{or}\ 7}$ alternatives found at the second glycosylation site in β_b are probably protected from glycosidic attack (Fig. 19-6). In no case would phosphate residues be expected because the action of phosphatases would have removed them from any site in the lysosome. Nevertheless, Hasilik and Neufeld (9), using ^{32}P labeling of fibroblasts, found that both the α- and β-subunits of hexosaminidase do contain phosphate. In the α chain there is only one eligible oligo-

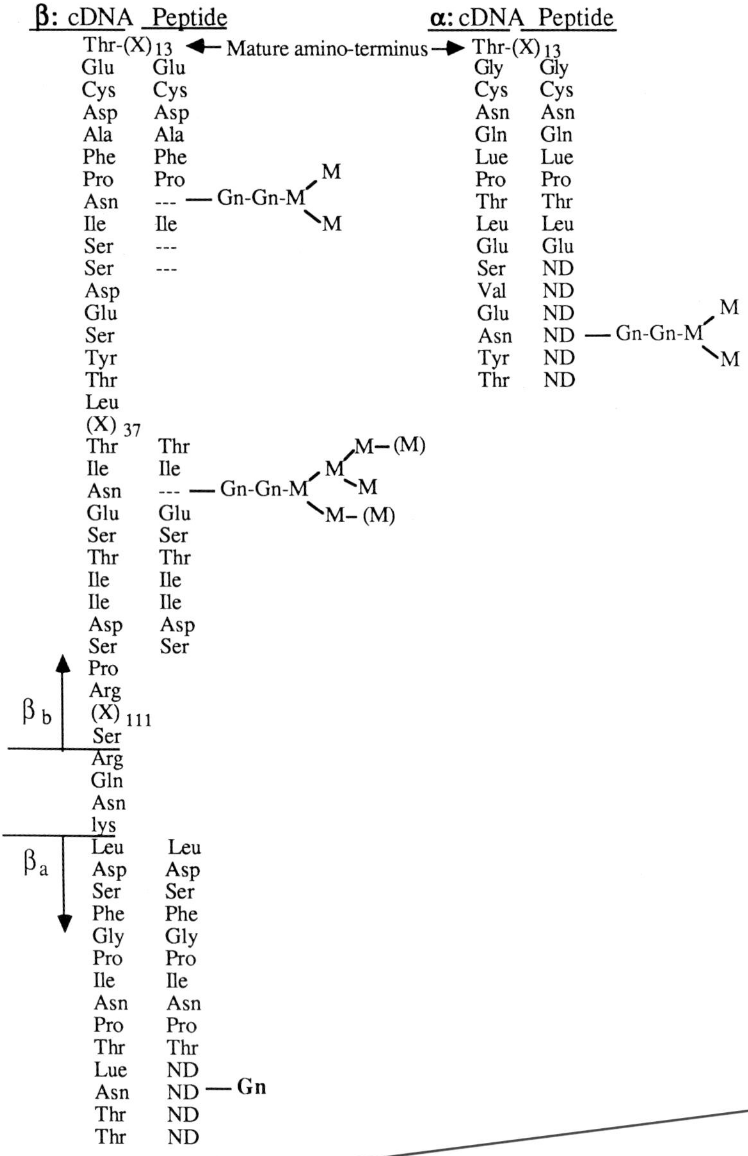

Fig. 19-6. Structures of the oligosaccharides and the amino acid sequences determined at their attachment sites with alignment to the deduced amino acid sequence from the α and β cDNA clones. Abbreviations are as in Fig. 19-5. (M) indicates the presence or absence of a mannose residue at the indicated site.

saccharide. In the β-subunit, it is not clear if only one or all of the sites might be phosphorylated.

Through the identification of proteolytic and gycosylation sites within the deduced amino acid sequences of the prepro-α and -β polypeptides, we have accounted for much of the ER, Golgi, and lysosomal processing of the hexosaminidase isozymes (Fig. 19-7).

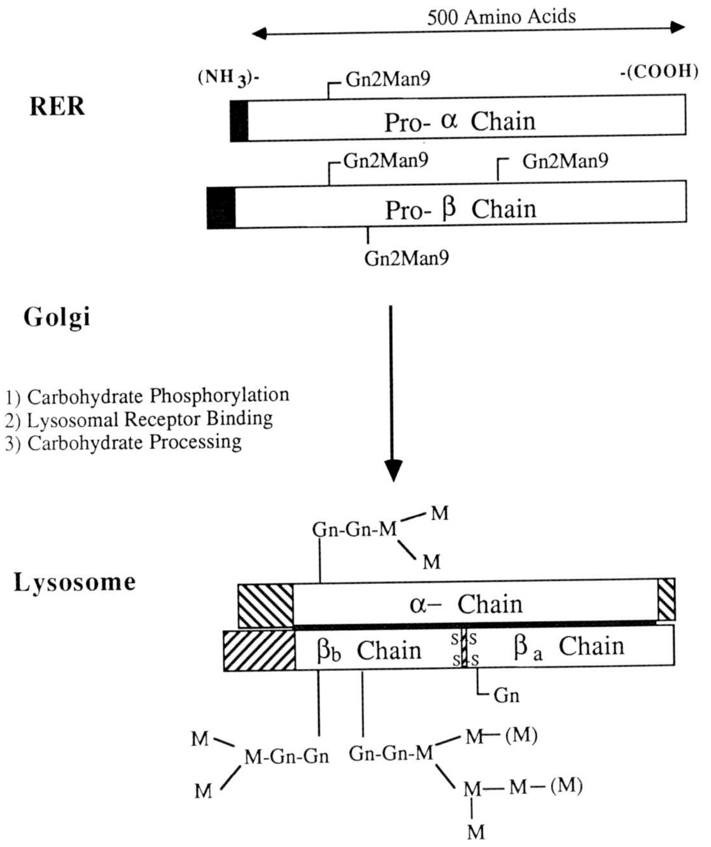

Fig. 19-7. Biosynthesis of hexosaminidase A. The steps of protein and oligosaccharide processing are shown along with their subcellular location. Abbreviations are as in Fig. 19-5.

Mutation in GM_2 Gangliosidosis

Initial investigations of mutation in Tay–Sachs and Sandhoff diseases have demonstrated considerable heteterogeneity at the level of mRNA and the gene. This is not surprising because of the known biochemical and clinical heterogeneity in these disorders. Myerowitz and Hogikyan (28) and we (14) found that patients with the classical, Ashkenazi form of Tay–Sachs disease have undetectable levels of prepro-α mRNA but other affected patients have some mRNA varying from none to normal levels. In the French-Canadian variant of Tay–Sachs disease, one patient showed a Southern blot pattern consistent with a deletion of the 5′ portion of the gene (28).

Similarly we found genetic heterogeneity to be characteristic of Sandhoff disease. We identified deletions of the 5′ half of the gene in two patients of different ethnic background, and there were different levels of mRNA in others (29). By combining residual levels of Hex A activity with the mRNA status of patient fibroblasts, we found that a correlation could be made with the clinical status of the patient (29). Detailed investigation of the distribution and haplotype origin of these mutations, especially in Ashkenazi Tay–Sachs disease, is now possible.

CONCLUSION

By combining conventional protein chemistry, biochemistry, and genetic analysis of patients with the use and sequence analysis of cDNA clones, we have been able to sort most of the processing steps in the biosynthesis of hexosaminidase. The availability of cDNA clones has allowed initial examination of the molecular basis of mutation in the GM_2 gangliosidoses. With the analysis of gene structure and the direct sequencing of mutant alleles, we will soon be able to define the precise nature of mutations in Ashkenazi Jews and other populations.

ADDENDUM

Recently, Little et al. (1988) [J. Biol. Chem. 263:4288–4392] and we [Stirling et al. (1988) FEBS LET. 231:47–50] experimentally determined the sites of signal peptide cleavage in the pre-α and pre-β chains, respectively. The cleavage site in the pre-α chain is located between residues 22 (Ala) and 23 (Leu), as predicted in Fig. 19-2. However, the cleavage site in the pre-β chain is located between residues 42 (Ala) and 43 (Ala) rather than residues 24 and 25, as we had predicted in Fig. 19-2. This suggests that the third (Met, residue 26 in Fig. 19-2) rather than the first in-frame ATG (Met, residue 1 in Fig. 19-2) is the initiation site for synthesis of the prepro-β polypeptide chain.

ACKNOWLEDGMENTS

We would like to thank C. Hew, M. Whiteside, and S. Joshi for determining the amino acid sequence of the glycopeptides. We also wish to thank J. Carver and A. Grey for their help in obtaining and interpreting the NMR spectra.

REFERENCES

1. Tay W. Symmetrical changes in the region of the yellow spot in each eye of an infant. Trans Ophthalmol Soc UK 1881; 1:155.

2. Sachs B. On arrested cerebral development with special reference to its cortical pathology. J Nervous Mental Dis 1890; 21:475–479.

3. Robinson D, Stirling JL. N-Acetyl-beta-glucosaminidases in human spleen. Biochem J 1968; 107:321–327.

4. Okada S, O'Brien JS. Tay–Sachs disease: Generalized absence of a beta-D-N-acetylhexosaminidase component. Science 1969; 165:698–700.

5. Conzelmann E, Sandhoff K. AB variant of infantile GM_2 gangliosidosis: Deficiency of a factor necessary for stimulation of hexosaminidase A-catalyzed degradation of ganglioside GM_2 and glycolipid GA_2. Proc Natl Acad Sci USA 1978; 75:3979–3983.

6. Bach G, Bargel R, Cantz M. I-cell Disease: Deficiency of extracellular hydrolase phosphorylation. Biochem Biophys Res Commun 1979; 91:976–981.

7. Geiger B, Ben-Yoseph Y, Arnon R. Purification of human hexosaminidase A and B by affinity chromatography. FEBS Lett 1974; 45:276–281.

8. Mahuran DJ, Tsui F, Gravel RA, Lowden JA. Evidence for two dissimilar polypeptide chains in the β_2 subunit of hexosaminidase. Proc Natl Acad Sci USA 1982; 79:1602–1605.

9. Hasilik A, Neufeld EF. Biosynthesis of lysosomal enzymes in fibroblasts: Synthesis as precursors of higher molecular weight. J Biol Chem 1980; 255:4937–4945.

10. Tsui F, Mahuran DJ, Lowden JA, Mosmann T, Gravel RA. Characterization of polypeptides serologically and structurally related to hexosaminidase in cultured fibroblasts. J Clin Invest 1983; 71:965–973.

11. Okayama H, Berg P. High-efficiency cloning of full-length cDNA. Mol Cell Biol 1982; 2:161–170.

12. Lamhonwah A-M, Barankiewicz TJ, Willard HF, Mahuran DJ, Quan F, Gravel RA. Isolation of cDNA clones coding for the α and beta chains of human propionyl-CoA carboxylase: Chromosomal assignments and DNA polymorphisms associated with PCCA and PCCB genes. Proc Natl Acad Sci USA 1986; 83:4864–4868.

13. O'Dowd B, Quan F, Willard H, et al. Isolation of cDNA clones coding for the β-subunit of human β-hexosaminidase. Proc Natl Acad Sci USA 1985; 82:1184–1188.

14. Korneluk RG, Mahuran DJ, Neote K, et al. Isolation of cDNA clones coding for the α-subunit of human β-hexosaminidase: Extensive homology between the α and β subunits and studies on Tay–Sachs disease. J Biol Chem 1986; 261:8407–8413.

15. Myerowitz R, Proia RL. cDNA clone for the α-chain of human beta-hexosaminidase: Deficiency of α-chain mRNA in Ashkenazi Tay–Sachs fibroblasts. Proc Natl Acad Sci USA 1984; 81:5394–5398.

16. Conzelmann E, Sandhoff K. Purification and characterization of an activator protein for the degradation of glycolipids GM_2 and GA_2 by hexosaminidase A. Hoppe-Seyler's Z Physiol Chem 1979; 360:1837–1849.

17. Myerowitz R, Piekarz R, Neufeld EF, Shows TB, Suzuki K. Human β-hexosaminidase α chain: Coding sequence and homology with the β chain. Proc Natl Acad Sci USA 1985; 82:7830–7834.

18. von Heijne G. Patterns of amino acids near signal-sequence cleavage sites. Eur J Biochem 1983; 133:17–27.

19. Mahuran DJ, Klavins MH, Neote K, Leung A, Gravel RA. Proteolytic processing of human pro-β hexosaminidase: Identification of the internal site of hydrolysis that produces the nonidentical β_a- and β_b-polypeptides in the mature β-subunit. J Biol Chem, submitted.

20. Hasilik A, von Figura K, Conzelmann E, Nehrkorn H, Sandhoff K. Lysosomal enzyme precursors in human fibroblasts. Eur J Biochem 1982; 125:317–321.

21. Schachter H. Glycoprotein biosynthesis and processing. In Callahan JW, Lowden JA, eds. Lysosomes and Lysosomal Storage Diseases. New York: Raven Press, 1981, pp 73–93.

22. Goldberg D, Gabel C, Kornfeld S. Processing of lysosomal enzyme oligosaccharide units. In Dingle JT, Dean RT, Sly W. eds. Lysosomes in Pathology and Biology. New York: Elsevier, North-Holland, 1984, pp 45–62.

23. Lang L, Reitman M, Tang J, Roberts RM, Kornfeld S. Lysosomal enzyme phosphorylation: Recognition of a protein-dependent determinant allows specific phosphorylation of oligosaccharides present on lysosomal enzymes. J Biol Chem 1984; 259:14663–14671.

24. Hoflack B, Kornfeld S. Lysosomal enzyme binding to mouse P388D1 macrophage membranes lacking the 215-kDa mannose 6-phosphate receptor: Evidence for the existence of a second mannose 6-phosphate receptor. Proc Natl Acad Sci USA 1985; 82:4428–4432.

25. Krentler C, Pohlmann R, Hasilik AK, von Figura K. Lysosomal membrane proteins do not bind to mannose-6-phosphate-specific receptors. Hoppe-Seyler's Z Physiol Chem 1986; 367:141–145.

26. O'Dowd BF, Mahuran D, Cumming D, Lowden JA. Characterization by nuclear magnetic resonance of the concanavalin A binding oligosaccharides on the βb chain of placental β-hexosaminidase B: Lectin binding to the separated polypeptide chains of hexosaminidase A and B. Can J Biochem Cell Biol 1985; 63:723–729.

27. O'Dowd BF, Cumming D, Gravel RA, Mahuran DJ. Characterization of the high mannose glycans and mapping of their attachment sites to the deduced amino acid sequence of cDNA clones coding for the preproα- and preproβ-polypeptide chains of β-hexosaminidase. J Biol Chem, submitted.

28. Myerowitz R, Hogikyan ND. Different mutations in Ashkenazi Jewish and non-Jewish French Canadians with Tay–Sachs disease. Science 1986; 232:1646–1648.

29. O'Dowd BF, Klavins MH, Willard HF, Gravel R, Lowden JA, Mahuran DJ. Molecular heterogeneity in the infantile and juvenile forms of Sandhoff disease (O-variant GM2 gangliosidosis). J Biol Chem 1986; 261:12680–12685.

20

Ornithine Transcarbamylase: Enzyme Deficiency and Biogenesis of Mitochondrial Proteins

WAYNE A. FENTON

In the years since electron microscopy was first used to observe the structure of cells, it has become clear that there is an intricate architecture within the living cell, an arrangement as complex as that by which cells themselves are integrated into an organism. The application of sophisticated biological and biochemical methods to the dissection of this structure has demonstrated that its establishment and maintenance are crucial to cellular homeostasis and that the localization of macromolecules within it plays a vital role in cellular physiology.

Simultaneously, the investigation of the pathogenesis of inherited human diseases has suggested that mutations in localization signals or in transport pathways may account for some heritable diseases. The techniques of modern molecular biology and genetics have permitted us to bring these two sets of observations together to understand both the importance of cellular structure and biogenesis to normal metabolism and the contribution of disruptions in these pathways to the pathophysiology of human disease. Much has been learned in the past few years about the transport of proteins into and out of cellular organelles, such as the nucleus and lysosomes, and the roles these pathways play in the cell. This chapter will focus on another organelle, the mitochondrion, and specifically on one protein, ornithine transcarbamylase, which we have used as a model to study the biogenesis of mitochondrial proteins involved in human disease.

ORNITHINE TRANSCARBAMYLASE

Ornithine transcarbamylase (OTC) is the second enzyme of the urea cycle, by which humans and other ureotelic organisms convert ammonia, the toxic waste product of nitrogen metabolism, to the excretable product, urea. OTC catalyzes the condensation of ornithine and carbamyl phosphate to form citrulline. OTC is also found in nonureoteles, including bacteria and yeast, where it plays an important role in the synthesis of arginine. In humans, OTC activity is found in significant amounts only in the liver, with a smaller amount in the gut. Because the first enzyme in the cycle, carbamyl-phosphate synthase I, is also liver specific, the liver is the only organ capable of urea synthesis (1).

OTC Deficiency

Deficiency of OTC activity was first described in girls who had hyperammonemia and protein intolerance. OTC activity in liver biopsy samples was 5–20% of normal. The discovery that the pedigrees of some of these girls included multiple neonatal male deaths led to the hypothesis that OTC deficiency was X linked and lethal in hemizygous males. This was borne out by the observation in these families of newborn boys who showed complete OTC deficiency on post-mortem assay of liver. Additional evidence of X linkage was the observation of cellular mosaicism of OTC activity in hepatocytes from an obligate heterozygous carrier. A cytochemical stain for OTC activity showed two populations of cells—some with normal activity and others with no activity—in accord with the Lyon hypothesis of random X-chromosome inactivation in females (2).

Most affected boys follow the same clinical course (3,4). Soon after the first protein feeding, they become irritable, then lethargic, with extremely high blood ammonia levels. If untreated, coma and death follow within 1 or 2 weeks of birth. Unfortunately, no treatment is completely effective in reversing or preventing this sequence of events. Peritoneal dialysis to remove serum ammonia, protein restriction to reduce the ammonia load, and administration of keto analogs of essential amino acids and ammonia-conjugating agents such as sodium benzoate or sodium phenylacetate to sequester ammonia biochemically have all been used with varying degrees of success. Currently, the favored therapy is a combination of protein intake restriction and sodium benzoate/phenylacetate administration, with a reported 1-year survival rate of ~70% (5). Whatever the treatment, the patients require careful monitoring and rapid intervention in the event of a crisis, but many still succumb to overwhelming hyperammonemia, often as a result of an intercurrent infection that produces a catabolic state. In addition, survivors are usually developmentally delayed and mentally retarded and have cerebral palsy and seizures that are attributed to the repeated neurologic insults of the episodic hyperammonemic crises.

Girls who are heterozygous for OTC deficiency show a wider range of symptoms, even within a family, because X-inactivation is random. Some are almost as severely affected as their brothers, with neonatal hyperammonemia, coma, and death. Others follow a milder course, with protein avoidance and episodic hyperammonemia that may lead to mental retardation or lethal crisis. Still others are clinically normal and are identified only by provocative tests or by the family history.

In addition to the severely affected boys, some boys have had a clinical picture similar to that of the heterozygous girls. These boys have partial OTC deficiency with mutant enzymes that show abnormal kinetic parameters or pH optima (6). The clinical course has generally correlated well with the amount of residual enzyme activity. These observations established that the structural locus for OTC is indeed on the X-chromosome and suggested that different mutations in the OTC gene could lead to functional impairment and the metabolic consequences of OTC deficiency.

Prenatal Diagnosis and Heterozygote Detection

Because OTC is expressed essentially only in liver and not in amniocytes or other readily accessible fetal tissue, prenatal diagnosis of OTC deficiency was not possible until recently. In a few cases, fetal liver biopsy was used, but that procedure carries a high risk to the fetus and also suffers from uncertainty because there is developmental regulation of OTC expression.

One major reason for cloning a human OTC cDNA was to provide the necessary reagents for DNA-based prenatal diagnosis and carrier detection of OTC deficiency, and for molecular analysis of naturally occurring mutations at the OTC locus. Once we had recovered a full-length OTC cDNA (see below), we used it as a probe for Southern blots of DNA prepared both from normal individuals and from OTC-deficient patients and members of their families (7). We surveyed the patterns generated following digestion with a number of different restriction endonucleases in an effort to detect restriction site alterations, deletions, or rearrangements at the OTC locus in patients and to identify restriction fragment length polymorphisms (RFLPs) in normal controls. In 40 affected individuals, we found no evidence of a change in a restriction site as a result of the mutation causing OTC deficiency. Nussbaum et al., however, described two families in which the mutation producing the OTC deficiency also resulted in the disappearance of a *Taq*I restriction site (8).

Four individuals of the 40 we examined showed changes in Southern blot patterns that could be attributed to deletions or rearrangements. One had a partial deletion of the 3′ end of the OTC gene, two had virtually complete deletions of the gene, and the fourth had a complex deletion/rearrangement. None of these alterations was seen cytogenetically. By examining Southern blot patterns from the mothers of the patients, we deduced that two of the deletions were inherited and the other two were new mutations in the patient.

Four RFLPs have been found at the OTC locus (Table 20-1). Two separate RFLPs have been identified with the enzyme *Msp*I, characterized by polymorphic bands at 6.6 and 6.2 and at 5.1 and 4.4 kilobase pairs (kbp), respectively. DNA from 35 unrelated control women was analyzed to determine the frequencies of these RFLPs in the general population. Forty-five percent of these women had bands at 6.6 and 5.1 kbp, 16% had bands at 6.6 and 4.4 kbp, 28% had bands at 6.2 and 5.1 kbp, and 11% had bands at 6.2 and 4.4 kbp. Thus, using RFLPs detected

Table 20-1 RFLPs at the OTC Locus

RFLP	Enzyme	Probe[a]	Polymorphic bands (kbp)	Constant bands (kbp)
1	*Msp*I	pHO731	6.6, 6.2	17.5, 5.4, 1.9
2	*Msp*I	pHO731	5.1, 4.4	17.5, 5.4, 1.9
3	*Bam*HI	p53H21	18.0, 5.2	—
4	*Taq*I	pHO731	3.7, 3.6	4.6, 2.4

[a]pHO731 is a plasmid containing a full-length human OTC cDNA. p53H21 is a plasmid containing a 1200-base pair fragment of single-copy human OTC genomic DNA.

with *Msp*I alone, it is possible to distinguish the two X-chromosomes in approximately 69% of control women.

An additional RFLP at the OTC locus was identified with *Taq*I, characterized by polymorphic bands at 3.7 and 3.6 kbp; it was detected in 11% of the control population. (This RFLP differs from the *Taq*I site mutation reported by Nussbaum.) Using the enzyme *Bam*HI, another RFLP, characterized by polymorphic bands at 18.0 and 5.2 kbp, was identified in 28% of the controls. No other RFLPs were detected when the DNA from 16 unrelated control women was digested with the endonucleases *Kpn*I, *Pst*I, *Sst*I, *Eco*RV, *Bgl*I, and *Xho*I. The polymorphisms identified with *Taq*I and *Bam*HI increase the number of women who are heterozygous for at least one of the four RFLPs to approximately 80%. Thus, these RFLPs are useful in many families for prenatal diagnosis of OTC deficiency, carrier detection, and carrier exclusion.

We have performed eight prenatal diagnoses based on RFLP analysis in pregnancies in which male fetuses were at risk for OTC deficiency. DNA was prepared either from chorionic villus biopsy samples or from cultured amniocytes obtained by amniocentesis. Seven of the fetuses were diagnosed as affected; in five cases, the parents elected to terminate the pregnancy. In two cases, we confirmed the diagnosis by assaying OTC activity in liver from the abortus; in both instances, OTC activity was undetectable. Of the three live-born infants, one was normal, as predicted; one affected boy died at about age 2 months in spite of treatment as outlined above; the other affected boy has been managed successfully for 18 months on a similar regimen.

An example of the diagnostic data is shown in Fig. 20-1 (9). In this family, amniocentesis provided material for fetal DNA preparation. The mother was an obligate carrier, based on family history, and she was heterozygous for the RFLP

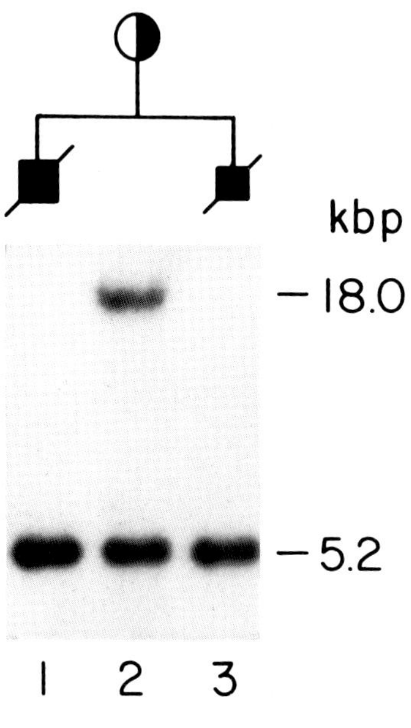

Fig. 20-1. RFLP analysis of DNA from a family at risk for a son with OTC deficiency. Lane 1, DNA from affected male sibling; lane 2, DNA from mother; lane 3, DNA from male fetus at risk. [From Fox et al. (9). Used with permission of the authors.]

identified with *Bam*HI (lane 2), characterized by bands at 18.0 and 5.2 kbp. DNA from the male fetus had a band at 5.2 kbp (lane 3). Because that band was also present in the previously affected child (lane 1), it identified the fetus as affected with OTC deficiency, and the pregnancy was terminated. The diagnosis was confirmed by direct assay of OTC activity from fetal liver obtained about 90 minutes after delivery of the abortus. The OTC activity was undetectable compared to the value obtained from an age-matched control fetal liver.

Detection of asymptomatic carriers of OTC deficiency is still a problem for families in which OTC deficiency has been diagnosed, because the fraction of OTC-deficient patients due to spontaneous mutation may be as high as one-third. Provocative tests in women who may be carriers of OTC deficiency, either by oral protein loading or by administration of allopurinol, have been used to identify carriers on the basis of increased orotic acid excretion, but uniformity and standardization of these tests are still lacking. On the other hand, women whose family histories indicate that they have a 50% chance of being a carrier of OTC deficiency may benefit from RFLP analysis. In these families, the X-chromosome associated with OTC deficiency can be traced in the family, and at risk women who bear that chromosome will be identified as obligate carriers. Prenatal diagnosis can then be offered to these women before they have an affected child. Similarly, DNA analysis can exclude carrier status in family members who do not have the affected chromosome.

For example, the woman in Fig. 20-1 had four sisters, each having a 50% chance of being a carrier. DNA from three sisters did not exhibit the 5.2-kbp band. These women were, therefore, not carriers and were at no greater risk of having a son with OTC deficiency than any other woman in the general population. One sister, however, did have the 5.2-kbp fragment associated with OTC deficiency in that family; she was designated a carrier and would have the option of prenatal diagnosis.

We are now analyzing some of these mutations at the molecular level, to provide general information about the mechanisms of mutation that result in human disease, and also to generate specific data on individual mutations to give some insight into normal OTC function and to suggest new approaches to diagnosis and therapy. For example, we have isolated the deletion junction fragment from the patient with the partial 3′ deletion described above. We are studying this fragment further to better understand the mechanism of the deletional event; the fragment itself provides a unique probe with which to assess the carrier status of women in the patient's family.

TRANSPORT OF MITOCHONDRIAL PROTEINS

About a decade ago, it became clear that most mitochondrial proteins share several features in biogenesis. They are encoded by nuclear genes, synthesized on cytoplasmic polyribosomes, and then imported by mitochondria. It was postulated that defective import of these proteins would probably be one of several mechanisms responsible for the deficiency of mitochondrial enzymes, in addition to more common abnormalities such as failure of synthesis, abnormal kinetics, or protein instability. Because OTC is a mitochondrial protein, and because of our interest in inherited human disease, we undertook the study of OTC biogenesis to understand the pathogenesis of OTC deficiency and to explore the normal pathways of mitochondrial protein transport.

We first purified OTC to homogeneity from human liver and found that it is a trimer of identical, 36-kDa subunits. We then raised specific, polyclonal antibodies to the purified protein in rabbits; the antibodies cross-reacted with other mammalian OTC species, particularly from rat and mouse. After demonstrating that intact mitochondria do not import homogeneous, assembled OTC, we decided to determine the nature of the primary translation product using cell-free protein synthesis in a rabbit reticulocyte lysate system programmed with rat liver polysomal RNA, immunoprecipitation with anti-OTC antiserum, and polyacrylamide gel electrophoresis.

We found that OTC was synthesized as a single polypeptide about 4 kDa larger than the purified OTC subunit (Fig. 20-2, lane 1) (10). Incubation of this larger putative precursor with isolated intact rat liver mitochondria led to its internalization and proteolytic cleavage to a size identical to that of the mature mitochondrial OTC subunit (36 kDa) (lane 3). We then found that this bonafide precursor polypeptide, designated pOTC, consisted of a cleavable 4-kDa leader peptide joined to the mature OTC sequence at the amino-terminus; the amino-terminal leader was absolutely required for mitochondrial uptake.

We next examined several key features of this import pathway (11). First, we found that some inhibitors of mitochondrial electron transport and uncouplers of oxidative phosphorylation prevented the posttranslational uptake and conversion of pOTC to mature OTC. The nature of these inhibitors suggested that maintenance of the inner membrane potential (rather than the supply of ATP) is the predominant source of the energy required for translocation. Second, we fractionated isolated mitochondria into a series of submitochondrial components and showed that the protease responsible for cleaving pOTC to its mature size is localized in the mitochondrial matrix and requires a divalent cation (preferably Zn^{2+}) for activity. Third, we found that assembly of mature OTC subunits into the active trimeric form occurred only after transport into the mitochondria and cleavage of the leader peptide.

Based on these findings and the results of others (12,13), we proposed a multistep model (Fig. 20-3) for the uptake of proteins destined for the mitochondrial matrix. These proteins are synthesized on cytoplasmic polyribosomes as larger precursors that bear amino-terminal leader peptides that function as address signals. Once the precursors are delivered to the mitochondria and recognized (presumably by receptors on the outer membrane), the precursors are translocated across both the outer and inner membranes by a process that requires an intact membrane potential. Then, the leader peptide is removed by a divalent cation-dependent matrix protease. Finally, the mature subunits assemble to the active conformation.

Similar models of import of mitochondrial proteins have been proposed by many groups working with proteins from *Saccharomyces cerevisiae* and *Neurospora crassa,* as well as from mammalian cells. It seems certain that the general features of the import of proteins to the internal subcompartments of the mitochondrion have been conserved throughout evolution.

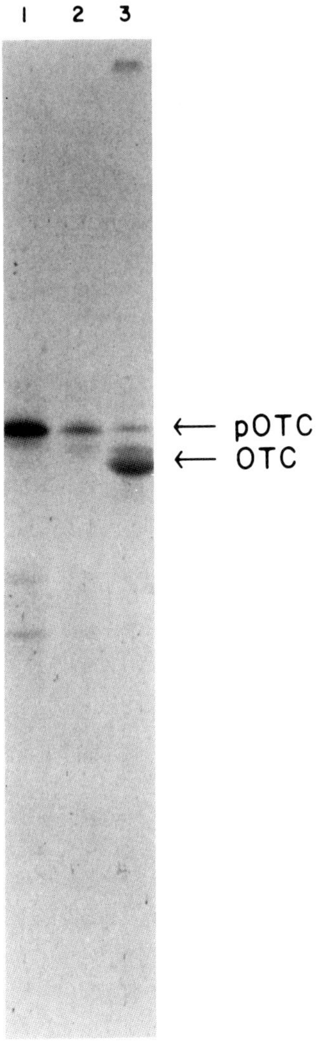

Fig. 20-2. *In vitro* translation and mitochondrial uptake and processing of OTC. Lane 1, the products of *in vitro* translation of rat liver mRNA with [^{35}S]methionine, immunoprecipitated with anti-OTC antiserum; lane 2, the supernatant recovered after incubation of the translation products with isolated intact rat liver mitochondria; lane 3, the mitochondrial pellet from such an incubation. [From Conboy and Rosenberg (10). Used by permission of the authors.]

Cloning of OTC cDNA

We decided that the molecular cloning of cDNAs for OTC would allow us to determine the structure of the amino-terminal leader peptide and to define the role it plays in directing mitochondrial uptake and processing. The crucial first step of the cloning strategy was to prepare a highly enriched OTC mRNA from rat liver, where the OTC message comprises about 0.1% of the total mRNA, using polysome immu-

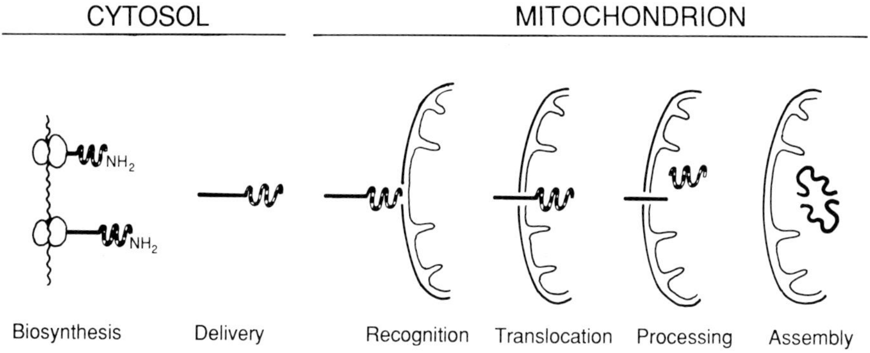

Fig. 20-3. Schematic model of the synthesis, uptake, and processing of a mitochondrial matrix protein. [From Rosenberg et al. (11). Used by permission of the authors.]

noadsorption (14). Polysomes were prepared from fresh rat liver homogenates and incubated with our polyclonal anti-OTC antibody. Those polysomes containing nascent OTC peptide chains reacted with the antibody and were recovered from the mixture by binding the attached antibody to a protein A-Sepharose affinity matrix.

After extensive washing to remove unbound material, the column was eluted with EDTA, which disrupted the bound polysomes and released the mRNA into the eluate. The mRNA was recovered by chromatography on oligo(dT) cellulose and, as shown in *in vitro* translation, comprised more than 90% OTC mRNA. Double-stranded cDNA was synthesized from that material, ligated into pBR322, and cloned in *Escherichia coli.* Colonies containing OTC sequences were identified by differential hybridization with radiolabeled cDNA probes produced from RNA of OTC-expressing and nonexpressing tissues and confirmed by the ability of the plasmid inserts to direct hybrid-selected translation of OTC. One was established as bonafide by determining that the amino acid sequence predicted from the DNA sequence of the insert corresponded exactly to the amino acid sequence of a peptide prepared from the purified rat OTC protein.

That insert was then used to screen a human liver cDNA library. A number of overlapping plasmid inserts were obtained that could be combined to yield an essentially full-length human OTC cDNA, with an open reading frame sufficient to encode pOTC (15). This was confirmed by constructing a eukaryotic expression vector, consisting of the OTC cDNA flanked by a viral control sequence and viral polyadenylation signal, and transfecting the vector into cultured HeLa cells, normally devoid of OTC protein, by calcium phosphate-mediated DNA transfer. The transfected cells now produced OTC protein, as judged by metabolic labeling with [^{35}S]methionine and immunoprecipitation, and, in the presence of an inhibitor of mitochondrial protein transport, accumulated pOTC. Stable cell lines derived from these transfectants contained measurable OTC activity in amounts up to 10% of that found in liver.

We deduced the complete primary structure of human pOTC from the nucleotide sequence of the cloned cDNA. The amino-terminal leader peptide contained 32 amino acid residues—in excellent agreement with prior work suggesting a mass of ~4 kDa (Fig. 20-4). The amino acid composition of the leader contains four

5 10

met leu phe asn leu arg ile leu leu asn asn ala

15 20

ala phe arg asn gly his asn phe met val arg asn

25 30

phe arg cys gly gln pro leu gln ↓ asn lys val....

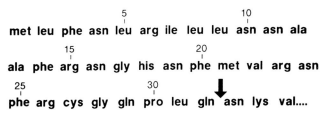

Fig. 20-4. Amino acid sequence of the leader peptide of human OTC. The arrow designates the site of cleavage between the leader peptide and the mature OTC subunit. [From Rosenberg et al. (11). Used by permission of the authors.]

arginines (at positions 6, 15, 23, and 26), but no glutamates or aspartates. This pattern suggests that the leader carries a net positive charge. The relative abundance of basic residues and absence of acidic ones are distinguishing features of most leader peptides that have been analyzed. Cleavage of the leader peptide from the mature portion occurs between glutamine and asparagine residues.

Molecular Analysis of OTC Transport

All our previous work had indicated that the amino-terminal leader peptides characteristic of mitochondrial precursors were necessary for import. To determine if they were also sufficient, we asked whether such a leader could "misdirect" a cytosolic protein to the mitochondria. We therefore fused the nucleotide sequence encoding the human OTC leader with a cDNA encoding the cytosolic enzyme, dihydrofolate reductase (DHFR). This fusion gene, along with required SV40 regulatory elements and a selectable marker, was used to transfect a mutant Chinese hamster ovary cell line devoid of endogenous DHFR. In transformants pulse-labeled with [^{35}S]methionine, the predicted chimeric protein was detected, as were the expected mitochondrial cleavage products. In addition, immunofluorescent staining of transformants with anti-DHFR antiserum revealed a pattern characteristic of mitochondrial localization. Finally, when the chimeric precursor was synthesized in a cell-free system and incubated posttranslationally with rat liver mitochondria, it was imported and cleaved. Thus, both in intact cells and *in vitro,* the human OTC leader sequence was sufficient to direct import of DHFR.

To take advantage of our ability to analyze OTC uptake and processing in the reconstituted *in vitro* system, we then transferred the OTC sequences to vectors containing the phage SP6 or T7 promoters. The appropriate phage RNA polymerase recognizes its promoter *in vitro* and, in the presence of nucleotides, synthesizes an RNA corresponding to the cDNA insert. This RNA can be translated *in vitro* in the rabbit reticulocyte lysate system to yield pOTC for incubation with mitochondria or mitochondrial fractions.

We have used these constructs as the starting point to generate a series of deletions and numerous site-directed and random amino acid substitutions in the human OTC leader peptide to obtain additional detailed information about functionally important features of this sequence (11). Analysis of the deletions revealed that, whereas neither the extreme amino-terminal nor carboxy-terminal portions of the leader is absolutely required for mitochondrial uptake and proteolytic processing, the mid-portion of the leader is critical. When codons 8–22 were deleted, no uptake or processing was noted. All other deletions resulted in much less extreme impairment of import.

To test the role of the arginine residues, those at positions 15, 23, and 26 were replaced at once by charge-neutral glycines. The triply substituted precursor failed to be taken up and cleaved by mitochondria *in vitro*. We then assessed the role of each of the four arginine residues individually. When the arginine residue at position 6 or position 26 was substituted by glycine, there was no significant effect on import and cleavage *in vitro*. Substitution of glycine for arginine at position 15 reduced processing by about 50%. The arginine at position 23 was, without question, the most crucial. When it was replaced by glycine, no processing of the precursor was observed. We next substituted three other amino acids—asparagine, alanine, and lysine—for arginine at position 23. The asparagine-substituted precursor was taken up and cleaved about 20% as efficiently as wild type, the alanine-substituted form about 50% as well, and the lysine-substituted molecule indistinguishably from wild type. Clearly, a positive charge at position 23 is favored (as shown by the results with arginine and lysine) but is not absolutely required (as shown by the ability of asparagine and alanine to retain appreciable function).

Given additional information from theoretical predictions that the order of increasing α-helix potential of the residues substituted here is $\mathrm{Gly} < \mathrm{Asn} \ll \mathrm{Arg} < \mathrm{Lys} < \mathrm{Ala}$, we proposed that residue 23 participates in (and perhaps is the center of) a short stretch of α-helix required in some, as yet unexplained, way for uptake and processing of the OTC precursor. This putative helix does not appear to be amphipathic because substitution of arginine for the residues at positions 21 or 24 (singly or in combination) does not interfere with uptake and cleavage. We do not know the extent of this helical region, how it promotes uptake and cleavage, or its relationship to the functional role of the other positively charged residues. It is clear, however, that a secondary structure component in this region, coupled with overall positive charge, plays a major role in directing OTC uptake by mammalian mitochondria.

We and others have attempted to simplify the models of transport to unify systems as diverse as those in yeast and humans, but data from many labs have emphasized the diversity and complexity of these pathways. For example, we have found evidence of an intermediate on the pathway from pOTC to mature OTC that has been partially translocated through the mitochondrial membranes and has had a portion of the leader peptide removed. A *Neurospora* protein shows similar behavior. In yeast, the amino-terminus of mitochondrial proteins is crucial for import, in contrast to our data for OTC above. Nevertheless, human pOTC can be taken up and processed by yeast mitochondria. Some investigations have suggested that translocation occurs through "pores" at contact points between the inner and outer membranes, whereas others have postulated that the leader peptides can diffuse directly through the lipid bilayer. Finally, groups working with yeast have claimed that no cytosolic factors, such as proteins or nucleic acids, are required for transport, and investigators of *Neurospora* seem to have demonstrated genetically that a protein factor is involved.

Thus, as we explore this complex system further and identify more components of the mitochondrial uptake apparatus, we uncover more potential sites for mutation. On the other hand, no well-established examples of human genetic disease can yet be attributed to defects in this pathway. Even mutations in individual proteins that might impair their own transport have been difficult to uncover. Perhaps most defects in this system are lethal prenatally, or perhaps we are not yet sophisticated enough to recognize their phenotype.

ACKNOWLEDGMENTS

The work described was carried out in the laboratories of Leon E. Rosenberg with the support of research grants from the National Institutes of Health (DK09527, GM32156). Special thanks are due Connie Woznick for her secretarial expertise.

REFERENCES

1. Grisolia S, Baguena R, Mayor F, eds. The Urea Cycle. New York: John Wiley, 1976.
2. Ricciuti FC, Gelehrter TD, Rosenberg LE. X-Chromosome inactivation in human liver: Confirmation of X-linkage of ornithine transcarbamylase. Am J Hum Genet 1976; 28:332–338.
3. Walser M. Urea cycle disorders and other hereditary hyperammonemic syndromes. In Stanbury JB, Wyngaarden JB, Fredrickson DS, Goldstein JL, Brown MS, eds. The Metabolic Basis of Inherited Disease, 5th ed. New York: McGraw-Hill, 1983, pp 402–438.
4. Rosenberg LE, Scriver CR. Disorders of amino acid metabolism. In Bondy PK, Rosenberg, LE, eds. Metabolic Control and Disease. Philadelphia: WB Saunders, 1980, pp 682–687.
5. Msall M, Batshaw ML, Suss R, Brusilow SW, Mellits ED. Neurologic outcome in children with inborn errors of urea synthesis: Outcome of urea-cycle enzymopathies. N Engl J Med 1984; 310:1500–1505.
6. Briand P, Francois B, Rabier D, Cathelineau L. Ornithine transcarbamylase deficiencies in human males: Kinetic and immunochemical classification. Biochim Biophys Acta 1982; 704:100–106.
7. Rozen R, Fox JE, Hack AM, Fenton WA, Horwich AL, Rosenberg LE. DNA analysis for ornithine transcarbamylase deficiency. J Inher Metab Dis 1986; 9:49–57.
8. Nussbaum RL, Boggs BA, Beaudet AL, Doyle S, Potter JL, O'Brien WE. New mutation and prenatal diagnosis in ornithine transcarbamylase deficiency. Am J Hum Genet 1986; 38:149–158.
9. Fox J, Hack AM, Fenton WA, Golbus MS, Winter S, Kalousek F, Rozen R, Brusilow SW, Rosenberg LE. Prenatal diagnosis of ornithine transcarbamylase deficiency with the use of DNA polymorphisms. New Engl J Med 1986; 315:1205–1206.
10. Conboy JG, Rosenberg LE. Posttranslational uptake and processing of *in vitro* synthesized ornithine transcarbamylase precursor by isolated rat liver mitochondria. Proc Natl Acad Sci USA 1981; 78:3073–3077.
11. Rosenberg LE, Fenton WA, Horwich AL, Kalousek F, Kraus JP. Targeting of nuclear-encoded proteins to the mitochrondrial matrix: Implications for human genetic defects. Ann New York Acad Sci 1986; 488:99–108.
12. Colman A, Robinson C. Protein import into organelles: Hierarchical targeting signals. Cell 1986; 46:321–322.
13. Douglas MG, McCammon MT, Vassarothi A. Targeting proteins into mitochondria. Microbiol Rev 1986; 50:166–178.

14. Kraus JP, Rosenberg LE. Purification of low-abundance messenger RNAs from rat liver by polysome immunoadsorption. Proc Natl Acad Sci USA 1982; 79:4015–4019.

15. Horwich AL, Fenton WA, Williams KR, et al. Structure and expression of a complementary DNA for the nuclear coded precursor of human mitochondrial ornithine transcarbamylase. Science 1984; 224:1068–1074.

21

Mitochondrial Diseases

SALVATORE DIMAURO, EDUARDO BONILLA, ERIC A. SCHON, MASSIMO ZEVIANI, SERENELL SERVIDEI, ARMAND F. MIRANDA, AND DARRYL C. DE VIVO

GENERAL CONSIDERATIONS

The mitochondria of eukaryotic cells are thought to be the progeny of a successful and long-lasting symbiotic relationship (endosymbiosis) between oxygen-breathing bacteria and the host cells (1). As evidence of this hypothesis, mitochondria have their own DNA (mtDNA) and protein synthesis machinery. Also, mitochondrial ribosomes resemble bacterial ribosomes in that they use formylmethionine as the initial amino acid of the nascent polypeptide, and they are sensitive to chloramphenicol (2–4).

Human mtDNA is a circular, double-helical DNA with tightly spaced (in some cases overlapping) genes, no introns, and a single promoter. Due, in part, to this compact structure, the mitochondrial genome is small (16.5 kb), contains only 13 structural genes, 2 tRNA genes, and 2 genes that encode the 16 S and 12 S rRNAs (Fig. 21-1). All structural proteins encoded by these genes are components of the respiratory chain, where they are subunits of oligomeric complexes that also include proteins imported from the cytoplasm (Table 21-1).

mtDNA from *Xenopus*, mouse, cow, and humans has been fully sequenced and shows extensive interspecies homologies. Although there may be some differences in the sequences of bovine liver and brain mtDNA (5), it is generally assumed that all mitochondria in different organs of the same individual have identical DNA molecules, in keeping with the view that mitochondria arise by duplication of preexisting organelles. Mitochondrial division seems to be coordinated with cell division; daughter cells have about the same number of mitochondria as the parent cell (2–4).

However, variation has been documented in human mitochondrial genes in a study of restriction fragment length polymorphisms (RFLP) in mtDNA from 145 individuals from different geographic regions (6). That analysis resulted in a genealogical tree in which all 134 identified human mtDNA variants arose from one woman who lived about 200,000 years ago, probably in Africa (7). That feat of genetic anthropology was facilitated by two features of mtDNA: lack of recombination and strictly maternal inheritance.

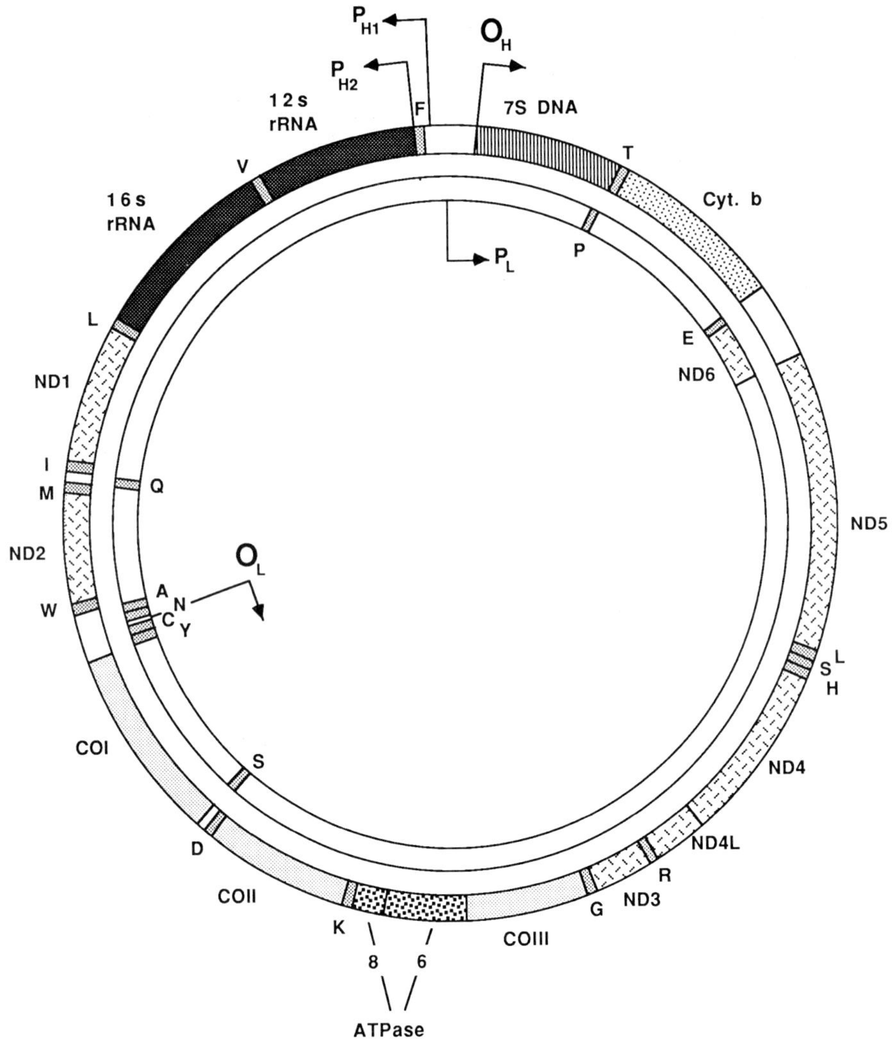

Fig. 21-1. Map of the human mitochondrial DNA. Genes are represented by differently shadowed areas. Individual patterns identify genes encoding the rRNAs, the tRNAs, or subunits of the same respiratory complex. The inner circle represents the light strand, and the outer circle the heavy strand. $\mathbf{O_H}$, $\mathbf{O_L}$: origin of replication of the heavy (**H**) or light (**L**) strand. $\mathbf{P_L}$, $\mathbf{PH_1}$, $\mathbf{PH_2}$: promoters of transcription for the light and heavy strands. Arrows indicate the direction of replication or transcription. **Cyt. b**: cytochrome *b;* **ND1** to **ND6**: subunits of NADH-dehydrogenase complex (complex I); **COI** to **COII**: subunits of cytochrome c oxidase (complex IV). **Single capital letters** indicate tRNA genes and identify the corresponding amino acids by the conventional single-letter code.

The concept of maternal inheritance is based on the observation that, in the formation of the zygote, virtually all mitochondria are contributed by the ovum. The concept has been documented in humans by RFLP studies of mtDNA (8). Characteristics that distinguish this mode of transmission from Mendelian inheritance include the following:

Table 21-1 Composition and Genetic Control of the Mitochondrial Respiratory Chain

		Polypeptides	
Complex	Prosthetic groups	Total number	mtDNA encoded
I	FMN; Fe–S clusters	25	6
II	FAD; Fe–S clusters b_{560} heme	4–5	0
III	b_{562}, b_{566}; c_1 heme [2Fe–2S] cluster	9–10	1
IV	aa_3 hemes; Cu_a; Cu_{a_3}	13	3
V	Mg^{++}	12–1	2

1. Subsequent generations show the marker trait, as in autosomal dominant diseases, but with many more affected individuals in each generation than in autosomal dominant transmission; theoretically, all children of an affected mother should be affected were it not for the threshold effect described below.
2. Inheritance of the trait is maternal, as in X-linked transmission, but children of both sexes are affected.
3. Because there are multiple copies of mtDNA in each cell, phenotypic expression of a mitochondrially encoded gene depends on the relative proportion of mutant and wild-type mtDNA within a cell; a pathological characteristic will be expressed only when the proportion of mutant DNA reaches a critical threshold (9,10).
4. At the time of cell division, the proportion of mutant and wild-type mtDNA in daughter cells may shift, changing the genotype and, possibly, the phenotype *(mitotic segregation)* (11).

Mitochondrial inheritance has not been documented conclusively in any human disease, but it has been suggested for chloramphenicol-induced blood dyscrasias (12) and for Leber's disease (optic atrophy) (13). A mitochondrially inherited factor, one that is still to be elucidated, may also be implicated in the exclusively maternal transmission of the congenital variant of myotonic dystrophy, although that disease is ordinarily inherited as an autosomal dominant trait and the mutant gene has been localized to chromosome 19 (14).

The best human example of maternal inheritance is a "mitochondrial encephalomyopathy," first described by Fukuhara (15) and later given the acronym "MERRF" (myoclonus epilepsy with ragged red fibers, the histochemical hallmark of structurally abnormal mitochondria in a muscle biopsy). Examination of a large American pedigree with that condition has confirmed all the basic features of maternal inheritance outlined above (10). The disease was expressed in four successive generations and affected both males and females. However, only women passed on the trait to their children; there were no instances of paternal transmission. Although clinical severity varied from case to case, all nine siblings in one generation were affected, which is virtually incompatible with autosomal dominant transmission.

In that family, there was marked variation of clinical expression: some patients had the full-fledged syndrome with myoclonus, ataxia, dementia, sensorineural deafness, and latic acidosis with hyperventilation; others had myoclonus, ataxia, and deafness, but no dementia; and still others had only hearing loss. Although MERRF is a progressive disorder, the remarkable clinical heterogeneity among patients could not be ascribed simply to different stages of progression. Rather, the involvement of different organs could have been related to the relative proportion of mutant and wild-type mtDNA in the stem cells of the various tissues.

Phenotypic expression was even more marked in another family (16) with a similar mitochondrial encephalomyopathy and evidence of maternal inheritance. Seven individuals in two generations showed different combinations of myoclonus epilepsy, ataxia, dementia, optic atrophy, spasticity, posterior column signs, lower motor neuron signs, deafness, myopathy, and cervical lipomas (16). Clinical expression varied from severe childhood encephalopathy to a mild disorder of adult onset.

Maternally inherited diseases should be caused by deleterious mutations in one or more of the few structural genes encoded by mtDNA. Cytochrome *c* oxidase (COX) deficiency has been found in two unrelated patients with maternally inherited MERRF (16,17). Presumably, the genetic error affects one of the three larger mitochondrial-encoded subunits of the complex (COX I, II, or III), but this remains to be documented.

Nevertheless, most mitochondrial proteins are encoded by nuclear DNA. The proteins are synthesized on cytoplasmic polyribosomes and imported into mitochondria. The complex posttranslational process of protein transport into mitochondria has been investigated extensively by Schatz and co-workers (18) and Rosenberg et al. (19) (see Chapter 20). The following general steps have been observed:

1. Mitochondrial polypeptides are synthesized as larger precursors, containing, in addition to the mature protein sequences, positively charged amino-terminal leader peptides.
2. The leader peptides function as address signals and probably recognize specific receptors on the outer mitochondrial membranes.
3. Interaction of the peptide and receptor somehow allows translocation of the precursor peptides across the outer and inner mitochondrial membranes. This process requires an intact electrochemical potential across the inner membrane.
4. Within the mitochondrial matrix, the leader peptides are cleaved by a divalent cation-dependent protease, and the mature polypeptides are assembled in their final intramitochondrial location.

Leader peptides are indispensable and sufficient signals for mitochondrial import, because mature mitochondrial proteins are not transported into mitochondria, but cytosolic proteins that have been artificially tagged with leader peptides are "erroneously" directed into mitochondria (18,19). Not only do leader peptides address proteins into mitochondria, but they also seem to contain the information necessary to direct imported proteins to the appropriate intramitochondrial compartment (intermembrane space, inner membrane, or matrix). Intramitochondrial targeting was demonstrated by experiments in which DNA fragments encoding leader peptides for mouse alcohol dehydrogenase III, a matrix enzyme, and cytochrome c_1, a protein of the inner membrane, were fused with the gene encoding the cytosolic enzyme dihydrofolate reductase. The corresponding fusion proteins were expressed in yeast cells and were localized in the submitochondrial compartment specified by the leader peptide (18). Although leader peptides differ in length and amino acid sequence, they share common features, such as lack of acidic amino acids, extended stretches of uncharged amino acids, and a relative abundance of basic residues.

Although considerable information has been collected about the mitochondrial import of nuclear-encoded proteins, little is known about the factors that coordinate the synthesis of mitochondrial-encoded and nuclear-encoded components of mitochondrial proteins. For example, unless nuclear genes are expressed more efficiently than mitochondrial genes, there are many more copies of mtDNA than nuclear DNA in a cell, so there should be a predominance of mitochondrial-encoded versus nuclear-encoded subunits of an enzyme that is under mixed genetic control, such as cytochrome *c* oxidase. An overriding control of the nuclear over the mitochondrial genome is suggested by nuclear DNA mutations that prevent the assembly of mitochondrial complexes without affecting individual subunits of the complex (4). Some nuclear-encoded factors seem to control assembly without being part of the assembled complex. However, by and large, the "cross-talk" between the nuclear and mitochondrial genomes remains to be elucidated.

The dual genetic control of mitochondrial proteins and the complexity of posttranslational events required for the import and correct assembly of mitochondrial proteins synthesized in the cytoplasm combine to explain how inherited mitochondrial diseases may be due to many different genetic errors. At least three mechanisms can be considered (Table 21-2): (1) alterations of transcription or translation of mtDNA-encoded polypeptides, (2) alterations of transcription or translation of nuclear-encoded polypeptides, and (3) alterations of the posttranslational processing of cytoplasmically synthesized proteins. Interaction with mitochondrial receptors, translocation, and proteolytic processing of precursor proteins are indispensable for correct assembly; genetic errors affecting any of these steps could result in enzyme deficiency.

Diseases due to mechanisms 2 or 3 ought to be transmitted by Mendelian inher-

Table 21-2 Possible Causes of Mitochondrial Enzyme Defects[a]

Mechanism	Example	Inheritance
1. Alteration of transcription or translation of mtDNA-encoded polypeptides	MERRF	Maternal
2. Alteration of transcription or translation of nuclear DNA-encoded polypeptides	CPT deficiency	Mendelian (A-R)
3. Alteration of posttranslational processing of nuclear DNA-encoded proteins	?	Mendelian
4. Alteration of prosthetic group	Trichopoliodystrophy (Menkes)	Mendelian (X-R)
5. Alteration of mitochondrial membrane	Adriamycin toxicity	Acquired

[a]A-R, autosomal recessive; X-R, X-linked recessive.

itance, but diseases due to mutations of mtDNA should be transmitted by non-Mendelian, maternal inheritance.

In addition to mutations that directly affect mitochondrial proteins, mitochondrial enzyme defects may also be caused by indirect mechanisms, such as alterations of prosthetic groups or alterations of the lipid milieu of membrane-bound enzymes (Table 21-2). Alterations of the prosthetic group could be due to defective synthesis or binding of a cofactor (e.g., biotin-dependent mitochondrial carboxylase deficiencies, due to biotinidase or to holocarboxylase synthetase deficiencies), or to a more general disorder in the homeostasis of a cofactor (e.g., Menkes disease, in which copper concentration is abnormally low and COX activity is decreased in certain organs). Alterations of the lipid composition of the inner mitochondrial membrane might result from defective mitochondrial synthesis of phospholipids such as cardiolipin, or to defective transport and incorporation of other phospholipids from the endoplasmic reticulum into mitochondria. The activity of some enzymes of the inner membrane, such as COX, is strongly influenced by the concentration of cardiolipin. Segregation of cardiolipin, which can be induced, for example, by administration of adriamycin, is accompanied by COX deficiency and may explain the cardiotoxicity of the drug (20).

MITOCHONDRIAL ENCEPHALOMYOPATHIES

This clinically heterogeneous group of disorders is still largely identified on the basis of morphological abnormalities of muscle mitochondria, usually first suggested by "ragged red fibers" in the modified Gomori trichrome stain of a muscle biopsy (Fig. 21-2). However, with the increase in biochemical investigations in the last decade, many specific errors of mitochondrial metabolism have been identified, and a biochemical classification of the mitochondrial encephalomyopathies is taking shape. In broad terms, these disorders can be divided into five groups:

1. Defects of mitochondrial transport, such as carnitine palmitoyltransferase (CPT) deficiency, which impairs transport of long-chain fatty acids across the inner mitochondrial membrane.
2. Defects of substrate utilization, such as pyruvate dehydrogenase complex (PDHC) deficiency or defects of β-oxidation.
3. Defects of the Krebs cycle, such as fumarase deficiency.
4. Defects of oxidation–phosphorylation coupling, exemplified by nonthyroidal hypermetabolism (Luft disease).
5. Defects of the respiratory chain.

We will confine this discussion to the defects of the respiratory chain for several reasons:

1. They account for the largest group of mitochondrial encephalomyopathies in which the biochemical errors are known.
2. The respiratory chain is the only area of mitochondrial metabolism that involves both nuclear and mitochondrial-encoded proteins. Therefore, genetic diseases of the respiratory chain can be transmitted by either Mendelian or maternal inheritance.

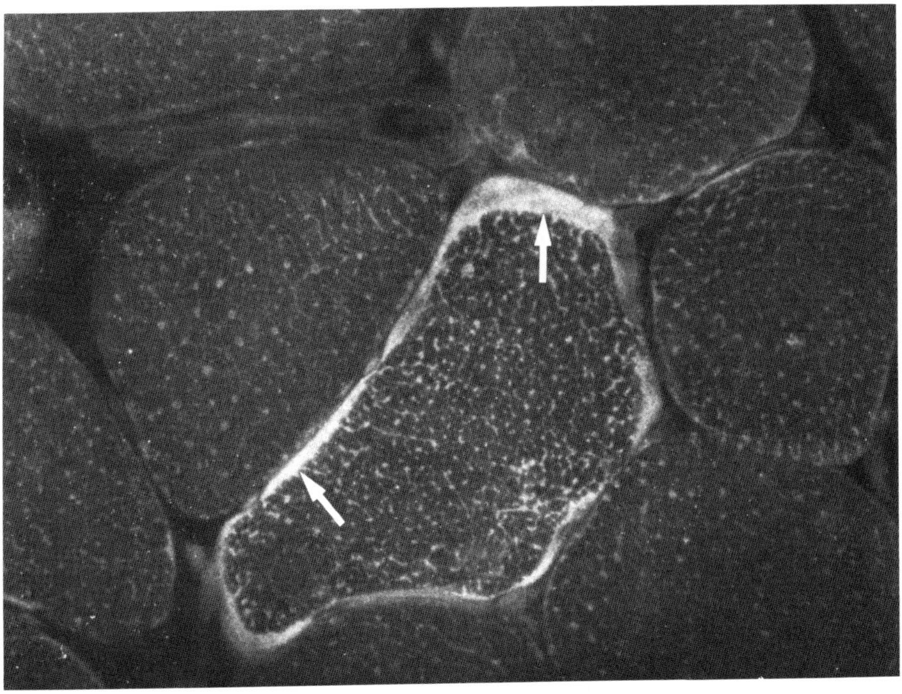

Fig. 21-2. Frozen muscle section from a patient with mitochondrial myopathy illustrating one ragged red fiber. The abnormal accumulations of mitochondria in subsarcolemmal regions of the fiber (arrows) were stained with the fluorescent dye rhodamine 123. ×450.

3. Biochemical abnormalites of this type have been studied in more detail than in other mitochondrial encephalomyopathies, and have been attributed to individual subunits in some cases.
4. DNA probes are available for all of the structural genes of mtDNA and for 4 of the 10 nuclear-encoded subunits of cytochrome *c* oxidase, so that hereditary genetic defects of the respiratory chain can now be studied at the molecular genetic level.

The respiratory chain is divided into five functional units or complexes that are embedded in the inner mitochondrial membrane (21) (Fig. 21-3).

Complex I (NADH–coenzyme Q reductase) transports hydrogen from NADH to coenzyme Q and consists of 25 polypeptides, 6 of which are encoded by mtDNA. The prosthetic group is composed of flavin mononucleotide and 6 or 7 nonheme iron–sulfur centers.

Complex II (succinate–coenzyme Q reductase) contains 4 or 5 polypeptides, including the flavin adenine dinucleotide-dependent enzyme succinate dehydrogenase, and a few nonheme iron–sulfur centers. This is the only complex that does not contain any mtDNA-encoded protein.

Complex III (reduced coenzyme Q–cytochrome *c* reductase) carries electrons from coenzyme Q to cytochrome *c*. It contains 9 or 10 proteins, 1 of which (the apoprotein of cytochrome *b*) is encoded by mtDNA.

Complex IV (cytochrome *c* oxidase) is composed of 2 cytochromes (*a* and a_3),

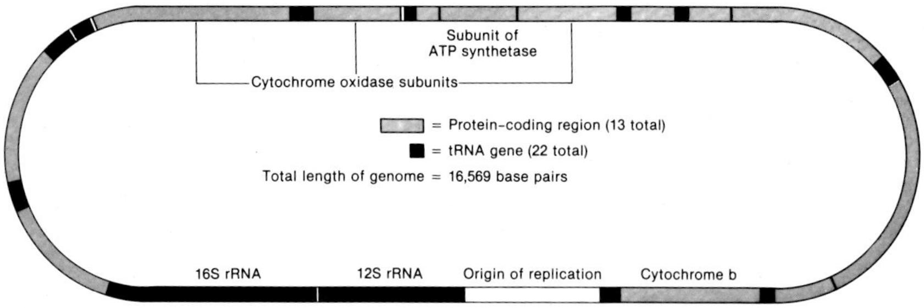

Fig. 21-3. Scheme of the respiratory chain. From DiMauro et al. (Am J Clin Genet).

2 copper atoms, and 13 protein subunits, 3 of which (subunits I to III) are encoded by mtDNA.

Complex V, or mitochondrial ATPase, consists of 12 to 14 subunits, of which 2 are encoded by mtDNA. Coenzyme Q and cytochrome *c* act as "shuttles" between complexes. Cytochrome *c* is a low-molecular-weight hemoprotein; coenzyme Q is a lipoidal quinone.

Documented human defects of the respiratory chain have affected each of the five complexes except complex II. From the *clinical* point of view, two major syndromes accompany defects of complex I or III (Fig. 21-4): (1) a pure myopathy with onset in childhood or adult life, characterized by increasing exercise intolerance with premature fatigue, myalgia, and limb weakness; and (2) different multisystem disorders dominated by muscle and brain manifestations (encephalomyopathies),

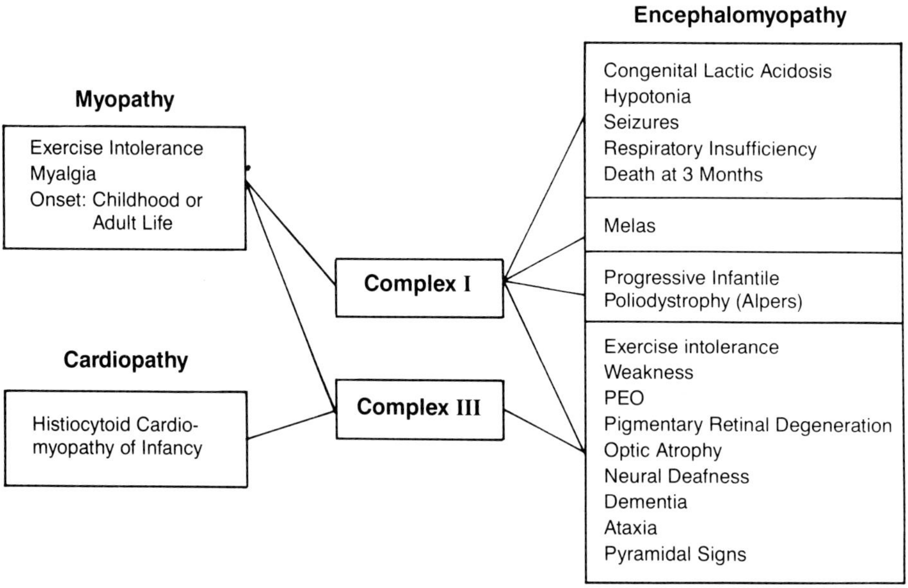

Fig. 21-4. Clinical presentations in complex I and complex III deficiencies. MELAS: mitochondrial encephalomyopathy with lactic acidosis and stroke-like episodes.

some causing congenital lactic acidosis and death in infancy and others manifesting later in life and running a chronic course. Complex III deficiency has also been associated with a rare cardiopathy of infancy (Fig. 21-4).

The clinical manifestations of cytochrome *c* oxidase (COX) deficiency (complex IV) are even more heterogeneous but, again, tend to fall into two groups, one dominated by muscle disease and the other by cerebral manifestations (encephalomyopathy) (Fig. 21-5). There are two major forms of COX-deficient myopathies, both evident at or soon after birth, with generalized hypotonia and weakness, lactic acidosis, and life-threatening respiratory insufficiency. However, although the "fatal" form has a downhill course with death usually occurring before 6 months of age, children with the "benign" form improve spontaneously and are normal by age 3 years. The clinical improvement is accompanied by a gradual recovery of COX activity in muscle, a unique phenomenon in human pathology (22).

The isolated involvement of muscle or heart in some patients and the generalized nature of the disease in others suggest that there are tissue-specific isoforms of the respiratory chain complexes. This has been documented biochemically in patients with complex III and complex IV (COX) deficiencies. In one baby with "histiocytoid cardiomyopathy of infancy," complex III activity and the amount of reducible cytochrome *b* were markedly decreased in cardiac muscle, although the enzyme activity was normal in skeletal muscle and liver (23). Conversely, in a patient with myopathy only (24), the complex was normal in cultured fibroblasts and transformed lymphoid cells (25). In the "fatal" infantile myopathy due to COX deficiency, enzyme activity was normal in liver, heart, and brain (26), and was demonstrated histochemically even in the intrafusal fibers of the muscle spindle (27). In contrast, COX deficiency was found in all tissues of several patients with subacute necrotizing encephalomyelopathy (Leigh syndrome) (28) and in one patient with MERRF (17).

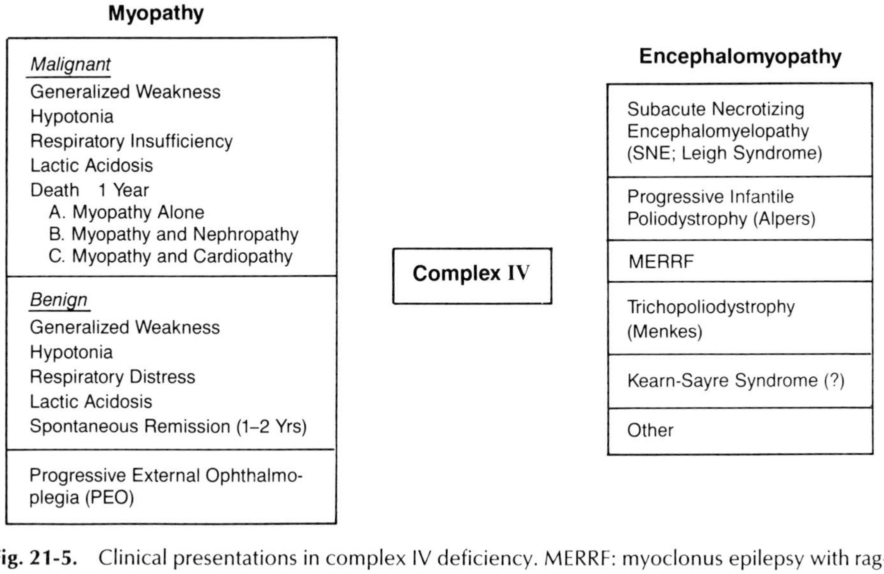
Myopathy
Malignant
Generalized Weakness
Hypotonia
Respiratory Insufficiency
Lactic Acidosis
Death 1 Year
A. Myopathy Alone
B. Myopathy and Nephropathy
C. Myopathy and Cardiopathy
Benign
Generalized Weakness
Hypotonia
Respiratory Distress
Lactic Acidosis
Spontaneous Remission (1–2 Yrs)
Progressive External Ophthalmoplegia (PEO)
Complex IV
Encephalomyopathy
Subacute Necrotizing Encephalomyelopathy (SNE; Leigh Syndrome)
Progressive Infantile Poliodystrophy (Alpers)
MERRF
Trichopoliodystrophy (Menkes)
Kearn-Sayre Syndrome (?)
Other

Fig. 21-5. Clinical presentations in complex IV deficiency. MERRF: myoclonus epilepsy with ragged-red fibers.

Because, as mentioned above, all mitochondria of one individual ought to have identical DNA molecules, the tissue specificity of respiratory chain complexes probably resides in one or more of the nuclear-encoded subunits. Biochemical and immunological studies in multiple rat tissues have shown that most nuclear-encoded COX subunits are actually tissue specific (29). On the other hand, generalized defects of respiratory chain complexes could be due to either mutation of nuclear-encoded subunits that are not tissue specific or to mutation of mtDNA-encoded subunits. Leigh syndrome, which seems to be transmitted as an autosomal recessive trait, and maternally inherited MERRF could be examples of these two mechanisms.

Identification of the different *biochemical blocks* in the respiratory chain was first made possible by polarographic studies showing differential impairment in the ability of isolated intact mitochondria to use different substrates. For example, defective respiration with NAD-dependent substrates, such as pyruvate or glutamate, but normal respiration with FAD-dependent substrates, such as succinate, suggested an isolated defect of complex I; on the other hand, defective respiration with both types of substrates in the presence of normal cytochrome *c* oxidase activity related the problem to complex III (Fig. 21-3). Polarographic studies were complemented by measurement of reduced-minus-oxidized spectra of cytochromes, showing decreased amounts of reducible cytochrome aa_3 in patients with COX deficiency and of cytochrome *b* in many (but not all) patients with complex III deficiency, and by direct measurement of electron transport through portions of the respiratory chain.

Only in the past 2 years, however, have attempts been made to go beyond the site of the lesion in the respiratory chain to a definition of molecular defects, using antibodies against the whole complexes or against individual subunits. In one infant with fatal infantile generalized complex I deficiency, electron paramagnetic resonance (EPR) spectroscopy of liver submitochondrial particles had shown lack of the iron–sulfur clusters of complex I, whereas the iron–sulfur clusters of complex II and III were normal (30). Immunological studies, using antibodies against the holoenzyme or against the four main subunits of the iron–protein fragments of complex I, showed selective absence of two polypeptides of 75 and 13 kDa (31).

In a young woman with complex III-deficient myopathy, immunoblots of muscle mitochondrial extracts with antibodies against beef heart complex III showed decreased staining of core proteins 1 and 2 and of peptide VI (32). In contrast, no subunit of complex IV seemed to be selectively involved when muscle mitochondria from a child with fatal infantile COX-deficient myopathy were subjected to immunoprecipitation followed by SDS–polyacrylamide gel electrophoresis (24). However, in that patient the amount of immunologically reactive enzyme protein was markedly decreased. Therefore, decreased synthesis or increased degradation of a muscle-specific COX subunit could have impaired assembly of the whole complex without affecting the subunit pattern of the residual holoenzyme.

Finally, muscle mitochondria from a patient with combined defects of complex III and IV (33) were studied by immunoblotting with antibodies against complex III and complex IV: the amounts of several components of both complexes were decreased (34). Because complex III and IV seem to be linked only at the level of mtDNA translation and processing of the polycystronic mRNA, it has been suggested that the defect in that child could have involved a nuclear-encoded protein controlling these events (34).

MOLECULAR GENETICS

The genes encoding several mammalian mitochondrial enzymes have been cloned and the primary structures of the proteins have been deduced from the nucleotide sequences of the cDNA. Molecular genetic techniques have been used to study the import of nuclear-encoded proteins into mitochondria. The crucial role of leader peptides both in directing these proteins to the mitochondria and in targeting them to different mitochondrial compartments has been elegantly demonstrated using fusion genes, combining nucleotide sequences for leader peptides and for different mitochondrial or cytosolic proteins (18,19). Furthermore, site-directed amino acid substitutions of the leader peptide for the matrix enzyme ornithine transcarbamylase (OTC) have demonstrated a critical functional role of the midportion of the leader as compared to the amino and carboxy terminals (19).

On the other hand, direct application of molecular genetics to the study of mitochondrial diseases has been limited. The clinical and biochemical heterogeneity of COX deficiency, illustrated above, suggests that there are probably different molecular defects. The selective involvement of one or more tissues in some forms of COX deficiency further suggests that some mutations affect the nuclear-encoded subunits that confer tissue specificity to the enzyme. We have therefore begun a long-range project aimed at elucidating the different genetic defects at the molecular level, using the tools of recombinant DNA technology. Our general strategy has been to use antibodies against individual COX subunits to screen human cDNA expression libraries. After isolation of full-length clones, the genes for the nuclear-encoded COX subunits are sequenced and the existence of tissue-specific isoforms is explored by Northern analysis of RNA from different normal human tissues. Using this approach, we have isolated the genes encoding human COX subunits IV, Vb, and VIII (35,36) [nomenclature of Kadenbach; for a comparison with the nomenclatures of Capaldi or Buse, see (37)]. Both COX IV and COX Vb genes are highly homologous to their bovine counterparts, and Northern analysis of RNA from different human tissues has shown transcript of identical size for each gene, suggesting that there are no tissue-specific isoforms of these subunits. In contrast, protein sequencing data indicate that many of the smaller nuclear-encoded COX subunits have tissue-specific isoforms, and Northern analysis of COX VIII transcripts in different tissues supports this observation.

These clones, in conjunction with already available clones containing the mitochondrially encoded genes for subunits I to III, are being used as molecular probes to analyze tissues from patients with different phenotypic expressions of COX deficiency. Once the other nuclear-encoded subunits genes have been cloned, application of Northern and Southern hybridization techniques as well as *in situ* hybridization should clarify the molecular basis for the clinical and biochemical heterogeneity of COX deficiency.

ADDENDUM

While this review was in press, Holt, Harding, and Morgan-Hughes (Nature 1988; 331:717–719) described deletions of muscle mtDNA in 9 of 25 patients with mitochondrial myopathies, including one with complex I deficiency and another with complex III deficiency. We have found large-scale deletions in muscle mtDNA in 7 of 7 patients with Kearns-Sayre syndrome, a mitochondrial encephalomyopathy characterized by ophthalmoplegia, pigmentary degeneration of the retina, onset before age 20, and at least one of the following: increased cerebrospinal fluid protein, heart block, or ataxia (Zeviani, Moraes, DiMauro, Nakase, Bonilla, Schon, Rowland [Neurology 1988; 38:1339–1346]). The relationship between site and size of the deletions and clinical manifestations, biochemical alterations, and mode of inheritance remains to be fully established.

ACKNOWLEDGMENTS

Some of the work discussed in this chapter was supported by Center grants NS11766 from the National Institute of Neurological and Communicative Disorders and Stroke and from the Muscular Dystrophy Association.

Dr. Zeviani was supported by a fellowship from the Italian Society for Neurological Research (ARIN) and Dr. Servidei by a fellowship from the Unione Italiana Lotta alla Distrofia Muscolare (UILDM), Sezione Laziale "Giulia Testore."

We are grateful to Dr. Lewis P. Rowland for his advice and encouragement and to Ms. Mary Tortorelis for typing the manuscript.

REFERENCES

1. Sagan D, Margulis L. Bacterial bedfellows. Natural Hist 1987; 3:26–33.
2. Tzagoloff A. Mitochondria. New York: Plenum Press, 1982.
3. Holtzmann E, Novikoff AB. Cells and Organelles, 3rd ed. New York: CBS College Publishing, 1984.
4. Alberts B, Bray D, Lewis J, Raff M, Roberts K, Watson JD. Molecular Biology of the Cell. New York: Garland, 1983.
5. Coote JL, Szabados G, Work TS. The heterogeneity of mitochondrial DNA in different tissues from the same animal. FEBS Lett 1979; 99:255–260.
6. Whittam TS, Clark AG, Stoneking M, Cann RL, Wilson AC. Allelic variation in human mitochondrial genes based on patterns of restriction site polymorphism. Proc Natl Acad Sci USA 1986; 83:9611–9615.
7. Cann RL, Stoneking M, Wilson AC. Mitochondrial DNA and human evolution. Nature (London) 1987; 325:31–36.
8. Giles RE, Blanc H, Cann HM, Wallace DC. Maternal inheritance of human mitochondrial DNA. Proc Natl Acad Sci USA 1980; 77:6715–6719.
9. Wallace DC. Assignment of the chloramphenicol resistance gene to mitochondrial deoxyribonucleic acid and analysis of its expression in cultured human cells. Mol Cell Biol 1981; 1:697–710.

10. Rosing HS, Hopkins LC, Wallace DC, Epstein CM, Weidenheim K. Maternally inherited mitochondrial myopathy and myoclonic epilepsy. Ann Neurol 1985; 17:228–237.

11. Wallace DC, Bunn CL, Eisenstadt JM. Mitotic segregation of cytoplasmic determinants for chloramphenicol resistance in mammalian cells. II. Fusion with human cell lines. Somatic Cell Genet 1977; 3:93–119.

12. Fine PEM. Mitochondrial inheritance and disease. Lancet 1978; 1:659–662.

13. Nikoskelainen EK, Savontaus ML, Wanne OP, Katila MJ, Nummelin KU. Leber's hereditary optic neuroretinopathy, a maternally inherited disease. Arch Ophthalmol 1987; 105:665–671.

14. Harper PS. Myotonic disorders, Chap. 40. In Engel AG, Banker BQ, eds. Myology, Vol 2. New York: McGraw-Hill, 1986, pp 1267–1296.

15. Fukuhara N. Myoclonus epilepsy and mitochondrial myopathy. In Scarlato G, Cerri C, eds. Mitochondrial Pathology in Muscle Diseases. Padova: Piccin, 1983, pp 88–110.

16. Berkovic SF, Carpenter S, Karpati G, et al. Cytochrome c oxidase deficiency: A remarkable spectrum of clinical and neuropathologic findings in a single family. Neurology 1987; 37:223 (abstr.).

17. Mendell JR, Barohn RJ, Yates AJ, et al. Autopsy case of myoclonic epilepsy and ragged-red fibers (MERRF) with cytochrome c oxidase deficiency. Evidence for a maternally inherited biochemical defect. Ann Neurol 1987; 22:128 (abstr.).

18. Van Loon APGM, Brandli AW, Schatz G. The presequences of two imported mitochondrial proteins contain information for intracellular and intramitochondrial sorting. Cell 1986; 44:801–812.

19. Rosenberg LE, Fenton UA, Horwich AL, Kalousck F, Kraus JP. Targeting of nuclear-encoded proteins to the mitochondrial matrix: Implications for human genetic defects. Ann New York Acad Sci 1986; 488:99–108.

20. Huart P, Brasseur R, Goormaghtigh E, Ruysschaert JM. Antimitotics induce cardiolipin cluster formation: Possible role in mitochondrial enzyme inactivation. Biochim Biophys Acta 1984; 799:199–202.

21. Hatefi Y. The mitochondrial electron transport and oxidative phosphorylation system. Annu Rev Biochem 1985; 54:1015–1069.

22. DiMauro S, Zeviani M, Servidei S, et al. Cytochrome oxidase deficiency: Clinical and biochemical heterogeneity. Ann New York Acad Sci 1986; 488:19–32.

23. Papadimitriou A, Neustein HB, DiMauro S, Stanton R, Bresolin N. Histiocytoid cardiomyopathy of infancy: Deficiency of reducible cytochrome b in heart mitochondria. Pediat Res 1984; 18:1023–1028.

24. Kennaway NG, Buist NR, Darley-Usmar VM, et al. Lactic acidosis and mitochondrial myopathy associated with deficiency of several components of complex III of the respiratory chain. Pediat Res 1984; 18:991–999.

25. Darley-Usmar V, Watanabe M, Uchiyama Y, et al. Mitochondrial myopathy: Tissue-specific expression of a defect in ubiquinol-cytochrome c reductase. Clin Chem Acta 1986; 158:253–261.

26. Bresolin N, Zeviani M, Bonilla E, et al. Fatal infantile cytochrome c oxidase deficiency: Decrease of immunologically detectable enzyme in muscle. Neurology 1985; 35:802–812.

27. Zeviani M, Nonaka I, Bonilla E, et al. Fatal infantile mitochondrial myopathy and renal dysfunction due to cytochrome c oxidase deficiency: Immunological studies in a new patient. Ann Neurol 1985; 8:672–675.

28. Moreadith RW, Batshaw ML, Ohnishi T, et al. Deficiency of the iron-clusters of mitochondrial reduced nicotinamide adenine dinucleotide-ubiquinone oxidoreductase (complex I) in an infant with congenital lactic acidosis. J. Clin Invest 1984; 74:685–697.

29. DiMauro S, Servidei S, Zeviani M, et al. Cytochrome c oxidase deficiency in Leigh syndrome. Ann Neurol 1987; 22:498–506.

30. Moreadith RW, Cleeter NWJ, Ragan CI, Batshaw ML, Lehninger AL. Congenital deficiency of two polypeptide subunits of the iron-protein fragment of mitochondrial complex I. J Clin Invest 1987; 79:463–467.

31. Darley-Usmar VM, Kennaway NG, Buist NMR, Capaldi RA. Deficiency in ubiquinone-cytochrome c reductase in a patient with mitochondrial myopathy and lactic acidosis. Proc Natl Acad Sci USA 1983; 80:5103–5106.

32. Sengers RCA, Trijbels JMF, Bakkeren JAJM, et al. Deficiency of cytochromes b and aa_3 in muscle from a floppy infant with cytochrome oxidase deficiency. Eur J Pediat 1984; 141:178–180.

33. Takamiya S, Yanamura W, Capaldi RA, et al. Mitochondrial myopathies involving the respiratory chain: A biochemical analysis. Ann New York Acad Sci 1986; 488:33–43.

34. Zeviani M, Nakagawa M, Herbert J, et al. Isolation of a cDNA clone encoding subunit IV of human cytochrome c oxidase. Gene, 1987; 55:205–217.

35. Zeviani M, Sakoda S, DiMauro S, Schon EA, Suske G. Isolation of three nuclear-encoded subunit cDNA clones of cytochrome c oxidase. Neurology 1987; 37 (Suppl 1):209 (abstr.).

36. Capaldi RA. Structure and functioning of cytochrome c oxidase. Chem Scripta, 1987; 278:59–66.

37. Kadenbach B, Merle P. On the function of multiple subunits of cytochrome *c* oxidase from higher eukaryotes. FEBS Lett 1981; 135:1–11.

22

Familial Amyloidotic Polyneuropathies

JOSEPH HERBERT

The familial amyloidotic polyneuropathies (FAPs) consitute a group of dominantly inherited disorders characterized by the deposition of amyloid fibrils in the peripheral nervous system. Although relatively rare in occurrence (except for several well-defined geographic foci), the FAPs are the first of the hereditary neuropathies, and one of only a handful of autosomal dominant diseases, to be defined at the molecular level.

NOMENCLATURE

In the past two decades the fundamental characteristics of amyloid fibrils have been elucidated. Until recently amyloids of different origin were generally considered similar in chemical composition; Glenner complained that amyloid was a "19th century term in search of a 20th century definition" (1). Great strides in protein chemistry have now provided that definition: amyloids may be formed by a wide variety of proteins that have in common only a specific conformation—the twisted β-pleated sheet fibril (2). This distinctive structure accounts for the unique tinctorial and optical properties of amyloid fibrils—the Congo Red birefringence.

The establishment of the biochemical nature of amyloid fibrils has made the traditional clinical nomenclature of primary and secondary amyloidoses obsolete. By convention, amyloids are now classified according to both biochemical composition and associated clinical pathology (3). Most FAP amyloids are derived from a plasma protein formerly known as prealbumin (or thyroxine-binding prealbumin). That term is a misnomer because prealbumin bears no relationship whatsoever to serum albumin and refers only to its electrophoretic migration in gels. To avoid confusion, the Joint Commission on Biochemical Nomenclature has recommended the term transthyretin (TTR) (4), which shall be used in this discussion. Unfortunately, since both terms are still in current usage, the confusion has not yet been resolved.

CLINICAL VARIANTS (Table 22-1)

Three comprehensive reviews of the clinical syndromes also provide detailed literature citations for the discussion that follows (5–7).

Table 22-1 Clinical Features of the Familial Amyloidotic Polyneuropathies[a]

Type	Upper limb neuropathy	Lower limb neuropathy	Cranial neuropathy	Autonomic neuropathy	Renal disease	Cardiac disease	Gastrointestinal disease	Ocular	Other
I	Late	Early	No	Early	Late	Prominent	Prominent	Scalloping Vitreous opacities	
II	Early	Late	No	No	No	Late	Rare	Vitreous opacities	Carpal tunnel
III	Early	Early	No		Nephrotic syndrome	No	Peptic ulcer	Cataracts	
IV	No	Late	Yes	No	?Yes	Yes	No	Lattice dystrophy Glaucoma	Cutis laxa
Jewish	Later	Early	No	Early	No	Mild	Peptic ulcer	Vitreous opacities	Malnutrition
Appalachian	Early Carpal tunnel	No	No	Yes	No	Yes	Yes	No	
Illinois/German	No	Yes	No	?	Yes		Yes		
Familial oculolepto-meningeal	No	Late	Yes	No	No	No	No	Vitreous opacities	Seizures, headache, dementia

[a]Modified from Glenner et al. (5).

FAP Type I (Portuguese, Andrade Type)

The prototype of the FAPs is the disorder first described in Portugal by Andrade (8). Since then several hundred families have been identified, centered around a fishing village in Northern Portugal called Povoa de Varzim. Onset of symptoms is in adult life, in the majority of cases in the third or fourth decade. In one-fifth of the cases there is first a sensory neuropathy of the lower extremities, followed within a few years by progressive autonomic dysfunction. In about a quarter of the cases, however, dysautonomia precedes the sensory symptoms. Motor neuropathy generally manifests later, first in the legs and progressing gradually to involve the trunk and upper limbs. Severe weight loss is prominent. The clinical manifestations are remarkably constant within and between affected families, with inexorable progression to death from inanition or intercurrent infection within 15 years.

Gastrointestinal disturbances are a prominent feature of FAP type I, and cardiac amyloidosis is not uncommonly the cause of death. Renal amyloidosis is variable and, when present, a late manifestation.

Ocular abnormalities occur frequently. Vitreous opacities due to amyloid deposition are seen in up to 75% of advanced cases. Scalloping of the pupils due to involvement of the ciliary bodies reportedly is seen in up to one-third of the cases and has been considered pathognomonic.

The central nervous system is spared. Amyloid deposition is limited to leptomeningeal vessels, the choroid plexuses, and subpial and subependymal locations. The spinal cord is unaffected, as are those portions of the proximal nerve roots within the subarachnoid space.

Pathologically, there is massive amyloid infiltration of the spinal and autonomic ganglia and the distal processes of the peripheral nerves. Systemic amyloid deposition is widespread; blood vessels of all size are involved, as are skin, smooth muscle, and occasionally striated muscle. Parenchymatous organs, such as kidneys, spleen, thyroid gland, and testes, are involved, whereas the liver is relatively spared. Cardiac amyloidosis may be marked and there may be deposits at various levels of the gastrointestinal tract.

In 1966, Nakao and collaborators first reported a Japanese pedigree bearing remarkable similarity to the Portuguese kindred (9). Since then, several hundred patients, located mainly in Ogawa village in the Nagano prefecture and Arao district in the Kumamoto prefecture, have been described, representing the largest focus of FAP type I outside of Portugal. In Northern Sweden, Andersson described 60 cases with a later age of onset than in the Portuguese pedigrees, but the clinical picture otherwise was identical (10). Other families have been identified in Greece, England, Germany, Brazil, Italy, and the United States.

The clinical picture is remarkably monomorphic, suggesting that the same mutation is responsible for the disease. This has now been established at the protein level for at least the Portuguese, Japanese, Swedish, Brazilian, Italian, and Greek varieties (11–15). It seems likely that the original mutation occurred in Povoa de Varzim sometime before the fifteenth century, and then spread worldwide with the Portuguese navigations. In Kyushu, the Portuguese had established trading posts in the sixteenth century, but Portuguese ancestry has not been established in any affected Japanese family.

FAP Type II (Swiss/Indiana, Rukavina Type)

Rukavina and collaborators described an autosomal dominant amyloidosis in a family of Swiss descent living in Indiana (16). The neuropathy involves predominantly the upper limbs, with the carpal tunnel syndrome a common early manifestation. Lower limb symptoms are either absent or manifest 5 to 10 years later. Vitreous opacities and bead-like deposits on the retinal vessels occur frequently. Gastrointestinal symptoms and hepatosplenomegaly occur occasionally. Death is often due to a restrictive cardiomyopathy.

Symptoms appear in the fourth or fifth decade and the progression is slow, with death occurring from 16 to 35 years after diagnosis. Pathologically, amyloid infiltration is seen in the myocardium, tongue, larynx, adrenals, pancreas, lungs, and prostate.

In Washington County, Maryland, 11 families of common German ancestry had a similar syndrome, except for a lower incidence of vitreous opacities (17).

FAP Type III (Iowa, Van Allen Type)

In 1969, Van Allen et al. (18) reported a variant form of FAP in an Iowa pedigree whose ancestors originated from the British Isles. The neuropathy was severe and involved all limbs, with no carpal tunnel syndrome. Vitreous opacities were absent but cataracts were prominent. Several patients suffered from inner ear deafness and peptic ulcer disease was common. Cardiac amyloidosis was not a feature but non-amyloidotic renal disease with nephrosis and renal failure was typical. At autopsy, amyloid was found in the liver, spleen, testes, adrenals, peripheral nerves, and sympathetic and dorsal root ganglia. The age at onset was in the third to fourth decade, and the average life span was 17 years after diagnosis. A family manifesting similar features has been described in Spain (19).

FAP Type IV (Finnish, Meretoja Type)

In 1969 Meretoja reported the association of corneal lattice dystrophy, cranial neuropathy, and cutis laxa as features of a generalized hereditary amyloidosis (20). Two large foci were identified in neighboring regions of Southern Finland. The disease was transmitted as an autosomal dominant trait, but two homozygotes were also documented.

In this variant, onset occurs in the third or fourth decade, with the appearance of filamentous corneal amyloid deposits, followed by corneal opacities, erosions, and anesthesia. Visual acuity is preserved until the seventh decade. A quarter of the cases develop chronic glaucoma. Vitreous opacities are notably absent. Pathologically, lattice filaments correspond to amyloid deposits around degenerated corneal nerve fibers.

Peripheral neuropathy is mild and occurs as a late manifestation. However, progressive facial weakness occurs in three-quarters of the patients, first involving the upper facial muscles and later the lower. The cochlear nerve is occasionally involved; other cranial nerves are rarely affected.

The cutaneous manifestations are due to widespread infiltration of the dermis by amyloid. The skin becomes atrophic and pendulous. Cardiac symptoms are common. Progression is slow, with most patients living to old age.

Pathologically, there is a moderate degree of amyloid deposition in peripheral nerves, leptomeninges, and the basal membranes of the sweat and sebaceous glands.

Other Variants

In addition to the four "classical" syndromes, several other clinical phenotypes warrant separate mention.

Pras et al. were the first to report a mutation in plasma TTR in a family of Ashkenazi Jewish origin (21). At the age of 25, the propositus in that family had vitreous opacities leading rapidly to blindness, followed 2 years later by sensorimotor neuropathy of the legs and dysautonomia. His father had had a similar clinical picture and died at age 36 from bleeding gastric erosions. At autopsy, perivascular amyloid affected blood vessels of all the organs of the body. Massive deposition was present in the spinal and autonomic ganglia and between fibers of the spinal nerve roots and peripheral nerves. Major deposits were found in skeletal and heart muscle of the father, and in the interstitium of the thyroid and prostate glands and seminal vesicles of the son (22).

Wallace et al. (23) described a pedigree from the Appalachian mountains presenting in the seventh decade with carpal tunnel syndrome, cardiomyopathy, and impotence (FAP Appalachian). There was no ocular disorder and little evidence of generalized polyneuropathy.

The same group described a kindred of German descent living in Illinois in which gastrointestinal and renal involvement is prominent (24).

Goren et al. (25) described an interesting disorder, known as familial oculoleptomeningeal amyloidosis, in a kinship of German origin residing in Ohio. This is the only variant in which cerebral symptoms were prominent. The age at onset was between 30 and 60 years and transmission was autosomal dominant. Headache, seizures, dementia, aphasia, and spasticity were salient features. Vitreous opacities were common and impaired hearing was occasionally present. Motor neuropathy was more prominent than distal sensory involvement. Death occurred 3–12 years after onset.

Pathologically, there was widespread amyloid deposition in the leptomeninges, around subarachnoid vessels and in the ocular vitreous. Deposition in cranial and spinal roots and ganglia, and in peripheral nerves, was present but less striking.

OTHER HEREDITARY AMYLOIDOSES

Table 22-2 places the FAPs in the context of the other hereditary amyloidoses. An encapsulated description of those disorders is given below. Table 22-3 enumerates those heredofamilial disorders for which the protein composition of the amyloid is known. This classification emphasizes the wide variety of proteins that may be associated with amyloid fibril formation.

Familial Mediterranean fever (FMF) is the only disorder in this group that is

Table 22-2 The Heredofamilial Amyloidoses[a]

I. Generalized Hereditary Amyloidosis
  A. Neuropathic
   1. Amyloidotic neuropathy type I (Andrade)
   2. Amyloidotic neuropathy type II (Rukavina)
   3. Amyloidotic neuropathy type III (Van Allen)
   4. Amyloidotic neuropathy type IV (Meretoja)
   5. Amyloidotic neuropathy Ashkenazi Jewish type
   6. Amyloidotic neuropathy Appalachian type
   7. Amyloidotic neuropathy Illinois/German type
   8. Familial oculoleptomeningeal amyloidosis
  B. Nonneuropathic
   1. Familial Mediterranean fever
   2. Familial amyloidotic cardiomyopathy (Danish)
   3. Amyloid nephropathy of Ostertag
   4. Muckle–Wells syndrome

II. Localized Hereditary Amyloidosis
   1. Heredofamilial cerebral hemorrhage (Icelandic)
   2. Familial Alzheimer disease
   3. Corneal lattice dystrophy
   4. Hereditary amyloid deposits of cornea
   5. Hereditary cutaneous amyloidoses
   6. Medullary carcinoma of thyroid (multiple endocrine adenomatosis type II)

[a]Modified from Glenner et al. (5).

transmitted in an autosomal recessive manner. It occurs among inhabitants of the Mediterranean basin and is characterized by episodic pyrexia, polyserositis, arthritis, and amyloid nephropathy. Peripheral neuropathy is not a feature of this condition. The amyloid fibril is derived from the acute-phase plasma protein serum amyloid A (SAA [26]).

Familial amyloidotic cardiomyopathy (FAC) was described in a Danish family in which 7 of 12 siblings suffered from progressive heart failure beginning in adult life (27). The amyloid in that condition may also be related to TTR (28), but the precise chemical composition has not yet been determined. Clinical reevaluation of patients many years after onset revealed no evidence of polyneuropathy.

Ostertag (29) described chronic nephropathy, hypertension, and hepatosplenomegaly in at least four members of a single family, spanning four generations. The disorder appears to be transmitted as an autosomal dominant trait and progresses

Table 22-3 Known Amyloid Fibril Proteins in the Heredofamilial Amyloidoses

Disease	Amyloid precursor
Familial amyloidotic polyneuropathies	Transthyretin variant
Familial amyloidotic cardiomyopathy	Transthyretin variant
Familial Mediterranean fever	SAA protein
Hereditary cerebrovascular amyloidosis	γ-trace protein
Familial Alzheimer disease	β protein precursor
Medullary carcinoma of thyroid (MEA II)	Procalcitonin

to death within 10 years. Pathologically, there are extensive amyloid deposits in kidneys, adrenal glands, and liver.

In 1962 Muckle and Wells (30) described childhood nerve deafness, urticaria, limb pains, and nephropathy in nine members of an English family, involving four generations. Transmission appears to be autosomal dominant. There are extensive renal amyloid deposits. The organ of Corti and the vestibular sensory epithelium are absent and the cochlear nerves atrophic, but there are no amyloid deposits in the auditory apparatus.

Hereditary cerebrovascular amyloidosis has been documented in kindreds of Icelandic extraction (31). These patients suffer from sudden, catastrophic, and often multifocal cerebral hemorrhages from intraparenchymal and leptomeningeal vessels. Arteries in both of these locations are thickened and contain perivascular amyloid deposits derived from γ-trace protein, a serum cysteine proteinase inhibitor (32).

The association of amyloid with the central cores of senile plaques and in perivascular deposits in Alzheimer disease is well known. A rare dominantly inherited form of the disorder with earlier onset has been described and a familial predisposition to the disorder has been documented even in "sporadic" cases. The major component of the amyloid is a 4.8-kDa peptide, β-protein, which is derived from a larger precursor encoded by a gene on chromosome 21 (33). Linkage between the FAD locus and the β-preamyloid locus has not been established (34, 35).

Finally, there is a familial syndrome characterized by the development of tumors of multiple endocrine glands, including the thyroid, parathyroid, and adrenal glands (multiple endocrine adenomatosis, type II). The thyroid neoplasms in this condition are of the rare medullary carcinoma variant, and contain stromal amyloid deposits derived from the calcitonin precursor molecule (36).

GENETICS

All of the amyloidotic polyneuropathies are transmitted as autosomal dominant traits with high penetrance. Males and females are equally affected with the notable exception of FAP type III, in which there is a striking preponderance of affected males (11 of 12 cases reported by Van Allen) (18). Father-to-son transmission was reported in three of their cases, ruling out an X-linked mode of transmission. The only female in their pedigree could have been a victim of uremic polyneuropathy, because her clinical course differed from that of her brethren and amyloidosis was not proven in her case. Recently two females with this variant have come to autopsy (M. Benson, personal communication). Gafni et al. (22) emphasized that the marked renal disease seen in this variant is not secondary to renal amyloidosis. This feature, they believed, along with peptic ulceration, cataracts, and deafness, should be considered independent phenotypic expression of a pleiotropic gene.

The prototypic clinical variants are consistent within families, with the possible exception of one reported family. These variants are therefore true examples of genetic heterogeneity rather than overlapping forms of the same disorder. This differentiation is now borne out at the protein level (11–15, 21–24).

As with most of the dominantly inherited disorders, onset of symptoms is in adult life, usually in the third or fourth decade, although expression of the mutant allele in childhood has been demonstrated. Within families, the age of onset is very predictable, but considerable variation exists between pedigrees expressing the same mutation. For example, the average age of onset in FAP type I of the Swedish variety is approximately 20 years later than in the Portuguese variety, although the subsequent clinical course and the documented mutation are identical. A Portuguese kindred manifesting with late-onset (in the sixth or seventh decade) but otherwise typical FAP type I carries the same mutation as the early-onset variety, and the mutant protein is detectable in the serum of asymptomatic heterozygotes in childhood (37).

Clearly, there are other genetic or epigenetic factors that modify the expression of the disease phenotype. The operation of these factors could account for apparently sporadic cases, in which there is no family history of the disease. Such cases have been noted particularly among Swedish patients with FAP type I.

ETIOLOGY OF THE FAPs

The critical contribution to the elucidation of the etiology of the familial amyloidotic polyneuropathies was made by Costa and associates in 1978, when they isolated type I amyloid fibril protein and found that it was related immunologically to serum TTR (38). In 1981 Merrill Benson reported partial amino acid identity between amyloid fibril protein and TTR (39), and soon thereafter several groups independently demonstrated a single amino acid substitution of methionine for valine at position 30 of the TTR monomer, in FAP Type I (11–13) (Fig. 22-1).

cDNA and genomic clones were isolated and sequenced by several investigators (40,41) and it was shown that the mutation is a result of a single base substitution of adenosine for guanosine. Affected individuals are invariably heterozygous for one mutant and one normal TTR allele, whereas unaffected family members are homozygous for the normal gene. No homozygous mutants have been reported, suggesting that this genotype may be lethal *in utero.*

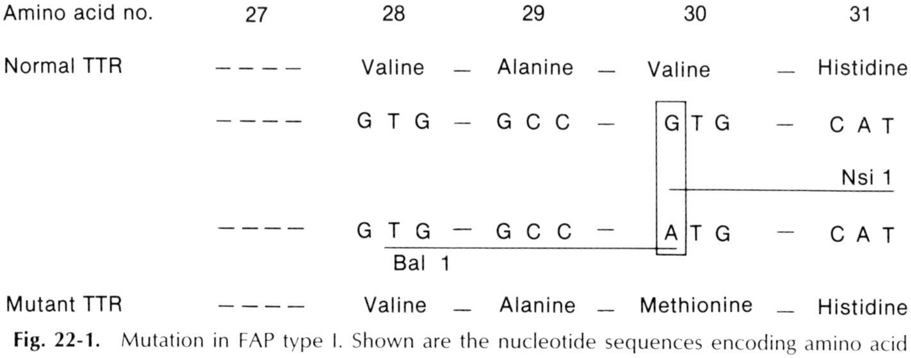

Fig. 22-1. Mutation in FAP type I. Shown are the nucleotide sequences encoding amino acid residues 28 to 31 of the normal and mutant TTR monomer. An adenosine for guanosine substitution (box) at the first position of the codon encoding amino acid 30 causes a methionine-for-valine substitution in the translated peptide. The mutation also creates new cleavage sites for two restriction enzymes, *Bal* and *NsiI*, which are under- and overlined, respectively.

The identical mutation at position 30 has now been described in Portuguese, Swedish, Japanese, Brazilian, and Greek pedigrees with FAP type I, strengthening the contention that this allele is indeed responsible for the disease and not just a cosegregating polymorphism. The normal and mutant TTR proteins circulate in the plasma in roughly equimolar amounts, indicating that the two TTR alleles are expressed codominantly. Opinions differ concerning the composition of the amyloid deposits; whereas one group of investigators found an overwhelming predominance of the mutant protein (11), another has reported a ratio of approximately 60–40% of mutant to normal TTR (42).

In FAP type I, the single base pair substitution results in the creation of a new restriction site for two restriction enzmes, *Bal*I and *Nsi*I, creating a simple test for heterozygosity (Fig. 22-2). Restriction fragment length polymorphism (RFLP) analysis has been used to detect presymptomatic carriers and could be employed in prenatal screening (40,41,43). Other relatively simple strategies for carrier detection have been based on the presence in the serum of the mutant protein (44–47).

A series of reports on the amino acid composition of TTR in several other FAP variants followed in rapid succession (Table 22-4). In type II there is a serine for isoleucine substitution at position 84 of the TTR monomer (48). In the Appalachian variant there is an alanine for threonine substitution at residue 60 (23), and in the Illinois/German variant, a tyrosine for serine substitution at residue 77 (24). A Scandinavian group has published a preliminary report of a methionine for leu-

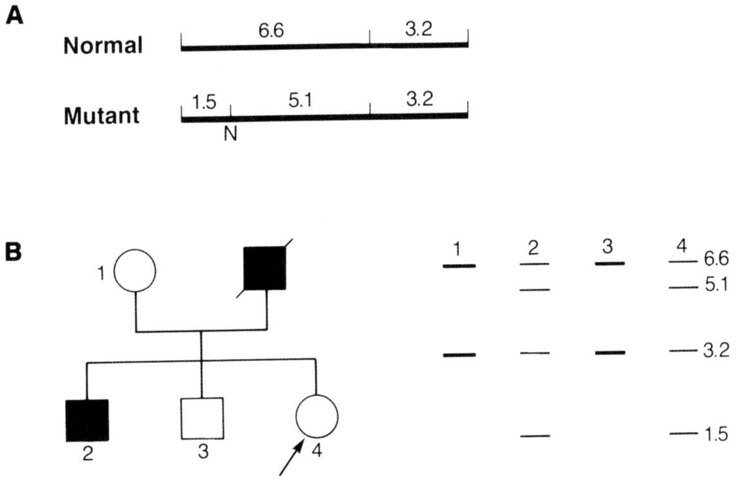

Fig. 22-2. Detection of TTR mutation by restriction analysis. (A) A restriction map of the human transthyretin gene and (B) a Southern blot analysis of DNA from four members of an FAP pedigree. (A) The adenosine for guanosine mutation at amino acid position 30 creates a new cleavage site for *Nsi*I (N). Digestion of normal DNA with *Nsi*I generates two fragments, measuring 3.2 and 6.6 kb long, which hybridize with a TTR cDNA probe. Digestion of the mutant DNA yields three hybridizing fragments: 3.2, 5.1, and 1.5 kb long. (B) All three hybridizing fragments are detected by Southern analysis of DNA from an affected boy (lane 2), but not in his unaffected mother (lane 1) or brother (lane 3). However, the proband (lane 4), who is clinically unaffected, shows the same pattern of hybridization as her affected brother. She is therefore heterozygous for the FAP mutation and will almost certainly develop the disease in the future. After Mita et al. (40).

Table 22-4 Characterized Mutations in Transthyretinopathies[a]

Mutation	Position	Phenotype	References
Val–Met	30	FAP type I	11–15
Ile–Ser	84	FAP type II	47
Phe–Ile ?	33	FAP Jewish	48
Thr–Gly ?	49		21
Thr–Ala	60	FAP Appalachian	23
Ser–Tyr	77	FAP Illinois/German	24
Leu–Met	111	FAC (Danish)	28

[a]Val, valine; Met, methionine; Ile, isoleucine; Tyr, tyrosine; Thr, threonine; Phe, phenylalanine; Gly, glycine; Ala, alanine; Ser, serine; Leu, leucine; FAP, familial amyloidotic polyneuropathy; FAC, familial amyloidotic cardiomyopathy.

cine substitution at position 110 in Danish patients with familial amyloidotic cardiomyopathy (28).

The situation appears more complex regarding the nature of the mutation in the Ashkenazi Jewish pedigree. Pras et al. (21) initially reported a glycine for threonine substitution at position 49, which would require a two-base change, which is unlikely; splenic amyloid from their patient (SKO) was also analyzed by Nakazato et al. (49), who found only an isoleucine for phenylalanine change at residue 33. Pras et al. reexamined amyloid components from both thyroid and spleen and corroborated their original report of a position 49 substitution. In addition, they confirmed the mutation at residue 33 in splenic amyloid, but could not identify the same substitution in amyloid fibrils derived from the thyroid (50).

All of the mutations described thus far are amenable to detection by restriction analysis as described above (M. Benson, personal communication) so that determination of heterozygosity is possible for asymptomatic family members or for purposes of prenatal screening.

BIOLOGY OF TRANSTHYRETIN

TTR [Goodman (51) has provided a comprehensive review and literature citations] is a 56-kDa protein composed of four identical subunits (52). Plasma TTR is synthesized in the liver and functions as a plasma transport protein. In humans, TTR is a minor transport protein for thyroxine, the major fraction being carried by thyroxine-binding globulin. However, in lower mammals such as the rat, TTR is the major throxine-binding protein. TTR binds thyroxine (T_4) with greater affinity than triiodothyronine (T_3). In contrast to the plasma, in the cerebrospinal fluid (CSF) TTR is the major carrier of thyroxine, with thyroxine-binding globulin contributing only about 15% to total bound thyroxine (53,54).

In addition, TTR serves an important role in the plasma transport of vitamin A. Vitamin A, in the form of retinyl esters, is transported in the chylomicrons from the intestine to the liver. In the hepatocytes, retinyl esters are hydrolyzed to the alcohol retinol, some of which is bound to a protein of hepatic origin, retinol-binding protein (RBP). The retinol–RBP complex (known as holo-RBP) is secreted into the plasma, where it is itself complexed to TTR in a 1:1 molar ratio. The size of the complex preserves RBP from rapid glomerular filtration. This complex transports retinol to vitamin A-requiring tissues where its uptake into target cells is mediated by a specific cell surface receptor. From the cell membrane, retinol is transferred to an intracellular retinol-binding protein different from plasma RBP. After releasing retinol, the resultant apo-RBP has a reduced affinity for TTR and the free uncomplexed RBP is excreted in the kidney. TTR filtered through the glomeruli is reabsorbed in the proximal tubule. TTR is catabolized mainly in the liver, muscle, and skin (55).

Binding of TTR to RBP is independent of its binding to thyroxine, and there is no evidence to suggest that the coincidence of these two binding functions is of any physiological significance. The development of two independent functions in the same molecule needs further study from the evolutionary standpoint. Since the total absence of TTR has not been reported, and since no homozygotes for the FAP mutations have yet been identified, it would appear that TTR is essential to life. However, it is not clear what contribution each of its separate functions makes in this regard.

Evolutionarily, both TTR and RBP are highly conserved proteins. RBP from three species of fish and from yellowtails binds retinol but lacks affinity for TTR, suggesting that the acquisition of a TTR-binding site was a later phylogenetic modification of an existing vitamin A transport system.

Plasma TTR concentrations decrease in malnutrition and liver disease and are highly correlated with RBP and retinol concentrations. TTR is a negative acute-phase reactant, with serum levels decreasing up to 50% in acute inflammatory conditions. In vitamin A deficiency, TTR synthesis in rat hepatocytes is unaffected, whereas RBP secretion from the liver declines markedly. Both RBP and TTR levels are reduced in hyperthyroid patients, but the RBP:TTR ratio remains constant. In FAP type I, the concentration of total TTR in the plasma is reduced (56).

The mutant TTR associated with FAP type I (methionine-30 variant) interacts normally with RBP (56), but both type I and type II variants show reduced affinity for thyroxine (57). In contrast, the Appalachian variant TTR shows normal affinity for thyroxine (57). Serum RBP levels are uniquely depressed in FAP type II, but it is unclear whether this is a primary effect or merely an accompaniment of the reduced plasma TTR levels seen in that condition (58).

TTR may exert a thymic hormone-like activity (59), and yet another transport function—of the small peptide thymuline—has been proposed for TTR (60). TTR is also present in pancreatic islet cells (61,62) and in isolated cells throughout the mucosa of the gastrointestinal tract (63). Plasma TTR levels are increased in some insulinomas and glucagonomas, and TTR is reportedly a highly reliable marker for carcinoid tumors (64) and neoplasms of the choroid plexus (65). There is a significant degree of homology between TTR, glucagon, secretin, and vasoactive intestinal peptide, suggesting that they may be related evolutionarily. High TTR mRNA levels have been detected in the visceral yolk sac of the rat (66) and mouse (67).

MOLECULAR BIOLOGY OF TRANSTHYRETIN

TTR has been studied extensively by X-ray crystallography (68). The following description of the three-dimensional structure and chemical properties of TTR is taken from the excellent summary by Felding (54). The monomeric subunit consists of 127 amino acids and has an extensive β-pleated structure composed of eight strands organized into two four-stranded sheets, and a single short α-helix. Almost half of the amino acids are organized into β conformation. The basic unit of structure is the dimer, formed by antiparallel hydrogen bond interactions that extend the two four-stranded sheets in the monomer to eight-stranded sheets in the dimer.

The tetrameric structure is formed by a face-to-face assembly involving both hydrophilic and hydrophobic interactions. There are no covalent bonds between subunits. The TTR subunits are arranged tetrahedrally in such a way that a central channel is formed that penetrates the entire molecule. The tetramer has three perpendicular axes of symmetry, so that rotation 180° around any one of the axes results in a molecular appearance identical to the original. Two symmetry-related binding sites for thyroxine are located in the central channel; however, due to negative cooperativity, high-affinity binding of only one molecule of thyroxine is observed. Two or four binding sites for RBP have been proposed, but it is not clear from the structural analysis where these sites are located. A curious feature of the structure is the presence of a cylindrical–helical impression on the external surface which is complementary to a DNA double helix. No function has yet been ascribed to this potential DNA-binding site.

The TTR tetramer is extremely stable. Its internal organization is unaffected by wide fluctuations in pH and it is resistant to a wide variety of denaturing conditions, requiring guanidine hydrochloride concentrations greater than 6 *M* to denature the molecule at a significant rate (69). Nevertheless, the tetramers do exchange subunits, albeit very slowly, under nondenaturing conditions (70). The binding of thyroxine increases the stability of the tetramer and reduces the subunit exchange rate.

Each monomeric subunit contains one cysteine residue at amino acid position 10, but no internal disulfide bonds are formed between monomers. The cysteines are located at the external margins of the central channel. Pettersson et al. (71) have noted microheterogeneity of TTR in different systems due to the status of the sulfhydryl groups and the equilibrium between tetramers and monomers. One of the observed monomeric forms is apparently an adduct with glutathione. A relationship appears to exist between the thiol groups, the binding of RBP and retinol, the tetramer–monomer equilibrium, and thyroxine binding. These complex interactions most likely contribute to the multifunctional properties of the molecule.

cDNA clones for human, rabbit, rat, and mouse TTR have been isolated and sequenced (72–76) (Fig. 22-3). TTR is encoded by a single gene in all of these species; in humans, the gene is located on chromosome 18 (77). There is a high degree of interspecies homology between the nucleotide coding and protein sequences, indicating that TTR is a highly conserved protein in evolutionary terms. Comparison of mouse and human sequences has revealed that regions corresponding to functional domains of the protein are highly conserved and that all but 1 of the 25 amino acid substitutions are located on the outer surface of the molecule (76). This suggests that those amino acids facing the interior of the molecule are critical to its function.

Genomic clones have been isolated and analyzed for human and mouse TTR (78–81). The human gene spans about 7 kb of DNA and consists of four exons and

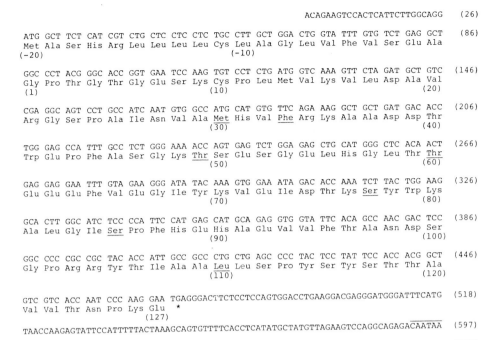

```
                                                       ACAGAAGTCCACTCATTCTTGGCAGG   (26)

 ATG GCT TCT CAT CGT CTG CTC CTC CTC TGC CTT GCT GGA CTG GTA TTT GTG TCT GAG GCT   (86)
 Met Ala Ser His Arg Leu Leu Leu Leu Cys Leu Ala Gly Leu Val Phe Val Ser Glu Ala
(-20)                                   (-10)

 GGC CCT ACG GGC ACC GGT GAA TCC AAG TGT CCT CTG ATG GTC AAA GTT CTA GAT GCT GTC  (146)
 Gly Pro Thr Gly Thr Gly Glu Ser Lys Cys Pro Leu Met Val Lys Val Leu Asp Ala Val
 (1)                                (10)                                    (20)

 CGA GGC AGT CCT GCC ATC AAT GTG GCC ATG CAT GTG TTC AGA AAG GCT GCT GAT GAC ACC  (206)
 Arg Gly Ser Pro Ala Ile Asn Val Ala Met His Val Phe Arg Lys Ala Ala Asp Asp Thr
                                    (30)                                    (40)

 TGG GAG CCA TTT GCC TCT GGG AAA ACC AGT GAG TCT GGA GAG CTG CAT GGG CTC ACA ACT  (266)
 Trp Glu Pro Phe Ala Ser Gly Lys Thr Ser Glu Ser Gly Glu Leu His Gly Leu Thr Thr
                                    (50)                                    (60)

 GAG GAG GAA TTT GTA GAA GGG ATA TAC AAA GTG GAA ATA GAC ACC AAA TCT TAC TGG AAG  (326)
 Glu Glu Glu Phe Val Glu Gly Ile Tyr Lys Val Glu Ile Asp Thr Lys Ser Tyr Trp Lys
                                    (70)                                    (80)

 GCA CTT GGC ATC TCC CCA TTC CAT GAG CAT GCA GAG GTG GTA TTC ACA GCC AAC GAC TCC  (386)
 Ala Leu Gly Ile Ser Pro Phe His Glu His Ala Glu Val Val Phe Thr Ala Asn Asp Ser
                                    (90)                                    (100)

 GGC CCC CGC CGC TAC ACC ATT GCC GCC CTG CTG AGC CCC TAC TCC TAT TCC ACC ACG GCT  (446)
 Gly Pro Arg Arg Tyr Thr Ile Ala Ala Leu Leu Ser Pro Tyr Ser Tyr Ser Thr Thr Ala
                                    (110)                                   (120)

 GTC GTC ACC AAT CCC AAG GAA TGAGGGACTTCTCCTCCAGTGGACCTGAAGGACGAGGGATGGGATTTCATG  (518)
 Val Val Thr Asn Pro Lys Glu  *
                        (127)
 TAACCAAGAGTATTCCATTTTTACTAAAGCAGTGTTTTCACCTCATATGCTATGTTAGAAGTCCAGGCAGAGACAATAA  (597)

 AACATTCCTGTGAAAGGC - poly(A)                                                      (615)
```

Fig. 22-3. Complete nucleotide and deduced amino acid sequences of a cDNA clone encoding human transthyretin. Nucleotides are shown in the upper register and the corresponding amino acid sequence in the lower register. Numbers at the end of each line refer to nucleotide position and numbers below each line refer to amino acid position. Nucleotides are numbered in the 5′ to 3′ direction, beginning with a short 5′ noncoding sequence. The prepeptide begins at position 27, which is the first nucleotide of the ATG codon triplet encoding the initiative methionine. Putative signal peptide residues are indicated by negative numbers. The mature polypeptide begins at the glycine residue numbered 1. The termination signal follows amino acid number 127 and is indicated by an asterisk. The polyadenylation signal is overlined. The seven amino acids implicated in substitutions causing human amyloidoses are underlined. After Mita et al. (73) and Soprano et al. (74).

three introns. An unusual feature of the gene in both species is the presence of two independent open reading frames, provided with the appropriate transcriptional regulatory sequences, within the first and third introns, respectively. However no mRNA corresponding to these sequences was detected in Northern analysis of liver, brain, and pancreatic RNA (S. Mita, personal communication).

In addition to the exon-coding regions, there are two highly conserved DNA regions in mouse and human TTR, one at the 3′ end of the first intron, and the other in the 5′ flanking region of the gene (78). The latter spans approximately 200 nucleotides and contains a liver cell-specific promoter located, in the mouse, between 108 and 151 nucleotides upstream of the transcription initiation site. In addition, a distal enhancer element is present between 1.6 and 2.15 kb upstream of the first exon; remarkably, this element can stimulate expression in hepatoma but not in HeLa cells, and is thus an example of a distal cell-specific enhancer, only the second of its kind to be described for a mammalian gene (79). Additionally, three sequences identical to the SV40 enhancer core sequence are present in the introns and the 3′ flanking region of the human gene (80,81).

Two glucocorticoid receptor (GCR)-binding site sequences are present in the 5′ flanking region of the human TTR gene (81), but similar sequences are not found in the homologous region of the mouse gene (78). However, two sequences homologous to the mouse mammary tumor virus GCR-binding site sequence are present in the 5′ flanking region in the mouse. Furthermore, in the highly conserved region of intron I, at least one GCR-binding site sequence is present in both species. In spite of the presence of such potential regulatory sequences, the glucocorticoid dexamethasone failed to induce TTR mRNA synthesis in the mouse liver (78). Further studies are needed to assess whether the TTR gene is responsive to steroidal modulation.

TRANSTHYRETIN IN THE CENTRAL NERVOUS SYSTEM

Although TTR is a minor component of the total plasma protein, its concentration is disproportionately elevated in CSF (82). This suggested to some authors selective uptake of TTR from the plasma across the blood–CSF barrier (83). However, the possibility of a separate intraaxial source of TTR was also entertained, and Aleshire et al. (84) presented ultrastructural evidence in support of TTR synthesis in the choroid plexus of the cerebral ventricles. Subsequently, several groups of investigators reported the presence of abundant TTR mRNA in rat and human brain, specifically within the choroid plexus (74,85–90). By *in situ* hybridization TTR mRNA is localized within the epithelial monolayer of the plexus, in agreement with the distribution of immunoreactive TTR (Fig. 22-4). TTR forms a major fraction of newly synthesized protein secreted into the medium by a rat choroid plexus explant, implying that, *in vivo,* TTR is also secreted into the CSF (89).

TTR transcripts and immunoreactivity have been detected in the primordial choroid plexus, the tela choroidea, as early as 10 days gestational age in the mouse (67) and 11 days in the rat (86). At these stages the embryonic ependymal lining has just begun to indent the ventricular cavity but the typical villous architecture has not yet developed. In other words, the biochemical specialization for TTR production precedes the morphological differentiation of the organ.

De novo TTR synthesis by the choroid plexus is of particular interest for several reasons: first, it establishes a major extrahepatic source of TTR synthesis, in addi-

Fig. 22-4. TTR synthesis in the choroid plexus. (A) Fourth ventricle space bounded by the cerebellum (cb) and medulla (me) in the rat. Note intense TTR immunostaining of the epithelium of the choroid plexus, but no staining of the ependymal lining of the ventricle. The arrow indicates a sharp point of transition at the tela choroidea. (Antiserum to rat TTR kindly provided by D. S. Goodman.) (B) *In situ* hybridization of ^{35}S-labeled TTR mRNA complementary probe to rat brain. After autoradiography, TTR transcripts are visualized as silver grains distributed uniformly within the cytoplasm of all choroid plexus epithelial cells (cp), but the ependymal lining (ep) of the cerebral ventricle (v) and the brain parenchyma (p) are unlabeled. (High-power view.) Reprinted with permission from *Neurology* (85).

tion to other possible minor sources noted in pancreas, testis, and chromaffin cells of the GI tract. Second, TTR is the first molecular marker within the central nervous system (CNS) assigned specifically to the choroid plexus. This has provided a diagnostic tool for the pathological assessment of choroid plexus neoplasms (65). Third, it raises the question of the function of TTR within the CNS.

The choroid plexuses are the major sites of CSF production and the choroidal

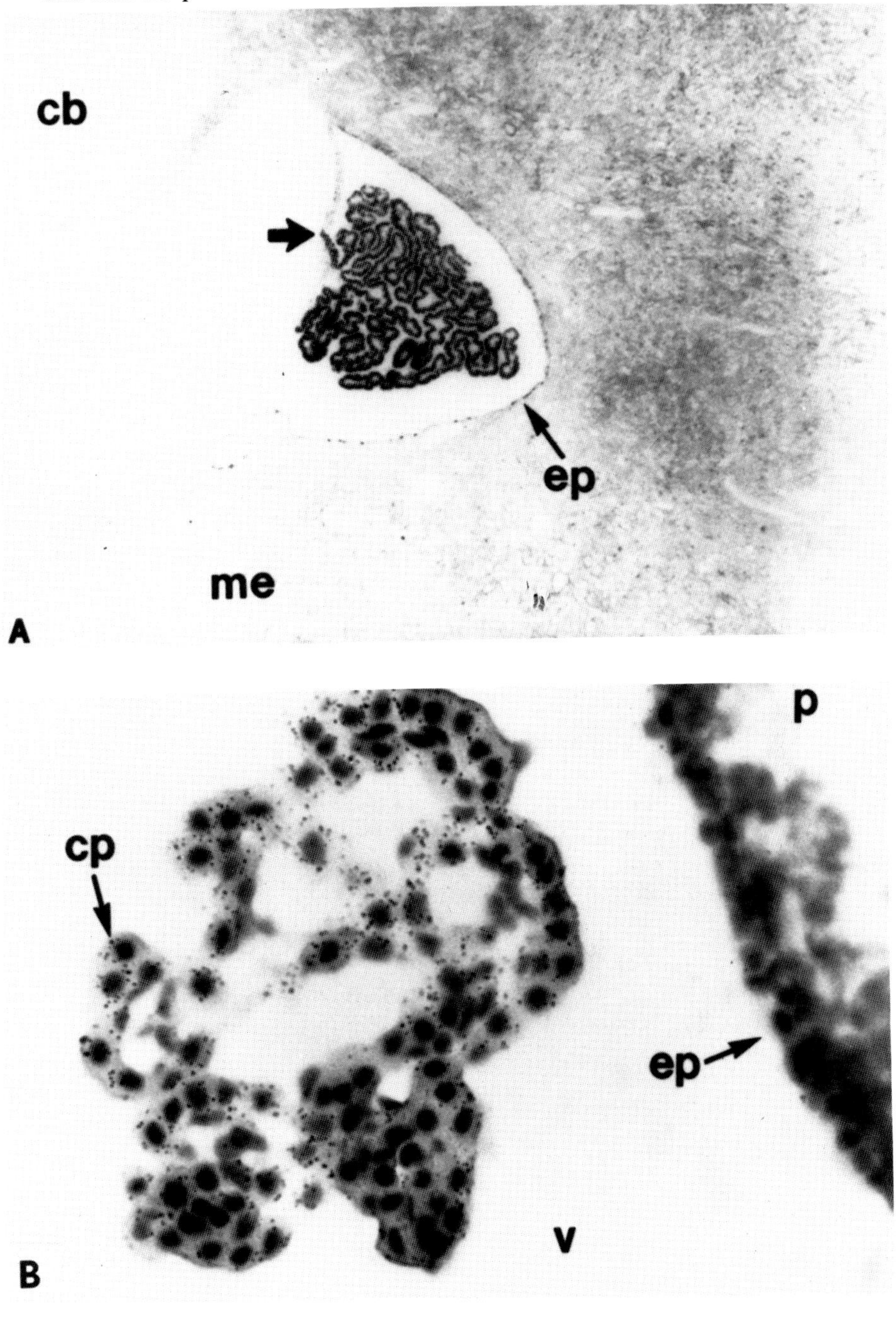

epithelium is the physical barrier to free diffusion of macromolecules from the plasma to CSF compartments. Thus TTR is ideally located to serve as a transchoroidal transport protein; in addition, its appearance in CSF suggests that TTR is involved in molecular translocation throughout the neuraxis. Little is known about vitamin A transport in the CNS, but there is little, if any, RBP in the nervous tissue (D. S. Goodman, personal communication) or CSF (91). Evidence has been presented that, in the rat, TTR is responsible for thyroxine transport into the CSF (92). Thyroxine-binding globulin plays only a minor role in CSF thyroxine transport compared to TTR (53,54). Since only about 3% of CSF TTR is complexed to thyroid hormone, there is, theoretically, sufficient protein available to fulfill other possible roles in CNS metabolism.

An interesting hypothesis has been proposed by Costa et al. (93). They suggested that TTR is trophic to peripheral nerve, and that the route of access is via the CSF. This implies specific sites for TTR recognition and uptake along the neuraxis, presumably at the level of the proximal nerve roots or spinal cord. No experimental evidence has yet been offered to support this hypothesis, but it could explain the predilection of TTR-amyloid for the peripheral nervous system and the fact that, unlike most other proteins, TTR concentration decreases with caudal progression along the subarachnoid space (94). The suggestion that TTR is trophic to peripheral nerve is also consistent with Andrade's observation that, in FAP type I, neuropathic features precede the deposition of amyloid, implying that the neuropathy is due to a metabolic derangement rather than to the physical presence of the amyloid (8).

In FAP type I, both mutant and normal TTR are found in spinal fluid. In contrast to the plasma, where the two species approximate equimolar concentrations (12) or normal TTR is in excess (11), in the CSF mutant TTR is the predominant species by a ratio of 4:1. However, the relative proportion of mutant and normal mRNA in postmortem choroid plexus homogenates has been reported to approximate 1:1 (90). In order to obtain a clear picture of TTR metabolism (both normal and mutant) within the CNS, further studies of this nature are required. Obviously, it is important to determine whether peripheral nerve amyloid is derived from plasma or CSF TTR, or both.

Since there is a separate source of TTR synthesis within the neuraxis, it is remarkable that, in most of the FAP variants, the brain is almost entirely free of amyloid deposits. In FAP type I, amyloid is found in the choroid plexus, the leptomeninges, and in subpial and subependymal locations, that is, only within structures interfacing directly with CSF. In other variants cranial neuropathy or leptomeningeal involvement may be more pronounced, but the cerebral parenchyma is always spared. This supports the notion that, whatever its function in the CNS, TTR is largely restricted to the CSF compartment.

TRANSTHYRETIN IN THE EYE

Ocular abnormalities are prominent in many of the clinical variants of FAP (Table 22-1). Ciliary body involvement with scalloping of the pupils is a feature of type I; vitreous opacities due to amyloid deposition are manifestations of types I and II, the Ashkenazi Jewish variant and familial oculoleptomeningeal amyloidosis; cataracts are common in type III; and corneal lattice dystrophy is the salient clinical sign in type IV. Additionally, localized heredofamilial amyloidosis restricted to the cornea is recognized. In our laboratory we have demonstrated, by Northern analysis, the presence of TTR mRNA in rat and bovine eyes, of similar relative abundance to that seen previously in the choroid plexus (95). By immunohistochemical analysis, we have localized TTR protein to the retinal pigment epithelium, the corneal endothelium, the iris, ciliary and lens epithelia, the ganglion cell layer and inner limiting membrane of the retina, and the vitreous humor (Fig. 22-5). The most intense staining is present in the retinal pigment epithelium, the developmental homologue of the choroid plexus. *In situ* hybridization studies demonstrated that the pigment epithelium is the unique site of TTR synthesis in the rat eye.

The role of TTR in ocular metabolism is obscure. Vitamin A is an integral component of the biochemistry of the visual process, but TTR has not been implicated in intraocular translocation of retinol (for review, see 96). Plasma RBP is thought to deliver retinol to the pigment epithelial surface, following which it is complexed to a series of cellular retinoid-binding proteins for further processing in the visual cycle. None of the latter binds to TTR, and RBP has not been reported in ocular structures. Therefore, it appears more likely that in the eye TTR is involved principally in the transport of thyroid hormone, as has been suggested for the choroid plexus (92).

It is remarkable that nature has designed two specialized systems, sequestered from the systemic circulation, for the production and secretion of TTR into their respective humors. The functional significance of this phenomenon is still largely unknown.

PATHOGENESIS

Most of the mutations involving TTR-amyloid are located on the external surface of the molecule, with the notable exception of the methionine-for-valine substitution at residue 30. Precisely how these amino acid changes alter protein structure is unknown. It is noteworthy that residues 30, 33, and 49, which have been implicated in FAP type I and the Jewish variant, respectively, lie in close stereochemical proximity to one another, suggesting that they may constitute a structural domain that, if altered, is important in the pathogenesis of FAP (15). Although some investigators have found that the amyloid deposits in FAP type I are composed predominantly of intact TTR (Met-30 variant) (11), others have reported that amyloid deposition may involve partial degradation of the TTR monomer or modifications such as the addition of a glycine residue at position 49 (42). It is not known whether these changes occur before or after amyloid deposition.

The FAPs are an extreme example of the dictum that knowing the etiology of a disease does not explain its pathogenesis. We have little, if any, understanding of the factors that determine amyloid deposition, tissue specificity, age at onset, clinical severity, and clinical course. Information is needed about the chemical features critical to structure and function of the TTR molecule, the role of TTR in neural

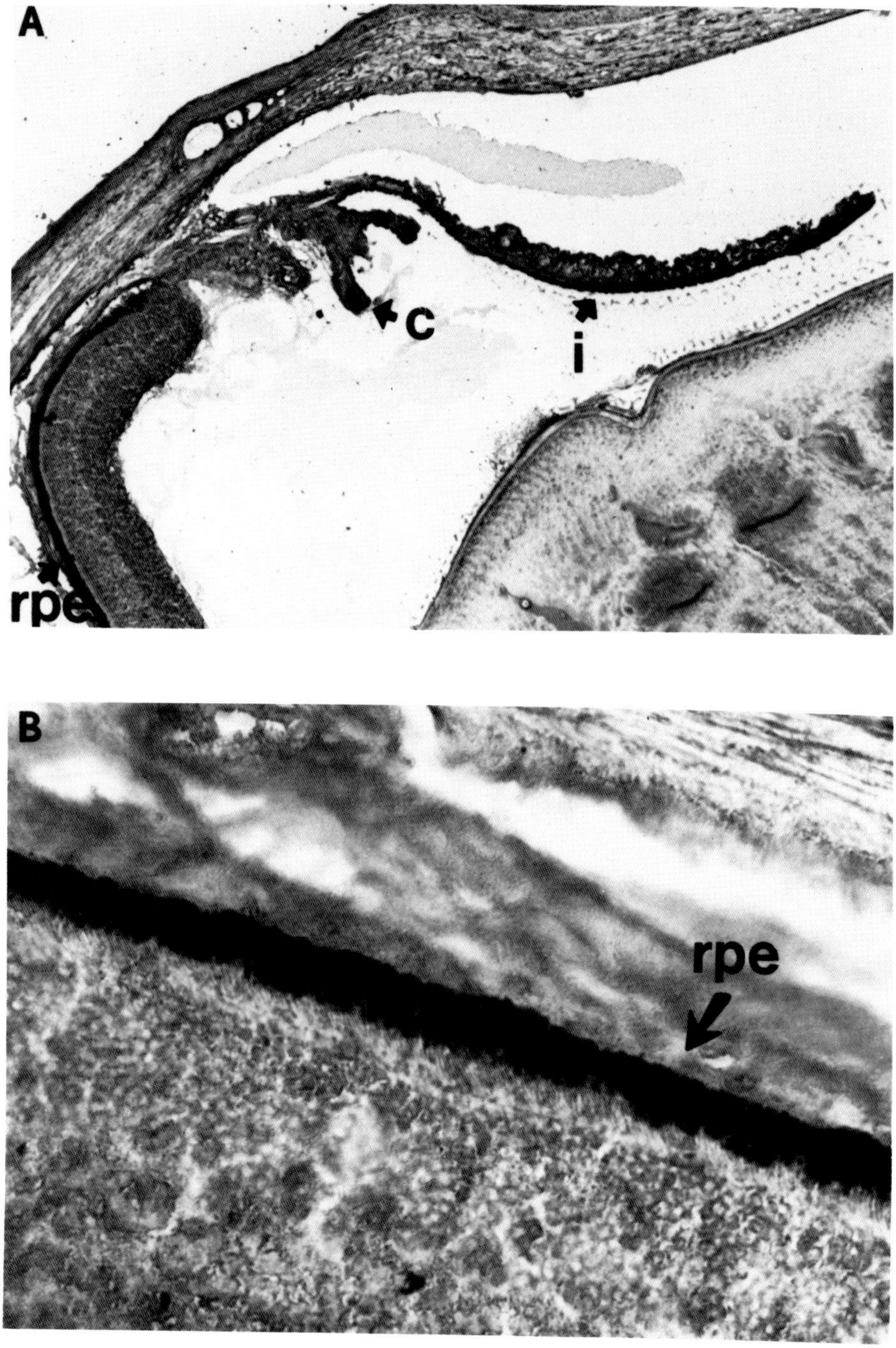

Fig. 22-5. Immunocytochemical demonstration of TTR in rat eye. (A) Anterior chamber, low-power view. (B) Retinal pigment epithelium, high-power view. Intense immunoperoxidase reaction product is present within the retinal pigment epithelium (rpe) whereas the iris (i) and ciliary (c) epithelia stain less intensely. (H and E counterstain. Antiserum to rat TTR kindly provided by D. S. Goodman.) Reprinted with permission from *Biochem Biophys Res Commun* (95).

metabolism, and the biochemistry of amyloid fibril formation. Protein chemists must determine which amino acid substitutions will favor amyloid fibril deposition. That process is determined not only by the nature of the protein but also by local factors in the microenvironment, and we need information about the nature of these tissue factors.

One promising approach is provided by molecular genetic technology, which can provide specific mutations in DNA (Fig. 22-6) (97). A short complementary oligonucleotide containing the desired mutated sequence can be created on a DNA synthesizer, then hybridized to the full-length wild-type gene contained in an appropriate vector. Synthesis of the remainder of the sequence complementary to the full-length gene can be completed by enzymatic reaction. The resulting DNA molecules are allowed to replicate in an appropriate host (e.g., *E. coli*) in culture; half of the progeny will express the wild-type and half the mutated gene. The mutated DNA can be isolated and amplified and the resulting DNA used for the synthesis, *in vitro*, of the mutant polypeptide.

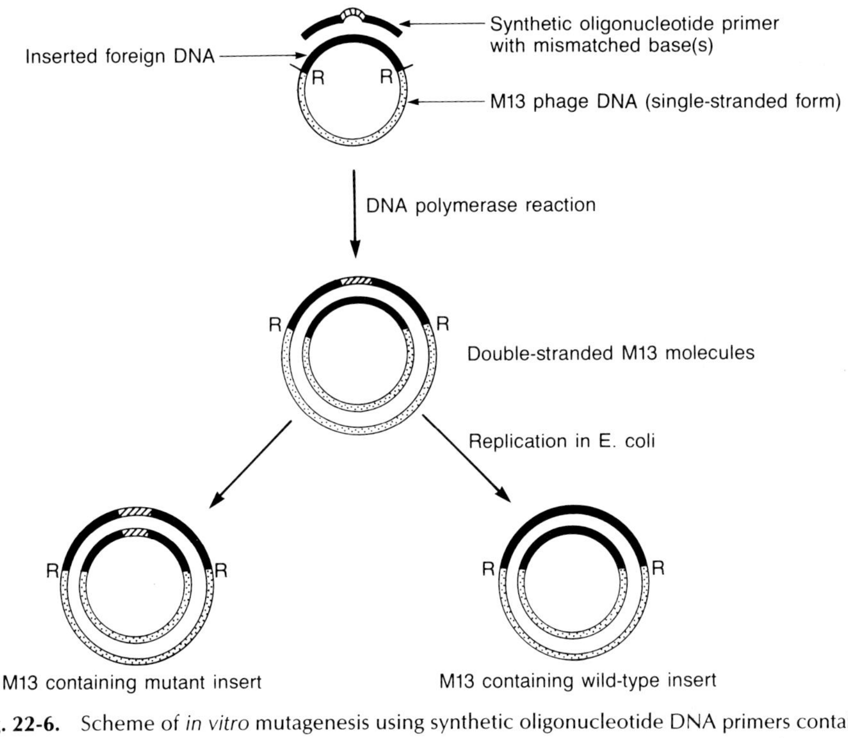

Fig. 22-6. Scheme of *in vitro* mutagenesis using synthetic oligonucleotide DNA primers containing mismatched base pairs. R indicates restriction enzyme sites in the vector M13 phage, into which the foreign DNA strand has been inserted. The synthetic oligonucleotide is hybridized to the single-stranded form of the recombinant phage containing the wild-type insert, and the complementary strand synthesized in the presence of the four deoxyribonucleotides and DNA polymerase. The resulting double-stranded recombinant M13 molecules are hybrids. These are then introduced into *E. coli,* where they replicate; one-half of the progeny will contain the wild type and one-half the mutant insert. Mutants can be identified by differential hybridization or by restriction analysis, if the mutation creates a new restriction site. Mutant inserted DNA can be recovered by cleavage at R with the appropriate restriction enzyme.

That approach could generate a panel of mutant TTR molecules and their biochemical characteristics could then be analyzed extensively in the test tube.

A variation on this theme allows for the expression of such artificially generated molecules in the natural biological setting. Mutant genes can be microinjected into the oocytes of mice and transmitted to progeny as stably inherited Mendelian traits (98). This approach is being applied to FAP in several laboratories (99). There is a high degree of homology between human and mouse TTR (Fig. 22-7). Each one of the amino acids involved in known FAP mutations (numbers 30, 33, 60, 77, 84, 111) is conserved between the two species, implying that these residues are required for functional or structural integrity of the molecule. It is theoretically possible to create transgenic mice that carry identical mutations to those identified in the human diseases. If mice behave like humans, the animals might develop a disease similar to FAP. That would be a true animal model of a human disease. The experimental disease could be an incisive tool for the study of pathogenesis and for possible therapy. The same approach can be used for other dominantly inherited disorders, such as Huntington disease, when the mutations are identified.

PROPOSED CLASSIFICATION

While this discussion is concerned with the FAPs, it should be emphasized that neuropathy and amyloidosis are not invariable manifestations of disorders associated with mutant TTR. For example, neuropathy is absent in the Danish form of

```
            -20                 1                                       40
Human       MASHRLLLLCLAGLVFVSEAGPTGTGESKCPLMVKVLDAVRGSPAINVAVHVFRKAADDT
Mouse       ---L--F---------------A-A--------------------VD---K--K-TSEGS
FAP                                                          M  I

                                                                       100
Human       WEPFASGKTSESGELHGLTTEEEFVEGIYKVEIDTKSYWKALGISPFHEHAEVVFTANDS
Mouse       ---------A----------D-K----V-R--L-------T------F-D----------
FAP                            A                Y      S

                                   127
Human       GPRRYTIAALLSPYSYSTTAVVTNPKE
Mouse       -H-H------------------S--QN
FAP                   M

CODE:  M=methionine   A=alanine     S=serine      H=histidine
       R=arginine     L=leucine     C=cysteine    G=glycine
       V=valine       T=threonine   E=glutamate   F=phenylalanine
       P=proline      K=lysine      I=isoleucine  D=aspartate
       Y=tyrosine     Q=glutamine
```

Fig. 22-7. Schematic representation of the homology between human (top register) and mouse (middle register) transthyretin amino acid sequences. Amino acids are designated by single letters as shown in the code. Numbers above the top register represent the presequence (−20 to 0) and active peptide sequence (1 to 127) of transthyretin. The C-terminus is at position 127. In the mouse sequence, a dash represents an amino acid residue that is identical to that shown in the human sequence in the register above. The lower register identifies the sites and identity of those amino acid substitutions known to be involved in human familial amyloidotic polyneuropathies. They are all located at conserved positions.

Table 22-5 The Transthyretinopathies, 1987

A. Known
  1. The familial amyloidotic polyneuropathies
     a. Type I (Andrade)
     b. Type II (Rukavina)
     c. Ashkenazi Jewish type
     d. Appalachian type
     e. Illinois/German type
  2. Familial amyloidotic cardiomyopathy
  3. Familial euthyroid hyperthyroxinemia

B. Suspected
  1. Familial amyloidotic polyneuropathies
     a. Type III (Van Allen)
     b. Type IV (Meretoja) (ref. 102)
     c. Familial oculoleptomeningeal amyloidosis
  2. Senile cardiac amyloidosis (ref. 103)

familial amyloidotic cardiomyopathy (27), in which a mutation at TTR amino acid residue number 110 has been reported. Familial euthyroid hyperthyroxinemia is caused by a variant TTR with an increased affinity for T_4, but neither amyloidosis nor neuropathy is a feature of that disorder (100,101). At least half a dozen other TTR mutants are currently being characterized (J. Saraiva, personal communication); not all are amyloidogenic, and several differ in their affinity for thyroxine.

Table 22-5 summarizes those clinical syndromes currently known or suspected to be associated with variant TTR. It is remarkable that single amino acid substitutions in the same molecule should be associated with such a diverse spectrum of clinical symptomatology. For this group of disorders I would therefore propose the term "transthyretinopathies"—which is not much worse than "hemoglobinopathies."

SUMMARY

The familial amyloidotic polyneuropathies (FAPs) comprise a group of dominantly inherited diseases characterized by the deposition of amyloid fibrils in the peripheral nervous system and in a variety of systemic organs. They are the first of the hereditary neuropathies to be characterized at the protein and DNA level. Single amino acid substitutions in the transthyretin (TTR) monomer result in the varying clinical phenotypes. Affected individuals are heterozygous for one normal and one mutant TTR allele. DNA restriction analysis provides accurate detection of presymptomatic carriers and prenatal determination of heterozygosity. TTR is a 56-kDa plasma protein, composed of four identical subunits, which plays a role in the transport of thyroxine and retinol (vitamin A). About 50% of the normal TTR monomer is in the β-pleated sheet conformation, but the factors that determine mutant TTR-amyloid deposition and the predilection for the peripheral nervous system are not understood. We have found separate sources of TTR synthesis in the choroid plexus of the cerebral ventricles and within the eye. These sites of synthesis may explain some of the clinical manifestations of the FAPs but the physiological significance of these findings is obscure. Understanding the pathogenesis of the FAPs will require more precise information about the structure and function of TTR, its physiological role in the nervous system, and the biochemistry of amyloid fibril deposition. Molecular genetic technology has contributed to the elucidation of the molecular basis of the FAPs and provides novel approaches for the elucidation of their pathophysiology.

ACKNOWLEDGMENTS

I am indebted to DeWitt S. Goodman, Maria Joao M. Saraiva, Shuji Mita, and Merrill D. Benson for critical review of the manuscript. This work was supported by a fellowship from the Muscular Dystrophy Association, NIH grant NSO1155 (Clinical Investigator Development Award), and grants from the Charles A. Dana and Aaron Diamond Foundations.

REFERENCES

1. Glenner GG. A retrospective and prospective overview of the investigations on amyloid and amyloidosis—the beta-fibrilloses. In Glenner GG, Costa PP, De Freitas, eds. Amyloid and Amyloidosis. Amsterdam: Excerpta Medica, 1980, pp 7–13.

2. Eanes ED, Glenner GG. X-Ray diffraction studies on amyloid filaments. J Histochem Cytochem 1968; 16:673–677.

3. Benditt EP, Cohen AS, Costa PP, et al. Guidelines for Nomenclature. In Glenner GG, Costa PP, De Freitas, eds. Amyloid and Amyloidosis. Amsterdam: Excerpta Medica, 1980, p XI.

4. Goodman DS. Statement regarding nomenclature for the protein known as prealbumin, which is also (recently) called transthyretin. In Glenner GG, Osserman EF, Benditt EP, Calkins E, Cohen AS, Zucker-Franklin D, eds. Amyloidosis. New York: Plenum Press, 1986, pp 287–288.

5. Glenner GG, Ignaczak TF, Page DL. The inherited systemic amyloidoses and localized amyloid deposits, chap 54. In Stanbury JB, Wyngaarden JB, Fredrickson DS, eds. The Metabolic Basis of Inherited Disease, 4th ed. New York: McGraw-Hill, 1978, pp 1308–1339.

6. Cohen AS, Rubinow A. Amyloid Neuropathy, chap 81. In Dyck PJ, Thomas PK, Lambert EH, Burge R, eds. Peripheral Neuropathy, Vol II. Philadelphia: Saunders, 1984, pp 1866–1898.

7. Andrade C. Amyloid neuropathy. In Vinken PJ, Bruyn GW, Myrianthopolous NC, eds. Handbook of Clinical Neurology, Vol 42: Neurogenetic directory, Part I. Amsterdam: Elsevier, 1981, pp 518–524.

8. Andrade C. A peculiar form of peripheral neuropathy. Familial atypical generalized amyloidosis with special involvement of peripheral nerves. Brain 1952; 75:408–427.

9. Nakao K, Togi H, Furukawa T. A pedigree of familial amyloid neuropathy. Clin Neurol (Tokyo) 1966; 6:369–370.

10. Andersson R. Hereditary amyloidosis with polyneuropathy. Acta Med Scand 1970; 188:85–94.

11. Saraiva MJM, Birken S, Costa PP, Goodman DS. Amyloid fibril protein in familial amyloidotic polyneuropathy, Portuguese type. J Clin Invest 1984; 74:104–119.

12. Nakazato M, Kangawa K, Minamino N, Tawara S, Matsuo H, Araki S. Identification of a prealbumin variant in the serum of a Japanese patient with familial amyloidotic polyneuropathy. Biochem Biophys Res Commun 1984; 122:712–718.

13. Dwulet FE, Benson M. Primary structure of an amyloid prealbumin and its plasma precursor in a heredofamilial polyneuropathy of Swedish origin. Proc Natl Acad Sci USA 1984; 81:694–698.

14. Benson M, Dwulet FE, Scheinberg MA, Greipp P. Chemical classification of hereditary amyloidosis in Brazilian families and identification of gene carriers. J Rheumat 1986; 13:927–931.

15. Saraiva MJM, Costa PP, Goodman DS. Transthyretin (prealbumin) in familial amyloidotic polyneuropathy: Genetic and functional aspects. In DiDonato S, DiMauro S, Mamoli A, Rowland LP, eds. Advances in Neurology, Vol 48: Molecular Genetics of Neurological and Neuromuscular Disease. New York: Raven Press, 1988, pp 189–200.

16. Rukavina JG, Block WD, Jackson CE, Falls HF, Carey JH, Curtis AC. Primary systemic amyloidosis: A review and an experimental genetic and clinical study of 29 cases with a particular emphasis on the familial form. Medicine 1956; 35:239–334.

17. Mahloudji M, Teasdell RD, Adamkiewicz JJ, Hartmann, WH, Lambird PA, McKusick VA. The genetic amyloidoses: With particular reference to hereditary neuropathic amyloidoses, Type II (Indiana or Rukavina type). Medicine 1969; 48:1–37.

18. Van Allen MW, Frolich J, Davis JR. Inherited predisposition to generalized amyloidosis. Neurology 1969; 19:10–25.

19. Gimeno A, Garcia-Alix C, de Arana JMS, Mateos F, Sotelo MT. Amyloidotic polyneuritis of type III (Iowa—Van Allen). Eur Neurol 1974; 11:46–57.

20. Meretoja J. Familial systemic paramyloidosis with lattice dystrophy of the cornea, progressive cranial neuropathy, skin changes and various internal symptoms. A previous unrecognized hereditable syndrome. Ann Clin Res 1969; 1:314–321.

21. Pras M, Prelli F, Franklin EC, Frangione B. Primary structure of an amyloid prealabumin variant in familial polyneuropathy of Jewish origin. Proc Natl Acad Sci USA 1983; 80:539–542.

22. Gafni J, Fischel F, Reif R, Yaron M, Pras M. Amyloidotic polyneuropathy in a Jewish family. Evidence for genetic heterogeneity of the lower limb familial amyloidotic polyneuropathies. J Med 1985; New Series 55 (216):33–43.

23. Wallace MR, Dwulet FE, Conneally PM, Benson MD. Biochemical and molecular genetic characterization of a new variant prealbumin associated with hereditary amyloidosis. J Clin Invest 1986; 78:6–12.

24. Wallace MR, Dwulet FE, Williams EC, Conneally PM, Benson MD. Identification of a new hereditary amyloidosis prealbumin variant, tyr-77, and detection of the gene by DNA analysis. J. Clin Invest 1988; 81:189–193.

25. Goren H, Steinberg MC, Farboody GH. Familial oculoleptomeningeal amyloidosis. Brain 1980; 103:473–495.

26. Levin M, Franklin EC, Frangione B, Pras M. The amino acid sequence of a major nonimmunoglobulin component of some amyloid fibrils. J Clin Invest 1972; 51:2773–2776.

27. Frederiksen TH, Gotzsche H, Harboe N, Kiaer W, Mellemgaard K. Familial primary amyloidosis with severe amyloid heart disease. Am J Med 1962; 33:328–348.

28. Husby G, Ranlow PJ, Sletten K, Marhaug G. The amyloid in familial amyloid cardiomyopathy of Danish origin is related to prealbumin. Clin Exp Immunol 1985; 60:207–216.

29. Ostertag B. Familiare Amyloid-erkrankung. Z Menschl Vererbungs Konstit Lehre 1962; 31:235–248.

30. Muckle TJ, Wells M. Urticaria, deafness and amyloidosis: A new heredofamilial syndrome. J Med 1962; 31:235–248.

31. Gudmundsson GJ, Hallgrimsson TA, Jonasson TA, Bjornason O. Hereditary cerebral hemorrhage with amyloidosis. Brain 1972; 95:387–404.

32. Cohen DH, Feiner H, Jensson O, Frangione AB. Amyloid fibril in hereditary cerebral

hemorrhage with amyloidosis (HCHWA) is related to the gastroenteropancreatic neuroendocrine protein, gamma trace. J Exp Med 1983; 158:623–628.

33. Goldgaber D, Lerman MI, McBride OW, Saffiotti U, Gadjusek DC. Characterization and chromosomal localization of a cDNA encoding brain amyloid of Alzheimer's disease. Science 1987; 235:877–880.

34. Tanzi RE, Gusella JF, Watkins PC, et al. Amyloid beta protein gene: cDNA, mRNA distribution, and genetic linkage near the Alzheimer locus. Science 1987; 235:880–884.

35. Tanzi RE, St George-Hyslop PH, Haines JL, et al. The genetic defect in familial Alzheimer's disease is not tightly linked to the amyloid-protein gene. Nature 1987; 329:156–157.

36. Sletten K, Westermark P, Natvig JB. Characterization of amyloid fibril protein from medullary carcinoma of the thyroid. J Exp Med 1976; 143:993–998.

37. Saraiva MJM, Costa PP, Goodman DS. Genetic expression of a transthyretin mutation in typical and late-onset families with familial amyloidotic polyneuropathy. Neurology 1986; 36:1413–1417.

38. Costa P, Figueira AS, Bravo FR. Amyloid fibril protein related to prealbumin in familial polyneuropathy. Proc Natl Acad Sci USA 1978: 75:4499–4503.

39. Benson MD. Partial amino acid sequence homology between heredofamilial amyloid protein and human plasma prealbumin. J Clin Invest 1981: 67:1035–1041.

40. Mita S, Maeda S, Ide M, Tsuzuki T, Shimada K, Araki S. Familial amyloidotic polyneuropathy diagnosed by cloned human prealbumin cDNA. Neurology 1986; 36:298–301.

41. Sasaki H, Sakaki Y, Matsuo H, et al. Diagnosis of familial amyloidotic polyneuropathy by recombinant DNA techniques. Biochem Biophys Res Commun 1984; 122:719–725.

42. Benson MD, Dwulet FE. Familial amyloidotic polyneuropathy type I: Characterization of the prealbumin amyloid subunit and precursor protein. In Glenner GG, Osserman EF, Benditt EP, Calkins E, Cohen AS, Zucker-Franklin D, eds. Amyloidosis. New York: Plenum Press, 1986, pp 367–373.

43. Sasaki H, Sakaki Y, Takagi Y, et al. Presymptomatic diagnosis of heterozygosity for familial amyloidotic polyneuropathy by recombinant DNA techniques. Lancet 1985; 1 (8420):100.

44. Saraiva MJM, Costa PP, Goodman DS. Biochemical marker in familial amyloidotic polyneuropathy, Portuguese type. Family studies on the transthyretin (prealbumin)-methionine-30 variant. J Clin Invest 1985; 76:2171–2177.

45. Nakazato M, Kangawa K, Minamino N, Tawara S, Matsuo H, Araki S. Radioimmunoassay for detecting abnormal prealbumin in the serum for diagnosis of familial amyloidotic polyneuropathy (Japanese type). Biochem Biophys Res Commun 1984; 122:719–725.

46. Benson MD, Dwulet FE. Identification of carriers of a variant plasma prealbumin (transthyretin) associated with familial amyloidotic polyneuropathy type I. J Clin Invest 1985; 75:71–75.

47. Altland K, Banzhoff A. Separation by isoelectric focusing of normal human plasma transthyretin (prealbumin) and a variant with a methionine for valine substitution associated with familial amyloidotic polyneuropathy. Electrophoresis 1986; 125:636–642.

48. Dwulet FE, Benson MD. Characterization of a transthyretin (prealbumin) variant associated with familial amyloidotic polyneuropathy type II (Indiana/Swiss). J Clin Invest 1986; 78:880–886.

49. Nakazato M, Kangawa K, Minamino N, Tawara S, Matsuo H, Araki S. Revised analysis of amino acid replacement in a prealbumin variant (SKO-111) associated with familial amyloidotic polyneuropathy of Jewish origin. Biochem Biophys Res Commun 1984; 123:921–928.

50. Pras M, Prelli F, Gafni J, Frangione B. Genetic heterogeneity of familial amyloidotic polyneuropathies of Jewish type. In Glenner GG, Osserman EF, Benditt EP, Calkins E,

Cohen AS, Zucker-Franklin D, eds. Amyloidosis. New York: Plenum Press, 1986, pp 385–389.

51. Goodman DS. Plasma retinol-binding protein, chap 8. In Sporn MB, Roberts AB, Goodman DS, eds. The Retinoids, Vol 2. Orlando, FL: Academic Press, 1984, pp 41–88.

52. Kanda Y, Goodman DS, Canfield RE, Morgan FJ. The amino acid sequence of human plasma prealbumin. J Biol Chem 1974; 249:6796–6805.

53. Hagen GA, Elliot W. Transport of thyroid hormones in serum and cerebrospinal fluid. J Clin Endocr Metab 1973; 37:415–422.

54. Felding P. Prealbumin: Metabolic and Chemical Studies. Thesis: Malmo, Sweden: University of Lund, 1984, p 7.

55. Makover A, Moriwaki H, Ramakrishnan R, Blaner WS, Goodman DS. Plasma transthyretin (TTR): Tissue sites of degradation and turnover in the rat. Fed Proc 1987; 46:1188 (abstr.).

56. Saraiva MJM, Costa PP, Goodman DS. Studies on plasma transthyretin (prealbumin) in familial amyloidotic polyneuropathy. J Lab Clin Med 1983; 102:590–603.

57. Refetoff S, Dwulet FE, Benson MD. Reduced affinity for thyroxine in two of three structural thyroxine-binding prealbumin variants associated with familial amyloidotic polyneuropathy. J Clin Endocr Metab 1986; 63:1432–1437.

58. Benson MD, Dwulet FE. Prealbumin and retinol binding protein serum concentrations in the Indiana type hereditary amyloidosis. Arthritis Rheum 1983; 26:1493–1498.

59. Burton P, Iden D, Mitchel K, White A. Thymic hormone-like restoration by human prealbumin of azothioprine sensitivity of spleen cells from thymectomized mice. Proc Natl Acad Sci USA 1978; 75:823–827.

60. Dardenne M, Pleau J-M, Bach JF. Evidence of the presence in normal serum of a carrier of the serum thymic factor (FTS). Eur J Immunol 1980; 10:83–86.

61. Jacobsson B, Pettersson T, Sanstedt B, Carlstrom A. Prealbumin in the islets of Langerhans. IRCS Med Sci 1979; 7:509.

62. Kato M, Kato K, Blaner WS, Chertow BS, Goodman DS. Plasma and cellular retinoid-binding proteins and transthyretin (prealbumin) are all localized in the islets of Langerhans in the rat. Proc Natl Acad Sci USA 1985; 82:2488–2492.

63. Liddle CN, Reid WA, Kennedy JS, Miller ID, Horne CHW. Immunolocalization of prealbumin: Distribution in normal human tissue. J Pathol 1985; 146:107–113.

64. Miller ID, Reid WA, Liddle CN, Horne CHW. Immunolocalization of prealbumin as a marker for carcinoid tumors. J Pathol 1984; 143:199–204.

65. Herbert J, Schon EA, Vallejos H, Dwork AJ. Transthyretin is a marker for choroid plexus neoplasms. Neurology 1988; 38 (suppl 1): 213 (abstr.).

66. Soprano DR, Soprano KJ, Goodman DS. Retinol binding protein and transthyretin mRNA levels in visceral yolk sac and liver during fetal development in the rat. Proc Natl Acad Sci USA 1986; 83:7330–7334.

67. Murakami T, Yasuda Y, Mita S, et al. Prealbumin gene expression during mouse development studied by in situ hybridization. Cell Diff 1987; 22:1–10.

68. Blake CCF, Geisow JJ, Oatley SJ, Rerat B, Rerat C. Structure of prealbumin: Secondary, tertiary and quaternary interactions determined by Fourier refinement at 1.8 Å. J Mol Biol 1978; 121:339–356.

69. Branch WT, Robbins J, Edelhoch H. Thyroxine-binding prealbumin. Conformation in urea and guanidine. Arch Biochem Biophys 1972; 152:144–151.

70. Fex G, Hansson B. Subunit exchange between human and dog prealbumins. Scand J Clin Lab Invest 1980; 40:71–78.

71. Pettersson T, Carlstrom A, Jornvall H. Different types of microheterogeneity of thyroxine-binding prealbumin. Biochemistry 1987; 26:4572–4583.

72. Dickson PW, Howlett GJ, Schreiber G. Rat transthyretin (prealbumin): Molecular

cloning, nucleotide sequencing, and gene expression in liver and brain. J Biol Chem 1985; 260:8214–8219.

73. Mita SS, Maeda S, Shimada K, Araki S. Cloning and sequence analysis of cDNA for human prealbumin. Biochem Biophys Res Commun 1984; 124:558–564.

74. Soprano DR, Herbert J, Soprano KJ, Schon EA, Goodman DS. Demonstration of transthyretin mRNA in the brain and other extrahepatic tissues in the rat. J Biol Chem 1985; 200:11793–11798.

75. Sundelin J, Melhus H, Shonit D, et al. The primary structure of rabbit and rat prealbumin and a comparison with the tertiary structure of human prealbumin. J Biol Chem 1984; 60:6480–6487.

76. Wakasugi S, Maeda S, Shimada K, Nakashima H, Migita S. Structural comparison between mouse and human prealbumin. J Biochem 1985; 98:1707–1714.

77. Wallace MR, Naylor SL, Kluve-Beckerman B, et al. Localization of the human prealbumin gene to chromosome 18. Biochem Biophys Res Commun 1985; 129:753–758.

78. Wakasugi S, Maeda S, Shimada K. Structure and expression of the mouse prealbumin gene. J Biochem 1986; 100:49–58.

79. Costa RH, Lai E, Darnell JE. Transcriptional control of the mouse prealbumin (transthyretin) gene: Both promoter sequences and a distant enhancer are cell specific. Mol Cell Biol 1986; 6:4699–4708.

80. Tsuzuki J, Mita S, Araki S, Shimada K. Structure of the human prealbumin gene. J Biol Chem 1985; 260:12224–12227.

81. Sasaki H, Yoshioka N, Takagi Y, Sakaki Y. Structure of the chromosomal gene for human serum prealbumin. Gene 1985; 37:191–197.

82. Weisner B, Roethig H-J. The concentration of prealbumin in cerebrospinal fluid, indicator of CSF circulation disorders. Eur Neurol 1983; 22:96–105.

83. Weisner B, Kauerz U. The influence of the choroid plexus on the concentration of prealbumin in cerebrospinal fluid. J Neurol Sci 1983; 61:27–35.

84. Aleshire SL, Bradley CA, Richardson LD, Parl FF. Localization of human prealbumin in choroid plexus epithelium. J Histochem Cytochem 1983; 81:608–612.

85. Herbert J, Wilcox JN, Pham KC, et al. Transthyretin: A choroid plexus-specific transport protein in human brain. The 1986 S. Weir Mitchell Award. Neurology 1986; 36:900–911.

86. Kato M, Soprano DR, Makover A, Kato K, Herbert J, Goodman DS. Localization of immunoreactive transthyretin (prealbumin) and of transthyretin mRNA in fetal and adult rat brain. Differentiation 1986; 311:228–235.

87. Stauder AJ, Dickson PW, Aldred AR, Schreiber G, Mendelsohn FA, Hudson P. Synthesis of transthyretin (pre-albumin) mRNA in choroid plexus epithelial cells localized by in situ hybridization in rat brain. J Histochem Cytochem 1986; 34:949–952.

88. Dickson PW, Aldred AR, Marley PD, Guo-Fen T, Howlett GJ, Schreiber G. High prealbumin and transferrin mRNA levels in the choroid plexus of rat brain. Biochem Biophys Res Commun 1985; 127:890–895.

89. Dickson PW, Aldred AR, Marley PD, Bannister D, Schreiber G. Rat choroid plexus specializes in the synthesis and secretion of transthyretin (prealbumin). Regulation of transthyretin synthesis in choroid plexus is independent from that in liver. J Biol Chem 1986; 261:3475–3478.

90. Mita S, Maeda S, Shimada K, Araki S. Analyses of prealbumin mRNAs in individuals with familial amyloidotic polyneuropathy. J Biochem 1986; 100:1215–1222.

91. Kleine TO. Different distribution pattern of retinol-binding protein (RBP) and prealbumin in cerebrospinal fluid (CSF) and blood serum. Prot Biol Fluids 1984; 32:219–229.

92. Schreiber G, Dickson PW, Aldred AR, Marley PD, Memting JGT, Sawyer WH.

Transthyretin and transferrin synthesis in choroid plexus for transport of thyroxine and iron between body and brain. Fed Proc 1987; 46:686 (abstr.).

93. Costa PP, Figueira AS, Bravo FR, Guimaraes A. Prealbumin and amyloid in the Portuguese type of familial amyloidotic polyneuropathy. In Glenner GG, Costa PP, De Freitas F, eds. Amyloid and Amyloidosis. Amsterdam: Excerpta Medica, 1980, pp 147–152.

94. Hill NC, Goldstein NP, McKenzie BF, McGuckin WF, Svien HJ. Cerebrospinal fluid proteins, glycoproteins and lipoproteins in obstructive lesions of the central nervous system. Brain 1959; 82:581–593.

95. Martone RM, Herbert J, Dwork AJ, Schon EA. Transthyretin is synthesized in the mammalian eye. Biochem Biophys Res Commun 1988; 151:905–911.

96. Bridges CDB. Retinoids in photosensitive systems, chap 10. In Sporn MB, Roberts AB, Goodman DS, eds. The Retinoids, Vol 2. Orlando, FL: Academic Press, 1984, pp 126–176.

97. Shortle DJ, DiMaio D, Nathans D. Directed mutagenesis. Annu Rev Gen 1981; 15:265–294.

98. Palmiter RD, Brinster RL. Germline transformation of mice. Annu Rev Genet 1986; 20:201–254.

99. Sasaki H, Tone S, Nakazato M, et al. Generation of transgenic mice producing a human transthyretin variant: A possible mouse model for familial amyloidotic polyneuropathy. Biochem Biophys Res Commun 1986; 139:794–799.

100. Moses AC, Lawlor J, Haddow J, Jackson IMD. Familial euthyroid hyperthyroxinemia resulting from increased thyroxine binding to thyroxine-binding prealbumin. New Engl J Med 1982; 306:966–969.

101. Lalloz MRA, Byfiels PGH, Himsworth RL. A prealbumin variant with an increased affinity for T4 and reverse-T3. Clin Endocrinol 1984; 21:331–338.

102. Maury CRJ, Teppo A-M, Karinemi A-L, Koeppen AH. Amyloid fibril protein in familial amyloidosis with cranial neuropathy and corneal lattice dystrophy (FAB Type IV) is related to prealbumin. Am J Clin Path 1988; 89:359–364.

103. Sletten K, Westermark P, Natvig JB. Senile cardiac amyloid is related to prealbumin. Scand J Immunol 1980; 12:503–506.

23

DNA Markers in Dominant Neurogenetic Diseases

JAMES F. GUSELLA AND T. CONRAD GILLIAM

In the majority of neurogenetic disorders, no specific biochemical defect has yet been identified as the fundamental cause of the clinical symptoms. Consequently, there is no straightforward approach for employing recombinant DNA techniques to directly clone a gene encoding the unknown defective protein. Fortunately, DNA methods can also be used, however, to facilitate and improve indirect methods of identifying the defect by first determining its chromosomal location, and then using this information to isolate the disease gene itself.

Traditionally, genetic disorders have been mapped to specific chromosomal locations by family studies demonstrating linked inheritance of the disease and a polymorphic marker. Until recently, the useful markers were effectively limited to two dozen or so loci defined by expressed polymorphic differences in antigenicity or electrophoretic mobility of plasma and red cell proteins. Since the value of a genetic marker is the capacity it provides to distinguish the two homologues of the chromosome on which it resides, high quality genetic markers can now be generated by methods that detect variations in the primary sequence of genomic DNA. Typically, these DNA differences are most easily detected when they affect the size of restriction fragment length polymorphisms (RFLPs) (1).

DNA markers have already been used successfully to identify the chromosomal positions of several neurogenetic disorders, including Huntington disease (2) (HD) and familial Alzheimer disease (3) (FAD). Similar approaches are currently being pursued for many other diseases. Linkage analysis with RFLPs is an expensive and labor-intensive strategy for investigating any particular disorder, but, since the genome size is finite, the method is guaranteed to be successful if adequate family pedigrees are available. This review will examine the application of DNA markers to HD, FAD, and neurofibromatosis, emphasizing the differences in approach that have been used in each case.

HUNTINGTON DISEASE

Huntington disease (HD), described by George Huntington (4) in 1873, is a neurodegenerative disorder that displays autosomal dominant inheritance with high penetrance (5). HD is characterized by apparently normal cerebral development, followed by premature neuronal cell death, with the striatum most prominently affected. Clinical symptoms include progressive motor disorder, usually chorea; psychological manifestations that may include mood shifts, personality changes, impulsive behavior, or chronic depression; and intellectual deterioration. Symptoms typically appear in the fourth or fifth decade of life, after the gene may have already been transmitted to progeny. The degenerative process progresses for 15 to 20 years, the HD victim becoming totally disabled physically and unable to communicate. Finally, death ensues, often due to secondary complications such as heart disease or pneumonia. There is no known therapy to cure this devastating disease, or even to slow the inexorable course.

In 1980, we started to explore the strategy of genetic linkage analysis using DNA markers in human disease. We chose HD as a model to assess the feasibility of identifying the chromosomal location of the HD gene based solely on the genetics of the disorder and without any knowledge of the nature of the defective gene. Family studies with expressed markers had already been carried out in HD without success, but those early studies had eliminated some 15% of autosomal regions as possible locations of the HD gene. There was no clue where, in the remaining autosomal regions, the HD gene might be located and, at the time, there was no abundant supply of markers from any one chromosomal region; we therefore pursued the strategy of typing randomly chosen anonymous DNA markers in large HD pedigrees, regardless of the map position of these loci. Given that about 200 evenly spaced markers would span the entire genome, 0.5% of arbitrary markers tested would be expected to display less than 10% recombination with the disease gene. In 1983, we found that the probe G8, which defines an RFLP locus termed *D4S10* on chromosome 4, was tightly linked to HD (2). This early success was due in large part to the availability of an unusually large Venezuelan pedigree, with dozens of affected individuals, ideally structured for proving linkage.

Mapping of HD to chromosome 4 narrowed the search for the defect to 6.5% of the genome. More precise localization of the *D4S10* marker on this autosome has reduced the region that could contain the HD gene to less than 0.3% of the genome, or to less than 300 genes (Fig. 23-1). G8 was mapped to the terminal 4p16 band of the short arm of chromosome 4 (Fig. 23-1) when we found the probe sequence in only a single copy in patients with Wolf–Hirschhorn syndrome, a birth defect caused by heterozygous deletion of that region (6). *In situ* hybridization of the probe to metaphase chromosomes of patients with cytogenetic abnormalities of chromosome 4 suggested that *D4S10* is located in the approximate center of the band (7,8). However, *in situ* hybridization to normal chromosomes, using a non-isotopically labeled probe, indicated that *D4S10* may be located in the terminal 4p16.3 subband (9). We have now constructed a panel of somatic cell hybrid lines for rapidly mapping new DNA fragments to the region of G8 (10). Our data using anonymous DNA fragments chosen from a chromosome 4 library suggest that 4p16.3 localization is more likely to be correct, and that G8 is probably located in the top 1–2% of the chromosome.

The precise location of G8 yields only an approximate location for HD because the two are not perfectly linked. After we found that G8 was linked to HD in the Venezuelan pedigree and also in one much smaller American pedigree (2), we and

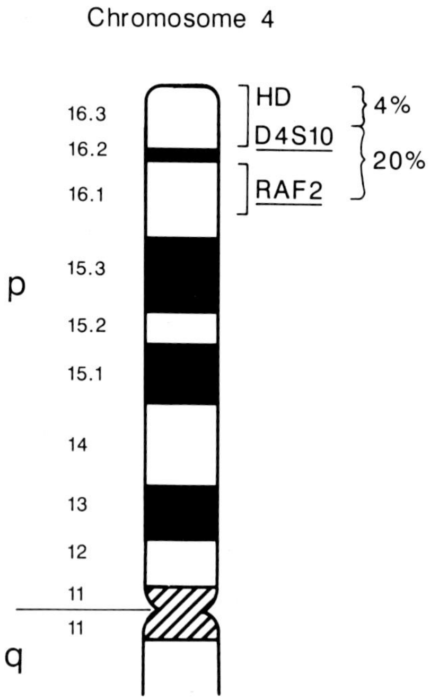

Fig. 23-1. Location of the Huntington disease gene. The HD gene is located in the terminal 4p16.3 subband of the short arm of chromosome 4, a region corresponding to perhaps 0.2% of the genome. The position of HD was defined by tracing the degree of coinheritance of the disease with the two mapped DNA markers, *D4S10* and *RAF2*, which show 4 and 20% recombination, respectively, with the disorder.

others typed many families from different parts of the world and different ethnic groups. The results indicate that approximately 4% recombination separates the disease gene from the marker locus (11). Furthermore, there was no evidence in over 50 families typed for nonallelic heterogeneity, indicating that the vast majority of HD cases are due to a defect mapping to the terminal region of chromosome 4.

When markers mapping below the region deleted in Wolf–Hirschhorn syndrome were typed in HD pedigrees, they were invariably farther away from HD than was *D4S10*. Three-point cross-analysis showed that the HD gene is located above *D4S10* on the chromosome, in the region of 4p16.3 closest to the telomere (12). This terminal location of the HD defect raises the formal possibility that the disease may be due to inheritance of an unbalanced translocation that is not detectable cytogenetically. In any event, the location of the HD defect has made it difficult to identify any marker flanking the disease gene with respect to G8. With a somatic cell hybrid mapping panel, however, an intensive effort is now underway to saturate this terminal region with cloned DNA segments.

Search for the HD gene itself has been narrowed tremendously by the identification of a linked marker. Numerous strategies are being used to identify and analyze DNA fragments in the region of the disease gene. RFLPs detected by cloned segments can be used to determine the genetic position of any candidate sequence relative to known recombination events between *D4S10* and HD. Similarly, a physical map of the terminus of chromosome 4 is being constructed using long range restriction mapping by pulsed field gel electrophoresis, and selective cloning strategies are being used to move rapidly along the chromosome. Ultimate cloning and characterization of the HD defect are now a matter of time.

While the search for the HD gene has progressed, it has also been possible to explore the nature of the defect by analyzing individuals who might be homozygous for the genetic abnormality. Using the *D4S10* marker to analyze 14 progeny of two affected parents in the Venezuela HD pedigree (14), we found that homozygosity for the HD gene is not a lethal event. In fact, HD homozygotes do not differ clinically in any way from HD heterozygotes. They do not show earlier onset and their symptoms are quantitatively and qualitatively similar to their heterozygous relatives. HD seems to be unusual among human genetic diseases in that the phenotype is completely dominant. A double dose of the disease allele does not increase the severity of manifestations and, similarly, the presence of the normal allele in HD heterozygotes does nothing to ameliorate the disease. These features are characteristic of known mutations in *Drosophila* that involve a gain rather than loss of function, so HD might involve increased or inappropriate expression of a normal protein, or production of an altered protein with a new functional property.

The HD gene will probably be isolated in the foreseeable future. In the meantime, linked DNA markers can be used for predictive testing of individuals with appropriate living relatives. The original discovery of linkage between *D4S10* and HD was achieved by finding two *Hin*dIII polymorphisms that were detected by G8. Subsequently, many additional RFLPs have been found with that probe and adjacent genomic sequences, making more than 95% of individuals heterozygous at the locus (15). Presymptomatic testing using the *D4S10* marker began on a trial basis in 1986, after enough data had been gathered to eliminate concern over the disastrous consequences that could result from premature predictions if there were significant nonallelic heterogeneity in HD. It is hoped that the information gained by pilot counseling programs will facilitate safe implementation of more widespread availability for the HD predictive test and will address concerns about possible detrimental effects of the test on "at risk" individuals, their friends, and their families. Ultimately, the trial programs will probably have an impact on delivery of similar predictive information in other untreatable disorders of late onset and for which presymptomatic tests are likely to be available.

NEUROFIBROMATOSIS

Neurofibromatosis is characterized by hyperplasia of cells of neural crest origin, particularly Schwann cells. Two distinct forms of neurofibromatosis have been recognized: Von Recklinghausen disease or peripheral neurofibromatosis (16) (NF1) and bilateral acoustic or central neurofibromatosis (17) (NF2). Both are autosomal dominant disorders that are ideally suited to investigation by linkage analysis. NF1 is one of the most common single gene disorders of the nervous system, affecting as many as 1 in 3000 births. The defect displays high penetrance, but remarkably variable phenotypic expression (18). Typical clinical features include neurofibromas (disordered growth of Schwann cells) and cafe-au-lait spots (dysregulated growth of melanocytes).

The defect in NF2 is highly penetrant, but much rarer than NF1, with a prevalence of perhaps 1 in 100,000. NF2 is also characterized by formation of benign Schwann cell tumors that infrequently become malignant sarcomas, a transformation attributed to secondary changes in individual cells. Some of the characteristics of NF2 overlap with those of NF1, but the two disorders seem to be genetically distinct. The hallmark of NF2 is the development of bilateral vestibular schwannomas, commonly called "acoustic neuromas." Similar tumors occur more frequently in a sporadic noninherited fashion, and less frequently in NF1, but are then almost invariably unilateral. The bilateral vestibular schwannomas of NF2 are slow-growing tumors that probably begin early in childhood, but symptoms do not appear until after puberty. Affected individuals are usually deaf by age 30. In addition to the vestibular schwannomas, cranial and spinal tumors are common and may be debilitating or even lethal.

Features other than inherited bilateral acoustic neuromas also distinguish NF2 and NF1:

1. Most important, each disease breeds true in a given pedigree.
2. A few cutaneous neurofibromas and cafe-au-lait spots may be seen in NF2, but these are much fewer than in NF1, where some patients have hundreds or thousands of skin lesions.
3. In NF1 more than 90% of cases have Lisch nodules (hamartomas) of the iris, but none are seen in NF2 (19).
4. The cafe-au-lait spots of NF1 have an abnormal melanosome, a "melanin macroglobule," that is not found in the cafe-au-lait spots of NF2 patients or normal individuals (20,21).
5. The vestibular schwannomas and peripheral tumors of NF2 contain high levels of glial growth factor that is present in much lower amounts in the tumors of NF1 (22).
6. Finally, NF1 is a relatively common disease with highly variable symptoms and an amazingly high mutation rate; about 50% of the cases seem to be new mutants. NF2 is much less prevalent and the phenotype is much less variable.

Faced with the prospect of searching for these defective loci throughout the genome by linkage analysis, some investigators have attempted to increase the chances for success by using as marker loci "candidate" genes that might play a primary role in these diseases. Tumor formation is involved in both conditions, so known oncogene loci and growth factor-related loci have been tested for linkage, particularly in NF1. The NF1 locus has recently been assigned to chromosome 17 by two independent studies, one linking the defect to a centromeric alpha satellite polymorphism (23), and the other demonstrating coinheritance with the gene encoding the nerve growth factor receptor (24). In the latter instance, although the nerve growth factor receptor was chosen as a candidate gene for the site of the mutation, measurable recombination between this marker and the NF1 locus indicates that the two are only neighbors in the subcentromeric region of chromosome 17.

The quest for the NF2 gene has taken a different route, based not on candidate genes, but on identifying a candidate region of the genome in which to concentrate linkage efforts. The relatively frequent occurrence of unilateral acoustic neuromas as noninherited sporadic tumors, compared to the familial formation of bilateral tumors in NF2, is reminiscent of the clinical presentation of retinoblastoma and Wilms tumor. Investigation of these tumors revealed chromosomal alterations in specific regions, 13q14 for retinoblastoma and 11p13 for Wilms tumor (25). The proposed mechanism of tumorigenesis in these cases could also be applied to acoustic neuromas (Fig. 23-2). The model invokes a single locus that normally suppresses formation of the tumor. For the occurrence of sporadic tumors, loss of both functional alleles at this locus must occur within a single cell and this requires coincidence of two rare events, hence the observation of a single unilateral tumor.

In familial cases, i.e., patients with NF2, one of the alleles at the locus is already defective when it is transmitted through the germline, accounting for the dominant pattern of inheritance. For tumor formation in NF2, only a single event need occur to destroy the remaining functional allele, leading to more frequent formation of tumors and the consequent observation of bilateral rather than unilateral acoustic neuromas. The functional allele might be lost in any number of ways, including point mutation, gene conversion, mitotic recombination, or deletion.

For retinoblastoma and Wilms tumor, the position of the locus implicated in tumor formation was first identified by cytogenetic observation of chromosome deletions. Since the Schwann cells of acoustic neuromas are difficult to culture, we used RFLP techniques to search directly for chromosomal aberrations in tumors. By comparing tumor DNA to DNA extracted from normal leukocytes of the patients with sporadic unilateral acoustic neuromas, we found specific loss of heterozygosity for DNA markers on chromosome 22 (26). Then, we found, as expected, a similar loss of heterozygosity for chromosome 22 markers in the acoustic neuromas of NF2 patients, suggesting that the primary defect of that disorder lies on chromosome 22. In support of this hypothesis, we identified loss of heterozygosity for chromosome 22 markers in three other tumor types from NF2 patients, indicating a role for this locus in many different cells of neural crest origin (27).

RFLP analysis has revealed that, in some cases, there is only a partial deletion of chromosome 22 material, narrowing the potential location of the defect to the region of the long arm below 22q11. The ultimate proof that the inherited defect in NF2 was at the same location as the chromosomal alteration in acoustic neuromas came with the demonstration of linkage between the disorder and two chromosome 22 DNA markers by Rouleau et al. (28).

Both NF1 and NF2 are serious diseases for which discovery of a DNA marker should have profound impact. Early diagnosis of NF2, in particular, is hampered by the lack of peripheral signs. A genetic marker would be invaluable for prenatal or presymptomatic diagnosis in both diseases. Presymptomatic diagnosis would be particularly important because there might be effective treatment. For example, in NF2, if surgery could be performed early, hearing and facial nerve functions might be saved. This is often not possible later in the disease.

It may be possible to identify progressively smaller deletions as increasing numbers of NF2 tumors are investigated. A similar approach to homing in on the NF1 gene could be used if the same mechanism acts on a locus on an as yet unidentified

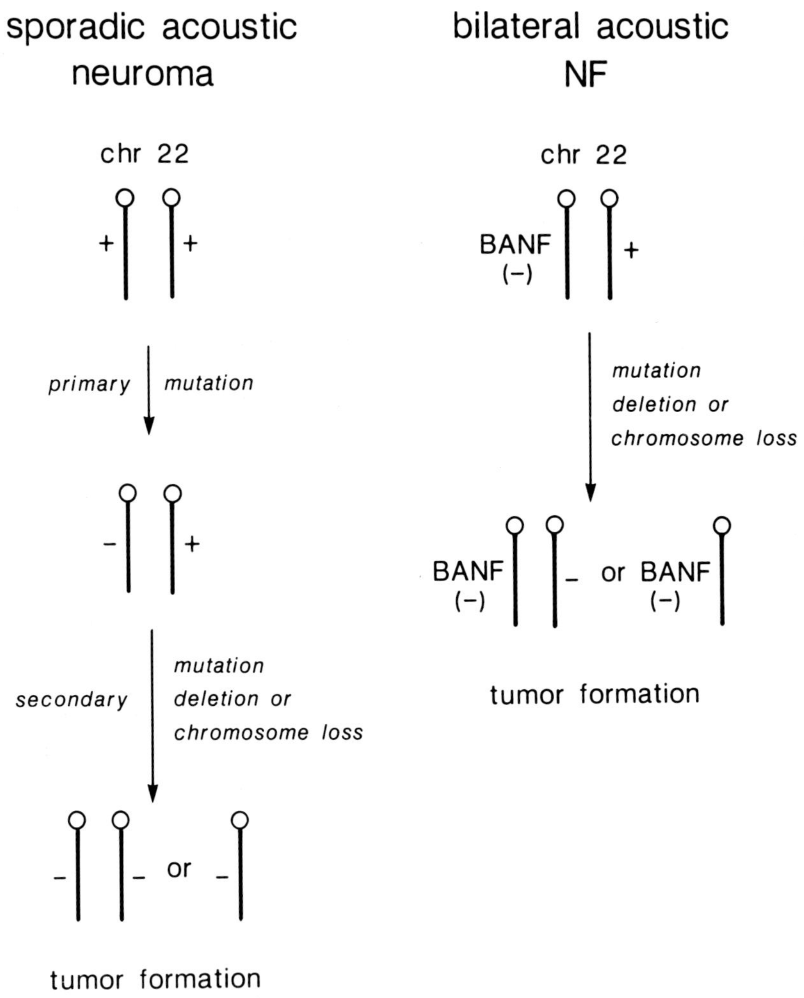

Fig. 23-2. Proposed model for tumorigenesis of acoustic neuroma. We postulate that acoustic neuromas are caused by the loss within a single cell of both copies of an as yet unidentified gene on chromosome 22 that normally suppresses tumor formation. In sporadic cases, both copies of the gene must be eliminated in a single cell by mutational or chromosomal mechanisms. In the inherited cases (NF2), one copy of the gene in question is already defective in the germline. Thus, elimination of only a single functional copy of the locus is required for tumor formation. This model predicts that the gene causing NF2 is on chromosome 22, and the mutation acts at the cellular level to cause a recessive loss of function. At the level of inheritance, however, the disorder displays dominant transmission since it is virtually certain that at least one cell in every NF2 gene carrier will suffer loss of the second "normal" allele, thereby resulting in tumor formation.

chromosome. To date, no strong evidence for loss of heterozygosity on chromosome 17 in NF1 tumors has been obtained, but the possibility that this phenomenon is masked by the presence of a mixed cell population in most tumors remains. Ultimately, whether by use of deletions or by moving along the chromosome from linked DNA markers, it should be possible to clone and compare both disease genes. In view of the similar effects, it is possible that the defective loci in NF2 and NF1 are related functionally or evolutionarily and that cloning of one defect might lead to direct isolation of the other.

FAMILIAL ALZHEIMER DISEASE (FAD)

Alzheimer disease (AD) is a degenerative neurological disorder of late onset, recognized by progressive impairment of memory and intellectual function (29). AD pursues a rapid course leading to profound mental and physical incapacity before death occurs 5 to 10 years after the onset of symptoms. There is no effective treatment for AD and the fundamental biochemical abnormality remains a mystery. It is thought that both environmental and genetic factors may play a role. However, in several large families, the disease seems to be caused by a genetic defect that is transmitted in an autosomal dominant fashion. The phenotype of the familial form of AD is indistinguishable from that of "sporadic" cases with respect to clinical and pathological characteristics, except for the evident pattern of inheritance and, typically, younger age at onset. Identification of the genetic defect causing FAD could provide a better understanding of both inherited and noninherited cases of AD.

The rapid course of FAD makes it difficult to identify large pedigrees with many living affected individuals. With the help of collaborators in several countries, we have investigated four FAD families suitable for genetic linkage analysis with DNA markers (3). We concentrated our initial choice of probes on loci for known genes that could be considered candidates for being the site of the primary defect because they are obviously related in some way to the pathogenic changes seen in AD brain. Although a number of markers were tested, none proved to be linked to the FAD defect. Furthermore, these early investigations emphasized how difficult it would be to find linkage with a randomly chosen DNA marker using the pedigrees available.

We concluded that the ideal method for approaching FAD might be to use the rapidly increasing capacity for multipoint analysis in the human genome. That technique, which resulted from the abundance of RFLP markers discovered in the last few years, involves simultaneous analysis of more than one marker for linkage to the disease locus, maximizing the informativeness of limited pedigree resources. The markers chosen, however, must have defined linkage relationships with respect to each other if multipoint analysis methods are to be useful.

A clue to the possible location of the FAD defect was provided by the observation of Alzheimer-like neuropathologic changes in older patients with Down syndrome (trisomy 21; DS) (30). We therefore attempted to generate RFLP markers for that autosome, using a chromosome-specific DNA library as a source of probes. We then constructed a linkage map of chromosome 21 by tracing the segregation of those markers in the Venezuela Reference Pedigree, a large section of the Venezuela HD pedigree (31). The reference pedigree comprises 17 large interrelated sibships that are ideally structured to determine the linkage relationships between individual DNA markers. With it, we constructed a linkage map that spanned the long arm of chromosome 21 with 12 DNA markers spread out from centromere to telomere.

With the linkage map in hand, we set out to type the DNA markers in FAD families and test for the presence of the FAD gene. Using multipoint analysis, we identified two markers showing linkage to the FAD defect in the upper portion of the long arm (3). We and others also found that the gene enclosing amyloid β protein, a main constituent of the amyloid plaques that are the hallmark of AD, is located on chromosome 21 (32–34) within the same region that seems to contain the FAD gene (Fig. 23-3). However, recent investigations using RFLPs for the amyloid β protein gene have shown significant recombination with FAD, demonstrating that the cause of this disorder is not an inherited defect in the amyloid β protein coding sequences (35,36).

Discovery of linkage markers for FAD opens the doors to isolating the defective gene based on chromosomal location, but it may take several years to find it by using currently known FAD pedigrees. In the meantime, the linked DNA markers could be useful for investigating many of the same issues, such as nonallelic heterogeneity, that are important in HD. The existence of a large proportion of apparently noninherited cases of AD also raises questions of whether and how often alleles at the FAD locus are involved in "sporadic" AD. Resolution of this issue might eventually lead to a DNA-based diagnostic test for FAD or AD, but there is

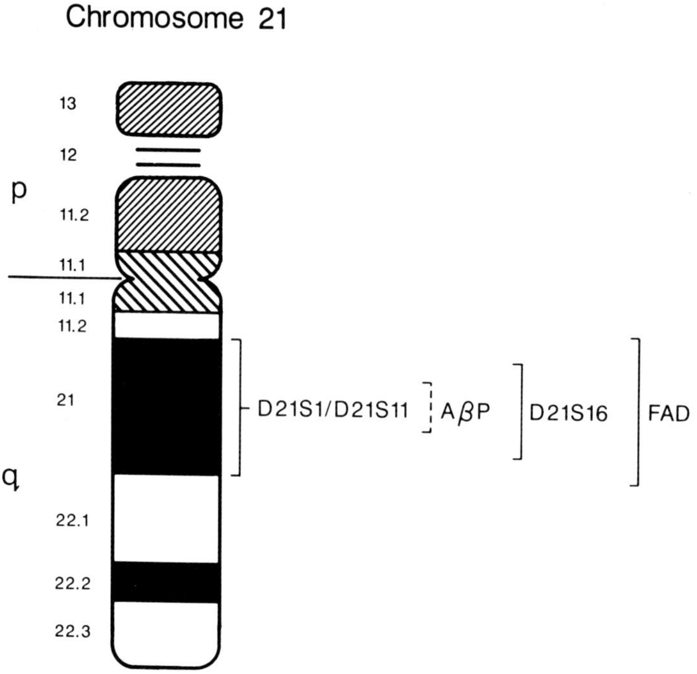

Fig. 23-3. Location of the familial Alzheimer disease gene. The gene causing FAD is genetically linked to *D21S1/D2S11*, a highly informative DNA locus in band q21 of chromosome 21. Also linked to the disease is *D21S16*, located approximately 8% recombination either above or below the other DNA marker. Similarly, the amyloid β protein gene (AβP) is located 4% recombination from *D21S1/D21S11*. Since FAD is linked to both DNA markers, the defect could be located anywhere within or near band q21, and need not necessarily coincide with the position of AβP in this region of approximately 10^6 base pairs.

no immediate prospect of applying the current markers to widespread predictive testing because there may be nonallelic heterogeneity, because of the uncertainty about the genetic nature of any individual AD case, and because of the imprecision in estimating the location of the gene defect relative to the markers.

SUMMARY

RFLP analysis has provided a powerful approach for investigating inherited disorders for which no defective protein has been identified. The chromosomal localization of inherited disease genes is now virtually guaranteed if adequate disease pedigrees can be studied. The strategy for approaching dominantly inherited single gene defects is straightforward, but can vary if some characteristic of the disease provides a potential shortcut to linkage by implicating a candidate gene or location. The use of multipoint analysis, with complete linkage maps that are being developed for each chromosome, will probably be the most efficient method to locate some defects, and may be essential for rare recessive diseases or multifactorial disorders in which several genes interact to produce the clinical phenotype. Once a defect has been mapped to a particular chromosomal region, the linked markers can be used for presymptomatic or prenatal diagnosis in appropriate families. The major impact of the approach, however, is that localization of the defect provides a way to clone and characterize the disease gene based only on chromosomal position, without any knowledge of the protein product of the gene. Imminent identification of the genes that cause Huntington disease, Alzheimer disease, and neurofibromatosis provides only three examples of an active and fruitful field of research: application of molecular genetic strategies to human neurological disease.

ACKNOWLEDGMENTS

The author's work discussed here is supported by NINCDS grants NS16367 (Huntington's Disease Center Without Walls), NS22031, NS20012, NS22224, NIA grant ADRC P50 AGO5134, and grants from the Hereditary Disease Foundation, the McKnight Foundation, the Massachusetts Chapter of the Huntington's Disease Foundation of America, and the Julieanne Dorn Fund for Neurological Research. JFG is a Searle Scholar of the Chicago Community Trust.

REFERENCES

1. Gusella JF. DNA polymorphism and human disease. Annu Rev Biochem 1986; 55:831–854.

2. Gusella JF, Wexler NS, Conneally PM, et al. A polymorphic DNA marker genetically linked to Huntington's disease. Nature (London) 1983; 306:234–238.

3. St. George-Hyslop PH, Tanzi RE, Polinsky RJ, et al. The genetic defect causing familial Alzheimer's disease map on chromosome 21. Science 1987; 235:885–890.

4. Huntington G. On chorea. Med Surg Rep 1872; 26:317–321.

5. Martin JB, Gusella JF. Huntington's disease: Pathogenesis and management. New Engl J Med 1986; 315:1267–1276.

6. Gusella JF, Tanzi RE, Bader PI, et al. Deletion of Huntington's disease linked G8 (D4S10) locus in Wolf-Hirschhorn syndrome. Nature (London) 1985; 318:75–78.

7. Magenis RE, Gusella JF, Weliky K, Olson S, Haight G, Toth-Fijel S, Sheehy R. Huntington disease-linked restriction fragment length polymorphism localized within band p16.1 of chromosome 4 by in situ hybridization. Am J Hum Genet 1986; 39:383–391.

8. Wang HS, Greenberg CR, Hewitt J, Kalousek D, Hayden MR. Subregional assignment of the linked marker G8 (D4S10) for Huntington's disease to chromosome 4p16.1–16.3. Am J Hum Genet 1986; 39:392–396.

9. Landegent JE, Jansen In de Wal N, Fisser-Groen YM, Bakker E, Van Der Ploeg M, Pearson PL. Fine mapping of the Huntington disease linked D4S10 locus by non-radioactive in situ hybridization. Hum Genet 1986; 73:354–357.

10. MacDonald ME, Anderson MA, Gilliam TC, et al. A somatic cell hybrid panel for localizing DNA segments near the Huntington's disease gene. Genomics 1987; 1:29–34.

11. Haines JL, Tanzi R, Wexler N, et al. Further evidence for the lack of heterogeneity of linkage of Huntington disease to D4S10. Cytogenet Cell Genet 1987; 46:625.

12. Gilliam TC, Tanzi RE, Haines JL, et al. Localization of the Huntington's disease gene to a small segment of chromosome 4 flanked by D4S10 and the telomere. Cell 1987; 50:565–571.

13. Gusella JF, Tanzi RE, Anderson MA, et al. DNA markers for nervous system disorders. Science 1984; 225:1320–1326.

14. Wexler NS, Young AB, Tanzi RE. Homozygotes for Huntington's disease. Nature (London) 1987; 326:194–197.

15. Gusella JF, Gilliam TC, Tanzi RE, et al. Molecular genetics of Huntington's disease. Cold Spring Harbor Symp Quant Biol 1986; 51:359–364.

16. Riccardi VM. Von Recklinghausen neurofibromatosis. New Engl J Med 1981; 305:1617–1627.

17. Martuza RL. Genetic neuroncology. Clin Neurosurg 1984; 31:417–440.

18. Seizinger BR, Tanzi RE, Gilliam RC, et al. Genetic linkage analysis of neurofibromatosis with DNA. Ann New York Acad Sci 1986; 486:304–311.

19. Martuza RL, Ojemann RG. Bilateral acoustic neuromas clinical aspects, pathogenesis treatment. Neurosurgery 1982; 10:1–12.

20. Martuza RL, Philippe I, Fitzpatrick TB, et al. Melanin macroblobules as a cellular marker of neurofibromatosis: A quantitative study. J Dermatol 1985; 85:347–350.

21. Kanter WR, Eldridge R, Fabricant J, et al. Central neurofibromatosis with bilateral acoustic neuroma: Genetic, clinical and biochemical distinctions from peripheral neurofibromatosis. Neurology 1980; 30:851–859.

22. Brockes JP, Breakefield X, Martuza RI. Glial growth factor like activity in Schwann cell tumor. Ann Neurol 1986; 20:317–322.

23. Barker D, Wright E, Nguyen, et al. Gene for von Recklinghausen neurofibromatosis is in the pericentromeric region of chromosome 17. Science 1987; 236:1101–1102.

24. Seizinger BR, Rouleau GA, Ozelius LJ, et al. Genetic linkage of von Recklinghausen neurofibromatosis to the nerve growth factor receptor gene. Cell 1987; 49:589–594.

25. Cavenee WK. The genetic basis of neoplasia: The retinoblastoma paradigm. Trends Genet 1986; 2:299–300.

26. Seizinger BR, Martuza RL, Gusella JF. Loss of genes on chromosome 22 in tumorigenesis of human acoustic neuroma. Nature (London) 1986; 322:644–647.

27. Seizinger BR, Rouleau G, Ozelius LJ, et al. Common pathogenetic mechanism for three tumor types in bilateral acoustic neurofibromatosis. Science 1987; 236:317–319.

28. Rouleau GA, Wertelecki W, Haines JL, et al. Genetic linkage of bilateral acoustic neurofibromatosis to a DNA marker on chromosome 22. Nature 1987; 329:246–248.

29. Katzman R. Alzheimer's disease. New Engl J Med 1986; 314:964.

30. Davies P. The genetics of Alzheimer's disease: A review and a discussion of the implications. Neurobiol Aging 1986; 7:459–466.

31. Tanzi RE, Haines JL, Watkins PC, et al. Genetic linkage map of human chromosome 21. Genomics 1988; in press.

32. Tanzi RE, Gusella JF, Watkins PC, et al. Amyloid B protein gene: cDNA, mRNA distribution, and genetic linkage near the Alzheimer locus. Science 1987; 235:880–884.

33. Goldgaber D, Lerman MI, McBride OW, et al. Characterization and chromosomal localization of a cDNA encoding brain amyloid of Alzheimer's disease. Science 1987; 235:877–880.

34. Kang J, Lemaire H-G, Unterbeck A, et al. The precursor of Alzheimer's disease amyloid A4 protein resembles a cell-surface receptor. Nature (London) 1987; 325:733–736.

35. Tanzi RE, St. George-Hyslop PH, Haines JL, et al. The genetic defect in familial Alzheimer disease is not tightly linked to the amyloid beta protein gene. Nature 1987; 329:156–157.

36. Van Broekhoven C, Genthe AM, Vandenberghe A, et al. Failure of familial Alzheimer's disease to segregate with the A4-amyloid gene in several European families. Nature 1987; 329:153–155.

24

Myotonic Muscular Dystrophy

ALLEN D. ROSES AND RICHARD J. BARTLETT

Myotonic dystrophy (DM) is the most common muscular dystrophy and is inherited as an autosomal dominant trait, with variable expressivity. Symptoms usually appear in adults but children may be affected (1,2). Clinical manifestations may be evident at birth, with severe limb hypotonia, difficulty sucking, and characteristic facial appearance—congenital DM; in contrast, there may be no symptoms throughout adult life. Heterozygotes in the same extended pedigree may demonstrate the entire range of variability.

Aspects of the clinical application of molecular genetic strategies to DM are similar to those of other late-onset diseases. However, some features are unique for DM. Many affected individuals already have children before they realize that they carry the gene. Some affected individuals seem to show little concern that they are gene carriers. The demand for genetic counseling by affected individuals is far less in DM families than the demand in other genetic diseases, such as Huntington disease, which has a roughly similar prevalence. Although not formally documented, we believe that secondary sex characteristics develop early in affected women, and if this is combined with lower than normal mental capacity it could account for youthful pregnancies. We have also found that large families may remain undiagnosed, because many individuals have only mild symptoms and others are misdiagnosed because of the wide variability of organ systems that may be symptomatic. These sociological factors may operate to perpetuate DM.

Using genetic linkage techniques and molecular genetic methods, the gene for DM has been mapped to the proximal long arm of chromosome 19 (3–5). The linkage grouping of DM, the Lutheran and Lewis blood groups, and the secretor loci, was one of the first linkage groups, described in 1954 (6). The first chromosomal localization for this linked group of loci came in 1980 when Whitehead et al. (3) linked the DNA polymorphisms for complement component 3 (C3) to DM and established that the locus for C3 is on chromosome 19. We have a strategy specifically directed to the definition of tightly linked probes on chromosome 19 that were close enough to use physical methods to identify the gene (7–9). Three areas of research activities have been involved: precise clinical diagnosis, genetic epidemiology, and recombinant DNA techniques.

We have examined and banked DNA (as lymphoblasts or fibroblasts) from more than 1500 family members of large DM pedigrees. Five families comprise more than 100 available individuals in at least three generations, three families contributed more than 200 tested members, and four generations have been banked (10). These large families, or smaller families with large and informative sibships, provided the bases for detecting linkage of restriction fragment length polymorphisms (RFLPs) that were developed from flow-sorted chromosome 19 genomic libraries (Fig. 24-1) (9).

Random clones were selected from libraries constructed from purified chromosome 19. Unique sequences were identified by screening the DNA inserts from selected clones with human repeat-sequence DNA. Unique sequences were then screened against a somatic cell hybrid panel that allows chromosome 19 probes to be localized to four different areas of the chromosome (Fig. 24-2). Previous data had established a close linkage between DM and the apolipoprotein C2 (apoC2) RFLPs (4,5). We used this information to choose unique sequences that showed the same pattern of hybridization on the hybrid screening panel as that of apoC2. This step eliminated the unique sequences from chromosomes other than chromosome 19 as well as those located outside of the area of 19cent-19q13.2. Thus, we identified unique sequences that were located in the "DM area" of chromosome 19; these probes could then be tested against a panel of DNA from DM heterozygotes that had been cut with several different restriction enzymes so that the chromosome 19 RFLPs would be useful in linkage studies (9,11,12). Sequences that detected polymorphisms on Southern blots (RFLPs) were then tested for linkage in the large DM families. The goal was to identify RFLPs with no crossovers.

As a technical point, to describe the magnitude of the screening project, 37 Southern blots of high quality were needed to represent over 400 family members in the screening families. Each RFLP that used a different endonuclease enzyme required a complete set of 37 blots containing the individual DNA samples cut with that enzyme. It was far easier to screen with multiple probes that detected RFLPs created with the same enzyme than it was to screen many probes that detected RFLPs created with different enzymes. The Southern blots could be stripped and reprobed several times; the same single set of blots created by one restriction enzyme could be screened rapidly with multiple RFLPs.

We initially found only one crossover between DM and apoC2 (5). Most of the known crossovers between DM and apoC2 had been reported from Prof. Harper's laboratory (4). Confirmation of the clinical diagnosis in his series is under way, but it is probable that the maximum recombination fraction is less than the reported value of 0.04. Our reported recombination fraction between DM and apoC2 is 0.01 (95% confidence limits, 0.00–0.02) with a lod score of 29.4 (5). Thus the probe for apoC2 is close to the gene for DM and is highly informative in families; it is useful for genetic counseling with better than 98% accuracy.

Another polymorphic probe was developed from our screening protocol and has demonstrated no crossovers to date. LDR152 (D19S19) is a 1.6-kbp fragment that was selected from a flow-sorted chromosome 19 library and detects polymorphisms with two enzymes. *Msp*I and *Pst*I (9,11,12). Neither RFLP is highly informative (allele frequency for each is approximately 0.15/0.85), but data from our large pedigrees and from Holland (Prof. H. H. Ropers) indicate a recombination frequency of 0.00 (95% confidence limits: 0.00–0.02), lod score of 21 (8,9). LDR152 (D19S19) has been used as a probe to screen genomic walking libraries; we have mapped 40

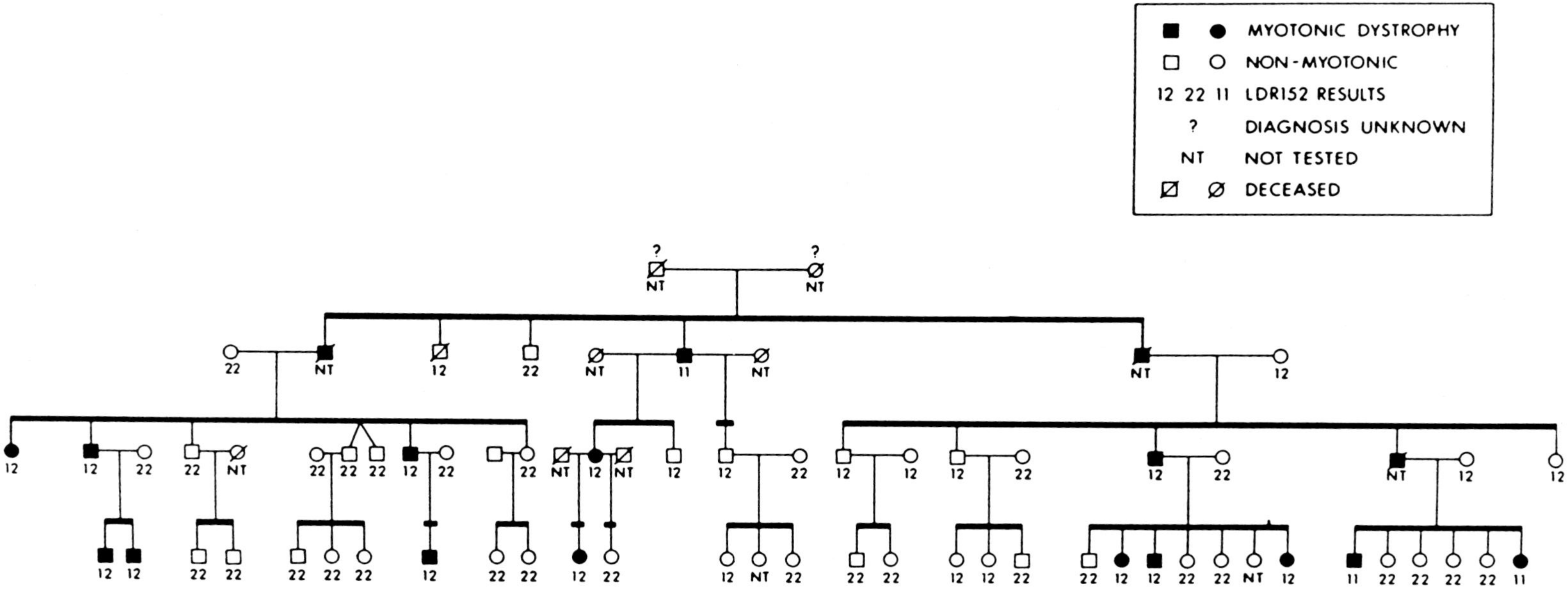

Fig. 24-1. Large DM pedigree segregating the D19S19 Msp Ipolymorphism. The larger MspI allele (1) is segregating with the DM allele in all affected individuals. Notice that this represents a four-generation pedigree with three generations of sampled families members, including fourteen members diagnosed with DM.

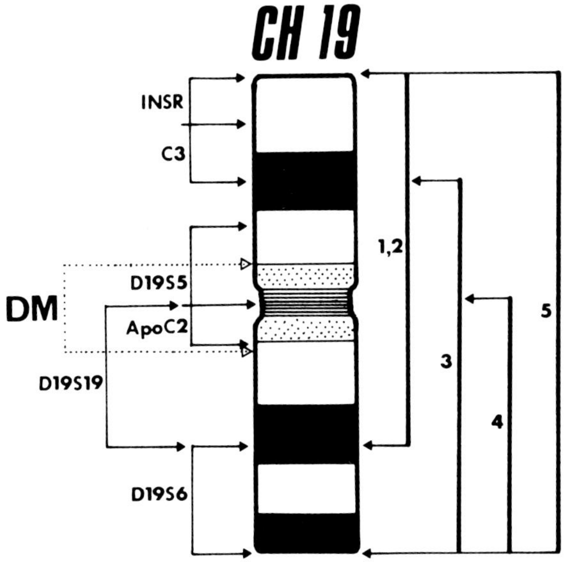

Fig. 24-2. Diagrammatic representation of a hybrid cell panel segregating CH19. Localization of translocation breakpoints are indicated on the right-hand side of the chromosome, while known or predicted localization of specific markers is as indicated on the left-hand side of the chromosome. CH19 can be segregated into at least four distinct portions using this panel.

kbp, about 20 kbp on either side of LDR152. Unique sequences from the walk are now being screened to identify additional RFLPs, particularly for those enzymes for which screening sets of Southern blots already exist: *Msp*I, *Pst*I, *Bgl*I, and *Sac*I from the screening of C3, apoC2, and LDR152.

Two other relatively new strategies are also being applied to define the DM gene: subtraction hybridization screening and the field inversion gel electrophoretic technique (FIGET) (13,14) as described by Smith and Cantor (Chapter 12). This is useful in mapping large DNA fragments and chromosome hopping (15).

Monaco and Kunkel and colleagues (Chapters 10 and 27) made use of a subtraction hybridization technique called PERT (phenol emulsion reassociation technique) (16) to compare genomic DNA from X-chromosomes from two sources: one was a normal boy and the other had a cytological deletion and several X-linked diseases that included Duchenne muscular dystrophy (DMD) and chronic granulomatous disease (17). They essentially matched all the fragments of the X-chromosome shared by the two starting materials and allowed the unmatched fragments from the normal X to define the fragments that were missing in the deleted X-chromosome from the patient. DNA/DNA screening led to identification of the gene for chronic granulomatous disease (18) and also for part of the large gene locus for DMD (19,20).

One specific aim in identifying and cloning the flanked segment of DNA that contains the DM gene is to be able to perform subtraction hybridization of that genomic segment against mRNA (cDNA) from tissues such as muscle, brain, liver, and testes, to determine which genes are expressed in that tissue and which should be tested as candidate genes. Thus, there is a rationale in delimiting the DM gene with flanking markers that may be found on a single FIGET fragment or two contiguous fragments. Linkage methods can theoretically define the 0.5–2 million bp within which the gene for DM must reside (Fig. 24-3). The theoretical possibilities must be brought down to manageable experiments. Testing each unique sequence

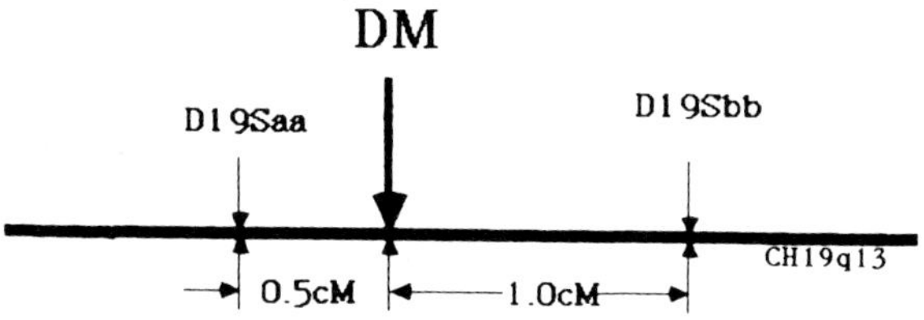

Fig. 24-3. Diagrammatic representation of mapping using flanking, tightly linked markers for DM. When these putative markers, D19Saa and D19Sbb, map as indicated, then the region within which the DM gene must lie is delimited by no more than 1.5% recombination. Assuming 10 6 pb per 1% recombination, the DM region would then be defined by these two markers as 1.5 × 10 6 bp.

A
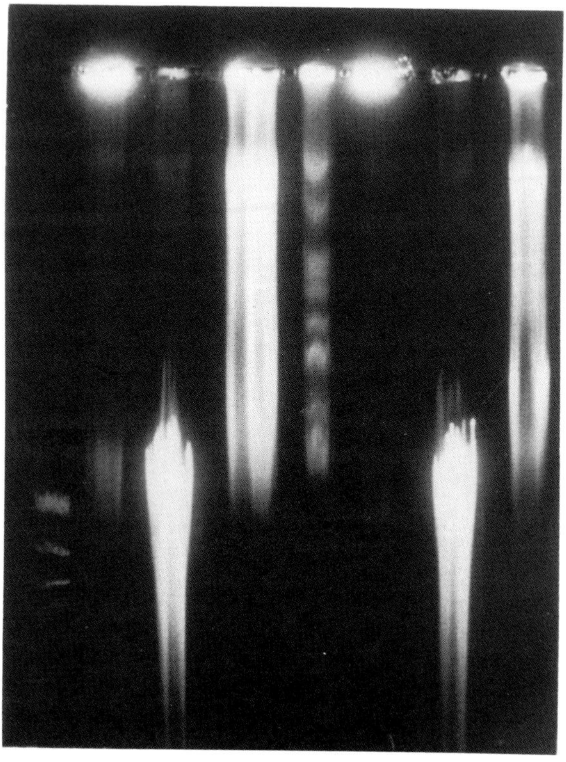

Fig. 24-4. Macro-restriction analaysis using the field inversion gel electrophoresis technique. Panel A shows the ethidium stained results of a FIGET electrophoresis which used six linear ramps of 6 hrs each with a switching ratio of 3:1 (forward:reverse) beginning with 3 sec, and ending with 90 sec in the forward direction. The *Hin*DIII lambda markers in lane 1 and yeast markers in lane 5 indicate that the resolution was from 20 to 1,500 kbp. The undigested human DNA samples (lanes 2 and 6) and *Spe*I and *Eag*I digested samples are in lanes 3 and 7, and 4 and 8, respectively. Panel B illustrates the results of hybridizing the D19S19 probe to this gel after blotting.

that is determined in a walk or a hop as an RFLP requires an extensive and lengthy input of time and resources; therefore our goal is to concentrate on the chromosome hopping techniques to define adjacent large segments of DNA that include apoC2 and LDR152, as well as additional fragments on both sides (15). Using RFLPs taken from the extremes of this FIGET map, we will confirm the linkage to DM and delineate the number of FIGET fragments of DNA that may contain the DM gene.

To pursue this strategy, chromosome 19 must be reproducibly cut into large fragments. Some restriction enzymes recognize rare sequences in genomic DNA and result in large DNA segments that can be separated by FIGET. For example, we made a Southern blot of a FIGET gel with DNA cut by several restriction enzymes (Fig. 24-4a). On this particular gel the larger fragments were produced by *Eag*I and the smaller by *Spe*I. In an autoradiograph of the same Southern blot probed with LDR152, a large 1100-kbp fragment was identified (Fig. 24-4b). By adjusting the field inversion ramping characteristics the band can be sharpened and

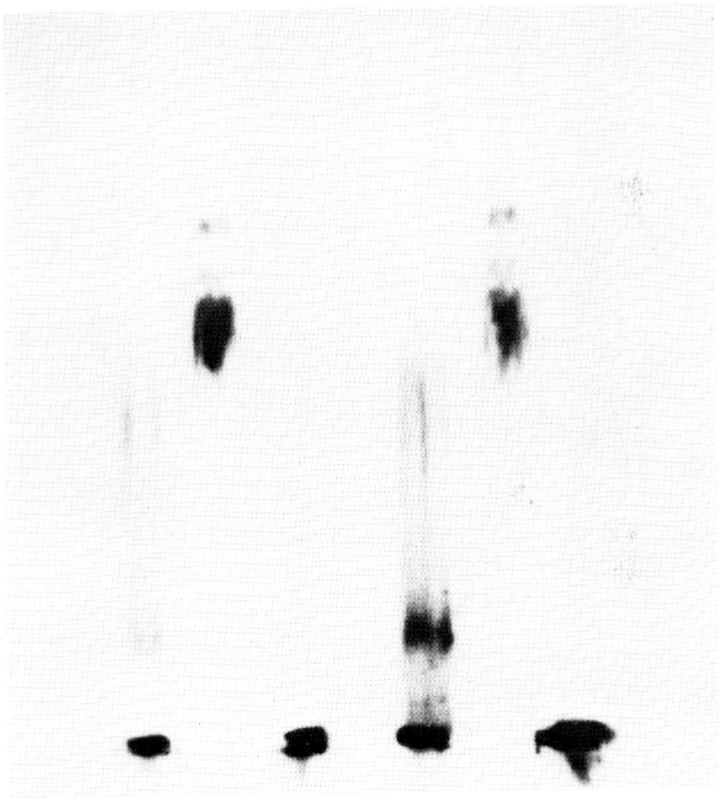

B

Fig. 24-4. ***(Continued)***

then isolated by cutting the gel. Small fragments can then be produced by incubating the fragment in a soft agar gel with a second restricton enzyme, then electroeluted and cloned. The terminal ends of the fragment will be the only ends cut with *Eag*I, and a selection vector that contains an *Eag*I cloning site can be used to select the terminal ends. Cloning the terminal ends of the large fragment will provide probes for screening a library to obtain the contiguous DNA that forms the terminal end of the adjacent large fragments (Fig. 24-5). Detecting both ApoC2 and LDR152 on one or two adjacent large fragments could identify the limits of the DNA that contain the DM gene. A flanking marker from the opposite end of the fragment could be sought directly from the mapped DNA rather than from our initial random screening strategy.

Screening the flanking RFLPs can be made rather straightforward by using DNA from family members who define known apoC2 crossovers (Fig. 24-6). We now have DNA from two individuals with confirmed recombination events, one from our laboratory and one from Prof. R. Williamson in England. The recombination events from Prof. P. Harper's laboratory are also being reexamined and sampled. Neither of the confirmed crossovers is informative for LDR152, so that three locus crosses cannot yet be analyzed. With more informative probes from the immediate vicinity of LDR152 we may be able to determine if we already have flanking markers.

This would be particularly exciting because preliminary FIGET analyses have suggested that both apoC2 and LDR152 may be localized on a single large fragment. It remains to be determined whether this fragment is a single fragment or one of two or more fragments that comigrate on the FIGET gels. If both probes are mapped to the same fragment, and if they can be shown to flank the DM gene by identifying more informative LDR152 probes, then the DM gene is contained within the delimited DNA fragment. We would then be in a position to clone the fragment and to employ subtraction hybridization techniques using muscle, liver brain, and thymus cDNA libraries that have already been constructed.

There are important experimental advantages to identifying a segment of DNA that contains the DM gene. Methods of subtraction hybridization are powerful tools for rapid screening of mRNA (cDNA) and DNA (18–20). Subtraction cloning of mRNA from muscle (or other tissues) and from chromosome 19 DNA can identify the genes coded in the limiting segment of DNA. These genes will be the candidates to be examined directly in DM tissues. DM may involve an intrinsic membrane protein (21–23) and candidate probes that have characteristics of channels or membrane-associated proteins may be rapidly identified.

Existing linkage and FIGET data suggest that the gene for DM may be of manageable size. This is in contrast to Duchenne dystrophy, in which the apparent size of the gene is too large to clone in available expression vectors. Once the DM gene is identified, currently available technology should be applicable to study the gene product. This may be achieved by expressing the normal and DM gene products to study physiological mechanisms in reconstituted membranes, if the gene turns out to be a channel protein or a component of an ATPase (23).

Treatment must also be considered in disease-oriented research. Since DM is inherited as an autosomal dominant trait, one of the gene copies is normal. As a late-onset disease somatic therapies might be directed at delaying expression of the

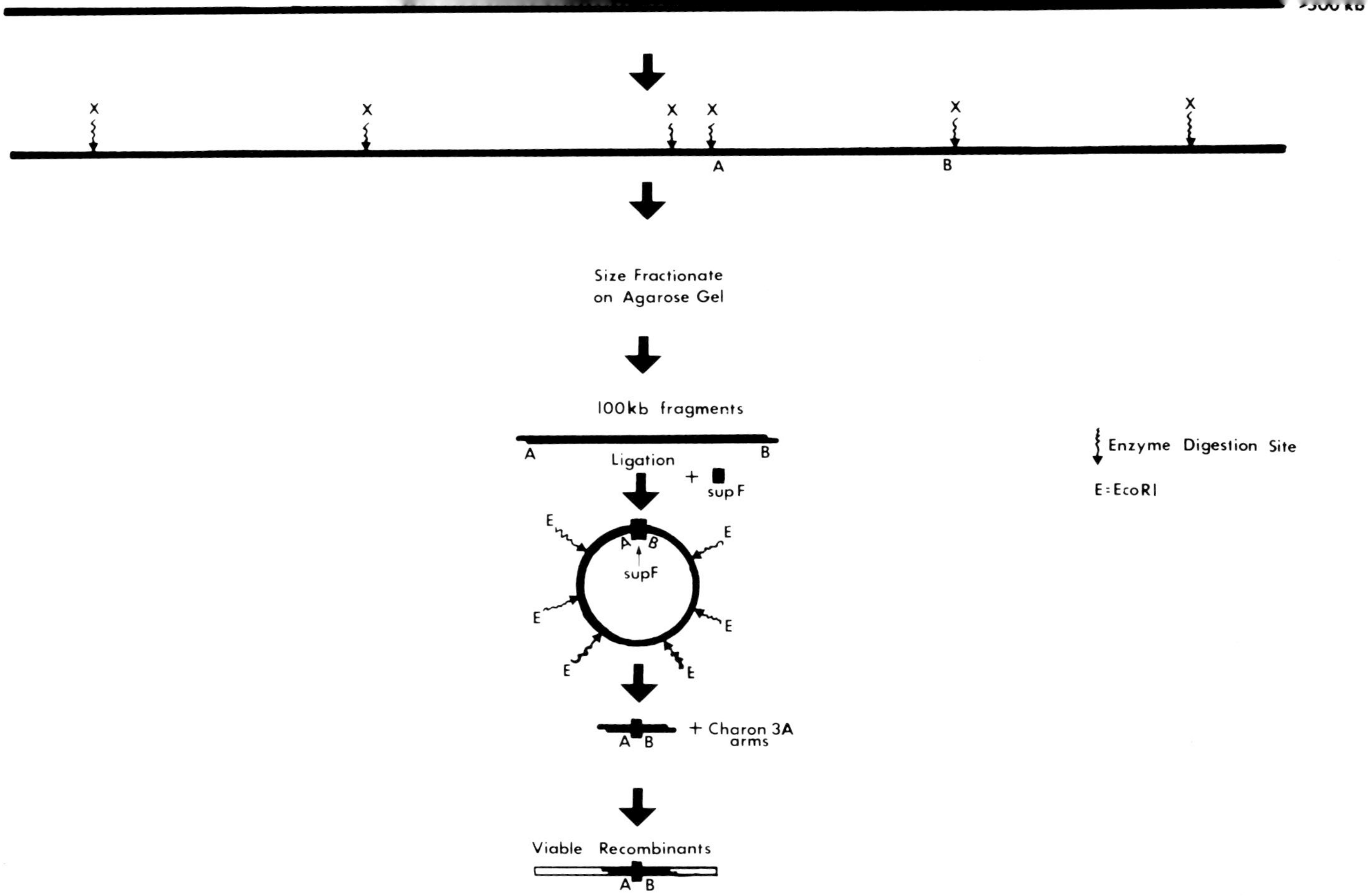

Fig. 24-5. Diagrammatic representation of the cloning of the end fragments of large restriction fragments from rare cutter digests. Restriction enzymes which contain CpG dinucleotides within their recognition sites will digest genomic DNA at a very low frequency partially because of the paucity of these sites within the genome. Therefore, using vectors which contain these sites, the termini from a particular fragment may be cloned to permit hopping from one end to the next, thus allowing movement along the chromosome in large increments defined by the length of the rare cutter fragments of interest.

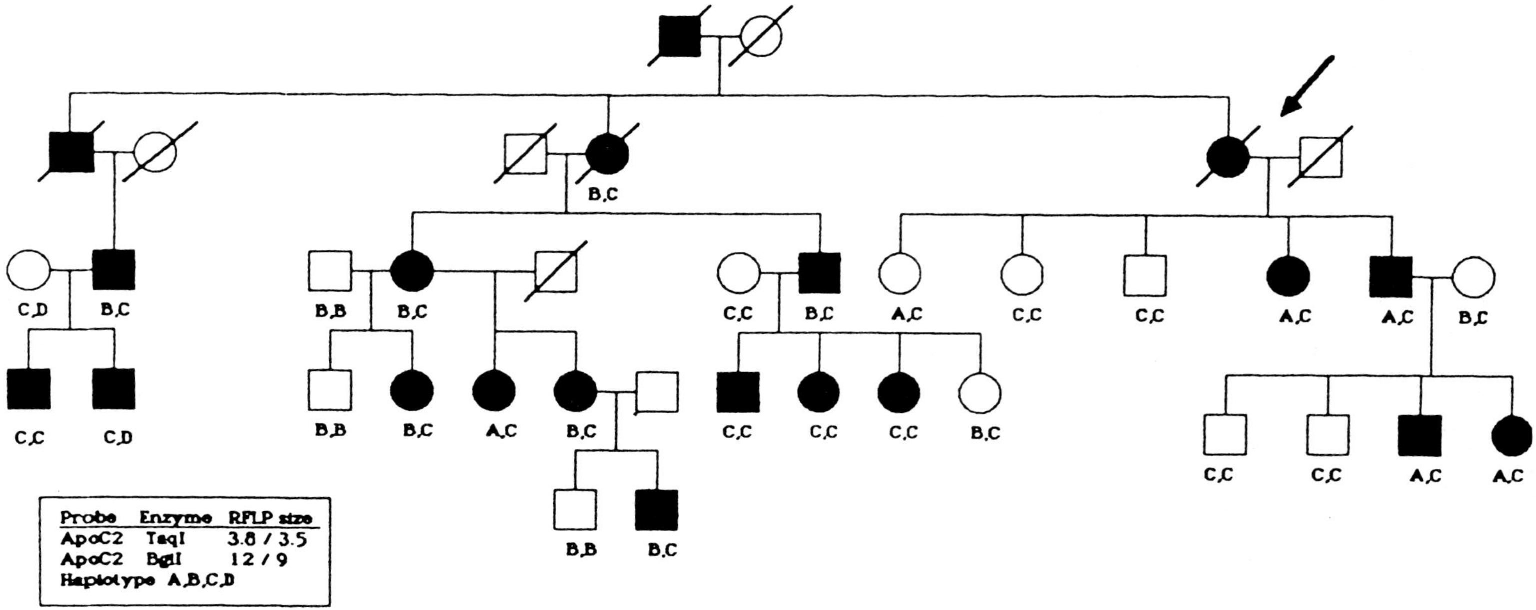

Fig. 24-6. A DM pedigree showing an apoC2 crossover event. On the left two branches of this pedigree the DM allele is segregating with the C haplotype. In contrast, the DM allele appears to travel with the A haplotype in the right branch. The arrow indicates the most likely candidate for the crossover, although the affected child with the A,C haplotype may be the original crossover.

disease. This possibility is also a hope for other diseases such as Huntington disease, inherited ataxias, and inherited neuropathies.

Chronic granulomatous disease was the first human genomic defect to be identified from the location of the gene rather than from a recognized protein mutation (18). Duchenne dystrophy exons have also been identified (20), and the DM gene has been identified (24). However, no other disease locus has yet been identified by a linkage approach without the benefit of a genomic deletion to focus chromosome mapping. Neurology has the reputation of dealing with a large number of nontreatable diseases. Somatic therapies based on control of gene expression may soon transform the practice of neurology. Molecular genetic probes now available have already redefined the standard of practice and will expand the responsibility of the diagnosing physician to other members of affected pedigrees. The MDA Clinic network is uniquely suited to disseminate new information to physicians as rapidly as possible, particularly in comparison with other neurogenetic diseases. This network has maturation pains because the information is coming fast and many practicing physicians have little background in the basic language of molecular genetics. Within a short time the MDA Clinic network should be able to test therapies based on rational design from the knowledge of the gene defects of inherited neuromuscular diseases.

ACKNOWLEDGMENTS

This work was supported by a grant from the Denver Fund for Health and Medical Research, Clinical Research Unit Grant RR-30 from the NIGMS, a clinical research grant from the Muscular Dystrophy Association, a grant from the Three Swallows Foundation, and NS-19999 from the NINCDS.

REFERENCES

1. Roses AD, Harper P, Bossen E. Myotonic muscular dystrophy (dystrophia myotonica, myotonia atrophy), chap. 13. In Vinken PJ, Bruyn CW, eds. Handbook of Clinical Neurology. New York: John Wiley, 1979, pp 485–532.

2. Harper PS. Myotonic Dystrophy. Philadelphia: Saunders, 1979.

3. Whitehead AS, Solomon E, Chambers S, Bodmer WF, Povey S, Fey G. Assignment of the structural gene for the third component of human complement to chromosome 19. Proc Natl Acad Sci USA 1982; 79:5021–5055.

4. Shaw DJ, Meredith AL, Sarfarazi M, et al. The apolioprotein CII gene: Subchromosomal localization and linkage to the myotonic dystrophy locus. Hum Genet 1985; 70:271.

5. Pericak-Vance MA, Yamaoka LH, Assinder RIF, et al. Tight linkage of apolioprotein C2 (ApoC2) to myotonic dystrophy (DM) on chromosome 19. Neurology 1986; 36:1418–1423.

6. Mohr J. A Study of Linkage in Man. Copenhagen: Munksgaard, 1954.

7. Yamaoka LH, Bartlett RJ, Ross DA, et al. Localization of cloned unique DNA to three different regions of chromosome 19: Screen for linkage probes for myotonic dystrophy. J. Neurogenet 1985; 2:403–412.

8. Roses AD, Pericak-Vance MA, Bartlett RJ, et al. Myotonic dystrophy—Update on progress to define the gene. Australia Pediat J, in press.

9. Bartlett RJ, Pericak-Vance MA, Yamaoka LH, et al. A new probe for the diagnosis of myotonic muscular dystrophy. Science 1987; 235:1648–1650.

10. Roses AD. Myotonic muscular dystrophy from clinical description to molecular genetics. Arch Intern Med 1985; 145(8):1487–1492.

11. Roses AD, Pericak-Vance MA, Ross DA, Yamaoka LH, Bartlett RJ. RFLPs at the D19S19 locus of human chromosome 19 linked to myotonic dystrophy (DM). Nucleic Acids Res RFLP Rep 1986; 14(13):5569.

12. Roses AD, Pericak-Vance MA, Yamaoka LH, et al. A new tightly linked DNA probe for myotonic dystrophy. (Letter) Neurology 1986; 36(8):1146.

13. Carle GF, Frank M, Olson MV. Electrophoretic separations of large DNA molecules by periodic inversion of the electric field. Science 1986; 232:65–67.

14. van Ommen GJB, et al. Restriction analysis of chromosomal DNA in a size range up to two million base pairs by pulsed field gradient electrophoresis. In Davies KE, ed. Human Genetic Diseases, A Practical Approach, Oxford: IRL Press, 1986, pp 113–133.

15. Collins FS, et al. Directional cloning of DNA fragments at a large distance from an initial probe: A circulation method. Proc Natl Acad Sci USA 1984; 81:6812–6816.

16. Kunkel LM, Monaco AP, Middlesworth W, Ochs HD, and Latt SA. Specific cloning of DNA fragments absent from the DNA of a male patient with an X chromosome deletion. Proc Natl Acad Sci USA 1985; 82:4778.

17. Francke U, Ochs HD, de Martinville B, et al. Minor Xp21 chromosome deletion in a male associated with expression of Duchenne muscular dystrophy, chronic granulomatous disease, retinitis pigmentosa, and McLeod syndrome. Am J Hum Genet 1985; 37:250.

18. Royer-Pokora B, Kunkel LM, Monaco AP, et al. Cloning the gene for an inherited human disorder—chronic granulomatous disease—on the basis of its chromosomal location. Nature (London) 1986; 332:32.

19. Monaco AP, Bertelson CJ, Middlesworth W, et al. Detection of deletions spanning the Duchenne muscular dystrophy locus using a tightly linked DNA segment. Nature (London) 1985; 316:842.

20. Monaco AP, et al. Isolation of candidate cDNAs for portions of the Duchenne muscular dystrophy gene. Nature (London) 1986; 323:646–650.

21. Wong P, Roses AD. Isolation of an abnormally phosphorylated erythrocyte membrane band 3 glycoprotein from patients with myotonic muscular dystrophy. J Membrane Biol 1979; 45:147–166.

22. Lipicky RJ. Studies in human myotonic dystrophy. In Rowland LP, ed. Pathogenesis of Human Muscular Dystrophies. Amsterdam: Excepta Medica, 1977, p 729.

23. Roses AD. The impact of molecular genetics on clinical neurology. Trends Neurosci 1986; 9(10):518–522.

24. Hoffman EP, Brown RH Jr, Kunkel LM. Dystrophin: The protein product of the duchenne muscular dystrophy locus. Cell 1987; 51:919–928.

25

Hereditary Motor and Sensory Neuropathies (Charcot-Marie-Tooth Disease)

PHILLIP F. CHANCE AND THOMAS D. BIRD

In 1886, Charcot and Marie (1) and Tooth (2) described families with inherited peripheral neuropathy. There had been many earlier reports of similar phenotypes, and many followed. Charcot-Marie-Tooth (CMT) disease is now recognized as a heterogeneous group of hereditary degenerative disorders of peripheral nerves, primarily affecting motor nerve functions. The clinical phenotype includes distal muscle weakness and atrophy, and diminished or absent tendon reflexes. Symptoms usually begin in the first or second decade of life; rare cases have been detected in infancy (3) and some patients note onset after age 20 years. Ninety-seven percent of affected individuals manifest either clinical symptoms or nerve conduction abnormalities by age 27 (4). There is usually slowly progressive loss of distal muscle mass and strength, but there is much variation in clinical manifestations. Some patients have severe distal muscle wasting, with marked hand and foot deformities, but the only clinical findings in other patients may be pes cavus or sluggish tendon reflexes.

Clinical variations in members of the same family, some severely affected and others obligate carriers with no clinical abnormality (5), suggest that variable expression is not due to genetic heterogeneity alone.

From the first description, CMT disease has been recognized as familial. Most families show autosomal dominant inheritance, but some are X-linked and rare families have an autosomal recessive mode. The prevalence of the several types of CMT as determined in Western Norway was $36/10^5$ for autosomal dominant, $3.6/10^5$ for X-linked, and $1.4/10^5$ for the autosomal recessive form (6). This genetic heterogeneity and numerous clinical variants have led to the complicated and confusing nosology of CMT (7,8). A popular classification is that of Dyck et al. (9), which designates this group of disorders as hereditary motor and sensory neuropathies (HMSN). Two major autosomal dominant types have been delineated, based on histological and electrophysiological data:

1. HMSN type I includes individuals with primarily demyelinating disease: very slow motor nerve conduction times and hypertrophied peripheral nerves.

2. HMSN type II includes individuals with axonal degeneration: mildly prolonged or normal motor nerve conduction times without peripheral nerve hypertrophy.

This categorization is helpful, but individuals with HMSN I or II are often indistinguishable by clinical examination or mode of inheritance, and electrophysiologic and histologic data are needed. Unfortunately, there are individuals with intermediate nerve conduction velocities or equivocal nerve histology and they are difficult to classify. Clearly, biologic markers of the CMT genes or tightly linked loci are needed to define the various HMSN subtypes and to find the gene product, still unknown.

GENETIC MAPPING OF HEREDITARY MOTOR AND SENSORY NEUROPATHY BY CLASSICAL METHODS

One of the first linkage observations regarding the CMT locus was the 1969 report of Swift and Horowitz (10) who described a multigeneration kindred in which CMT cosegregated with familial jaw cysts. Heimler et al. (11) restudied the same family in 1978 and found that the jaw cysts were due to the basal cell nevus syndrome, a disorder with loose linkage to the RH blood group locus (12), on chromosome 1. However, inspection of that pedigree reveals no instance of male-to-male transmission of the neuropathy, and there could have been the chance cosegregation of an X-linked neuropathy with an autosomal disorder, the basal cell nevus syndrome.

Bird et al. (13a,13b) used red blood cell markers and polymorphic serum markers to study two large HMSN kindreds. Autosomal dominant inheritance was confirmed by male-to-male transmission. Affected family members had very slow nerve conduction times and hypertrophy on sural nerve biopsy, compatible with HMSN I. Linkage to the Duffy locus on chromosome 1 was strongly suggested by a maximum lod score of 2.3 at $\theta = 0.1$. This suggested that the HMSN I and Duffy loci exhibited about 10% recombination with odds for linkage of about 200 to 1.

Additional families meeting the same clinical and electrophysiological criteria were studied by Guiloff et al. (14), with further evidence (lod = 0.72, $\theta = 0.10$) for linkage of HMSN I to the Duffy locus. The data of Dyck et al. (lod = 1.02, $\theta = 0.1$) (15) and Stebbins and Conneally (lod = 3.03, $\theta = 0.1$) (16) agreed with or supported the observed linkage between HMSN I and Duffy.

Genetic heterogeneity of HMSN I has been established by studying families who met clinical and electrophysiologic criteria of HMSN I but showed no linkage to Duffy. Bird et al. (17) first reported linkage data that excluded linkage between HMSN I and Duffy (lod = −1.675, $\theta = 0.10$). Dyck et al. (15) (lod = −6.74, $\theta = 0.10$), Rossi et al. (18) (lod = −3.81, $\theta = 0.12$), and Marazita et al. (19) (lod = − 0.55, $\theta = 0.1$) confirmed those observations. The discovery of families with HMSN I, which is not linked to Duffy, suggested that there are at least two or more clinically indistinguishable disorders within the category of HMSN I. Families with HMSN I not linked to Duffy are designated HMSN Ia; those linked to Duffy are designated HMSN Ib (15).

There are two possible explanations for the observation of families with a wide range of positive and negative lod scores between HMSN I and Duffy. First, there could be random segregation of the two genes, with no linkage. On the other hand, there might be genetic heterogeneity within the HMSN I phenotype, with one form linked to Duffy, and the other (or others) not linked. We have applied a statistical test for genetic heterogeneity (HOMOG) (20) to the linkage data from 12 HMSN I families (including 10 published pedigrees). The results strongly favor heterogeneity $\chi^2 = 5.601$, $p < 0.009$). Three families have a high probability of linkage to Duffy ($p > 0.93$), four families are highly unlikely to be linked to Duffy ($p < 0.26$), and five families show an indeterminant likelihood of linkage (see Table 25-1). Pedigrees with HMSN I linked to Duffy (HMSN Ib) may be less common than the unlinked variety (HMSN Ia).

There is no compelling published information on the chromosomal assignments of HMSN Ia or HMSN II. Spaans et al. (21) found an association of a hypertrophic neuropathy with slow nerve conduction velocities with myotonic dystrophy linked to the Lewis-secretor region of chromosome 19 in one family. Rabbiosi et al. (22) described a family in which HMSN II and palmoplantar keratoderma cosegregated. Unfortunately, the locus of palmoplantar keratoderma is unknown.

GENETIC MAPPING OF HEREDITARY MOTOR AND SENSORY NEUROPATHY BY RESTRICTION FRAGMENT LENGTH POLYMORPHISMS

Although linkage to Duffy places the HMSN Ib locus on chromosome 1, the regional assignment of the Duffy and HMSN Ib loci is broad (1p2 > q2) (23). The Duffy blood group locus is linked to the uncoiler locus (1qh), a cytologic marker, mapped to the region of 1q12 (24) (Fig. 25-1). Furthermore, combined linkage studies have suggested that Duffy is loosely linked to the blood coagulation protein antithrombin III (AT3) (25); that locus has been assigned to 1q23 > q25 by molecular hybridization (26–28).

These data suggest that the most plausible location of the Duffy locus is on the long arm of chromosome 1, between the centromere and the AT3 locus (29). There-

Table 25-1 Charcot-Marie-Tooth (HMSN I) Heterogeneity Analysis (HOMOG)

Author (family)	Reference	Posterior probability of linkage to Duffy
Stebbins and Conneally (Indiana)	16	0.9992
Bird (A)	13a,b	0.9869
Dyck (L)	15	0.9275
Guiloff (8,10)	14	0.8678
Bird (B)	13a,b	0.7982
Bird (Lo)	41	0.6698
Bird (J)	41	0.4811
UCLA (N005)	19	0.2562
Bird (D)	17	0.0249
Dyck (P)	15	0.0027
Rossi	18	0.0002

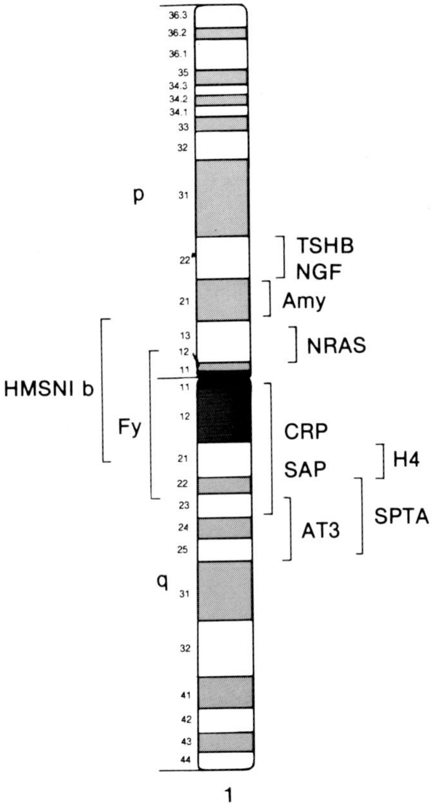

Fig. 25-1. Idiogram of human chromosome 1, depicting loci in the region of the hereditary motor and sensory neuropathy (HMSN Ib) and Duffy (Fy) loci. TSHB, β-subunit thyroid stimulating hormone; NGF, nerve growth factor, Amy, amylase; CRP, C-reactive protein; SAP, serum amyloid P; AT3, antithrombin III; H4, histone 4; SPTA, α-spectrin; NRAS, N-*ras* oncogene.

fore, the HMSN Ib locus should be located on the proximal long arm of chromosome 1 in the region of 1 cen > q2. However, there is also evidence that the Duffy locus may map to the proximal short arm. These studies include those with amylase (1p21) that demonstrated linkage with Duffy (lod = 4.18, θ = 0.04) (30). Also, Duffy may be loosely linked (lod = 2.18, θ = 0.15) to the nerve growth factor (NGF) locus (31), which maps to 1p22 (32).

The linkage relationship of AT3 and HMSN Ib has been examined by Chance et al. (33) who studied the same two families earlier reported by Bird et al. (13a,13b). Since the combined lod score was −2.337, θ = 0.05, close linkage between the AT3 and HMSN Ib loci was excluded. Additionally, lod scores were calculated using AT3 and Duffy as markers. At θ = 0.10 the lod score was −0.919 and at θ = 0.04 linkage of AT3 and Duffy was rejected by a lod score of −2.049. Additionally, Lebo et al. (34), in a study of the family reported by Stebbins and Conneally (16), found that the AT3 marker recombined 30% of the time with the HMSN Ib gene. These two studies do not support the hypothesis that the HMSN Ib and AT3 loci are closely linked (21), but rather suggest the HMSN Ib locus is more proximally located on the long arm of chromosome 1 closer to the centromere, or possibly even on the proximal short arm.

To identify loci tightly linked to CMT, it is necessary to use clones that are closer to the centromere and the proximal short arm of chromosome 1. NGF has been assigned to the region 1p22 (32). Examination of the two families of Bird et al. (13a) with the NGF clone detected recombination (lod = -3.54, $\theta = 0.1$), suggesting that the HMSN Ib locus is proximal to band p22 of chromosome 1 (35). Clearly, the linkage relationships of loci flanking the centromere of chromosome 1 (Fig. 25-1) merit further study to identify the ones that show tight linkage to the HMSN Ib locus.

Human chromosome 1 is thought to encompass a genetic distance of about 250 cM (36). The region 1p2 > 1q2 thought to contain the HMSN Ib locus is a genetic distance of approximately 80 cM (8×10^7 DNA base pairs). This is a large genetic distance, and it is unlikely that any of the 8–10 genes already assigned to this region will demonstrate tight linkage to the HMSN Ib locus. Therefore, additional random unique sequences, as derived from flow-sorted chromosome specific libraries (37), will be needed for linkage studies in HMSN Ib families for precise localization of the HMSN Ib gene.

The X-linked form of HMSN (38–40) has been mapped to the region Xq13 > Xq21 (41–43), as described in Chapter 30 by Fischbeck.

ACKNOWLEDGMENTS

This work was supported in part by a research grant from the Muscular Dystrophy Association and an NIH Clinical Investigator Development Award (NSO1144) to PFC, and NIH grants GMO7454 and GM15253, and Veterans Administration Medical Research Funds to TDB.

Michael Boehnke, Ph.D., provided valuable help with the statistical test of heterogeneity. The authors also thank John A. Phillips, III, M.D., for his advice and Anita Lewis for expert preparation of the manuscript.

REFERENCES

1. Charcot JM, Marie P. Sur une forme particuliere d'atrophie musculaire progressive souvent familiale debutant par les pieds et les jambes et atteignant plus tard les mains. Rev Med 1886; 6:97–138.

2. Tooth HH. The Peroneal Type of Progressive Muscular Atrophy. London: Lewis, 1886.

3. Vanasse M, Dubowitz V. Dominantly inherited peroneal muscular atrophy (hereditary motor and sensory neuropathy type I) in infancy and childhood. Muscle Nerve 1981; 4:26–30.

4. Bird TD, Kraft GH. Charcot-Marie-Tooth disease. Data for genetic counseling relating age to risk. Clin Genet 1978; 14:43–49.

5. Dyck PJ, Lambert EH, Mulder DW. Charcot-Marie-Tooth disease: Nerve conduction and clinical studies of a large kinship. Neurology 1963; 13:1–11.

6. Skre H. Genetic and clinical aspects of Charcot-Marie-Tooth disease. Clin Genet 1974; 6:98–118.

7. Dyck PJ, Lambert EH. Lower motor and primary sensory neuron diseases with peroneal muscular atrophy (I. Neurologic, genetic, and electrophysiologic findings in hereditary polyneuropathies). Arch Neurol 1968; 18:603–618.

8. Dyck PJ, Lambert EH. Lower motor and primary sensory neuron diseases with peroneal muscular atrophy (II. Neurologic, genetic, and electrophysiologic findings in various neuronal degenerations). Arch Neurol 1968; 18:619–625.

9. Dyck PJ, Thomas PK, Lambert EH, Bunge R. Peripheral Neuropathy. Philadelphia: W.B. Saunders, 1984, pp 1600–1655.

10. Swift MJ, Horowitz SL. Familial jaw cysts in Charcot-Marie-Tooth disease. J. Med Genet 1969; 6:193–195.

11. Heimler A, Freidman E, Rosenthal AD. Naevoid basal cell carcinoma syndrome and Charcot-Marie-Tooth disease. J Med Genet 1978; 15:288–291.

12. Anderson DE. Linkage analysis of the nevoid basal cell carcinoma syndrome. Ann Hum Genet 1968; 32:113–123.

13a. Bird TD, Ott J, Giblett E. Evidence for linkage of Charcot-Marie-Tooth neuropathy to the Duffy locus on chromosome 1. Am J Hum Genet 1980; 32:99A.

13b. Bird TD, Ott J, Giblett ER. Evidence for linkage of Charcot-Marie-Tooth neuropathy to the Duffy locus on chromosome 1. Hum Genet 1982; 34:388–394.

14. Guiloff RJ, Thomas PK, Contreras M, Armitage S, Schwarz G, Sedgwick EM. Linkage of autosomal dominant type I hereditary motor and sensory neuropathy to the Duffy locus on chromosome 1. J Neurol Neurosurg Psych 1982; 45:699–674.

15. Dyck PJ, Ott J, Moore SB, Swanson CJ, Lambert EH. Linkage evidence for genetic heterogeneity among kinships with hereditary motor and sensory neuropathy, type I. Mayo Clin Proc 1983; 58:430–435.

16. Stebbins NB, Conneally PM. Linkage of dominantly inherited Charcot-Marie-Tooth neuropathy to the Duffy locus in an Indiana family. Am J Hum Genet 1982; 34:195A.

17. Bird TD, Ott J, Giblett ER, Chance PF, Sumi SM, Kraft GH. Genetic linkage evidence for heterogeneity in Charcot-Marie-Tooth neuropathy (HMSN type I). Ann Neurol 1983; 14:679–684.

18. Rossi A, Paradiso C, Cioni R, Rizzuto N, Guazzi G. Charcot-Marie-Tooth disease: Study of a large kinship with an intermediate form. J Neurol 1985; 232:91–98.

19. Marazita ML, Cederbaum SD, Spence MA, et al. Possible heterogeneity in the linkage between autosomal dominant CMTI and Fy. Cytogenet Cell Genet 1985; 40:688–689.

20. Ott J. Analysis of Human Genetic Linkage. Baltimore: Johns Hopkins University Press, 1985, pp 200–203.

21. Spaans F, Jennekens FGI, Mirandolle JF, Bijlsma JB, de Gast GC. Myotonic dystrophy associated with hereditary motor and sensory neuropathy. Brain 1986; 109:1149–1168.

22. Rabbiosi G, Borroni G, Pinelli P, Cosi V. Palmoplantar keratoderma and Charcot-Marie-Tooth disease. Arch Dermatol 1980; 116:789–790.

23. Povey S, Morton NE, Sherman SL. Report of the committee on the genetic constitution of chromosome 1 and 2. Cytogenet Cell Genet 1985; 40:67–106.

24. Donahue RP, Bias WB, Renwick JH, McKusick VA. Probable assignment of the Duffy blood group locus to chromosome 1 in man. Proc Natl Acad Sci USA 1968; 61:949–955.

25. Winter JH, Bennett B, Watt JL, et al. Confirmation of linkage between antithrombin III and Duffy blood group and assignment of AT3 to 1q22 > q25. Ann Hum Genet 1982; 46:29–34.

26. Bock SC, Wion KL, Vehar GA, Lawn RM. Cloning and expression of the cDNA for human antithrombin III. Nucleic Acids Res 1982; 10:8113–8125.

27. Kao FT, Morse HG, Law ML, Lidsky A, Chandra T, Woo SLC. Genetic mapping of

the structural gene for antithrombin III to human chromosome 1. Hum Genet 1984; 64:34-36.

28. Bock SC, Harris JF, Balazs I, Trent JM. Assignment of the human antithrombin III structural gene to chromosome 1q23 > 25. Cytogenet Cell Genet 1985; 39:67–69.

29. Hamerton, JL, Povey S, Morton NE. Report of the committee on the genetic constitution of chromosome 1. Cytogenet Cell Genet 1984; 37:3–21.

30. Hill CJ, Rowe SI, Lovrien EW. Probable genetic linkage between human serum amylase (Amy 2) and Duffy blood group. Nature (London) 1972; 235:162–163.

31. Darby JK, Kidd JR, Pakstis AJ, et al. Linkage relationships of the gene for the beta subunit of nerve growth factor (NGFB) with other chromosome 1 loci. Cytogenet Cell Genet 1985; 39:158–160.

32. Francke U, Coussens L, Ullrich A. The human gene for the beta subunit of nerve growth factor is located on the proximal short arm of chromosome 1. Science 1983; 222:1248–1250.

33. Chance PF, Murray JC, Bird TD, Kochin RS. Genetic linkage relationships of Charcot-Marie-Tooth disease (HMSN Ib) to chromosome 1 markers. Neurology 1987; 37:325-329.

34. Lebo R, Anderson L, Lau C, Carver V, Conneally PM. Linkage analysis of Charcot-Marie-Tooth syndrome. Muscle Nerve 1986; 9:235.

35. Chance PF, Dracopoli N, Bird TD. Unpublished data.

36. Southern EM. Application of DNA analysis to mapping the human genome. Cytogenet Cell Genet 1982; 32:52–57.

37. Carrano AV, Gray JW, Langlois RJ, Burkhart-Schultz KJ, Van Dilla MA. Measurement and purification of human chromosomes by flow cytometry and sorting. Proc Natl Acad Sci USA 1979; 76:1382–1384.

38. Allan W. Relation of hereditary pattern to clinical severity as illustrated by peroneal atrophy. Arch Intern Med 1939; 63:1123–1131.

39. Erwin WG. A pedigree of sex-linked recessive peroneal atrophy. J Hered 1944; 35:24-26.

40. Fryns JP, Van Den Berghe H. Sex-linked recessive inheritance in Charcot-Marie-Tooth disease with partial clinical manifestations in female carriers. Hum Genet 1980; 55:413–415.

41. Gal A, Mucke J, Theile H, et al. X-Linked dominant Charcot-Marie-Tooth disease: Suggestion of linkage with a cloned DNA sequence from the proximal Xq. Hum Genet 1985; 70:38–42.

42. Fischbeck KH, ar-Rushdi N, Pericak-Vance M, et al. X-Linked neuropathy: Gene localization with DNA probes. Ann Neurol 1986; 20:527–532.

43. Bird TD. Unpublished data.

26

Hereditary Ataxias

ROGER N. ROSENBERG AND ABRAHAM GROSSMAN

The hereditary ataxias, also called the spinocerebellar degenerations, comprise a series of clinical manifestations that include ataxia and dysmetria, resulting from the predominant involvement of the cerebellum and its afferent and efferent pathways. These disorders are system degenerations; many of them are specific entities that are clearly inherited as autosomal dominant or autosomal recessive and X-linked traits. Although the clinical manifestations and neuropathologic findings of cerebellar disease dominate the spinocerebellar degenerations, there may also be characteristic changes in the basal ganglia, optic atrophy, retinitis pigmentosa, or peripheral neuropathy (1). There are many gradations from pure cerebellar manifestations to mixed cerebellar and brainstem disorders, cerebellar and basal ganglia syndromes, and spinal syndromes or peripheral neuropathy. The clinical picture may be consistent in one family, but sometimes there is a characteristic syndrome in most family members and an entirely different disorder in one or several members.

The typical clinical picture and the age at onset of symptoms and signs are used to classify the inherited spinocerebellar diseases. Major categories of disease are included in this designation, but the divisions are arbitrary and there are gradations between the several entities. The important and commonly inherited spinocerebellar degenerations include (1) Friedreich syndrome, the spinal form of spinocerebellar degeneration, (2) Roussy–Levy syndrome, (3) Refsum syndrome, (4) Bassen–Kornzweig syndrome, (5) olivopontocerebellar degeneration, (6) Joseph disease, (7) dyssynergia cerebellaris myoclonica, (8) ataxia telangiectasia, (9) Marinesco–Sjögren syndrome, (10) hereditary spastic paraplegia, and (11) Charcot-Marie-Tooth disease.

These entities, as classified by Greenfield (2), can be grouped into predominantly spinal forms, spinocerebellar forms, and pure cerebellar forms. The olivopontocerebellar degenerations (OPCD) were subclassified by Konigsmark and Weiner (3) into at least five subgroups with both autosomal dominant and autosomal recessive forms of inheritance. The many minor variants of OPCD described, for example, by Brown in 1892 (4), Marie in 1893 (5), Dejerine and Thomas in 1900 (6), Holmes in 1907 (7), and Schut in 1950 (8), as listed by Konigsmark and Weiner (3), could be examples of phenotypic variability due to the combination of a single gene mutation that is transmitted as an autosomal recessive trait and another single gene mutation that is transmitted as an autosomal dominant trait in which many other host genes modify expression and penetrance of the mutant gene.

Insights into the molecular causes of these diseases are beginning to be described in some of the spinocerebellar diseases, including the Freidreich syndrome, Refsum disease, and Bassen–Kornzweig syndrome. In the other spinocerebellar degenerations, although the disorders have been well described both clinically and pathologically, the specific biochemical disorder remains elusive. The spinocerebellar system seems to be highly vulnerable to different molecular abnormalities, as evidenced by different molecular defects in the Bassen–Kornzweig and Friedreich disorders, yet the system can respond only in a limited manner and without a great deal of pathologic variation, as evidenced by the similarity of neuropathologic findings in these disorders. Common neuropathologic features, from the peripheral nerve through the spinal cord and up to the cerebellum and cerebellar connections, are seen in the broad spectrum of spinocerebellar degenerations. These generalized and nonspecific changes lead to specific clinical syndromes that are caused by different molecular lesions inherited in a characteristic autosomal recessive or autosomal dominant manner.

SPINOCEREBELLAR ATAXIAS

A collection of spinocerebellar degenerations occurs in childhood sporadically, in siblings or clearly transmitted as a genetic autosomal recessive or dominant disorder. These disorders include several specific entities that share common clinical features and pathologic changes. The syndrome includes several recognized inborn errors of metabolism, including several disorders of lipids (phytanic acid storage disease, a-β-lipoproteinemia, moderate β-galactosidase deficiencies, and juvenile arylsulfatase deficiencies); diseases of oxidative metabolism [deficiencies of the pyruvate dehydrogenase complex, defect of mitochondrial malic enzyme, neuromuscular disorders with "ragged red" fibers, and abnormalities of cytochrome *b* or of nicotinamide adenine dinucleotide (NADH) (oxidation), aminoacidurias (intermittent maple syrup urine disease, γ-glutamyl-cysteinyl transferase deficiencies, and Hartnup disease)]; and partial deficiency of hypoxanthine guanine phosphoribosyltransferase (HGPRT) deficiency. Clinical expression of these inborn errors includes disordered cerebellar and corticospinal functions that are progressive and symmetrical.

In addition to those already mentioned inborn errors producing spinocerebellar degeneration, deficiencies of enzymes of the pyruvate dehydrogenase complex that catalyze the conversion of pyruvate to acetyl-CoA and carbon dioxide have also been identified in patients showing both ataxia and peripheral neuropathy. Pyruvate oxidation defects have also been described in patients with ataxia and peripheral neuropathy caused by nongenetic, acquired conditions such as thiamine deficiency, alkylmercury poisoning, and elemental mercury poisoning.

The activities of several enzymes of oxidative metabolism in cell-free extracts of fibroblasts from four patients with Friedreich ataxia and from a fifth patient with a variant of this syndrome were assayed by Blass and co-workers (9,10) and were reduced in activity.

In several patients Stumpf et al. (11) found that mitochondrial malic enzyme of fibroblast cultures was reduced by at least 50%, but other patients had normal levels in fibroblast cultures (12–14). Patients with a demonstrated defect in oxidative metabolism might be referred to as having Friedreich disease. Those with the same phenotype but no oxidative defect could be referred to by the nonspecific designation, Friedreich syndrome.

HEREDITARY ATAXIA WITH MUSCULAR ATROPHY (ROUSSY–LEVY SYNDROME)

Roussy–Levy syndrome, originally described in 1926 (15), is an intermediate form between Friedreich syndrome and Charcot-Marie-Tooth syndrome, with a combination of minor cerebellar signs and peripheral neuropathies, especially in a peroneal distribution. Harding and Thomas (16) include it as a form of both hereditary motor and sensory neuropathy type 1 (slow nerve conduction velocities) and type 2 (normal conduction velocities). In reports of Friedreich syndrome in large families, the presence of the Roussy–Levy variant is seen in some family members (17).

OLIVOPONTOCEREBELLAR DEGENERATIONS

The olivopontocerebellar syndromes are disorders characterized by impaired cerebellar function and impairment or reduction in neurons in the inferior olivary nuclei of the medulla, in the basis pontis, in the cerebellar cortex, and in the deep cerebellar nuclei (18).

Progressive ataxia, dysarthria, dysmetria, dysadiodochokinesia, nystagmus, and loss of fast saccadic eye movements are combined with spasticity, optic atrophy, distal sensory loss, and late intellectual dysfunction as the essential clinical features (3,19). There may be modifications in different families, with either autosomal recessive or dominant inheritance (3). The variants described by Holmes, Sanger-Brown, and Marie, among others, are clinically similar, separated by arbitrary and minor differences in clinical features. It can be argued that the olivopontocerebellar degenerations, autosomal dominant or autosomal recessive, comprise only a few unique diseases, with penetrance and expression of the primary gene mutation altered clinically by modifying genes.

In the second or third decade of life, there is ataxia of limbs and trunk, with dysmetria and dysarthria. Upper motor neuron signs and spasticity follow. Nystagmus, optic nerve atrophy, and loss of fast saccadic eye movements may occur. Muscle atrophy and fasciculations may be seen in facial muscles, muscles of mastication, and tongue. Sensory loss may be seen in distal limbs. Intellectual deterioration and dementia may occur late in the course. Ophthalmoplegia, extrapryamidal signs, and optic atrophy with visual loss may be encountered.

The syndrome of dyssynergia cerebellaris myoclonica of Ramsay–Hunt (20) is another rare variant beginning in childhood with myoclonus, seizures, and progressive ataxia inherited in an autosomal dominant manner (21).

The development early in life of progressive symmetrical impairment of cerebellar functions, followed by progressive and symmetrical spasticity, is characteristic of the olivopontocerebellar degenerations. Abnormalities of eye movement, intellectual impairment, and muscle atrophy with sensory distal loss may complete the clinical picture. Cerebellar atrophy, pontine atrophy with minor cerebral atrophy, and large lateral ventricles may be seen on brain CT. There may be slowing of motor nerve conduction velocities with signs of denervation on electromyography. CSF protein and cell counts are normal. In the Schut–Swier kindred (22,23), there was evidence of linkage between the gene for ataxia and the human lymphocyte antigen (HLA) complex on chromosome 6.

Plaitakis et al. (24,25) and Duvoisin et al. (26) found a 50% reduction in glutamate dehydrogenase (GDH) activity in nonneural tissues from patients with recessively inherited disorders or dominantly inherited syndromes with incomplete penetrance that could result in toxic levels of glutamate in the cerebellum, possibly causing excitotoxic degeneration of cerebellar neurons. Dominantly inherited patients have been reported as having low GDH activity in leukocyte homogenates (27), but it is not clear if this enzyme abnormality is a primary or secondary change. We studied GDH activities in brain homogenates from patients with dominantly inherited olivopontocerebellar degeneration, Joseph disease, and control subjects and found no significant difference in the three groups. It is our view that GDH is not involved in the pathogenesis of disease in these dominantly inherited syndromes (28).

JOSEPH DISEASE

In 1972, two papers described dominantly inherited neurologic disease in two Azorean-Portuguese families residing in Massachusetts. The first report (29) referred to "Machado disease" to acknowledge the proband family in which there was a progressive cerebellar disease with distal limb atrophy and sensory loss beginning in the fifth decade. A sural nerve biopsy from one patient documented segmental demyelination.

The second report (30) described "nigro-spinal-dentatal degeneration with nuclear ophthalmoplegia" in the Thomas family, presenting in the third to fifth decades with gait ataxia, external ophthalmoplegia, facial–lingual fasciculations, and elements of spasticity or rigidity. The neuropathologic features included neuronal loss in the substantia nigra, dentate nuclei, and Clark's column in the spinal cord. Demyelination was seen in the posterior columns, lateral corticospinal, and spinothalamic tracts of the spinal cord, most evident in the thoracic portion.

In 1976, Rosenberg et al. (31) described a progressive motor system disease beginning in the second and third decades in another Portuguese family, the Joseph family of California, consisting of 329 persons in nine generations at that time. The clinical picture differed from that of the two previous reports because affected persons in the Joseph family had progressive spasticity, lurching unsteadiness of gait due to spasticity (without cerebellar signs), spastic dysarthria, loss of fast saccadic eye movements, ophthalmoparesis for upward gaze, and facial–lingual fasciculations, with prominent dystonia of head, face, extremities, and trunk in some individuals. The neuropathology consisted of neuronal loss in the striatum, substantia nigra, basis pontis, dentate nucleus, cerebellar cortex, and anterior horns of the spinal cord. CSF homovanillic acid levels were reduced, correlating with neuronal loss in the zona compacta of the substantia nigra. The Joseph family disorder differed from the Massachusetts cases because (1) there was striatonigral degeneration, (2) dystonia was severe in some patients, and (3) cerebellar signs were not evident in any family member examined in 1975. The disease was also different from olivopontocerebellar degeneration because the medullary inferior olives were unaffected. It was unlike Huntington disease (HD) because intelligence remained normal in all Joseph patients, and the cerebral cortex was histologically normal.

Epidemiologic studies were necessary to establish the spectrum of clinical and neuropathologic findings. The logical place to proceed was in the Azores Islands, particularly on the island of Flores, from which both the Joseph and Thomas families originated. Other suggestions of variable clinical and neuropathologic features were made as personal communications to Dr. Rosenberg from Dr. Coutinho and Dr. Andrade of the Neurology Service, Hospital Santo Antonio, Porto, Portugal. They traveled to Flores in December 1976, where they found patients with prominent spasticity and extrapyramidal signs and some with cerebellar disorders or peripheral neuropathy. Coutinho and Andrade (32) emphasized types I, II, and III as variants of the autosomal dominant disorder in affected individuals on Flores. Type I corresponded to the typical Joseph phenotype, with only extrapyramidal and pyramidal signs, including dystonia, athetosis, and rigidity with associated spasticity. Type II included patients with cerebellar signs and spasticity. Type III was reserved for patients with the Machado phenotype—patients with a dominantly inherited form of motor polyneuropathy and true cerebellar disorder (32,33).

Romanul et al. (34) described another Massachusetts family of Portuguese ancestry with a similar clinical spectrum of gait ataxia, parkinsonian rigidity, ophthalmoparesis, muscle fasciculation, areflexia and associated nystagmus, cerebellar tremor, and extensor plantar responses. Postmortem study of one patient showed striatonigral degeneration with severe neuronal loss and gliosis in the putamen; in another patient, the basal ganglia were normal. Their observations supported the concept of a clinical and neuropathologic variation. It seemed likely that the disorders in all the reported families of similar ancestry could be a single genetic entity with variable expression, and that a better understanding of this disease might accrue from a visit to Flores and San Miguel, from which the family reported by Romanul et al. (34) migrated.

AZOREAN NEUROEPIDEMIOLOGY

In 1977, Rosenberg and Nyhan joined Coutinho in the Azores to study the disease, particularly on the island of Flores. Rosenberg, Nyhan, and Coutinho encountered the typical syndromes described in the Joseph and Thomas families on Flores in the Sousa family (33).

Synthesizing the data from the Joseph, Thomas, and Sousa families (33), the data of Coutinho and Andrade (32), and the data of Romanul et al. (34), we believe we are witnessing the expression of a single gene that causes (1) early-onset disease with pyramidal and extrapyramidal signs (type I), (2) intermediate age of onset and progression, with cerebellar, pyramidal, and extrapyramidal findings (type II), and (3) later onset, with peripheral neuropathy and cerebellar disorder (type III). Dominantly inherited parkinsonism with distal atrophy in one family could be a type IV disease. It took a systematic evaluation of the families in California, Massachusetts, and Flores, however, to document the clinical variations. The neuropathologic findings of Romanul et al. (34), indicating the wider pathologic involvement with striatonigral degeneration in only some patients, also implied varied expression of this mutant gene.

A similar disorder has been found in two families, one in northeastern Portugal (35) and one in New York (36)—a black family traced to North Carolina, but neither family had any identifiable linkage to the Azores. Neuropathologic studies are required to prove that these families have the same disease.

Joseph disease has also been evaluated in Japan (37,38). The clinical features are the same as in the American and Protuguese families and the neuropathologic features are also quite similar but also have an accentuation of a dentatorubroluysian degeneration (39). Joseph disease has also been described in India (40). The disease may have been brought to India and Japan, possibly in the sixteenth century, by Portuguese navigators and clergy.

Molecular Genetics of Joseph Disease

The primary molecular defect in autosomal dominant spinocerebellar degeneration is unknown. In general, the molecular defect in most dominantly inherited diseases has not been characterized, in contrast to the known enzyme abnormalities of recessively inherited disorders. DNA polymorphism studies have not yet been reported in human spinocerebellar degenerations, so we now describe preliminary restriction fragment length polymorphism (RFLP) findings in Joseph disease, using leukocyte or brain genomic DNA. The data suggest that the primary mutation for Joseph disease may reside on chromosome 1, near the amylase gene locus (1p21), from loose linkage data with the Rh blood group factor and pancreatic amylase polymorphism (amy-2) (41,42). The evidence of linkage between 1p loci and the Joseph disease gene is weak, and efforts are being made to look elsewhere for linkage (42).

A modifier gene, if it cosegregated with the primary mutant gene in an individual, could explain the variation in gene expression and penetrance, as well as occasional skipped generations, that are seen so often in dominantly inherited neurological diseases. An electrophoretic variant of erythrocyte acid phosphatase (ACP-1) was associated with absence of neurological disorder in one family with Joseph disease and there could be a modifier gene close to the ACP-1 locus on chromosome 2p23 (43).

Hybridization Probes

We used two probes for the short arm of human chromosome 1 and two probes for the short arm of chromosome 2. The probe for the amylase (Amy) gene, located on the p1 band of chromsome 1 was used (42). The probe, pcXP-38, was obtained from Dr. R. MacDonald, UTHSC at Dallas. This probe contained 1.2 kilobases (kb) of 5′ end of rat cDNA, and has been described in detail (44). RFLP was reported when human genomic DNA was digested with *Msp*I and *Pst*I restriction enzymes and hybridized with Amy probe (45).

Short Arm of Chromosome 1

We used a rat probe, pcXP-38, for an amylase (Amy) gene sequence. RFLPs of Amy sequences have been reported when human genomic DNA was digested with *Msp*I or *Pst*I restriction enzymes (45). We also found RFLP in JD families after DNA digestion with these enzymes. However, only two of five JD families showed polymorphism in the Amy gene.

Members of one family (W15) had either two or one *Msp*I polymorphic fragments (approximately 4.5 and 4.6 kb) that hybridized with the Amy probe (Fig. 26-1). One nonaffected, not at risk member (2) had only one 4.6-kb fragment. Another nonaffected but at risk member (1) had another 4.5-kb fragment affected members (3 and 4) had both, 4.5- and 4.6-kb fragments of DNA. RFLPs were not found with *Pst*I.

However, we found RFLPs after hybridization of Amy probes with *Pst*I-digested genomic DNA from the second JD family (W2.3–7) (Fig. 26-2). In general,

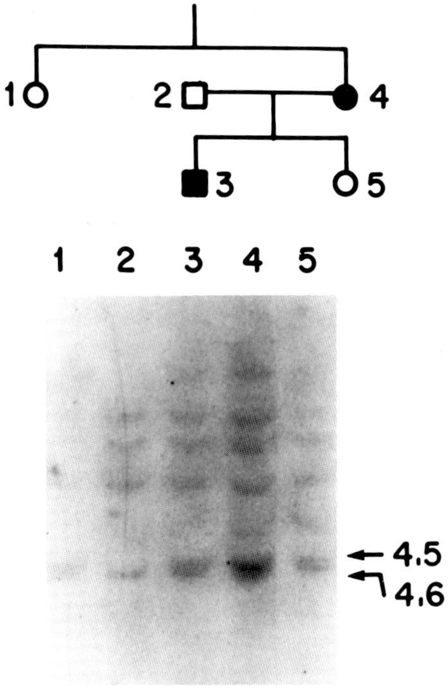

Fig. 26-1. RFLP in JD (W15) family are shown. Solid square and circle indicate persons affected with Joseph disease. DNAs of patients have been digested with *Msp*I restriction enzyme and hybridized with the pcHPA nick-translated probe, characteristic for human Amy gene. Arrows and numbers indicate sizes of hybridizing polymorphic genomic DNA fragments in kilobases (see text for details).

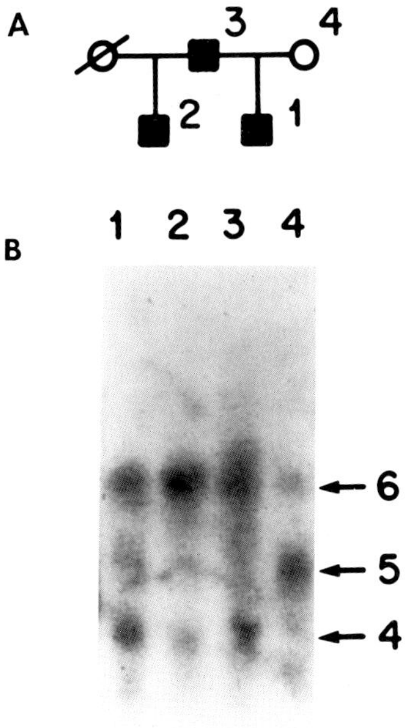

Fig. 26-2. RFLPs in JD (2.3-7) family are shown. Solid squares indicate affected Joseph disease patients. The circle with cross-line (∅) indicates a deceased person. DNAs of patients have been digested with *Pst*I restriction enzyme and hybridized with (A) the pcxp38 and with (B) the pcHPA nick-translated probes characteristic for the human Amy gene. Arrows and numbers indicate sizes of hybridizing polymorphic genomic DNA fragments in kilobases (see text for details).

three bands of approximate size 6.0, 5.0, and 4.0 kb showed visible polymorphism. Patient 4, not affected and not at risk, had only one 5.0-kb band. Her affected husband (patient 3) had two bands of 6.0 and 4.0 kb. Their affected son (patient 1) has three bands of 6.0, 5.0, and 4.0 kb, similarly to his affected half-brother (patient 2). As the human amylase gene maps to 1p, these RFLPs support the conclusion that the Joseph disease gene is linked to the amylase locus at 1p.

Short Arm of Chromosome 2

Two probes, POMC and N-*myc*, are known to be on the short arm of chromosome 2 (45). These probes were used in hybridization experiments for one large family (E-2) to find possible cosegregation of RFLP with the absence of disease.

The POMC probe showed RFLPs after *Rsa*I digestion of control human genomic DNA and E-2 family DNA (45). Two DNA fragments of approximate size 2.5 (type 1) and 2.2 kb (type 2) were polymorphic (Fig. 26-3). Two members of the family not at risk and nonaffected, patient 3 and her at risk daughter (patient 15), were homozygous for the 2.5-kb band; three affected children of affected mother (5, 6, and 8) were homozygous for the 2.2-kb band; and 10 family members were heterozygous for both bands. Four members (1, 2, 7, and 9) among these 10 were affected and the others were at risk (Fig. 26-3). The 2.5-kb (type 1) and 2.6-kb (type 2) DNA fragments cosegregated with the A and B type of acid phosphatase-1 (ACP-

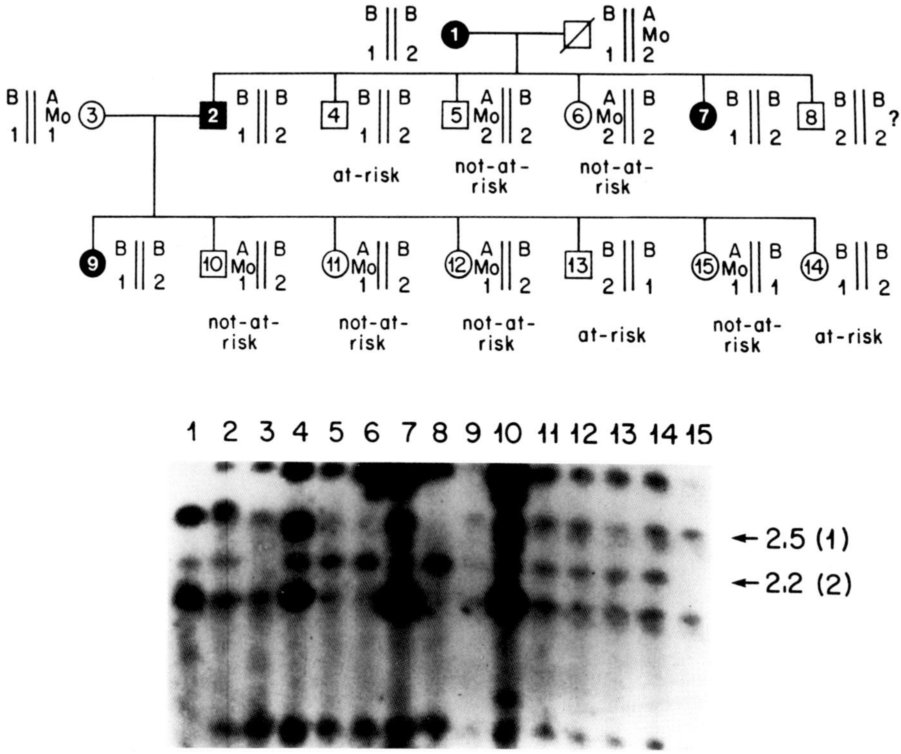

Fig. 26-3. Erythocyte acid phosphatase type and RFLPs are shown for Corderio family members. A portion of this family's pedigree is shown here. Solid circles and squares indicate patients with Joseph disease. Patient numbers correspond to the patient's position on the Southern blot. DNAs were digested with *Rsa*I restriction enzyme and hybridized with POMC nick-translated probe. Two DNA fragments, 2.5 kb (1) and 2.2 kb (2), are polymorphic. Combinations of alleles for ACP-1 (A and B) and RFLP (1 and 2) and their association with the short arm of the chromosome 2 are shown for each individual. Genotypes of ACP-1 and RFLPs and the combination of alleles for the deceased person (⍁) have been deduced from the segregation of these alleles among his children and grandchildren (see text for details).

1). POMC and ACP-1 genes are closely linked and situated in the same region (2p23) of chromosome 2.

We followed cosegregation of these two genes with the appearance of JD in that family. Four affected members of the family (1, 2, 7, and 9) had identical genotypes for ACP-1 (BB) and for POMC (1,2) genes (Fig. 26-3).

We did not find RFLP after digestion of genomic DNA with either *Eco*RI or *Taq*I enzymes and hybridization with the N-*myc* probe (Fig. 26-4A and B). Only one fragment of 1.9 kb hybridized with the N-*myc* probe after digestion of genomic DNA with *Eco*RI enzyme (Fig. 26-4A). Two *Tag*I fragments, approximate size 4 and 3.8 kb, hybridized with the N-*myc* probe (Fig. 26-4B).

These data support the conclusion that the ACP-1 locus is linked to the POMC locus at 2p23. Further, the ACP-1 locus may be linked to a modifier locus at 2p23 as patients with the BA isozyme type of ACP-1 had a reduced incidence of disease.

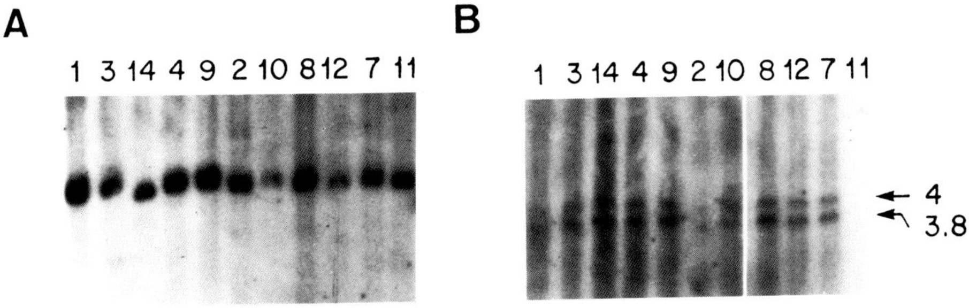

Fig. 26-4. DNA of Cordeiro family members has been digested with (A) *Eco*RI or with (B) *Tag*I restriction enzymes and hybridized with the nick-translated N-*Myc* probe. Numbers on the top of Southern blots correspond to the numbers of family members in Fig. 26-3. Arrows and numbers indicate the size of DNA fragments hybridized with the probe.

Modifier *(Mo)* Gene

Joseph disease is transmitted as an autosomal dominant trait (Figs. 26-1–26-3). The proportion of affected children of heterozygous (affected) and homozygous for normal allele (nonaffected) parents should be close to 50%. However, in the 164 families with 694 descendants at risk, only 266 descendants (38.3%) older than age 40 years were clinically affected, significantly less ($p < 0.001$) than the expected number (347) of affected descendants, and implying reduced penetrance in about 10% of the families, including the Cordeiro family. Using the curve of cumulative distribution for age at onset, we calculated that 6 children of 16 under risk should have developed the disease. However, only 3 were seen (Fig. 26-3).

Erythrocyte acid phosphatase-1 (ACP-1) was investigated for its three alleles (A, B, C) in the Cordeiro family. Two ACP-1 alleles, A and B, were found in the family (Fig. 26-3). Four affected persons had the BB phenotype of ACP-1. Nine children with a BA phenotype have not yet developed JD, nor have four with the BB genotype. The absence of affected descendants with the BA phenotype of EAP-1 apparently is not a matter of chance ($0.1 > p > 0.05$); it led us to the conclusion that the BA phenotype of EAP-1 is linked to the absence of JD expression.

DNA Polymorphisms

We found DNA polymorphisms in leukocyte or brain genomic DNA from two families with JD, using the rat amylase cDNA probe. It was suggested by finding linkage between the JD locus and the Rh factor mapped to chromosome 1p32-1pter and the Amy-2 locus at 1p21 (41,42). With the rat amylase probe, assigned to the short arm of chromosome 1 (1p21), we studied two JD families and found segregation of fragments in those affected but not in family members who were not at risk. In family 1 (W-15), two affected persons showed amylase cDNA hybridization to 4.5- and 4.6-kb fragments and a not at risk person showed only a 4.6-kb fragment hybridization signal (Fig. 26-1). That pattern implied that the JD gene is on the chromosome which yielded the 4.5 *Msp*I fragment.

In family 2 (W2.3-7), three affected persons had amylase cDNA probe hybridization as follows: (1) 6.0- and 4.0-kb fragments, (2) 6.0-, 5.0-, and 4.0-kb fragments, and (3) 6.0- and 4.0-kb fragments. A not at risk family member had only a 4.0-kb hybridization signal (Fig. 26-2). The data suggest that, in this family, the mutant JD gene may be on a chromosome that yielded 6.0- and 4.0-kb *Pst*I fragments. In both families, at risk persons and affected patients had the same patterns. If there is a linkage, the at risk individuals may develop the disease in the future, and the JD locus may be linked to the amylase locus. These studies will be extended to additional families and additional chromosome 1 probes. One family with another spinocerebellar degeneration, autosomal dominant Charcot-Marie-Tooth disease, has also been mapped to chromosome 1 with linkage with the Duffy locus (46).

One electrophoretic isoform of ACP-1 was associated with absence of clinical disease in one JD family (Fig. 26-3), suggesting that the postulated modifier gene may be linked to the ACP-1 locus. The modifier gene product, it is presumed, corrects or stabilizes the mutant JD locus product to prevent clinical expression. The gene map location for ACP-1 is 2p23, which is also the map location for the proopiomelanocortin locus (POMC gene). At the 2p23 region in JD patients, there was polymorphism for both ACP-1 and POMC. However, the N-*myc* DNA probe gave no polymorphism, an important internal control because the N-*myc* gene is also assigned to the 2p23 region, so emphasizing the specificity of the ACP-1 and POMC polymorphisms.

The Cordeiro family (Fig. 26-3) is a JD family with fewer than expected affected descendants. We studied segregation of three inherited traits (JD, ACP-1, and POMC). Alleles of ACP-1 are designated as A and B, and 2.5- and 2.3-kb *Rsa*I fragments of DNA hybridizing with POMC probe are designated as 1 and 2.

If the *Mo* gene is located on chromosome 2, then the affected patient #9 (Fig. 26-3) inherited the $^{B}_{2}$ chromosome from her father (patient #2) who is affected and apparently does not have a modifier gene, and the $^{B}_{2}$ chromosome without the *Mo* gene from her healthy mother (patient #3). Similarily, the paternal chromosome $^{B}_{1}$ of the affected patient #2 does not carry the *Mo* gene. Therefore, the homologous second chromosome $^{A}_{2}$ of his deceased healthy father is the chromosome that would carry the *Mo* gene. His brother (#4) and sister (#7) with paternal $^{B}_{1}$ chromosomes, which is without the *Mo* gene, are either at risk or affected. His brother (#5) and sister (#6) with $^{A}_{2}$ paternal chromosomes are not at risk for the disease expression, since the $^{A}_{2}$ chromosome might carry an *Mo* gene, although they could be carriers of JD gene. Patient #8 carries the $^{B}_{2}$ maternal chromosome and a crossover paternal chromosome $^{B}_{2}$. The clinical status of this patient depends on where the hypothetical *Mo* gene is located, i.e., closer to EAP or to the POMC gene.

As we indicated above, the $^{B}_{1}$ chromosome of #3, a healthy patient, does not carry the *Mo* gene since her daughter who inherits $^{B}_{1}$ chromosome (patient #9) is affected. However, if patient #3 does carry an *Mo* gene, it should be situated on the chromosome $^{A}_{1}$. Therefore, all her children with the $^{B}_{1}$ maternal chromosome (# 13 and #14) are at risk for Joseph disease or affected (#9). At the same time, her children #10, #11, #12, and #15 are not at risk to develop disease, although they might be carriers of the JD gene. We will follow descendants not affected at the present time in this family to see who will develop JD.

We believe that the primary mutant JD gene product encoded on chromosome 1 interacts with genes at the 2p23 region. The product of a modifier gene linked to ACP-1 and POMC may interact with the JD gene product to correct the expected biological phenotype and minimize expression. The possibility of a modifier gene in JD to explain variation in penetrance and expressivity of dominantly inherited diseases is a concept that needs further study. That postulate can be tested directly, rather than attributing clinical variation to genetic heterogeneity involving many genes.

The primary mutation for JD may reside on chromosome 1 and the gene for some families with olivopontocerebellar degeneration may be on chromosome 6 near the HLA complex. The JD locus may be influenced by a modifier gene locus at 2p23 to account for variation in clinical expression.

REFERENCES

1. Siekert RG, et al. Symposium on ataxia in childhood. Proc Mayo Clin 1985; 34:659.
2. Greenfield JG. The Spino-Cerebellar Degenerations. Springfield, IL: Charles C Thomas, 1954.
3. Konigsmark BW, Weiner LP. The olivopontocerebellar atrophies: A review. Medicine 1970; 49:227–241.
4. Brown S. Hereditary ataxia. Brain 1892; 15:250.
5. Marie P. Sur l'hérédoataxie cérébelleuse. Sem Med (Paris) 1893; 13:444–447.
6. Dejerine J, Thomas A. L'atrophie olivo-pontocerebelleuse. N Iconog Salpetriere 1900; 12:330.
7. Holmes G. A form of familial degeneration of cerebellum. Brain 1907; 30:466.
8. Schut J. Hereditary ataxia: Clinical study through six generations. Arch Neurol Psychiat 1950; 63:535–568.
9. Blass JP. Disorders of pyruvate metabolism. Neurology 1979; 29:280–286.
10. Blass JP, Kark RAP, Menon NK. Low activities of the pyruvate and oxoglutarate dehydrogenase complexes in five patients with Friedreich's ataxia. New Engl J Med 1976; 295:62–76.
11. Stumpf DA, Parks JK, E'quren LA, Haas R. Friedreich's ataxia. III. Mitochondrial malic enzyme deficiency. Neurology 1982; 32:221–227.
12. Chamberlain S, Lewis PD. Normal mitochondrial malic enzyme levels in Friedreich's ataxia fibroblasts. J Neurol Neurosurg Psychiat 1983; 46:1050–1051.
13. Gray R, Kumar D. Mitochondrial malic enzyme in Friedreich's ataxia: Failure to demonstrate reduced activity in cultured fibroblasts. J Neurol Neurosurg Psychiat 1985; 48:70–74.
14. Fernandez RJ, Civantos F, Tress E, Maltese W, DeVivo D. Normal fibroblast mitochondrial malic enzyme activity in Friedreich's ataxia. Neurology 1986; 36:869–872.
15. Roussy G, Levy G. Sept cas d'une maladie familiale particuliere: Troubles de la marche, pieds bots et areflexie tendineuse generalisee, avec accessoirement legere maladresse des mains. Rev Neurol 1926; 33:427.
16. Harding AE, Thomas PK. The clinical features of hereditary motor and sensory neuropathy: Type I and II. Brain 1980; 103:259–280.
17. Oelschlager R, White HH, Schinike RN. Levy-Roussy syndrome: Report of a kindred and discussion of the nosology. Acta Neurol Scand 1971; 47:80.
18. Geary JF, Earle KM, Rose AS. Olivopontocerebellar atrophy. Neurology 1956; 6:218–224.
19. Hassin GB, Harris TH. Olivopontocerebellar atrophy. Arch Neurol Psychiat 1936; 35:43–63.
20. Hunt JR. Dyssynergia cerebellaris myoclonica. Brain 1921; 44:490–538.

21. Skre H, Loken AC. Myoclonus epilepsy and subacute presenile dementia in heredo-ataxia. Acta Neurol Scand 1970; 46:42.

22. Jackson J, Currier R, Terasaki P, Morton N. Spinocerebellar ataxis and HLA linkage: Risk predictions by HLA typing. New Engl J Med 1970; 296:1138–1141.

23. Haines J, Schut L, Weitkam PL, Thayer M. Spinocerebellar ataxis in a large kindred: Age at onset, reproduction, and genetics linkage studies. Neurology 1984; 34:1542–1548.

24. Plaitakis A. Abnormal metabolism of neuroexcitatory amino acids. In Duvoisin RC, Plaitakis A, eds. Olivopontocerebellar Atrophies. New York: Raven Press, 1984, pp 225–243.

25. Plaitakis A, Nicklas WJ, Desnick RJ. Glutamate dehydrogenase deficiency in 3 patients with spinocerebellar syndrome. Ann Neurol 1980; 7:297–303.

26. Duvoisin RC, Chokroverty S, Lepore F, Nicklas W. Glutamate dehydrogenase deficiency in patients with olivopontocerebellar atrophy. Neurology 1983; 33:1322–1326.

27. Finocchiaro G, Taroni F, DeDonato S. Glutamate dehydrogenase activity in leukocytes and muscle mitochondria in olivopontocerebellar atrophies. Neurology (Suppl. 1) 1985; 35:193.

28. Grossman A, Rosenberg R, Warmoth L. Glutamate and malate dehydrogenase activities in Joseph disease and olivopontocerebellar atrophy. Neurology 1987; 37:106–111.

29. Nakano K, Dawson D, Spence A. Machado disease: A hereditary ataxia in Portuguese immigrants to Massachusetts. Neurology 1972; 22:49–55.

30. Woods BT, Schaumburg H. Nigro-spino-dentatal degeneration with nuclear ophthalmoplegia. J Neurol Sci 1972; 17:149–166.

31. Rosenberg RN, Nyhan WL, Bay C, Shore P. Autosomal dominant striatonigral degeneration. Neurology 1976; 26:703–714.

32. Coutinho P, Andrade C. Autosomal dominant system degeneration in Portuguese families of the Azores Islands. Neurology 1978; 28:703–709.

33. Rosenberg RN, Nyhan WL, Coutinho P, Bay C. Joseph disease: An autosomal dominant neurological disease in the Portuguese of the United States and the Azores Islands. In Kark P, Rosenberg RN, Schut L, eds. The Inherited Ataxias. New York: Raven Press, 1978, pp 33–57.

34. Romanul F, Fowler H, Radvany J et al. Azorean disease of the nervous system. New Engl J Med 1977; 296:1505–1508.

35. Lima L, Coutinho P. Clinical criteria for diagnosis of Machado-Joseph disease: Report of a non-Azorean Portuguese family. Neurology 1980; 30:319–322.

36. Healton E, Brust J, Kerr D, Resor S, Penn A. Presumably Azorean disease in a presumably non-Portuguese family. Neurology 1980; 30:1084–1089.

37. Goto I, Tobimatsu S, Ohta M, Hosokawa S, Shibasaki H, Kuroiwa Y. (1982): Denatatorubropallidoluysian degeneration: Clinical, neuro-ophthalmologic, biochemical, and pathologic studies on autosomal dominant form. Neurology 1982; 32:1395–1399.

38. Yuasa T, Ohama E, Harayama H, Yamade M, Kawase Y, Wakabayashi M, Atsumi T, Miyatake T. Joseph's disease: Clinical and pathological studies in a Japanese family. Ann Neurol 1986; 19:152–157.

39. Sakai T, Ohta M, Ishino H. Joseph disease in a non-Portuguese family. Neurology 1983; 33:74–80.

40. Bharucha N, Bharucha E, Bhabha S. Machado-Joseph Azorean disease in India. Arch Neurol 1986; 43:142–144.

41. Forster-Gibson C, Myers S, Simpson N, Rosenberg R, Sequeiros J, MacLeod P. Investigation of linkage in 12 kindreds with Machado-Joseph's disease. Am J Human Genet 1984; 36:9S.

42. Myers S, MacLeod P, Forse R, Forster-Gibson C, Simpson N. Machado-Joseph disease: Linkage analysis between the loci for the disease and 18 protein markers. Cytogenet Cell Genet, in press.

43. McKusick V. Mendelian Inheritance in Man, 6th ed. Baltimore: Johns Hopkins University Press, 1983.

44. MacDonald R, Crevar M, Swain W, Pictet R, Thomas G, Rutter W. Structure of a family of rat amylase genes. Nature (London) 1980; 287:1–5.

45. Skolnick M, Willard H, Menlove L. Report of the committee on human gene mapping by recombinant DNA techniques. Human gene mapping 7. Cytogenet Cell Genet 1984; 37:210–273.

46. Bird T, Oh J, Giblett E. Evidence for linkage of Charcot-Marie-Tooth neuropathy to the Duffy locus on Chromosome 1. Am J Hum Genet 1982; 34:388–394.

27

Duchenne Muscular Dystrophy: Identification of the Gene

LOUIS M. KUNKEL

Most human genetic disorders have been characterized clinically, but the primary biochemical defects remain unknown (1). Duchenne (DMD) and Becker (BMD) muscular dystrophies are two examples of the many biochemically uncharacterized disorders. Each has a distinctive clinical course (2), but the primary abnormality remains elusive despite extensive research (3). Both DMD and BMD are candidate disorders for "reverse genetics": identification of the disrupted genetic locus that gives rise to the disorder, followed by characterization of the product of the locus in patients (4). Reverse genetics bypasses the extensive effort of identifying the primary altered protein. Instead, it involves the tools of molecular biology, together with genetic information, to isolate the actual primary genetic determinants that are responsible for the disorder (Fig. 27-1).

This chapter will examine these steps with reference to the DMD and BMD phenotypes. The hope is that the application of reverse genetics to BMD and DMD will lead to an understanding of the primary defects, and perhaps even alter the course of the disorder.

1. Chromosomal Map Position
2. Random DNA Fragment Within Locus
3. Search for Genomic "Exons"
4. Identify RNA Transcript
5. Isolate cDNA Clones
6. Detect Specific Mutation in Patient DNA
7. Sequence cDNA—Predict Amino Acid Sequence
8. Study Protein in Patients and Normals

Fig. 27-1. Outline of the general steps for reverse genetics.

MAPPING OF THE DMD AND BMD LOCI TO Xp21

The evidence that DMD and BMD reside in Xp21 was obtained from two lines of research. The first was the description of cytological alterations in rare female patients with clinical manifestations of muscular dystrophy (5). Each one was heterozygous for a balanced X-autosome translocation, all with the breakpoint on the short arm of the X-chromosome at band Xp21. The normal X-chromosome was found to be late replicating and presumably inactive; the translocation chromosome was active, giving evidence that the DMD locus had been disrupted by the translocation break. In addition to the females with translocations, several males have had cytologically detected interstitial deletions of Xp21 (6–8) associated with DMD as well as other concomitant disorders.

Second, the structural evidence for an Xp21 location of DMD and BMD was supported by linkage analysis of restriction fragment length polymorphisms (RFLP) (9), detecting cloned DNA segments in families with DMD or BMD. Cloned DNA segments that mapped physically close to Xp21 (8,10) were also found to be tightly linked to DMD and BMD mutations (11,12). Both structural and genetic mapping therefore indicated that the locus altered in both BMD and DMD was in Xp21.

SPECIFIC CLONING OF DNA SEGMENTS FROM Xp21

Once DMD and BMD were located in Xp21, in principle, the loci could be identified and isolated by cloning Xp21 in its entirety. That effort would be laborious but would eventually yield a segment of DNA that encoded the DMD/BMD gene product. My colleagues and I tried to streamline the effort (Fig. 27-2). If DMD could be physically and genetically localized to a defined segment of 2–3 million basepairs of DNA, then random cloned segments from within that region could be used to identify structural alterations in DNA isolated from patients.

In other cloned disease loci, 5–10% of patients have had detectable deletions of coding sequences (13,14). We therefore assumed that similar deletions would be found in DMD and BMD. If the structural alterations were not detected by the first analysis, then chromosome walking (see Chapter 10) could be initiated at each cloned segment (expanding cones in Fig. 27-2) until walking encountered the structural alteration. The random cloned segment that detected a primary mutation (such as a deletion) would be the cloned DNA segment closest to the DMD/BMD locus.

A crucial aspect of this strategy is the acquisition of random cloned segments from a defined region of DNA. In a preliminary analysis of more than 30 Xp cloned segments, it was demonstrated that at least one cloned DNA fragment (754; 15) was absent from the DNA of BB, a patient with a cytologically detected delection at Xp21 (8). He also had three X-linked disorders; the most important for this discussion was DMD. The locus for DMD was therefore defined as the region of DNA absent from this patient. A *competition–reassociation* and cloning strategy (16) was used to enrich for additional segments of DNA that might be missing from the DNA of BB. The DNA of BB was sheared to 1500 bp in length and mixed in a

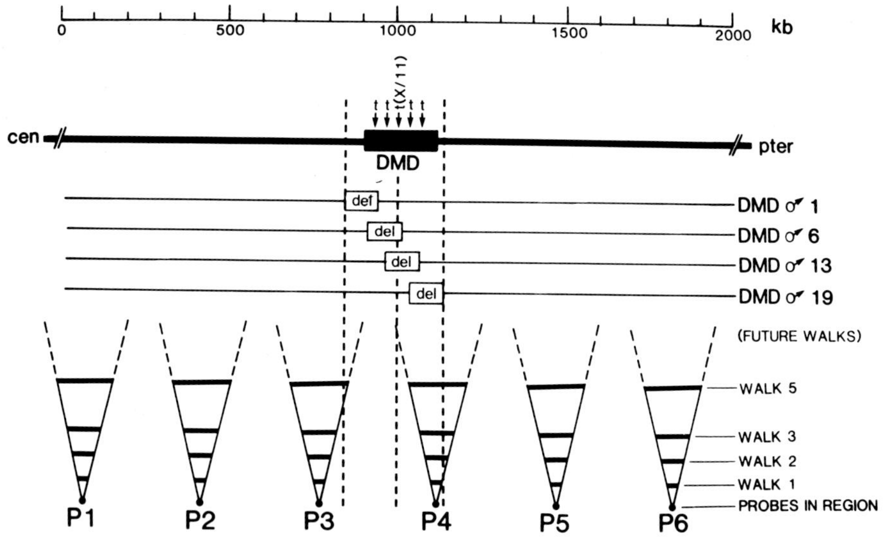

Fig. 27-2. Schematic outline of the strategy to saturate Xp21 and identify the DMD locus. If by both physical and genetic methods DMD can be delimited to a region of approximatley 2000 kb as indicated in the kilobase scale, then one of the random probes P1–P6 might detect an alteration in a patient. Both translocations (indicated with a t) and hypothetical deletions (DMD; 1,6,13,19) are indicated in the figure. The expanding cones represent successive chromosome walking steps with the amount of DNA obtained after each walk drawn to scale. The DMD locus is indicated as a dark block arbitrarily placed in the center, although it could be to either side or even much larger. The dotted vertical lines indicate the extents either centromeric or terminal in which disruptions might be detected.

200-fold excess with *Mbo*I-cleaved DNA isolated from a 49,XXXXY cell line. The mixture was dissociated by heating to 100°C and allowed to reassociate under conditions that dramatically enhance the rates of reassociation (phenol emulsion reassociation technique, or PERT; 17).

Three types of reassociated molecules were expected. Most of the reassociated molecules would be those both strands of DNA with irregularly sheared ends. A minor portion would be molecules with one strand sheared and one strand with a "straight" *Mbo*I termini formed by *Mbo*I-ended molecules that found a complementary strand in the excess of sheared DNA. Another minor component would be molecules with straight *Mbo*I termini on both reassociated strands. These last molecules would be *Mbo*I ended molecules that found no complementary strand in the sheared deletion DNA sample and had self-reassociated.

The DNA molecules with *Mbo*I ends on both termini were compatible in ligation with *Bam*HI-cleaved plasmid vector DNA. Transformation of bacteria with that ligation mixture yielded a library of recombinant DNA segments (PERT library; 18) that, in principle, should have been enriched for cloned segments absent from the excess deletion DNA sample. The degree of enrichment of the PERT library was established by analysis of each individual clone on Southern blots (19) for absence or presence in the DNA of the original deletion patient. More than 350 individual clones were analyzed, and of the unique sequence-containing clones, nine (3.6%) were found to be absent from the DNA of BB. Seven of the nine PERT clones and the cloned segment 754 (15)(Fig. 27-3) were partially subdivided to subregions of Xp21. The subregions were defined by physical mapping of an individual clone relative to different translocation breakpoints. Each of the eight cloned segments was a unique point in Xp21 that might, in principle, detect a primary structural alteration in the DNA of a DMD or BMD patient.

PROXIMITY OF PERT CLONES TO DMD

Originally, the DNA of 57 boys with BMD or DMD was isolated and tested by hybridization with each of the 8 clones absent from the DNA of BB. For example, (Fig. 27-4) identical filters with DNA samples isolated from 6 DMD boys were hybridized with two different PERT library cloned segments. The DNA fragment (PERT379) was present in all 6 DNA samples, but PERT87 was absent from the DNA sample in lane 2. The PERT87 clone was absent from the DNA isolated from 5 patients with DMD, but no other PERT clone or clone 754 was absent from those 5 DNA samples (20). Detection of deletions in the DNA of DMD patients by clone PERT87 indicated that the DNA surrounding PERT87 must be within or near the DMD locus. To obtain more DNA surrounding PERT87, we used chromosome walking in human recombinant phage libraries starting with the original 200 bp clone (20). The human DNA inserts obtained were then subcloned in plasmid vectors and tested for hybridization to unique DNA segments from Xp21. Each unique

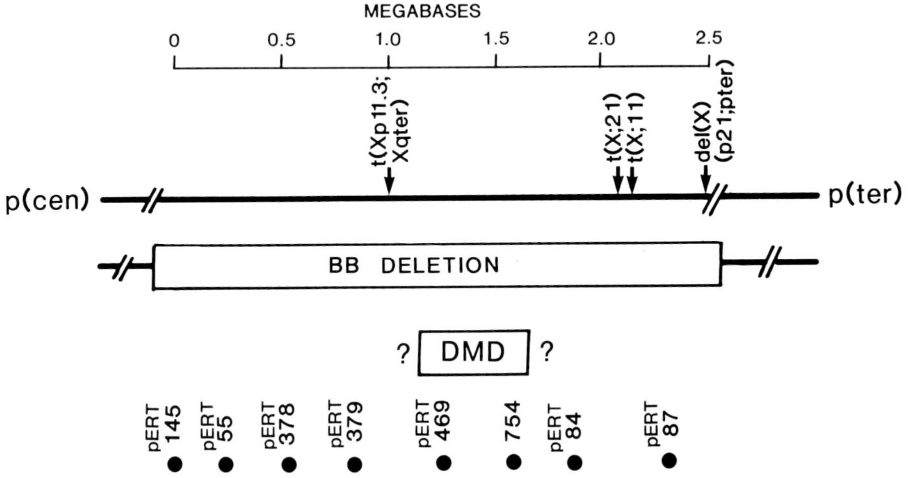

Fig. 27-3. Schematic representation of the 7 PERT clones and 754 within the large BB deletion. An approximate kilobase scale is given above a schematic of various translocation breakpoints. Below is shown the position of eight independent cloned DNA segments, each absent from the DNA isolated from BB. Each of the eight was mapped relative to the translocation breakpoints (described in 18), and the positions are indicated. The block highlighting DMD implies the question being asked: namely, which of the eight clones will detect a structural alteration in the DNA isolated from a DMD male.

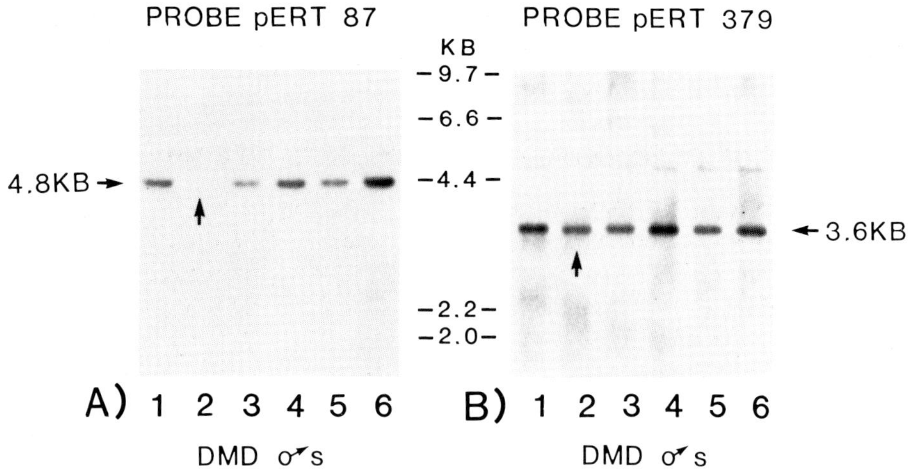

Fig. 27-4. Autoradiographs of the hybridization results for two PERT clones with six DMD males. Identical nitrocellulose filters were prepared (see Chapter 10), and each was hybridized with a radiolabeled cloned segment absent from BB. The filter in A was hybridized with PERT87 and that in B with PERT379. One microgram of DNA isolated from six unrelated DMD males was cleaved with *Pst*I, separated by electrophoresis, and blotted. The males are indicated as 1–6. The size of the hybridizing *Pst*I fragments for each probe is given on the sides of the figure. The arrow in lane 2 indicates the DNA sample that exhibits deletions for PERT87 but not PERT379.

sequence-containing subclone was tested for detection of RFLP alleles (20,21). The segregation pattern of the unique subclones that detect RFLP alleles were studied in DMD families. No recombinants were observed between PERT87 subclones and DMD mutations in the first set of families studied (20). Later, as the number of families increased, recombinants were found (22,23). Despite these recombinants between the PERT87 subclones and the DMD locus, the overall segregation results have indicated that the PERT87 subclones are genetically close to DMD mutations.

In addition to segregation experiments, several subclones (among many obtained in the chromosome walk) from the newly named DXS164 locus (PERT87) were made available to other investigators for analysis of deletions in other DMD patients. In an international cooperative effort, 1346 boys with BMD or DMD were studied for deletion of DNA surrounding the DXS164 locus. Eighty-eight (6.5%) were found to have complete or partial deletions of the distributed subclones (22). No DNA sample isolated from a normal male had a deletion of any PERT87 subclone. The DNA samples deleted for the DXS164 locus were further tested with additional subclones from the contiguous block of walked DNA. Many deletions demonstrated complete absence of all 137 kb of contiguous cloned DNA, but approximately 40% were partially deleted (some subclones were present, others absent) (22).

When analyzed in terms of the DXS164 restriction map, the deletion breakpoints were scattered throughout the DXS164 locus, and the deletions extended in either direction, toward the centromere or toward the terminus. This pattern implied that the DXS164 locus was central to deletion mutations found in the DMD males. Segregation results with RFLP-detecting probes in DMD families gave the same indication (22). In addition, the translocation breakpoints in DMD females were found to be heterogeneous and flanked the DXS164 locus (24).

IDENTIFICATION OF POTENTIAL CODING SEQUENCES FOR DMD

The central location of DXS164 subclones to DMD mutations implied that they might contain a portion of the coding sequences for the DMD gene. Outlined in Fig. 27-5 is a schematic of the extensively walked DXS164 locus and possible means to identify coding sequences of the gene. The nucleotide sequences of most coding segments of any gene (exons) have been partially conserved between species in evolution. In contrast, noncoding sequences (introns) have generally not been conserved between species. Based on this knowledge, more than 50 unique sequence subclones from the DXS164 locus were tested for sequence conservation by hybridization to Southern blots of DNA from different mammals. Two subclones hybridized to all mammalian DNA samples tested and one, PERT87-25, was also found in birds (25). The region of conserved nucleotides in humans was narrowed to a small segment that was used to isolate the equivalent region from a mouse genomic DNA recombinant library. The primary nucleotide sequence was determined for each region from both mouse and humans (25).

With the primary sequence in hand, we searched within the sequences for specific splicing signals (splice sites) surrounding a potential exon of a gene (Fig. 27-6) (26,27). By comparing many exon boundaries for other cloned DNA segments, we found a consensus nucleotide sequence at both the 3′ and 5′ sides of an exon (26,27). These nucleotide segments are essential for the proper splicing of the mature mRNA from its large precursor. Sequences that indicated a potential exon of a gene were found in the two conserved regions of human DNA. The equivalent regions were also found in the mouse sequences (25). The central location of these

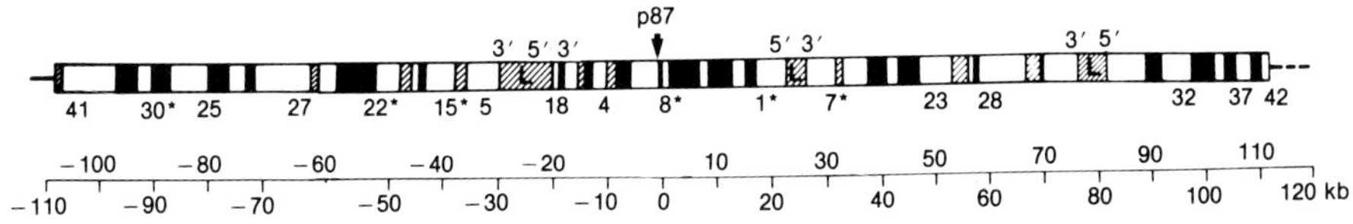

HOW ARE POTENTIAL CODING SEQUENCES IDENTIFIED?

1. DIRECT ANALYSIS FOR TRANSCRIPTION
2. SEQUENCE CONSERVATION DURING EVOLUTION
3. DETECTION OF PRIMARY MUTATION IN PATIENT DNA

Fig. 27-5. A schematic map of the DXS164 locus and the potential ways that coding sequences might be identified. The DXS164 locus is presented with a kilobase scale below, along with various subclone nubmers (RFLP detecting clones are marked with an asterisk). The block diagram depicts regions that are unique sequence (open areas), moderately repeated (hatched lines), and highly repeated (dark areas). The original 200-bp p87 (PERT87) clone marks the center of the nine bidirectional walks undertaken to clone the DXS164 locus. The question addressed is, Where would coding sequences for the gene reside?

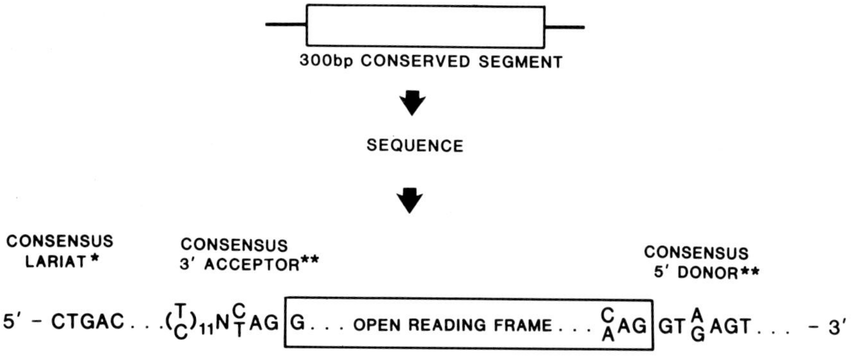

Fig. 27-6. Schematic of how to search for an exon. A region of DNA is first demonstrated to be conserved during evolution, e.g., the 300-bp region from the DXS164 locus. The DNA sequence is determined and a search is made for the designated sequence signals that surround an exon. An uninterrupted stretch of amino acids is indicated as an open reading frame. The consensus sequences are given and were taken from Mount (26, **) and Keller and Noon (27, *). These sequences are not invariant but are those which are most commonly found surrounding an exon.

two regions to DMD mutations, their conservation, and the presence of nucleotide consensus sequences that correspond to those surrounding an exon indicated that these regions were likely to be a portion of the DMD gene.

IDENTIFICATION OF MUSCLE TRANSCRIPTS AND CLONING OF cDNAs

For a particular segment of DNA to be identified as part of a gene, it should be transcribed as an RNA in some cell type during development or in adult life. A search was made for the presence of the two potential exons of the gene for DMD in cellular RNA. RNA was isolated from several human cell lines and fetal and adult human tissues. The RNA was separated by molecular weight on denaturing agarose gels (see Chapter 8) and transferred to nylon membranes. The membrane was hybridized with each potential exon as a radiolabeled DNA segment. Only the RNA from fetal skeletal muscle exhibited a faintly hybridizing 16-kb mRNA, with a smear of smaller hybridizing fragments (25). No other RNA sample tested exhibited the large transcript. The hybridizing RNA was demonstrated to be RNase sensitive, and, later, more extensive analysis in other mouse tissues indicated that fetal or adult skeletal or cardiac muscle showed the same large transcript.

The detection of the 16-kb muscle transcript by potential exons near or within DMD mutations clearly implied the importance of this transcript to the phenotype of DMD. To obtain the entire representation of this transcript, cDNA libraries were constructed from human skeletal muscle RNA samples. The libraries were screened with the same conserved human exons that detected the transcript. We plaque purified phage that contained hybridizing human cDNA clones, and the DNA inserts were hybridized back to human genomic DNA to demonstrate X-chromosomal localization of the cDNA clones. Initially, a 1.0-kb human cDNA (a small part of the 16-kb transcript) hybridized to 7 *Hin*dIII fragments of X-chromosome origin.

When it was hybridized to the entire DXS164 locus, that same DNA clone detected 7 exons of a gene spread over at least 110 kb of genomic DNA. By extrapolation, if the remainder of the cDNA representative of the 16-kb transcript (now in the process of being isolated) were also spread over such large genomic distances, then the DMD gene may exceed 2 million basepairs. The large size of the locus could explain the high frequency of new mutations, the heterogeneity of translocation breakpoints, and the large size and heterogeneous breakpoints of the deletions of Xp21.

PROSPECTS

The large size of the transcript and the muscle-specific expression imply that the altered product in DMD is a large muscle-specific protein. One such protein (nebulin) has been described to be absent or altered in DMD patients (28), but later work showed that nebulin is not the DMD gene product. When the entire cDNA representation of the 16-kb transcript is cloned, the primary nucleotide sequence will be determined for the DMD gene. Translation of the triplet code into an uninterrupted amino acid sequence will indicate the primary protein sequence of the DMD gene. Once the entire protein sequence is determined, then predictions may be made on the nature and function of the DMD protein. Techniques (described elsewhere) are also available and will be used to produce large quantities of the cloned DMD protein to serve as antigens for the production of antibodies to the DMD protein. The availability of antibodies to the DMD protein will not only aid in the understanding of the relationship between nebulin and DMD, but also will ultimately assign an overall function to this protein.

ADDENDUM

Kunkel and his associates have identified the DMD gene product as dystrophin. As discussed in Chapter 2, they found that the protein is usually absent in Duchenne patients. In Becker patients, however, dystrophin is altered in size, reduced in amount, or both.

ACKNOWLEDGMENTS

I am grateful to C. Bertelson and A. Monaco for critical reading of the manuscript. This study was supported by the Muscular Dystrophy Association of America and grants from the NIH (RO1 NS 23740-01 and HD18658).

REFERENCES

1. McKusick V. Mendelian Inheritance in Man. Baltimore: Johns Hopkins University Press, 1984.

2. Moser H. Duchenne muscular dystrophy: Pathogenic aspects and genetic prevention. Human Genet 1983; 66:17–40.

3. Rowland LP. Biochemistry of muscle membranes in Duchenne muscular dystrophy. Muscle Nerve 1980; 3:3–20.

4. Orkin SH. Reverse genetics and human disease. Cell 1986; 47:845–850.

5. Boyd Y, Buckle VJ. Cytogenetic heterogeneity of translocations associated with Duchenne muscular dystrophy. Clin Genet 1986; 29:108–115.

6. Baehner RL, Kunkel LM, Monaco AP, et al. DNA linkage analysis of X-chromosome-linked chronic granulomatous disease. Proc Natl Acad Sci USA 1986; 83:3398–3401.

7. Bartley J, Patil S, Davenport S, et al. Duchenne muscular dystrophy, glycerol kinase deficiency, and adrenal insufficiency associated with an Xp21 interstitial deletion. J Pediat 1986; 108:189–192.

8. Francke U, Ochs HD, de Martinville B, et al. Minor Xp21 chromosome deletion in a male associated with expression of Duchenne muscular dystrophy, chronic granulomatous disease, retinitis pigmentosa, and McLeod syndrome. Am J Hum Genet 1985; 37:250–267.

9. Botstein D, White RL, Skolnick M, Davis RW. Construction of a genetic linkage map in man using restriction fragment length polymorphisms. Am J Hum Genet 1980; 32:314–331.

10. de Martinville B, Kunkel LM, Bruns G, et al. Localization of DNA sequences in the region Xp21 of the human X chromosome: Search for molecular markers close to the Duchenne muscular dystrophy locus. Am J Hum Genet 1985; 37:235–249.

11. Davies KE, Pearson PL, Harper PS, et al. Linkage analysis of two cloned DNA sequences flanking the Duchenne muscular dystrophy locus on the short arm of the human X chromosome. Nucleic Acids Res 1983; 11:2303–2312.

12. Goodfellow PN, Davies KE, Ropers HH. Report of the committee on the genetic constitution of the X and Y chromosomes. Cytogenet Cell Genet 1985; 40:296–352.

13. Rosen R, Fox J, Fenton WA, Horwich AL, Rosenburg LE. Gene deletion and restriction fragment length polymorphisms at the human ornithine transcarbamylase locus. Nature (London) 1985; 313:815–887.

14. Yang TP, Patel PI, Chinault AC, et al. Molecular evidence for new mutation at the hprt locus in Lesch–Nyhan patients. Nature (London) 1984; 310:412–414.

15. Hofker MH, Wapenaar MC, Coor N, Bakker E, van Ommen GJB, Pearson PL. Isolation of probes detecting restriction fragment length polymorphisms from X-chromosome specific libraries: Potential use for diagnosis of Duchenne muscular dystrophy. Human Genet 1985; 70:148–156.

16. Lamar EE, Palmer E. Y-encoded, species-specific DNA in mice: Evidence that the Y chromosome exists in two polymorphic forms in inbred strains. Cell 1984; 37:171–177.

17. Kohne DE, Levinson SA, Byers MJ. Room temperature method for increasing the rate of DNA reassociation by many thousandfold: The phenol emulsion reassociation technique. Biochemistry 1977; 16:5329–5341.

18. Kunkel LM, Monaco AP, Middlesworth W, Ochs HD, Latt SA. Specific cloning of DNA fragments absent from the DNA of a male patient with an X chromosome deletion. Proc Natl Acad Sci USA 1985; 82:4778–4782.

19. Southern EM. Detection of specific sequences among DNA fragments separated by gel electrophoresis. J Mol Biol 1975; 98:503–517.

20. Monaco AP, Bertelson CJ, Middlesworth W, et al. Detection of deletions spanning the Duchenne muscular dystrophy locus using a tightly linked DNA segments. Nature (London) 1985; 316:842–845.

21. Aldridge J, Kunkel L, Bruns G, et al. A strategy to reveal high frequency RFLPs along the human X chromosome. Am J Hum Genet 1984; 36:546–564.

22. Kunkel LM, Hejtmancik JF, Caskey CT, et al. Analysis of deletions in DNA from patients with Becker and Duchenne muscular dystrophy. Nature (London) 1986; 322:73–77.

23. Fischbeck KH, Ritter AW, Tirschwell DL, et al. Recombination with pERT87 (DXS164) in families with X-linked muscular dystrophy. Lancet 1986; ii:104.

24. Boyd Y, Munro E, Ray P, et al. Molecular heterogeneity of translocations associated with muscular dystrophy. Clin Genet 1987; 31:265–272.

25. Monaco AP, Neve R, Colletti-Feener C, Bertelson CJ, Kurnit DM, Kunkel LM. Isolation of candidate cDNAs for portions of the Duchenne muscular dystrophy gene. Nature (London) 1986; 323:646–650.

26. Mount SM. A catalogue of splice junction sequences. Nucleic Acids Res 1982; 10:459–472.

27. Keller EB, Noon WA. Intron splicing: A conserved internal signal in introns of animal pre-mRNAs. Proc Natl Acad Sci USA 1984; 81:7417–7420.

28. Wood DS, Zeviani M, Prelle A, Bonilla E, Salviati G, Miranda AF, diMauro S, Rowland LP. Is nebulin the defective gene product in Duchenne muscular dystrophy? New Eng J Med 1987; 316:107–108.

28

Glycerol Kinase Deficiency: Association with Duchenne Muscular Dystrophy, Adrenal Insufficiency, and Mental Retardation

WILLIAM K. SELTZER AND EDWARD R. B. McCABE

Glycerol kinase deficiency (GKD) is an X-linked recessive inborn error of metabolism in which the activity of this enzyme is diminished or absent from those tissues normally responsible for maintaining blood glycerol homeostasis (1,2). GKD may occur alone or as part of a syndrome consisting of Duchenne muscular dystrophy, congenital adrenal hypoplasia, and developmental delay.

GKD was first reported in 1977 almost simultaneously by two groups who described two very different clinical manifestations. In the family first reported, two boys aged 2 and 5 years had psychomotor retardation, dystrophic myopathy, poor somatic growth, and osteoporosis (3,4). Subsequently, adrenal hypoplasia and adrenocortical insufficiency were also recognized in the same children (1,5). A urinary screen for organic acids by gas chromatography revealed an abnormal peak that was identified as glycerol by mass spectrometry. Plasma glycerol levels in these children were 2.5–7.8 mmol/liter (normal range: 0.01–0.20 mmol/liter). Leukocyte and fibroblast glycerol kinase activity was less than 10% of normal values.

In the second report of GKD, a 70-year-old man was incidentally identified as having GKD when a report of elevated serum triglyceride concentration was found to be erroneous, due to high levels of glycerol instead (6). Subsequently, a brother and a son of the proband's daughter were also identified as having GKD. However, none of the three affected individuals had any clinical symptoms at all. The pattern of inheritance of GKD was consistent with X linkage. The discrepancy between the clinical findings in these two families raised the question of the relationship between glycerol kinase deficiency and the other phenotypic features.

BIOCHEMISTRY OF GLYCEROL KINASE

Glycerol kinase is responsible for the first step in glycerol metabolism. Glycerol arises from dietary sources either as free glycerol or as glycerolipids (primarily triglycerides and phospholipids) that are hydrolyzed in the intestinal mucosa or liver to liberate glycerol and free fatty acids or acyl phosphates. Glycerol is also produced in tissues as a normal consequence of triglyceride and phospholipid turnover. Free glycerol is phosphorylated in the ATP-dependent glycerol kinase reaction to yield glycerol 3-phosphate, a key intermediate of carbohydrate and lipid metabolic pathways (Fig. 28-1).

The glycerol 3-phosphate produced may enter the glycolytic pathway to be oxidized for the production of energy or, alternatively, may be used for gluconeogenesis. Glycerol 3-phosphate is also the precursor for the synthesis of phosphatidic acid in the biosynthesis of triglycerides and phospholipids.

Occupying a pivotal point between the pathways of carbohydrate and lipid metabolism, glycerol kinase is apparently present in all tissues. Enzymatic activities vary over a wide range (7). Highest activities are found in the liver, kidneys and leukocytes; these organs are presumably responsible for homeostasis of blood levels of glycerol (0.01–0.20 mmol/liter). Within cells, glycerol kinase is found both free in the cytosol and membrane bound in mitochondria and endoplasmic reticulum (8,9). The association of glycerol kinase with these organelles involves reversible binding to a membrane pore-forming protein, porin. Membrane binding is tissue specific and it is regulated both developmentally and metabolically (9–12). The function of this property is unknown, but it could be a way to modulate enzymatic activity or coordinate different metabolic pathways (5,11).

METHODS FOR IDENTIFYING GLYCEROL KINASE DEFICIENCY

Generally, two criteria are used to diagnose GKD. First, there is biochemical evidence of the consequences of the functional deficiency: hyperglycerolemia and glyceroluria; second, glycerol kinase enzyme activity is absent or greatly reduced in leukocytes and fibroblasts. Hyperglycerolemia or glycerol in the urine is usually the first indication of abnormal glycerol metabolism.

Glyceroluria is conveniently detected by a urine screen for organic acids using gas chromatography (GC) and identified by mass spectrometry (MS). In the organic acid extraction procedure, much of the glycerol normally present is lost; consequently, in unaffected individuals, there is rarely a glycerol peak. Quantitation of urinary glycerol excretion requires a direct assay. Free glycerol is measured by a radiochemical procedure using added [U-^{14}C]glycerol and commercially available purified glycerol kinase (13). The free glycerol in the specimen dilutes the labeled

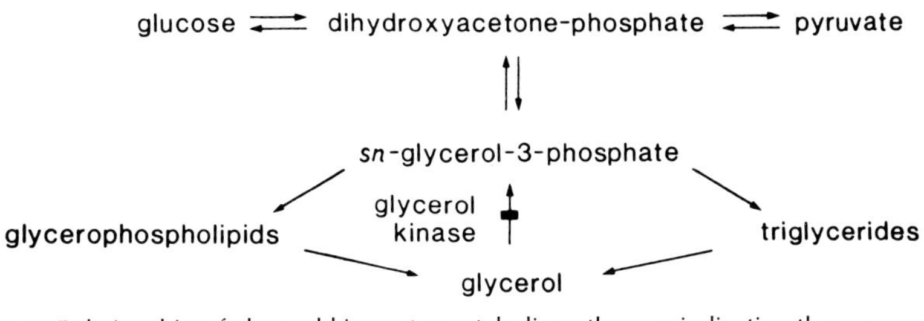

Fig. 28-1. Relationship of glycerol kinase to metabolic pathways, indicating the enzymatic deficiency in GKD.

glycerol of the assay and therefore decreases the specific activity of the substrate and hence the amount of [U-^{14}C]glycerol 3-phosphate formed in the reaction period. The [U-^{14}C]glycerol 3-phosphate produced is measured by liquid scintillation and compared to a standard curve generated by addition of known amounts of unlabeled glycerol to assay reactions. Normal urine glycerol levels are 0.01–0.25 mmol/liter; in GKD patients, levels are 40–350 mmol glycerol/liter (14). On the historical grounds that normal urines reveal no glycerol peak by GC–MS, the high urine glycerol level in GKD patients is called "glyceroluria" rather than "hyperglyceroluria." However, abnormally high glycerol levels in urine (by either the GC–MS method or by radiochemical assay procedure) are not necessarily indicative of GKD because elevated levels may be induced iatrogenically. The most common external source of urinary glycerol is contamination of the urine specimen with skin lotions; glycerol is also used in suppositories, medicinal additives, and as an osmotic agent in treating cerebral edema, all of which may elevate urine glycerol levels.

Positive, elevated glycerol screening tests should be followed up with another urine specimen and a blood sample for determination of leukocyte glycerol kinase activity and plasma glycerol concentrations. Plasma glycerol content is quantitated by the same method used for urine. Normal plasma glycerol levels are between 0.01 and 0.20 mmol/liter (14). Plasma from GKD individuals contains 1.5–7.7 mmol glycerol/liter (14). High plasma glycerol levels may also be detected by the standard serum triglyceride assay that measures free glycerol after hydrolysis of triglycerides. Abnormally high plasma glycerol content is therefore mistaken for an apparently elevated triglyceride level, or pseudohypertriglyceridemia (15). Abnormally high triglyceride values should be evaluated by direct assay for free glycerol.

For diagnostic purposes, glycerol kinase activity is most easily measured in leukocytes, transformed lymphocytes, or cultured skin fibroblasts. Glycerol kinase activity is low in these tissues, so sensitive radiochemical methods have been developed using [U-^{14}C]glycerol as the substrate for the glycerol kinase reaction in three different ways (Table 28-1):

1. Direct measurement of glycerol kinase activity by the quantitation of phosphorylated [U-^{14}C]glycerol, e.g., L-[U-^{14}C]glycerol 3-phosphate (16,17).
2. Indirect assessment of glycerol kinase activity by quantitation of [U-^{14}C]glycerol-derived ^{14}C incorporation into protein (18). The rate of incorporation of glycerol-derived label into protein depends on the conversion of glycerol to amino acids, and the first step of this pathway is the phosphorylation of glycerol. This is compared to the incorporation rate of L-[4,5-^{3}H]leucine into the protein, a pathway that does not involve glycerol kinase.
3. Indirect assessment of glycerol kinase activity by the complete oxidation of [U-^{14}C]glycerol to $^{14}CO_2$ (2). That rate is compared to the oxidation of L-[U-^{14}C]glycerol 3-phosphate, a route that is independent of glycerol kinase.

The indirect assay methods assess the ability of glycerol kinase to interact with complex, multistep pathways, whereas the direct assay determines glycerol kinase activity independent of other processes. For this reason, the direct assay is the preferred method for diagnostic purposes, whereas the indirect methods are useful for studying the relationship of glycerol kinase activity to cellular metabolism.

Table 28-1 Comparison of Methods for Determining Glycerol Kinase Activity in Diagnostic Tissues[a]

Tissues	Glycerol kinase specific activity (μU/mg protein)	[^{14}C]Glycerol/[^{3}H]leucine incorporation into protein (dpm/μg protein/24 hours) $\times 10^3$	Glycerol oxidation (pmol $^{14}CO_2$/mg protein/hour)
Leukocytes			
GKD	51 (n = 5) (15–78)		
Controls	524 (n = 13) (415–631)		
Fibroblasts			
GKD	16 (n = 11) (9–20)	8 (n = 11) (0–39)	2.1 (n = 5) (0.8–2.5)
Controls	141 (n = 12) (76–258)	229 (n = 12) (100–388)	21 (n = 11) (19–32)
Amniocytes			
GKD	N.D.	N.D.	
Controls	86 (n = 9) (50–119)	27 (n = 9) (10–45)	

[a]Data presented in this table represent the results from this laboratory (W. K. S.) and those cited in the literature (2,24). Values are expressed as the mean with the range in parentheses. All three subgroups of glycerol kinase deficiency are represented in each assay method except in the glycerol oxidation method in which only data for the complex syndrome are presented. The methods and experimental conditions used to determine these activities are referenced in the text. GKD, glycerol kinase-deficient patients; N.D., none done.

Together, low cellular glycerol kinase activity and high urine and plasma glycerol levels are diagnostic of GKD.

GKD, congenital adrenal hypoplasia, and Duchenne muscular dystrophy can each be diagnosed prenatally in families at risk. Glycerol kinase activity is normally present and can be measured in amniocytes obtained by amniocentesis, but GKD has not yet been documented in an affected fetus (19). Similarly, if the family is informative, the possibility that a boy has Duchenne dystrophy can be assessed by restriction fragment length polymorphism (RFLP) analysis, using Xp21 genomic DNA probes for DNA obtained at amniocentesis (20). The presence of congenital adrenal hypoplasia may be determined by the level of maternal estriol (21). Dehydroepiandrosterone, produced by the fetal adrenal gland, is the major precursor for placental synthesis of estriol. If the fetus has congenital adrenal hypoplasia, maternal estriol levels will be abnormally low (21). Since all of the phenotypic components of the complex GKD syndrome are heritable as single X-linked disorders, evaluation of any component would be diagnostic. Prompt treatment of infants with both GKD and congenital adrenal hypoplasia is required to prevent fatal adrenal crisis.

CLINICAL PHENOTYPES OF GLYCEROL KINASE DEFICIENCY

As of April 1987, GKD had been identified in 32 individuals (31 males and 1 normal karyotype female) in 23 families (22). Residual enzymatic activities measured in leukocytes, transformed lymphocytes, or fibroblasts range from less than 1 to

20% of normal values (Table 28-1). All affected individuals have had glyceroluria and hyperglycerolemia, but clinical phenotypes have varied and can be divided into three groups: the complex syndrome, the juvenile form, and the benign form (22).

The *complex form* of GKD is characterized by lack of glycerol kinase enzyme activity in the diagnostic tissues, hyperglycerolemia and glyceroluria, and one or more of the following: adrenal insufficiency (usually generalized cortical insufficiency), muscular dystrophy, and mental retardation. The dystrophic myopathy varies in severity, from classical Duchenne dystrophy to a nonprogressive and mild myopathy, although all show high values of creatine kinase (CK) in serum (22).

The degree of mental impairment, when present, also shows marked variation. Poor somatic growth, osteoporosis, and cryptorchidism with absence of testes in association with gonadotropin deficiency have been reported inconsistently (23). Twenty-two individuals in 17 families had this complex GKD syndrome.

The *juvenile form* of GKD is characterized by hyperglycerolemia and glyceroluria, episodes of vomiting, metabolic acidosis, stupor that may lead to coma, and possible glycerol intolerance (22). There is no myopathy and serum CK values are normal. Two unrelated individuals have been documented with this form of GKD.

The *benign form* of GKD is not associated with any clinical symptoms but glycerol kinase activity is lacking in the diagnostic tissues and hyperglycerolemia and glyceroluria are present (22). Eight individuals in four families have been described with benign GKD.

BIOCHEMISTRY OF THE ENZYMATIC DEFICIENCY OF GLYCEROL KINASE

Biochemically, the deficiency of glycerol kinase is evidenced as either diminished or total absence of activity in the tissues routinely examined. Fibroblasts and leukocytes generally have less than 10% of normal glycerol kinase activity (Table 28-1). These cells all oxidize glycerol at markedly reduced rates and show an impaired ability to incorporate glycerol into phospholipids, triglycerides, and protein (2,18,24,25). However, the same cells in culture do not show altered growth or gross morphology.

Enzyme activity was studied at autopsy from a patient with complex GKD and a myopathy that was milder than Duchenne dystrophy. The enzymatic deficiency appeared to be generalized, with equally low levels in liver, kidney, and intestine as well as leukocytes and skin fibroblasts (1). Kinetic analysis of the residual glycerol kinase activity in liver indicated impaired ability to bind and phosphorylate glycerol (Fig. 28-2A) (17). In light of the recent molecular genetic findings of extended deletions spanning the glycerol kinase locus in some patients with the complex GKD syndrome, the residual glycerol kinase activity in these patients might be the effect of some other protein with glycerol kinase activity.

Muscle from that child was assayed for glycerol kinase activity and compared to normal muscle and muscle from individuals with typical Duchenne dystrophy (26). Glycerol kinase activity in muscle was normal in both the Duchenne muscle as well as in the patient with the complex GKD syndrome, including the dystrophic myopathy (Fig. 28-2B). Kinetic analysis of the glycerol kinase activity in muscle

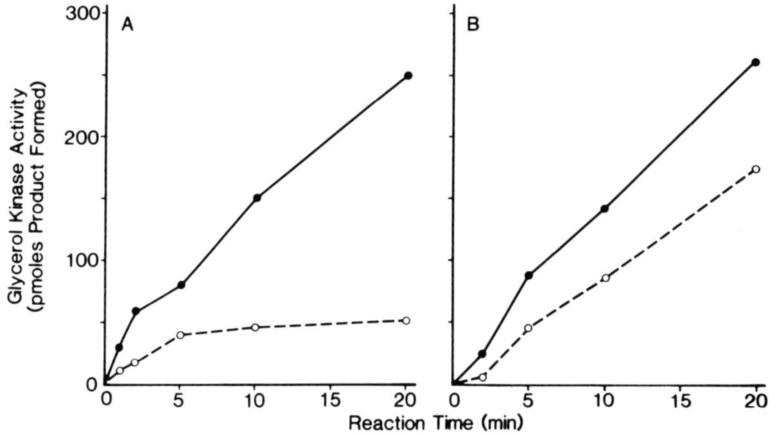

Fig. 28-2. Glycerol kinase activity progress curves for liver (A) and muscle (B) from control (●) and GKD (○) individuals. Glycerol kinase reactions were carried out at 37°C and contained liver (10–15 μg protein) or muscle (70–80 μg protein) homogenate and 100 m*M* Tris–HCl (pH 7.4 at 37°C), 5 m*M* $MgSO_4$, 6.25 m*M* NaF, 20 m*M* 2-mercaptoethanol, 6 m*M* ATP, and 250 μ*M* [U-^{14}C]glycerol. Details of the experimental procedure are referenced in the text.

from the patient with complex GKD showed normal glycerol binding and phosphorylation (26). Therefore, the Duchenne-like myopathy of the complex form of GKD cannot be attributed to the lack of glycerol kinase activity in muscle. Moreover, muscle glycerol kinase activity is probably genetically distinct from the enzyme in liver.

The biochemical studies on GKD explain the hyperglycerolemia and glyceroluria in these individuals. It is more difficult to understand how functional deficiency or absence of glycerol kinase activity could lead to the complex phenotype. Although glycerol kinase is involved in several metabolic pathways, it is clear that the lack of glycerol kinase activity, per se, does not invariably lead to severe metabolic disorder. For example, there is little abnormality in the juvenile form and none at all in the benign form of GKD. The lack of metabolic aberration is probably the result of compensating pathways that also provide glycerol 3-phosphate and phosphatidic acid for synthesis of triglycerides and glycerophospholipids and pathways that provide alternative substrates for gluconeogenesis and oxidative metabolism. Some aspects of the variation in clinical phenotype may be resolved by further biochemical studies, but molecular genetic approaches have provided the framework to understand this diversity.

MOLECULAR GENETICS OF GLYCEROL KINASE DEFICIENCY

The early recognition that each of the components of the complex GKD syndrome could be inherited as separate X-linked disorders prompted suggestions that an extended deletion could affect multiple loci to cause the complex syndrome (2,11). This theory was supported by the identification of small cytogenetic deletions of the Xp21 region of males with the complex GKD syndrome (24,27). This region on the short arm of the X-chromosome had been subject to intense investigation in relation to Duchenne and Becker dystrophies (28,29). The coincidence of GKD and dystrophic myopathy in individuals with small Xp21 deletions and the known X-linked inheritance of the cytomegalic form of adrenal hypoplasia implied that small deletions affected closely linked but separate loci for Duchenne dystrophy, congenital adrenal hypoplasia, and GKD to account for the complex GKD syndrome (24,27,30). The clinical heterogeneity, ranging from the benign form with no symptoms to the complex syndrome could be due to differences in the mutations, ranging from point mutations within the glycerol kinase gene to large deletions of varying size that affect one or more of the closely linked loci. Support for this hypothesis has grown with the increased ability to perform detailed mapping studies of the Xp21 region as more Xp21-specific probes have been generated. The X-specific probes used for the molecular genetic studies on GKD are those routinely used for the investigation of Duchenne muscular dystrophy (28,29,31).

Data accumulated from the molecular genetic investigations on all known glycerol kinase-deficient individuals have been complied (Fig. 28-3). The relationship of these deletions to a map of X-specific genomic DNA and cDNA probes spanning the region Xp21 is also shown. Because of the importance of the correlation of clinical phenotype with the molecular genetic studies, a brief description of the clin-

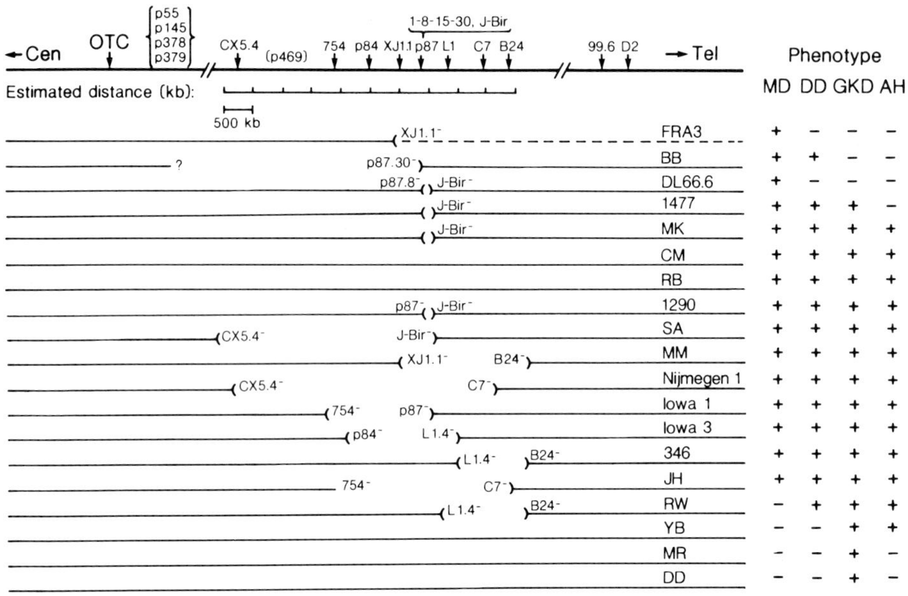

Fig. 28-3. Relationship of Xp21 deletions to phenotypes associated with GKD. Top line: map of region Xp21 with derived positions of probes used in these studies. Lower lines: deletion maps of patients previously reported. Deletions, when present, are indicated by breaks in the solid lines with end points denoted by parentheses. The half dotted line of FRA3 represents the X:21 translocation. Phenotype: the presence (+) or absence (−) of Duchenne muscular dystrophy (DMD), developmental delay (DD), glycerol kinase deficiency (GKD), and adrenal hypoplasia (AH) in the corresponding patients. Case reports, where available, and references for DNA analysis are cited in the appendix. [Adapted from Francke et al. (31) and van Ommen et al. (33).]

ical presentation of the individuals described in Fig. 28-3 is given in the appendix, in addition to the molecular genetic data. The individuals presented here with clinical phenotypes ranging from Duchenne dystrophy alone to isolated GKD have been instrumental in localizing the loci responsible for the separate disorders of this region.

Duchenne and Becker dystrophies may follow disruption of the genome anywhere in the region spanning sequences complementary to probes XJ1.1 and J-Bir (Fig. 28-3) (28,29). For Duchenne dystrophy to be a component of the complex GKD syndrome, at least a portion of the region from J-Bir to L1.4 must be involved (patients MK and RW, Fig. 28-3) (31). This suggests that the Duchenne locus terminates just distal to the J-Bir sequence and proximal to the glycerol kinase gene, which may reside as far distally as the L1.4 sequence (patients Iowa 3 and RW, Fig. 28-3). Absence of the J-Bir sequence in patient DL66.6, who does not have GKD, supports the contention that the glycerol kinase gene lies distal to this sequence. The considerable heterogeneity of the molecular genetic deletions in the patients with the complex GKD syndrome is similar to that observed in patients with Duchenne or Becker muscular dystrophy.

It is generally agreed that there is an increased incidence of mental retardation in patients with Duchenne dystrophy, but the basis for the association is not known. A genetic basis has been considered (32). If there is a specific locus for mental retardation in this region, it seems to reside between the Duchenne muscular dystrophy locus and the glycerol kinase gene. Individuals with Duchenne muscular dystrophy and mental retardation (patient BB, for example) would define the proximal end of the mental retardation locus, whereas individuals with apparently normal development and either isolated GKD or the complex GKD syndrome with adrenal hypoplasia (patient YB, Fig. 28-3) would define the distal end. Inconsistencies in this formulation (patient DL66.6, for example), however, indicate the speculative nature of this hypothesis.

Individuals with the benign or juvenile forms of GKD have consistently failed to show any deletions (Fig. 28-3), and may be due to mutations limited to the glycerol kinase locus (31).

The occurrence of normal adrenal function in patient 1477 with the complex GKD syndrome including Duchenne muscular dystrophy implies that the congenital adrenal hypoplasia locus is distal to the glycerol kinase gene. Both this patient and patient MK are missing only the J-Bir sequences. Since patient MK has adrenal insufficiency as a component of the complex syndrome, the congenital adrenal hypoplasia locus must extend into the region between the L1 and J-Bir sequences.

Taken as a whole, the data presented here suggest that the order of these loci from centromere to telomere is: Duchenne muscular dystrophy–glycerol kinase deficiency–congenital adrenal hypoplasia. Deletions that involve two or more of these loci could explain the complex phenotypes. However, deletions have not yet been found in some individuals with the complex GKD syndrome. It is anticipated that these individuals do indeed have deletions that affect two or more loci, but the vast expanse of the region, spanning the J-Bir to L1 sequences ($\geq$300–500 kb), is too large for available probes to delineate the small deletions (31,33). These individuals should be the most informative for mapping the individual loci.

CONCLUSIONS

GKD is a clinical disorder that has been important in stimulating the biochemical investigations on glycerol kinase and glycerol metabolism. GKD has also become a model for microdeletions and the analysis of multilocus genetic mutations in a complex disorder.

APPENDIX

The clinical phenotypes of the individuals presented in Fig. 28-3 are briefly described below. The molecular genetic data relevant to definition of the deletion, when detectable, are also presented.

Individuals without Glycerol Kinase Deficiency

Individual FRA3 has an X:21 translocation that resulted in Duchenne muscular dystrophy (33,34). The translocation produced abnormal RFLPs that hybridized to probes p84 and XJ1.1.

Individual BB was the first reported case of Duchenne dystrophy with cytogenetic evidence of a small deletion in the Xp21 region (35). The syndrome included developmental delay, chronic granulomatous disease, McLeod red cell phenotype, retinitis pigmentosa, and idiopathic intestinal pseudoobstruction. The proximal deletion end point was defined by the presence of sequences complementary to the OTC probe and the absence of a fragment hybridizing to probe 754. The region between probes OTC and 754 was not analyzed. The distal deletion end point was defined by the absence of a probe p87-80 hybridizing fragment and the presence of probe J-Bir hybridizing fragments.

Patient DL66.6 had only Duchenne muscular dystrophy (33). There was no report of glycerol kinase deficiency in this individual. The proximal end of this deletion was defined by the abnormal RFLP hybridizing to probe p87.1 and the absence of a fragment hybridizing to probe p87.15. The distal end of the deletion was defined by the absence of a probe J-Bir hybridizing fragment and the presence of sequences complementary to probe L1.

Individuals with Glycerol Kinase Deficiency

Patient 1477 was one of two brothers with the complex GKD syndrome. He also had severe Duchenne muscular dystrophy with elevated CK levels. Adrenal function was normal. This patient was mentally retarded. The deletion was defined only by the absence of the J-Bir sequences. The p87 sequences proximal to the J-Bir region and the L1 sequences distal were present (36).

Patient MK had the complex GKD syndrome, consisting of adrenocortical insufficiency with extremely hypoplastic adrenal glands, severe retardation of psychomotor development, and myopathic changes resembling Duchenne muscular dystrophy. CK levels were consistently elevated (37). This deletion was defined only by the absence of sequences complementary to the probe J-Bir. The p87.30 sequence and the L1.4 sequence were present (31).

Patient CM (Colorado, patient 2) was the surviving brother of the two originally reported brothers with the complex GKD syndrome consisting of adrenocortical insufficiency, psychomotor retardation, myopathy, poor somatic growth, and osteoporosis (1,4). CK levels (MM type) were greatly elevated. The myopathy has not been progressive, there was no calf hypertrophy, and the patient was still walking at age 14 years. No deletion was detected in his DNA (31).

Patient RB had the complex GKD syndrome consisting of adrenocortical insufficiency, severe retardation of psychomotor development, myopathy, and high CK levels (MM type). At 8 months he had hypotonia and decreased muscle mass (37). No deletion was found (31).

Patient 1290 had the complex GKD syndrome consisting of adrenal insufficiency, mental retardation, and Duchenne dystrophy. The deletion was defined by the absence of p87 and J-Bir sequences. There were DNA sequences complementary to probes XJ1.1 proximally and C7 distally (36).

Patient SA had the complex GKD syndrome consisting of adrenal insufficiency, mental retardation, and Duchenne dystrophy (38). The proximal end of the deletion was defined by the presence of the OTC gene and absence of the fragment hybridizing to probe CX5.4. The distal end of the deletion was defined by the absence of the J-Bir region and the presence of sequences complementary to the L1.4 probe (31).

Patient MM (patient from Iowa 4 family) had the complex GKD syndrome, with growth and developmental retardation, adrenal insufficiency, and myopathy. CK levels were elevated. He had cryptorchidism and no testes were found on laparotomy (23). The proximal end of the deletion was defined by the presence of the probe p84 RFLP and the absence of the XJ1.1 region. The distal end of the deletion was defined by the absence of the probe B24 hybridizing fragment and the presence of sequences complementary to probe 99.6 (31).

Patient Nijmegen 1 had the complex GKD syndrome, with severe psychomotor and mental retardation, adrenal insufficiency (younger sibling with documented congenital adrenal hypoplasia of the cytomegalic type), and Duchenne dystrophy. Serum CK levels were elevated. Serum gonadotropin levels were low. The right testicle was not completely descended and the left one was not palpable (39). The deletion was defined at the proximal end by the presence of an abnormal RFLP hybridizing to the probe CX5.4. The distal end of the deletion was defined by the absence of complementary sequences to probe C7 and the presence of the fragment hybridizing to probe B24 (30,33).

Affected individuals in the Iowa 1 family had the complex GKD syndrome consisting of adrenocortical insufficiency, developmental delay, and high serum CK levels. Muscle biopsies have not been obtained (2). The proximal end of the deletion was defined by the presence of probe p469 and the absence of probe 754 hybridizing fragments. The distal end of the deletion was defined by the absence of sequences hybridizing to probe series p87 and the presence of the probe L1 hybridizing fragment (33,40).

Affected individuals in the Iowa 3 family had the complex GKD syndrome. The proband had high CK levels and muscle histopathology characteristic of Duchenne muscular dystrophy. At age 6 years he had never walked. He had adrenocortical insufficiency, developmental delay, and generalized spasticity (24). The proximal end of the deletion was defined by the presence of sequences complementary to the probe 754 and the absence of the p84 hybridizing fragment. The distal end of the deletion was defined by the absence of the probe L1.4 hybridizing fragment and the presence of the probe C7 fragment (33,40).

Patient 346 had the complex GKD syndrome consisting of adrenal hypoplasia, developmental delay, and Duchenne dystrophy. The proximal end of the deletion was defined by the presence of the p87.30 fragment and the absence of the L1 fragment. The distal end of the deletion was defined by the absence of the B24 fragment and the presence of the 1aE3 fragment. The DNA sequences complementary to probe 1aE3 lie proximal to the probe 99.6 sequences (41).

Patient JH had the complex GKD syndrome. Serum CK levels were high and he was in a wheelchair at age 10 years. He showed poor somatic growth, developmental delay (functional IQ of about 60), and congenital adrenal hypoplasia requiring mineralocorticoid therapy. The proximal end of the deletion was defined by the absence of sequences complementary to probe 754. Complementary sequences did hybridize to probe OTC. The region between probes OTC and 754 was not analyzed. The distal end of the deletion was defined by the absence of the C7 fragment. Probe B24 was not used, but the probe 99.6 hybridizing fragment was present (42).

Patient RW (patient from Iowa 2 family) had the complex form of GKD but no clinical evidence of myopathy and serum CK levels were normal. He had adrenocortical insufficiency and mild developmental delay (2). The proximal end of the deletion was defined by the presence of probe J-Bir hybridizing fragments and the absence of the L1.4 fragment. The distal end of the deletion was defined by the absence of sequences complementary to probe B24 and the presence of the probe 99.6 fragment (31,40).

Patient YB had the complex GKD syndrome with adrenal insufficiency but no developmental delay or myopathy. He was studied by J.-P. Harpey and colleagues. CK levels were normal. He was of Algerian heritage and his parents were second cousins. Hypoglycemia and apnea were seen in the neonatal period. No deletion was found (31).

Patient MR had the juvenile form of GKD. At age 4 years, he had episodic vomiting, metabolic acidosis, and stupor that may have responded to a low glycerol diet. Adrenal function was normal and there was no evidence of myopathy on muscle biopsy. Serum CK was normal (43). No deletion was found (31).

Patient DD (patient from Ontario family) had the benign form of GKD. At age 21 years, he had normal height and intellectual development. There was no history of metabolic disorder (6). No deletion was found (31).

REFERENCES

1. Guggenheim MA, McCabe ERB, Roig M, et al. Glycerol kinase deficiency with neuromuscular, skeletal, and adrenal abnormalities. Ann Neurol 1980; 7:441–449.

2. Bartley JA, Miller DK, Hayford JT, McCabe ERB. Concordance of X-linked glycerol kinase deficiency with X-linked congenital adrenal hypoplasia. Lancet 1982; ii:733–736.

3. McCabe ERB, Guggenheim MA, Fennessey PV, O'Brian D, Miles B, Goodman SI. Glyceroluria, psychomotor retardation, spasticity, dystrophic myopathy, and osteoporosis in a sibship. Pediat Res 1977; 11:527.

4. McCabe ERB, Fennessey PV, Guggenheim MA, et al. Human glycerol kinase deficiency with hyperglycerolemia and glyceroluria. Biochem Biophys Res Commun 1977; 78:1327–1333.

5. Seltzer WK, Firminger H, Klein J, Pike A, Fennessey P, McCabe ERB. Adrenal dysfunction in glycerol kinase deficiency. Biochem Med 1985; 33:189–199.

6. Rose CI, Haines DS. Familial hyperglycerolemia. J Clin Invest 1978; 61:163–170.

7. Thorner JW, Paulas H. Glycerol and glycerate kinases. In Boyer PD, ed. The Enzymes, Vol 8. New York: Academic Press, 1973, pp 487–508.

8. Seltzer WK, McCabe ERB. Subcellular distribution and kinetic properties of soluble and particulate-associated bovine adrenal glycerol kinase. Mol Cell Biochem 1984; 64:51–61.

9. Ostlund A-K, Gorhing U, Krause J, Brdiczka D. The binding of glycerokinase to the outer membrane of rat liver mitochondria. Its importance in metabolic regulation. Biochem Med 1983; 30:231–245.

10. Fiek C, Benz R, Roos N, Brdiczka D. Evidence for the identity between the hexokinase binding protein and the mitochondrial porin in the outer membrane of rat liver mitochondria. Biochim Biophys Acta 1982; 688:429–440.

11. McCabe ERB. Human glycerol kinase deficiency: An inborn error of compartmental metabolism. Biochem Med 1983; 30:215–230.

12. Sadava D, Depper M, Gilbert M, Bernard B, McCabe ERB. Development of enzymes of glycerol metabolism in human fetal liver. Biol Neonate 1987; 52:26–32.

13. Newsholme EA. Radiochemical assay for glycerol. In Bergmeyer HU, ed. Methods in Enzymatic Analysis, Vol 3. New York: Academic Press, 1974, pp 1409–1415.

14. McCabe ERB. Disorders of glycerol metabolism, Chap 36. In Scriver CR, Beaudet AL, Sly W, Valle D, eds. Metabolic Basis of Inherited Disease. New York: McGraw-Hill, 1988, in press.

15. Goussault Y, Turpin E, Neel D, et al. "Pseudohypertriglyceridemia" caused by hyperglycerolemia due to congenital enzyme deficiency. Clin Chem Acta 1982; 123:269–274.

16. Seltzer WK, McCabe ERB. Human and rat adrenal glycerol kinase: Subcellular distribution and bisubstrate kinetics. Mol Cell Biochem 1984; 62:43–50.

17. McCabe ERB, Sadava D, Bullen WW, McKelvey HA, Seltzer WK. Human glycerol kinase deficiency: Enzyme kinetics and fibroblast hybridization. J Inher Metab Dis 1982; 5:177–182.

18. Seltzer WK, Dhariwal G, McKelvey HA, McCabe ERB. 1-Thioglycerol: Inhibitor of glycerol kinase activity in vitro and in situ. Life Sci 1986; 39:1417–1424.

19. Williamson R, Patil S, Bartley J, Greenburg F. Prenatal evaluation for glycerol kinase deficiency (GKD) associated with congenital adrenal hypoplasia. Am Soc Hum Gen 1984; 36:200S.

20. Bakker E, Bonten EJ, De Lange LF, et al. DNA probe analysis for carrier detection and prenatal diagnosis of Duchenne muscular dystrophy: A standard diagnostic procedure. J Med Genet 1986; 23:573–580.

21. Hensleigh PA, Moore WV, Wilson K, Tulchinsky D. Congenital X-linked adrenal hypoplasia. Obstet Gynecol 1978; 52:228–232.

22. McCabe ERB. Glycerol kinase deficiency. In Buise ML, ed. Birth Defects Encyclopedia. Dover, MA: Birth Defects Information Services, 1988, in press.

23. Wise JE, Matalon R, Morgan AM, McCabe ERB. Phenotypic features of patients with congenital adrenal hypoplasia and glycerol kinase deficiency. Am J Dis Child 1987; 141:744–747.

24. Bartley JA, Patil S, Davenport S, Goldstein D, Pickens J. Duchenne muscular dystrophy, glycerol kinase deficiency, and adrenal insufficiency associated with Xp21 interstitial deletion. J Pediat 1986; 108:189–192.

25. Bartley JA, Ward R. Glycerol kinase deficiency inhibits glycerol utilization in phosphoglyceride and triacylglycerol biosynthesis. Pediat Res 1985; 19:313–314.

26. Seltzer WK, Angelini C, Dhariwal G, Ringel SP, McCabe ERB. Muscle glycerol kinase in Duchenne dystrophy and glycerol kinase deficiency. 1988; submitted.

27. Hammond J, Howard NJ, Brookwell R, Purvis-Smith S, Wilcken B, Hoogenraad N. Proposed assignment of loci for X-linked adrenal hypoplasia and glycerol kinase genes. Lancet 1985; i:54.

28. Kunkel LM, Hejtmancik JF, Caskey CT, et al. Analysis of deletions in DNA from patients with Becker and Duchenne muscular dystrophy. Nature (London) 1986; 322:73–77.

29. Thomas NST, Ray PN, Worton RG, Harper PS. Molecular deletion analysis in Duchenne muscular dystrophy. J Med Genet 1986; 23:509–515.

30. Wieringa B, Hustinx Th, Scheres J, Renier W, ter Haar B. Complex glycerol kinase deficiency syndrome explained as X-chromosomal deletion. Clin Genet 1985; 27:522–523.

31. Francke U, Harper JF, Darras BT, et al. Congenital adrenal hypoplasia, myopathy and glycerol kinase deficiency: Molecular genetic evidence for deletions. Am J Hum Genet 1987; 40:212–227.

32. Rossman NP, Kakulas BA. Mental deficiency associated with muscular dystrophy. Brain 1966; 89:769–787.

33. van Ommen GJB, Verkerk JMH, Hofker MH, et al. A physical map of 4 million bp around the Duchenne muscular dystrophy gene on the human X-chromosome. Cell 1986; 47:499–504.

34. Verellen-Dumoulin Ch, Freund M, De Meyer R, et al. Expression of an X-linked muscular dystrophy in a female due to translocation involving Xp21 and non-random inactivation of the normal X chromosome. Hum Genet 1984; 67:115–119.

35. Francke U, Ochs HD, de Martinville B, et al. Minor Xp21 chromosome deletion in a male associated with expression of Duchenne muscular dystrophy, chronic granulomatous disease, retinitis pigmentosa and McLeod syndrome. Am J Hum Genet 1985; 37:250–267.

36. Davies KE, Patterson MN, Kenwrick SJ, et al. Fine mapping of glycerol kinase deficiency and congenital adrenal hypoplasia within Xp21 on the short arm of the human X chromosome. Am J Med Genet 1987; 29:557–564.

37. Kohlschutter A, Willig RP, Schlamp D, et al. Glycerol kinase deficiency—A condition requiring prompt identification—Clinical, biochemical and morphological findings in two cases. Europ J Pediat 1987; 146:575–581.

38. Saito F, Goto J, Kakinuma H, et al. Inherited Xp21 deletion in a boy with complex glycerol kinase deficiency syndrome. Clin Genet 1986; 29:92–93.

39. Renier WO, Nabben FAE, Hustinx TWJ, et al. Congenital adrenal hypoplasia, progressive muscular dystrophy, and severe mental retardation, in association with glycerol kinase deficiency, in male sibs. Clin Genet 1983; 24:243–251.

40. Patil SR, Bartley JA, Murray JC, Ionasescu VV, Pearson PL. X-Linked glycerol kinase, adrenal hypoplasia and myopathy maps at Xp21. Cytogenet Cell Genet 1985; 40:720–721.

41. Kenwrick S, Patterson M, Speer A, Fischbeck K, Davies K. Molecular analysis of the Duchenne muscular dystrophy region using pulsed field gel electrophoresis. Cell 1987; 48:351–357.

42. Clarke A, Roberts SH, Thomas NST, Whitfield A, Williams J, Harper PS. Duchenne muscular dystrophy with adrenal insufficiency and glycerol kinase deficiency: High resolution cytogenetic analysis with molecular, biochemical, and clinical studies. J Med Genet 1986; 23:501–508.

43. Ginns EJ, Barranger JA, McClean SW, et al. A juvenile form of glycerol kinase deficiency with episodic vomiting, acidemia and stupor. J Pediat 1984; 104:736–739.

29

X-Linked Mental Retardation Syndromes

KAY E. DAVIES

The observation that there is an excess of mentally retarded males over females led clinicians over 50 years ago to propose that sex-linked inheritance was a major cause of mental deficiency. Clinical and family studies 30 years later identified specific syndromes. There are now almost 60 X-linked traits listed in McKusick's catalogue (1) that result in, or are associated with, mental retardation (Table 29-1). In a few mental retardation syndromes, family studies suggest that there is a single gene defect on the X chromosome. However, there are also X-linked diseases in which mental retardation affects only a small percentage of patients. This is the pattern in both Duchenne and Becker muscular dystrophy; only a minority are mentally retarded. It is by no means clear whether this is due to a combination of factors associated with these diseases or whether there is a mutation in a specific gene within the same region of Xp21.

By far the most common form of X-linked mental retardation is that associated with a cytogenetically detectable fragile site on the long arm of the human X chromosome. This has now become the focus of attention for both clinicians and molecular biologists because it is prevalent in many populations and because there is no reliable carrier detection or antenatal diagnosis.

CLINICAL ASPECTS

Most X-linked mental retardation syndromes have not been localized to regions of the X chromosome because there are too few families for study and few large pedigrees in which relevant members are alive. Only two have been mapped along the chromosome: one form shows linkage to the blood group marker Xg on the short arm; the other is the fragile X syndrome, associated with a fragile site near the telomere of the long arm.

Newly ascertained cases of mental retardation have been termed "nonspecific" (1). In a study of 24 pedigrees in British Columbia, Herbst and Miller (2) calculated that the incidence of mental retardation was 1.83 per 1000 live male births, with a carrier frequency of 2.44 per 1000 live female births. Assuming a mutation rate for X-linked loci of $3–9 \times 10^{-5}$ and a fitness of zero for affected males, they estimated that between 7 and 19 genes on the X chromosome are responsible for nonspecific retardation.

Table 29-1 X-Linked Mental Retardation Syndromes

Type (McKusick number)	Alternate names
1. Allan syndrome (30960)	Mental retardation, X linked with hypotonia Mental retardation and muscular atrophy Mental retardation with muscle weakness and cerebellar atrophy Allan–Herndon–Dudley syndrome
2. Atkin syndrome	Atkin–Flaitz syndrome Mental retardation, X linked, with macrocephaly and macroorchidism
3. Davis syndrome (30964)	Mental retardation with spastic paraplegia Mental retardation with spastic diplegia Mental retardation, X linked, with progressive spastic quadriparesis
4. FitzSimmons syndrome (30956)	X-linked mental retardation, spastic paraplegia, and palmoplantar hyperkeratosis
5. Fragile (X)-linked mental retardation	Fragile X syndrome Marker X syndrome Martin–Bell syndrome
6. Gareis syndrome (none)	Gareis–Mason syndrome Mental retardation, X linked with clasped thumbs (absence of extensor pollicis brevis)
7. Glycerol kinase deficiency (30703)	Hyperglycerolemia
8. Golabi syndrome (none)	Mental retardation, X linked, with multiple congenital anomalies (microcephaly, postnatal growth retardation, eye anomalies, and heart defect) Golabi–Ito–Hall syndrome
9. Homes syndrome (none)	Mental retardation, X linked, with microcephaly, renal, foot, and other anomalies
10. Juberg syndrome	Mental retardation, X linked, with growth retardation, deafness, and microgenitalism Juberg–Marsidi syndrome Renier syndrome Mental deficiency, epilepsy, spasticity, and deafness syndrome
11. Lujan syndrome (none)	Mental retardation, X linked, with marfanoid habitus
12. Renpenning syndrome (30950)	X-linked mental retardation with microcephaly and short stature
13. Schimke syndrome (none)	X-linked mental retardation, progressive basal ganglion, dysfunction, mental and growth retardation, external ophthalmoplegia, postnatal microcephaly, and deafness
14. Seemanova syndrome	X-linked microcephaly with epilepsy, spastic tetraplegia, and absent abdominal reflexes
15. Vasquez syndrome (none)	Mental retardation, X linked, with hypogonadism, gynecomastia, short stature, and obesity
16. X-linked mental retardation with macroorchidism but without fragile X (30953)	Fragile X-negative Martin–Bell syndrome
17. Nonspecific mental retardation	Nonspecific X-linked mental retardation Fragile (X)-negative X-linked mental retardation, macroorchidism

In a survey of school children in Coventry, Webb et al. (3) estimated that nearly 1 in 1100 males and 1 in 700 females carry the gene for the fragile X syndrome. The fragile X syndrome is the most common cause of mental retardation after Down syndrome.

Male patients suffering from mental retardation associated with a fragile site on the X chromosome have a characteristic phenotype that becomes more evident in adults. The phenotype is called the Martin–Bell syndrome, after the family study in which the characteristics were first described. Affected boys have extraordinarily large ears, long foreheads, prominent jaws, and large testes (macroorchidism). IQ ranges from 30 to 65, and many of the boys are in institutions. Affected females have an IQ in the range 60–85 and have no obvious specific clinical features, but more studies are required because some retarded girls do show dysmorphic features.

In view of the limited data available on most X-linked mental retardation syndromes, this chapter will be confined to a discussion of the techniques that can be used for the analysis of the fragile X syndrome.

CYTOGENETIC DETECTION OF THE FRAGILE X

The fragile site on the X chromosome is the only one so far reported in the human genome that is associated with a phenotypic effect (4). The term "fragile site" refers to a heritable gap in the chromosome that is seen in cytogenetic studies after the cells have been cultured under specific conditions. The fragile X is seen after culturing cells in folate-deficient media or by the addition of the folic acid antagonist, methotrexate (Fig. 29-1). If thymidine is excluded from the medium, 5-fluorodeoxyuridine (FUdR) enhances the expression of the fragile X by blocking the remaining pathway to deoxythymidine triphosphate and DNA synthesis. Folic acid deficiency and methotrexate act through different effects on the same pathway.

Cytogenetic studies, using R- and G-banding, suggest that the fragile site is in the band Xq27 (Fig. 29-2). Scanning electron microscopy studies have refined this localization to the subband Xq27.3 (5).

Cytogenetic detection of the fragile site is relatively straightforward in affected

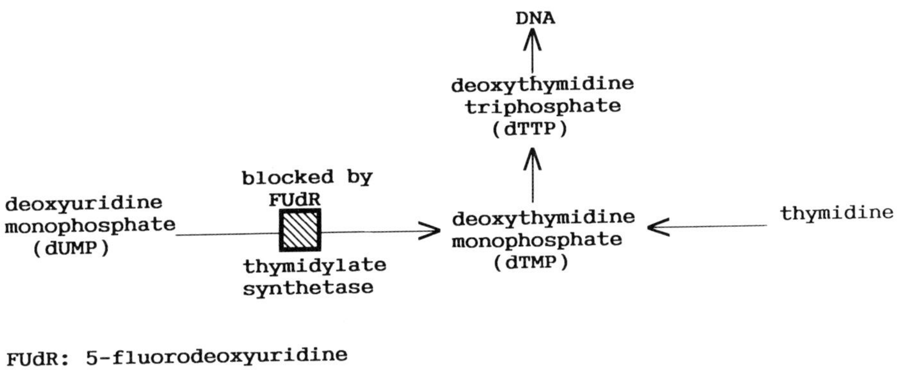

Fig. 29-1. Pathway leading to the expression of the fragile X. FUdR, 5-fluorodeoxyuridine.

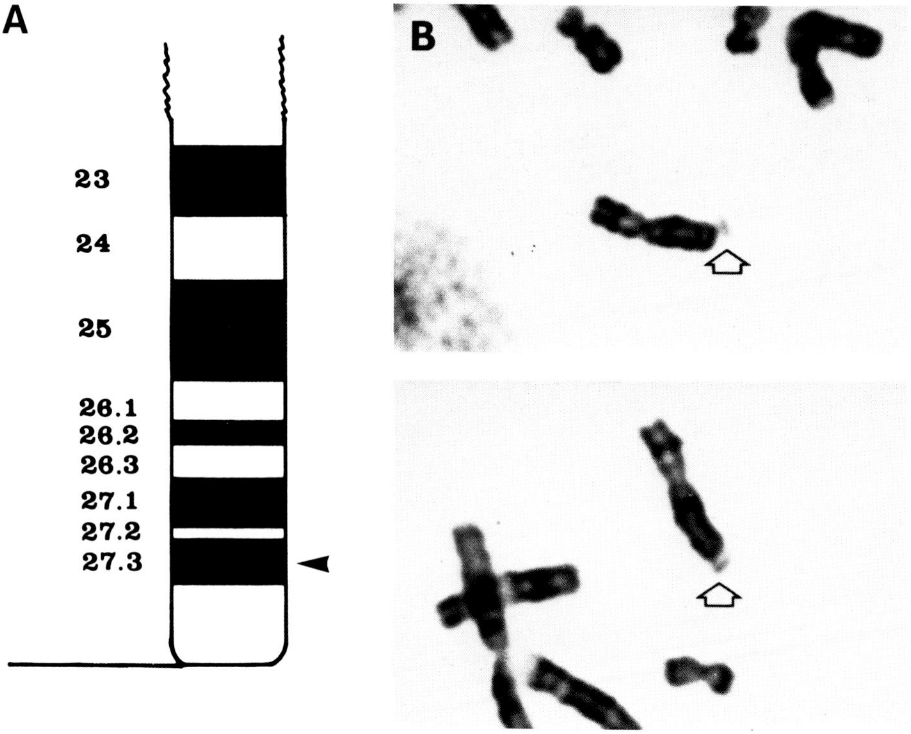

Fig. 29-2. (A) An ideogram of the long arm of the X-chromosome. (B: Partial metaphases from lymphocyte cultures of a patient with the fragile X syndrome; GTG-banded chromosomes. The arrow indicates the fragile site.

males but the number of lymphocyte cells that display it varies from 2 to 50%. It is not known why the fragile site is generally not observed in more than 50% of cells. The situation for females is rather more complicated because daughters of normal transmitting males are almost never mentally retarded and show the fragile site at low frequency, whereas their daughters are often mentally retarded and show high levels of expression of the fragile site.

Although there seems to be no correlation between the proportion of fragile X-positive cells and mental impairment, there is a correlation between expression of the fragile site and intelligence. Females expressing the fragile site tend to be mentally retarded, whereas those who do not express the fragile site are usually mentally normal. The relationship between X inactivation and the intellectual status of females is unclear, but the incidence of manifesting heterozygotes (33–50%) is unusually high for an X-linked recessive disorder.

Culture conditions that induce the fragile site such as the starving of cultures of deoxythymidine triphosphate (dTTP) also increase DNA breakage and mutation rates. High levels of thymidine induce the fragile site, perhaps by depriving the cell of dCTP by feedback inhibition. In view of these observations, the most favored model for fragile site expression is the presence of a region of DNA that is incompletely replicated or repaired. This would alter the local protein and DNA interactions, loosening the chromosome structure and so expressing the fragile site. The fragile site has also been seen under specific conditions of culture in hybrids derived from normal males from families that segregate for the disease and also in unrelated individuals (6). These observations suggest that all X chromosomes contain sequences that could induce the fragile site; these sequences might be altered in affected individuals, increasing expression of the fragile site.

SEGREGATION ANALYSIS OF THE FRAGILE X

The fragile X syndrome differs from the other X-linked disorders in that it does not show the typical pattern of inheritance due to an X-linked recessive mutation. The mutation can be passed on by phenotypically normal males; the degree of penetrance of the disease is thought to be approximately 80% (7). It is the only X-linked disorder that shows this phenomenon, which obviously complicates genetic counseling.

For example, a pedigree (Fig. 29-3A) may include normal transmitting males. In that example the three males in generation II are normal transmitters because at least one of their daughters gave birth to a mentally retarded male.

Other factors of importance in the genetic counseling of these families have become apparent from the extensive segregation studies of Sherman et al. (7). One of the most startling features is that the risk of mental retardation in children of

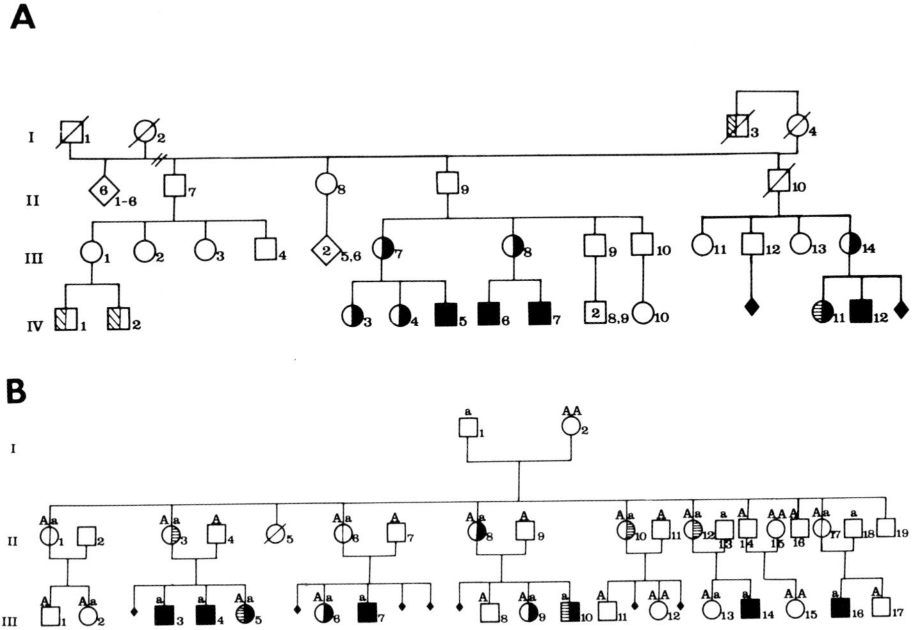

Fig. 29-3. (A) Pedigree segregating for the fragile X syndrome with normal transmitting males [taken from Froster-Iskenius et al. (18)]. (B) Segregation analysis of the *F9* locus (alleles *A* and *a*) with the disease [taken from Camerino et al (8)]. (▧, ◍) Mentally retarded, unknown cause; (◧, ◐) mentally retarded; (◫, ◍) dull; (◨, ◑) fragile X positive > 5% cells; (◫, ◍) fragile X positive < 2% cells; (□, ○) normal.

mothers of normal transmitting males is significantly lower than in the children of daughters of normal transmitting males.

Sometimes it is not possible from pedigree analysis alone to determine whether the mother or the father is transmitting the disease. In some families this dilemma, which can have important implications for genetic counseling, can be solved by examining the segregation of the linked DNA markers (8). A restriction fragment length polymorphism (RFLP) detected by the factor IX gene sequence (alleles *A* and *a*) has been localized proximal, but close to, the fragile site by *in situ* hybridization and somatic cell hybridization (Fig. 29-3B). The factor IX gene is defined on the Human Gene Map as the *F9* locus. If the male I-1 in Fig. 29-3B is a normal transmitting carrier of the mutation, then all his daughters are obligate carriers and have all inherited allele *a* at the *F9* locus together with the disease.

In the third generation, the affected boys have also inherited allele *a*, whereas their normal brothers inherited allele *A*. However, if the female I-2 is the carrier, then her daughters must have inherited allele *A* with the mutation. For this to be the case, the offspring in generation III must all be recombinants, which is highly unlikely. It can therefore be concluded that individual I-1 has a high probability of being a normal transmitting male. If so, female II-1 is also a carrier even though she has not yet had any affected offspring and does not express the fragile site.

In many cases, such an extensive family history is not available and it may be impossible to give reliable information. Figure 29-4A gives an example of this. The daughter, II-1, and her obligatory carrier sister, II-2, have inherited different alleles of a hypothetical closely linked marker from their mother. No fragile sites were detected in the cells of either daughter. Since II-2 is homozygous, it is not possible to determine which allele (*A* or *a*) of the DNA marker is being coinherited with the mutation. If the mutation were inherited from the mother I-2 then the risk for the sister II-1 is small. In contrast, if I-1 is a normal transmitting male, then II-1 is an obligatory carrier. Figure 29-4B gives the risks of mental impairment in a model pedigree.

RFLP STUDIES IN X-LINKED MENTAL RETARDATION

One of the first studies of the fragile X syndrome using RFLPs created excitement because it suggested close linkage between the *F9* locus and the fragile X mutation (8). No recombinants were observed in 12 meioses (Fig. 29-3B). However, further studies of the same marker in other families showed a rather different pattern, in which the *F9* locus was only loosely linked to the mutation (9). Markers on the distal side of the fragile site, St14 and DX13, were found to be linked at a recombination fraction of approximately 0.12. Unlike the proximal markers, this genetic distance seems to be roughly the same in all families studied. A map of these markers is given in Fig. 29-5.

The apparent heterogeneity of the linkage between *F9* and the fragile site was first thought to be due to the mode of transmission (10). Families with normal transmitting males were thought to display less recombination than those with only female carriers. This was later found to be erroneous because some male transmission families fit the "high recombination class." The genetic studies suggest that

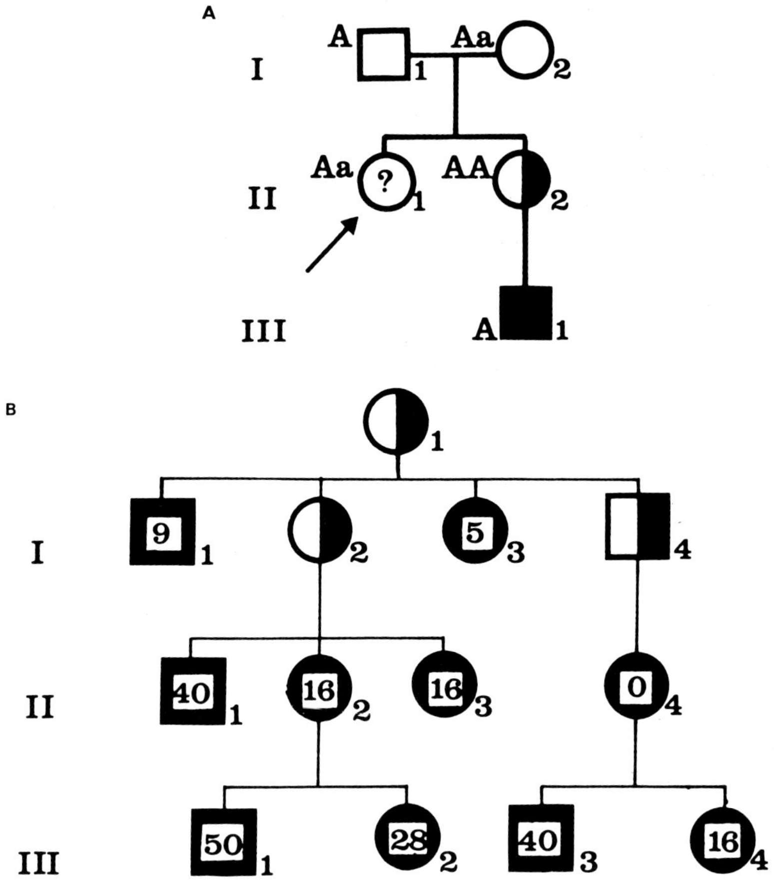

Fig. 29-4. (A) Pedigree where DNA and cytogenetic analysis are uninformative. (B) Risks of mental impairment in a model pedigree.

there is genetic heterogeneity although the data base is still small. The apparent heterogeneity may be related to the degree of expression of the fragile site in the affected males (11), an observation that remains to be substantiated.

The order of the markers and genetic distance from the fragile site (Fig. 29-6) have been deduced from studies in many laboratories. The closest proximal probes are cx55.7 and 4D8, which are linked at a recombination fraction of 0.05 (12). In addition, a third marker, probe λ135, was found to be close to the fragile site by *in situ* hybridization (13). The distal markers St14 and DX13 are linked at a recombination fraction of 0.12.

PRENATAL DIAGNOSIS AND CARRIER DETECTION

Even with very closely linked DNA markers the dilemma of the carrier status of II-1 illustrated in Fig. 29-4A cannot be resolved. These situations must await direct detection of the mutation. However, prenatal diagnosis is now possible in families

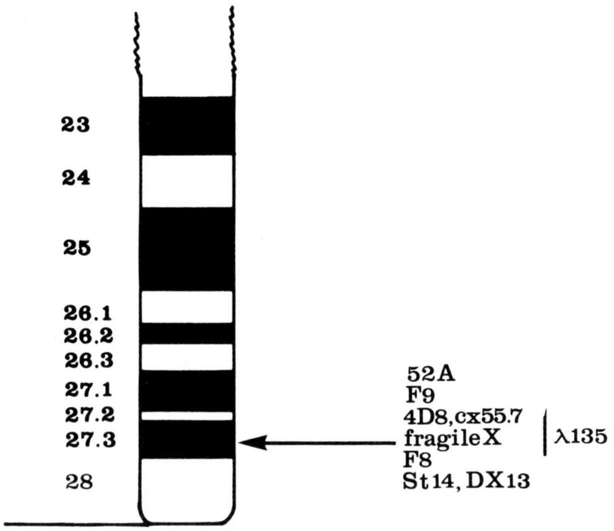

Fig. 29-5. Map showing the relative positions of RFLPs used for the genetic analysis of the fragile X syndrome.

in which the mode of transmission of the mutation is clear because, although the individual markers are not sufficient for accurate prediction, the use of bridging markers greatly enhances accuracy.

The use of either 4D8, cx55.7, or λ135 on the proximal side together with St14 or DX13 on the distal side should provide a diagnostic accuracy of greater than 95% if the female is heterozygous for the bridging markers. If cytogenetic evidence is also available, RFLP data can be combined to give good risk estimates. The principle of these calculations is similar to that used for bridging markers and serum creatine kinase levels in the diagnosis of Duchenne muscular dystrophy.

SOMATIC CELL HYBRIDS AND THE FRAGILE X

Somatic cell hybrids (see Bruns, Chapter 11) play an important role in the elucidation of the molecular basis of the fragile X syndrome. For example, one of the theories for the expression of the fragile site is the presence or absence of an autosomal repressor sequence. That theory could be tested by attempting to express the

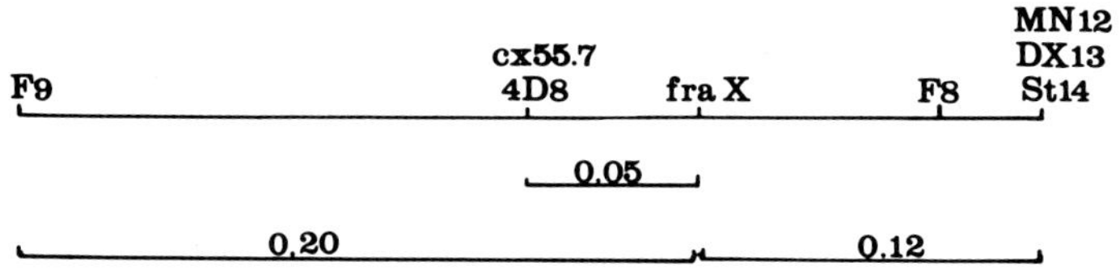

Fig. 29-6. Genetic map around the fragile X locus. The numbers refer to the recombination fraction between markers.

fragile site in hybrids with only the human X chromosome present in the absence of human autosomes. Such human–Chinese hamster hybrids have been generated from the X chromosome of an affected male (6). No other human chromosomes were present and yet the cells could be induced to express the fragile site. This experiment suggests that expression of the fragile X does not depend on other sequences in the human genome.

The same authors repeated the experiment with the X chromosome of a normal transmitting male and found little expression in the hybrid. Therefore, it is unlikely that there is an autosomal suppressor sequence because removal of other chromosomes did not increase the level of expression of the fragile site.

Somatic cell hybrids have also been used as a source of DNA probes from the region. Hybrids have been generated that contain reduced amounts of the human X chromosome translocated onto Chinese hamster chromosomes. One hybrid of this type contains Xq24-Xqter (14). The human component of these hybrids can be isolated by making a library of the total DNA and then screening the resultant clones with a human repetitive sequence (Fig. 29-7). A repetitive sequence, Alu, is often used because it is present in about 300,000 copies per genome, or once every 10 kb on average. Since each clone in a phage library contains 15–25 kb of human sequence, most of the human clones contain an Alu sequence. Positive clones can be selected from among the negative Chinese hamster clones; these should be derived from the Xq24-Xqter region. This can be verified by Southern blot analysis against hybrid cell lines that contain the distal part of the X chromosome.

Once DNA probes from the region have been identified, they can be used to evaluate RFLPs for genetic linkage analysis in affected families.

Somatic hybrids have also been used to demonstrate that the fragile site results

SOMATIC CELL HYBRID CONTAINING THE HUMAN X CHROMOSOME

▼

IRRADIATE

▼

CELLS WITH FRAGMENTED CHROMOSOMES

▼

FUSE WITH HGPRT– CHINESE HAMSTER CELL LINE

▼

SELECT FOR HYBRIDS EXPRESSING HGPRT

▼

MAKE DNA AND GENOMIC PHAGE LIBRARY

▼

SCREEN RECOMBINANTS WITH ALU REPEAT SEQUENCE

▼

PICK POSITIVE CLONES FROM TERMINAL REGION OF THE HUMAN X CHROMOSOME

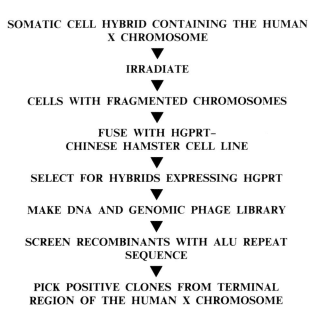

Fig. 29-7. Schematic diagram of the cloning of hybrid DNA to yield human sequences from a specific chromosomal region.

in chromosome breakage. Warren (15) constructed hybrids by fusion of lymphoblasts from normal and affected males with Chinese hamster cell lines that were hypoxanthine-guanine phosphoribosyltransferase (HGPRT) positive and glucose-6-phosphate dehydrogenase (G6PD) positive; The HGPRT gene is proximal to the fragile X locus and G6PD is distal. He then induced the hybrids for expression of the fragile X and examined the resulting hybrids. He found that a larger proportion of the hybrids expressing the fragile site had retained the HGPRT gene but lost the G6PD locus from their X chromosomes, whereas the normal X chromosome hybrids had retained both loci. Since G6PD lies distal to the fragile site, the pattern implies chromosome breakage in the region of the fragile site.

MICRODISSECTION

An alternative method of generating DNA sequences close to the fragile site is to microdissect the telomeric region of the long arm of the X chromosome (16). A micromanipulator is used to dissect the chromosomes in a metaphase spread on a coverslip. These are then collected under an oil drop where they are extracted with phenol and the DNA is cut with an appropriate restriction enzyme for cloning. Although only very small amounts of DNA (10 pg) are prepared in this way, several hundred clones can be obtained from the specific region of interest. The only limitation is that the chromosome must be recognizable in the karyotype without staining, because staining would damage the integrity of the DNA for cloning. For the fragile X region, it is possible to use metaphase spreads from patients with large dicentric X chromosomes; these chromosomes give two Xqter regions every time the chromosome is recognized. For other parts of the genome, there are now so many rearrangements in cell repositories that most regions would be accessible with this technology.

Once several hundred clones have been generated from the region Xq27-Xqter they can be used to construct a fine pulsed field gel electrophoresis map (often called a macrorestriction map) of the region (see Smith and Cantor, Chapter 12). Assuming that Xq27-Xqter contains about 15,000 kb of DNA, then 100 clones from dissection experiments should permit construction of a map with markers every 150 kb if they are randomly distributed. Such a normal map can then be used to determine the order of probes in fragile X patients. Female carriers would be the best individuals to study because they should show the normal pattern along with any changes due to the mutation.

DIRECT DETECTION OF THE MUTATION

In view of the expression of a fragile site closely associated with mental retardation, it is reasonable to assume that they are caused by the same, or closely linked, mutations. The mutation might be expected to be due to a deletion in some boys because a small percentage of patients suffering from X-linked disease show deletions of part or all of the relevant gene. Alternatively, the fragile X expression might be the result of the amplification of specific DNA sequences (17). Such changes will be detectable on conventional gels if the DNA probe lies close to the site of the mutation. However, even if close markers are not identified, it should be possible to detect such chromosomal changes by pulsed field gel electrophoresis (see Smith and Cantor, Chapter 12). If the disease turns out to be highly heterogeneous and to cover a large locus, like Duchenne muscular dystrophy (see Kunkel, Chapter 27), it might be necessary to use pulsed field gel electrophoresis as a diagnostic technique.

ACKNOWLEDGMENTS

I would like to thank Dr. Patterson for reading the manuscript and Dr. Ursula Froster-Iskenius for Fig. 29-2. I am very grateful for support for the study of X-linked diseases from the Medical Research Council of Great Britain, The Muscular Dystrophy Association, and the Muscular Dystrophy Group of Great Britain. I thank Rachel Kitt for typing the manuscript.

REFERENCES

1. McKusick V. Mendelian Inheritance in Man. Baltimore: Johns Hopkins University Press, 1986.
2. Herbst DS, Miller JR. Non-specific X-linked mental retardation. II. The frequency in British Columbia. Am J Med Genet 1980; 7:461–469.
3. Webb TP, Bundey S, Thake A, Todd J. The frequency of the fragile X chromosome among schoolchildren in Coventry. J Med Genet 1986; 23:396–399.
4. Sutherland GR, Hecht F. Fragile Sites on Human Chromosomes. New York: Oxford University Press, 1985.
5. Harrison CJ, Jack EM, Allen TD, Harris R. The fragile X: A scanning electron microscope study. J Med Genet 1983; 20:280–285.
6. Ledbetter DH, Airhart SD, Nussbaum RL. Somatic cell hybrid studies of fragile (X) expression in a carrier female and transmitting male. Am J Med Genet 1986; 23:429–444.
7. Sherman SL, Jacobs PA, Morton NE, et al. Further segregation analysis of the fragile X syndrome with special reference to transmitting males. Hum Genet 1985; 69:289–299.
8. Camerino G, Mattei MG, Mattei JF, Jaye M, Mandel JL. Close linkage of fragile X mental retardation syndrome to hemophilia B and transmission through a normal male. Nature (London) 1983; 306:701.
9. Goodfellow PN, Davies KE, Ropers H-H. Report of the committee on the genetic constitution of the X and Y chromosomes. Cytogenet Cell Genet 1985; 40:296–352.
10. Brown WT, Gross AC, Chan CB, Jenkins EC. Genetic linkage heterogeneity in the fragile X syndrome. Hum Genet 1985; 71:11–18.
11. Brown WT, Jenkins, EC, Gross AC, et al. Further evidence for genetic heterogeneity in the fragile X syndrome. Hum Genet 1987; 75:311–321.
12. Brown WT, Wu Y, Gross AC, Chan CB, Dobkin CS, Jenkins EC. RFLP for linkage analysis of fragile X syndrome. Lancet 1987; 1:280.
13. Mulligan LM, Holden JJA, Duncan AMV, Poon R, White BN. Isolation of a DNA segment from the fragile region, at Xq27.3, involved in the fragile X form of mental retardation. Science 1986; in press.
14. Nussbaum RL, Airhart SD, Ledbetter DH. A rodent-human hybrid containing Xq24-qter translocated to a hamster chromosome expresses the Xq27 folate-sensitive fragile site. Am J Med Genet 1986; 23:457–466.

15. Warren ST. Genetic evidence for chromosome breakage at the fragile X site. 7th Int Cong Hum Genet, Berlin, 7th. 1986; 217 (Abstr.).

16. Brown SDM. Mapping mammalian chromosomes: New techniques for old problems. Trends Genet 1985; 1:219–220.

17. Nussbaum RL, Airhart SD, Ledbetter DH. Recombination and amplification of pyrimidine-rich sequences may be responsible for initation and progression of the Xq27 fragile site. Am J Med Genet 1986; 23:715–721.

18. Froster-Uskenius U, Schulze A, Schwinger E. Transmission of the marker X syndrome trait by unaffected males: Conclusions from studies of large families. Human Genet 1984; 67:419–427.

FURTHER READING

Sutherland GR, Hecht F. Fragile Sites on Human Chromosomes. New York: Oxford University Press, 1985.

Nussbaum RL, Ledbetter DH. Fragile X syndrome: A unique mutation in man. Ann Rev Genet 1986; 20:109–145.

X-Linked mental retardation 2. Am J Med Genet 1986; 23 (special issue).

30

Other X-Linked Neuromuscular Diseases

KENNETH H. FISCHBECK

Research in Duchenne muscular dystrophy has shown what can be accomplished with a systematic approach to the isolation and characterization of the genetic defect responsible for an X-linked neuromuscular disease. Gene isolation offers at least three advantages in the understanding and management of a disease. First, it can help to identify the disease unambiguously (although we must be able to define a disease clinically to some extent *before* the gene can be isolated). Second, isolation of the gene makes it possible to identify the gene defect in asymptomatic or presymptomatic carriers, and to provide genetic counseling in time to prevent the disease from being passed to later generations. Finally, gene isolation offers the opportunity to determine the pathogenesis of the disease and to develop rational treatment.

Several neuromuscular diseases are suitable for the application of the same techniques that have been successful in pursuing the Duchenne gene. Neuromuscular diseases cause progressive weakness due to degeneration of skeletal muscle, peripheral nerve, or motor nerve cells. Each neuromuscular disease affects a specific and highly specialized population of cells. The anatomy, biochemistry, and physiology of these cells have been thoroughly studied both *in situ* and in tissue culture, so that we are in a good position to understand the role of an identified genetic defect in cellular metabolism.

It is worthwhile to consider *X-linked* neuromuscular diseases because the characteristic pattern of inheritance solves the first problem in gene isolation, determining the chromosome on which the gene is located. Recessive disorders due to defects on the X-chromosome are expressed much more commonly than disorders on other chromosomes because males may be *hemizygous;* they have only one X-chromosome and, thus, no normal allele to compensate for the defective gene.

Because of the work on Duchenne dystrophy and other X-linked disorders, the X-chromosome has been covered with probes for DNA polymorphisms that can be used in genetic linkage studies. About 200 such markers were listed as of the human gene mapping conference in 1986 (1). Each X-linked disease gene that is isolated and characterized, such as the genes for Lesch–Nyhan disease, ornithine transcarbamylase deficiency, hemophilia, male color blindness, chronic granulomatous disease, and Duchenne dystrophy, brings us closer to having a detailed map of the entire human X-chromosome, which is probably about 200 million

DNA base pairs long and may contain more than 1000 genes. The Duchenne gene alone will fill in nearly 1% of this map. The goal of completely mapping the human genome will certainly be accomplished first for the X-chromosome.

Finally, understanding and identifying genetic defects in X-linked varieties of human neuromuscular disease may help us understand the autosomal and sporadic varieties of these diseases that may be more common. For example, characterizing the gene defect in X-linked neuropathy may help in understanding the more common autosomal dominant Charcot-Marie-Tooth disease. Isolating the gene for X-linked spinal muscular atrophy may give clues to the pathogenesis of amyotrophic lateral sclerosis, a sporadic form of motor neuron disease of much greater prevalence.

Here I will discuss three other X-linked neuromuscular diseases in which progress has been made in gene localization: Emery–Dreifuss muscular dystrophy (a disease of muscle), X-linked neuropathy (a disease of peripheral nerve), and X-linked, adult-onset spinal muscular atrophy (a disorder of motor nerve cells).

EMERY-DREIFUSS MUSCULAR DYSTROPHY

Duchenne muscular dystrophy has a typical clinical presentation and course that have been recognized for almost 120 years. Only boys are affected. Symptomatic weakness of proximal limb muscles starts in early childhood. The weakness progresses until the boy is wheelchair bound around age 10 and dies about age 20. Becker dystrophy, first described in the 1950s, is a milder X-linked disease, but apart from the difference in age at onset and severity the clinical picture is similar to Duchenne dystrophy: the weakness affects the proximal muscles of the legs first and then the arms. Both diseases cause hypertrophy of calf muscles, marked elevation of serum creatine kinase levels, and, on muscle biopsies, muscle fiber degeneration and fibrosis.

In 1966 Emery and Dreifuss reported a Virginia family with a different form of X-linked muscular dystrophy (2). That family, reported several years previously as a variant of Duchenne, actually showed features that are not seen in Duchenne dystrophy. Subsequent reports of additional families from the United States and Europe clearly defined the disease (3–6).

The course of Emery–Dreifuss muscular dystrophy is milder than Duchenne and more like Becker dystrophy. Weakness begins in childhood but progresses so slowly that the affected men can still walk at age 50. Contractures are a prominent feature, particularly at the elbows, ankles, and neck. The contractures are out of proportion to the weakness and are often the first symptom. The pattern of weakness differs from the proximal muscle accentuation of Duchenne or Becker dystrophy; here the muscles of the upper arms and lower legs are most affected, a so-called humeroperoneal pattern. There is also no calf hypertrophy. Nearly every affected patient has a severe cardiac arrhythmia, usually bradycardia due to atrial failure, and this may be life threatening.

Laboratory evaluations of patients with Emery–Dreifuss muscular dystrophy show much less elevation of serum creatine kinase than the levels in Duchenne dystrophy. Electromyography and muscle biopsies usually show signs of myopathy

Table 30-1 Pooled Linkage Results for Emery–Dreifuss Muscular Dystrophy (Lod Scores)[a]

Probe (marker)	0.00	0.05	0.10	0.20	0.30	0.40
pERT87 (DXS164)	—	−6.47	−3.59	−1.21	−0.28	0.01
754 (DXS84)	—	−1.01	−0.34	0.09	0.14	0.08
F9	—	−4.07	−2.46	−1.06	−0.42	−0.10
F8	7.75	7.15	6.52	5.15	3.63	1.90
DX13 (DXS15)	—	5.82	6.09	5.29	3.76	1.84
St14	—	9.52	9.45	8.02	5.86	3.19

[a]Data from Hodgson et al. (9), Yates et al. (10), and Thomas et al. (11).

but without the active muscle fiber degeneration and regeneration of Duchenne dystrophy, and there may be signs of denervation such as fibrillations and muscle fiber atrophy. Preferential atrophy of type I muscle fibers has been reported.

At least 10 families have been reported with Emery–Dreifuss muscular dystrophy in a characteristic X-linked recessive pattern of inheritance (2–6). Obligate heterozygous females in these families have been asymptomatic, and clinical evaluations of the cardiac and skeletal muscles of the carriers have been negative. However, autosomal dominant transmission of a similar disease has also been reported (7,8); so more than one gene defect may lead to the same disease phenotype.

In one family, X-linked Emery–Dreifuss muscular dystrophy and color blindness segregated together: every male with Emery–Dreifuss was also color blind (3). That pattern indicated that the gene for Emery–Dreifuss muscular dystrophy is close to the gene for male color blindness; that gene has been isolated and localized to the distal long arm of the X-chromosome. Tight linkage of the Emery–Dreifuss gene has been found with DNA polymorphisms identified by three probes: clotting factor VIII (the gene that is defective in hemophilia A), DX13 (DXS15), and St14 (DXS52) (9–11). The loci identified by DX13 and St14 are at or near the tip of the long arm of the X-chromosome (Fig. 30-1A). The analysis has also excluded close linkage to clotting factor IX and short arm markers L1.28 (DXS7), 754 (DXS84), and pERT87 (DXS164) (Table 30-1). Therefore, Emery–Dreifuss muscular dystrophy is genetically as well as clinically distinct from Duchenne and Becker dystrophies.

X-linked muscular dystrophies in animals might be models of Duchenne dystrophy (12,13). If these disorders are not genetically homologous with Duchenne, they might turn out to be homologous with Emery–Dreifuss muscular dystrophy. Because there are so few human families with Emery–Dreifuss, identification of an equivalent animal disease could be the most productive way to isolate and characterize the defective gene.

X-LINKED NEUROPATHY

About 100 years ago Jean Charcot and Pierre Marie, in France, and Howard Tooth in England independently described the hereditary neuropathy that bears their names. Progressive degeneration of peripheral nerve leads to weakness, muscle atrophy, and sensory loss in the lower legs and hands. The weakness often causes

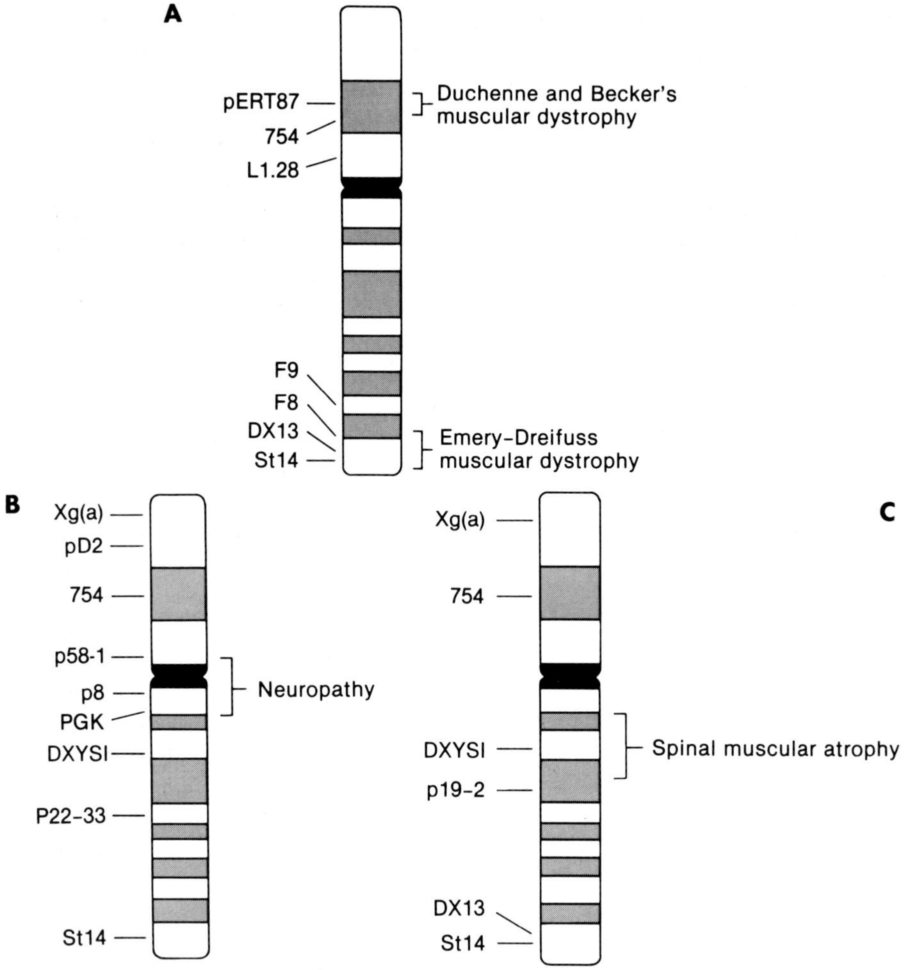

Fig. 30-1. Diagrams (A,B,C) of the human X-chromosome, showing the relative locations of genetic markers and the genes for the X-linked neuromuscular diseases discussed in this chapter.

progressive foot deformity, with high arches and hammertoes. Tendon reflexes are lost. The disease generally affects members of each generation in a family, and is inherited in an autosomal dominant pattern.

Shortly after the original description of Charcot-Marie-Tooth disease, Herringham described a family with the same clinical manifestations but a different pattern of inheritance (14). That family included 19 affected males in five generations and no affected females. Nine women who had affected fathers and affected sons were themselves asymptomatic. As Herringham said:

> . . . that the women of this family, themselves even uncommonly buxom and healthy, should be able to . . . transmit to the males alone tissues unlike their own, and endowed with a regular form of weakness which they do not themselves possess, is . . . marvellous. It seems as if the daughter of the diseased father carried from the beginning of her life ova of two sexes, the female healthy, the male containing within it the representation of the father's disease.

We have since come to recognize that pattern of inheritance as X-linked recessive, indicative of a gene defect on the X-chromosome. The pedigree reported by Herringham differed from X-linked inheritance in one detail, however, because there was apparently father-to-son transmission in a cousin of the proband. An X-linked disease should not be passed from father to son because the father donates only a Y-chromosome, not an X-chromosome to his sons. In Herringham's family, the discrepancy could have been due to a different disease in the affected son, such as poliomyelitis, or to an allele inherited from his mother; male-to-male transmission did not occur in 10 other opportunities in that family.

Although there have been other reports of X-linked neuropathy since Herringham's paper (15–22), it was uncertain whether these families had an *X-linked* disease or *sex-limited* expression of an autosomal dominant disorder (23). Rozear et al. (24) provided convincing evidence that the gene defect was located on the X-chromosome in a 50-year follow-up study of a North Carolina family that had originally been described by Allan (15). Using Allan's records, the investigators identified more than 800 members of an extended pedigree covering 9 generations. They found no male-to-male transmission in 28 chances, making autosomal dominant transmission exceedingly unlikely (about 1 chance in 250 million).

X-linked neuropathy families have been variously reported as X-linked "recessive" (15,16,18), X-linked "recessive with partial manifestations in female carriers" (20), X-linked "incompletely dominant" (21), or X-linked "dominant" (17,19,22); it is likely that all these terms refer to the same disease, in which heterozygous females may or may not be symptomatic. Affected males are uniformly symptomatic, with distal weakness and sensory loss beginning in late childhood or adolescence and leading to moderate disability in the third decade. Nerve conduction studies show moderate to marked slowing of conduction in affected males, but only mild slowing or normal conduction in heterozygous females (22,24). In general, nerve conduction values have been 25–40 m/second; the values lie between the marked slowing of the hypertrophic demyelinating (type I) hereditary neuropathy and the mild slowing of the neuronal (type II) form. A sural nerve biopsy in one patient showed onion-bulb formation, consistent with primary demyelination (24).

It is difficult to determine the proportion of families with hereditary neuropathy that have the X-linked form of the disease. Skre estimated that in Western Norway X-linked recessive neuropathy had a prevalence of 3.6/100,000, one-tenth as common as autosomal dominant Charcot-Marie-Tooth disease (25). In Ontario a family history indicated X-linked inheritance in one-third of the hereditary neuropathy families and the pattern was consistent with X-linked inheritance in another one-third (26). The largest X-linked families reported have been of Northern and Western European ancestory; the disease may be more common in these ethnic groups.

The genetic defect in X-linked hereditary neuropathy has been localized to the region of the centromere of the X-chromosome (Fig. 30-1B). Linkage studies done before DNA probes were available indicated that the disorder is not linked to the Xg(a) blood type, on the distal short arm (19), or to color blindness, on the distal long arm (21). Studies in three laboratories, including our own, using probes for DNA polymorphisms on the X-chromosome, disclosed significant linkage to the marker DXYS1 on the proximal long arm (27–29). We also found linkage to p58-1 (DXS14) on the proximal short arm. We excluded close linkage to pD2 (DXS43) and 754 (DXS84) on the distal short arm and p22-33 (DXS11) and St14 (DXS52) on the distal long arm (Table 30-2). We have also done multipoint linkage analysis, which placed the neuropathy gene defect between p58-1 and DXYS1, near the X-chromosome centromere (30). Further analysis showed close linkage to p8 (DXS1) and PGK, which are between DXYS1 and the centromere. None of these markers showed perfect linkage to the neuropathy gene in the families we have been studying, however.

Actually isolating the neuropathy gene will probably require finding a patient or family with a major structural alteration in this region, such as a chromosomal deletion or translocation. A family with neuropathy, deafness, and mental retardation (31) may be useful for that purpose. Although the neuropathy in this family is axonal rather than demyelinating, our linkage analysis placed the gene defect in the same region of the chromosome. As with Duchenne dystrophy patient "BB" (32), the association of more than one X-linked disease in the family may indicate a microdeletion that could be used to isolate the gene for X-linked neuropathy.

X-LINKED, ADULT-ONSET SPINAL MUSCULAR ATROPHY

In 1968, Kennedy et al. described two familes with adult-onset spinal muscular atrophy in an X-linked recessive pattern (33). The disease had been described previously, including one family with three affected brothers (34), but Kennedy et al. were the first to emphasize the X-linked pattern of inheritance. The disorder has now been described in more than 20 families (33–45), and a characteristic clinical picture has emerged.

Symptoms begin much later in other hereditary spinal muscular atrophies, such as Werdnig–Hoffmann and Kugelberg–Welander diseases. Affected men usually note proximal arm and leg weakness at about age 30. Examination shows dysarthria, limb weakness, muscle atrophy, and fasciculations, but there are no upper motor neuron signs. Many affected men have gynecomastia, which is sometimes the earliest manifestation of the disease. Decreased fertility has also been reported. The weakness progresses for many years. Most patients have increasing dysphagia and breathing difficulty; they may die by age 55 of recurrent aspiration. In some

Table 30-2 Pooled Linkage Results for X-Linked Neuropathy (Lod Scores)[a]

Probe (marker)	0.00	0.05	0.10	0.20	0.30	0.40
Xg(a)	—	−4.09	−2.16	−0.70	−0.19	−0.04
pD2 (DXS43)	—	−2.94	−1.85	−0.86	−0.39	−0.13
754 (DXS84)	—	−4.11	−2.47	−1.00	−0.33	−0.04
p58-1 (DXS14)	—	3.61	3.42	2.69	1.77	0.80
PGK	—	4.09	3.99	3.19	2.15	1.10
p8 (DXS1)	—	2.35	2.65	2.30	1.58	0.78
DXYS1	—	6.14	6.68	5.77	3.91	1.82
p22-33 (DXS11)	—	−3.39	−2.13	−1.02	−0.49	−0.18
St14 (DXS52)	—	−2.28	−1.23	−0.34	0.00	0.09

[a]Data from Gal et al. (27), Beckett et al. (28), Fischbeck et al. (29), and additional unpublished results.

families the disease is milder, with onset after age 40 or 50 and survival after age 60 or 70.

Unlike X-linked neuropathy, this motor neuron disease follows a true X-linked *recessive* pattern; women who are obligate carriers of the disease gene (mothers and daughters of affected men) are asymptomatic and show no abnormalities on examination. Because the onset is late and because one or more generations may seem to be spared because carriers are asymptomatic, the family history may be remote or unobtainable (41). There has been no father-to-son transmission of the disease, although there have been few opportunities for this to occur.

We have carried out linkage studies in nine families with X-linked, adult-onset spinal muscular atrophy (45). We found significant linkage to DXYS1, the same proximal long arm marker that is linked to the neuropathy gene. We could exclude close linkage to short arm markers Xg(a) and 754 (DXS84). Another proximal long arm marker, p19-2 (DXS3), gave a weakly positive score for linkage (Table 30-3). This analysis places the spinal muscular atrophy gene near DXYS1 on the proximal long arm of the X-chromosome (Fig. 30-1C).

There may be a clue to the nature of the gene defect in the endocrine manifestations (gynecomastia and decreased fertility) that are often associated with this disease. There have been at least five endocrinologic studies (38,42–44,46). Testosterone levels have been normal or only mildly decreased. Follicle-stimulating hormone and leutinizing hormone levels are normal, and the response to human chorionic gonadotropin may be blunted. Testicular biopsies showed failure of germinal cell maturation with variable Sertoli cell involvement (42,43,46). Although different interpretations have been offered for these findings, the consensus is that the gynecomastia and testicular failure are probably caused by end-organ unresponsiveness to androgen, perhaps due to a defect in the androgen receptor (43).

Androgen receptors are concentrated in motor neurons of the brainstem and spinal cord (47) and an abnormality of androgen receptors has been postulated to play a role in the pathogenesis of amyotrophic lateral sclerosis (48). The testicular feminization syndrome *(Tfm)* of humans and mice is an X-linked disorder with a demonstrable defect in androgen receptors. Migeon et al. (49) used hybrid *Tfm* cells to localize the gene for the human androgen receptor (49); they placed the gene near the X-chromosome centromere, in the same region as the gene for X-linked adult-onset spinal muscular atrophy.

An androgen receptor derangement could account for both the motor neuron degeneration and the gynecomastia and testicular failure seen in X-linked spinal

Table 30-3 Pooled Linkage Results for X-Linked Spinal Muscular Atrophy (Lod Scores)[a]

Probe (marker)	0.00	0.05	0.10	0.20	0.30	0.40
Xg(a)	—	−2.59	−1.54	−0.64	−0.26	−0.07
p754 (DXS84)	—	−1.21	−0.51	−0.01	0.12	0.10
DXYS1	—	3.12	2.83	1.91	0.88	0.10
p19-2(DXS3)	—	0.97	1.08	0.90	0.62	0.31
DX13 (DXS15)	—	−0.96	−0.64	−0.33	−0.16	−0.06
St14 (DXS52)	—	−0.28	−0.10	−0.01	0.00	−0.01

[a]Data from Fischbeck et al. (45) and additional unpublished results.

muscular atrophy. Studies are now in progress to determine whether receptor abnormality can be identified.

The work presented here demonstrates the value of genetic linkage studies in neuromuscular disease. In Emery–Dreifuss muscular dystrophy, genetic linkage has helped to identify the disease and to confirm that it differs from Duchenne and Becker dystrophies. In hereditary neuropathy, linkage helped to prove that there is an X-linked form of the disease. In X-linked spinal muscular atrophy, gene localization by linkage analysis has suggested a possible pathogenesis for the disease. In each of these disorders, the linkage studies have provided markers that can be used for genetic counseling and have also brought us closer to identifying the primary genetic defect.

ACKNOWLEDGMENTS

The author's work is supported in part by grants from the Muscular Dystrophy Association, the March of Dimes Birth Defects Foundation, and the National Institutes of Health (NS08075 and NS00695).

REFERENCES

1. Goodfellow PN, Davies KE, Ropers HH. Report of the committee on the genetic constitution of the X and Y chromosomes. Cytogenet Cell Genet 1985; 40:296–352.

2. Emery AEH, Dreifuss FE. Unusual type of benign X-linked muscular dystrophy. J Neurol Neurosurg Psychiat 1966; 29:338–342.

3. Thomas PK, Calne DB, Elliott CF. X-Linked scapuloperoneal syndrome. J Neurol Neurosurg Psychiat 1972; 35:208–215.

4. Rowland LP, Fetell M, Olarte M, Hays A, Singh N, Wanat FE. Emery-Dreifuss muscular dystrophy. Ann Neurol 1979; 5:111–117.

5. Hopkins LC, Jackson JA, Elsas LJ. Emery-Dreifuss humeroperoneal muscular dystrophy: An X-linked myopathy with unusual contractures and bradycardia. Ann Neurol 1981; 10:230–237.

6. Merlini L, Granata C, Dominici P, Bonfiglioli S. Emery-Dreifuss muscular dystrophy: Report of five cases in a family and review of the literature. Muscle Nerve 1986; 9:481–485.

7. Miller RG, Layzer RB, Mellenthin MA, Golabi M, Francoz RA, Mall JC. Emery-Dreifuss muscular dystrophy with autosomal dominant transmission. Neurology 1985; 35:1230–1233.

8. Gilchrist JM, Leshner RT. Autosomal dominant humeroperoneal myopathy. Arch Neurol 1986; 43:734–735.

9. Hodgson S, Boswinkel E, Cole C, et al. A linkage study of Emery-Dreifuss muscular dystrophy. Hum Genet 1986; 74:409–416.

10. Yates JRW, Affara NA, Jamieson DM, et al. Emery-Dreifuss muscular dystrophy: Localization to Xq27.3-qter confirmed by linkage to the factor VIII gene. J Med Genet 1986; 23:587–590.

11. Thomas NST, Williams H, Elsas LJ, Hopkins LC, Sarfarazi M, Harper PS. Localization of the gene for Emery-Dreifuss muscular dystrophy to the distal long arm of the X chromosome. J Med Genet 1986; 23:596–598.

12. Bulfield G, Siller WG, Wight PAL, Moore KJ. X chromosome-linked muscular dystrophy (mdx) in the mouse. Proc Natl Acad Sci USA 1984; 81:1189–1192.

13. Valentine BA, Cooper BJ, Cummings JF, deLahunta A. Progressive muscular dystrophy in a golden retriever dog: Light microscope and ultrastructural features at 4 and 8 months. Acta Neuropathol 1986; 71:301–310.

14. Herringham WP. Muscular atrophy of the peroneal type affecting many members of a family. Brain 1888; 11:230–236.

15. Allan W. Relation of hereditary pattern to clinical severity as illustrated by peroneal atrophy. Arch Intern Med 1939; 63:1123–1131.

16. Erwin WG. A pedigree of sex-linked recessive peroneal atrophy. J Hered 1944; 35:24–26.

17. Woratz G. Neurale Muskelatrophie mit dominantem X-chromosomalem Erbgang. Abh Dtsch Akad Wissensch Berl 1964; Nr 2.

18. Campeanu E, Morariu M. Les relations entre génotype et phénotype dans la maladie de Charcot-Marie-Tooth. Rev Roum Neurol 1970; 7:47–56.

19. de Weerdt CJ. Charcot-Marie-Tooth disease with sex-linked inheritance, linkage studies and abnormal serum alkaline phosphatase levels. Eur Neurol 1978; 17:336–344.

20. Fryns JP, Ven Den Berghe H. Sex-linked recessive inheritance in Charcot-Marie-Tooth disease with partial clinical manifestations in female carriers. Hum Genet 1980; 55:413–415.

21. Iselius L, Grimby L. A family with Charcot-Marie-Tooth's disease, showing a probable X-linked incompletely dominant inheritance. Heriditas 1982; 97:157–158.

22. Phillips LH, Kelly TE, Schnatterly P, Parker D. Hereditary motor-sensory neuropathy (HMSN): Possible X-linked dominant inheritance. Neurology 1985; 35:498–502.

23. Harding AE, Thomas PK. Genetic aspects of hereditary motor and sensory neuropathy (types I and II). J Med Genet 1980; 17:329–336.

24. Rozear MP, Pericak-Vance MA, Fischbeck KH, et al. Hereditary motor and sensory neuropathy, X-linked. Neurology 1987; 37:1460–1465.

25. Skre H. Genetic and clinical aspects of Charcot-Marie-Tooth's disease. Clin Genet 1974; 6:98–118.

26. Beckett J, personal communication.

27. Gal A, Mucke J, Theile H, Ropers HH, Wienker TF. X-Linked dominant Charcot-Marie-Tooth disease: Suggestion of linkage with a cloned DNA sequence from the proximal Xq. Hum Genet 1985; 70:38–42.

28. Beckett J, Holden JJA, Simpson NE, White BN, MacLeod PM. Localization of X-linked dominant Charcot-Marie-Tooth disease (CMT 2) to Xq13. J Neurogenet 1986; 3:225–231.

29. Fischbeck KH, ar-Rushdi N, Pericak-Vance M, Rozear M, Roses AD, Fryns JP. X-Linked neuropathy: Gene localization with DNA probes. Ann Neurol 1986; 20:527–532.

30. Fischbeck KH, Tirschwell D, Ritter A, Pericak-Vance M, Rozear M. Multipoint linkage analysis of X-linked neuropathy and markers near the X chromosome centromere. Am J Hum Genet 1986; 39:A154 (abstr.).

31. Cowchock FS, Duckett SW, Streletz LJ, Graziani LJ, Jackson LG. X-Linked motor-sensory neuropathy type II with deafness and mental retardation: A new disorder. Am J Med Genet 1985; 20:307–315.

32. Francke U, Ochs HD, deMartinville B, et al. Minor Xp21 chromosome deletion in a male associated with expression of Duchenne muscular dystrophy, chronic granulomatous disease, retinitis pigmentosa, and McLeod syndrome. Am J. Hum Genet 1985; 37:250–267.

33. Kennedy WR, Alter M, Sung JH. Progressive proximal spinal and bulbar muscular atrophy of late onset: A sex-linked recessive trait. Neurology 1968; 18:671–680.

34. Magee KR. Familial progressive bulbar-spinal muscular atrophy. Neurology 1960; 10:295–305.

35. Tsukagoshi H, Shoji H, Furukawa T. Proximal neurogenic muscular atrophy in adolescence and adulthood with X-linked recessive inheritance: Kugelberg-Welander disease and its variant of late onset in one pedigree. Neurology 1970; 20:1183–1193.

36. Stefanis C. Papapetropoulos T, Scarpalezos S, Lygidakis G, Panayiotopoulos CP. X-Linked spinal and bulbar muscular atrophy of late onset: A separate type of motor neuron disease? J Neurol Sci 1975; 24:493–503.

37. Ringel SP, Lava NS, Treihaft MM, Lubs ML, Lubs HA. Late-onset X-linked recessive spinal and bulbar muscular atrophy. Muscle Nerve 1978; 1:297–307.

38. Schoenen J, Delwaide PJ, Legros JJ, Franchimont P. Motorneuropathie héréditaire: La forme proximale de l'adulte liée au sexe (ou maladie de Kennedy). J Neurol Sci 1979; 41:343–357.

39. Paulson GW, Liss L, Sweeney PJ. Late onset spinal muscular atrophy: A sex linked variant of Kugelberg-Welander. Acta Neurol Scand 1980; 61:49–55.

40. Barkhaus PE, Kennedy WR, Stern LZ, Harrington RB. Hereditary proximal spinal and bulbar motor neuron disease of late onset: A report of six cases. Arch Neurol 1982; 39:112–116.

41. Harding AE, Thomas PK, Baraitser M, Bradbury PG, Morgan-Hughes JA, Ponsford JR. X-Linked recessive bulbospinal neuronopathy: A report of ten cases. J Neurol Neurosurg Psychiat 1982; 45:1012–1019.

42. Hausmanowa-Petrusewicz I, Borkowska J, Janczewski Z. X-Linked adult form of spinal muscular atrophy. J Neurol 1983; 229:175–188.

43. Arbizu T, Santamaria J, Gomex JM, Quilez A, Serra JP. A family with adult spinal and bulbar muscular atrophy, X-linked inheritance and associated testicular failure. J Neurol Sci 1983; 59:371–382.

44. Guidetti D, Motti L, Marcello N, et al. Kennedy disease in an Italian kindred. Eur Neurol 1986; 25:188–196.

45. Fischbeck KH, Ionasescu V, Ritter AW, et al. Localization of the gene for X-linked spinal muscular atrophy. Neurology 1986; 36:1595–1598.

46. Imai H, Beppu H, Uono M. Endocrinological investigation in patients with progressive proximal spinal and bulbar muscular atrophy of late onset. (Kennedy-Alter-Sung type). Clin Neurol 1980; 20:704–712.

47. Sar M, Stumpf WE. Androgen concentration in motor neurons of cranial nerves and spinal cord. Science 1977; 197:77–80.

48. Weiner LP. Possible role of androgen receptors in amyotrophic lateral sclerosis: A hypotheis. Arch Neurol 1980; 37:129–131.

49. Migeon BR, Brown TR, Axelman J, Migeon CJ. Studies of the locus for androgen receptor: Localization on the human X chromosome and evidence for homology with the *Tfm* locus in the mouse. Proc Natl Acad Sci USA 1981; 78:6339–6343.

IV

SOCIAL POLICY AND MOLECULAR GENETICS

31

Gene Replacement Therapy: The Example of Lesch–Nyhan Syndrome

GRANT R. MACGREGOR, DAVID L. NELSON,
STEPHEN M. W. CHANG, AND C. THOMAS CASKEY

Hypoxanthine-guanine phosphoribosyltransferase (HPRT; EC 2.4.2.8) catalyzes the transfer of 5′-ribose phosphate from 5′-phosphoribosyl-1-pyrophosphate (PRPP) to the free bases hypoxanthine or guanine to form the nucleotides 5′-IMP or 5′-GMP, respectively. HPRT cDNAs for the human (1), mouse (2), and Chinese hamster (3) genes have been cloned and the fine structure of the human genomic locus has been determined (4). In humans, lack of HPRT results in different clinical disorders; the severity of each condition is related to the extent of the enzyme deficiency.

Complete absence of HPRT activity results in the Lesch–Nyhan (L–N) syndrome (5), an X-linked disease that usually causes symptoms in the first year of life: mental retardation, self-injurious behavior, spasticity, and choreoathetosis. Less severe enzyme deficiency leads to gouty arthritis, a common disease of adults (6). The adults develop hyperuricemia and uricosuria but are of normal intelligence and account for 5% of people with *de novo* purine overproduction (7).

The pathogenesis of neural dysfunction in Lesch–Nyhan syndrome is poorly understood. Watts et al. found elevated levels of HPRT activity in the cerebral cortex, basal ganglia, cerebellum, and medulla but not in the spinal cord of 8-week-old rats (8). *De novo* purine biosynthesis is lower in brain than in liver (9) and levels of amidophosphoribosyltransferase (AMPRT), the rate-limiting enzyme of the *de novo* pathway, are lower in brain than in other organs and the ratio of salvage pathway/*de novo* synthesis is higher in brain than in other tissues (9). This pattern suggests that brain and perhaps basal ganglia cells in particular are exceptionally dependent on hypoxanthine "salvage" via HPRT. In Lesch–Nyhan patients oxypurine levels are elevated 2- to 3-fold in serum and CSF but oxypurines alone are probably not responsible for the cerebral disorder because the neurologic symptoms are unaffected by allopurinol (a xanthine oxidase inhibitor). Moreover, patients with xanthine oxidase deficiency also have high serum levels of oxypurines, but they have no neurological symptoms (10).

We are attempting to develop a novel treatment for HPRT deficiency by introducing new genetic information into somatic tissues. This would complement the

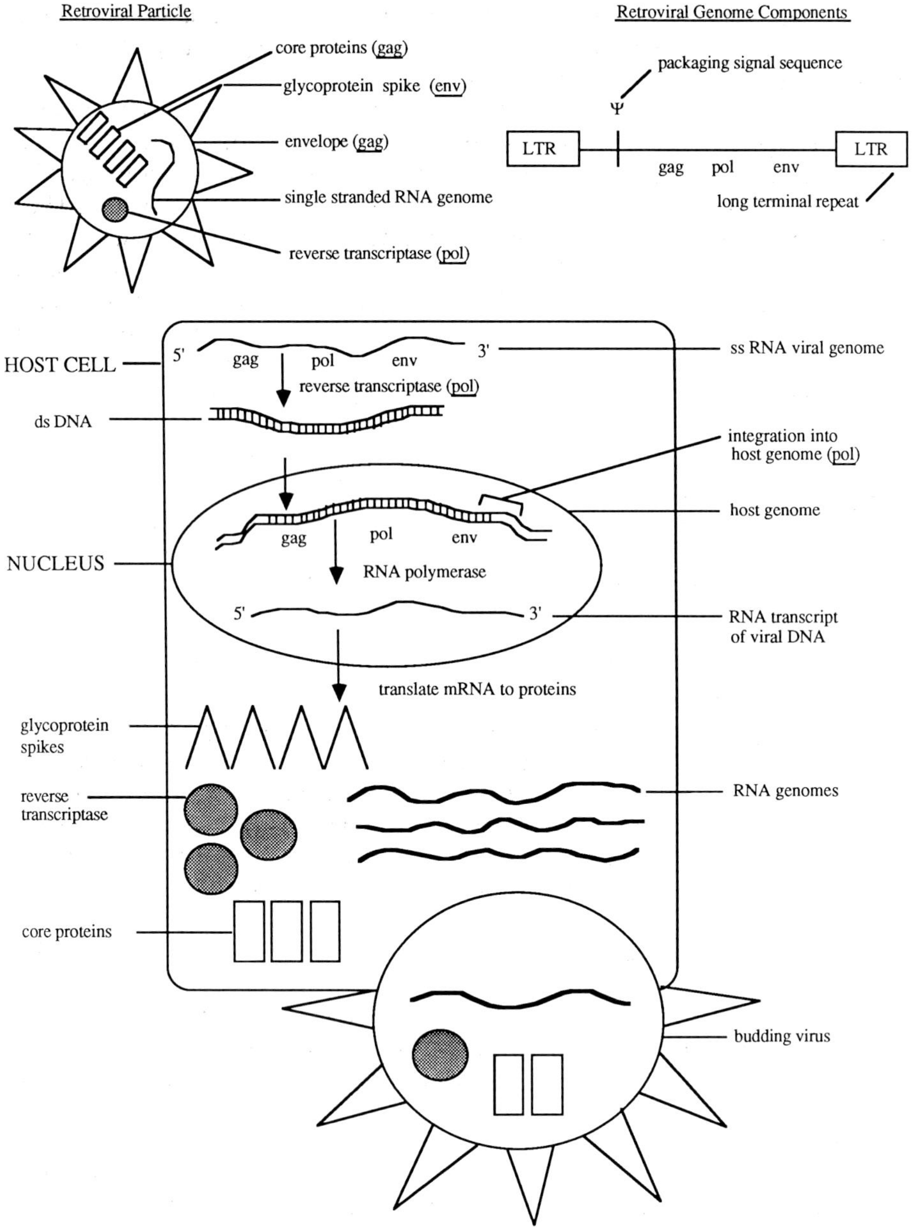

Fig. 31-1. Basic structure and life cycle of a retrovirus. The components of a retroviral particle are illustrated at the top left of the figure. The gene that codes for each component is indicated in parentheses after the component in question. The basic components of the retroviral genome are the *long terminal repeats* (LTR) that contain an RNA polymerase transcriptional element (promoter), the psi (Ψ) sequence required for packaging of the viral RNA genome, and the *gag, pol,* and *env* genes that code for the core and envelope protein, the reverse transcriptase, and the glycoprotein spikes, respectively. The retroviral particle binds to the cell membrane and is brought into the cytoplasm by pinocytosis where linear viral DNA is synthesized from the RNA template.

endogenous, defective gene if the introduced genetic material could be expressed at a level sufficient to ameliorate the deficiency. These are general goals on gene therapy, also applicable to other diseases.

The study of viruses has contributed to the development of methods for introducing foreign genetic material into cells (Fig. 31-1). The modification of both DNA and RNA viruses to include additional cloned genes has permitted high efficiency transfer and expression of these genes in many different cell types (11). Retroviruses are excellent vectors for gene replacement therapy for several reasons. Target cells can be infected with high efficiency—up to 100%. Many cells can be treated simultaneously. The retroviral DNA integrates as a single copy at a random chromosomal site. The infection and maintenance of retroviral DNA do not seem to harm cells (except in the rare cases when insertion of the retroviral DNA into a specific locus disrupts normal cellular function). Finally, many different types and species of cells can be infected.

Progress in understanding the retroviral life cycle has facilitated the process of modifying viral genomes to generate recombinant viral vectors that can transduce foreign genes into cultured cells or whole animals. Much of this pioneering work was carried out in the laboratories of R. Mulligan (12) and I. Verma (13). The basic principles of using defective retroviruses to transduce foreign genes (Fig. 31-2) can be stated briefly:

1. A cloned retroviral DNA genome is modified to accept a gene "cartridge" by removing the viral gene sequences.
2. The *gag, pol,* and *env* genes of the virus are removed by digestion of the viral DNA with appropriate restriction endonucleases to leave only the viral *long terminal repeat* (LTR) transcriptional elements and the packaging Ψ signal.
3. The DNA sequence encoding the gene to be introduced into the target cells is cloned into the modified vector. If the gene is being introduced with its own transcription-promoting element, the cassette may be inserted in either orientation. However, if the gene lacks its own promoter, the insertion position must be controlled so that the 5′ end of the message is proximal to the 5′ LTR element.
4. After the defective retrovirus is constructed, RNA genomes must be produced from the DNA form and then packaged to generate intact retrovirus particles. Mann et al. (14) developed cultured cell lines that harbor retroviral genomes from which the packaging (Ψ) sequences had been deleted. These integrated viral DNA genomes (proviruses) transcribe a retroviral message that in turn is translated to generate the viral *gag, pol,* and *env* proteins. However, this RNA message lacks the Ψ sequence and consequently cannot be packaged to form a mature viral particle.

Subsequently, the linear DNA migrates to the nucleus and is incorporated into the host genome by an enzyme encoded by the *pol* gene. This *proviral* form of the virus is the template from which both virion RNA and viral messages are synthesized. Normal cellular transcriptional machinery transcribes and processes the mRNAs. The various components required to generate progeny viral particles aggregate in the cytoplasm and intact viral particles are budded off from the cell membrane.

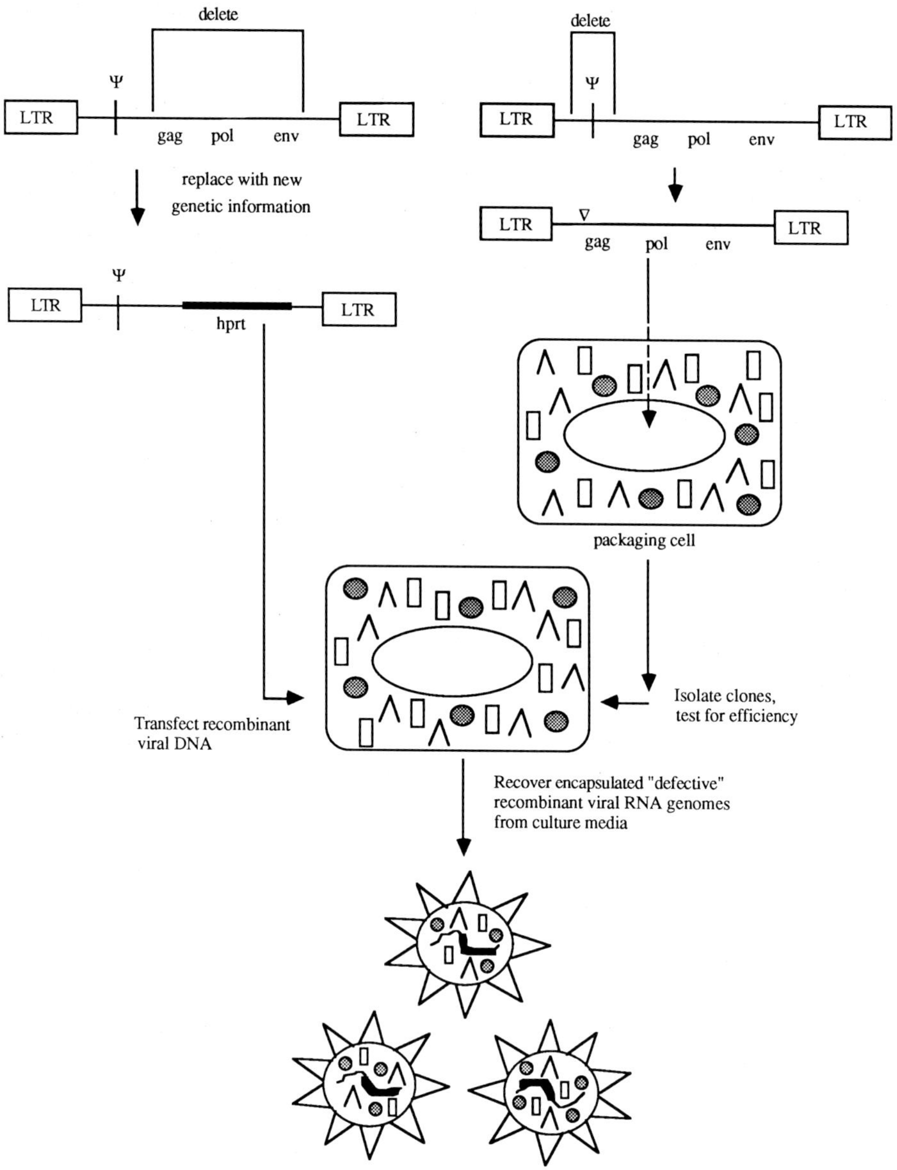

Fig. 31-2. Construction of retroviral packaging cell lines and defective retroviral genomes. Symbols are similar to those in Fig. 31-1.

5. These so-called "packaging lines" or "helper cells" are transfected with the defective retroviral construction that contains the LTR elements, the missing packaging sequence (Ψ), and the gene to be transduced. The helper cell provides all the components necessary for packaging viral RNA transcripts from the transfected retroviral DNA construction.
6. Subsequently, mature viral particles (containing RNA copies of the transfected retroviral construct) are produced and released from the helper cells

directly into the culture medium. The viruses can now be used to introduce the genetic information into recipient cells.

Most planned treatments of genetic disorders concern the use of retroviral vectors to introduce genes into bone marrow cells. Among bone marrow cells, approximately 0.01% are pluripotent hematopoietic stem cells with capacity for self-renewal. Primary bone marrow cells from donor mice can be cultured and then infected with viral supernatants produced by the packaging cell lines. Typically, the medium is filtered before it is added to the bone marrow culture to remove any nonadherent helper cells. Alternatively, bone marrow cells can be cocultivated with the virus-producing cell line. Twenty-four hours later, the infected bone marrow cells are injected into the tail vein of recipient mice that have been prepared by γ-irradiation to destroy the endogenous bone marrow.

Success of the procedure is determined by analyzing colonies that appear on the surface of the spleen 2 weeks after injection of the treated bone marrow. The cells that give rise to these colonies are termed "colony-forming units" or CFU-S (14). These colonies can be assayed directly to determine whether the retrovirally transduced gene is present within the colony cells by Southern (DNA) analysis, and whether it is expressed by Northern (RNA) analysis and protein tests.

Using this technique, successfully transduced genes have included those for (neo^r) (15,16), adenosine deaminase (ADA) (17), and purine nucleoside phosphorylase (PNP) (17). It was possible to generate ample stocks of defective retrovirus particles that contained the gene to be transferred and the gene product was identified in infected tissue culture cells. However, except in the case of *neo,* the gene has not been expressed in spleen colonies.

Using another method, Miller et al. (13) found that a human HPRT cDNA could be expressed in whole spleens of recipient mice. However, these experiments were performed before the development of retroviral packaging lines and they used a replication competent helper virus (13). They constructed a defective retroviral genome that could express a cDNA for HPRT via transcription from a viral LTR element. After transfection of a recombinant HPRT/retroviral DNA construction into a HPRT-deficient cell line they superinfected the transfected cells with non-defective helper retroviruses. In doing so, they recovered a mixture of intact viral particles that contained either a HPRT transducing genome or the wild-type retroviral genome.

Cultured rodent or human HPRT-deficient cells were infected with the mixture of wild-type (helper) and HPRT transducing virus (13), and an HPRT-proficient phenotype was observed. The same mixed stock of HPRT transducing viruses was then used to generate production of human HPRT protein in mice (18). However, there were other problems. Transfer was inefficient, being clearly observed in only one of several animals. Also, the use of mixed stocks of defective and wild-type helper virus to transduce the HPRT gene was an unsatisfactory factor as spread of wild-type virus to other tissues can lead to undesirable features such as thymic lymphoma. Recently, more promising progress has been made by the same group. Hock and Miller generated high titers of replication-defective retroviral vectors that transduced *neo* or a mutant dehydrofolate reductase (dhfr) gene into cultured human hematopoietic progenitor cells. These genes were expressed as judged by resistance of the infected bone marrow cells to either the neomycin analog G418 or methotrexate, respectively (19). Using similar methods, we have transferred defective retroviral genomes that contain human HPRT (20) or human ADA (21) genes into cultured murine bone marrow cells, with subsequent expression of the human gene products.

A major hurdle in gene therapy for Lesch–Nyhan patients involves the method of viral delivery. The target cells are in the brain, in particular the basal ganglia where production of HPRT seems essential for normal neurological development and function. Unlike the hematopoietic system, there is no hope that brain tissue can be removed for *in vitro* transfer of the genes with subsequent return to the organ. Therefore, any treatment of the brain must be carried out in a living person. The patients would have to be exposed to helper-free stocks of HPRT-transducing defective retroviruses that would have to infect neurons at high efficiency. Each defective retroviral particle can infect only one target cell. Thus, it would be necessary to generate large quantities of defective HPRT-transducing particles to infect adequate target cells to ameliorate the clinical L–N syndrome. One of the major stumbling blocks to gene therapy is the inability to generate, in a routine fashion, cell lines that produce sufficient amounts of defective retroviral particles. Generation of 10^5–10^6 particles/ml of culture media would be a reasonable goal, but 10^1–10^3/ml are usually found. The reason for this is not clear and, as such, many different defective retroviral constructs have to be tested before a suitable viral construct/cell line combination is found.

One of the major determinants of the host cell range of the retrovirus is the *env* gene that encodes the viral surface glycoproteins that recognize target cell receptors, a step needed for infection of the cell by the virus. We are currently trying to transduce and obtain expression of defective retroviral genomes in neural tissues of mice.

In wild mice, retroviruses cause a late-onset disease that generates hind limb paralysis and lymphoma. These "neurotropic" viruses infect brain and spinal cord, and disrupt normal functions of lower motor neurons. We are exploring the potential of one of these viruses [Cas-Br-E (22)] to deliver HPRT to CNS cells in newborn mice. We have found that the wild-type virus can infect many organs, including brain. Experiments that included a defective HPRT-containing retrovirus in the Cas-Br-E viral stock used to infect the mice have been unsuccessful. We are now using a competent Cas-Br-E virus that contains the human HPRT gene to transfer HPRT to neural tissues of mice.

Another problem is the ability of retroviruses to infect nondividing cells, a process that has been shown to be extremely inefficient. This is a major hurdle because neuronal cell division ceases at or around birth, and delivery of an HPRT gene directly to neurons appears unlikely (23). However, glial cells exhibit metabolic cooperation (24) and continue to divide throughout the life span of mice or humans. Therefore, production of HPRT might be induced in glial cells or other actively dividing brain cells.

Considerable research effort by many laboratories has been concerned with the generation of an HPRT-deficient mouse that could prove of value as an animal model for the correction of L–N disease. Recently two laboratories published reports describing the successful generation of such an animal (25,26). However, as yet the male HPRT-deficient mice have shown no manifestation of a neurological dysfunction. In contrast to human L–N patients, there is presumably no build-up of uric acid seen in these mice due to the presence of uricase, an enzyme not found in man. In this regard, the usefulness of the HPRT-deficient mouse as an animal model for correction of the neurological deficiency by gene transfer is limited although it will be of interest to see whether an HPRT/uricase-deficient mouse remains apparently normal in neurologic function.

We must also ask whether gene therapy for L–N syndrome is justified. The connection between HPRT deficiency and the neurological disorder implies that the enzyme must function throughout the entire prenatal period. However, there is no evidence that gene therapy can be effected *in utero*. On the other hand, prenatal diagnosis of L–N syndrome is effective, so that termination of the gestation of an affected fetus may be most appropriate for parents who can accept that course.

An alternative potential therapy for L–N concerns the transplantation of fetal neurons into the brains of L–N patients. Recently, functional recovery in a rat model of Parkinson disease following transplantation of cultured human sympathetic neurons was demonstrated (27). Improved techniques of stereotaxic injection and the relative immunotolerance of the CNS make this alternative worthy of investigation although the availability of material for such treatments may be scarce.

HPRT deficiency is a candidate for gene therapy because there is so much information about the molecular biology of the disease and because the illness is grave and current treatments are inadequate. Moreover, it would appear that gene expression of HPRT can be relaxed with regard to tissue distribution, levels of activity, and developmental expression without any detrimental effects arising. Through HPRT gene replacement, ameliorative effects are expected in the systemic manifestations including reversal of the arthropathy and renal disease. Whether the cerebral disorder can be ameliorated is an open question.

ACKNOWLEDGMENTS

G.R.M. is the recipient of an Arthritis Foundation Postdoctoral fellowship. C.T.C. is an Investigator of the HHMI at Baylor College of Medicine. S.M.W.C. is the recipient of Individual Research Service Award GM10149 from the NIH. D.L.N. is a Senior Research Associate of the HHMI.

REFERENCES

1. Jolly DJ, Okayama H, Berg P, et al. Isolation and characterization of a full-length expressible cDNA for human hypoxanthine phosphoribosyltransferase. Proc Natl Acad Sci USA 1983; 80:477–481.

2. Brennand J, Chinault AC, Konecki DS, Melton DW, Caskey CT. Cloned cDNA sequences of the hypoxanthine phosphoribosyltransferase gene from a mouse neuroblastoma cell line found to have amplified genomic sequences. Proc Natl Acad Sci USA 1982; 79:1950–1954.

3. Konecki DS, Brennand J, Fuscoe JC, Caskey CT, Chinault AC. Hypoxanthine guanine phosphoribosyltransferase genes of mouse and Chinese hamster: Construction and sequence analysis of cDNA recombinants. Nucleic Acids Res 1982; 10:6763–6775.

4. Patel PI, Framson PE, Caskey CT, Chinault AC. Fine structure of the human hypoxanthine phosphoribosyltransferase gene. Mol Cell Biol 1986; 6:393–403.

5. Lesch M, Nyhan WL. A familial disorder of uric acid metabolism and central nervous system function. Am J Med 1964; 36:561–570.

6. Kelley WN, Rosenbloom RM, Henderson JR, Seegmiller JE. A specific enzyme defect in gout associated with an over-production of uric acid. Proc Natl Acad Sci USA 1967; 57:1735–1739.

7. Seegmiller J. In Bondry A, Rosenberg F, eds. Metabolic Control and Disease. Philadelphia: W.B. Saunders, 1980, pp 777–937.

8. Watts RW, Spellacy E, Gibbs DA, Allsop J, McKeran RO, Slavin GE. Clinical postmortem, biochemical and therapeutic observation on the Lesch–Nyhan syndrome. Q J Med 1982; 201:47–48.

9. Howard WJ, Kerson LA, Appel SH. Synthesis *de novo* of purines in slices of rat brain and liver. J Neurochem 1970; 17:121–123.

10. Sweetman L. Urinary and cerebrospinal fluid oxypurine levels and allopurinol metabolism in the Lesch–Nyhan syndrome. Fed Proc 1967; 27:1055–1058.

11. Gluzman Y. In Gluzman Y, ed. Eukaryotic Viral Vectors. Cold Spring Harbor, NY. Cold Spring Harbor Laboratory, 1972.

12. Mulligan RC. In Inouye I, ed. Experimental Manipulation of Gene Expression. New York: Academic Press, 1984.

13. Miller AD, Jolly DJ, Friedmann T, Verma IM. Transmissible retrovirus expressing human hypoxanthine phosphoribosyltransferase: Gene transfer into cells obtained from humans deficient in HPRT. Proc Natl Acad Sci 1983; 80:4709–4713.

14. Mann R, Mulligan RC, Baltimore D. Construction of a retrovirus packaging mutant and its use to produce helper-free defective retrovirus. Cell 1983; 33:153–159.

15. Whetton AD, Dexter TM. Haemopoietic growth factors. Trends Biochem Sci 1986; 11:207–211.

16. Eglitis MA, Kantoff P, Gilboa E, French-Anderson W. Gene expression in mice after high efficiency retroviral mediated gene transfer. Science 1985; 230:1395–1398.

17. Magli M-C, Dick JE, Huszar D, Bernstein A, Phillips RA. Modulation of gene expression in multiple haematopoietic cell lineages following retroviral vector gene transfer. Proc Natl Acad Sci USA 1987; 84:789–793.

18. McIvor RS, Johnson MJ, Miller AD, et al. Human purine nucleoside phosphorylase and adenosine deaminase: Gene transfer into cultured cells and murine hematopoietic stem cells by using recombinant amphotropic retroviruses. Mol Cell Biol 1987; 7:838–846.

19. Miller AD, Eckner RJ, Jolly DJ, Friedmann T, Verma IM. Expression of a retrovirus encoding human HPRT in mice. Science 1984; 225:630–632.

20. Hock RA, Miller AD. Retrovirus mediated transfer and expression of drug resistant genes in human haematopoietic progenitor cells. Nature (London) 1986; 320:275–277.

21. Chang SMW, Wager-Smith K, Tsao TY, Henkel-Tigges J, Vaishnav S, Caskey CT. Construction of a defective retrovirus containing the human HPRT cDNA and its expression cultured cells and mouse bone marrow. Mol Cell Biol 1987; 7:854–863.

22. Belmont JW, Henkel-Tigges J, Chang SMW, et al. Expression of human adenosine deaminase in murine haematopoietic progenitor cells following retroviral transfer. Nature (London) 1986; 322:385–388.

23. Jolicoeur P, Nicolaiew N, DesGroseillers L, Rassart E. Molecular cloning of infectious viral DNA from ecotropic neurotropic wild mouse retrovirus. J Virol 1983; 45:1159–1163.

24. Gruber HE, Koenker R, Lutchman LA, Willis RC, Seegmiller JE. Glial cells metabolically cooperate: A potential requirement for gene replacement therapy. Proc Natl Acad Sci USA 1985; 82:6662–6666.

25. Hooper M, Hardy K, Handyside A, Hunter S, Monk M. HPRT deficient (Lesch–Nyhan) mouse embryos derived from germline colonization by cultured cells. Nature (London) 1987; 326:292–295.

26. Kuehn MR, Bradley A, Robertson EJ, Evans MJ. A potential animal model for Lesch–Nyhan syndrome through introduction of HPRT mutations into mice. Nature (London) 1987; 326:295–298.

27. Kamo H, Kim SU, McGeer PL, Shin DM. Functional recovery in a rat model of Parkinson's disease following transplantation of cultured human sympathetic neurons. Brain Res 1986; 397:372–376.

32

The Ethics of Gene Therapy and DNA Diagnosis

R. RODNEY HOWELL

This volume has defined in some detail the current state of information about molecular genetics and the potential application of these technologies to human genetic disease. My concern now is the diagnosis and treatment of genetic diseases that are debilitating or lethal.

For decades, on a regular basis, we have been able to treat genetic diseases that are due to a deficiency of circulating proteins. Some of the proteins and hormones used in these therapies are now produced by recombinant DNA technology. The best examples are hemophilia, in which Factor VIII is supplied, and human growth hormone deficiency where growth hormone is synthesized *in vitro* and provided to the patient. There seem to be no substantive ethical concerns about this form of treatment that supplies a circulating protein or hormone and in no known way could alter the genetic constitution of the person being treated.

Other than to supply soluble proteins, as in hemophilia, the most broadly applicable use of modern molecular genetics is for clinical diagnosis. Several genes have been isolated and informative restriction fragment length polymorphisms (RFLPs) are used in the diagnosis of major genetic diseases from the hemoglobinopathies to phenylketonuria, Duchenne muscular dystrophy, and Huntington disease. That information can be used for family counseling, accurate proband diagnosis, and prenatal diagnosis.

However, it is my impression that, with appropriate and carefully planned safeguards, there is popular support for somatic cell gene treatment under the proper circumstances. Religious leaders have tried to make a distinction between two perceptions: whether medicine is in the service of humanity and human dignity, or if the physician is considered primarily an agent of the community, serving the interest of the healthy at the expense of the care of the sick (1). I believe that, for many people, the term "genetic manipulation" is ambiguous, suggests adventuresome efforts, and conjures up distrust. The term "genetic surgery" might be acceptable because it connotes an attempt to modify nature rather than trying to make a fundamental change (1).

To permit the rapid application of new technology to human genetics and, at the same time, to protect the human gene pool as well as the ethical and moral considerations, a national oversight committee has been formed, the National Institutes of Health Recombinant DNA Advisory Committee. The responsibilities of the Recombinant DNA Advisory Committee include both technical and ethical issues. Additionally, procedures have been developing for the coordination of experiments that involve the use of recombinant DNA technology in humans by the Institutional Review Boards of research institutions, the Food and Drug Administration, the Institutional Biosafety Committees, the Office of Recombinant DNA Activities, and the Recombinant DNA Advisory Committee. A working group on social and ethical issues reviews proposals for experiments that involve human subjects and are submitted to the Recombinant DNA Advisory Committee. The working group on human gene therapy comprises three laboratory scientists, three clinicians, three ethicists, three lawyers, two specialists in public policy, and a representative of the public.

Anderson (2) has outlined the four levels of possible application of genetic modification in humans:

1. *Somatic cell gene therapy* implies the introduction of genes into the body cells of patients.
2. *Germline therapy* denotes the insertion of genetic material into the reproductive tissue of the patient, so that the genetic abnormality would also be corrected in his or her children.
3. *Enhancement genetic engineering* is the term that implies insertion of a gene that would "enhance" normal production of an essential substance. For example, an additional growth hormone gene might be put into a child to enhance growth.
4. *Eugenic genetic engineering* is defined as the attempt to improve complex genetic traits. These genetic traits primarily involve the function of many genes and would affect widely diverse characteristics such as personality, character, or intelligence.

The ethical and moral concerns about these levels of genetic intervention vary enormously. At present, there are no reasonable approaches to "eugenic genetic engineering," and even if there were efforts to that kind, there could be strong opposition. Opposition has also been expressed about "enhancement" genetic engineering and also about any interference with the germline.

Therefore, the host of ethical issues of human genetic intervention are being evaluated. To abide by traditional ethical guidelines, it will be critical to choose conditions for the application of genetic techniques to serious genetic disease for which there is no effective treatment. As mentioned in these proceedings, Lesch–Nyhan disease, adenosine deaminase deficiency, and Duchenne muscular dystrophy would be good examples.

However, we should not launch an early gene treatment project in conditions such as phenylketonuria, for which there is already effective, but not curative, treatment.

Somatic cell therapy would be warranted in a situation such as adenosine deaminase deficiency. The gene could be introduced *in vivo* (for example, in autologous bone marrow cells that have been aspirated and then reintroduced into the person). Alternately, the gene could be introduced systemically by a retrovirus vector so the gene could be expressed in all tissues. Carefully constructed guidelines must be used for these trials and safety tests (ideally in animals) would be needed before any human application.

The Recombinant DNA Advisory Committee of the National Institutes of Health has recommended that the risks and benefits to the patients must be considered. Also, long-term effects and social and ethical effects must be considered. The committee insists upon evidence that geneticists have examined the issue thoroughly to be absolutely certain that the disease selected is a good candidate. It is necessary to know the natural history of the disease, the variability, and whether the course can be predicted. If the disease is variable, can we know whether treatment has influenced the cause of the disease or whether there are fortuitous changes. What is the effect of the therapy proposed? Finally, does the gene therapy provide real advantages over other therapy. Experimental studies in cultured cells and animals, particularly primates, must be available before gene therapy is attempted in humans. Shapiro et al. (3) have also considered the risk of being the first patient to receive gene therapy because that patient will attract tremendous publicity (3).

Anderson (2) has stated that if we can show that we have an effective delivery system, that we have sufficient expression in bone marrow cells, and that the probability is low for infection or malignancy, it would be *unethical to delay* human trials with serious, otherwise untreatable, diseases. There is already public pressure for gene therapy, at least somatic cell therapy, if the experimental background is adequate.

Isolation of the gene for Duchenne muscular dystrophy has generated new hopes. Perhaps the gene can be introduced into mammalian experimental cells, either animal model or cultured cells, to demonstrate that the gene can actually function in these new conditions. A major problem for muscle disease is the identification of a specific vector that could introduce the gene into muscle cells. These similar problems exist for other diseases that require a specific protein for a specific organ including neurons.

There will also be problems of "scientific ethics" (3). Who owns the technology or the biologic material (the genes that have been cloned)? There may be problems of secrecy, dealing with patents and profitability, in conflict with the quick dissemination of research findings.

In spite of all of these ethical and technical questions about genetic therapy, the characterization of the genetic defect in Duchenne muscular dystrophy and other untreatable diseases now provides the real possibility of a specific treatment for currently fatal diseases.

REFERENCES

1. Pope John Paul II: The physician and the rights of mankind. J Am Med Assoc 1984; 251:1037–1038.

2. Anderson WF. Human gene therapy: Scientific and ethical considerations. J Med Philos 1985; 10:275–291.

3. Shapiro LJ, moderator. New frontiers in genetic medicine. Ann Intern Med 1986; 104:529–539.

4. Recombinant DNA research: Request for public comment on "Points to Consider in the Design and Submission of Human Somatic Cell Gene Therapy Protocols." Fed Reg 1985; 50(14)2940–2945.

33

The Oracle of DNA

NANCY S. WEXLER

From the beginning of human history people have sought to tame the unpredictable through reading the future. Bones, stones, tea leaves, dice, cards, entrails—all manner of animate and inanimate objects have been used to describe the shape of the future by the shape of their castings. Fortune-tellers with crystal balls or Tarot cards, shamans high on hallucinogens, wizards, witches, and soothsayers have been revered in all cultures and epochs.

As vaunted as the clairvoyants may be, their messages were often cryptic and hard to decipher. The Greeks refused to act on any important mission without first consulting the Oracle of Delphi. And yet tragedy often befell them. Macbeth's misunderstanding of the witches' prophesies to him led to his overweening arrogance and downfall. Those who open the doors to the future do not necessarily also provide their consultees with the wisdom to use foreknowledge constructively.

The critical question placed most often before the oracles, ancient and modern, is, at some elemental level, "Am I going to live or die on this mission? When and how will I die? Will I suffer?" People are mesmerized by the question and terrified of the answer. And yet many people have an internal sense of when and how they will die based on what they see happening to their relatives, particularly their parents and grandparents. If you ask people how old they will be when they die, they will readily answer, "Very old, because all of my relatives lived into their nineties," or "I feel I must live in a frenzy now because all my relatives died of heart attacks before age forty." People use genetic prophesy on a daily basis.

And so it would appear that, in the elegant organization of nucleotide bases along the double helical spine, the die is really cast. But perhaps not forever.

PROPHESY WITHOUT PREVENTION

The recent spate of astonishing genetic discoveries in this era of recombinant DNA now finds us in a unique quandry. We are in the problematic position of being able to indentify those who carry genes for fatal or crippling conditions without being able to prevent or treat the diseases.

Since 1983 genes have been located for Huntington disease, Duchenne muscular dystrophy, polycystic kidney disease, cystic fibrosis, phenylketonuria (PKU), manic depression, Alzheimer disease, and two forms of neurofibromatosis. The same strategies are being brought to bear to search for other genes such as those for schizophrenia, familial amyotrophic lateral sclerosis, dystonia, some cancers, alcoholism, Tourette syndrome, and others—all disorders for which there is either no treatment or inadequate treatment. The expectation is that once the genetic lesion is characterized, effective treatments can be developed and genetic techniques can be used to intervene in the gene itself to ameliorate or cure the disease. Genetic research must follow a linear order from chromosomal localization of the gene to therapy. The anticipated benefits would also follow a certain order, from diagnosis to eradication of the disease. For most genetic investigators, the capacity to diagnose is merely a way-station along the road to a cure. And yet now, for patients and families, for physicians, for employers and insurance companies, and for the community at large, we find ourselves in the terrifying position of being able to predict severe disability or certain death without being able to prevent or treat it.

For some conditions, early diagnostic information can bring substantial gains. Children on controlled diets to prevent PKU thrive with an intact IQ. Physicians, apprised before or just after a birth that a baby has cystic fibrosis or sickle cell anemia, are likely to intervene vigorously with antibiotics or other medication in attempts to stave off some of the severe sequelae of those diseases. Even familial hypercholesterolemia is more manageable with new drug treatments that were developed as a direct result of basic work in molecular biology to understand the primary gene lesions.

For the most part, however, genetic diagnostic tests can be used in many more situations than we have recourse to treatment. They can be used prenatally with the aim of terminating a pregnancy in which the health of the fetus is seriously compromised, or to predict future disability in an asymptomatic individual. The most disquieting dilemma provoked by genetic advances is the capacity to prophesy before we can prevent. During this painful hiatus in science, when one breakthrough has been accomplished, making it possible to detect the abnormal gene, but the second breakthrough leading to treatment is still in the offing, we are challenged with all manner of psychological, social, ethical, and moral questions.

Genetic prediction has consequences for the individual, the family, and society. The gravity of these consequences and the choices they provoke differ depending upon (1) whether the inheritance pattern is dominant or recessive, (2) age at onset of symptoms, (3) if the disease is treatable, (4) the gravity, burden, and subjective perception of the disease, and (5) the symptoms, particularly if there is loss of mental capacities.

DOMINANT VERSUS RECESSIVE

If a disorder is recessive, such as Tay–Sachs disease or cystic fibrosis, the capacity for early diagnosis is relevant primarily for prenatal or perinatal diagnosis. Carrier parents can choose not to conceive, knowing they are carriers, or to complete only pregnancies in which the fetus does not carry two copies of the disease-producing allele.

Our ability to use DNA markers to detect carriers of cystic fibrosis is newly acquired. Many more people are confronted with the possibility of prenatal diagnosis for a different kind of disease than we are accustomed to detecting prenatally. Children with cystic fibrosis can now live moderately normal lives into the twenties and thirties. And parents are having to make choices not previously available. Nevertheless, the parents are only carriers and they will never be diagnosed with the disease. Their concern about cystic fibrosis is for their children.

If the illness is inherited in a dominant fashion, the "healthy," or asymptomatic individual, is being diagnosed with an illness that will appear in the future, sometimes many years later. If the diagnostic test is used prenatally, a positive diagnosis in the fetus immediately diagnoses the parent.

Huntington disease is the first autosomal dominant, late-onset, severe disorder for which a DNA marker, tightly linked to the disease gene (on chromosome 4) now permits presymptomatic diagnosis (1). There also seems to be little or no significant genetic heterogeneity that might preclude widespread use of the test. About 75 families worldwide have been tested and share the same Huntington disease locus (2).

In 1987, markers localizing genes for familial Alzheimer disease (3) and two forms of manic-depressive disorder were discovered (one on chromosome 11 and one X-linked) (4, 5). In these conditions, genetic heterogeneity is likely to play a major role, complicating diagnostic testing by requiring probes that are specific for a particular subset of families. Variable penetrance, particularly with respect to manic depression, gives counseling a unique twist. Should people be told they are most likely carrying a gene for predisposition to manic-depressive illness, when (at least in the form found on chromosome 11) there is only a 63% penetrance rate (4)? How will people interpret a 63/37 risk—perhaps as 50/50 or 100%? Will anticipating the illness precipitate it? Should prophylactic lithium therapy be given even though there is a large chance that the person will never become symptomatic?

In some conditions, knowledge of vulnerability to a late-onset illness may help a person to receive immediately appropriate treatment once symptoms appear, and may even preclude some complications that might be induced iatrogenically. For example, bipolar illness may begin with unipolar symptoms that are treated with tricyclic antidepressants. There is evidence that these medications may actually exacerbate the cycling pattern once the bipolar phase begins (6). It might be advisable to avoid these drugs in people who are genetically vulnerable to bipolar manic-depressive disorder.

HUNTINGTON DISEASE AS A MODEL FOR PRESYMPTOMATIC DIAGNOSIS

Huntington disease and familial Alzheimer disease are both autosomal dominant, neurodegenerative disorders of late onset, invariably fatal and untreatable. The presymptomatic test for Huntington disease, using DNA probes approximately 4 cM (four million base pairs) from the defective gene, is now being offered under strict research protocols in several locales (Harvard, Johns Hopkins, and Columbia Universities, University of British Columbia, Canada, and Universities of Manchester and Cardiff, England). In the 2 years since the test became clinically available, about 100 individuals worldwide have received diagnostic information. Some people were already symptomatic when they came for the test and they were diagnosed clinically. Others changed their minds and postponed testing until treatment becomes available, or the test is 100% accurate. Some decided they would prefer not to receive any predictive information.

Centers offering the test have been impressed by the low response rate of persons at risk to avail themselves of the test. In all three centers in the United States, serving an at risk population in the thousands, fewer than 200 individuals have actually enrolled in a testing program (7). These findings are in sharp contrast to the enthusiastic response to testing that was found before the test was actually available. In several surveys over the past decade, between 63% and 79% of the at risk respondents said they would take a presymptomatic test (8–11). Obviously, the reality of the test, particularly a linkage test that involves other family members and is not entirely accurate, is less appealing than the prospect.

Psychologists Kahneman and Tversky offer some insight into this seeming change of opinion (although it is probably still too early to tell) (12,13). They argue that people are not adverse to taking risks but are rather adverse to taking losses. If they perceive that they have an advantage, however small, they tend to be conservative rather than gambling to achieve a greater gain at the risk of losing everything—the "bird in hand" philosophy. On the other hand, if people perceive themselves to be already in a loss situation, they will gamble to escape the loss, even if it means they may lose more. If they are already down, the argument goes, what does a little more mean if it is possible not to lose anything. The same situations, described either in terms of losses or gains, can provoke opposite responses in the same people. Genetic counselors know this phenomenon well, taking pains to emphasize balanced phraseologies for risk, for example, "You have 3 in 4 chances to have a normal baby" (gain) and "You have a 1 in 4 chance to have a genetically defective baby" (loss).

If people coming for the presymptomatic test for Huntington disease perceive themselves to be in a loss position because they are at risk, if being at risk has hampered their lives tremendously, they will be willing to risk a terrible loss—hearing that they most likely have Huntington disease—for the gain of learning they are gene free. If, in contrast, people have been relatively happy when they come for testing, they may realize that the increment of satisfaction or relief brought by a negative test would not outweigh the disastrousness of presymptomatic diagnosis. People who want the test are more likely to perceive themselves already in a loss position and are willing to gamble. Those who see themselves already in a gain position fear that they have a lot to lose with a bad outcome of the test. A large part of the counseling is in helping people to realize more clearly what they stand to gain or lose with different test outcomes.

For many who receive a high probability of being gene free, the good news is tempered with concern for siblings and other relatives. For those who learn that they have a high probability of having the Huntington disease gene and are likely to become ill at some indeterminate time in the future, the information is devastating. Some have said that no matter how prepared they thought they were, they could not imagine beforehand how it would be as a reality. And it was much worse than they had thought (14). The only suicide attempt was made by a person who came for presymptomatic testing and, instead, was diagnosed clinically (15).

Nevertheless, a presymptomatic test for Huntington disease does provide some obvious benefits. Individuals "at risk" can plan for marriage, career, finances, and, most particularly, children, knowing in advance if all is likely to be jeopardized. Those with a lowered risk could stop being hypervigilant for symptoms, as many are, whereas those carrying the gene would undoubtedly find themselves constantly screening themselves for the first signs. The test gives no indication when symptoms will appear, so the insidious onset of minute disturbances in movement, memory, and mood can bring wracking uncertainty even to probable gene carriers. Many who wish to take the test want to resolve the ambiguity of being at risk, for better or worse. They cannot abide not knowing how to interpret minor adventitious movements, moodiness, and depression, or difficulties at work or in personal relationships. There is a danger that these people are seeking desperately to learn that they do not have the illness but are not truly prepared to learn that they do carry the gene. If they have a positive test, the ambiguity of waiting for disease onset may be unbearable and they could decide to become patients as a kind of riddance phenomenon, foreclosing on possibly years of health.

The advent of a presymptomatic test for Huntington disease using DNA markers has led to a new category of individual: the presymptomatic person with the Huntington disease gene. Not yet a patient, no longer at risk, how does this person live? Many Huntington disease patients have had long and productive lives, families, and careers in all areas of endeavor, including medicine, law, academics, the military, construction, and others. Will a young person, age 21, with a 96% probability of carrying the gene, be willing to expend the time, money, and energy on developing a career? Will future patients shy away from taxing professions, be intimidated by the failures that are inevitable in any pursuit, doubt their own abilities, and abandon pathways that could make their lives rewarding while they are still young and healthy? Will physicians, parents, friends, and relatives discourage them from pursuing careers they could not sustain once they become ill? Will universities and employers refuse to accept the presymptomatic individual if this status is known?

What is the diagnosed individual's social and ethical responsibility? What if a neurosurgeon, a pilot, or a stockbroker tests positively for the likely presence of the gene. Should they stop working immediately? How do they monitor themselves adequately? Should those around presymptomatic persons in potentially high risk jobs, to themselves or others, try to dissuade them from continuing these activities? Although neurosurgery under the knife of someone at risk may be acceptable, knowing the diabolic gene is present—a gene that causes the tiniest uncontrollable movements of fingers, the most fleeting lapses of memory and judgment—should someone intervene and prevent the surgeon from operating? Should the surgeon's continuation with his or her job be considered such a failure of good judgment that the disease must certainly be starting? Many of these same concerns will be true when a test for familial Alzheimer disease is also available.

Prenatal testing for Huntington disease offers some relief from a few of the quandries posed by presymptomatic testing but introduces other twists. At risk parents might want to ensure that their children are free from the Huntington disease gene without learning their own genotype. If so, they can test fetuses. In three of four instances, the fetus should be negative for the haplotype that indicates the presence of the disease gene. The at risk parent's genotype is unknown—unless the fetus tests positively.

If an at risk parent wants to avoid any chance of learning his or her genotype inadvertently through testing the child, or perhaps there are too few family members to determine his or her genotype, a "nondisclosing" prenatal test can be offered. If the fetus inherits a chromosome 4 from an unaffected grandparent, spouse of the symptomatic patient, the risk of the fetus is 2%. If the fetus inherits a chromosome 4 from the affected grandparent, the risk is 50%, or the same risk as the parent. It is not known if the chromosome 4 from the affected grandparent is carrying the healthy or lethal allele. The genotype of the at risk parent is not required for this test.

All centers that offer tests for Huntington disease make the test available only to persons age 18 or older who can give an informed consent. Current opinion is that parents should not be privy to this knowledge for their minor children. A disturbing ethical problem is raised by prenatal testing. Sometimes, a fetus is tested, found to be at 50% risk, and the parents decide not to terminate the pregnancy. If the at risk parent develops Huntington disease, the genotype of the tested child is also known because they share the same chromosome 4 (barring recombinations). A minor child has thus been de facto tested presymptomatically. And yet, programs cannot insist that couples abort children who are found to have a 50% risk. Sage counseling is needed to determine the intentions of prospective parents who want prenatal testing and to apprise them of these dilemmas.

Some argue that presymptomatic tests of this nature should not be used until treatment is also available. The potency of knowledge, for benefit or harm, has been an issue since Adam and Eve and the apple. Will genetic tests once more expunge us from the blissful ignorance of the Garden of Eden?

Sophocles grappled with the dreadful consequences of acts committed knowingly or unwittingly in *Oedipus Rex.* As the blind seer, Teiresias (many of those who see the future are blind to the present—as if what they see is so powerful it robs them of ordinary sight), is summoned by Oedipus to explain why Thebes is sinking into putrefaction and decline; Teiresias realizes what devastating message he must deliver to Oedipus about his own actions: Teiresias: "Ah me! It is but sorrow to be wise when wisdom profits not." And later, after refusing to speak: "Though I hide all in silence, all will come" (16).

Finally Teiresias is goaded into revealing the truth, which propels Oedipus' wife Jocasta to suicide, and sends Oedipus, after stabbing his eyes out with Jocasta's broaches, reeling broken and alone into exile and perpetual night.

In all likelihood, revelations of genetic truths will not have such dire consequences. Appropriate counseling before and after receipt of genetic information is absolutely essential. In our society there is another hazard in making these tests available: there is little or no reimbursement from insurance companies for either genetic or psychological counseling. Genetic counselors do not usually see clients on a long-term basis and most are not trained as psychotherapists. Admission to a psychiatric hospital is almost the only mental health service for which there is insurance coverage. The person tested must pay for psychotherapeutic help for decision making before testing or follow-up after test results are delivered.

PRENATAL PREDICTION: COMPASSION OR CONVENIENCE?

Our increasing capacity to test prenatally for hereditary disease provokes some of the most trenchant questions. With prenatal detection increasingly possible, larger numbers of couples will be offered the possibility to know, before the birth of their child, if it is carrying a genetic defect. One immediate impediment in taking advantage of this information is the lack of understanding in the general public of genetic concepts and probability theory. Sadly, the public is joined in this ignorance by many physicians because of the woeful paucity of genetics teaching in medical school (17). As of 1985, only 18% of United States medical schools had genetics courses (17). Two years later, 24% were considered to have good or excellent instruction (18). There is little in the general medical school curriculum on genetics, small emphasis in practice, and a resulting inability among practitioners to interpret genetic tests or provide adequate counseling (17). In one survey, only half of the pediatricians and family practitioners could properly interpret a PKU test result (19). Medical malpractice suits may eventually make doctors more aware of their obligations to provide adequate genetic information to their patients, but we can hope that genetic awareness will not be born in the courts.

Some individuals will not wish to avail themselves of new genetic prenatal information because they are opposed to abortion under any circumstances. In the United States, however, over 80% of the population approves the use of abortion if the child will suffer deformities or death (20). In this new genetic terrain, prospective parents are mapping fresh landscapes. What if prenatal testing is done for a disease of late onset? Should there be an abortion for a fetus destined to become ill at age 40? age 80? What is the quality of the years "at risk" for someone knowing that this is their destiny? Children with cystic fibrosis are leading increasingly longer lives. Is this sufficient reason not to abort?

Some suggest that the new genetic and reproductive technologies being developed are also leading to the "commodification" of parenthood (21). Sperm, eggs, uteruses, bodies, and children can be bought and sold, rented and returned like so many dry goods. But the tremendous efforts expended, the expense, and the time and energy devoted to becoming parents also speak to a yearning for parenthood that is no longer impeded by biological constraints.

One fear, expressed cogently by Barbara Katz Rothman in her book *The Tentative Pregnancy,* is that with our increased capacity to screen prenatally for genetic defects, couples will insist on "the perfect baby"—which still cannot be delivered.

> An irony in all this is that the technology still cannot guarantee a "blue ribbon baby." A fetus can pass all of the tests and still be far from perfect at birth. A child can be born or become retarded, disabled, disfigured from thousands of causes. One can rationally decide to abort a fetus with spina bifida because life in a wheelchair is not acceptable—and then have a baby's back broken in a car accident. One can choose not to carry to term a fetus with Down syndrome because the quality of life of the retarded is not acceptable—and then have the baby suffer permanent brain damage from some illness. There are limits to control, and our children are always "hostages to fortune." Does the conscious, deliberate emphasis on control and "standards of acceptability" prepare us for the *reality* of parenthood? (22)

There is a crucial element missing in this rather materialistic analysis of parental motivations—the parents' empathy with their handicapped or ill child. Many parents realize that life is so full of unexpected, uncontrollable wounds, such as auto accidents or disease, that they would prefer their children to start out as best prepared, body and mind, as possible. Frequently, prospective parents are quite familiar with the diseases for which they are testing their fetuses. They have seen these diseases face to face and are terrified of them. A previous child may have died a horrifying, deteriorative death—a parent, a brother, or an uncle may have succumbed. Parents cannot bear to inflict this same torture on children they love. They cannot tolerate the idea that their own wish to have a child could cause untold suffering for that child, even if the child makes a good adaptation to its plight. Parents who use genetic services are not necessarily grocery store "aficionados," shopping for the perfect tomato. Rather, they are desperate to protect their children from harm; this may mean terminating a pregnancy before the fetus is viable to prevent later trauma for that child. Some would even argue that not terminating the pregnancy of a genetically impaired fetus, insisting knowingly that the child be born handicapped, is tantamount to child abuse.

In this light, the question of how long a life must be before it is worth bringing it into this world is irrelevant. The amount of time before death is not the issue; rather, it is the prospective parent's personal idiosyncratic view of the disease, how devastating it is, how much their child might suffer—as a child or an adult. Just as parents want to give their children everything that they never had, for narcissistic as well as generous motivations, so too they want to spare them from the pain they have suffered.

A potential hazard in our expanding capacity to diagnose more disorders in the carrier state and prenatally is the parents' loss of innocence. Parents will find it increasingly difficult to plunge ahead with a pregnancy and take the consequences. It will soon become mandatory for physicians to offer the option of testing, at least for certain conditions, or if there is a known risk. A new law in California, for example, now requires physicians to screen for maternal serum α-fetoprotein to detect neural tube abnormalities. Pregnant women can refuse the test only by signing a waiver of liability. Doctors who do not offer amniocentesis to pregnant women over age 35 are liable for malpractice suits because amniocentesis is now standard medical practice.

With court-ordered loss of innocence, parents will be confronted with increasingly more complex choices and culpability. If they choose not to avail themselves of services offered, they are responsible for any abnormalities the child might bear. In electing not to detect prenatally, they are more actively visiting potential suffering on their children.

Rothman foresees that subtle or overt pressures may be brought to bear on parents who do not avail themselves of these services or do not terminate an abnormal fetus. Knowing of a family history of a particular disorder, insurance companies might require such testing and refuse to pay the medical costs for a defective baby. Physicians may bring increasing pressure on parents, citing the immense emotional and financial burdens on the family, suffering of other siblings, and costs to society for the care of abnormal children. Family and friends may also weigh in to persuade a couple to terminate "defective" fetuses and try again for a better baby. Rothman cites the compulsion to choose the socially acceptable choice: "Amniocentesis and selective abortion, like embryo transplants, surrogate motherhood, and other new reproductive technology, are all being used to give the illusion of choice" (23). Rothman believes that society constrains the choices to be made by making certain choices more easily assimilable culturally.

Some prospective parents find the ambiguity of their circumstances particularly perplexing. If they are carriers for Duchenne muscular dystrophy, for example, they may feel as if aborting a child with this disease is tantamount to aborting a beloved brother or other relative. People at risk for Huntington disease sometimes feel "How can I abort someone who has the same risk as me?" To abort a child with the same genetic identity—in one respect—is confusingly close to denying the validity of your own identity. This argument is often voiced by members of handicapped rights groups who object to the promotion of abortion for conditions from which they suffer.

One of the more specious arguments against using new genetic technology is that we, as a culture, may lose certain heroes known to be afflicted by particular diseases. No Woody Guthrie, who died of Huntington disease, no Sylvia Plath, who died a suicide. Or perhaps we lose a category of people such as artists like Van Gogh, whose wellsprings of inspiration may be the same DNA flaws that cause them searing psychological distress. Yet, myriads of gametes are never fertilized, never born, and never mourned.

THE EUGENIC PROTEST

The most strident cautionary voice on the perils of genetic research and testing is that of Jeremy Rifkin, Director of the Center for Economic Trends (24). Rifkin carries the "commodification" argument to the extreme, arguing that modern parents are no more interested in the individuality of their children than are buyers of automobiles. In his 1985 book, *Declaration of a Heretic* (25), Rifkin explains that Galileo was a heretic not because of his radical scientific views but because he opposed the established church of his day. Today's "church" is our society's complacent acceptance of scientific technology, Rifkin argues; our "faith system," the scientific world view; and Nobel laureates, the new prelates of the scientific establishment Church. In titling his book, Rifkin modestly announces himself as a modern day Galileo, with important truths about the universe and new insights into the nature of reality and existence.

And what are these truths? Rifkin equates the atomic age with the genetic age, the bomb with the gene. Splitting the atom and deciphering the double helix are discoveries of equal import to him and equal potential for destruction. Rifkin sees all of scientific progress as a perilous descent down a rain soaked, muddy, "slippery slope" where footing can be lost at any moment, plunging us all into the new eugenics (25):

> Once we decide to begin the process of human genetic engineering, there is really no logical place to stop. If diabetes, sickle cell anemia, and cancer are to be cured by altering the genetic make-up of an individual, why not proceed to other "disorders": myopia,

> color blindness, left-handedness? Indeed, what is to preclude a society from deciding that a certain skin color is a disorder? In fact, why would we ever say no to any alteration of the genetic code that might enhance the well-being of the individual or the species? It would be difficult to even imagine society rejecting any genetic modification that promised to improve, in some way, the performance of the human race. (26)

In one sentence Rifkin travels from left handedness (hardly a disorder) to presumably the ravages of cancer, swift and fatal heart attacks, and the lingering decrepitude of Huntington and Alzheimer disease. He asks if "guaranteeing our health is worth trading away our humanity" (27). How is our humanity enhanced by permitting the perpetuation of diseases that devastate body and mind? He sees it as our moral duty to carry our faulty genes into the future so that our children, too, may suffer. Rifkin also seems never to have heard of, or at least taken seriously, the civil rights movement in this country. Nor does he feel we are genuinely capable of moving beyond the flirtation of our society in the early 1900s with eugenics when Charles B. Davenport was in residence at Cold Spring Harbor running the Eugenics Records Office, instead of James D. Watson with a team of creative molecular geneticists. Rifkin assumes a unanimity of definitions of perfection that are unlikely to exist, such as the choice of the ideal skin color.

Most troublesome, Rifkin is particularly stingy in his willingness to concede that society as a whole and individuals within it are capable of making discriminatory choices, capable of expending major efforts to treat severely debilitating and life-threatening diseases, with whatever means, genetic or otherwise, and leaving minor irritations alone. They will have enough to do to cure the myriad diseases to which humans are subject; if society does not wish minor genetic tinkering, it should not support it financially.

Ironically, gene therapy is probably the most natural of modern therapeutics. Scientists are attempting to use the body's healthy genes, the natural genes, to fix nature gone awry. By placing a normal hemoglobin gene in bone marrow stem cells, scientists are mimicking the normal state. If a way can be found to replace lost cancer suppressor genes or switch off genes undergoing abnormal amplification, children may not lose eyes to retinoblastoma, nor would other cancer patients be subjected to the devastating chemicals and blasts of radiation that are today's therapy but kill more than the intended targets. Furthermore, all current gene therapy research for humans is aimed at somatic therapy rather than germline therapy, curing patients rather than their progeny.

LEGAL ACCOUNTABILITY: WRONGFUL LIFE SUITS

Support for amniocentesis and selective abortion, in the absence of gene therapy, may come from an unexpected direction. Using a tort action known as wrongful life, children born with genetic defects have sometimes sued physicians whose duty it was to warn parents of potential genetic conditions. At first these cases were not accepted for legal action because the courts could not measure the value of a life lived or unlived. However, some cases are now being heard, although none has been won yet by the plaintiff (28).

These cases differ from typical malpractice cases because they presume that a person's life should never have existed at all, if the defendant had done his or her duty. To date, a wrongful life action has not been brought by a child against parents. In the past, children were constrained from suing parents, but courts now permit cases that involve property and finance (29). In 1987, an Illinois appeals court ruled that a 5-year-old girl, injured in a car accident while still in her mother's womb, could sue her mother for negligence (30). If a child born with severe deformities or a genetic defect decided that the parents could have detected the disorder prenatally, a suit against the parents might be based on wrongful life or negligence. Justice Oliver Wendell Holmes wrote a legal opinion in the late 1880s stating that there is "a conditional, prospective liability for one not yet in being" (31). Later courts have argued that every child has a right to begin life with physical and mental health (32). Marjorie Shaw, M.D., J.D., an expert in law as it pertains to genetics, has concluded that

> knowingly, capriciously, or negligently transmitting a defective gene that causes pain and suffering and an agonizing death to an offspring is certainly a moral wrong if not a legal wrong. Thus, if reproduction is contemplated (or not consciously prevented) there is an ethical obligation not to harm the offspring and one's genotype should be determined so that appropriate steps can be taken to avert the disease in future generations. (33)

Pressures on parents to use genetic services will certainly, in part, be considerations of ethical duty to the child and responsibility to society. Parents might also choose to do what is most convenient for them, feeling themselves incapable of or unwilling to raise a handicapped child. But the moral and ethical responsibilities of the parents to do no harm to their children may yet be reinforced by court actions. After all, the children suffer the handicaps, not the parents.

CONFIDENTIALITY AND OTHER CONSIDERATIONS

The noisy debate over eugenics threatens to obscure some difficult and immediate problems. Many individuals do not wish to have third party carriers pay for their genetic tests because the carriers cannot guarantee confidentiality of the results. If insurance companies know that someone has had a positive test for having a high probability of developing Huntington disease, Alzheimer disease, or some other devastating disorder of late onset, there is a grave risk that the insurance policy will be canceled (34). Insurance companies have not yet acted to make a predisposing gene equivalent to a preexisting condition that would nullify a contract. If third-party carriers do not pay for the expensive genetic tests, however, few people will be able to afford them. Genetic testing will be the prerogative of only the wealthy.

The use of genetic tests for screening in the workplace is another potential source of abuse. One biotechnology company, Focus Technologies, is developing a panel of genetic tests that could be used by employers to screen employees (35). Employers might altruistically be saving employees from potentially hazardous work environments, but they may also be trying to reduce the company's health care costs or to avoid spending the money necessary to make an environment safe for the genetically vulnerable.

Similarly, a predisposing gene might be seen as grounds for dismissal rather than grounds for receiving improved medical care. For example, an employee with a gene for familial hypercholesterolemia should not be taken off the job for fear of a sudden heart attack, but rather should have periodic tests of blood cholesterol levels. More difficult to contend with is the employer's response to people with genes predisposing to psychiatric disorders. Routine health checks are not easy in these circumstances. And the company, like the family, cannot be seen as hovering over the genetically vulnerable person, waiting "for the other shoe to drop." In some civil service or tenured positions, it is difficult to force employees to leave even if job performance is slipping. Employers can legally require genetic tests as a condition of employment; the employee is free to take the tests or leave—a choice that is not viable in many small towns.

Proposed legislation has been designed to protect citizens from mandatory screening for the human immunosuppressive virus (HIV), for the maintenance of confidential records if tests are made, and for protection in employment. HIV legislation could also have unexpected benefits for people with genetic diseases (36). At the moment, the Rehabilitation Act of 1974 provides protection from discrimination against the handicapped, or those who "give the appearance of being handicapped." Persons who test positively for genes that predispose to particular disorders may be discriminated against even though they were not handicapped at the time of employment. The HIV and genetic disease constituencies should recognize that legislation beneficial or harmful for one group may have the same implications for the other group.

There are certainly many parallels between the testing programs for persons at risk for Huntington disease and HIV (37–40). Both disorders are sexually transmitted, although HIV poses a far greater risk to the community at large. Neither test is 100% accurate and both tell those with positive outcomes only that a disease is likely to appear at some unspecified time in the future. Both disorders produce dementia, are marginally treatable, and invariably fatal. There is a risk of discrimination and stigmatization to both groups (although far greater for HIV carriers) and both groups must struggle continuously with terrifying ambiguities and certainties. Both groups are in need of quick access to counseling for which they must pay personally.

MAPPING THE FUTURE

For the first time in history, the time is ripe for making major inroads on diseases that wreak havoc not only on one generation but on each subsequent generation. Through learning to understand and work with DNA, scientists can become healers at the most fundamental levels. Once we have the capacity to move ahead, however, if we choose not to, we are consigning some people to certain death as actively as if we were to withhold antibiotics or oxygen. The road is before us; we no longer have the excuse that there is nothing to be done. If we act recklessly on this road, we will also rue the consequences. At each step we must weigh and reaffirm our commitments—particularly to the Hippocratic oath and to Galen's oath to do no harm. Acting in a responsible, thoughtful manner can bring freedom from suffering, pain, and fear in ways not yet imagined. The oracular vision is only as good as its human interpretation.

REFERENCES

1. Gusella JF, Wexler NS, Conneally PM, et al. A polymorphic DNA marker genetically linked to Huntington's disease. Nature (London) 1983; 306:234–238.

2. Conneally PM, Wexler NW, Gusella JF. Genetic heterogeneity in Huntington's disease. 1988; in preparation.

3. St. George-Hyslop PH, Tanzi RE, Polinsky RJ, et al. The genetic defect causing familial Alzheimer's disease maps on chromosome 21. Science 1987; 235:885–887.

4. Egeland JA, Gerhard DS, Pauls D, et al. Linkage between a gene conferring predisposition to bipolar affective disorders and DNA markers on chromosome 11. Nature (London) 1987; 325:783–787.

5. Baron M, Risch N, Hamburger R, Mandel B, Kushner S, Newman M, Drumer D, Belmaker RH. Genetic linkage between X-chromosome markers and bipolar affective illness. Nature 1987; 326:289–292.

6. Potter D. Personal Communication (1986).

7. Myers R. Personal Communication (1987).

8. Kessler S, Field T, Worth L, Mosbarger H. Attitudes of persons at risk for Huntington disease toward predictive testing. Am J Med Genet 1987; 26:259–270.

9. Markel DS, Young AB, Penney JB. At-risk person's attitudes toward presymptomatic and prenatal testing of Huntington disease in Michigan. Am J Med Genet 1987; 26:295–305.

10. Mastromauro C, Myers RH, Berkman B. Attitudes toward presymptomatic testing in Huntington disease. Am J Med Genet 1987; 26:271–282.

11. Meissen GJ, Berchek RL. Intended use of predictive testing by those at risk for Huntington disease. Am J Med Genet 1987; 26:283–293.

12. Kahneman D, Tversky A. The psychology of preference. Sci Am 1982; 246:161–171.

13. Tversky A, Kahneman D. The framing of decisions and the psychology of choice. Science 1981; 211:453–458.

14. Myers R. Personal Communication (1987).

15. Hayden M. Personal Communication (1987).

16. Sophocles. Oedipus the King. In Eight Great Tragedies. New York: Mentor Books, 1957, pp 65–66.

17. Childs B, Huether CA, Murphy EA. Human genetics teaching in U.S. medical schools. Am J Hum Genet 1981; 33:1–10.

18. Schmickel R, et al. Unpublished paper prepared for the American Society of Human Genetics (Courtesy of NA Holtzman) (1987).

19. Holtzman NA, in preparation (1988).

20. Rothman BK. The Tentative Pregnancy. New York: Viking Press, 1986, p 3.

21. Ibid, p 2.

22. Ibid, p 7.

23. Ibid, p 14.

24. Rifkin J. Algeny: A New Word—A New World. New York: Viking Press, 1983.

25. Rifkin J. Declaration of a Heretic. Boston: Routledge & Kegan Paul, 1987, pp 4–6.

26. Ibid, pp 66–67.

27. Ibid, p 111.

28. Wexler NS. Will the circle be unbroken? Sterilizing the genetically impaired. In Milunsky A, ed. Genetics and the Law, Vol. II. New York: Plenum Press, 1980.

29. Ibid, p 318.

30. Hotz RL. Life, death & DNA. Atlanta J Atlanta Const 1987; 4/26–4/29:3.

31. *Dietrick vs. Inhabitants of North Hampton,* 138 Mass. 14, 52 Am. Rep. 242 1884.

32. Wexler, p 316.

33. Shaw MW. Invited editorial comment: Testing for the Huntington gene: A right to know, a right not to know, or a duty to know. Am J Med Genet 1987; 26:243–246.

34. Davis J. Director, New York State Genetics Task Force, Chair, Committee on Reimbursement. Personal Communication (1987).

35. Report from Representatives of Focus Technologies, Meeting of the Office of Technology Assessment, Panel on Gene Linkage (1987).

36. Waxman H. Congressional Representative from California. Personal Communication (1987).

37. Farrer LA. Suicide and attempted suicide in Huntington disease: Implications for preclinical testing for persons at risk. Am J Med Genet 1986; 24:305–311.

38. Lamport AT. Presymptomatic testing for Huntington chorea: Ethical and legal issues. Am J Med Genet 1987; 26:307–314.

39. Overall C. Ethics and Human Reproduction: A Feminist Analysis. Winchester MA: Allen & Unwin, 1987.

40. Smurl JF, Weaver DD. Presymptomatic testing for Huntington chorea: Guidelines for moral and social accountability. Am J Med Genet 1987; 26:247–257.

Glossary

ERIC A. SCHON

Acceptor Splice Site. The AG dinucleotide at the extreme 3′ end of an intron, which is one signal for intron excision (see Donor Splice Site).

Allele. One form of a pair of genes at the same locus; an individual is homozygous if the alleles are the same, heterozygous if the alleles differ. If there are more than two alleles for the locus in the population, they are called multiple alleles, or the set is an allelic series (see also Genetic Compound.)

Allele-Specific Oligonucleotides (ASO). Synthetic DNA probes used to detect single base mutations.

Allelic Exclusion. Assembly of functional immunoglobulin heavy or light chain genes on only one of the two homologous chromosomes for each chain, thus ensuring that a given B cell makes only one unique functional antibody after DNA rearrangement; the alleles encoding the Ig genes on the "unused" chromosome are excluded from the rearrangement process.

Allosterism. Modulation or regulation of the activity of a protein through reversible changes in the physical conformation of the protein.

Alternative Splicing. The production of multiple mRNAs from a single transcription unit.

Alu Family. A set of highly repetitive DNA elements dispersed throughout the human genome; each member is about 300 bp in length.

Amphiphilic Helix. A protein α-helix with a net spatial charge distribution, e.g., positively charged amino acids (lysine, arginine) concentrated on one face of the helix, and noncharged (alanine, valine) or negatively charged (glutamate, aspartate) amino acids concentrated on the opposite face. Signal sequences of polypeptides destined for transport to mitochondria are thought to consist of amphiphilic helices.

Amyloid. Protein aggregates consisting of twisted β-pleated sheet fibrils. Amyloid fibrils stain with Congo red and exhibit optical birefringence. Several different proteins can form amyloid.

Annealing. Hybridization of nucleic acids.

Antimessage Sense. The strand of DNA or RNA that reads in the opposite (complementary) 5′–3′ orientation to that of the mRNA (see Antiparallel and Antisense).

Antiparallel. The orientation of DNA strands in a duplex. One strand runs 5′ → 3′ in one direction whereas the complementary strand runs 5′ → 3′ in the opposite direction. Thus, by virtue of the polarity of DNA, the two strands are antiparallel.

Antisense RNA. An RNA transcript having the polarity opposite (i.e., complementary) that of normally transcribed RNA. The term is used in studies of gene regulation and in the RNase cleavage assay.

Antitermination. Reduction or prevention of transcription termination at a terminator site due to the action of specific protein factors that either modify RNA polymerase or "sequester" the termination signal.

Attenuation. A mode of regulation in prokaryotes that links transcription with translation; a translation feedback mechanism switches between termination and readthrough in the upstream region of a transcript.

Autosome. Any chromosome in a eukaryotic organism other than the sex chromosomes.

Auxotroph. A cell that requires a particular nutrient, without which it would die; auxotrophy is useful as a selectible marker of cells that lack specific enzymes.

Base Pair (bp). A pairing of two nucleotide bases, each one associated with one of the two strands in a nucleic acid duplex, usually in accordance with the Watson–Crick pairing rules: A with T (or U in RNA) and G with C.

Blot. The transfer of macromolecules (DNA, RNA, protein) from a gel to a solid support, such as nitrocellulose or nylon, by passing high-salt buffers through the gel. The transfer can be performed "passively" (by capillary action through a stack of towels), under vacuum, or electrolytically (see Dot, Northern, Southern, Western, and Spot Blot).

Branch Point. The site near the 3′ end of an intron to which the 5′ end of the intron is ligated, to form a lariat during splicing.

CCAAT Box. A eukaryotic upstream promoter DNA element that affects the efficiency of the initiation of transcription.

Cap. The structure added after transcription to the 5′ end of eukaryotic messenger RNAs, containing an unusual methylated guanine nucleotide.

cDNA. Complementary DNA, i.e., single-stranded DNA complementary to an RNA; it is synthesized *in vitro* by using the RNA as a template for the action of reverse transcriptase.

cDNA Clone. A double-stranded cDNA carried in a cloning vector. See Vector.

Centimorgan (cM). 0.01 Morgan. The unit of distance in a genetic map, defined by the frequency of recombination. One cM is equivalent to 1% recombination frequency, i.e., the distance along the genetic map that would experience a recombination event in 1% of meiotic events. In terms of the physical map, 1cM is equivalent to about 1 million bp.

Central Dogma. The core postulate of molecular biology: Information flows from gene (DNA) to message (RNA) to protein.

Centromere. The constricted region of a chromosome that is the site of attachment of the mitotic spindle; it is also the point of joining of the two sister chromatids (visible in metaphase).

Chambon Rule. Correct splicing of eukaryotic mRNAs requires that the DNA contain GT at the 5′ splice site and AG at the 3′ splice site. Other nucleotides do not substitute.

Chiasm. A cross, formed in meiosis by the junction of chromatid strands of homologous chromosomes; it is evidence of interchange of DNA (crossover) between the two members of the pair.

Chimera. An organism (or person) composed of cells that arose from different zygotes. In human genetics, chimeras result from the exchange of stem cells *in utero* by dizygotic twins.

Chloramphenicol. An antibiotic that inhibits translation by bacterial and mitochondrial, but not by eukaryotic, ribosomes.

Chromatid. One of the two daughter copies of a chromosome after replication.

Chromatin. The complex of genomic DNA and associated proteins.

Chromosome. A discrete, large-scale unit of the genome, consisting of packaged chromatin, usually visible macroscopically only during cell division.

Chromosome Hopping. Screening a library of many clones with unique probes to isolate clones that arise from different segments of the same chromosome, without having to clone the intervening DNA region; also called chromosome jumping.

Chromosome-Specific Libraries. Genomic libraries consisting of a vector plus DNA inserts derived from a single chromosome. The DNA is obtained from flow-sorted chromosomes (see Fluorescence-Activated Cell Sorting).

Chromosome Walking. Screening a library of clones that contains overlapping inserts, to isolate clones containing insert fragments that are adjacent to the original insert from which the probe was derived.

Cis-Acting Factor. A regulatory element or factor that operates only when it is physically contiguous with the gene it regulates (see also Trans-Acting Factor).

Cistron. A genetic unit of DNA that encodes a protein; it is synonymous with a "gene."

Clonal Selection. Proliferation and maturation of lymphocytes as a result of interaction with a foreign antigen.

Clone. (1) A large number of molecules or cells, derived from, and identical to, an initial ancestral molecule or cell. (2) A segment of DNA that has been amplified as a recombinant molecule in bacteria. The word can also be used as a verb ("The gene has been cloned").

Coding Strand. The strand of DNA having the same sequence as its cognate RNA.

Codon. A ribonucleotide triplet that either encodes a specific amino acid or provides the signal to terminate translation.

Colony-Forming Unit (CFU). A colony of transformed cells, usually from spleen (CFU-S); the transformation is due to infection with a retroviral construction, and is used to detect cells that contain a DNA sequence of interest.

Colony Hybridization. A method for using probes to detect individual bacterial colonies containing recombinant DNA sequences homologous to that probe.

Complementary Strands. DNA or DNA–RNA duplexes in which the nucleotide sequence on one strand is represented on the other strand by the sequence of nucleotides to which it would naturally hybridize, i.e., A with T or U; G with C.

Complexity. The total length of different DNA sequences in a genome or a subset of the genome; simple sequences (e.g., ATATATAT . . .) are of low complexity, whereas most coding regions are of high complexity.

Compound Heterozygote. See Genetic Compound.

Concordance. In chromosome assignment, the simultaneous presence or absence of a gene marker and a particular human chromosome in a somatic cell hybrid. Also used to describe the appearance of a disease in human siblings or twins who are then "concordant" (or "discordant," if only one twin has the disease).

Consanguinity. Marriage between related individuals who have one or more common ancestors, most commonly but not always cousins.

Consensus Sequence. The sequence of nucleotides that serves a particular function. The consensus or "ideal" sequence is derived by comparing nearly identical sequences from different sources. Also called the "canonical sequence."

Constant Region. The invariant portion of an antibody immunoglobulin that determines binding to antigens or cells.

Corepressor. An effector that combines with a nonfunctional repressor to allow it to bind at the operator site.

Cos Site. Cohesive ends of phage λ arms that are required for *in vitro* packaging of DNA fragments in forming a cloning vector.

Cosmid. A plasmid containing the phage λ cos site, thus enabling the plasmid to be packaged *in vitro* in the phage coat. Cosmids can be used as vectors that contain large (about 35–45 kb) recombinant DNA inserts.

CpG. A dinucleotide (cytidine–phosphodiester bond–guanidine) associated with both DNA methylation and high frequency of mutation.

CRM. Acronym for "cross-reacting material"; the form of a protein that has lost enzymatic activity but retains antigenic activity and reacts with antibodies to the native enzyme.

Crossing-Over. The process by which homologous segments of paired maternal and paternal chromosomes are exchanged.

C-Terminus. The carboxy terminus of a polypeptide; the "back" end.

C Value. The total amount of DNA in the haploid genome; C stands for content or complexity.

Cycloheximide. An antibiotic that inhibits eukaryotic protein synthesis.

Cytoplasmic Inheritance. Inheritance of genetic material that is located not in the nucleus (as in Mendelian inheritance) but in the cytoplasm, e.g., mitochondrial and chloroplast DNA (see Maternal Inheritance).

Dalton (Da). The unit of molecular mass, based on the molecular mass of carbon as 12.000. Some protein molecular masses are expressed in kilodaltons (kDa).

Deletion. The removal of a sequence of DNA, with ligation of the remaining flanking DNA.

Denaturation. Dissociation of double-stranded DNA or RNA into the single-stranded state, usually performed by heating ("melting") or chemically (use of alkali for DNA).

Denaturing Gradient Gel Electrophoresis. A type of electrophoresis in which DNA duplexes are separated according to melting or denaturing characteristics. Also called Lerman gels.

Diagonal Transmission. Expression of an inherited disorder in an affected son of a clinically unaffected mother who carries the gene; seen in the inheritance of X-linked recessive genes.

Differentiation. The process by which primitive cell types proliferate and develop specialized functions.

Diploid. The state of a cell in which each autosome is present in two copies, along with one copy of each of the two sex chromosomes.

Discordance. In chromosome assignment, the case in which a gene marker is present but a particular human chromosome is absent in a somatic cell hybrid, or, conversely, the gene is absent but the chromosome is present. Also used to describe twins when a disease appears in only one of them.

Disjunction. Movement of sister chromatids or sister chromatid pairs to opposite poles during cell division.

Divergence. The percentage difference between amino acid sequences in the same protein from two different organisms, or between nucleotide sequences in two related nucleic acids; divergence is a measure of evolutionary distance.

DNA. Deoxyribonucleic acid, the genetic material or the nucleic acid constituent of the genome.

DNA Fingerprinting. RFLP analysis of the hypervariable regions of genomic DNA. The pattern for each individual is unique, hence the term "fingerprint."

DNA Ligase. The enzyme responsible for joining two DNA fragments covalently; used to make recombinatnt DNA molecules.

DNA-mediated Transformation Introduction of foreign DNA into a eukaryotic cell, for the purpose of studying or altering gene expression in the host cell.

Dominant. The phenotype of a heterozygote, resulting from expression of only one allele (the dominant allele), independent of the expression of the second allele.

Donor Splice Site. The GT dinucleotide at the extreme 5′ end of an intron, which is one signal for intron excision (see Acceptor Splice Site).

Dot Blot. See Spot Blot.

Double Minute (DM). An extrachromosomal, but chromosome-like self-replicating structure, consisting of an array of linked amplified genes; visible in the nucleus at metaphase. ("Minute" in the sense of "small.")

Downstream. Further in the direction of expression from 5′ to 3′ in DNA; in transcription, downstream is more "3′-ward."

Effector. A protein that modulates the binding affinity of a repressor at the operator site.

Electrophoresis. Separation of molecules (DNA, RNA, protein) that migrate in an electric field, now almost always carried out in gels of agarose or polyacrylamide.

Endonuclease. An enzyme that cleaves DNA or RNA within the polynucleotide chain.

Enhancer. A regulatory element that increases the transcriptional efficiency of promoter; it is cis acting, but can function in any location relative to the promoter.

Episome. A plasmid that can be integrated into the bacterial chromosome.

Epistasis. An effect of one (nonallelic) gene on another; also, the effect of a "modifier" gene on a second gene locus.

Ethidium Bromide. A stain used to visualize DNA in ultraviolet light; therefore, a reagent for analysis of DNA after separation by gel electrophoresis.

Eukaryote. An organism in which the cells contain a nucleus.

Exon. Any portion of a split gene that encodes part of the mature messenger RNA (see also Intron).

Exonuclease. An enzyme that cleaves DNA or RNA processively from either the 5′ or 3′ end.

Expression Vector. A cloning vector (e.g., λgt11) that, in a suitable bacterial host, produces a protein that is encoded by the inserted exogenous DNA; colonies of bacteria that contain the clone of interest can be isolated by using antibodies to the protein instead of with DNA probes.

Expressivity. The clinical manifestations in an individual with an abnormal gene; "variable expressivity," seen mostly in autosomal dominant conditions, describes the mild syndromes in some members of a family and the more severe or even different clinical manifestations in others.

FIGET. Acronym for "field inversion gel electrophoretic technique." A method to separate large pieces of DNA, except that the alternating field is placed at 180° rather than at 90°, as in pulsed field gel electrophoresis.

Fingerprinting. (1) Cutting cloned DNA with a set of restriction enzymes and running the cleavage products on an electrophoretic gel. The pattern is unique for each clone, and can be used to link clones that contain overlapping inserts. (2) Using hypervariable region RFLPs as unique identifiers of an individual's DNA. (3) Running two-dimensional gels of either RNA that has been cleaved by RNases or of peptides that result from the cleavage of a protein by proteases. The method is used to identify the composition or presence of the RNA or protein.

Fluorescence-Activated Flow Sorting. A method for purifying individual cells or chromosomes one by one. The method relies on the optical detection of particles of different sizes or fluorescence intensities.

Fragile Site. A heritable "gap" in the chromosome that is seen in cytogenetic studies after the cells have been cultured under specific conditions; one example is X-linked mental retardation associated with a particular fragile site on X.

Frameshift Mutation. A mutation caused by insertion or deletion of single DNA nucleotides in a coding region, so that the triplet-based translational reading frame is thrown out of register. Some frameshift mutations result in premature "reading" of a termination codon not too far downstream from the point of the frameshift; others produce long stretches of altered amino acid sequence, usually converting the polypeptide into a "nonsense" protein.

Fusion Peptide. A recombinant protein arising from insert DNA sequences that are fused (or ligated) to expression vector sequences; this recombinant produces a translatable mRNA (see also Expression Vector).

G-Banding. A method of visualizing metaphase chromosomes with the Giemsa stain, which stains alternating dark (stained) and light (unstained) bands. The G-banding pattern is unique for each chromosome, and helps to identify it.

Gene. An integral unit of usable genetic information; most often, a segment of DNA encoding a polypeptide or a structural RNA as well as the associated regulatory elements.

Gene Amplification. The production in the genome of many copies of a specific gene, usually as a result of environmental influences.

Gene Replacement Therapy. The introduction of new genetic information into somatic tissues (not necessarily including the germline) to complement or replace the action of an endogenous defective gene and, if expressed at sufficiently high levels, to ameliorate an inherited genetic error.

Genetic Code. The 64 possible triplets of codons, and the amino acids that they specify; three of the codons are termination codons.

Genetic Compound. An individual carrying two different alleles, both of which are abnormal; the individual lacks a normal allele, but is not homozygous; also known as a compound heterozygote.

Genetic Map. The arrangement of genes or genetic markers along the chromosome; distances are expressed in recombination units (morgans), in contrast to the physical map.

Genome. The total heritable information content of an organism, as contained in the DNA.

Genotype. The description of an organism in terms of the genome; often used to describe the alleles at a particular locus.

Goldberg–Hogness Box. The TATA box.

Haploid. The state of a cell in which each autosome is present only once, along with one of the two sex chromosomes; the haploid state is characteristic of the gametes.

Haplotype. (1) A group of alleles from closely linked loci, usually inherited as a unit, as in the histocompatibility complex. (2) A set of restriction fragment sites closely linked to one another and to a gene of interest.

HAT Medium. A selective medium containing hypoxanthine, aminopterin, and thymidine. Useful in selecting for cells deficient in nucleotide biosynthesis.

Helper. (1) A prokaryotic or eukaryotic virus that enables a "defective" virus to be packaged and released from the host cell. (2) A eukaryotic cell that provides all the components necessary for packaging defective viral sequences.

Hemizygote. An individual carrying only one allele of a particular gene, rather than the usual two. Males are hemizygous for X- and Y-linked genes.

Heteroduplex. Double-stranded DNA, RNA, or DNA–RNA hybrids in which the two strands are derived from two different sources. Heteroduplexes are usually not perfectly complementary.

Heterogeneity. If a phenotype can be produced by different genetic mechanisms, the phenotype is genetically heterogeneous.

Heterogeneous Disease. A single disease phenotype caused by different genetic mechanisms in different individuals, i.e., there are multiple disease alleles that affect the same gene in different ways. Alternatively, abnormalities of different genes (and different gene products) might cause the same clinical manifestations.

Heterogeneous Nuclear RNA (hnRNA). The precursor to mRNA within the nucleus.

Heterozygote. An individual with two different alleles for a particular gene, one on each homologous autosome.

High-Resolution Chromosome Banding. Banding of chromosomes in prophase rather than metaphase, which enables visualization of about 10 times as many bands; karyotypic anomalies are thus easier to detect.

Histones. Highly basic proteins that bind DNA in the nucleosome.

Holoenzyme. The complete enzyme, made up of all constituent subunits and prosthetic groups.

Homogeneously Staining Region (HSR). A karyotypic abnormality in which the normal banding pattern is disrupted, due to the integration of an array of linked amplified genes.

Homozygosity Mapping. A method used to map genes that cause recessive traits when the gene product is not known; based on the assumption that in children of consanguineous marriages, RFLPs in regions adjacent to the disease locus will be homozygous preferentially.

Homozygote. An individual carrying two identical alleles for a particular gene, one on each homologous autosome.

Horizontal Expression. Evidence of an inherited disorder in the same generation of siblings or cousins (although the parents may be unaffected); seen in the inheritance of autosomal recessive genes.

Host. Recipient of a recombinant DNA molecule (usually a bacterium); an organism that permits replication of recombinant DNA.

Hotspot. A region of DNA that is unusually prone to recombination, rearrangement or deletion.

Housekeeping Genes. Genes that are active in all cells because the products are necessary for normal, "nonspecial" functioning of the cell. Organ-specific genes are outside the realm of housekeeping.

HTF Island. Acronym for "*Hpa*II tiny fragment island"; nonrandom clustered regions of the genome containing multiple recognition sequences for the restriction endonuclease *Hpa*II (CCGG). HTF islands have been associated with loci containing structural genes.

Hybridization. The pairing of two single strands of DNA, or one of DNA with one of RNA, to yield a stable duplex.

Hypervariable Region. (1) A region of the chromosome containing variable numbers of tandemly repeated short DNA sequences; useful in RFLP analysis. (2) The portion of an antibody that determines antigen-binding specificity.

Illegitimate Recombination. Nonhomologous recombination; recombination between DNA strands that have no nucleotide sequences in common.

Inducer. An effector that decreases the binding affinity of a repressor.

Inducible Enhancer. An enhancer that responds to an activating factor.

Informative. A family is informative if both parents are heterozygous for an RFLP marker, so that the two parental chromosomes can be distinguished.

In Frame With. The proper positioning of a codon triplets to produce a continuous translation product, such as a fusion peptide.

Initiation Codon. The codon specifying the first amino acid in a polypeptide chain; usually AUG, which specifies methionine.

Insert. An exogenous DNA (restriction) fragment that is combined (ligated) with a vector to produce a recombinant DNA molecule.

Insertion. An additional DNA segment ligated into a stretch of DNA.

***In Situ* Hybridization.** Hybridization of a DNA or RNA probe directly in cells, tissues, or chromosomes mounted on slides. The technique is used to locate RNA messages for specific proteins, or viruses, within cells.

Interference. The tendency for an already established crossover event to suppress (negative interference) or enhance (positive interference) the formation of other crossovers in nearby regions.

Interphase. The period of the cell cycle when it is not in mitosis.

Intervening Sequence (IVS). An intron.

Intron A transcribed segment of DNA that does not appear in the final mature RNA; the mRNA segment encoded by an intron is excised by splicing to form a "mature" mRNA.

***In Vitro* Mutagenesis.** Modifying DNA regions *in vitro*. Both deletions/insertions and single-point mutations can be created. This type of "reverse genetics" is useful in studying gene regulation.

***In Vitro* Packaging.** The assembly of infectious phage (e.g., λ) particles when extracts that contain viral proteins are mixed with recombinant DNA molecules that have cos sites.

***In Vitro* Translation.** Synthesis of polypeptide chains *in vitro* in a synthesizing system, such as rabbit reticulocyte or wheat germ lysates, using exogenously added mRNA.

Isoschizomers. (1) Restriction enzymes isolated from different strains of bacteria, but which have the same recognition sequences. (2) Restriction enzymes cleaving at different positions within the same recognition sequence.

Isozymes. Two or more different, but related, enzymes, each catalyzing the same reaction. Isozymes are often tissue specific.

Karyotype. The overall display of chromosomes at mitosis; usually visualized in metaphase.

Kilobase Pair (kb). 1000 base pairs of DNA or 1000 bases of RNA.

Klenow Fragment. The large fragment obtained from the tryptic digestion of DNA polymerase I; Klenow retains the DNA polymerase activity, but no longer has the 5′-to-3′ exonuclease activity. It is used in many recombinant DNA techniques, including end labeling and Sanger sequencing.

Lariat. The topologically closed splicing intermediate formed by the ligation of the 5′ end of an intron to an internal branch point just prior to the 3′ end of the intron.

Leader. The 5′ untranslated segment of the transcription unit immediately preceding the initiation codon.

Legitimate Recombination. Homologous recombination between DNA strands that have regions of similar or identical nucleotide sequence.

Library. A collection of recombinant DNA molecules that includes most of the genomic or cDNA sequences from an organ or organism.

Ligation. Covalent joining of two or more DNA molecules by DNA ligase enzyme.

Linkage. The tendency of two genes to be inherited together, by virtue of proximity to each other on the chromosome.

Linkage Disequilibrium. If the frequency of a specific gene haplotype is not equal to the product of the individual gene frequencies, two linked markers occur together on the same chromosome more frequently than expected by chance.

Linked Clone Library. A library of clones constructed with overlapping inserts, so that a single probe will hybridize to more than one clone, thus linking them up.

Linked Clone Map. A set of isolated and characterized cloned DNA fragments that together span the entire DNA region of interest.

Linker. A small piece of synthetic DNA used in constructing libraries, usually for the attachment of inserts to vector sequences.

Linking Probes. Cloned small pieces of DNA that contain the recognition sequence for a restriction endonuclease (usually a "rare cutter"); useful for long-range chromosome walking.

LIPED. The acronym, a contraction of linkage and pedigree, of the most commonly used computer program for linkage analysis and for calculating lod scores.

Locus. The position of a gene on the chromosome; it may be occupied by any of the alleles for that gene.

Lod Score. Acronym for the "log of the odds" that two markers are genetically linked; a score of 3 (i.e., 1000:1 odds) is usually taken as the minimum lod score for significant linkage.

Long Terminal Repeat (LTR). A segment of repeated DNA that contains transcriptional control elements, and is located at each end of retroviral DNA.

Lyonization. Inactivation of one of the two X-chromosomes in the somatic (not germline) cells of a female embryo. The choice as to which particular X-chromosome is inactivated in each cell is random.

Lysis. The death of a bacterium due to the action of a phage, with release of progeny phage that can then infect other bacteria. Infection of bacteria with lytic phages on an agar plate gives rise to clear plaques on a background "lawn" of uninfected cells; each circular, clear plaque results from a single infective center.

Lysogen. A bacteriophage that has integrated its DNA into the host chromosome.

Lysosome. A subcellular organelle in which molecules containing complex sugars (e.g., glycoproteins and glycolipids) are broken down by acid-stable enzymes, such as hexosaminidase, fucosidase, or glucosidase. Similarly, proteins may be degraded by lysomal proteases.

M13. A filamentous bacteriophage in which the DNA goes through either single- or double-stranded forms in different portions of the life cycle. This property makes it useful in Sanger sequencing and *in vitro* mutagenesis.

Main Band DNA. The bulk of DNA that forms a broad peak on a density gradient.

Map. A representation of positions on the chromosome. The map may be either genetic or physical (see Genetic Map, Physical Map).

Marker. (1) A molecule of known size used to calibrate sizes on an electrophoretic gel. (2) Any allele of interest. (3) A genetic characteristic (either genotype or phenotype) used to track a second less well-defined characteristic, due to the genetic linkage of the two. Using an agreed-upon frame of reference, markers can be considered to be *proximal to, distal to,* or *bridging* the locus of interest.

Maternal Inheritance. Inheritance of genetic material through the mother; cytoplasmic inheritance. Virtually no human mitochondria pass from the sperm to the zygote, so all mitochondria inherited by the child arise from the egg (see Cytoplasmic Inheritance).

Maturase. An intron-encoded enzyme that is responsible for excising the segment of pre-mRNA encoded by that intron, as part of the splicing of the pre-mRNA into a mature message; found in some yeast mitochondrial genes.

Mature Transcript. The RNA transcript after all processing reactions (e.g., capping, polyadenylation, splicing) have been completed; the RNA is now ready for use as a messenger.

Maxam–Gilbert Sequencing. Sequencing of DNA using base-specific chemicals to modify and then cleave DNA at each nucleotide in the DNA.

MB. Megabase pairs, or million base pairs.

McKusick. Victor McKusick, author of *Mendelian Inheritance in Man* (Baltimore: Johns Hopkins University Press, 8th Ed., 1988), the standard reference compendium of genetic disorders. It also contains a valuable, up-to-date summary of the human chromosomal gene map.

Meiosis. (1) Division of diploid sex cells to produce the haploid gametes. (2) In RFLP analysis, an event in which DNA recombination may occur.

Melting. The denaturation of duplex DNA (or of DNA–RNA hybrids) into the single-stranded constituents.

Mendelian Inheritance. Inheritance of genetic material that is located in the nucleus, on the chromosomes. Genes are inherited from both parents, with the parental haploid sets combining in the zygote to form the diploid set of the child.

Message Sense. The strand of DNA or RNA that reads in the same 5′–3′ orientation as the mRNA.

Messenger RNA (mRNA). RNA encoding a protein or polypeptide.

Metabolic Interference. The hypothesis that two alleles at a locus, or two alleles of genes at different loci, cause a harmful effect only when they are present together in the same individual.

Metaphase. A stage in cell division in which the condensed chromosomes are stationary and aligned in the middle of the cell; the stage most often used for visualizing chromosomes and for karyotype analysis.

Methylase. An enzyme that puts methyl groups on specific (restriction site) sequences of DNA; methylated sequences are usually refractory to digestion by restriction enzymes.

Missense Mutation. A mutation in a codon that converts it to one encoding a different amino acid; the change may or may not affect the function of the protein.

Mitochondrion. An intracellular organelle found in all eukaryotes, responsible for oxidative energy production.

Mitosis. Division of a eukaryotic somatic cell.

Mitotic Segregation. Unequal distribution of mitochondria from the parent cell to the two daughter cells after mitosis.

Modifier Gene. A gene that modulates the expression or activity of a second (nonallelic) gene (see Epistasis).

Morgan (M). The unit of the genetic map, expressed as the frequency of recombination between two genes.

Mosaic. An individual or tissue with at least two cell lines that differs in genotype or karyotype, derived from a single zygote.

Mosaicism. Unequal distribution of alleles among different cells of the same organism (a phenomemon often encountered in transgenic mice).

Multifactorial Threshold Disease. A disorder caused by the interaction of genetic and environmental factors, expressed only above a threshold of genetic susceptibility.

Multipoint Mapping. Linkage analysis with more than two loci.

Mutation. Any inheritable change in the sequence of DNA.

Negative Control. Regulation by a factor that represses transcription in the absence of any derepressing factors.

N-Terminus. The amino terminus of a polypeptide; the "front" end.

Nick Translation. A method of incorporating radioactive nucleotides into DNA through the combined actions of DNase I (an endonuclease) and *E. coli* DNA polymerase I; useful for making radiolabeled hybridization probes.

Nondisjunction. The failure of sister chromatids or sister chromatid pairs to go to opposite poles during cell division; the main cause of trisomy 21 in Down's syndrome.

Nonsense Mutation. Any mutation resulting in the creation of an in-frame codon that does not encode an amino acid; it is therefore a termination codon and results in premature termination of the polypeptide.

Northern Blot. A method of detecting and sizing RNA species by transferring them from an electrophoretic gel to a nitrocellulose or nylon filter by "blotting," and hybridizing the RNA on the filter with specific labeled DNA or RNA probes.

Nucleus. The large organelle of the cell containing the chromosomes.

Nucleosome. The fundamental structural unit of the chromosome, consisting of about 140 bp of DNA wrapped around a core of histones, plus 60 bp of DNA linking it to the next nucleosome.

Nucleotide (nt). The fundamental unit of DNA or RNA, consisting of three elements: a cyclic base [adenine (a), cytosine (c), guanine (g), thymine (t), or uracil (U)], a five-membered sugar ring (ribose), and a phosphate group.

Okayama–Berg Vector. A plasmid cloning vector specifically designed to enhance the isolation of full-length cDNAs.

Oligo(dT). A synthetic oligonucleotide of deoxythymidine, 10–30 nt long, that can hybridize to the poly(A) tails of mRNA; used for isolating mRNA and for priming cDNA synthesis with reverse transcriptase.

Oligonucleotide. A short piece of single-stranded DNA (10–100 nt long), usually synthesized *in vitro* chemically.

Oncogene. In eukaryotes, a normal cellular counterpart to a transforming gene of an RNA tumor virus.

Open Reading Frame. A stretch of DNA triplets or RNA codons uninterrupted by a termination codon; it may encode a polypeptide.

Operator. The site of binding of a repressor to DNA.

Operon. A complete unit of gene expression, including all the structural genes plus any regulator genes and control sequences.

Origin. The point on DNA at which replication commences.

p. Designation of the short arm of a chromosome (from the French, petit).

Palindrome. A sequence of DNA that is identical on both strands when each is read 5′ to 3′; palindromic sequences are the recognition sites of most restriction endonucleases.

Partial Digestion. Digestion of DNA with a suboptimal amount of enzyme, so that some, but not all the sites for that enzyme are cleaved; useful in long-range restriction mapping.

Penetrance. The fraction of individuals carrying an abnormal allele who actually express the abnormal phenotype.

Peptide. A small segment of a polypeptide or protein.

pERT. An acronym for "phenol-enhanced reassociation technique". A method of reassociating (i.e., hybridizing after denaturation) low abundance DNA species at greatly enhanced rates. pERT was used most notably in the isolation of the region of the X-chromosome containing a large deletion in the DNA of a patient with Duchenne muscular dystrophy (DMD), by enriching for DNA in the normal X-chromosome that had no "reassociation partners" in the DMD locus.

pERT Library. A library of clones of double-stranded reassociated DNA derived from the pERT method.

Phage. A bacterial virus.

Phase. (1) The RFLP allele associated with the disease mutation. (2) The association of alleles for two genes on the same chromosome. In a double heterozygote (AaBb; A and B are dominant; a and b are recessive), *coupling phase* occurs when A and B are on one chromosome and a and b are on the other; *repulsion phase* occurs when A and b are on one chromosome and a and B are on the other. (3) In family studies, the phase is "known" when the disease site is linked to a marker locus so that maternal and paternal haplotypes can be identified.

Phenotype. The description of an organism in terms of observable physical characteristics, enzymes, or proteins (see also Genotype).

Physical Map. The physical structure of genomic DNA; distances are measured in numbers of base pairs.

Plaque. A clear, circular area of bacteria lysed by phage, as visualized against a background "lawn" of uninfected bacteria, representing one infective center, and used to detect clones that contain a specific recombinant DNA.

Plaque Hybridization. A method for using probes to detect individual plaques containing recombinant DNA sequences homologous to that probe.

Plaque Purity. Isolation of a clone of a single plaque containing a recombinant sequence of interest, which is not contaminated by other, unwanted, clones.

Plasmid. An autonomous, self-replicating, piece of extrachromosomal circular DNA found in many bacteria; it need not be an episomal element.

Plus–Minus Screening. Examination of two different library populations of cDNAs in order to isolate clones that are in one library but not the other (see Subtractive Hybridization).

Point Mutation. A mutation (substitution) at a single base or base pair.

Polyadenylation. The addition of a stretch of polyadenylic acid to the 3′ end of a messenger RNA. The amount of poly(A) is related to the stability of the message.

Polyadenylation Signal. The sequence AAUAAA, located about 25 nt upstream from the polyadenylation site, which is a signal for polyadenylation.

Poly(A) Polymerase. The enzyme that catalyzes the addition of poly(A) to the 3′ end of a eukaryotic mRNA.

Polycistronic Message. A single transcript that contains the information encoded by contiguous genes, often due to readthrough by RNA polymerase.

Polylinker. A short piece of DNA containing multiple unique restriction enzyme recognition sites, inserted into a plasmid. The polylinker makes it relatively easy to insert exogenous DNA fragments into the plasmid.

Polymerase Chain Reaction (PCR). An enzymatic method of amplifying the amount of a specific segment of DNA of interest; conceptually, it is a type of cloning that does not require a vector.

Polymorphism. The presence of two or more alleles in a population at the same time, defined by differences in either phenotype or genotype; now often used to describe RFLP patterns. A trait is "polymorphic" if there is more than one form and if the less common form is found in at least 2% of individuals. The greater the number of different forms and the higher the proportion of individuals affected, the more informative the RFLP or other trait.

Polymorphism Information Content (PIC). A scale to determine how polymorphic a locus must be, and how many pedigrees would be required to establish linkage; ranges from zero for a totally uninformative locus to 1.0 for a highly informative locus.

Polyribosome; Polysome. A string of ribosomes bound to a messenger RNA.

Positive Control. Regulation by a factor that increases transcription (see Negative Control).

Pribnow Box. A promoter region in prokaryotes located about 10 bp upstream of the transcription start site; it has a consensus sequence of TATAAT on the coding strand.

Primary Structure. The linear sequence of nucleotides in DNA or RNA, or of amino acids in a polypeptide.

Primary Transcript. The original unprocessed RNA derived from transcription of the DNA, encoded in the transcription unit.

Primer. A short segment of single-stranded DNA used to initiate DNA polymerization; often used in construction of cDNA libraries.

Probability Ratio. In lod score analysis, the probability (or likelihood) of the data, given a specific value of the recombination frequency (θ_i) divided by the probability of the data if there were independent assortment ($\theta = ½$).

Proband. The family member through whom the family is ascertained; also called the propositus or index case (if affected).

Probe. A labeled molecule (usually DNA, but sometimes RNA) used in hybridization and blotting experiments to detect sequences complementary to that molecule.

Processed Gene. A retroposed gene that is a double-stranded cDNA copy of a mature, processed, mRNA message, which is reinserted into the genome at a site removed from the initial gene. Because it is processed, the gene has few or no introns, contains poly(dA) at the 3′ end, and is flanked by duplicated repeats at the site of insertion. Most processed genes are pseudogenes.

Prokaryote. An organism that lacks a nucleus (e.g., bacteria).

Promoter. A region of DNA associated with initiation of transcription by RNA polymerase.

Protease. An enzyme that catalyzes the cleavage of proteins into smaller polypeptides or into free amino acids.

Protein. The polypeptide chains that make up a biologically active molecule.

Provirus. A viral DNA genome integrated into a host genome.

Pseudogene. An apparently inactive component of the genome derived by mutation of an ancestral functional gene.

Pulsed Field Gel Electrophoresis (PFGE). Electrophoretic separation of very large pieces of DNA (50–1000 kb or more) using alternating electrical fields that are oriented perpendicular to each other.

q. Designation of the long arm of a chromosome.

Q-Banding. A method of visualizing metaphase chromosomes with quinacrine dye, which stains them in alternating dark (stained) and light (unstained) bands. The Q-banding pattern is unique for each chromosome, and helps to identify it.

Reading Frame. One of three possible ways to translate a nucleotide sequence as a series of triplets; an "open" reading frame contains no termination codons, and could be translated into a protein.

Readthrough. Continuation of RNA polymerase past a termination point.

Recessive. The phenotype displayed only when both alleles are identical (i.e., in a homozygote); in a heterozygote, the phenotype resulting from a recessive allele may be masked by the dominant allele. Often, the gene product of a recessive allele is either absent or inactive.

Recombinant. (1) The DNA produced by ligating two or more DNA molecules (insert and vector), placed in a host cell to yield a clone. (2) An individual who has a new combination of genes that were present together in either parent, as a result of crossing-over.

Recombination. (1) Formation of new combinations of linked genes by crossing-over between two loci. (2) The interconversion of two DNA duplexes, usually by crossing-over. (3) In RFLP analysis, the situation in which the RFLP and the disease phenotype may become unlinked, due to a recombination event during meiosis.

Recombination Frequency. The proportion of DNA strands that experience an odd number of crossover events between two loci. Two markers are likely to be inherited together if they are close to each other, so the recombination frequency increases as the two markers (e.g., the RFLP and the disease gene) become physically farther apart.

Regulon. A set of widely scattered genes that is regulated coordinately as one group.

Repetitive DNA. Identical or nearly identical DNA sequences that are represented multiple times (from a few thousand to a million copies) in the genome.

Replication. The process of duplication of parental DNA into two daughter molecules, each one of which segregates into a daughter cell after cell division.

Repressor. A negative-acting protein that blocks or supresses transcription.

Respiratory Chain. The series of enzyme complexes in bacterial and mitochondrial membranes responsible for transferring electrons to oxygen (producing water) and, in the process, providing the driving force for producing cellular energy, as ATP. The chain, also called the electron transport chain, consists of five complexes: Complex I (NADH–coenzyme Q reductase), II (succinate–coenzyme Q reductase), III (reduced coenzyme Q–cytochrome *c* reductase), IV (cytochrome *c* oxidase), and V (mitochondrial ATPase).

Restriction Endonuclease. An enzyme that recognizes specific short (usually palindromic) sequences in the DNA and cleaves the DNA at or near that point; used extensively in all facets of recombinant DNA technology.

Restriction Map. A physical map in which unique restriction endonuclease sites are ordered along a stretch of DNA.

Retroposon. A mobile genetic element which goes through an RNA intermediate before it is reinserted, as DNA, into the genome.

Retrovirus. An RNA virus that has a DNA intermediate in its life cycle.

Reverse Genetics. Use of recombinant DNA technology to identify the DNA mutation that gives rise to a disorder, and then to characterize the protein that the gene encodes. The method is the reverse of traditional biochemical genetics, in which the abnormal protein is identified first, and the protein is then used as a tool to isolate the defective gene.

Reverse Transcriptase. An enzyme that can polymerize single-stranded DNA from a template of RNA; useful in synthesizing cDNAs during the creation of cDNA libraries.

RFLP. Acronym for "restriction fragment length polymorphism"; changes in the size of DNA fragments due to the absence or presence of a particular restriction endonuclease site in the genome of some individuals. RFLPs are used as linkage markers, and are used in pedigree analysis (see Polymorphism).

RFLP Haplotype. A combination of individual RFLP patterns, each created by a single restriction enzyme (see also Haplotype).

Rho Factor. A protein required for transcription termination of many prokaryotic genes.

Ribonucleoprotein Particle (RNP). Protein–RNA (mainly snRNA) complexes located within the nucleus, and which seem to be associated with hnRNA splicing.

Ribosomal RNA (rRNA). RNA that is a major component of the ribosome. Two major eukaryotic rRNAs are called 18 S and 28 S (S, Svedberg unit, a measure of size as determined by centrifugation).

Ribosome. The cytoplasmic organelle, composed of ribosomal proteins and rRNA, on which translation takes place.

Ribosome-Binding Site (RBS). A site in the 5′ untranslated region of mRNA to which the ribosome binds before translation commences.

Ribozyme. RNA that acts as a true catalytic molecule, usually as an enzyme to catalyze splicing.

RNA. Ribonucleic acid, the nucleic acid constituent of the cell used in moving the information inhered in the DNA throughout the cell; consists of three major categories—messenger, ribosomal, and transfer RNA.

RNA Polymerase. An enzyme that reads a DNA template and transcribes a strand of isomorphic RNA reading, by definition, in the "sense" orientation.

RNase A Cleaveage Assay. Detection of single-base mismatches in sense–antisense RNA heteroduplexes; used in searching for point mutations.

Sanger Seqeuncing. Sequencing of DNA using base-specific terminators of DNA polymerase or reverse transcriptase. Also called dideoxy sequencing.

Satellite DNA. DNA consisting of many tandem repeats of a short (less than 10 bp) simple sequence unit, and which is observed as a subsidiary "satellite" band on density gradients relative to main band DNA.

SDS–PAGE. Acronym for "sodium dodecyl sulfate–polyacrylamide gel electrophoresis." A common method of separating polypeptides electrophoretically.

Secondary Structure. The internal folding pattern of a polynucleotide or polypeptide chain.

Segregation. (1) The clean separation of members of a pair of alleles when an individual forms germ cells. (2) The appearance of two or more markers that always "travel together" in genetic analyses.

Self-Splicing. Splicing of an RNA by the RNA itself, acting as a ribozyme.

Sense and Nonsense. (1) A codon that specifies an amino acid (e.g., UGG, tryptophan) as compared to one that does not (e.g., UGA, chain termination). (2) The coding strand of DNA (sense) as opposed to the complementary, noncoding, strand of the double helix (nonsense or antisense).

Shine–Dalgarno Sequence. A site in the 5′ untranslated region of prokaryotic messages implicated in binding of the mRNA to the ribosome to initiate translation.

Shocked Cells. Host bacteria that are made "competent" to take up DNA molecules by treating them with cold solutions of $CaCl_2$ followed by heating to elevated temperatures (e.g., shift from 37°C to 42°C); used in transforming cells with DNA, e.g., in subcloning and transformation experiments.

S1 Nuclease. An enzyme that digests single-stranded but not double-stranded nucleic acid.

Sigma Factor. A subunit of prokaryotic RNA polymerase that enables it to recognize promoter sequences and initiate transcription at the correct site.

Silent Mutation. A point mutation at a redundant codon position, which does not alter the amino acid specified by the codon.

Sister Chromatids. The two copies of a chromosome produced after replication.

Small Nuclear RNA (snRNA). A group of RNAs located only within the nucleus, many of which appear to be involved in hnRNA splicing.

Somatic Cell Hybrid. A hybrid cell containing chromosomes from two different species; often a single human chromosome in a background of the full complement of rodent (usually mouse or hamster) chromosomes. Somatic cell hybrids are useful in assigning genes to individual human chromosomes.

Southern Blot. A method of detecting and sizing DNA species by transferring them from an electrophoretic gel to a nitrocellulose or nylon filter by "blotting," and hybridizing the DNA on the filter with specific labeled probes. Named after E. Southern, who devised the method.

Sp1. The first identified trans-acting transcription factor that binds to an enhancer region.

Spliceosome. The ribonucleoprotein complex associated with intron splicing.

Splicing. The removal of introns from a primary mRNA transcript during the process of RNA maturation; the RNA encoded by introns is spliced out and the flanking exon sequences are ligated together.

Spot Blot. A Southern or Northern hybridization in which DNA (or RNA) is spotted directly on a filter without prior electrophoretic separation, and hybridized with a probe; useful in quantitation of autoradiographic signals, and in the assignment of genes to flow-sorted chromosomes.

Sticky Ends. The single-stranded nucleotides at the ends of DNA molecules, produced by restriction enzyme digestion, that have affinity for and bind the complementary ends of other DNA molecules.

Stop Codon. One of the three codons that do not specify an amino acid and that cause termination of translation: UAG ("amber"), UAA ("ochre"), and UGA ("opal").

Stringency. The combination of temperature, salt concentration, and buffer composition that determines the degree of hybridization of two nucleotide sequences. High-stringency conditions select for hybridization of pairs of complementary single-stranded sequences that contain very few nucleotide mismatches.

Subtractive Hybridization. A method of plus–minus screening in which cDNAs unique to cell type A are isolated by removing all cDNAs held in common between cell types A and B. One way of doing this is by hybridizing total cDNA from cell type A with total mRNA from cell type B, and discarding the cDNA:mRNA hybrids; the unhybridizable fraction thus represents the "unique" cDNA population of cell type A, which can then be cloned.

Synteny. (1) Different genes present on the same chromosome. (2) The same gene present on the analogous chromosome in two species.

Tandem Repeats. Multiple copies of the same sequence repeated in series, as opposed to dispersed repeats.

TATA Box. A eukaryotic AT-rich promoter element associated with the accuracy of transcription initiation; it is located about 30 nt upstream of the cap site, and has a typical sequence of ATATAAA.

Telomere. Either end of a chromosome.

Terminal Deoxynucleotidyltransferase. An enzyme that adds deoxynucleotide residues covalently to the 3′ end of a DNA molecule.

Termination Codon. Any of the three stop codons (UAA, UAG, or UGA) that result in termination of protein synthesis; also called a nonsense codon.

Termination Factors. Factors required to terminate transcription (see Rho Factor).

Tertiary Structure. The overall three-dimensional structure of a polynucleotide or polypeptide chain (i.e., all the internal secondary structures taken together).

Theta (θ). In lod score analysis, the recombination frequency; ranges from $\theta = 0$ (no recombination; markers are tightly linked) to $\theta = \frac{1}{2}$ (markers are totally unlinked, i.e., they are on different chromosomes).

Threshold Effect. In mitochondrial diseases, the requirement that a cell or tissue must contain a certain minimum number of defective mitochondria (i.e., mutant mitochondrial DNA) before a pathological phenotype is expressed. The term is also used to describe the effects of multifactorial inheritance in which one or more genes creates "liability" that becomes manifest only under the influence of other genes or environmental factors.

Trans-Acting Factor. A regulatory element or factor that is derived from a locus physically unlinked from the gene it regulates.

Transcription. The synthesis of RNA from DNA templates, achieved by the action of RNA polymerase.

Transcription Unit. The distance in the DNA between the sites of initiation and termination of a single RNA transcript; sometimes the transcript includes more than one gene.

Transcription Vector. A cloning vector in which the exogenous DNA is inserted downstream from a transcriptional promoter; on addition of RNA polymerase, either sense or antisense RNA can be transcribed, depending on the orientation of the insert.

Transfection. The uptake of recombinant DNA molecules by shocked host bacterial cells.

Transfer RNA (tRNA). Small (about 100 nt) RNAs to which amino acids are attached, and which are used as "adaptors" on the ribosome to transfer amino acids to the growing polypeptide chain.

Transgenic. The incorporation of the genes of one animal into the germline of another. Most often associated with the insertion of human genes into mouse embryos (transgenic mice).

Transition. A point mutation in which one purine (A or G) is substituted for another, or one pyrimidine (T or C) for another.

Translation. The synthesis of a polypeptide based on the codons present in the mRNA.

Translocation. A chromosomal rearrangement in which the DNA from one chromosome is moved to another; often the translocation is reciprocal, or balanced—two chromosomes exchange information.

Transposon. A mobile DNA element that can replicate and insert a copy of itself at a new chromosomal location.

Trans Splicing. Intermolecular splicing of two RNAs to form one message. Trans splicing is the rule, not the exception, in transcription in trypanosomes.

Transversion. A point mutation in which a purine (A or G) is substituted for a pyrimidine (T or C) or vice versa.

Upstream. Away from the direction of expression; when speaking of transcription, it is more "5′-ward."

Upstream Promoter Element (UPE). Any DNA sequence upstream of a gene that regulates transcription of that gene.

Vector. A self-replicating DNA molecule (virus or plasmid); insert DNA molecules can be ligated to it at unique restriction sites.

Vertical Transmission. Expression of an inherited disorder in successive generations (parent to child to grandchild), as in the inheritance of autosomal dominant genes.

VNTR. Acronym for "variable-number tandem repeat." A highly polymorphic DNA region; also known as a hypervariable region.

Western Blot. A method of detecting and sizing specific polypeptides or proteins by transferring them from an electrophoretic gel (usually run by SDS–PAGE) to a nitrocellulose or nylon filter by "blotting," and detecting the protein bound to the filter with specific antibodies.

Wild Type. The normal or reference phenotype or genotype.

Index

Page numbers in italics refer to figures, and *t* after a page number refers to tables.

VINCENT VEE-TWINS

OSPREY COLLECTOR'S LIBRARY

VINCENT VEE-TWINS

The famous 1000 Series, plus 500 Singles

Roy Harper

Published in 1982 by Osprey Publishing Limited,
12–14 Long Acre, London WC2E 9LP
Member company of the George Philip Group

British Library Cataloguing in Publication Data
Harper, Roy
Vincent vee-twins: the famous
1000 series, plus 500 singles.—(Osprey
collector's library)
1. Vincent vee-twin motorcycles
I. Title
629.2'275 TL448.V/
ISBN 0-85045-435-2

Editor Tim Parker

Filmset and printed in England by
BAS Printers Limited, Over Wallop, Hampshire

Contents

Acknowledgements

For his considerable help with the text of this book I must thank Phil Irving. Derek Harper and John Mellor assisted with the appendices. For several of the illustrations used I must also thank various magazines, notably *Motor Cycle News* (editor, Bob Berry) and *Motor Cycle Weekly* (editor, Mick Woollett) for their contributions. Others who did their very best were Mirco Decet, Roy Bacon, Richard Renstrom, Denis Jenkinson, Brian Terry and Tim Parker. Then there were I. W. Buckden, Bruce Photography, M. M. Evans, Brian Holder, Keig, Keystone Press Agency, *Motorcyclist* magazine, Planet News, K. Simmons, Elder Smith & Co. Ltd., R. Stafford and Len Thorpe.

A number of illustrations come from brochures, parts lists and rider's handbooks issued by the Vincent factory and now made available by the Vincent HRD Owners' Club. I must thank them.

When I saw the first title in the Osprey Collector's Library series, *BSA Twins & Triples* by Roy Bacon, I wanted to see a Vincent title to fit into that series. I am, therefore, grateful to Osprey Publishing for giving me that opportunity. I hope this edition will add something to the history of the magnificent Vincent motorcycle and encourage further enthusiasm for the marque.

Roy Harper
Chelsea, London
November 1981

To Phil Irving

Preface by Phil Irving

When Philip Vincent and I were collaborating in the creation of the postwar Vincent-HRD in 1945, we were doing so in circumstances which were unusual to say the least. The War was by no means finished: V2 rockets made life a trifle uncertain, food and petrol were scarce, and strictly rationed, and most material we would need was either in short supply or grudgingly obtained by permits from the Ministry of Supply on condition that 80 per cent of our eventual output would be exported. The factory contained plenty of machine tools which had been installed for manufacturing munitions, but only a few of them were suitable for making motorcycles.

Nevertheless, with Government contracts nearing their ends, it was imperative to push ahead with the design of our peacetime product with all speed. As a result of many spare-time conversations it had already been decided that we could not hope to compete with the makers of medium-weight models which were already in production, or could be easily re-introduced, and our salvation would lie in exploiting the reputation of the pre-war 'Series A' Rapide by retaining its performance and moderate weight, but generally tidying up the appearance, and making the complete engine and gear-box ourselves in one unit.

At the outset, there was no thought of making the 'Series B' Rapide, as it was called, into a racing machine, as our aim was to provide a model which was not only the fastest thing one could buy on two wheels, but would also be flexible, economical, and have a service life of around 100,000 miles. As most of the output would be going to foreign countries, it was essential that routine maintenance should be easily understood, and provision should be made for attaching a sidecar on either side, which would require being able to fit the kick-starter on either side—something nobody had thought of before.

One regrettable defect of the pre-war model had been the clutch which not only lacked enough grip, but needed far too much effort to lift it; this was eventually remedied with a servo design (often miscalled centrifugal) which had the ability to transmit 100 horse power but could be lifted with one finger—a great boon to ex-servicemen who had had their hands injured during the war.

Of course, all this time we proceeded as if the factory would exist forever: we had no inkling that it would be closed down in less than a decade, yet two decades later on many machines would still be in use, and their market value would have risen to several times their original cost. Values are still rising, thanks in part to the continued availability of spare parts from the Vincent Owners' Club Spares Company.

To go back to the original design, our pre-war wheels which had been extremely reliable could still be used, as could the Brampton girder forks. We refused to follow the current trend towards telescopic forks which we considered to be lacking in torsional rigidity, besides 'diving' very badly during brake application. In the preliminary discussions it had been decided to shorten the wheelbase to 55 inches and to simplify the whole construction by making the power unit the heart of the construction and to which the rear forks would be attached directly.

The headlug would be attached to the front cylinder by a bracket and long through-bolts, thus eliminating the front down-tube in the fashion utilized on the P & M (Panther) for many years. The headlug would also be attached to a hollow sheet-steel backbone which in turn was bolted to the rear cylinderhead and also forming the oil tank. The usual pair of springs was attached to the rear of this backbone which also contained a sidecar attachment point, the idea being that the front frame assembly and forks could be detached from the engine by the removal of four major bolts and wheeled away when a top overhaul was in progress.

This 'boneless wonder' construction as someone scoffingly termed it, proved to be eminently satisfactory, one or two severe mishaps early in the testing period failing to show up any weakness—a promise which was later borne out in the heaviest sidecar racing.

Although steel was hard to obtain, there was plenty of aluminium available from melting-down crashed or obsolete aircraft; partly for this reason, and partly to reduce weight and avoid excessive use of chromium plating, aluminium in various alloys was used extensively throughout. Other parts such as brakerods or banjo unions and axle tommy bars, which were usually plated, were made of stainless steel, that being one reason why old Vincents do not degenerate into rusty heaps. The exhaust pipes could have been stainless too, except that we already knew that they turned a repulsive yellow colour when heated to running temperature.

There was neither time, money, or manpower to initiate an expensive test programme in the modern manner. Instead, a great deal of forward thinking was put into the design at the paper stage. The one and *only* experimental model was fired up on April 27, 1946, and was tried out the same day by *The Motor Cycle* and anon *Motor Cycling*. Their reports were so eulogistic that the trickle of advance orders became almost a flood, but filling them was a different matter until more machine tools arrived and the staff could be increased. The latter difficulty was solved by advertizing for men in the motorcycle press, which resulted in a rush of applicants from whom the most promising were selected. They include the unforgettable Paul Richardson. The late C.J. Williams was already on the staff, and most of the long range testing was carried out by him.

Although not designed for racing, and there were no events in the UK for large-capacity machines, it was different overseas, where the first imports were eagerly acquired for competition, and thereby started a long run of successes. The Argentines and Brazilians were particularly keen on solo races which were mostly held on public roads through populous cities!

A severe setback occurred early in 1947 when Government ineptitude caused all industry to shut down for over three weeks in freezing weather. Vincent promptly enlisted all our idle personnel to move the machine tools and stores into a much larger building, and by the time electric power was again available we had two factories both in working order at much less expense than if an outside contractor had done the work.

A real break came from an unexpected source—the Clubmans TT Races which commenced in 1947 but without any Vincent entries. The Senior class included one-litre machines and in 1948, 11 Vincent HRD's were entered, rather to the amusement of many people who considered them too cumbersome for the difficult Island circuit. Amusement turned to indignant cries of unfairness when Vincent HRD's not only won, but made the fastest lap, while nine out of the ten which left the grid finished the course—an unprecedented demonstration of reliability which effectively silenced the scoffers and filled our order book to overflowing, although the production at only about 30 per week lagged well behind the demand.

The cancellation in 1948 by Argentina of a long-standing trade agreement effectively prevented all purchases from England. As we had been selling 20 machines a week in that area and had a batch of 50 models ready for dispatch to the San Rosario Police Force, the sudden cessation of this lucrative market struck a deadly blow at our fortunes, and undoubtedly assisted in the downfall of the Company.

However, the introduction of the Series D and diversification into other products were not enough to stave off the fate which was about to befall other manufacturers as well as us. The company continued to exist and operate a spare parts and repair service for several years until it was sold in 1960, and P.C. Vincent ceased to be a part of the organization. The Spares supply was, and is, continued by the Spares Company,

Left **Howard R. Davies used** 'Mercury' **in his HRD transfers and Philip C. Vincent followed suit when he bought HRD Motors Ltd in 1928**

Right **Howard R. Davies won two Senior TT Replicas in 1925—one as the winning rider and the other (shown) as the manufacturer of the HRD motorcycle**

H·R·D
THE VINCENT
H.R.D. Co. Ltd
STEVENAGE
HERTS

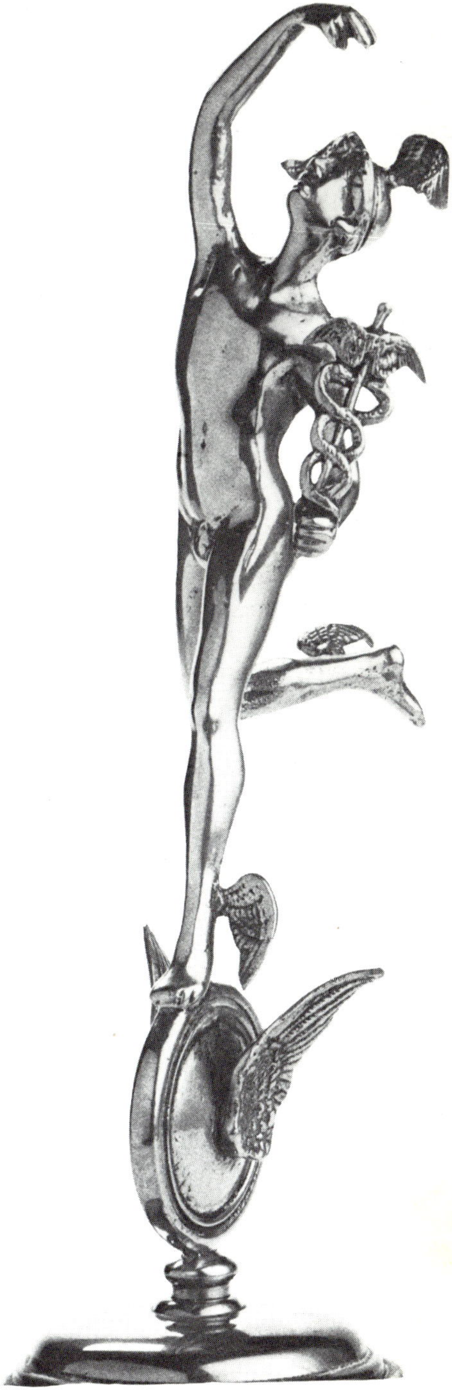

founded by the Vincent HRD Owners' Club, and largely thanks to that world-wide band of enthusiasts, the name and fame of the Vincent lives on and will do so for a long time yet.

Phil Irving
Melbourne, Australia
December, 1981

PCV (left) and HRD at the 1969 Vincent HRD Owners' Club dinner. 'Of course you know each other,' said the chairman. 'Never seen him before in my life,' replied HRD

Phil Irving writing *Vincent H.R.D. Development 1931–1936* **in author Roy Harper's study in Chelsea**

Proem

A Cambridge undergraduate with a passion for fast motorcycles sat astride a chum's McEvoy Anzani 1000 cc vee-twin. He retarded the ignition, kick-started the machine, revved the engine and revelled in the power at his command. Nervously, for this was his first ride on a 'thousand', he disengaged the clutch, slipped the hand-change mechanism into first gear and slowly released the clutch lever. Cautiously, he moved off in first gear and advanced the ignition. Acceleration was rapid, exhilarating, exciting as he shifted through the gears into top. The wind stung his face, blew his hair awry, clawed his eyes until they watered as the powerful machine thundered through the quiet Cambridgeshire countryside.

The twenty-year-old rider, Philip Vincent, had already built his own motorcycle—a 350 cc special with rear sprung frame—and his designer's mind appreciated the torque of the engine while the off-beat exhaust note was akin to the heartbeat of a faithful steed. The handlebars could be lower, of course, and the rear end was distinctly agricultural, but as he rode the bike back to George Lomonossoff, its anxious owner, Philip Vincent dreamed of producing a vee-twin that would be faster than any other motorcycle in the world.

The only picture I've seen of PCV aboard a pre-war Vincent. High level pipes were *de rigueur* **in the mid-thirties**

DAR23

Another Cambridge chum, John Cockshott, introduced Philip to the joys of Rudge ownership. Philip was impressed with the four-valve-engined motorcycle's ability to cover 200-mile trips reliably, despite 'a tendency to shake things off'.

To further the courtship of his other object of hero-worship—a beautiful girl who lived near Leicester—John attached a sidecar to the Rudge. The lady was impressed. She would be able to sit in comfort out of the stream of the usual stuff that went a pillion passenger's way—wind (and sometimes rain) laced with a dash of oil or petrol.

John proudly helped her aboard the combination for the first time and roared off happily.

One would like to be able to record that the acquisition of the chair helped John's suit to such an extent that the pair became man and wife and lived blissfully ever after. But this idyll was not to be. The bolts that attached the chair to the motorcycle grew tired and suddenly John was riding solo. The chair had quietly detached itself and deposited its sacred cargo safely by the roadside. The girl's enthusiasm for sidecars waned and John was left with a broken heart and a bent chair.

No doubt influenced by this little mishap, Philip always built a substantial safety margin into his frame members when he commenced manufacture; and several of his models were to incorporate Rudge engines.

This shows the rear sprung frame of Philip Vincent's first motorcycle; trail could be adjusted for experimental purposes

Philip and Elfrida Vincent in their flat (1977). Columbine, the cat, is a Siamese seal-point. I was so taken by her that I promptly bought a Siamese kitten from Harrods to prevent mice eating my books

1 Early Vincents (1927–1934)

Before founding his motorcycle company, Philip sought advice from Arthur Bourne, editor of *The Motor Cycle*, who said that motorcyclists were a conservative lot and would prefer to buy an established name rather than a new one. HRD Motors was for sale but would pater come across with the wherewithal? Yes, he would, and Philip purchased the company for £400. The HRD logo was retained with Vincent's name above within a modest scroll. The HRD logo always appeared, interestingly, as H·R·D, with full-points either side of the 'R', never with a third after the 'D'. For the sake of clarity and typesetting ease no full points appear in this text.

Founded in 1924 by Howard Raymond Davies, the successful TT rider, the HRD company enjoyed immediate acclaim when Howard won the 1925 Senior TT race with his JAP-engined motorcycle (after finishing second in the Junior TT).

Freddie Dixon, who customized his mount with great running boards and a backrest (Fred liked his comforts), won the Junior TT in 1927 but the company was faltering due to a lack of adequate control on expenditure (for example, entertainment of customers and others left a bill of £3000). Howard went into voluntary liquidation and when Philip enquired about the company it was owned by Ernie Humphries of OK Supreme.

Philip's father, Conrad, was a wealthy rancher in Argentina and he put up £30,000 to finance the company. There was one condition—the purse-strings were to be held by an associate, Frank Walker, who became the managing director of the Vincent HRD Company, Ltd., upon its formation.

Frank found a garage in Stevenage (with a listed Tudor building as a workshop) that allowed room for expansion; it was also reasonably close to London and Birmingham, a main line railway station was near by and the Great North Road on the firm's doorstep provided a ready-made test track.

Included in the deal were 'two or three good mechanics' who worked in the garage. Thus, in June 1928, with a famous monogram, an assortment of spare parts, a handful of personnel, suitable premises, substantial financial backing and one motorcycle (the Vincent Special), Philip Vincent announced to an uninterested world that he was open for business.

This first Vincent motorcycle, the forerunner of the famous series of vee-twins, was unique only in the construction of its frame, all other items being supplied by proprietary firms. But what a frame! Originally designed in 1925, when Philip was a 16-year-old pupil at Harrow School, the frame incorporated a top tube that could be lengthened or shortened, thereby enabling Philip to experiment with different rake angles of the steering head.

A 1926 HRD with 350 cc ohv JAP engine. Howard was one of the first manufacturers to fit a saddle tank. How sleek and fleet the bike looks!

The most innovative feature was the rear sprung frame. There had been rear sprung systems before, of course, some of them hefty contraptions that had gained a reputation for giving their riders some exciting moments culminating in a close-up of the rough macadam roads. Philip's frame, in contrast, was safe, strong and so neat that it made any other system look like something left over from a combine harvester.

The rear fork consisted of two triangles, one on either side of the wheel rigidly braced together by massive lugs at two corners, and the wheel spindle at the third. The fork was pivoted at its base on a large taper roller bearing, which could be adjusted to take up any wear. It was also protected from the weather. The apex of the fork thrusted against two coil springs while damping was achieved by friction plates.

This basic design, later modified, was used on all Vincent motorcycles until production ceased. Philip never made a rigid-framed model and was scornful of companies like BSA, who were still producing rigid-framed motorcycles two decades after the first Vincent appeared in 1927. When BSA finally produced a sprung frame, they chose the less efficient plunger system (and later the swinging arm), which left the chainstays protruding into space. (The Japanese, at least, have learned about rear springing and many of their models feature the Vincent rear sprung frame system but without the same quality bearings.)

The Vincent frame was built around a MAG 350 cc engine imported from Switzerland and Philip claimed a top speed of 82 mph, which was going some for the twenties.

However, JAP engines were used for the models exhibited at the Motor Cycle Show held later in the year at Olympia. I asked Philip which model was the most popular. 'There wasn't a popular model,' he replied, 'we didn't book a single order.'

The new 'HRD' models were so different from

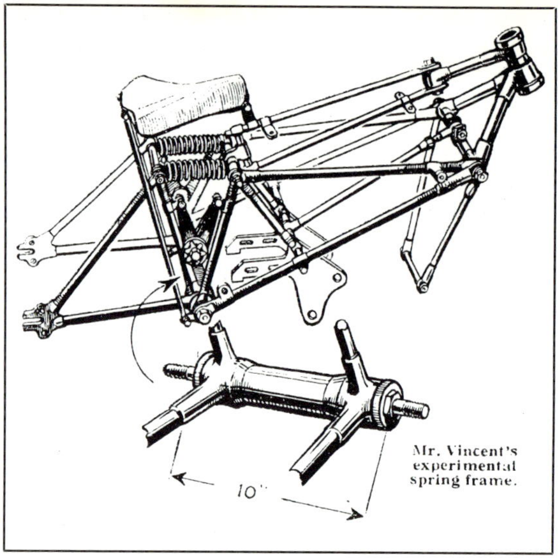

The original Vincent sprung frame (1927). Note the friction dampers and profusion of nuts and bolts

Howard Davies' sleek motorcycles that the original HRD enthusiasts were deterred. Howard's rigid-frame machine had won last year's Junior, all right, so who needed a sprung frame? Vincent's bike had hefty tubes sprouting from every direction and the petrol tank appeared to owe its genesis to a rugger ball. To a man, they shunned PCV's creations, politely described by a restrained Arthur Bourne as 'hardly pretty' with such a 'maze of tubing'.

Howard himself was astonished and couldn't understand why anyone should buy a name and then make something so utterly different.

Philip was unrepentant. His attitude, arrogant in its way but without it there would never have been a Vincent vee-twin, could be summed up as, 'My motorcycle, with its rear sprung frame to prevent you being bumped about, is good for you. Take it or leave it.'

The public left it.

After the abortive show was over, Philip reminded his old Cambridge chums that they had been at college together and one or two sales were effected. The chums were delighted with the bicycles, some of them competing successfully in trials, but only 24 machines were sold in the first year of trading.

This first season, 1929, saw the Wall Street crash in America and the recession spread throughout the world, bankrupting multinational companies and private investors alike. The little company in Stevenage would have been all right with Conrad Vincent's money behind it but due to the Argentine Government's restriction on the outflow of capital, the money was never forthcoming.

However, two events helped the firm to survive. The first was a world tour by Gill, who left England on a JAP-engined Vincent HRD combination in 1929. After an eventful trip (on which he took firearms to quell the natives), having been chased by marauding tribesmen and later locked in gaol in Istanbul as a spy, he reached Australia with Stephens (his long-suffering pillion passenger).

The 'pillion' was a metal frame and Stephens promptly vacated it. Phil Irving took up the option and kept the show on the road with some nifty spanner work, the boys reaching England in

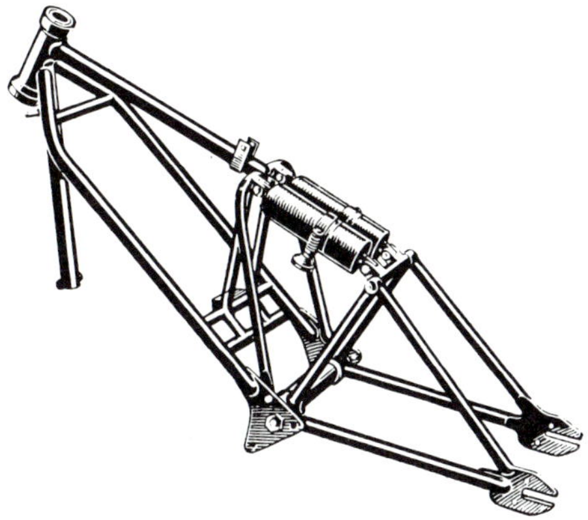

THE VINCENT "H.R.D." SPRING FRAME.

Left **The production (mainly welded) spring frame used from 1928 until 1932. Spring boxes are fitted for the first time**

Below **One of the new Vincents exhibited at Olympia in 1928, fitted with a 350 cc JAP ohv engine. Arthur Bourne thought the frame 'a trifle complicated'**

1930 and gaining the young firm some useful publicity.

The second was the appearance on the scene of Bill Clarke, an ebullient character who loved fast motorcycles, cars and women. He bought a Vincent HRD and was so impressed that he encouraged his father, Captain Clarke of the Imperial Tobacco Company, to invest in the Vincent company. The Captain became chairman and his son also served on the board. (To minimise expenditure the directors forewent their salaries.)

One problem for PCV, the technical director, of having a board of directors was that approval had to be obtained for new projects. Not that there was much friction in the early days. The firm was basically an assembly plant using bought-in parts but the frame and the stainless steel petrol tank were very different from anything produced elsewhere.

Late in 1931, the triangulated frame shed its side tubes and a prototype diamond frame was built. Phil Irving, who joined the company six weeks before the 1931 Motor Cycle Show at Olympia, was given the job of tidying up the frame and preparing drawings, etc for the new models to be exhibited.

Irving, a craggy, blunt, practical Australian with an encyclopaedic knowledge of motorcycle design, complemented Philip Vincent, a charming Englishman—educated at Harrow and Cambridge—possessing an inventive, sometimes fanciful, mind. Both men were keen motorcyclists and

Gill's 600 cc JAP-engined Vincent that he took around the world in 1929, returning in 1930 with Phil Irving

VINCENT "H.R.D." WORLD TOUR COMBINATION.

A 500 cc ohv JAP-engined Racing Model exhibited at Olympia in 1928. It was finished in green (including the leather saddle)

therefore a rider's satisfaction was a prerequisite to any new invention. Both men were perfectionists and the words 'that will do' was a heresy at the works.

Vincent had the ability to pick loyal colleagues who would literally work all night for him. Time was so short before the Show opened that the team actually worked a 74-hour shift and the machines were ready in time. There were little deficiences—like the gleaming stainless steel petrol tank with an aperture cut underneath to accommodate an unexpectedly tall JAP engine's valve gear. Throughout the Show the petrol cap was kept tightly screwed down and guarded lest an awkward customer asked sarcastic questions like, 'If the petrol is held up in the tank by the wind pressure generated by this bike's fantastic speed, what keeps it up before you start?'

Vincent enjoyed being first. He was the first to use stainless steel for a motorcycle's petrol tank and offered it until 1939. Quality was the keynote. Not for Vincent a chromium-plated tank peeling with rust.

The JAP engines, however, were lacking in quality and the road testers were strewing the countryside with an assortment of big-ends.

Rudge 'Python' and 'Sports Python' engines were offered as an alternative, the latter being good for over 80 mph. The new models were well-received at the Show but sales were still very slow, despite the new frame. When I asked Philip Vincent about sales figures, he remembered that his firm sold, '24 in 1929, 36 in 1930, 48 in 1931 and 60 in 1932', a consistent increase each year with a continual fall in the percentage increase, which shows how absurd figures can be.

These sales excluded a three-wheel van for shopkeepers, which was popular with the delivery boys who could whizz round town matching the roadholding against rivals driving Ariels and Beesas.

Sales topped 70 in 1933 as the new design team continued to improve the Vincent HRD motorcycles. One of the most important inventions was the 'duo' braking system. Two drums were fitted to each wheel, the drums and spoke flanges being separate so that the drums could not be pulled out of truth by the spokes. Each drum on the rear wheel was fitted with a different size sprocket, enabling the gear ratio to be altered simply by reversing the wheel. Front and rear wheels were interchangeable (spacers being used in the rear frame member). The prototype duo brakes (with spoke flanges integral with the drums) were used on a sidecar outfit built for the TT bristling with ingenuity, like the passenger having a steering wheel that could be operated independently of the rider's. This was fine until rider and passenger chose different lines through a bend on the Island circuit. Regrettably, the combination was never raced in the Island due to the cancellation of the 1933 sidecar TT race.

Later in the year, at the Motor Cycle Show at Olympia, new bicycles included the 'Model W', a water-cooled 250cc two-stroke Villiers-engined

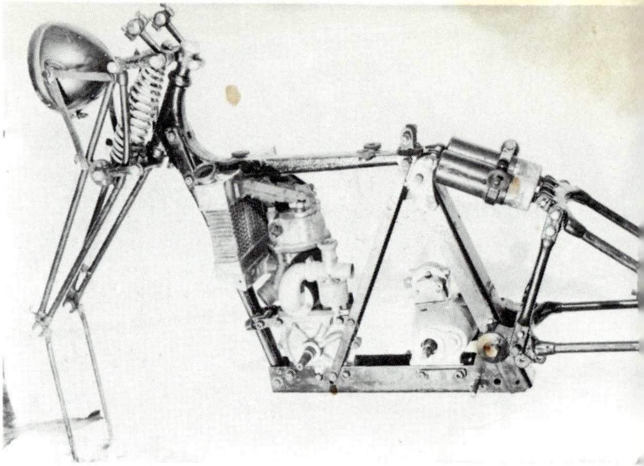

Above **The Model L (1932) was the forerunner of the Model W (1933), whose radiator was partially hidden when the metal panels were fitted**

Below **PCV's first venture into (partial) enclosure—the Model L with 250 cc Villiers engine. (Only one was made)**

Above **The 600 cc water-cooled Model JW shown at Olympia in 1933. (Note 'duo' brakes introduced that year)**

Below **Various bodies were available with this prewar chassis**

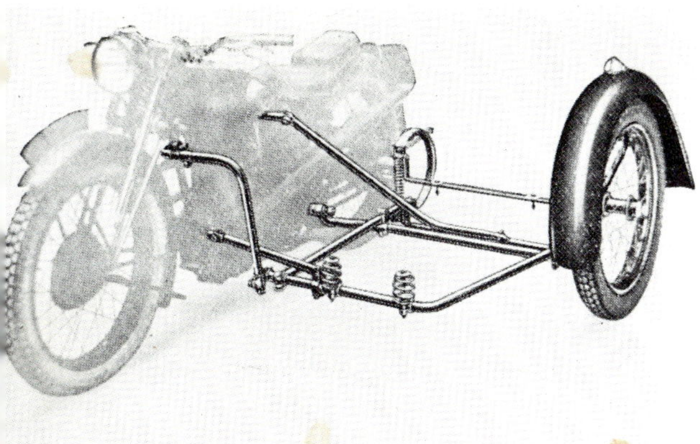

Sidecar Chassis £8 : 19 : 6

machine (considered to be too expensive by the ride-to-work brigade) and the Model JW, a water-cooled 600 cc JAP-engined combination subsequently used by a customer until he complained that it overheated. Where did he ride it? In London's rush-hour traffic.

During the Show, Ted Prestwich sold PCV the idea of entering the 1934 Senior TT using a marvellous new JAP engine of his. It was the very last word in overhead-valve engine development and was sure to be a winner. 'Can't lose, old boy.'

Weeks and then months passed before this fantastic engine reached the Vincent team. There was just enough time to discover, in tests at Brooklands, how bad the engine was before the team left for the Island. The engines supplied by JAP continually broke down during practice as big-ends, pistons, valve guides, and a host of other components failed. During the actual race, all three Vincent entries retired with engine trouble. As a result of the inadequacies of JAP engines coupled with the increasing difficulty of

H·R·D

obtaining other proprietary engines, there was only one solution—Vincents must make their own engine. The new engine had to be ready in time for the 1934 Show and that was less than four months away. Could it possibly be designed, manufactured and tested in time?

This is probably the earliest Vincent extant. It was made in 1931 and has been owned for half-a-century by Jack Dale. The Rudge engine's output provides a maximum speed of about 85 mph

2 The first Vincent engine

Pre-war directors. Bill Clarke (left), Frank Walker (centre) and Philip Vincent. (Bill's father, Captain Clarke, was chairman)

The first question to be decided was long stroke or short stroke? Norton's overhead cam long stroke engine (79 mm × 100 mm) was sweeping aside all opposition on the racetrack and was the obvious example to follow but Vincent and Irving settled for a short-stroke engine (84 mm × 90 mm) for various reasons—there would be less flexing and possibility of fracture of the connecting-rod; engine height would be reduced, push-rods would be shorter and when it came to cleaning the bore, your horny handed mechanic could swish around with ease!

The camshaft was positioned as high as possible, thereby reducing reciprocating weight at high revolutions and, unusually, the valve gear was splayed at an angle from the camshaft, the push-rod, rocker and valve being located in the same plane. Each valve was carried in two guides instead of the usual single one, the fork-ended valve rocker bearing on a hardened collar midway between the guides instead of the top end of the valve. This arrangement further reduced the over-all height of the engine and also the length of the push-rods, and meant less wear and better seating of the valves. Hairpin springs were *de rigueur* at the time and could be replaced without disturbing other components.

The home mechanic was considered—the tappet clearance was nil when the engine was cold. The system was described by PCV as 'semi-overhead cam', which is inaccurate but eye-catching. I prefer to describe the engine as overhead valve.

The cam pinion was driven by a large phosphor bronze idler meshed with the half-time pinion on the mainshaft. All pinions were marked to indicate valve timing—one simply lined up the dots.

The flywheel assembly was carried on four bearings, one roller and one ball bearing on each side and the detachable driveside main bearings could be removed without the need to take the engine from its frame.

The 1934 TT Model with 500 cc JAP engine, which was so unreliable Vincent decided to make its own.

Contrary to usual practice, the timing gear cover was cast integrally with the alloy crankcase. A screwed-on cover plate gave access to the self-withdrawing magneto pinion, so the magneto could be removed and replaced without disturbing anything else.

Access to the camshaft and cam follower was provided by a screwed-on cover plate and additional oil was fed to them via two thin external pipes. These were condemned by the critics for their ugliness but as a disgruntled Philip Vincent pointed out to me, 'Other manufacturers didn't have any ugly pipes because they didn't

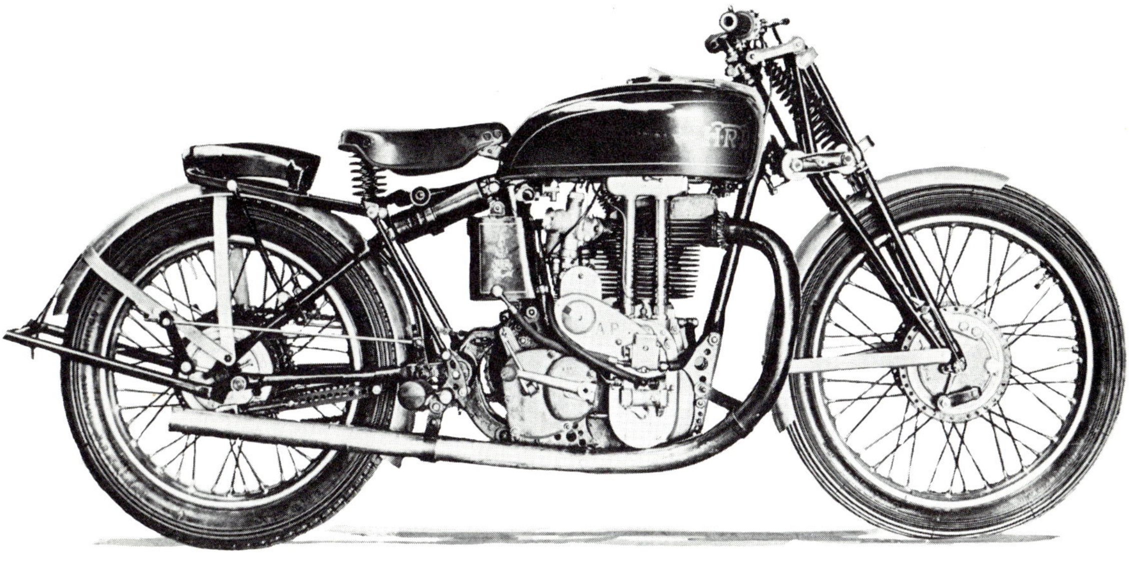

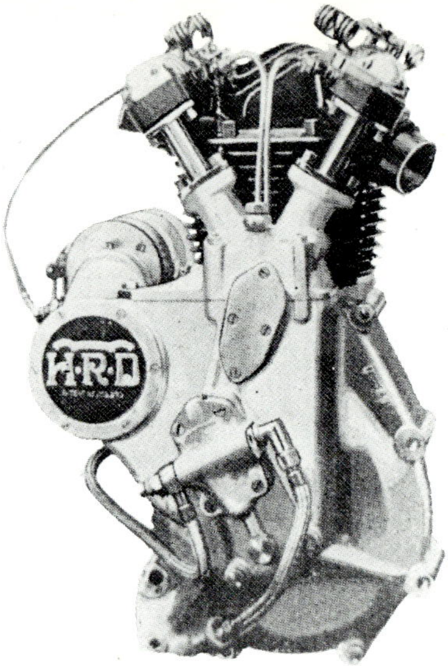

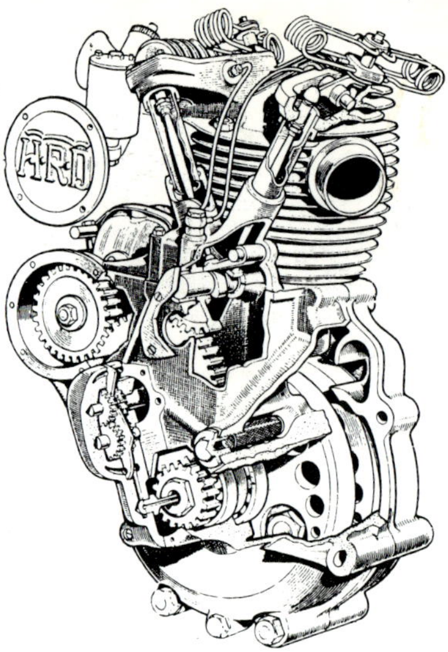

bother to give additional lubrication to these essential components.'

The barrel and head were of iron and with a compression ratio of 6·8:1, the engine produced 25 bhp at 5300 rpm. Power was transmitted via a Burman gearbox and checking the primary chain for the correct tension was easy for our home mechanic—a tiny lever on the chaincase was pulled clear from its recess and turned anti-clockwise. The tension was correct if only a slight effort was needed to push the internal feeler arm over the chain. If too much effort was required, the chain was too tight; if no resistance was felt the chain was too slack. Simple!

Carburation was by $1\frac{1}{16}$ in Amal and both lighting and ignition were provided by a BTH combined magneto and dynamo.

With the introduction of the new engine, Vincent HRD models were given names instead of just letters. The above engine was used for the Meteor and a slightly tuned version powered the Comet. With a $1\frac{1}{8}$ in carburettor and 7·3:1 compression ratio, the latter produced 26 bhp at 5600 rpm.

Petrol tanks were painted with black enamel

Left **The new Vincent engine (500 cc single)**

Centre **The new engine was modified in 1935 with a different sump position and bolts at the base of the push-rod tubes**

Above **Cutaway of the early engine—easily recognizable by the absence of bolts at the base of the push-rod tubes and crankcase design (1934)**

Below **1935 Comet engine. The unique primary chain adjuster is clearly seen on the cover**

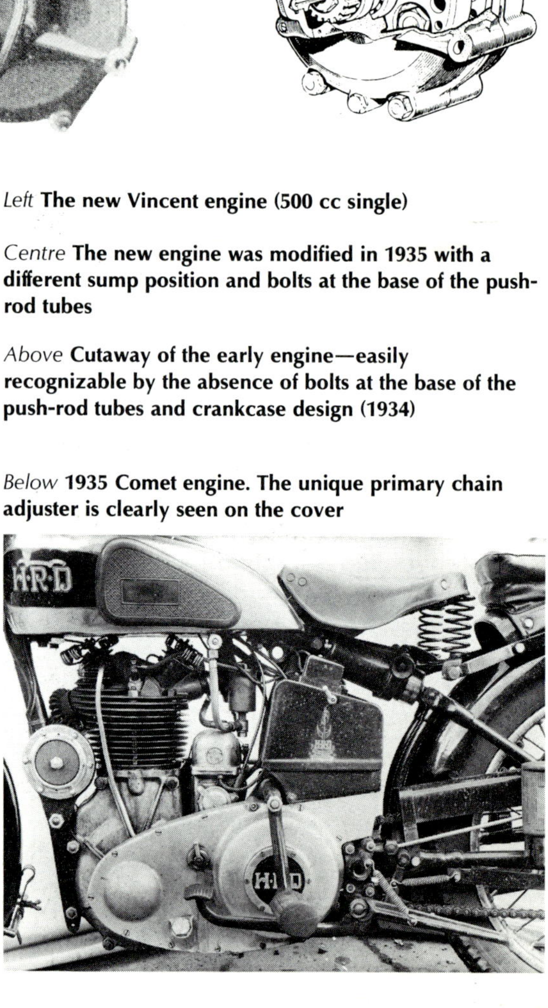

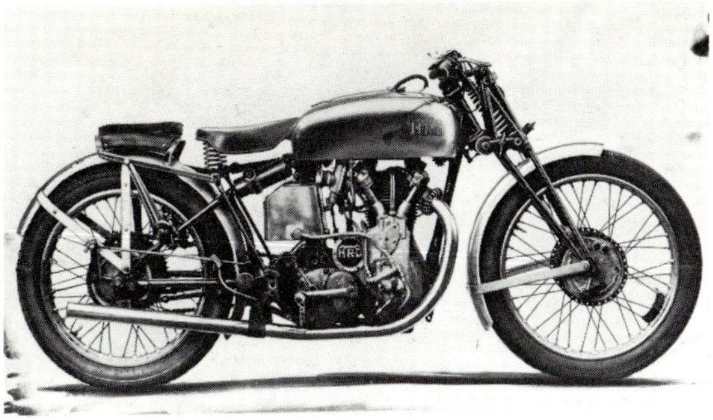

Above **1935 Racing Model. Note the elektron brake plates (efficient) and the stainless steel oil tank (not very efficient!)**

Below **The earliest Vincent-engined Vincent motorcycle extant. It was displayed at Olympia in 1934. (Eng. no. M208, frame no. D903)**

with gold lines and logo, after the original HRD motorcycles. A stainless steel tank was available for the connoisseur.

The new models looked very smart and fashionable with their high-level exhaust pipes but could they motor? Long before one had been started, let alone road tested, PCV, always the optimist, was cheerfully promising the press and any one else around top speeds of 80 mph for the Meteor, 90 mph for the Comet and 100 mph for the TT Model.

The bikes were taken to the 1934 Motor Cycle Show at Olympia but even then no engine had been started! However, during the Show one of the road testers achieved 90 mph on a Comet and there were great sighs of relief all round.

Top **TT Model built for the 1935 season**

Left **Right side of a TT Model built for the 1935 season**

The bikes were fast but were they reliable? Vincent HRD machines fitted with the new engine were entered in the 1935 Senior TT race in the Isle of Man and finished seventh, ninth and twelfth, which was a satisfactory demonstration of reliability. Sales consequently rose to around 160 for 1935 (from just over 100 in 1934) and in 1936 nearly 200 were produced, including a sporting version of the Comet—the Comet Special. Aided by a bronze cylinder head, 8:1 compression ratio and $1\frac{5}{32}$ in carburettor, the Comet Special was the first road-going Vincent to achieve 'the ton', and thanks to George Brown and company the local constabulary found the pace a trifle hot.

George would be chased back to the works and the slow, gentle family saloons driven by sober citizens were usually avoided. By the time

1935 Comet with stainless steel petrol tank

The 1937 version of the Comet

he had dashed into the works like a fox reaching its lair, George usually had enough seconds in hand to hide his hot bike in a shed and his pursuers would find him working away on a stone-cold engine.

At last he was caught and booked for riding at a dangerous speed (110 mph). The magistrate, reared in the age of the horse, listened to the evidence with incredulity. '110 miles per hour, officer? Impossible! Case dismissed.'

To control these excesses, the authorities erected 30 mph speed limit signs around Stevenage. But they made no difference.

If a 500 cc Single could achieve 100 mph, what could a Twin do? And did the small firm on the Great North Road in Stevenage have the resources to design and manufacture a 1000 cc vee-twin motorcycle?

Left **Right side of the supercharged TT Replica whose output was about 37 bhp—no more than the unsupercharged machines**

Below **One of the two 1936 supercharged TT Replicas. The blower was not a success and was removed after practice**

Left **Jack Williams aboard a works TT Model for the 1935 Senior TT. After the war, Jack joined the staff of PCV**

Below **Jack Williams (under contract to Vincent) at the start of the 1936 TT riding a works TT Replica**

Above **Jack Williams in the Island for Vincent. He failed to finish**

Above **Jack Williams (18) and Phil Irving (right) in the Island (1935). Jack finished seventh**

Below **PCV, Jock West, Norman Croft, Jock Forbes and Jack Williams in the Isle of Man (1936)**

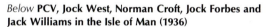

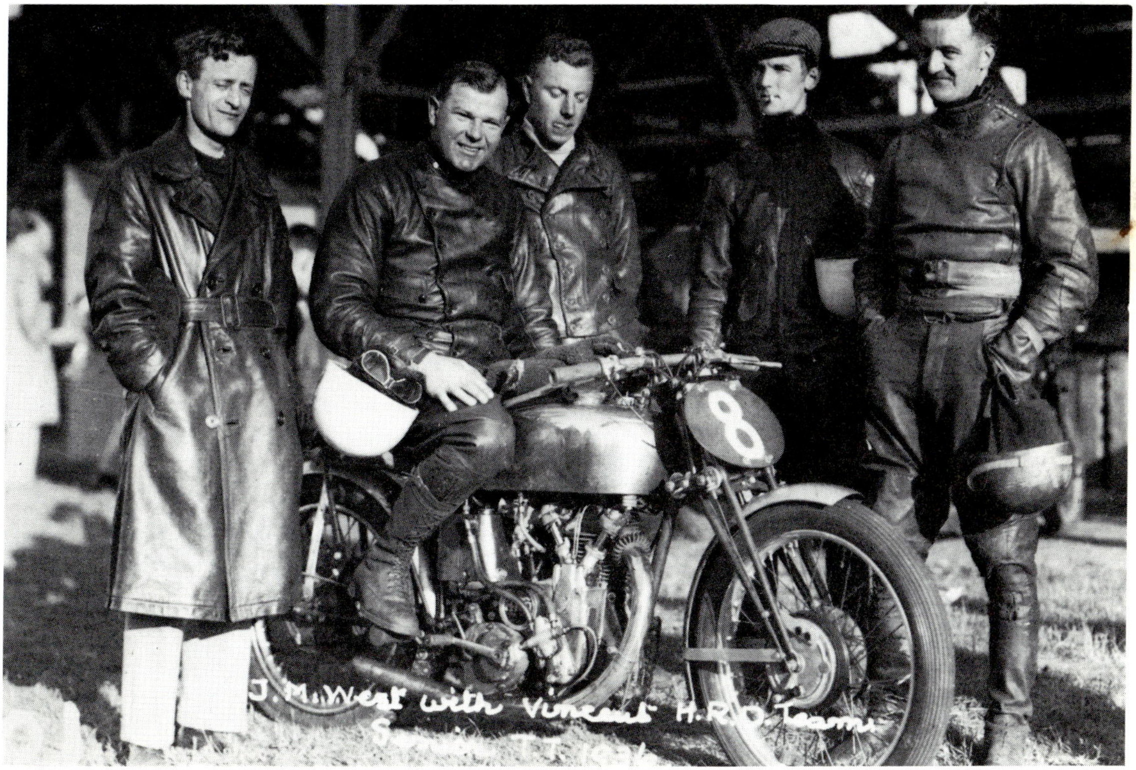

3 The plumber's nightmare

Nothing concentrates a designer's mind so wonderfully as a slender budget. The Vincent HRD Company never had the resources to spend vast amounts of money on research, thus the new Vincent thousand, that would be the world's fastest production vehicle on two, three or four wheels, had to be based on existing designs made in Stevenage instead of being a totally new concept. There was no magical moment of divine inspiration, nobody leapt out of a bath screaming, 'Eureka, I've got the layout for a new vee-twin.'

One day, Phil Irving noticed a drawing of the Vincent 500 cc engine lying on top of another similar drawing. and overlapped the two crankcases. The resultant vee-twin formation had distinct possibilities. A new crankcase was required, of course, but existing barrels, heads, valves, springs, connecting rods, carburettor, timing gear, gearbox and other parts could probably be used without alteration. The rear sprung frame would cope all right but the diamond frame would be too small. The dual-brakes were strong enough to stop the bike, especially if the weight was kept down.

It just so happened that Eric Fernihough had ordered from Vincents a special frame 3 in longer than standard (resulting in a $58\frac{1}{2}$ in wheelbase) to take a JAP vee-twin engine. The frame would not be required after all, and the first Vincent ($47\frac{1}{2}$ degree) vee-twin engine was made and fitted into this frame.

Both exhaust ports on the engine faced

forwards, the rear cylinder being offset $1\frac{1}{4}$ in towards the timing side relative to the front cylinder, thereby enabling the exhaust port and valve to receive a direct draught of cool air to assist prolonged high speeds without overheating.

Timing gear was the same as the single engine's except that the rear camshaft pinion took the place of the magneto drive pinion and the magneto was driven by chain at half engine speed off the front camshaft. The large phosphor-bronze idler gear was inserted through a slot in the front of the timing case so that it could be removed without disturbing the camshaft covers.

The cylinder heads and barrels were the same design as the Meteor's (but were later made of alloy instead of cast iron to reduce weight).

Lubrication was on the dry sump principle by a double gear pump running at quarter engine speed delivering oil direct under pressure to the big-ends, each valve rocker bush, the outer camshaft bushes and to the rear of each cylinder. Adjustments were provided to regulate the flow to the rockers and the cylinder, this flow being separately variable to each of six points.

The prototype Vincent vee-twin (Series A Rapide) that was described by a pressman as a 'plumber's nightmare', (1936)

With a compression ratio of only 6·8:1 and with two $1\frac{1}{16}$ in Amal carburettors, bhp was 45 at 5500 rpm.

The four-pint oil tank was incorporated in the stainless steel and black enamel petrol tank. (A stainless steel and red enamel tank was later offered to order.)

A pillion seat and eight-day clock were included in the price of £138—nearly 40 per cent above the price of the cheapest family saloon car.

Total weight was kept down to 430 lb and the dual braking system would enable rider and machine travelling at 30 mph to stop within 27 feet.

The motorcycle was ready by October 1936 and Philip Vincent had no doubt that it would be faster than George Brough's SS100. As Philip was always pleased to point out, George laboured under the disadvantage of using JAP engines in his Brough Superiors. And faster it was. No less than 108 mph was clocked on the Rapide's first speed test and Howard Davies' famous monogram was to gain added lustre during the next two decades as record after record fell to the thundering twins from Stevenage.

The prototype vee-twin took its honoured place at the Motor Cycle Show at Olympia later in 1936. The firm designated their new offspring the Series A Rapide, Bill Clarke nicknamed it 'The Snarling Beast' and a reporter, amazed at the plethora of petrol and oil pipes, dubbed it 'the plumber's nightmare'.

The public disbelieved the performance figures and only a handful of machines were sold during the 1937 season. Judicious use of the

Another view of the Vincent 1000. When I asked PCV why it had a studded tyre on the front wheel and a ribbed tyre on the back he said that the handling was much the same and anyway the wheels were interchangeable!

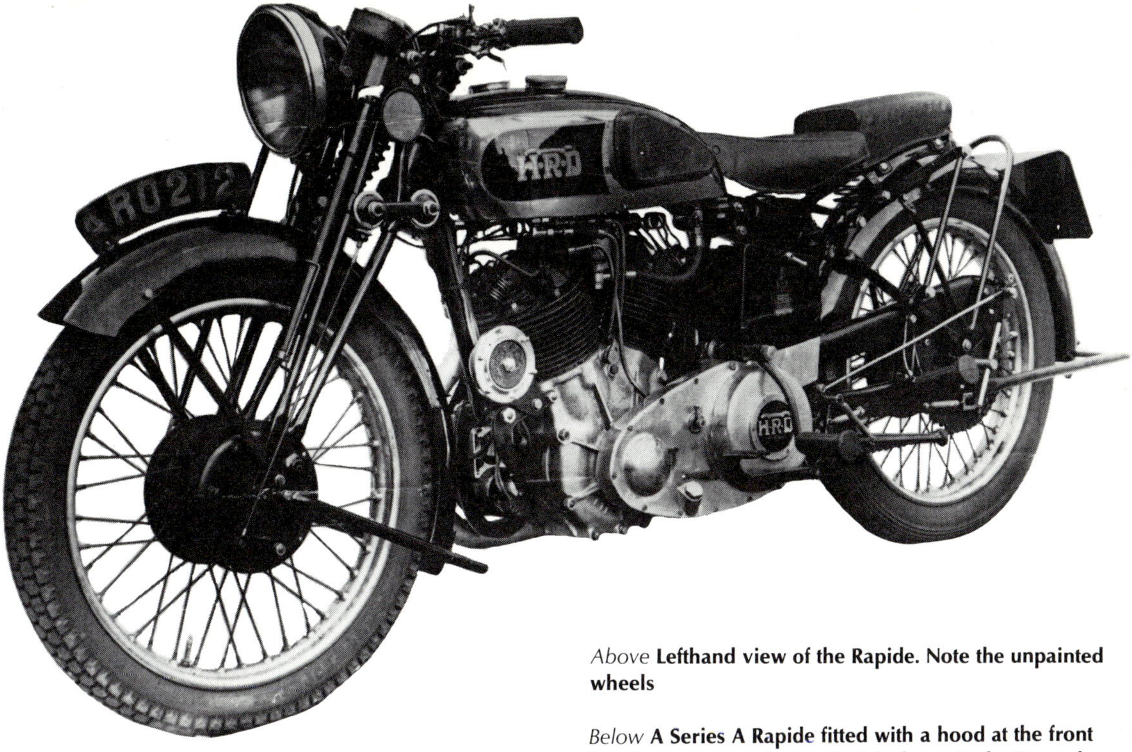

Above **Lefthand view of the Rapide. Note the unpainted wheels**

Below **A Series A Rapide fitted with a hood at the front headlamp for use during the black-out in the Second World War**

throttle was required to reach safe maximum speeds in the intermediate gears (80 mph in second, 98 in third) due to the inability of the Burman clutch and gearbox to cope with the unprecedented torque. If the clutch did not burn out the gearbox would get you!

The pre-war Rapide was the wild one of the Vincent vee-twin range. Compared with its postwar brothers it was noisier, lighter, longer and oilier. (If you ride in convoy with a Series A man, make sure he's behind you.)

Where the postwar Rapide is well-mannered and docile, the plumber's nightmare is impatient. Sit astride a Series A Rapide and it clatters and vibrates, 'straining in the slips', as it were, an impression confirmed as you squeeze the clutch lever against the handlebars and engage first gear. The clutch drags and the brute creeps forward. Feed in the clutch, tweak the throttle and the bike kicks like a spurred stallion. Within eight seconds you're into the seventies and into third gear . . . the brute cleaves aside the air and as you click into top and pass the ton the impressive thing is that the Vinnie keeps accelerating.

There's more weight on the off-side of the engine than the near-side, so right-hand bends can be 'ear-holed' but left-handers have to be fought.

Meanwhile, oil is rushing about, busily lubricating bearings, big-ends, connecting-rods, pistons, bores, valves, guides, cams, followers, tappets, push-rods and the right leg of your trousers.

About 50 vee-twins survive out of the 78 made between October 1936 and July 1939, by which time war was imminent and many employees downed tools to sort out the 'BMW boys'.

The most powerful of the Series A vee-twins was the works racing special built in June 1938 and entered for the Dublin 100 handicap race. This incorporated a magnificent five-gallon stainless steel petrol tank, close-ratio Burman gearbox, modified flywheels, racing cams, stainless steel oil tank (not very efficient for heat dissipation), high level exhaust pipes with megaphones, BTH racing magneto and elektron brake plates with cast finned drums.

This thousand cubic centimetres of naked brutality was ridden by Manliffe Barrington at Dublin and he promptly shattered the lap record by a staggering 6 mph (86·67 mph) and eventually finished second. Over 90 mph was apparently possible in bottom gear.

The unprecedented torque from the record eater posed problems for the rear tyre. Nothing less than a 4 in cover had to be used to keep wheelspin within acceptable limits.

To lighten their bikes, entrants trimmed weight from the front end. This was unnecessary with the racing Rapide—acceleration was so brisk that the front end became uncommonly light! Furthermore, fuel rushed to the rear of the petrol tank and left the tap without liquor and the front cylinder starved unless at least two gallons remained in the tank.

Vincents chief tester, 'Ginger' Wood, took the

This Series A Rapide was registered on April 7, 1938 (engine no. V1022: frame no. DV1524). It was road tested on March 1, 1939 by *Motor Cycling*—**top speed was 110 mph**

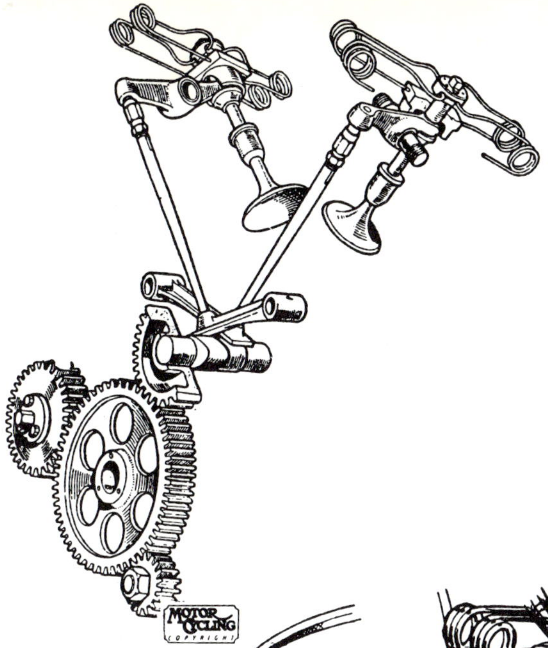

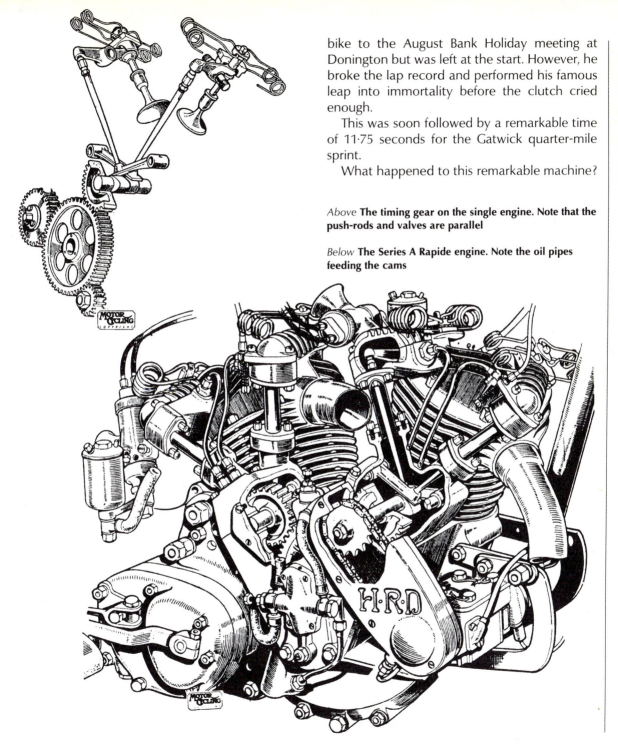

bike to the August Bank Holiday meeting at Donington but was left at the start. However, he broke the lap record and performed his famous leap into immortality before the clutch cried enough.

This was soon followed by a remarkable time of 11·75 seconds for the Gatwick quarter-mile sprint.

What happened to this remarkable machine?

Above **The timing gear on the single engine. Note that the push-rods and valves are parallel**

Below **The Series A Rapide engine. Note the oil pipes feeding the cams**

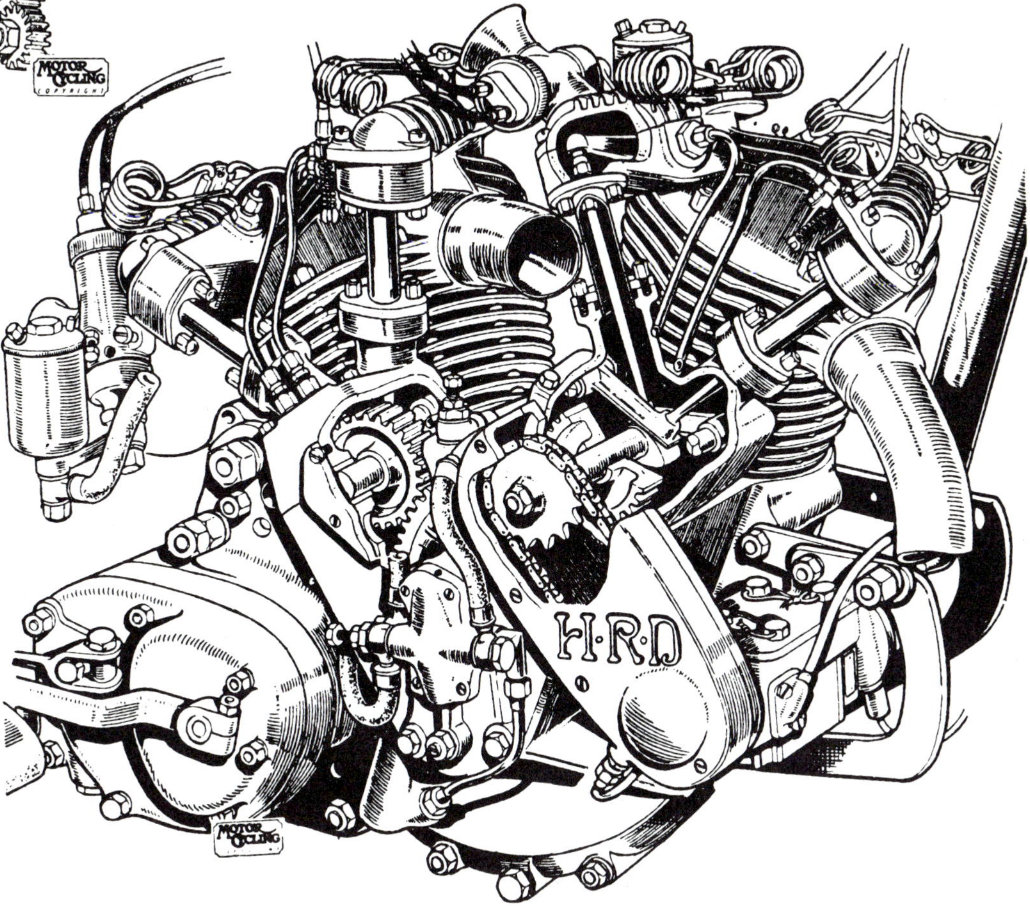

The Vincent company sold it in 1947 but the original crankcases (number TT1030) had been replaced with crankcases V1071 in the original frame. The bike has been restored to a really superb standard by Tony Wilson, a dedicated Series A enthusiast, and is regularly seen at Vincent HRD Owners' Club meetings.

The man who had done so much to develop the Series A range, Phil Irving, had left Vincent for Velocette in 1936. However, during the war, PCV invited him to return as the works chief engineer and the co-designer of a new vee-twin he planned to produce when the fighting was over. They had an efficient frame and a powerful engine; but could they harness this power to offer riders those incompatible virtues—speed and reliability?

Above **'Ginger' Wood with the works Rapide that broke the course record at Donington in 1938**

Below **An example of high-quality work by two Vincent HRD Owners' Club members. John Lumley photographed this pre-war Rapide rebuilt by Alan Edwards**

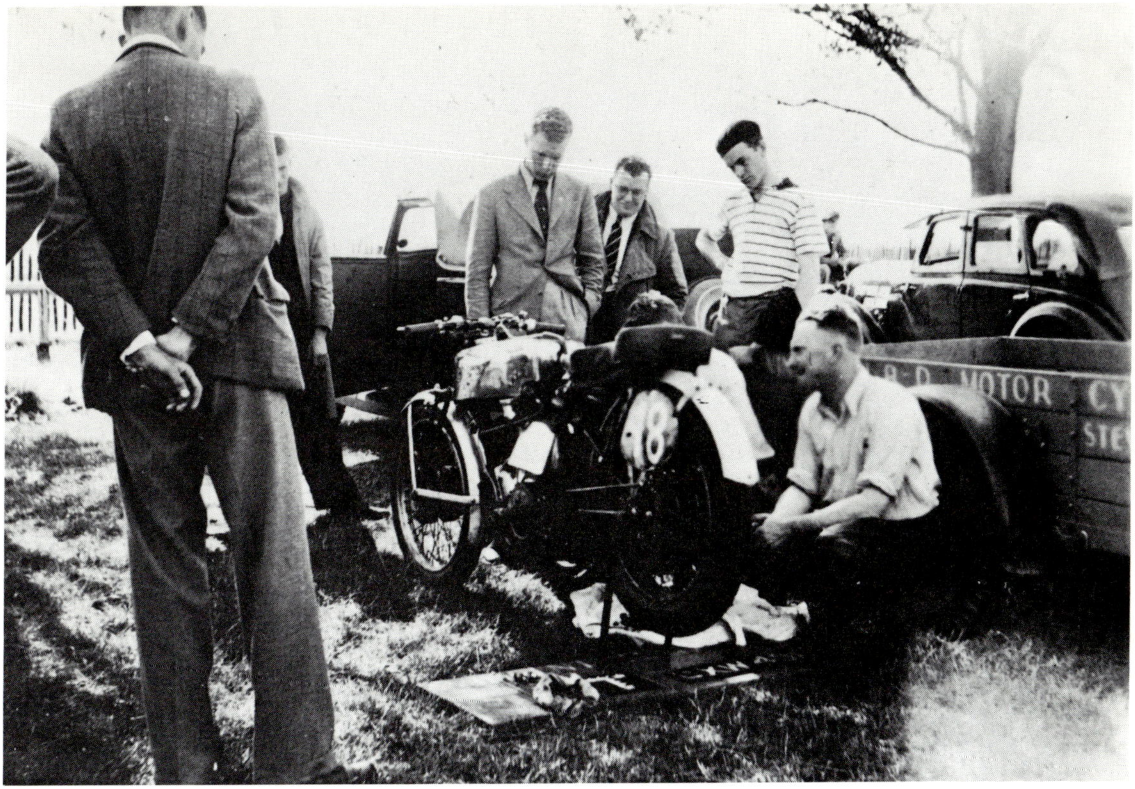

Above **'Ginger' Wood (right) crouching by the works Rapide**

Below **The war-time engine developed for the War Ministry but never put into mass production**

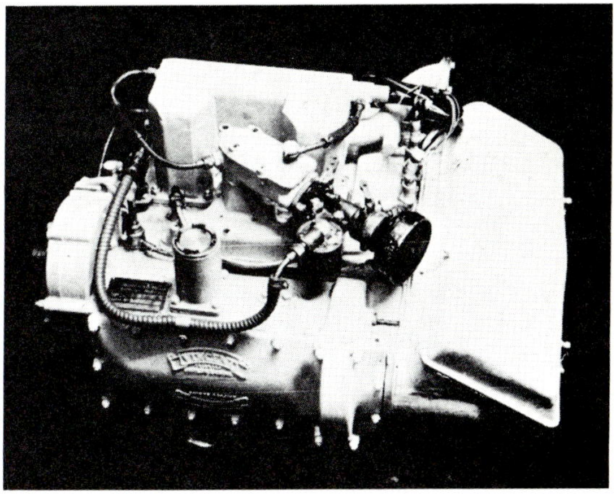

Above **The marine engine made for lifeboat use. PCV considered installing it in a motorcycle but the project died at the drawing board**

4 The Series B Rapide

As we have seen, the materials used for multi-plate clutches in the thirties provided insufficient grip to cope with the Series A Rapide's torque. An entirely new design of clutch had to be evolved first of all.

A ribbed, cast-iron drum was bolted to the primary chain sprocket and inside the drum were two leading shoes with a set of clutch linings on each. The shoes were forced against the drum by the action of a single-plate clutch which turned a ring coupled to the ends of the shoes by toggle links. The single-plate clutch—via tongues mating with slots in the edge of the drum—also turned the drum and would be able to cope with power output up to about 60 mph. The clutch would, in fact, be able to handle 120 bhp, so it would have no difficulty in coping with the Series A Rapide's output.

To enable the clutch to be as close-in to the engine as possible, Vincents chose cross-over drive with the rear chain on the right-hand side of the machine.

The clutch was tested in the Rapide and worked satisfactorily running in the oil circulating in the primary chaincase, so Vincent and Irving then considered various arrangements for the new vee-twin engine layout. For a while, they favoured a vertical rear cylinder with the front cylinder almost horizontal and the front down tube joined to the crankcase between the two barrels.

This would have meant different castings for the pair of heads and barrels; it would also have resulted in the lower cylinder fins becoming badly clogged with dirt on most roads.

Instead, they simply increased the vee-twin angle on their pre-war engine from $47\frac{1}{2}$ to 50 degrees, to accommodate the Lucas magneto then available, and by eliminating the front down tube, PCV was able to reduce the wheelbase from $58\frac{1}{2}$ in to 56.

The top tube was also eliminated and was replaced by a one-piece all-welded 16-gauge steel oil tank incorporated into a backbone that ran from the steering column to the dualseat and the two rear springs attached to the rear frame member.

Phelan and Moore had used a similar construction for decades, their inclined single-cylinder Panther engine acting as a frame member. Vincent, like P & M, attached a bracket

One of the five prewar Series B TT Replicas produced. (Restored by VOC member Tony King). The saddle is non-standard

to the cylinder head by four long bolts but were concerned about the head joints being gas-tight under the action of the frame stresses. Each head was therefore held down by four tubular bolts through which passed four $\frac{3}{8}$ in high-tensile bolts that retained the head brackets.

One problem was thermal expansion. Aluminium was used wherever possible due to a steel shortage and the engine, with its aluminium heads, barrels and crankcase, would expand more than the steel upper frame member. The rear fixing holes were therefore slotted and the rear bolt was not fully tightened, thus allowing for expansion of the massively strong power unit. (A locknut prevented the bolt from becoming too loose.)

Incorporated in the oil line feed from the rear of the oil tank to the Pilgrim rotary reciprocating plunger oil pump was an automatic cut-off valve. When the oil pipe was disconnected, the oil remained in the tank and upon re-assembly, oil flowed again. This is a better system than the oil tap usually fitted by lesser manufacturers—absent-minded people would forget to turn on the tap after re-assembly.

The oil return pipe was run below the petrol tank and a feed was taken to each rocker to provide adequate lubrication. This was a much neater arrangement than the four separate 5 in long oil pipes used on the Series A Rapide.

The cylinder heads were made from Y-alloy or RR53B and were different, the front head being more suitable if racing enthusiasts wished to open out the inlet port to accept larger carburettors.

High thermal conductivity was essential for the exhaust valve seats, so they were made of aluminium bronze while austenitic cast-iron was chosen for the inlets because of its work-hardening property.

The valve springs were neatly enclosed by threaded inspection caps sporting the Vincent HRD logo.

The barrels were finned aluminium jackets

One of the few 1946 B Rapides. Note the circular clutch cover and the 8 in. headlamp

Above **A 1948 B Rapide meticulously restored by Allan Mallinson, a former editor of** *MPH*

Below **Right-hand view of a 1946 Rapide. Note the stove-enamelled spring boxes under the saddle**

shrunk on to cast-iron liners, the rear cylinder being offset $1\frac{1}{4}$ in (as pre-war) with the exhaust facing forward and benefiting from being in the cold air stream.

The liners were held in crankcases and cast in aircraft specification alloy (DTD424).

To allow for the expansion and contraction of the aluminium barrels and heads (which weighed 13 lb less than their pre-war counterparts), the large idler that drove the two camshafts on their fixed spindles was made of phosphor-bronze and they were supported by a steady plate made of aluminium to keep them parallel at different temperatures.

The timing chain used on the A Rapide was

Above **The prototype Series B Rapide first road tested on 27 April 1946. With chair attached, it was ridden as a works hack and is still used by its present owner**

Above **Access points of the lubrication system. To drain the oil, simply remove the feed pipe banjo bolt**

Below **Exploded view of the postwar 1000 cc power unit**

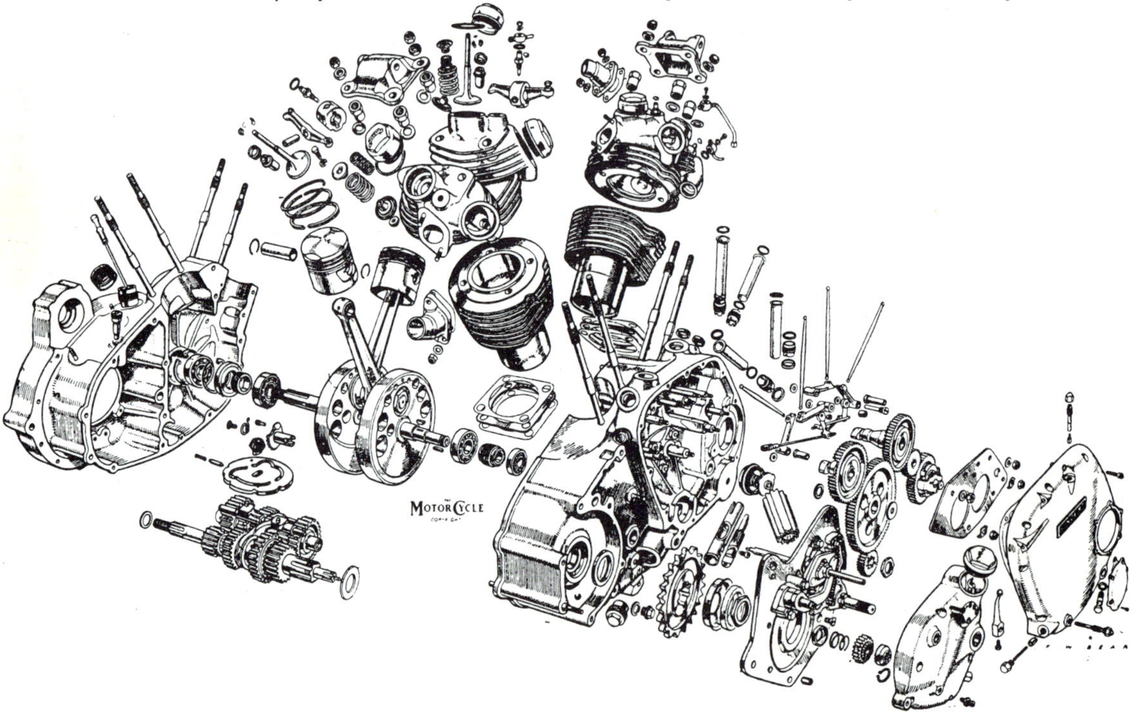

discarded, the magneto being driven by a timed crankcase breather. This in turn was driven by the front camwheel. The Lucas magneto was fitted with a centrifugal device that automatically advanced the ignition as revolutions increased and therefore one handlebar lever and cable could also be omitted.

To enable zero tooth backlash to be attained on the camshafts, the large idler gear spindle was adjustable and various sizes of half-time pinions were available to mesh with it.

The pivots for the camfollowers were located below the camshafts (to eliminate the need for separate camshaft covers) and this meant that the cams had to be redesigned.

Phil Irving consequently made the first cams by hand for the prototype engine, which retained the useful feature of zero clearance for the tappets with the engine cold.

On the opposite side of the engine, the usual engine-shaft shock absorber consisting of a single, large, square-sectioned spring was passed over in favour of 18 small springs because they were more readily available and narrower.

The engine shaft sprocket was linked to the clutch sprocket by an endless triplex chain which

Ted Davis aboard his B Rapide entered in the 1948 Clubman's TT. He finished fifth; George Brown broke the lap record

could be adjusted for tightness by screwing in a bolt underneath the primary cover, thereby pressing a curved blade against the chain. This chain also drove a Miller dynamo as per the system used many years previously on the Indian motorcycle. In use, the dynamo sprocket emitted a dreadful noise like a howling cat on hot bricks and was promptly redesigned to be not unpleasant to the trained Vincent ear.

The engine shafts were made of EN24, a high strength nickel-chrome molybdenum steel, and pressed into the unfinished flywheels; the outside diameter and one raised face flange on each flywheel were then ground off the mainshaft centres to bring them exactly true to the shafts themselves. These ground locations were then used to position the flywheel for the boring and facing of the crankpin holes. This method of construction ensured vibration-free running throughout the revolution range.

The two big-ends each consisted of three rows of 45 rollers (3 × 5 mm) running on a case-hardened EN36 nickel chrome steel crankpin.

Connecting rods were forged in EN16 steel and heat-treated to give 55/65 tons per square inch.

Aluminium pistons—first used successfully by W.O. Bentley nearly three decades earlier—with a compression ratio of 6·35:1 were fitted and with $1\frac{1}{16}$ in carburettors, bhp was 45 at 5300 rpm.

To handle this output the four-speed gearbox, an integral part of the immensely strong power unit, was designed to provide 300,000 miles of safe usage. For example, the pinions were reputedly larger than those fitted to the type 57 Bugatti gearbox.

The new box was very robust but would it be strong enough to cope with those heavy footed riders notorious for their ability to inflict damage? There were a lot of heavy feet around in the forties. All over Europe, Asia and Africa, the 'Don R boys' were knocking the stuffing out of their WD boxes and now they and their mates were leaving the Army, Navy and Air Force in their thousands, all wearing great hefty boots shod with steel tips, heels and studs.

The home mechanic was considered as usual—to ease assembly after an overhaul, a small spiral spring was fitted to hold the bevel in

The Series B Meteor was introduced in 1948 before going into production in spring 1949. This is the Show model at Earl's Court (1948). Note the footbrake and front stand

When PCV changed the logo from *HRD* to *Vincent* in 1949, the new motif was painted over the old on the press pictures

first gear position as the gear cluster was inserted into the box.

The rear frame member was attached to the rear of the gearbox (a bolt ran through the pivot bearing) while the apex was bolted to the two spring boxes attached to the upper frame member containing the oil tank.

Brazed tubular construction was again used for the rear forks and the tubes running from the apex to the pivot bearing were curved to follow the circumference of the rear wheel. The pivot bearing, as pre-war, consisted of two taper-roller bearings that would only need to be moved round once every few thousand miles.

Lugs fitted to each side of the rear frame housed friction dampers adjustable by means of an alloy knob. Adjustable, also, was the height of the dualseat—stay bolts in the shanks running from the friction dampers to the rear of the saddle could be unscrewed and therefore lengthened after disconnection from the lugs in the dualseat. Vincent, incidentally, were the first manufacturers to fit a dualseat as standard (to the chagrin of the Series A enthusiasts who preferred a separate saddle and pillion seat).

Other items that could be adjusted to suit the individual rider included the position of the fold-up footrests, gear lever and its knob, and position of the rear brake pedal.

The designers originally intended the footrest automatically to move out of the way when the long, loping kickstart lever was depressed but the idea was dropped, as was the intention to fit a twist-grip dipswitch on the handlebars, a BTH magneto, a felt-lined tool tray mounted inside a maroon dualseat and both stainless steel and maroon wheels and petrol tank.

Perhaps more than any other feature, the petrol tank gives a motorcycle its distinctive line. Its very prominence ensures that the tank will carry the company logo. Philip Vincent wanted to continue with a tank similar to the pre-war design but stainless steel was in short supply.

It was therefore decided to use the Pinchin Johnson rubber undercoat and black stoving enamel for the pressed steel petrol tank (as used so successfully on the John Marston Sunbeams) and gold leaf for the lines and logo.

A petrol tap was fitted on each side of the tank, the left-hand side containing a reserve of petrol that would carry the rider 5–10 miles, and at the rear of the tank a small recess below the position of the tap collected any dirt or water that would otherwise have entered the carburettor. Seen from either side, the tank drooped at the rear and Philip gloomily recalled that far from being hailed as a good idea, the recess was criticised by a reporter who thought it an eyesore that completely ruined the appearance of the motorcycle.

However, in the spring of 1946, as the prototype B Rapide took shape, no reporter had yet cast either aspersions or praise and PCV invited Arthur Bourne and Graham Walker, respectively the editors of *The Motor Cycle* and *Motor Cycling*, to test the new model on Saturday, April 27, 1946.

The bike wasn't quite finished—the petrol tank, for example, had yet to be enamelled—but Arthur wasn't bothered about trifles like that and published an enthusiastic report. Both Arthur and Graham likened the 1000 cc Rapide to a five-hundred job . . . a comment more true nearly four decades later when some Japanese five-hundreds are close to being heavier than the 455 lbs of the Series B Rapide.

Top speed wasn't ascertained but third gear nearly took the rider to the magic ton. Price was nearly double the cost of the pre-war Rapide at £265 5*s* 5*d* (including £54 5*s* 5*d* purchase tax, a 'temporary' tax introduced during the war and retained until it was replaced by the obnoxious Value Added Tax).

Five months after these road tests, on September 19, 1946, the first production B Rapide was finally ready and road tested. The next day it was air freighted to Buenos Aires to be exhibited at a trade fair by Nestor Vila Moret, director of Cimic Ltd., Vincent's agent for the Argentine.

PCV chose Buenos Aires for the first production model because of family links. His father had emigrated to Argentina in the 1880's and Philip's sister still lived there.

Several orders came from the Argentine but not all could be met due to production difficulties in the United Kingdom and restrictions on the movement of capital by the Argentine Government.

One problem was a lack of space due to utilising the whole factory for the war effort. During hostilities, the firm had ceased manufacturing motorcycles and produced various parts for shells, fuses, mines and aircraft.

The Air Ministry were interested in a marine engine, light in weight and economical in operation, for an airborne lifeboat to be used for air-sea rescue.

Vincent patented a 500 cc two-stroke marine engine in which two pistons adjoined a crankshaft with two further pistons adjoining the opposing crankshaft. The specification called for 15 bhp at 3000 rpm and tests showed that the boat would travel over 1000 miles at an average

The 500 cc Comet engine. The plain inspection caps suggest that the date is 1949

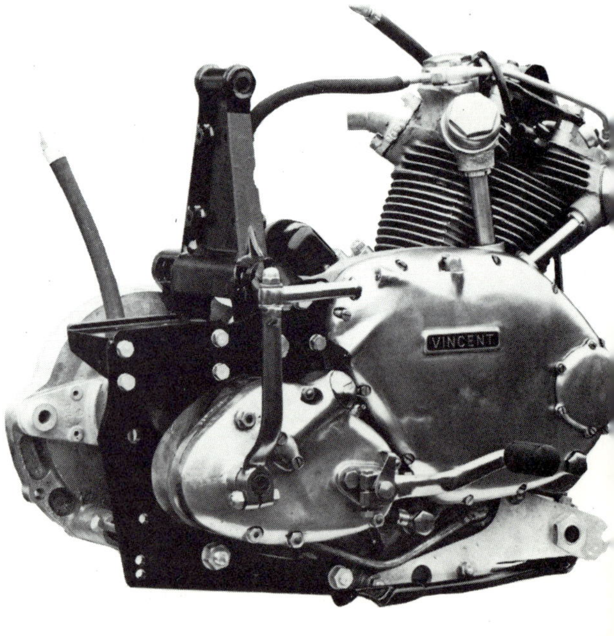

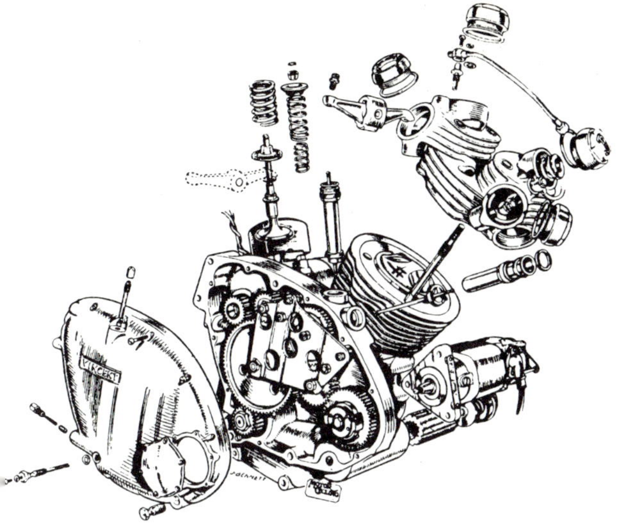

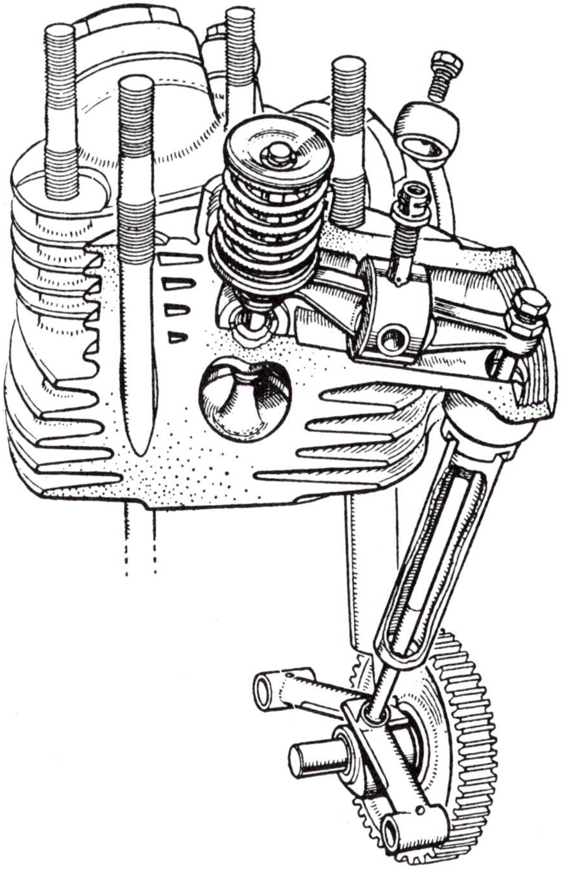

speed of $5\frac{1}{4}$ knots on 50 gallons of petrol. The Air Ministry were interested enough to place an order for 50 engines and PCV couldn't resist considering the engine for a small postwar motorcycle. A drawing was made showing the engine enclosed in metal fairings. Top speed was estimated at 75 mph with a fuel consumption of 180 mpg at a cruising speed of 50 mph. Unfortunately, the project was dropped due to the high retail cost of the machine.

Vincent also used the marine engine in a prototype car whose plywood body was made by de Havilland. The engine was mounted transversely which meant a very short bonnet and an unusual shape. As an MG and Rolls-Royce enthusiast, I had difficulty in appreciating the marine-engined car's beauty.

So did the man at the works who chopped it up for firewood.

A doubled-up version of the marine engine (the SDRE) was produced for a Ministry of Supply contract to supply engines for the Signals Research and Development Department; but the idea was dropped by the Ministry so the engine never went into production.

With no substantial Ministry of Defence contract to fulfil once the war was over, the firm pressed on with the design and manufacture of the Series B Rapide with little opposition from any quarter of the globe.

Before production of the Rapide began in the autumn of 1946, PCV looked for larger premises to cope with his envisaged output of 50 motorcycles per week.

The old factory was suitable enough when the firm was mainly an assembly shop for bought-in proprietary parts but as more and more items were manufactured on the premises, the pressure on available space continually increased.

Above **Exploded view of the Grey Flash engine, the tuned Single that appeared in 1949**

Below **Postwar valve mechanism. Adjustment was easy. Nil clearance**

Inside the shrine! The 'production line' for Series B Rapides

For example, the machine shop expanded threefold during the war and the general disruption meant that by 1946 the firm had the wrong plant for mass-production of motorcycles.

Philip was therefore delighted to gain the use of a modern factory about half-a-mile away with a main building 200 by 80 feet. These new works at Fishers Green were eventually used for the manufacture and assembly of new machines while the old works were retained and increasingly used for spares and service to keep the customers happy.

The service manager, the late Paul Richardson, whose linguistic skills enabled him to cope should 'pain and anguish wring the brow' of any esteemed customer (foreign or British), had to be in top form on one occasion just after the war.

An Italian had been interned and had asked the firm if he could store his Series A Comet at the works 'for the duration' as he knew he could leave it there with complete confidence. He was assured that the firm would look after it.

The war dragged on. Other Comets had to be kept on the road. Spares were in short supply and customers frequently arrived asking for replacement mudguards, footrests, etc. If none was available it would not do any harm to borrow just one or two minor parts from the Italian's Comet.

The war ended and as Paul sat in his office one day the Italian arrived and said how grateful he was that the firm had taken such good care of his Comet. Throughout his internment, he had continually dreamed of collecting his Vincent and riding it again as it had given him so much pleasure. He couldn't wait to get back in the saddle so could he please have his Comet immediately. 'That presented a little difficulty,' said Paul as he drew on his pipe and blew smoke into the air. 'Only the frame was left.'

Paul received several complaints about the new clutch. This unorthodox mechanism, on which so much time and energy had been spent, had worked well when tested in the Series A Rapide but when fitted to the Series B prototype in 1946 continually slipped due to running in oil. The reason was probably that the new clutch linings in the B were not impervious to oil, unlike the earlier linings.

The chaincase was re-designed so that the clutch was separated from the oil in the primary chaincase. Unfortunately, oil often seeped along the splines of the gearbox shaft past the little oil seal and ruined the Ferodo MR41 linings. The problem was eventually cured when Duron P28B moulded shoe linings, which were impervious to oil, were introduced.

Once set up properly, the Vincent clutch gave several thousand miles of trouble-free service, as demonstrated by Eddie Stevens, the author of *Know thy Beast*.

The circular clutch cover was ground down when the bike was 'ear-holed' round left-handers and was replaced by one with a cutaway at the bottom.

Another problem was the lightness of the clutch spring pressure. (The rider could squeeze the clutch lever against the bars with his little finger.) If any dirt or rust fouled the cable, the springs were too weak to return the clutch plates and handlebar lever to their correct positions. Six strong springs were therefore fitted as standard and the problem was cured.

The ubiquitous oil, so necessary for the lubrication of the valve guides, ran down them into the combustion chamber and fouled the sparking plugs, especially at low engine revolutions.

PCV admitted that the front plug oiled up

Part of the machine shop at the works (postwar)

A Wickman multi-spindle machine at the Fishers Green works spewing flywheels

because the oil level in the rocker tunnel was too high relative to the guide and the front cylinder inlet tunnel was almost horizontal; thus it lacked much of the drainage possessed by the rear cylinder inlet tunnel.

To improve matters, grooves were cut along the underside of the rocker bearings and metering wires were inserted in the oil feedbolts to reduce the oil flow.

A rocker tunnel sometimes became oversize due to the action of the rocker bearing and the wear could have been eliminated had the original design specified shrunk-in iron liners for the rocker tunnels.

In the event, some owners rode along happily unaware that worn bushes were thumping away at the tunnels . . . and these worn tunnels gave ham-fisted bodgers endless opportunities. To take up the gap between bush and tunnel, some of the boys 're-shaped' the bush with a large hammer; others stuffed in wooden wedges or bashed nails into the gap.

Professionals effected a permanent cure for tunnel wear by sleeving the tunnel and fitting a special feedbolt marketed by the Vincent HRD Owners' Club.

Oil usually found its way into the Miller dynamo and ruined it. However, a cynical pal suggested that the Miller dynamo was probably chosen by the designers because of its ability to run in oil longer than any other.

Another fault not anticipated was splitting (and therefore leaking) of the petrol tank. This was cured by fitting an alloy distance piece and steel tie bolt through a pair of tabs at the rear of the tank. Nevertheless, the bodger came into his own—he made up his own distance piece too long and hammered it into position between the tabs with inevitable consequences.

Many people outside the Vincent 'movement' thought that the engine was sealed to prevent enthusiastic bodgers from ruining the firm's reputation. In fact, no big motorcycle was ever easier to service. From the big tommy bar holding in the front wheel to the little tommy bar in the rear mudguard flap keeping the rear stand in place, the Rapide was tailor made for the home mechanic.

To dismantle the machine one unscrewed the little tommy bar, heaved the bike onto its rear stand and unscrewed a bolt (beneath the front section of the crankcase) holding a pair of front stands. The lifting of the front end enabled this front stand to spring down and keep the engine well off the floor. (These stands doubled up as prop stands that swung out on either side for parking the Vinnie at the kerbside.)

After the removal of petrol tank and cables, the complete front end assembly, incorporating upper frame member, Brampton forks, etc. could then be wheeled away leaving the engine completely exposed and devoid of impeding frame tubes. The complete operation could be done within 15 minutes. The front and rear wheels could be taken out and replaced (including brake adjustment) without recourse to a spanner.

The early rear frame was too short to permit the safe use of an oversize or racing tyre and was replaced by a modified frame $\frac{1}{2}$ in longer.

The rear wheel was held in the fork end lug by a tommy bar axle of 40-ton steel, the tommy bar itself being stainless steel.

On the opposite side, a lipped axle nut made of stainless steel hexagonal bar could not, of course, turn in its slot when the tommy bar was tightened by hand (with a final prod from one's boot). A lipped round nut would have been better; the hexagonal nut proved irrisistible to the bodger, who immediately attacked it with a ring spanner in a vain attempt to tighten the nut. He merely forced the lip on the nut to damage the slot; if he was particularly determined, he succeeded in fracturing the lug.

The axle abutted against the chain adjuster shank—to tighten the rear chain one simply screwed in the chain adjuster by hand (when the axle was loose, of course); a spring loaded tapered alloy body located in a groove in the lug to prevent the adjuster becoming loose when the bike was being ridden.

These various and unusual fittings gave the Vincent its unconventional appearance and limited appeal. By 1947, only 10 bikes a week were being produced for a seller's market.

Servicemen were being demobbed with their gratuities and many blew the lot on a motorcycle. Ex-WD berets, goggles, overcoats and boots provided weather protection for the boys as they rode to work or toured the countryside.

Motorcycling was fun. There were no speed limits on the open roads, very few cars, few big trucks, no MOT tests, double-white lines, parking wardens, and no compulsory wearing of crash helmets.

After all, the war had been fought over the question of freedom, had it not? The Allies hadn't freed Europe from the German onslaught for the British to lose their own freedom to the bureaucrats in Whitehall.

Motorcyclists took this freedom for granted and there was no organised opposition to Government legislation or many clubs in existence for the various marques.

Nevertheless, there was a great camaraderie in the motorcycle movement and if a rider broke down he was soon assisted by another.

THE SERIES "B"
THE VINCENT
H·R·D
Rapide
The World's fastest standard Motorcycle!

Cover of the catalogue produced for the 1947 season

Instead of being enviously told 'Thasworf money' (as today), a Vincent owner was asked how fast the bike would go and many a friendly duel with Triumph, Norton or BSA was settled without mishap.

Sometimes the camaraderie became a little strained between Rapide and Triumph owners. A Vincent rider regarded the vertical twin as a mobile vibro-massage machine and the Triumph rider thought Vincent owners were aloof (always waving to each other, never to any one else).

For example, when a pal of mine courted a girl who lived in Newcastle, some 300 miles away, he left his London office at 5 pm every Friday, rode to his girl friend's house and arrived before 11 pm. This was 'really tramping' in pre-motorway days and a colleague disbelieved the time, so Brian gave him the Newcastle telephone number and said, 'Ring at 11 pm next Friday and I'll answer the phone.' Which he did!

In the cold winter months he restored his circulation by getting off the Rapide occasionally and pushing it forward at a smart pace. One evening, a passing Triumph owner noticed Brian trotting along, assumed the Vincent had broken down and called out, 'What's the trouble?'

'No trouble,' replied Brian. 'I'm doing this to keep warm.'

Whereupon the Triumph owner roared off muttering, 'Sarcastic blighter.'

The Rapide was ideal for such long trips but uncomfortable around town. At low speeds, the steering was heavy, the turning circle was large and the plugs oiled up. The Rapide needed to be ridden for 12 miles before its oil warmed up to working temperature so anyone who used the bike mostly around town might have experienced an unusually high rate of wear.

As the number of Rapides in use around the world increased, so the legend of the Vincent grew.

In Australia, a Rapide owner chased by a police motorcyclist went so fast that he overtook another police motorcyclist riding flat out in pursuit of another speeding miscreant.

In the USA, a Rapide owner and his wife went for a ride and came across a hillclimb in progress. Unfortunately, no competitor had a motorcycle capable of taking him to the summit. The Rapide owner rode to the top of the hill and down again. Then his wife repeated the performance and they continued on their way.

Over the border in Canada, a Rapide owner was the only rider to reach the top at the 200-feet Mount Kuhn, Heidelber, hillclimb in October 1947.

Meanwhile, back in the United Kingdom, George Brown—backed up by his beautiful wife Ada and assisted by his skilled brother, Cliff—was winning short-circuit races, sprints and hillclimbs using either his 500 cc or 1000 cc Vincent special.

The 1000 cc special, christened *Gunga Din* by the late George Markham, became the most famous Rapide of all. It was so powerful that some owners of Ajays, Nortons and Triumphs refused to race if George and *Gunga* were among the entries and instead sulked at home.

The achievements of George Brown riding *Gunga Din* encouraged Philip Vincent to produce his ultimate road-going motorcycle—the Black Shadow. His long-cherished ambition to manufacture a 120 mph roadster was about to come true. Or was it?

5 The Black Shadow

When Philip drew up the specification for the Black Shadow and presented the case for making a sports version of the Rapide, his plan was vetoed by Frank Walker, the managing director.

In vain, Philip argued the necessity for continual improvement of the vee-twin—the firm could not stand still. The lead established over all other manufacturers in the world had to be maintained.

Walker remained unmoved.

Philip argued the economic benefits. Rapide owners would change to Black Shadows. Already there was a clamour for something faster, namely, the number of enquiries received for tuning the standard Rapide. The faster model would have great success in competition and enhance the firm's reputation. The model would cost more but profits would be greater. Walker remained unconvinced.

Fortunately, Philip refused to give up his ambition—but for his steely determination to produce the Black Shadow, the most famous model ever made by the company would never have appeared.

Certainly there was a great potential in Britain for a sports version of the Rapide. In the forties, Britain had a great engineering tradition and made the world's finest aeroplanes, cars and motorcycles. A new product had to be of exceptional quality to satisfy the discerning motorcyclist.

The Vincent was the fastest and most expensive (some 50 per cent dearer than a Norton, for example) and created the most expectations . . . perhaps that is why it received so much more analysis and criticism than any other make.

The Rapide was superbly made: the Black Shadow even more so. Unbeknown to Frank Walker, Phil Irving and George Brown covertly assembled a brace of Black Shadows. Ports were smoothed and stream-lined, connecting-rods were polished, Comet cams were fitted, the carburettor size was increased from $1\frac{1}{16}$ to $1\frac{1}{8}$ in. Only 72 octane petrol was available and the compression ratio was limited to 7·3:1. As a result of the modifications, bhp rose to 55 at 5700 rpm.

Greater speed required better brakes. Ribbed drums were therefore fitted to the brakes and the increased output was feted with a 5 in jumbo speedometer that inspired delusions of grandeur in the rider before he had even kicked over the machine.

Inevitably, of course, Frank Walker discovered the clandestine assemblies and expressed himself somewhat forcibly. However, Philip had presented him with a *fait accompli* so there was little he could do about it and the model was

The prototype Black Shadow (1948). Note the speedometer case (production models' speedometers were curved)

announced to an interested world on February 24, 1948.

To distinguish the Black Shadow from the Rapide, Philip informed his staff that the engine would be all black. The boys were aghast. Cover that beautiful, shiny Rapide engine in black? Nobody would buy the new model. Once again criticism came his way but PCV insisted that the colour must be black and events proved that his decision was correct, the Black Shadow eventually outselling the Rapide.

Optimistic as always, Philip announced that the new 'Rapide Black Shadow' would have a top speed of 125 mph (but only machines in tip-top condition could attain it).

Some 50 of these early Series B Black Shadows were reputedly made, the most famous being the one sent to the late Rollie Free for an attempt on the American speed record. The Black Shadow was ordered by John Edgar, a wealthy American journalist whose ambition was to own the fastest road machine in the United States. What better than to buy a Black Shadow that had been 'breathed on' by the boys back in England to enable it to beat the existing Harley-Davidson record of 136 mph.

How satisfying, once the bike had taken the record, to return it to standard specification and then be able to boast that the Vinnie was the quickest bike on the road; and if anyone wagered otherwise, the official record certificates would enable John Edgar 'to clean up'.

John put his requirement to Philip Vincent, then visiting California, and was assured that not only would the Black Shadow break the existing record, it would achieve 150 mph.

Basing his confidence on the performance of *Gunga Din*, PCV wrote to Phil Irving towards the end of April 1948, setting out in detail his requirements.

The bike had to land in Los Angeles before September, so time was short, especially as a new cam had to be produced. Nowadays, a new cam form could quickly be calculated with the use of a computer but Irving had no such help in 1948. He designed the cam contours, made the master and then ground three complete sets of cams. These new 'Mark II' cams were fitted to a Black Shadow and enabled George Brown to achieve 143 mph at Gransden Aerodrome before he began to run out of airstrip.

This performance was satisfactory and the bike was shipped to the United States for an attempt on the record on September 13, John

A Series B Black Shadow in the despatch department ready for action during the 1949 season

The first Black Shadow being introduced to the press early in 1948

Edgar stating that 'the thirteenth was the only possible date for the record attempts'.

The rider, Rollie Free, was decidely unorthodox. He replaced the comfortable saddle with a 4 in pad mounted on the rear mudguard and extended himself along the machine, his legs protruding two to three feet beyond the rear wheel. Rollie set a new record of 148 mph but wanted to achieve the magic figure of 150 mph. He removed his leathers and blue racing helmet and attacked the record wearing nothing but swimming trunks and running shoes. He told me that it was against the rules to attempt records without a crash helmet but his bald head would probably turn blue at the chilly time of the high-speed runs (around 6·00 to 8·00 am), so the absence of a helmet might not be noticed.

Rollie was not a tall man and the running shoes (size 12), which were borrowed from a friend, hung from his feet as he set a new record of 150·313 mph. He swore that the extra drag from those great shoes reduced his top speed by half-a-mile!

This speed was the fastest ever recorded by anyone riding an unsupercharged motorcycle and whetted Rollie's appetite for an attack on the absolute world record of 173·625 mph, of which more anon.

Vincents decided to market a production racing motorcycle of similar specification to Rollie's mount and called the new model the Series C Black Lightning. This appeared at the 1948 Earls Court Motor Cycle Show priced at £400 (excluding purchase tax), £85 more than a Black Shadow. The extra money went on the Mark II cams, steel idler timing gear, $1\frac{5}{32}$ in Amal TT carburettors, Lucas racing manual magneto, a pair of 2 in straight through exhaust pipes, alloy

The 1948 Series C Black Shadow at the Motor Cycle Show at Earls Court. Note Girdraulics and 7 in. headlamp but B rear frame member

wheel rims, rev-counter and a choice of pistons up to 13:1 compression ratio. The gearbox was modified to give quicker changes, some dogs having alternate teeth cut away to allow freer engagement.

This model was designated the Series C because it sported a new set-up at the rear and a very unconventional front end—the Girdraulics.

The rear frame member incorporated curved instead of straight lugs for the seat stays and a new damper designed and made by Vincents was fitted between the two spring boxes.

The New "Girdraulic" Fork

British Registered Trade Mark No. 675,834.

Staunch believers always in the necessity of preserving absolute rigidity against all forms of lateral whip or twist in all components connecting the two wheels of a motor cycle, if perfect handling is to be expected at high speeds, we refused to join the general post-war stampede into new fork designs—we wanted time to consider all possible methods, weigh the pros and cons and strike the best balance—always with firm insistence on *lateral rigidity* for perfect steering and cornering.

As makers of *standard* models capable of 150 m.p.h., we have to exercise great care in fork design.

Fortunately we started with the considerable advantage of far greater experience in motor cycle springing, for we are the *only* well known makers *who have never built a rigid frame*.

During the last twenty-one years we have perfected our fully triangulated and completely whip-free spring frame and we did not intend to throw away this great advantage by using a parallel membered front fork, depending largely on the wheel spindle for any rigidity.

We therefore decided to retain the well proved girder principle for its inherent rigidity. We obtained a new high level in rigidity-weight ratio by using forged light alloy blades of tapered oval section and preserved the maximum of this rigidity by using one-piece forged links. The new fork is at least twice as rigid as the earlier tubular design.

The long soft action so popular in these days has been provided by mounting very long springs in telescopic cases between the bottom of the crown lug and the fork ends, but we have kept this action within the limits demanded by safety at high speeds. An ingenious eccentric action provides a variation of spring loading and fork trail to suit solo or sidecar use ; this adjustment can be made in a few minutes without using any extra or different parts.

The other great improvement of modern fork design is also provided in the form of a two-way hydraulic shock-absorber, with hydraulic limit stops in both directions to eliminate metallic "bottoming," but with these important differences, vital differences to the thousands who live great distances from the factory or the dealer :—

1. The shock-absorber can be detached easily.
2. It weighs less than two pounds and is only eight inches long, so it can be sent by parcel post.
3. The machine can be ridden at reasonable speeds without the shock-absorber.

Now that simplifies service, doesn't it ?

To ensure long life and freedom from attention, large diameter ground high tensile steel spindles are used in conjunction with oil retaining bronze brushes. No oiling is required.

Girdraulic Forks have now been in the hands of the public for well over a year and have given extremely reliable service. They have been used most successfully for breaking world and national speed records and for arduous racing and trials work. They have received the highest praise, alike from expert riders, the press and the public. We have no hesitation in saying that the Girdraulic is the finest motor cycle fork ever made.

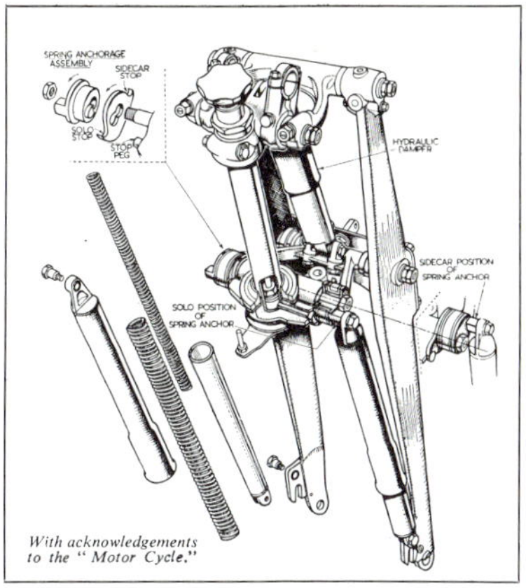

With acknowledgements to the "Motor Cycle."

THE VINCENT

Above **The new Girdraulics (1948) showing solo or sidecar trail**

Below **The Series B Black Shadow (1948) with** *Vincent* **logo painted over** *HRD*

Rollie Free precariously perched on John Edgar's Black Shadow before becoming the first man to ride an unsupercharged motorcyle at 150 mph (1948, USA)

The Girdraulics were remarkable and a tremendous improvement over the popular telescopic forks widely used at the time by many other manufacturers.

The fork blades were made of heat-treated forgings of L40 aluminium alloy by the Bristol Aircraft Company and were therefore immensely strong and almost bodger-proof (the tapped threads could be vulnerable).

Rigidity of the front forks was ensured by the one-piece links, the top one being a forging of heat-treated light alloy, the bottom link fashioned from forged steel—pivotting on 40 ton ground steel spindles running in Oilite flanged bushes.

Long springs mounted adjacent to each blade ran from the fork ends to the eccentrics mounted in the bottom fork links. As an enthusiastic sidecarist, Phil Irving designed the forks so that spring strength and trail could be quickly adjusted for chair use simply by altering the position of the eccentrics with a spanner. (It was inadvisable to ride solo with the eccentrics in the sidecar position.)

The firm designed and manufactured its own hydraulic damper (with 3 in movement), there being nothing suitable on the market at the time, and this was fitted in the position occupied by the fork spring on the Bramptons.

Unfortunately, the damper tended to leak after working very efficiently for a short time; preventing the oil from escaping was a problem never satisfactorily solved during production.

The Girdraulics were a great improvement over the Bramptons for road work but George Brown told me he preferred the latter for racing, the heavier Girdraulics requiring a lot of heaving and hoing through the bends.

The new forks were certainly tougher. Where the Bramptons buckled in a crash, the Girdraulics remained unimpaired, along with the upper frame member.

I once had an illustration of their strength while riding a Black Shadow combination. I took a left-hander too quickly, the chair wheel lifted, I shot across the road and had the choice of hitting a telegraph pole or a squat, ugly pillar box. I rushed between the two and rapidly entered a neighbour's garden without having time to open the wrought-iron gates first. On impact, the gates disintegrated and a strut flew the length of that neat, suburban garden, decapitating roses on the way, and banged on the front door. Meanwhile, the Vincent headed for the only tree in sight—a rose tree—and wrapped itself around the trunk. As I sat there on the petrol tank covered in rose petals the owner of the house emerged. I was grinning from nerves, relief and some amusement but he failed to see the funny side. His new wrought-iron gates with a special weather-proof finish had only recently been installed. I was relieved to see that the bike was undamaged apart from a bent front wheel and then managed to remove the combination from the garden.

Stan Duddington (right) with the two engines he built for Reg Deardon's supercharged Black Lightning (1949)

While I was getting a replacement front wheel from another Vincent, the aggrieved party rang the police and a constable consequently asked me what was I doing crashing a vehicle with a road fund licence applicable to an Austin 7. The owner of the Black Shadow, a happy Irishman, just happened to know the Chief Constable and we heard no more. I told the Irishman that I wasn't too happy about buying his Vincent. 'Why not?' he asked. 'It's got a bent front wheel.' I was honour-bound to buy the bike (which cost £120), and I enjoyed many miles of pleasurable riding before selling it to the jazz musician, Jim Bray.

One problem when selling a Black Shadow was starting the Beast in front of the prospective customer. The Lucas magneto—fitted with platinum points—provided a good enough spark but turning over 1000 cc required a hefty boot, especially in pre-multi-grade oil days and the secret starting technique was handed down owner unto owner.

Starting could be a dangerous business, especially if the ignition or cam timing was incorrect, and various friends of mine suffered indignities, including that of being rapidly propelled over the handlebars and hurled against the garage wall.

As the road tester for *The Motor Cycle* was obliged to admit, when he rode a Black Shadow for the first time, 'Engine starting from cold was found difficult. . . .'

Speeds were remarkable in that road tester's report: 87 mph in second gear, 110 mph in third, while the reference to top speed was a terse 'Not obtained'. (The staff of *The Motor Cycle* were unable to find a road or airfield which would allow the Black Shadow to reach its maximum speed.)

Motor Cycling's road tester achieved 122 mph while *Moto Revue* of Paris and *Cycles* of Los Angeles clocked the Black shadow at 128 mph.

Philip Vincent was absolutely delighted with these performances and cited with pleasure the American who lived in Los Angeles but worked in El Paso in Texas. At weekends he rode home on his Black Shadow to see his girl friend—a round trip of 2000 miles.

The achievements of the model were splashed across the pages of the world's press but behind the glamour the Vincent HRD Company was in dire economic straits. In my ignorance, I asked PCV why he didn't raise money from the banks, using his assets as collateral to pay his creditors. He replied that the company had already borrowed to the limit of its resources. To obtain enough capital to finance the postwar expansion, the firm had let in a group of trust companies. These were more interested in dividends than motorcycles, the relationship did not work, and the firm went bankrupt in 1949.

Fortunately, instead of forcing Vincents into liquidation, the banks appointed a receiver on

A beautifully restored Series C Black Shadow. (No tooltray—who said Vincents need lots of attention)

VINCENT

August 4. He was E.C. Baillie (a chartered accountant) and he kept the company afloat for a few more years.

There were very few design changes during the production period of the Series C range (1948 to 1954). This was partly due to the firm's shaky financial position, partly because Phil Irving went back to Australia in 1949, and partly because the conservative motorcycling public did not appreciate change for its own sake. The benefit was that dealers could stock the standard range of spares to fit any vee-twin produced between 1946 and 1954. (Unlike the position today when some of the latest Japanese machines are off the road for weeks due to the difficulty of obtaining spares.) As Oscar Wilde once remarked, 'One should never be too modern or one suddenly finds oneself old-fashioned.'

One controversial modification was the replacement of the phosphor-bronze large idler with a forging made from the tough aluminium alloy RR77 (from engine number 3815 on the singles: engine number 4548 on the twins).

This component gained a reputation for unreliability and when it began to break up, shards of aluminium spread throughout the engine and damaged bearings, especially the big-ends. However, a correctly manufactured aluminium idler properly meshed with its half-time pinion will give up to 100,000 miles of use. Unfortunately, unscrupulous or ignorant people have produced inferior pattern spares for the Vincent over the years to the detriment of the motorcycle's reputation for reliability.

A member of the Vincent HRD Owners' Club, Tony Rose, decided to test the reliability of his new 1951 Black Shadow—*Rumplecrankshaft*. He asked PCV point blank if a standard Black Shadow would last 100,000 miles without requiring decarbonisation or overhaul. 'Yes', replied the optimist, blissfully unaware of Tony's reputation for wearing out motorcycles at an alarming rate. Tony began the test in November 1951 and covered 20,000 high-speed miles in two months before inadvertently dropping the bike on a patch of ice. Blast! He would have to fit a sidecar. Tony did not like chairs very much—they slowed a man down. Nevertheless, he attached a Blacknell Bullet to *Rumplecrankshaft* and rushed about at over 90 mph to the surprise of various solo riders and car drivers who thought that they were quicker.

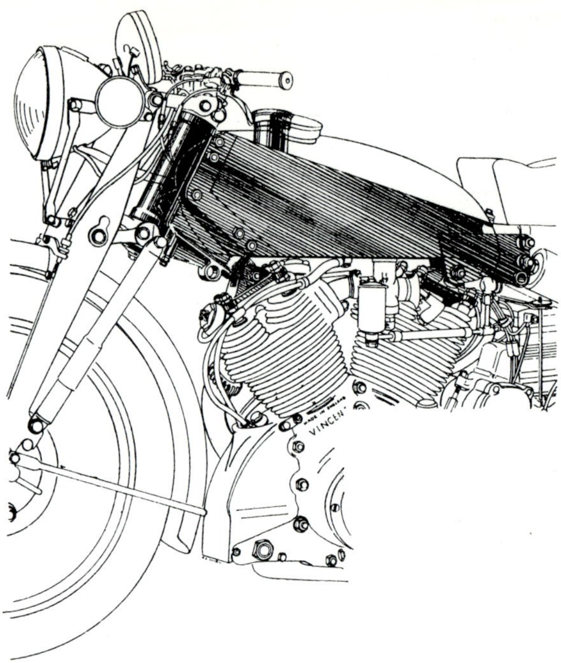

The upper frame member. The headlug and oil tank should not be parted during overhaul

Tony piled up the miles, but only the Bullet wilted under the strain and was replaced with a Swallow Commando.

On Tuesday, February 2, 1953, with over 100,000 miles on the clock, Tony rode into the yard at Stevenage and handed the outfit to Paul Richardson. Under press supervision, the engine was stripped by two mechanics, Jack Godfrey and Bill Bruce. They had the flywheels apart within 42 minutes and after the large mileage (including 80,000 with a sidecar), wear was as follows:

Rocker pins and rocker bores (third set)	0·003 in
G9 layshaft—first gear, bore wear	0·0006 in
G8/1 layshaft—second gear, bore wear	0·00025 in
G5 layshaft—journal wear	0·0005 in
Cylinder barrels, bore wear, front	0·0035 in to 0·004 in
Cylinder barrels, bore wear, rear	0·0025 in to 0·0035 in
Crankpin, roller tracks wear	0·0003 in
Big-end liner	0·0004 in

These figures speak for themselves. The Black Shadow was invariably left in the open during the 15 months of the test and Tony wore out numerous tyres and chains and three sets of exhaust pipes and silencers. The clutch shoes and linings were replaced after 50,000 miles but the aluminium idler survived the ordeal.

Philip Vincent (absent with flu) was delighted with this achievement and told me that Tony Rose was the type of rider he had in mind when he produced the Black Shadow.

The Vincent motorcycle therefore gave reliability as well as speed; the public obviously appreciated its looks, for Philip reckoned that maximum output was achieved in 1952.

Yet I must confess that when I saw a Vincent vee-twin for the first time, its lines did not appeal—only its performance. Speed has always been fashionable. However, after being inculcated in Vincent lore and riding the beast, I

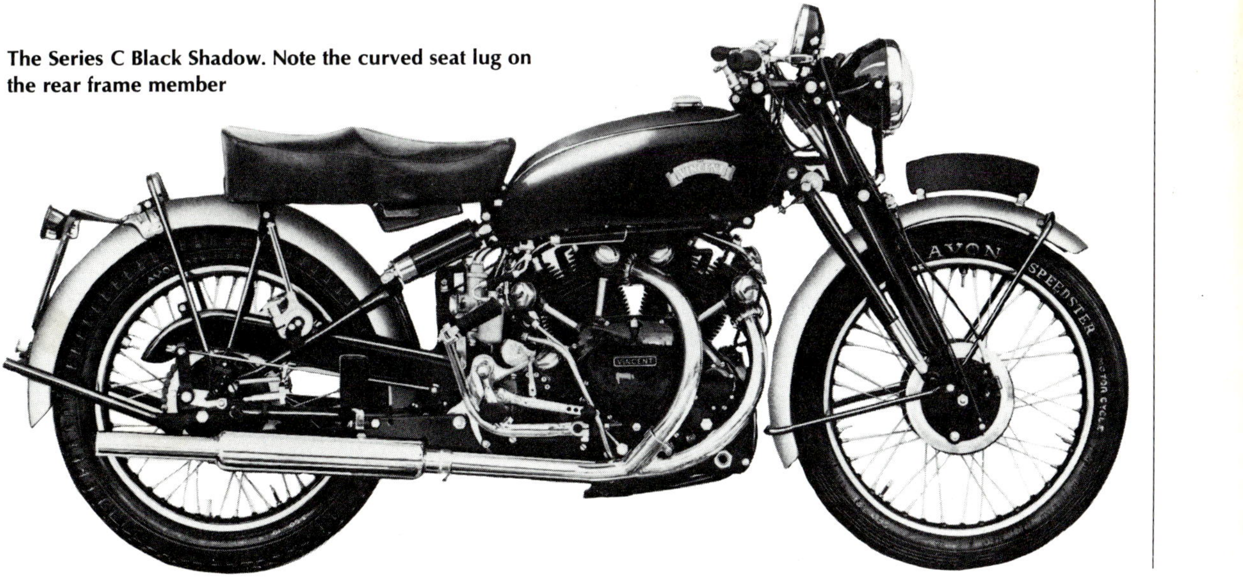

The Series C Black Shadow. Note the curved seat lug on the rear frame member

grew to love it despite its foibles. The Series C Black Shadow became the ultimate roadster for looks and performance and at last I was able to buy one, the 1952 works show model, no less.

This beautiful motorcycle, its chromium plated nuts and bolts sparkling against the black engine like the night hung with stars, was good for 105 mph in third gear. I never did find a road long enough to reach maximum in top gear.

This followed ownership of a Vincent Comet, a model advertised in 1948 to help the firm's turnover and to reduce production costs of the many items common to singles and twins.

The major parts that differed included the crankcase, the Burman gearbox, a new oilbath for the primary chain, a single exhaust pipe, new engine plates and a cast aluminium frame brace in place of the twin's rear cylinder to support the upper frame member.

The petrol available (72 octane 'pool') was inferior to the pre-war grade but the new Comet achieved nearly 90 mph—a speed good enough to see off most police cars of the era.

I felt that the wheelbase could have been shorter with advantage but that would have

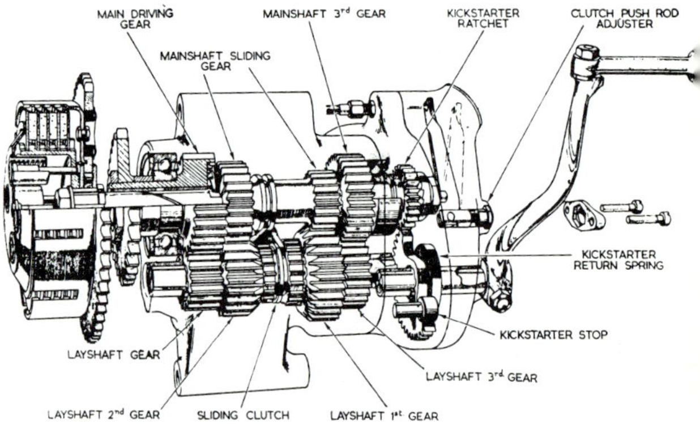

Right **The Burman gearbox used on the 500 cc models**

Below **This is a 1948 Series C 'Rapide Black Shadow' photograph with the Vincent logo painted over the HRD motif for use in 1949–50 catalogues**

meant reducing the length of the upper frame member thus involving the firm in unwanted extra costs.

With $1\frac{1}{8}$ in carburettor and 6·8:1 compression ratio, the Series C Comet had a cruising speed of 70 mph while fuel consumption for all-round work was about 70 mpg.

Girdraulics were scarce in 1948 and 1949 and were usually fitted to Black Shadows. Some years ago, I bought the 1949 Rapide first owned by George Formby (the second owner was Eddie Stevens) and learnt that this motorcycle left the works fitted with Girdraulics.

When I mentioned this to PCV he replied that very few early 1949 Rapides sported Girdraulics, due to the difficulty of getting supplies, and he instructed his colleagues to give priority to export orders and machines purchased by famous personalities. So my Rapide (*Pommers*) probably owed its Girdraulics and therefore Series C designation to George Formby.

Another 500 cc single, the Series B Meteor, was offered concurrently with the Comet and was fitted with Bramptons, a $1\frac{1}{16}$ in Amal carburettor, 6·45:1 compression ratio and instead of the unique front prop stands fitted to the Comet, it used the lower front mudguard stay as a stand for front wheel removal.

Production proper of the Meteor and Comet began in spring 1949, the former being dropped from the range a year later but the latter continued to be made until 1954.

To cater for the different taste of the Americans, 'Touring' versions of the Comet, Rapide and Black Shadow were produced from 1948 until 1954 at no extra cost. They incorporated curved 'cowhorn' handlebars, heavy steel mudguards and larger tyres (3·50 × 19 in front and 4·00 × 18 in rear instead of 3·00 × 20 in and 3·50 × 19 in).

During a visit to Canada and the USA in 1977, when I had the pleasant experience of riding my Vincent on North American roads, I asked several Vincent owners there why a different specification had been preferred. The concensus of opinion was that the larger tyres were more suitable for the mediocre but straight roads that ran for vast distances across those vigorous,

This is probably the first Black Lightning. It was on display at the 1948 Motor Cycle Show

The Series C Rapide Touring model with steel mudguards that added 7 lb. to the weight of the machine

hospitable countries. Interestingly the Canadians and Americans had gone to great lengths to restore their Vincents to original specification and there were very few 'Touring' machines to be seen. By 1977, the main highways were first class roads.

Some of the early machines exported to North America were finished in red or blue instead of black for the petrol tank, rear frame member and Girdraulics, etc.

When Girdraulics became readily available in 1950, the Series B Rapides and Black Shadows were phased out, by which time the famous HRD monogram had also been dropped to be replaced by the Vincent scroll. A former works employee assured me that Vincent was taking an egotrip but when I asked Philip for his reasons for the change he said that many Americans, who had never heard of Howard Davies, confused 'HRD' with Harley-Davidson and that would never do!

The changeover in 1949 has caused a considerable amount of confuson among historians and judges of *concours d'elegance* because several machines left the works a mixture of 'B' and 'C' parts—some with the HRD logo, some with 'THE VINCENT' and not a few with a blank space where the logo had once been. Confusion was compounded in 1949 when the works issued pictures of their new models. It was obvious that 'HRD' had been painted over and 'THE VINCENT' substituted.

When I asked PCV why he didn't take new pictures with the different logo in position he defended the action on the grounds of cost.

Early motorcycles in 1949 retained the 'HRD' transfers on the steering lug and sides of the petrol tank but some machines left the works with nothing at all cast on the crankcases (alongside the engine number), the inspection caps or the timing cover, the logo having been polished off. This action was surely unnecessary. (My George Formby Rapide has nothing on the crankcase but 'HRD' on the timing cover.)

Some machines were fitted with Series B rear frame members (straight lug for the seatstays) and Series C front end (Girdraulics). Early tooltrays (installed under the saddle) had a handle each side; later tooltrays had one handle only (fitted on the right-hand side).

The small ('HRD') transfer on top of the petrol tank featuring Mercury atop a winged wheel was identical to the one on the steering lug. The monogram was replaced by 'THE VINCENT' while below it the company name and address was retained—THE VINCENT "HRD" CO. LTD. STEVENAGE, HERTS.

In November 1952, the firm changed its name to the Vincent Engineering Company Limited, but another manufacturer had already registered that title, so in December the company registered the name of Vincent Engineers Limited and the transfers were changed accordingly.

However, many people continued to call the bike the HRD long after the changeover to Vincent. (After stopping my 1956-registered Vincent—which has no reference anywhere to HRD—outside a Lincolnshire farm during a ride in the seventies, I was surprised when the farmer emerged and asked, 'What year be the HRD?')

Philip reckoned that the production of the Series C range reached its peak that year (1952) and this increased output was partly due to the publicity gained from participation in the 1950 Senior TT in the Isle of Man.

The TT venture was Baillie's idea. He did not

Left-hand view of a Series C Rapide. (More Series C motorcycles were made than A, B and D combined)

KNK 610

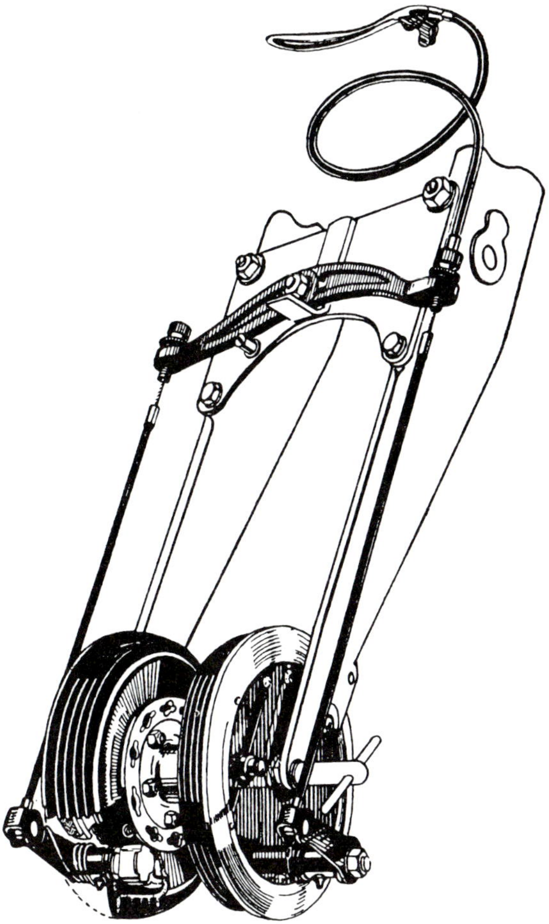

expect to win any prizes but at least the firm's presence would emphasis that it was still in business.

Phil Irving was in Australia and Matt Wright had left to rejoin the racing staff of AMC but PCV still had men of the calibre of George and Cliff Brown and Jack Williams.

Time for radical design changes to the 500 cc single was short. Philip had but seven weeks from Baillie's decision to his disembarkation for New York on April 15th and decided to concentrate on improving engine breathing and designing a new big-end that could withstand high revolutions.

Fortunately, the previous year, a racing version of the single had been developed and later displayed at Earls Court. George and Cliff had tuned a Comet and Vincent had marketed the new model as a Series C Grey Flash. It was available in three versions—as a stripped racing motorcycle for £275 plus purchase tax, as a roadster with lights, horn, and other components for £290 plus purchase tax, or as a roadster supplied with racing equipment for easy conversion into a racer (£300 plus purchase tax).

The engine was tuned to Black Lightning specification and produced 35 bhp at 6200 rpm. The petrol tank and frame members were finished in grey-green to distinguish them from the Comet's black enamel.

Grooves were cut on the inside of each Girdraulic blade to reduce weight and several holes were drilled in frame members to serve the same end. As a result of modifications, etc., weight was kept down to 330 lb.

Top **Black Shadow rear-brake torque arms and cross shaft. (Rapides were fitted with unribbed brake drums)**

Bottom **Series C Black Shadow front brakes. (Early models did not have the balance beam stop. Rapide, Comet and Meteor models had unribbed drums)**

Far left **Denis Jenkinson with a Vincent outfit (and French skier!) on the Ballon d'Alsace in 1952**

The Grey Flash was raced impressively by John Surtees, who was an apprentice at one Vincent factory for a time, before he moved on to faster machines.

Philip Vincent therefore had a proven machine as the basis for his TT racers and set to work on his modifications to the heads and big-ends.

Luckily the single's cylinder head was the same as the vee-twin's front head with its inlet port's greater potential for being opened out safely (whereas on the rear head there was the risk of going through into the rocker tunnel). Nevertheless, PCV calculated that by building up the thickness around the inlet port and altering slightly the angle between the valves, it would be possible to increase the diameter of the inlet port

Top **The Black Shadow used by DSJ and Cyril Quantrill for a week-end trip to Vosges mountains (1952)**

Below **The Series C Meteor (Girdraulics instead of Bramptons). The HRD logo has been painted over. Only one was made (1950)**

Above **The Series C Comet prototype displayed at Earls Court (1948). Production began spring 1949 with combined footrest and rear brake lever instead of the set-up shown**

Bottom **The reliable Series C Comet which can cruise all day at 70 mph and give 70 mpg**

and valve to work in conjunction with a larger exhaust valve.

The outside foundry produced a batch of big-port heads within a fortnight and a TT Amal racing carburettor was matched to the $1\frac{1}{4}$ in inlet port.

Meanwhile, Philip had designed a new big-end made, of all materials, aluminium! The problem

Top **The prototype Series C Comet with HRD logo painted over**

Bottom **The Series C production Comet. Note combined footrest and rear brake lever**

was that the aluminium crankpin would expand more than the steel connecting-rod. The theory was that the RR77 crankpin would run at a lower temperature than the connecting-rod and anyway there would be ample clearance allowed in the design when the engine was cold to enable expansion to take up the play without seizing up.

Top **The 1949 Grey Flash capable of 115 mph. (The HRD logo on the crankcase and clutch cover has been painted over)**

Below **Cyril Julian balancing above the Montlhéry Black Shadow (1952). Ted Davis hovers behind while John Surtees examines his gloves**

A tall order. PCV instructed his tool-room operators to make crankpins with outer diameters ranging from two to four thousands of an inch less than the big-end eye.

Tests on the brake suggested that 23 thou' was the most suitable diameter for the crankpin; a complete Grey Flash was assembled and taken to the old test track, Gransden Aerodrome, where George Brown clocked 119 mph, a big improvement over the usual maximum speed of a Grey Flash—110 mph.

The basic weakness of the aluminium big-end was its susceptibility to any grit that circulated

Overleaf **Cover of a catalogue used in 1949 and 1950**

THE
VINCENT

One of the vee-twins used at Montlhéry in May 1952

through the engine. Where the tough standard steel big-end would resist the attentions of a wayward minute shard of metal, the aluminium big-end would be ploughed up and seize, a mishap that ruined three new big-ends during practice in the Island.

The Burman gearboxes played up and Albion boxes had to be substituted at the eleventh hour but the works team of C.A. Stevens, K. Bills and M. Barrington was ready in time for the start; Ken Bills deputised at the last minute for George Brown, who had not fully recovered from a car crash. Another Grey Flash ran in the hands of a notable private entrant, Johnny Hodgkin.

However, by the fifth lap, only Ken Bills was still in the race, the others having retired with engine failure. He finished twelfth at an average speed of 83·79 mph and collected a second-class replica.

The importance of competing in the Senior TT in the Isle of Man—the Blue Riband of motorcycle racing—was that the public and traders were re-assured about the company's financial position.

When a company is in difficulties, the word soon gets round and credit, for example, is hard to obtain. The public are reluctant to purchase motorcycles that may be useless within weeks through lack of spare parts.

The TT venture in June stimulated confidence and the firm enjoyed record sales and output during the following two years, enabling it to discharge the receiver and revert to normal trading conditions on July 22, 1952. The receiver, Baillie, became chairman of the new board of directors.

The ambitious firm that had started so modestly in 1928 with a handful of enthusiasts now employed between 300 and 400 people to cope with production, sales and service of more than 60 motorcycles per week.

The company's improved financial position enabled it to chase a few records at Montlhéry earlier in 1952 in partnership with the Wakefield Oil Company with the Castrol brand.

Vincent endeavoured to produce the first motorcycle to average 100 mph for 24 hours.

Philip had a thing about using a standard Black Shadow 'Same as you can buy' and retaining the standard big-end for the record attempts. His colleagues advised him that the standard big-end would not last the 24 hours and that a racing big-end should be fitted instead. Vincent took no notice but altered the specification of the Black Shadow with a higher compression ratio (8:1), Black Lightning cams, Amal $1\frac{5}{32}$ in carburettors and 2 in open exhaust pipes. The front brakes and one rear brake were removed for lightness and a special 5 gallon petrol tank reduced the number of stops needed to refuel.

Rab Cook, the Editor of the Vincent HRD Owners' Club journal, *MPH*, felt obliged to point out that the specification was hardly that of a standard Black Shadow, so if other parts had been modified, why not the big-ends?

The long-distance record was attacked on May 14 and 6 hours later the Black Shadow claimed new records for the 6 Hours and 1000 Kilometres—originally set up 17 years earlier by Milhoux and Charlier riding an FN.

Other records fell before the big-ends failed after 11 hours, probably due to the intense heat (80°F in the shade).

A well-known Vincent and rider. Eddie Stevens (author of *Know thy Beast*) **aboard his Series C Rapide**

VINCENT
JER 161

Would a racing big-end have fared any better? Ted Davis thought so; so did a journalist, Vic Willoughby, who was riding the Black Shadow when the big-end seized. Vic told me that PCV later agreed that he was mistaken in using the standard big-end for the attempt. However, the firm now held the following records:

Right **A close-up of Malcolm Attrill's** *The Fast Lady*

Below The Fast Lady **with Steib. Malcolm raced a Vincent 1000 kneeler outfit successfully for many years**

Denis Jenkinson's Series C Rapide at Odiham (1960)

Record	*mph*	*Previous holder*	*mph*
6 hours	100·60	FN	96·72
1000 km.	100·80	FN	96·32
7 hours	99·73	Gnome et Rhone	91·90
8 hours	99·48	Gnome et Rhone	91·30
9 hours	99·40	Gnome et Rhone	91·70
10 hours	99·17	Gnome et Rhone	91·54
1000 miles	99·20	Gnome et Rhone	91·36
11 hours	92·50	Gnome et Rhone	91·37

Gunga Din was used by John Surtees and Ted Davis for attacking some short distance speed records but the tyres were unable to stand up to the strain of high-speed travel—some 150 mph—and shed their treads to the detriment of the riders' posteriors. John and Ted must have wondered what they had done to receive such undeserved flagellation.

It was indicative of the high standard that the team had set itself that the Montlhéry effort—which after all had yielded several records—was regarded as something of a failure. Nevertheless, these latest records further increased interest in the Beast and the factory was kept busy in 1952 and 1953 satisfying the demand for its products.

They were heady days for Philip Vincent. All over the globe his motorcycles were breaking records in private hands, the company was profitable (he was earning enough to pay surtax—19*s* 6*d* in the pound) and to compound his happiness he married (in 1953) Elfrida, a beautiful Irish girl who had recently joined the firm.

Display by Kings of Oxford (1950). Top of the heap is Reg Dearden's supercharged Black Lightning. No. 15 is *Gunga Din* **fitted with Girdraulics**

'Gunga Din'
HOLDS
VINCENT
BLACK SHADOW
Kings of Oxford
Kings of Oxford
Kings of Oxford
'BLACK LIGHTNING'
for EXPORT to BELGRADE
SIMILAR
AVAILABLE ...
HOME SALES NOW

6 *Polyphemus*, the Vincent three-wheeler

Although the factory was enjoying a boom in the early fifties, Philip was well aware that the postwar demand for motorcycles would soon diminish. He therefore sought an additional product that would utilise the Rapide engine in standard or tuned form and finally decided to design and manufacture a sports three-wheeler.

The firm had previously produced a four-wheeled car (not the amazingly ugly thing with a plywood body that was chopped up for fire-wood). A standard Cooper body had been mounted on a specially modified Cooper F3 chassis housing a Black Lightning engine and the late Eric Winterbottom had raced the Cooper-Vincent with some success. (He was once timed at 132 mph at Gransden—fast for the era.) Some other Cooper-Vincents were made and incorporated Black Shadow engines.

Philip thought that a fast three-wheeler would be popular and decided to manufacture one using an independently sprung front axle, incorporating rack and pinion car-type steering and hydraulic car-type brakes, fastened by a triangular form of chassis to the power unit, which was located along the centre line just behind the seats of the driver and passenger, and carrying the rear wheel in one of the standard Rapide triangulated rear forks.

Philip discussed the project with Dick Shattock, who consequently produced the bodywork made from hand-beaten 16-gauge aluminium. This was bolted on to a chassis constructed from 4 in steel tubing. The engine was cooled by a

The production prototype of the Cooper-Vincent touring car which regrettably never went into production

duct that ran from the orifice at the front of the car directly on to the front cylinder and further cooling was provided by vents on either side of the body. Fuel was supplied by gravity feed from a 2 gallon petrol tank mounted behind the passenger seat, while the standard Series C oil tank, in the conventional position above the engine, provided the frame member into which a Series D Armstrong rear suspension unit fitted. The rear frame was pivoted in the usual motorcycle position and housed an 18 in wheel fitted with a 4 in Avon sidecar tyre. 14 in Morris Minor wheels with 5 in tyres were used with the RGS front assembly; 8 in hydraulic brakes were fitted at the front with 7 in drums, cable operated, at the back.

The car retained a popular feature on Vincent motorcycles of a reversible rear wheel, which could incorporate two different sizes of sprocket.

Other details included Lucas magneto, Miller dynamo, wheelbase 11 ft 2 in; track 4 ft 3 in; length 11 ft 2 in. (Fuel consumption with $1\frac{1}{16}$ in carburettors would be 40–45 mpg.)

When the three-wheeler was ready for its first road test, the mechanic who had been working on the car climbed a little awkwardly into the driving seat and after receiving a push-start shot up the Great North Road at an impressive speed. He returned at a similar velocity and rushed through the factory gates, removing a drain pipe as he swerved round the yard, before coming to rest when the engine stalled. When a director demanded an explanation the driver assured him that although, as a motorcyclist, he had found all the pedals a little confusing, he had enjoyed his first ride in a car.

The prototype three-wheeler was locked away for a few weeks while the directors considered whether or not to put it into production. Many components were standard motorcycle parts (e.g. Rapide engine, clutch, gearbox, transmission, rear forks, suspension, oil tank, rear wheel and brakes) so the assembly operations could be run alongside the motorcycle production line, diverging near the last stages of completion.

It was decided to entrust the three-wheeler to Ted Davis, the chief tester, for further design and development. Ted had been racing his Vincent combination very successfully and was more interested in improving the car's performance than increasing its appeal as a standard touring vehicle. With a Rapide engine the three-wheeler was capable of 90 mph which Ted regarded as totally inadequate, so he installed *Gunga Din*'s 'Lightningized' engine.

With Paul Richardson, Vincent's service manager, as passenger, Ted made his first run in the three-wheeler with the new power unit. He accelerated through the gears, clicking smoothly into top at about 90 mph as he enjoyed the ride up the Great North Road. As he drove over the brow of a hill Ted's tendency to heave on the steering wheel as the car dropped down the other side pulled the wheel off its column, where it couldn't be replaced, and Paul's inclination to stutter increased considerably when Ted handed him the wheel.

The amazingly ugly plywood-bodied, marine-engined car, which fortunately never went into production

The steering wheel behaved itself subsequently and Ted once recorded 117 mph in the three-wheeler; he drove it in various sprints and races and once a plug lead came adrift. To replace the lead it was, of course, necessary to lift the large rear cowl that covered the engine and rear end. Ted and his passenger were in a hurry, the cowl was quickly flicked open whereupon it gathered momentum, broke the brackets holding it to the top frame behind the occupants' seats, and carried on until it came to rest on the track.

However, Ted usually enjoyed his sojourns in the car and would have liked more time to develop the potential of the three-wheeler but the project was curtailed because of the financial plight of the Vincent HRD Company.

'We were not able to manufacture the three-wheeler,' Philip told me, 'because we would have had to make substantial modifications to the works to incorporate the body-building department, but the company's finances were too run down to cover that requirement, so unfortunately the three-wheeler was allowed to die a natural death. This was a great pity because, of all the many models we made, I'm pretty sure the public would have acknowledged the three-wheeler to have been the most outstanding of all, especially in its super sports form, which would have incorporated a Black Lightning engine with all the many improvements effected in consequence of the Picador development programme. This unit would have provided a maximum speed of 120 mph at least, and with a Rapide ratio gearbox the acceleration would have been simply terrific. Road-holding and handling were excellent and neither front wheel ever gave any indication of lifting. Weight distribution was also excellent in the three-wheeler. The heavy units, represented by the engine and gearbox, were located in the position along the wheelbase that has been adopted by grand prix car designers, i.e. just in front of the rear wheel.

'In this model the front springing system, which was carried out by means of four elliptic springs, was naturally rather weighty, this weight

The Vincent three-wheeler on test. It was capable of 117 mph with *Gunga Din*'s engine fitted

Polyphemus, the Vincent three-wheeler

serving to maintain stability if the inside wheel tended to lift on corners.'

The three-wheeler was not a very practical vehicle for use in town, lacking a self-starter and reverse gear. 'Excuse me a moment,' says the driver to his passenger as he hops out, removes the engine panel and endeavours to kick-start the engine without the aid of de-compressors. Sometimes the engine is surprised and fires: usually a mighty leap on the kickstarter produces no result whatsoever and the sight and sound of a sweating, cursing driver leaves the passenger unimpressed. (No wonder Ted Davis used to keep going until he found a suitable hill.)

The absence of a reverse gear is a problem however. The first time I reversed the car (by standing and pushing alongside the driver's door and turning the steering wheel), the offside front wheel ran over my right foot. The car was very heavy for a tricycle, weighing 1100 lb and thereby qualifying for the full rate of road fund tax, but Philip Vincent told me that he would have reduced the weight by using a fibre glass body had the vehicle gone into production. He had not decided on any colour schemes for the body and the prototype left the works in its natural alloy finish.

Shortly after the Vincent company was taken

Left **Philip Vincent (left) describing the three-wheeler while Ted Davis (right) takes care of the muscle work**

Below **Engine room of the three-wheeler (Series B crankcases, Series C rear frame member, Series D Armstrong damper/spring unit!)**

over by Harpers in 1959, the three-wheeler was sold to a Mr. Gibson of London. A year later it was back at Harpers, still with its original finish of polished alloy.

In 1961, a German living in Sussex, Josef Karasek, read about the three-wheeler in a magazine advertising tricycles and purchased it. Karasek, who surprisingly had never heard of Vincent motorcycles before then, had the car sprayed British Racing Green and raced it at Silverstone a couple of times. He didn't get on too well. 'It was all right on the straights,' he said, 'but when I turned the wheel to take a bend the car went straight on.' Joe kept this rare vehicle in a shed on his farm but I did not know the three-wheeler's whereabouts until I began the research for a previous book.

Now, classic bike enthusiasts fall into two basic categories; the first consists of a small privileged minority who can drive through a village and smell a rusting wreck blushing unseen and wasting its sweet fragrance among the chickens; the second consists of the vast majority who pass through happily oblivious to the treasures hidden all around.

My brother Derek belongs to the former group: I to the latter. So when I wanted to know the whereabouts of the Vincent three-wheeler I didn't ask him if he knew the location, I merely suggested that he lead me to it. After driving through miles of tiny lanes we arrived at Joe's farm and there, stored for years in a barn, was the tricycle, the Vincent three-wheeler, the *prototype* three-wheeler. I had only seen photographs before of the car in pristine condition, and as I gazed at the vehicle with its airless tyres, rusty wheels, wires trailing on the floor and scratched body, I fell in love with it. A dealer would advertise it as 'ideal for restoration.' 'I'll make Joe an offer,' I said to Derek. 'What's a fair price?' 'There isn't one,' he replied. 'It's the only Vincent three-wheeler so it's worth what some one is prepared to pay. I know what I'd give but Joe won't sell. Not even for ready money.' However,

Top **Early shot of the three-wheeler before its windscreen was fitted**

Right **I bought the car in 1974, named it** *Polyphemus*, **and had this photograph taken outside my Chelsea flat**

Joe did sell. And *Polyphemus* formerly *Locomotion* during Joe's ownership) is now being restored to his old glory. (What I did not know was that all over the country many a Vincent enthusiast was sitting on a heap of cash waiting to move in when the whisper came that Joe was selling.)

Oddly enough, the engine is an early one with 'HRD' on the crankcases (number 3155) so the works obviously dropped in an old Rapide unit before selling the car. (The frame number in the log book is RC/1A/5928 TT.) I've bought a Picador engine with the intention of installing it in the three-wheeler complete with reverse gear and starter motor.

7 The Picador

The Picador power unit was a derivation of the Vincent vee-twin and was produced concurrently with the motorcycle engine in the early fifties.

The Air Ministry was interested in a powered target aeroplane to replace the aged DH Tiger Moth whose performance was restricted to towing sleeve targets in straight and level flight at a limited range of speed and altitude.

A major disadvantage of using the Tiger Moth, as trainee gunners blasted away with more enthusiasm than skill at the sleeve target, was the aircrew's psychological discomfort.

The pilot-less aeroplane—the U120D—was designed by the ML Aviation Company of White Waltham and was radio controlled from the ground to carry out vertical and horizontal manoeuvres.

When it was time to terminate the flight a radio signal switched off the fuel and magneto and released a parachute while simultaneously inflating a shock absorbing cushion and flotation bag, thereby enabling the aeroplane to descend horizontally touching down on land or water.

Should radio contact be lost, the engine fail, or the aeroplane descend to a pre-determined height, its radio equipment automatically switched off the fuel and released the parachute.

Life expectancy of the U120D was 12 hours but as it was constructed in three sections it would be possible to extend the life of the aeroplane by replacing any damaged section quickly.

The Picador engine on test in the fifties. It produced 65 bhp at 5000 rpm

PCV hoped that a substantial order from the Air Ministry for the Picador engine would enable improvements to be made to his vee-twin motorcycle and extend its production life.

The specification called for a tuned Black Shadow engine—running practically flat out—propelling the aeroplane at a speed of over 180 mph.

The design originally submitted by Vincent incorporated the Picador mounted transversely in the aeroplane with heads and barrels projecting above the fuselage.

ML Aviation's designers objected to the

engine protuberances spoiling the line of their beautiful bodywork and instead favoured installing the engine back to front which would, of course, mean less cooling for the hot exhaust ports. However, this would not be a problem because the air would be pretty nippy at the altitude flown by the 'kite'.

The propellor shaft was therefore mounted behind the engine and driven by a primary chain (in an oil bath) via a bevel gear to provide the necessary right-angled drive, the overall reduction ratio being 2:1.

The aeroplane had to be capable of inverted flight, which would present carburation difficulties, so fuel injection was used and Jack Williams made the necessary modifications to the standard CAV twin-cylinder diesel-type injectors.

Output of the pump was related to the throttle opening while a barometric aneroid control ensured that the air-fuel ratio remained constant at different heights.

A common shaft operated both throttles which were kept fully open by a spring. For starting, the throttles were held in a slow-running position by a peg, which was removed immediately the Picador engine had been turned over by an electric starter motor on the launching ramp engaging a starting dog on the drive side end of the crankshaft. When the engine fired the rockets were automatically ignited and the U120D accelerated into the air. (The injectors were located in the wall of the long induction pipes, which provided ram effects.) Development work continued for several months, a major problem being the final bevel gear drive for the airscrew. The teeth failed after three or four hours' running instead of lasting the expected 25. Heavier teeth solved the problem.

The shafts pressed into the flywheel tended to turn because the standard EN8 steel used was unable to grip them sufficiently, so a tougher steel was used for the flywheels, the shaft size was increased to $1\frac{1}{8}$ in while the crankpins were a press fit in the flywheels at each end.

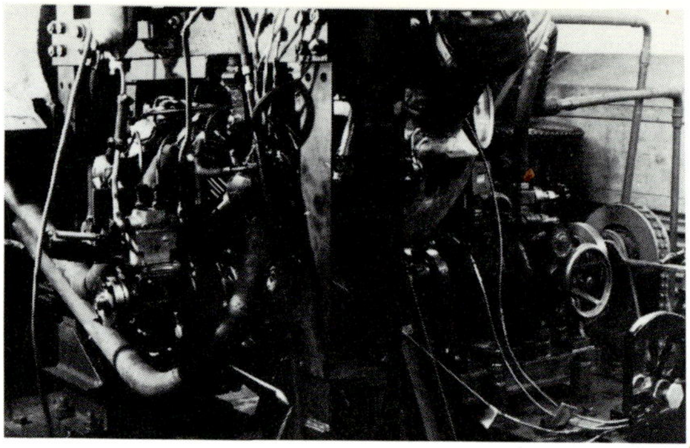

Top **The Picador engine (developed from the standard vee-twin engine) on test**

Bottom **The Picador engine developed in the fifties for use in a target aircraft built for the Air Ministry**

100 octane fuel was provided for the tests some months before it would be available in Britain for motorists; and Vincents were able to raise the compression ratio to 10:1.

During early tests, the pistons cracked but the manufacturers sorted that one out for Vincents. Another problem was failure of the drive side big-end bearings due to oil starvation but this was cured by using a steeper pitched double-start worm, which doubled the speed of reciprocation of the pump.

When ready for installation in the U120D, the engine weighed just 200 lb and incorporated

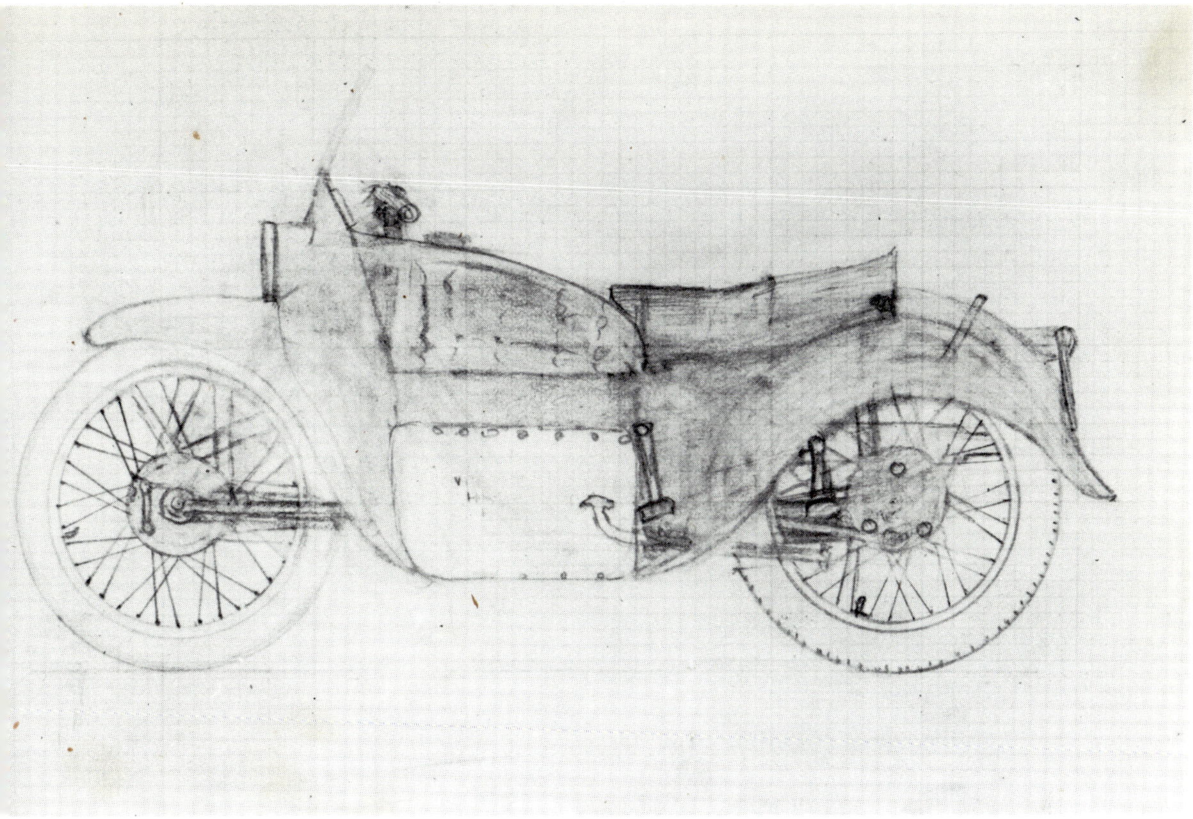

War-time drawing of a proposed marine-engined motorcycle that would have had a top speed of 75 mph (fuel consumption 180 mpg at a cruising speed of 50 mph)

Mk II cams, Scintella magneto and KLG FE102 sparking plugs. Its output was about 70 bhp at 5000 rpm.

The test flight of the aeroplane was made at Tenby, Glamorganshire, and Jack Williams was there to watch over the fortunes of the Picador. Unfortunately, there was a low cloud base and the aeroplane could be heard but not seen as it buzzed somewhere overhead. According to Jack, the chap operating the radio controls carefully guided the aeroplane into the sea.

The men from the Air Ministry were accordingly unimpressed. They couldn't have wayward aeroplanes homing into unauthorised targets and banned any more U120D flights in England. The other aeroplanes were subsequently taken to Castel Benito, Libya, which provided a big enough space for the technicians to play harmlessly. Two of the aircraft flew off into the sun and were never seen again.

8 Fireflys and Foxes

Had the Picador project been successful—at one stage the Air Ministry was thinking of buying 1000 engines—the production life of the Vincent vee-twin motorcycle would have been extended.

About 30 Picador engines were reputedly made in the fifties and some were sold off by a London dealer for about £65 each after the termination of the U120D project by the Air Ministry.

This cancellation was a major setback to PCV, who therefore concentrated his efforts on distributing various mopeds and lightweight motorcycles to augment the faltering sales of his vee-twin.

Philip told me that he became interested in handling a cyclemaster in 1953 because it was becoming obvious that sales of big motorcycles were bound to taper off after the postwar boom. The seller's market was turning into a buyer's market and it would be very difficult to maintain sales of an expensive machine like the Vincent thousand.

Philip had been approached by 'Millers', whose equipment he already used, to take over from them the manufacture and marketing of a 48 cc two-stroke unit that could be fitted to a pedal cycle.

Philip appreciated that the position of an engine below the bottom bracket of the cycle frame would give good weight distribution (unlike many rival units that hovered over the front wheel or vibrated away on one side or the

The Firefly (1953) which had a claimed cruising speed of 18 mph until it expired

other of the back wheel).

Vincents sold the 'Firefly' unit only in 1953 for £25 and two years later offered the complete machine for £38 19*s* 5*d*. By 1956 sales had already fallen and the bicycle was discontinued but the unit only was sold until 1958. Philip estimated that nearly 3000 units were produced (but not all were sold).

Several members of the Vincent HRD Owners' Club were dismayed when the Firefly made its appearance under the mighty Vincent banner and arguments ensued as to whether or not owners of these 'pop-pops' should be allowed to join the club. In fact, the person who bought the Firefly usually had no interest in motorcycling anyway and had he wanted to attend a Vincent rally his mount's ability to get him there was always in doubt.

The drive to the rear wheel was via a rubber roller pressed into contact with the rear tyre tread. (This could be disengaged by use of the clutch. Sometimes it disengaged itself.) Unfortunately, early Fireflys suffered trouble with the bonding of the rubber mounting on its shaft and all over Britain disgruntled, sweating owners were pushing or pedalling home their silent machines. (Some just abandoned the lumber by the wayside.)

If the Fireflys managed to survive in dry weather the rain usually got them. The roller slipped against the wet rear tyre as the engine revved while the motorcyclist-turned-cyclist pedalled optimistically. All this tested the patience of the long-suffering works road testers—even if the thing kept going its cruising speed was just 18 mph. To boost his mileage, one tester rigged up his Firefly on a set of rollers every night; then he switched on the rollers and went to bed. On one occasion, PCV—his unsuspecting boss—congratulated him on his high test mileage.

Although the problem with the rubber mounting was eventually cured, the Firefly was never popular, especially with Vincent HRD Owners' Club members, but as *One Track* pointed out, it was very useful as hardcore.

The Firefly was eclipsed by the vastly superior NSU Quickly, a moped imported from Germany ready assembled for distribution. Vincents handled the machine for a year and sold about 1000 per month. The Quickly deal was part of an agreement between Vincent and NSU, whereby the English company fitted British proprietary

The elegant NSU Quickly, which was distributed very successfully by Vincent until NSU formed its own distribution company

parts to the range of small-capacity motorcycles produced by the German company.

At that time, the Germans had some difficulty penetrating Commonwealth markets because the Commonwealth Preferential Tariff was in force. Under this agreement, the United Kingdom imported Commonwealth foodstuffs duty free and in return other Commonwealth countries reduced their external tariffs on manufactured goods from the UK. (This arrangement was phased out when the British Government took the UK into the Common Market. Food from the Commonwealth countries became subjected to import taxes and the UK lost its preferential treatment when it exported manufactured goods to other Commonwealth countries.)

To qualify as being 'Made in Britain' each NSU motorcycle had to include at least 51 per cent of British components. It could then be exported to other Commonwealth countries and qualify for the lower tariff extended to British goods.

Four interesting models were produced—

The NSU-Vincent Fox. This two-stroke version is recognizable from the overhead sparking plug

98 cc four-stroke, 123 cc two-stroke, 200 cc two-stroke and 250 cc overhead cam.

NSU claimed that the 98 cc ohv 'Fox' produced 6 bhp at 6500 rpm with a compression ratio of 7·2:1 but Philip thought that the claim was exaggerated.

The lubrication system was primitive. A pouch screwed to the wall received a supply of oil from the periphery of the clutch unit and fed the main bearings via a bore in the crankcase. Surplus oil collected in cavities inside the flywheel and subsequently lubricated the big-end under centrifugal force. The oil pouch also acted as a sludge trap. There was no provision for reboring the barrel; once it became worn it had to be replaced.

40 ohv Foxes were produced at a retail price of £124 16*s* (including £20 16*s* purchase tax). One languished in a dealer's window for a few years

until I bought it at a reduced price. It was ideal for travelling around London in the rush hour and every working day I used to give a lift to a colleague en route to my office. The little bike used to cope satisfactorily with two adults but one disadvantage was the gearchange mounted on the left instead of on the right like the Vincent twins and singles. I invariably braked when I wanted to change gear and vice versa.

I experienced problems with the tiny valve gear. A ball bearing situated in a cup on the end of the rocker arm located on a cap fitted over the top of the inlet valve. This invariably became loose in the cup and sometimes fell out. I never did solve that one, but I still have the bike which, like many other motorcycles all over the world, is undergoing long-term restoration.

Top speed was 46 mph and that was too fast for the miniature brakes, which proved that one didn't have to travel quickly to find excitement. The rear springing consisted of a miniature 'monoshock' and was very efficient.

The 123 cc two-stroke Fox reputedly developed 5 bhp at 5500 rpm with a compression ratio of 6·1:1. It gave a smoother ride than the four-stroke and top speed was about the same.

The bikes were smartly finished in black with gold lines—mini-Vincents—but they failed to induce many people to part with their money at Earls Court in 1953 and only 160 two-strokes were made. A handful have survived.

The prototype of the overhead cam 250 cc Max was displayed at Earls Court but again never went into production due to the high cost of the German components.

The third NSU-Vincent advertised in the catalogue—the 200 cc two-stroke Lux—was not shown at Earls Court and again the prohibitive costs of the German components prevented it from going into production.

Right **A 1954 hand-out advertising the unsuccessful lightweight wares**

Below **The NSU-Vincent 250 cc ohc Max. Only one was made**

VINCENT **Firefly** ALL-WEATHER CYCLE MOTOR

The safest all-weather cycle motor which becomes an integral part of your machine. Positioned below the bottom bracket, there is no splash, your clothes keep clean, and back wheel wobble is avoided. The larger capacity engine eats the hills and a $\frac{5}{8}$-gallon petrol tank gives you a long cruising range. The large flexible low geared roller drive ensures smooth, shockless transmission, overcomes slip in wet weather and provides a long tyre life of 5,000 miles plus. Running costs are in the region of 3 miles a penny.

NSU **Quickly!**

Safety, comfort, silence, complete reliability, light weight and ease of operation render this beautifully finished autocycle the obvious choice for all who require personal transport at minimum cost.

The machine has powerful motor cycle brakes, sprung front forks, " silenced air " carburation, a 2-speed twist-grip operated gearbox and a clutch which enables the machine to be used as a motor cycle. The standard specification includes full lighting set, electric horn, swinging saddle (adjustable for height), luggage carrier, safety lock, central stand and balloon tyres.

NSU VINCENT

Fox 98 c.c. o.h.v.

Luxurious in conception and of proved stamina, the 98 c.c. o.h.v. Fox appeals to discerning riders who require Vincent performance and quality combined with minimum running costs. This high-quality lightweight motor cycle is of most up-to-date design and incorporates such refinements as bottom link front fork, swinging arm rear suspension, helical gear primary drive and 4-speed gearbox.

NSU VINCENT

Fox 123 c.c. Two-Stroke

Similar in specification to the 98 c.c. model, the 123 c.c. Fox offers the advantages of a slightly larger engine with the simplicity inherent in two-stroke engine design. This machine, as well as the 98 c.c. model, is available with dual seat equipment and chromium plated rims as extras.

9 *Knights of the Road*

The failure of the NSU project encouraged Philip Vincent to see what modifications could be made to make his vee-twin more saleable.

By 1954, the postwar boom in motorcycle sales began to taper off and Philip Vincent pondered on the future development of his famous vee-twin. Despite the many racing and record-breaking successes of the motorcycle, Vincent still regarded his creation as a high-speed roadster able to carry its rider safely over long distances. The engine had proven reliability and power and the strong frame members gave the machine unequalled lateral rigidity and safety.

100 octane fuel was at last readily available in Britain so Philip had the opportunity to increase the performance of his road-going motorcycles by amongst other changes raising the compression ratio, opening out the ports and fitting larger carburettors.

In fact, he went in the opposite direction and decided to enclose the Black Shadow, Rapide and Comet to give the rider weather protection (which resulted in a diminution in speed).

He had always seen the motorcycle developing into a vehicle that could be used by a firm's managing director for travelling to the office clad in smart suit, gloves, shoes and bowler hat instead of looking like something from outer space.

The Series D range was therefore going to be the gentleman's motorcycle—the vee-twin with easier starting and fitted with a centre stand that

would be simple to operate compared with the balancing act required to get a Series C vee-twin on to its rear stand.

This could be quite a tricky manoeuvre, especially if panniers full of luggage were fitted. After resting the bike on one of its side-stands, the rider proceeded to the rear of the machine, unscrewed the little tommy bar and released the rear stand which was then lowered to the ground to a position to the rear of the wheel. Putting a toe behind the stand to prevent it from sliding, the rider then endeavoured to balance the bike upright while simultaneously heaving on the lifting handle located atop a mudguard stay just behind the dualseat. If all went well, the bike was pulled backwards, the rear stand attained a vertical position before settling just beyond it, and the bike was safely parked resting on its stand and front wheel, leaving the rear wheel free of the ground. Not everyone got it right first time and they soon discovered how much new headlamps and exhaust pipes cost.

As an admirer of the Rudge, Vincent was well aware of the virtues of its centre stand operated by a long lever on the left-hand side. (The rider could raise or lower the stand without leaving the saddle—impossible to the average rider of a 'D' due to the extra weight of the machine.)

The Series 'D' stand was made from $\frac{7}{8}$ in tube and mild steel strip bronze welded and was liable

The only 500 cc enclosed Series D Model, the Victor, made in 1954 but never to go into production

to fail, especially if the rider habitually started the motorcycle when it was on the stand.

A design fault resulted in the rear brake cable being trapped between the lifting handle and rear cowl when the motorcycle was on the stand.

Regrettably, the excellent Series C prop-stands were discarded so the rider had no provision for parking the bike on a cambered road. Nevertheless, it was much easier to park the bike on the centre stand instead of the rear stand system used on its predecessor.

The front mudguard stay doubled as the front stand. It was bent to curve round the cam boss when in position keeping the wheel off the ground but it was essential to have some clearance between the stay and the boss before removing the wheel, otherwise it was almost impossible to replace the wheel because the stand got in the way.

The front and rear wheel tommy bars were replaced by plain hexagon axles, the reason being an aesthetic one—Philip Vincent thought the tommy bars spoilt the appearance of the machine and there was always the danger of the rear tommy bar fouling the rear cowling.

Once again a lipped nut was used and some Series D rear frame members were to take their places in the service bay alongside C rear frames with ruined lugs.

The unique rear chain adjusters were replaced by bolts and locknuts which resulted in a more precise adjustment of the chain. To save weight and expense, the left-hand rear brake was omitted and the remaining brake was operated by cable. The stop light was activated by application of the front brake (instead of the rear brake as on the C).

The wheels and tyres on the C Black Shadow, Rapide and Comet (Avon 3·00 × 20 in front with 3·50 × 19 in rear) gave precision to the steering

One day in 1954, Vic Willoughby put in 500 miles before breakfast on the D prototype. Pump attendants were unimpressed

Mobilgas
SPECIAL
NATIONAL
BENZOLE MIXTURE
LRO 992

but again for aesthetic reasons the D models were fitted with different wheels and tyres—Avon $3{\cdot}50 \times 19$ in front with $4{\cdot}00 \times 18$ in rear. The unfortunate result was a worsening of the handling of the Vincent vee-twin, especially at low speed when the bike tended to roll.

The enclosed versions of the Black Shadow, Rapide and Comet—the Black Prince, Black Knight and Victor—were clad in fibreglass covers. Philip Vincent was the first manufacturer to use this material to enclose a high speed motorcycle and one advantage was that weight was kept down—the enclosed vee-twin was only 13 lb heavier than the 'naked' model.

Philip had marketed an enclosed motorcycle before, of course, in 1933 when the model 'W' and model 'J.W.' failed to excite many buyers and were quietly withdrawn.

Black Shadows and Rapides had been successfully lugging heavily laden sidecars around the country but that market was being captured by the various mass-produced family saloons.

The weather protection offered by the Series D range might appeal to the dyed-in-the-wool

Sir Nigel, **my standard Black Knight (with sidecar footwear), posing at a Vincent HRD Owners' rally at Stanford Hall**

sidecarist who could transport the wife at speed without having to listen to her.

However, the general public preferred the comfort and sociability of a small saloon car in which it was possible to communicate without resorting to sign language.

There was no evidence that the world was eagerly awaiting the appearance of an enclosed high-speed motorcycle but Philip proceeded with his concept of protecting the rider from the weather—not, note, offering him streamlining.

The era was a romantic one. The British had accepted several necessary restrictions on freedom during the war as well as rationing, many items from food to motorcycles being difficult to obtain.

During the fifties, under Churchill's Government, rationing of food and various bureaucratic controls were ended and people experienced greater freedom again. For many, this freedom was personified by a fast motorcycle on which bare-headed couples could ride off to the country or seaside. Romantic music matched this mood and ballads sung by Nat 'King' Cole, Frank Sinatra and Bing Crosby and others dominated the 'Top Twenty' played each weekend on Radio Luxembourg.

Philip Vincent himself was a romantic and in the spirit of the time gave his vee-twins suitable names with appropriate transfers to match. The Black Prince sported a large transfer fore and aft displaying a large Vincent scroll against a silver mediaeval helmet and visor above the name of the model. The Black Knight's battle motif was an axe while the only enclosed Comet produced—the Victor—carried a logo of a steel gauntlet. Smaller versions of these transfers were affixed on the instrument panel of each respective model between the speedometer and steering damper knob.

There were no transfers on the petrol tanks of the enclosed models but the 'open' Black Shadow—unlike the Rapide—had its name 'writ large' across the tank. Bereft of his magnificent 5 in speedometer, which had been replaced with a conventional 3 in unit, the Black Shadow owner needed something to boost his ego!

The practical problems of making fibreglass panels suitable for fast motorcycles were immense. Always the perfectionist, Philip insisted that the panels be stove-enamelled and able to withstand a considerable weight dropped from on high. Needless to say, several panels failed the test and the dustbins became crowded.

The rear cowl was fixed to a metal frame that also carried the new oil tank, whose capacity was one pint less than its predecessor mounted above the cylinder heads on the C range.

The D upper frame member consisted of a tube with a casting at each end and was both lighter and weaker than the C oil tank that it replaced. (The rear casting sometimes failed when the bike was used for sidecar work.)

Philip Vincent was not a sidecar enthusiast and in fact he omitted to incorporate a sidecar mounting hole at the rear of the upper frame member. (The C upper frame member incorporated a mounting hole beneath the rear spring bolts.)

Exploded view of the fairings. (The one item I need for *Sir Nigel*, **a bracket, is not shown!)**

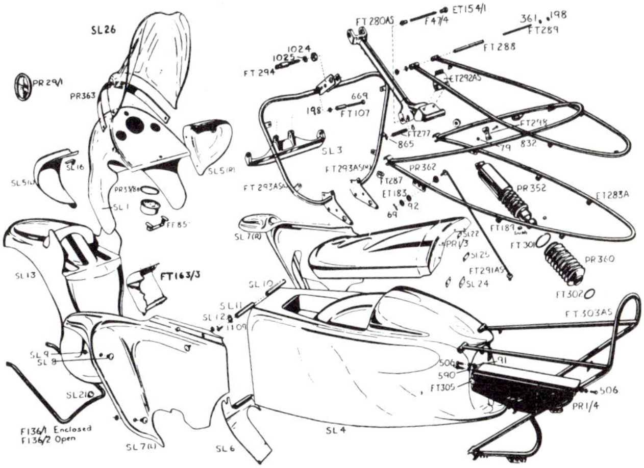

However, the pivot bolt and sleeve of the rear cowl could be removed and replaced by an eye bolt to provide a pick-up point.

The rear wheel was readily accessible; the saddle, which pivotted at the petrol tank end, was lifted and then rested upside down on top of the petrol tank and the complete rear cowl unit—including rear light and number plate—could be raised once the flanged nuts screwed on to the ends of the rear fork pivot had been loosened with a plug spanner.

A prop clipped inside the cowl was then used to hold it up while work on the rear end proceeded.

Carrying luggage on the enclosed models was a problem. Suitable fibreglass panniers had been designed but were never produced. However I had a frame made up to bolt into the tapped lugs inside the cowl of *Sir Nigel*, my Black Knight. For most trips, a bag strapped to the tank was adequate. The capacity of the tank had been increased to 4 gallons as a result of moving the oil tank.

The side panels were easily removed—each one was retained by three alloy thumb screws. At last, both spark plugs were on the same (right-hand) side, so only one panel had to be removed

Engine room of the Victor. (The enclosed models should not have a transfer on the tank cover)

Left side of the Victor with the panel in position—easily detachable by unscrewing three alloy thumb screws

to change the plugs each time they oiled up in London.

Each side panel incorporated a 'beak' at its upper leading corner to assist in protecting the rider from draughts and road dirt.

To my mind, the best-looking section of the enclosures was the front mudguard. This large fairing reduced considerably the amount of rain thrown at the rider. Above it, another fibre glass section housed the adjustable headlamp and instrument dashboard on to which was bolted the superbly efficient windscreen. An adjustable deflector was bolted to the top of the windscreen to take any rain over the rider's head.

Fibre glass muffs covered the brake and clutch levers, the final shape of all the fibre glass panels being determined after exhaustive riding tests by Ted Davis. Ted rode around the country to gauge the handling in various wind conditions in all weathers until the development on the prototype enclosed vee-twin—a Black Knight—was completed by October 1954.

The firm producing the original fibre glass mouldings was based in Northumberland. One day, Ted—accompanied by his wife, Ann, on the pillion—was approaching Newcastle at a fast rate of knots during a gale. A strong gust of wind blew a motorcycle in front of Ted across the road but the Black Knight held its course and Ted then knew that the handling in cross winds was satisfactory.

Ann appreciated the comfort of the Black Knight's rear sprung system, which was a great improvement over the Series C set-up.

The two spring boxes and damper on the C had been replaced by a single combined 300 lb spring and damper ('monoshock') which gave no less than 7 in travel. The dualseat was supported by the frame which was bolted to the upper frame member and attached to the rear fork pivot; it was therefore unaffected by the rear wheel bobbing up and down.

Tucked away under the left-hand side panel was a 6 volt Lucas coil wired to a distributor mounted at the front of the engine. This coil ignition system, which provided a good spark at low engine revolutions, enabled the rider to start a cold machine with a push of the kickstarter instead of the hefty boot previously required to persuade a vee-twin fitted with a magneto to show a spark or two. (Nevertheless the Vincent HRD Owners' Club has always had competent riders who could start their well-maintained 'thousands' with ease, magneto or not.)

As a result of this thoughful action by PCV, the Beast was easy to start and many Vincent owners continued to lead useful lives instead of sitting around at home nursing their hernias or broken ankles.

Other modifications included the use of Amal Monobloc carburettors which had the advantage of not dripping when the bike was leaning at an angle on a prop stand. In fact, they were fitted to the prototype Black Knight before the prop stands were discarded and replaced by the centre stand which only worked on a level

surface anyway, thus negating the advantage of the non-leaking carburettors.

However, a great improvement was effected in the timing chest, where a one-piece all-steel idler shaft replaced the two-part assembly used previously (the idler gear spindle had sometimes worked loose).

Crankcase breathing was improved by the fitting of a long pipe which ran from the inlet rocker box of the front cylinder.

The stainless steel pushrod tubes were replaced by inferior light-alloy ones but PCV justified their use on the grounds of lower cost.

The shock absorber on the 'C' engine shaft incorporated 18 pairs of small springs and some of these tended to break under stress. The recesses for them in the end plate did not allow the springs to settle down freely and knowledgeable enthusiasts like Eddie Stevens therefore fitted the end plate the wrong way round so that the springs abutted against a flat surface.

PCV increased the number of inner and outer springs to 22 pairs and fitted a strong spring plate and a tab washer to lock the $\frac{3}{4}$ in nut that held the lot together.

Although these various alterations were incorporated into the models in time for the Motor Cycle Show at Earls Court in November 1954, the bikes were not really ready for public display—the finish of the fibre glass panels was dreadful. The mouldings were uneven and patchy but there was nothing else available for the Show models. Nevertheless, pictures of them were issued to the press for publicity.

Many advertisements of the era featured girls with tight blouses and exposed thighs astride a manufacturer's bike of the future. Enthusiasts were not always impressed and some complained that they could not see the carburettor because a thigh was in the way.

Brian Terry (in blazer), a former owner of the Victor, outside the offices of *Motor Cycling*

The girl was usually accompanied by a smart, handsome Brylcreemed male grinning inanely into the distance. For some reason, route maps were particularly amusing and the happy couple were sometimes seen astride the machine pointing at the map and laughing insanely.

Philip Vincent was lucky in that he was married to a very beautiful girl who only had to stand near his motorcycle to attract attention to it. Philip Vincent was a keen photographer and his many motorcycles were therefore well-photographed for posterity.

During the Show, Vic Willoughby road-tested a Black Knight and published his opinions in *The Motor Cycle* on November 25, 1954. A myth had already originated from the armchair critics that an enclosed Vincent would be unstable in cross-winds but Vic reported that handling was unaffected by the enclosures.

The finish of these enclosures might be all the rage among other customers of the fibre glass manufacturer—boat owners, for example—but Philip terminated his dealings with the firm and entered into a contract with another, Microplas, who already supplied sports car bodies and in time were to produce excellent panels for the Vincent motorcycle.

Unfortunately, there was to be a four-month wait before the new panels could be made and delivered. It was difficult to explain to customers eager for their new mounts that there was nothing wrong with the shape of the mouldings—only the finish was faulty and the firm refused to release any motorcycle until it was fitted with panels of acceptable quality.

To keep the factory working Philip reluctantly decided to produce 'open' versions of the Black Prince, Black Knight and Victor which were, of course, the Series D Black Shadow, Rapide and Comet.

These motorcycles appeared in March 1955 and were followed a month later by the enclosed models.

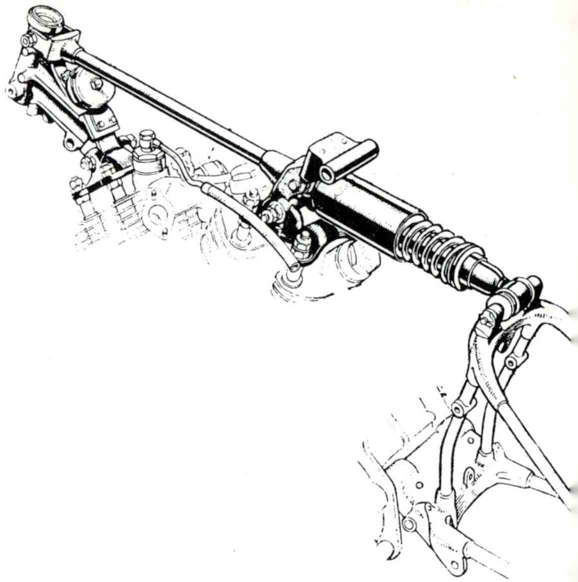

The Series D upper frame member which replaced the oil tank of the C. The headlug and spring-damper unit are also new

Regrettably, producing the D without its enclosures resulted in some ugly features—the high position of the steering damper knob, for example. However, the Series D Black Shadow was Philip's favourite model out of the many he produced over nearly three decades; when I asked him why he replied that it was the only Black Shadow he had designed entirely himself.

Sadly, he could never ride one—a crash in 1947 had left him unable to balance a solo motorcycle and I often wonder what changes he might have made had he ridden a Black Shadow.

Philip loved his motorcycles but once again the public thought they were ugly. Covering that superb vee-twin engine with black plastic was regarded as particularly heinous and a pal of mine likened the Black Knight to 'a bat out of hell'.

The 'beaks' were no doubt efficient but they attracted particular criticism at a time when streamlining was the vogue.

Only 460 motorcycles were produced during

1955, including some 200 enclosed machines. (Many were not registered for the road until 1956.)

In addition to the one (500 cc) Victor, a D Comet was produced and enclosures were fitted for the 1954 Show.

Officially, the last Vincent motorcycle—a Black Prince—left the production line on December 16, 1955 but a few motorcycles were made up later from spare parts.

Manufacture of these high-quality motorcycles was no longer profitable; in fact, the firm was losing money on each machine sold, which is difficult to understand. (If demand for quality goods is low, reduce output and increase the price.)

The market is like a pyramid. A motorcycle manufacturer can cater for the quality end of the market at the tip of the pyramid and make superb machines that only a small number of people can buy; he can cater for the average buyer who wants something functional that will be of good quality but costing less than the best; finally, he can go for the mass market where people buy the cheapest product and throw it away after a short time or when it dies on them.

Philip Vincent was artistic and just as a writer wants everyone to read his books, Philip wanted every motorcyclist to ride his machines. Hence his concept of producing a machine with built-in

I took this picture at the Shadow Lake Rally of the VOC in 1977, the first time since 1955 that so many enclosed models were together

Paul Richardson (with famous pipe) and Rollie Free stand behind *Sir Nigel* at Shadow Lake, Ontario, in 1977

weather protection. He hoped it would appeal to those who had shunned motorcycles because the rider was so exposed to the inclement English climate.

Unfortunately, Philip fell between two stools—the people who wanted enclosed motorcycles bought scooters. They were fashionable and socially okay. Motorcycles were not all right in England in the fifties and even a director of BSA was suspect for committing the social crime of riding a motorcycle, albeit a BSA, to a board meeting of the company.

I suggested to PCV that he might have done better to have kept production down in the early fifties instead of buying expensive high-production machines for the tool-room that could never be used to capacity because the firm was really catering for the small top end of the market. (Not everyone who could afford a Vincent wanted one, of course.)

Philip regarded his vee-twin as the Continental Bentley of motorcycles but just as a 'mass-produced Bentley' was a contradiction in terms, so it was impossible to mass-produce a Vincent and retain the quality that made the motorcycle famous in the first place.

However, Philip argued that increased production meant decreased costs per unit, which was true, but the Series D range was beginning to suffer a diminution of quality because of this desire for mass-production. For example, the alloy push-rod tubes quickly became scored and distorted when the bodgers did an 'overhaul' and attacked them with pliers. As a result, they leaked oil and owners replaced them with the proven tougher stainless steel tubes.

Crankcases were die-cast from 1954 and some appear to have been bored with too large an aperture for the lower end of the push-rod tubes, for the gap between tube and crankcase orifice was too large for the seal to prevent oil escaping. Quite a few Vincent enthusiasts felt that it was better for the firm to cease production if quality was going to be sacrificed.

However, the motorcycle was so well made and reliable that it was often kept for years as part of the family with its own pet name. This was fine for the owner but not for the manufacturer; as Philip once explained to me, 'It's a very thankless task to make a high-quality product. A high-quality motorcycle doesn't wear out and so the customers do not need to replace it. I knew many people who wouldn't buy a new one because they were satisfied with the bike they had.'

I've been satisfied with my Black Knight for more than a decade. The myth about the enclosed models being difficult to handle in cross-winds still persisted when I bought the bike in 1972. I cautiously rode along the East coast of England, which often suffers from gales straight

Works picture of the Series D Rapide which went into production in the spring of 1955. Note the Lucas coil, which enables the rider to start the bike while seated on the saddle

off the sea, but found the handling all right at speed. As already recorded, it rolled somewhat at low speed but the handling was satisfactory after the fitting of today's Avon Roadrunner tyres.

Proof of *Sir Nigel*'s ability to cope with cross-winds was demonstrated in 1977, when I rode the Vinnie from Niagara Falls to Toronto in rain with a strong wind blowing off Lake Eyrie. The wind blew at an angle, coming over my right shoulder, and hit the windscreen and deflected the front end. I screwed down the steering damper knob and continually corrected the steering yet never felt in any danger.

The bike has always behaved impeccably at speeds around 100 mph and this is partly due to the forward position of the centre of gravity compared with many of today's motorcycles, several of which can 'pop a wheelie' despite their relatively small power output because not much weight is toward the front end. The faster one goes, the greater the tendency for the front-end to lift and therefore the greater the need to design a fast bike with the centre of gravity well forward.

The enclosures unfortunately reduce the top speed by about 10 per cent but they certainly enable me to maintain higher cruising speeds for longer periods than when I ride a Series C, especially during wet and windy weather. The improved rear springing allows me to travel over bumpy roads more quickly and more safely.

Sir Nigel has been brought up to Black Prince specification with ribbed brake drums, 150 mph speedometer, Black Prince transfers, 7·3:1 compression ratio and $1\frac{1}{8}$ in carburettors, the only visual difference being the engine number (Black Prince and Shadow motorcycles have two capital Bs in their engine symbols).

Various items need replacing from time to time but the bike itself never wears out. PCV reckoned the frame members would last for a million miles so they will not require replacement during my lifetime.

Top **The Series C Rapide gave a harder ride to passengers than the D Rapide (*left*) but the girls (bless 'em) stayed with it, at least until they had married the rider. I nearly proposed to one of my girl friends when she suffered the C suspension for weeks without complaint, then she spoilt everything by asking, 'Would you sell your Vincent to get married?'**

Above **The Series C Touring Rapide gave greater comfort than the standard Rapide (larger tyres, higher handlebars and better weather protection due to the bigger mud-guards) but the handling was inferior**

Overleaf **The Series D Comet. The gaiters, rear chain adjusters and extension to the chainguard are non-standard**

UNK 575
VINCENT

VINCENT

Top **Right side of the Series D Comet (owned from new by the late Tom Reeve). The silencer is non-standard**

Right **Officially, only one D Comet was made but other D models were assembled from spares during 1956**

Engine parts can be renewed and stainless steel items are now available in place of many nuts and bolts, formerly supplied plated with chrome or cadmium.

Modern tyres, koni suspension units and the latest ignition and lighting systems keep the bike up-to-date. (I don't believe in keeping the bike as an original piece fit for a museum but no modification is irrevocable—the machine can be reverted back to its original specification at any time.)

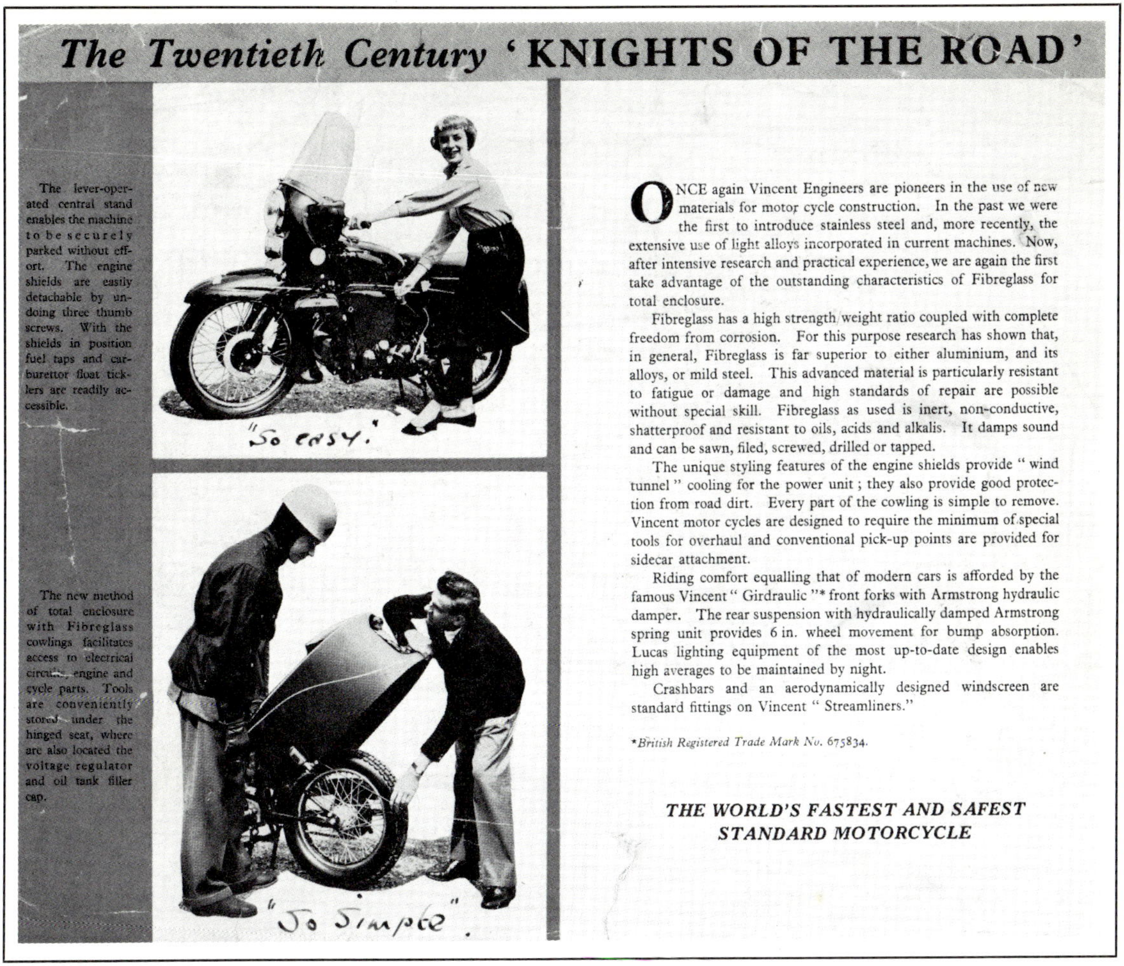

The Twentieth Century 'KNIGHTS OF THE ROAD'

The lever-operated central stand enables the machine to be securely parked without effort. The engine shields are easily detachable by undoing three thumb screws. With the shields in position fuel taps and carburettor float ticklers are readily accessible.

"So easy."

The new method of total enclosure with Fibreglass cowlings facilitates access to electrical circuits, engine and cycle parts. Tools are conveniently stored under the hinged seat, where are also located the voltage regulator and oil tank filler cap.

"So Simple".

ONCE again Vincent Engineers are pioneers in the use of new materials for motor cycle construction. In the past we were the first to introduce stainless steel and, more recently, the extensive use of light alloys incorporated in current machines. Now, after intensive research and practical experience, we are again the first take advantage of the outstanding characteristics of Fibreglass for total enclosure.

Fibreglass has a high strength/weight ratio coupled with complete freedom from corrosion. For this purpose research has shown that, in general, Fibreglass is far superior to either aluminium, and its alloys, or mild steel. This advanced material is particularly resistant to fatigue or damage and high standards of repair are possible without special skill. Fibreglass as used is inert, non-conductive, shatterproof and resistant to oils, acids and alkalis. It damps sound and can be sawn, filed, screwed, drilled or tapped.

The unique styling features of the engine shields provide " wind tunnel " cooling for the power unit ; they also provide good protection from road dirt. Every part of the cowling is simple to remove. Vincent motor cycles are designed to require the minimum of special tools for overhaul and conventional pick-up points are provided for sidecar attachment.

Riding comfort equalling that of modern cars is afforded by the famous Vincent " Girdraulic "* front forks with Armstrong hydraulic damper. The rear suspension with hydraulically damped Armstrong spring unit provides 6 in. wheel movement for bump absorption. Lucas lighting equipment of the most up-to-date design enables high averages to be maintained by night.

Crashbars and an aerodynamically designed windscreen are standard fittings on Vincent " Streamliners."

**British Registered Trade Mark No. 675834.*

THE WORLD'S FASTEST AND SAFEST STANDARD MOTORCYCLE

The proof copy of the Series D catalogue. (Captions have been written in ink.) The catalogue was never printed due to the firm's financial problems

Enclosed models in good condition fetch more than the cost of a brand new Hesketh 1000. When I asked Philip what he thought of the high prices for his old motorcycle he was pleased to point out that a Series D was a far better machine than anything else available on the market.

Less than 11,000 Vincent motorcycles were produced between 1927 and 1955 and probably 50 per cent of them survive, hence their rarity.

After production of the Vincent vee-twins ceased, the company continued in business designing and manufacturing an efficient little two-stroke engine.

10 Industrial two-stroke engine

The two-stroke industrial engine which could be used for a variety of applications

Philip recalled that the idea of making a two-stroke engine originated in February 1954 when Qualcast enquired about the possibility of Vincent producing a small power unit for their lawn mowers.

Just four weeks later Vincent arrived at the Derby works of Qualcast and proudly demonstrated his prototype two-stroke which promptly broke one of the transmission mounting brackets of the test lawnmower. As Philip put it, 'The mower was unaccustomed to such torque.'

Qualcast never did place an order (concerned, perhaps, at the prospect of engines popping out of their mountings all over the country) but Vincent now had an efficient industrial engine that could be used for a variety of applications—lawn mowers, cultivators, milking machines, pumps, compressors, paint spraying equipment, generating sets and outboard engines, etc. and the hardy road testers had suddenly turned into gardeners, sailors, and milkmaids.

The 73 cc (43 × 50 mm) engine developed 1·3 bhp at 2500 rpm with a compression ratio of 6:1. Aluminium alloy was used for the cylinderhead, piston, crankcase and barrel, which incorporated a replaceable cast iron liner. The big-end ran on a single track of 15 $\frac{3}{16} \times \frac{3}{16}$ in roller bearings on a crankpin pressed into the flywheel. Ignition was provided by a Vincent rotating magnet magneto (timed at 25 degrees before TDC) with a KLG F70 sparking plug. The fuel tank held two pints of 'pool' (70–75 octane) petrol mixed with Castrol XL oil in the ratio of 24:1.

A 99 cc version had also been developed with the cylinder bored out to 50 mm; with a compression ratio of 7:1 and the ignition advanced to 30 degrees before TDC, this engine produced 2·1 bhp at 2500 rpm. By 1957, its output had become 4 bhp at 4500 rpm with a compression ratio of 10:1.

In 1956, a firm became interested in fitting the 100 cc engine into a water scooter (named the Amanda after a director's daughter) and for several weeks the scooter was tested at Ruislip Lido and elsewhere. An American company then became interested and their representative was so impressed after a demonstration with the Amanda fitted with a developed version of the engine (4 bhp at 4500 rpm with a compression ratio of 10:1) that he placed an order for about 6000 water scooters.

Vincents subsequently assembled the craft at a new production site at Llanwrog, near Caernarfon, North Wales.

The engines were produced at Stevenage and fitted to the hulls manufactured at Llanwrog by another company. The propeller shaft (with its unique centrifugal clutch) was bonded into the fibre glass hull and the complete Amanda was tethered in a test tank constructed from bricks and cement and filled with water. The engine was then started and run continuously for a while before being packed and shipped to Chicago.

The Amanda was coloured red from around and below the waterline and white above, making a cheerful picture on the sea as it was taken out for tests.

One day, an unexpected tragedy occurred when Johnny Penn was drowned while testing a scooter along the coast.

Production reached over 100 craft per week and when nearly 1000 scooters had been made

A three-wheel delivery van. ('Not Vincent', said Phil Irving.) Presumably the seating arrangements have been removed otherwise the driver would need pretty strong trousers

by May 1958, news from Chicago shook Philip Vincent and his colleagues as they worked away at Stevenage and Llanwrog.

The American company had given free rein to its PR men and they had built up the image of the Amanda and laid on a well-publicised public demonstration to show how efficiently the little water scooter could skim across the water. Scantily clad nymphs were ready to do their stuff at the drop of a cheque; reporters, photographers and camera crews clustered round with their equipment as the demonstration proceeded.

It was a hot sunny day as the Amanda water scooter purred along, its rider beaming happily at the cameras. Under the effect of heat from the sun above and the engine below, the fibre glass

Below **The two-stroke engine ran on a mixture of oil and petrol in the ratio of 1 to 24. The tank held two pints**

Above **Sectioned two-stroke engine. Aluminium was extensively used (head, cylinder muff and cases)**

hull began to soften and the rider slowly sank into the engine compartment.

The Americans promptly cancelled the order, leaving the Vincent HRD Company in a truly desperate financial position. Vincent had set down in detail the specification of the fibre glass and he accused the manufacturer of cheeseparing when making the hull to save money. To the end of his life he maintained that had the hulls been made according to the original specification, the Amanda would have been a huge commercial success and the company would have continued to prosper.

Now, after three decades of brilliant achievement, the company would soon be bankrupt. It is never good policy for a firm to rely for its survival on one project but perhaps there was no option after the failure of other schemes.

Once again Baillie was called in as receiver, only this time he sold the company (to Harper). Selling scores of well-made motorcycles in 1949 was one thing; shifting hundreds of unseaworthy scooters in 1958 was something else.

A British Member of Parliament had taken one down the Thames but the hull melted on his craft, too. All very embarrassing.

Harper used a pair of industrial engines to power a Harper-Vincent 'Go-kart' but the project was a flop and Harper eventually sold the company to Cope-Allman.

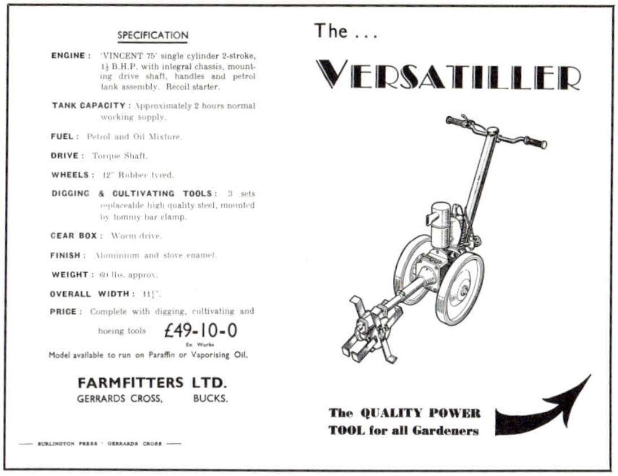

Above **The Versatiller . . . for hoeing and digging. The high speed road testers suddenly discovered they were gardeners**

Below **Vincent inboard marine engine for bonding into plastic hulls. Not a commercial success**

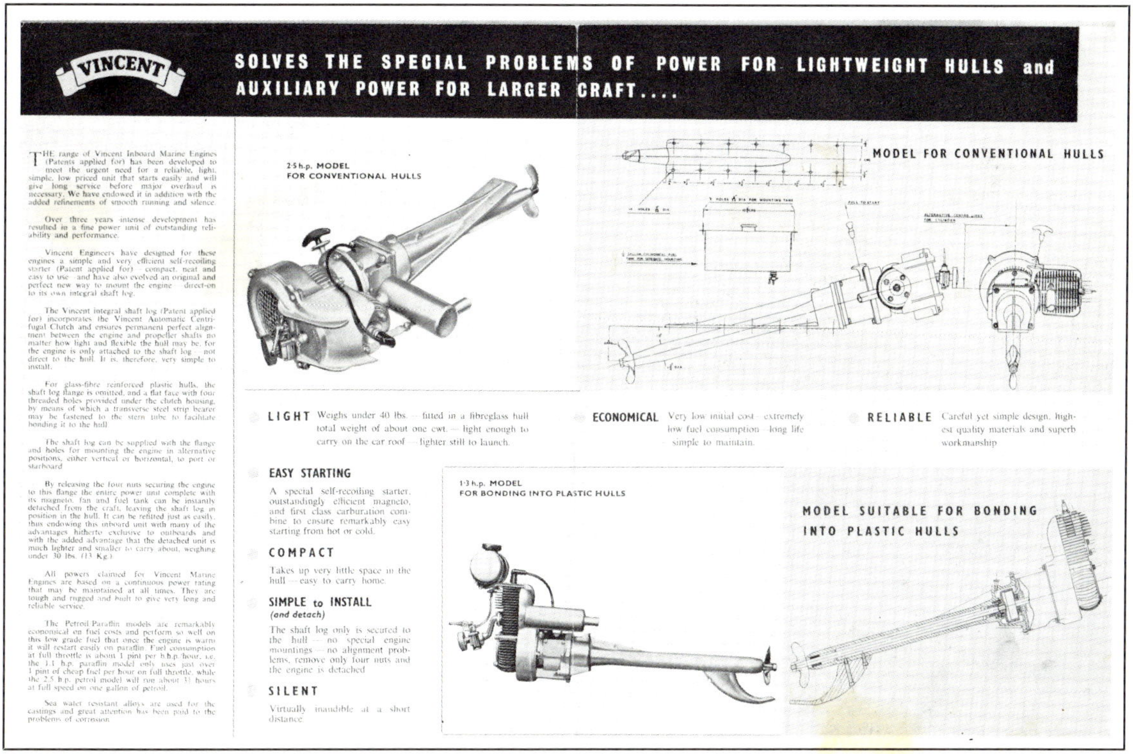

THE
First
ALL BRITISH
Kart Engine
For Performance
and Quality
THE
HARPER/VINCENT
MARK
1 and 2
A 4 H.P. WINNER
FOR CLASSES 1 AND 2
HARPER ENGINES LIMITED
H
G
A Member of the Harper Group

Above **A couple of two-stroke (99 cc) engines were fitted to a Harper/Vincent go-kart in the sixties after PCV sold his firm to the Harper Group. Not a success**

Right **A scantily clad nymph skimming along the Thames in the Amanda water scooter. A British Member of Parliament tried it but sank outside the House of Commons**

PCV did not fit into the new set-up but before he left he designed a scooter using a larger version of the industrial engine. This prototype was acquired by one of the employees, who, finding it somewhat cumbersome to ride, disposed of it.

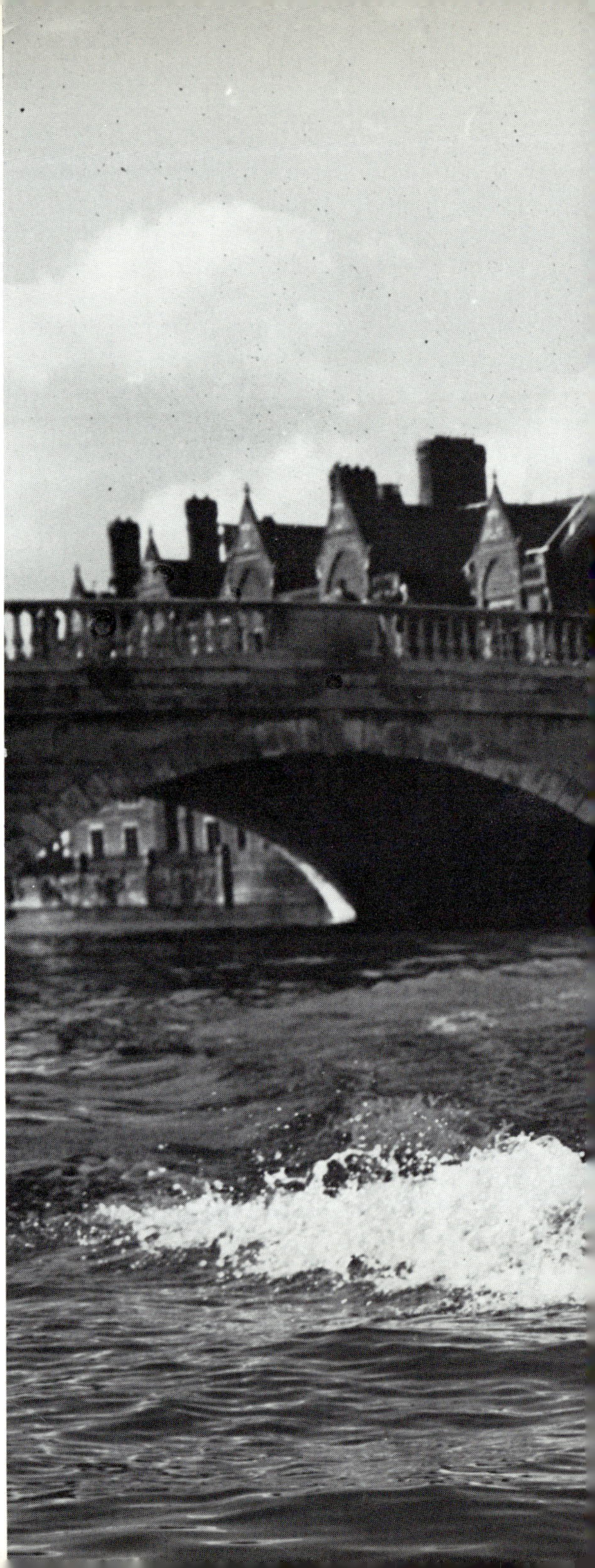

SUPER
TWIN

11 The record breakers

As shown, the Vincent HRD Company was always short of capital while the vee-twin motorcycle was in production. It was therefore impossible for the firm to give financial support to private owners who competed in races or attacked records but works advice was always readily given. Employees who knew their way around a race track included George Brown, John Surtees, Ted Davis and C.J. Williams and they themselves benefited from Phil Irving's technical knowledge. (Phil had published his classic work *Tuning for Speed* in 1948 and three decades later it is still in print.)

Anyone wishing to learn how to tune the Beast therefore had plenty of information readily available and the late forties saw much feverish activity as enthusiasts around the globe drilled, hacked and filed away.

Not everyone's engineering ability matched his intention—clutch drums burst, cylinder heads were ruined as runaway drills widening the ports went through into the rocker tunnels, frame members were drilled with holes 'for lightness' until, looking like black gorgonzola cheese, they failed.

However, in the early postwar days a standard, untuned Rapide was so far ahead of anything else available that provided the rider remained in the saddle he usually won a race against other makes.

Philip Vincent was always delighted to hear of any racing successes and he kept a scrapbook recording results as they came in to his office

Ted Davis's fast 1000 cc outfit which he raced successfully in the early fifties

from every continent.

I asked him which particular result gave him the most satisfaction and he recalled that 'In a 12-hour road race in Argentina in 1948 on an oval, dusty track, Rapides, which were normal touring models, came first, second, third, fourth and fifth and the winner averaged 76.86 mph. 922 miles in 12 hours on the road.' (The winner had bought his mount, second-hand, the previous day before taking part in the race at Rafaela.)

The Argentines loved fast motorcycles and a

George Brown racing the early version of *Gunga Din*. Note the Series B petrol tank, rear frame member and timing chest cover; also the Brampton forks. Series C components were later fitted in place of the B parts

large proportion of Vincent's early postwar production was bought by them. Comets, Rapides and one or two Grey Flashes and Black Shadows were to be seen all over the country, their owners riding them as competitively on the road as on the race track.

The police clamoured for 'a slice of the action' and Peron duly purchased several Rapides for his force and was invariably escorted by Vincent-mounted patrolmen.

The road-going champion, Jose Cruz, won his title on the first production Vincent (shipped to Buenos Aires in 1947 as previously mentioned); the success of the Rapide resulted in several orders being placed with Vincents but exports to Argentina diminished when import restrictions were imposed because of that country's shortage of foreign exchange.

Paul Richardson told me that the British Government had an arrangement whereby the sale of the beef imported from Argentina balanced the value of manufactured goods exported from Britain. One man's meat was another man's Vincent.

There were no such problems exporting to Australia, New Zealand, South Africa and Rhodesia, for example, and standard Rapides swept aside the opposition in solo and sidecar races. The Vincent vee-twin was not only fast—it was reliable. This was proved in the Isle of Man Clubman's TT race in 1948 when 10 out of 11 Rapides finished the race, one crashing from rider's error. George Brown led the race until he ran out of petrol but not before setting a new lap record for the event.

It was the first road race of the winner, J. Daniels, whose time would have given him second place in the Senior. He was followed home by another Vincent rider, Phil Heath, while Ted Davis finished fifth ahead of George, who fought off exhaustion to push his mount home to clinch sixth position.

The speedway engine (used by George Brown in his 500 cc Vincent special on which he won several short-circuit races after the Second World War)

The following year, the organisers imposed fuel restrictions on the entrants and George forcefully pointed out that he went to the Island to race and not to take part in a farcical test of fuel consumption.

Riding *Gunga Din*, so named by the late Charles Markham, George won several races, sprints and hillclimbs during the postwar decade. Particularly outstanding was his record time of 37·13 seconds at Shelsley Walsh hillclimb in September 1949 when he beat the best times recorded by the car-racing experts such as Raymond Mays and Joe Fry.

Gunga had been used at Jabbeke, Belgium, by Rene Milhoux in 1948 to set new world records of 83·5 mph for the standing start kilometre and 94 mph for the standing start mile. Rene also broke the Belgian National Solo record with a speed of 143 mph. A year later Milhoux broke the world sidecar record for the flying 5 kilometres with a speed of 126 mph.

George also competed successfully in 500 cc races riding his 'Speedway Special'. This was a mixture of Series A and B parts, the basis being a light engine developed after the war for use on the speedway tracks.

This sport had an enormous following in the forties and fifties, especially at Wembley. (The Wembley Lions had a supporters' club with more than 50,000 members.)

There was obviously a large potential market for a new speedway engine but J. A. Prestwich had the monopoly and Vincent never broke it, despite George's successes with the speedway engine.

PCV gave up trying to get a foothold in the lucrative speedway market and told George to leave his special at home and concentrate instead on racing a tuned Comet to help the sales of this new model.

While George was sorting out the opposition

The Dearden supercharged Black Lightning at Shadow Lake, Ontario (1977). *Right* I've never seen the fairing for the frame at the rear. Performance was 100 mph in first but maximum speeds in other gears were never established

in the UK, Frank Platt began a long series of wins on the first Rapide to reach Australia. Frank imported the Vincent towards the end of 1946 and during the first half of 1947, he won all nine races entered, including the Victorian Grand Prix, to the delight of another Victorian, Phil Irving.

Other Australians quickly ordered new Rapides and many of them found the big vee-twin engines ideal for pulling a chair round the racing circuits.

The Brampton front forks sometimes wilted under the stresses on the corners but the 1948 Girdraulics solved the problem and have since remained superior for this application to any other front fork made.

Back in the United Kingdom, Jack Surtees, father of John, rode his Vincent combination successfully on the circuits, then Ted Davis took over the role of the leading Vincent sidecarist,

Far right **Les Wharton set a new Australian sidecar record of 122.6 mph and a new solo record of 139.8 mph with his Black Lightning in 1953**

winning 20 races in his first season (1952) aboard his Black Lightning/Watsonian outfit. He won another 20 races before feeling it was necessary to lift the heads.

Ted used a caged roller big-end and told me that he had never heard of this racing big-end failing during a race. He therefore preferred it to PCV's big-end with its aluminium crankpin. 'The trouble with the aluminium big-end,' recalled Ted, 'was that it was unpredictable. It might last the race but often it failed almost immediately. The record for it failing on a Grey Flash I rode was 50 feet from the start.'

Inevitably, riders seeking fame as their country's fastest motorcyclist tuned up their Rapides and attacked their national records, a brave few going all out to be the world's fastest solo motorcyclist.

Some weeks after Rollie Free's sensational runs at Utah in 1948, *The Motor Cycle* offered a handsome trophy and £500 (later raised to £1000) to the first British rider of an all-British motorcycle to beat Henne's world record of 173·625 mph (which he set in 1937 aboard a supercharged BMW).

The race was on.

The first in the field was the late Reg Deardon, who purchased a modified Black Lightning from the works. Modifications included lengthening the wheelbase by 6 in to incorporate a Shorrock's blower and the fitting of fairings to provide streamlining.

With an estimated 90 bhp at 6500 rpm to hand, the monster was ridden by George Brown at over 100 mph in first gear and was ready for Reg to attack the world record on a German autobahn. The proposed date was early in 1951 but Deardon never made the world record attempt.

Cover picture of a postwar catalogue showing Rollie Free with his Black Lightning and Marty Dickerson (Rapide)

VINCENT
THE WORLD'S FASTEST STANDARD MOTORCYCLE
THIS IS A FACT, NOT A SLOGAN—THESE TWO PHOTOGRAPHS OF TYPICAL PRIVATE OWNERS PROVE IT
ROLLIE FREE
holder American Class A Speed Record
156·58 M·P·H Certified by A·M·A
MARTY DICKERSON
holder American Class C Speed Record
141·72 M·P·H Certified by A·M·A

Russell Wright aboard his record-breaking Black Lightning on which he raised the world solo record to 185 mph on a wet road! near Christchurch, New Zealand (1955)

This Black Lightning, developed at a cost of £3000, languished for several years; in 1973, the machine was purchased for £1000 by an American, Michael Manning of Philadelphia. In 1977, he took it in a van to a Vincent HRD Owners' Club rally at Shadow Lake, Ontario, and persuaded the engine to fire; then he rode it along a public road—the first time, apparently, that the bike had been running since the fifties. The sound was beautiful. The Black Lightning was so far ahead of its rivals that various national records were broken by colossal margins.

After Rollie Free and Rene Milhoux had led the way in 1948 by setting new American and Belgian national records, other private owners around the world acquired Vincent vee-twins in various mechanical condition and set about tuning them prior to record breaking.

Most record-breaking Vincent challengers were usually unsupercharged and each potential machine had its own individual modifications, as there was nothing much published to provide

After Russ Wright's record runs, Bob Burns set a new sidecar record of 162 mph on the same machine, seen here at Pendine Sands later in the year

guidance for travelling at high speeds in excess of two and a half miles a minute.

However, PCV and Phil Irving had learned much from Rollie's records and were able to provide some sound advice, although neither Rollie nor any other record breaker disclosed all his secrets.

The first rider to attack Henne's world record was Rollie himself. He purchased a Black Lightning in 1950 and fitted a fibre glass shell. On September 7, he made the attempt but crashed at 144 mph. Undaunted, he discarded the shell and four days later attacked various records from 1 kilometre to 10 miles.

Speeds set by Roland Free September 11, 1950

1 km	160·10 mph	153·58 mph	156·77 mph
1 mile	160·05 mph	153.26 mph	156.58 mph
5 km	156·55 mph	153·13 mph	154·82 mph
5 miles	156·15 mph	152·81 mph	154·46 mph
10 km	154·96 mph	152·29 mph	153·61 mph
10 miles	153·92 mph	150·75 mph	152·32 mph

After Joe Simpson—riding another Vincent—stole the mile record with a speed of 160·69 mph in 1953, Rollie, then aged 52, took the record back within a week with a speed of 160·73 mph.

Since then, Vincent owners have attacked various class records at Utah, the most recent being Dave Matson. In 1980, riding a fuel-injected Vincent bored-out to 1360 cc, he raised the Modified Production Special Partial Streamliner record (the MPSAG 2000 class) from 173·832 mph—set by Harley-Davidson in 1971—to 178·247 mph.

On changing to nitro, Dave attained 195·016 mph during a qualifying run but on a later ride the special flywheels broke.

Meanwhile, in the Southern Hemisphere, Vic Proctor in South Africa bought a Black Lightning and soon set a new National Record with a speed of 136·26 mph for the flying mile, a speed he increased in 1951 to 137·48 mph followed by another record run a year later—this time achieving 148·89 mph.

While Vic was doing his stuff in South Africa, Les Lamb in New Zealand had raised the National Record from 120·8 mph to no less than 139·54 mph riding a second-hand 1947 Series B Rapide. Activity in the 'Colonies' was really humming. In Australia, Les Warton broke the National Record for the flying mile with a solo speed of 139·8 mph and with a chair fitted he raised the sidecar record to 122·6 mph. However, Colin Crothers recorded the best Vincent speed in Australia in the fifties, clocking 147·5 mph in 1954, the year that Bob Burns, across the water in New Zealand, began record attempts with Russell Wright that would go down in the record books as some of the finest achievements ever recorded by private owners.

Early in 1954 Bob, a Scot domiciled in New Zealand, decided to attack Les Warton's speed of 122 mph using his tuned second-hand Rapide pulling a third wheel. Bob was such a skilled tuner that he easily passed Les's speed on an unofficial early morning run along a main highway!

George Brown's Vincent Special *Nero* in 1961

Left **Fitted with a polished aluminium fairing to reduce wind resistance,** *Nero* **just starting**

Top **With a third wheel attached, George broke several world sidecar records and national records in the sixties**

A streamlined shell around the Vincent would help to take him within reach of Bohm's world sidecar record of 154 mph. The shape of the shell was copied from Bohm's streamlined NSU that had recently raised the world solo record to 180 mph in addition to taking the world sidecar record.

Bob thought that the streamlined outfit would have broken Bohm's record all right but before his planned attempt in September 1954, he was warned by Avon that their tyres might not be safe at 'ultra-high speeds'. Nevertheless, Bob couldn't resist tweaking the throttle and he set a new British Empire record of 145·8 mph, which wasn't a bad effort for a second-hand Vincent.

The same day he met another motorcyclist who loved Vincent vee-twins—Russell Wright—who had bought a Black Lightning that had been displayed on the Vincent company's stand at the 1953 Earls Court Motor Cycle Show.

George and *Super Nero* close to 200 mph at Elvington (1970) during a record-breaking week-end

Russ offered Bob the use of the Black Lightning for his next record attempts near Christchurch on December 17 and 18 and PCV rushed out a pair of the big port ($1\frac{1}{4}$ in) heads developed for the 1950 TT races.

After sorting out some carburation problems, Bob just beat the record with an average of 155·2 mph for the two runs along the measured mile and the Vincent name was entered in the absolute world speed record book for the first time.

Seven months later, Russ and Bob went for the double—the coveted ultimate speed records for solo and sidecar motorcycles.

On July 2, 1955, on a wet highway near Christchurch, Russ averaged 185 mph for the flying mile after riding along the course in each direction, thus improving on the record set four years earlier by Hertz riding a streamlined NSU entered by the factory.

Then Bob fixed a third wheel to the Vincent and raised the sidecar record to 162 mph. Britain had gone nap! British machines and riders now held the world land speed records for vehicles powered by two, three and four wheels in addition to holding the world air speed record and the world water speed record.

Congratulations poured in from all over the world and the record holders went to the United Kingdom with the Vincent later in the year to receive the £1000 cheque and to attend the Motor Cycle Show at Earls Court. (During the trip, Bob consulted experts at Amal, Avon and Vincent about further improvements to the bike for another attempt.)

Sadly, Vincent no longer exhibited at the show due to their financial difficulties but Avon were delighted to display the record-breaking outfit on their stand.

Right **George and** *Super Nero* **resting awhile at Kirkstown Airfield during the Belfast and District Club's short circuit meeting**

Below **Starting** *Super Nero*'**s powerful engine by spinning the back wheel on rollers driven by a Bedford van**

ASSOCIA
Castrol
RENOLD CHAINS
EC 67
CONWAYS FOR VINCENTS
Brian Chapman
MIGHTY MOUSE.

In 1956, Burns and Wright went to Utah and Bob achieved 176 mph to regain the sidecar record, raised in 1955 to 174 mph by Noll riding a works-prepared supercharged BMW.

Wright's record had gone to Hertz whose cigar-shaped projectile entered by NSU was timed at 210 mph at Utah. Russ's best one-way speed was 198·3 mph, not quite enough but it was to stand for many years as the fastest speed recorded by anyone riding a conventional, unsupercharged motorcycle (subsequent contenders were to use an engine mounted in a cradle housed in a low, streamlined shell).

Before leaving the USA, Burns and Wright learned that Johnny Allen, an American riding a cigar-shaped device using a supercharged 650 cc Triumph engine, had raised the solo record to 214 mph.

This really upset the Germans, who lodged a protest. What was the point of spending thousands of pounds to prove the superiority of German riders and machines if all the effort was to be nullified by amateurs having a go on their private machines?

The FIM refused to ratify the times recorded by Burns and Allen because it did not approve of the timing apparatus used, yet the timing equipment had been developed by the Institute of Technology, California, one of the FIM's approved horological laboratories.

As far as motorcyclists around the world were concerned, Allen was the fastest man on two wheels and Burns was the fastest on three. Burns and Wright had kept the Vincent name alive after production of the Beast had ceased but at considerable financial cost to themselves despite sponsorship. Reluctantly, they sold their machines to meet expenses incurred in the USA and returned to New Zealand.

Brian Chapman shook the sprinting world in the seventies with his 25-year-old Vincent 500 cc engine mounted in a special frame (*Mighty Mouse*). Terminal speed for the standing quarter was over 150 mph

Meanwhile, other enthusiasts, impressed by the torque of the vee-twin engine, used it as the basis of various record-breaking devices. One, Roy Charlton, using his tuned vee-twin, *Rumblegutz*, actually beat George Brown in a sprint at the Brighton Speed Trials.

George was the most successful of the Vincent sprinters from the forties till the sixties. After leaving the works in the early fifties, he captured several trophies and records riding *Nero*. Later, he built a supercharged Vincent special, *Super Nero*, and took several more records before heart trouble ended his career in the sixties.

Brian Chapman took over with his remarkable supercharged 500 cc Vincent-engined special, *Mighty Mouse*. A carpenter by trade, Brian learnt about tuning Vincent engines firstly by reading Phil Irving's books and secondly by trial and error in actual competition.

In 1980, he ran the standing quarter in less than 9 seconds with a terminal speed of nearly 160 mph. In 1981, riding a supercharged Vincent thousand-engined special, *Super Mouse*, Brian recorded 8·5 seconds for the standing quarter with a terminal speed of 169·7 mph.

In 1981, Brian competed with his supercharged Vincent vee-twin (*Super Mouse*)

ASPEN
RACING JACKETS
Castrol
SANTA POD

12 The Vincent HRD Owners' Club

Owners are still able to use their 30-year-old Vincent engines for record-breaking, racing and road use because spare parts are readily available. When production of spares ceased in 1975, the Vincent HRD Owners' Club formed the VOC Spares Company Limited, a £50,000 firm mainly financed by members purchasing shares.

The result is that now, in 1982, spares are more easily obtainable for a Vincent than for many machines produced during the past decade. The majority of Vincents owned by members are in standard trim and restorers interested in competing in concours d'elegance events take great care to achieve originality.

The club was formed in 1948 and the founder, Alan Jackson, invited PCV to be the club's honorary president. Philip was delighted to accept the presidency, which he held for 31 years until his death in 1979. (Then Phil Irving—formerly a vice-president—was asked to become the new president.)

Both Philip and Alan agreed that the club should be completely independent—financially and administratively—from the factory but several of the Vincent HRD Company employees were to become members. Some works personnel—George Brown, Paul Richardson, Ted Davis and E. C. Baillie—became honorary members and the result was close co-operation

Bryan Phillips, secretary of the VOC, signs up the 100th new member booked at the Motor Cycle Show at Earls Court, 1960

H·R·D

Top **Vincent gathering snapped by Paul Richardson at his Stevenage home (1974). L to R: Ted Davis, George Rose, C. J. 'Jack' Williams, John Edwards, myself, Phil Irving and Edith. Front: Maurice Brierley and Herbert Tucker-Peake**

Right **Custom tank at Shadow Lake, Ontario, 1977. Several export Vincents were finished in Chinese red (including frame)**

between factory and club to the benefit of both.

Before Philip designed the Series D range in 1954, he invited members to give their views and he consequently incorporated some of the modifications suggested by riders.

Today's membership totals nearly 2000 devotees around the world linked by the club magazine, *M.P.H.* The first issue was produced in January 1949, a modest publication of eight pages and a flimsy cover. Nowadays, the magazine runs to 52–60 pages every month and is perused by a wider readership than the VOC.

The club has several regional sections in various parts of the globe and its diverse activities include rallies, sprints, highspeed trials, dinner dances, camping week-ends, publishing and technical assistance for owners with that rare complaint—a recalcitrant Vincent.

Philip Vincent always showed a keen interest in the club's fortunes and he was flattered when the founders of the VOC Spares Company invited him to be the firm's honorary president in 1975.

GEORGE BROWN

Top **VOC 21st Annual Dinner, 1969. L to R: Paul Richardson, Ann Rose, Tony Rose and Ann Phillips**

Left **Eddie Stevens taking part in a George Brown Memorial Run held in Stevenage after George's death in 1979**

Although he thought of his motorcycles as 'rather old-fashioned' he still regarded them as superior to any motorcycle made since his firm ceased production; and the knowledge that the VOC Spares Company would keep the Beast on the road for the foreseeable future was a source of contentment to him. As he once said to me, 'The Beast died a death many years ago but it won't lie down!'

Appendix

Specifications

Series A (from 1936)

Model	**Meteor**	**Comet**	**Comet Special**	**Rapide**
Year from	1934	1934	1935	1936
Year to	1939	1939	1937	1939
Bore (mm)	84	84	84	84
Stroke (mm)	90	90	90	90
Capacity (cc)	499	499	499	998
Comp. ratio (to 1)	6·8	7·3	8	6·8
Valve position	ohv	ohv	ohv	ohv
inlet opens BTDC	40	44	48	40
inlet closes ABDC	52	56	60	52
exhaust opens BBDC	65	68	71	65
exhaust closes ATDC	33	38	42	33
Valve cl/ce (cold) in. (in.)	Nil	Nil	Nil	Nil
Valve cl/ce (cold) ex. (in.)	Nil	Nil	Nil	Nil
Ignition timing degrees	42	42	40	42
Points gap (in.)	0·014	0·019	0·019	0·012
Primary drive chain (in.)	duplex 0·375	duplex 0·375	duplex 0·375	duplex 0·375
Rear chain (in.)	$\frac{5}{8} \times \frac{3}{8}$	$\frac{5}{8} \times \frac{3}{8}$	$\frac{5}{8} \times \frac{3}{8}$	$\frac{5}{8} \times \frac{1}{4}$
Sprockets: engine (T)	30	30	30	32
Sprockets: clutch (T)	56	56	56	56
Sprockets: gearbox (T)	19	19	19	22
Sprockets: rear (T)	46	46	46	45
O/A ratio: top	4·6	4·6	4·3	3·58
O/A ratio: 3rd	5·8	5·8	5·0	4·51

N.B. The Series A designation began in 1936.

Phil Irving aboard the Vindian, a Vincent vee-twin installed in an Indian frame finished in blue enamel (1949) which never went into production

Indian

Model	Meteor	Comet	Comet Special	Rapide
Year from	1934	1934	1935	1936
Year to	1939	1939	1937	1939
O/A ratio: 2nd	7·4	7·4	6·2	6·49
O/A ratio: 1st	12·4	12·4	7·8	9·55
Front tyre (in.)	3·00 × 20	3·00 × 20	3·00 × 20	3·00 × 20
Rear tyre (in.)	3·25 × 20	3·25 × 20	3·25 × 20	3·50 × 20
Rim front	WM1	WM1	WM1	WM1
Rim rear	WM2	WM2	WM2	WM2
Brake front dia. (in.)	7	7	7	7
Brake front width (in.)	$\frac{7}{8}$	$\frac{7}{8}$	$\frac{7}{8}$	$\frac{7}{8}$
Brake rear dia. (in.)	7	7	7	7
Brake rear width (in.)	$\frac{7}{8}$	$\frac{7}{8}$	$\frac{7}{8}$	$\frac{7}{8}$
Front suspension	Brampton	Brampton	Brampton	Brampton
Front movement (in.)	3 approx.	3 approx.	3 approx.	3 approx.
Rear type	pivoted fork	pivoted fork	pivoted fork	pivoted fork
Rear movement (in.)	4 approx.	4 approx.	4 approx.	4 approx.
Petrol tank (Imp. gal.)	$3\frac{1}{4}$	$3\frac{1}{4}$	$3\frac{1}{4}$	$3\frac{1}{2}$
Oil tank (Imp. pint)	$3\frac{1}{2}$	$3\frac{1}{2}$	$3\frac{1}{2}$	4
Box capacity (Imp. pint)	$1\frac{1}{2}$	$1\frac{1}{2}$	$1\frac{1}{2}$	$1\frac{1}{2}$
Ignition system	Miller mag-dyno	Miller mag-dyno	Miller mag-dyno	Miller mag-dyno
Output (watts)	36	36	36	36
Battery (volt)	6	6	6	6
Wheelbase (in.)	55	55	55	56
Dry weight (lb.)	385	385	385	430
Power: bhp	25	26	28	45
@ rpm	5300	5600	5600	5500

Series A

Model	TT Replica	TT Model
Year from	1935	1934
Year to	1937	1935
Bore (mm)	84	84
Stroke (mm)	90	90
Capacity (cc)	499	499
Comp. ratio (to 1)	8	8
Valve position	ohv	ohv
inlet opens BTDC	48	48
inlet closes ABDC	60	60
exhaust opens BBDC	71	71
exhaust closes ATDC	42	42

Model	TT Replica	TT Model
Year from	1935	1934
Year to	1937	1935
Valve cl/ce (cold) in. (in.)	Nil	Nil
Valve cl/ce (cold) ex. (in.)	Nil	Nil
Igntition timing degrees	42	42
Points gap (in.)	0·010	0·010
Primary drive chain (in.)	duplex 0·375	duplex 0·375
Rear chain	$\frac{5}{8} \times \frac{1}{4}$	$\frac{5}{8} \times \frac{1}{4}$
Sprockets: engine (T)	30	30
Sprockets: clutch (T)	56	56
Sprockets: gearbox (T)	19	19

Phil Irving described the Vindian as 'a monstrous device weighing some 580 lb, and riding it made one feel rather like a yachtsman at the helm of the *Queen Elizabeth*'

In the sixties, a dealer hoped to produce a production Viscount motorcycle incorporating a Vincent vee-twin engine mounted in a Norton featherbed frame but only a few were made

Model	TT Replica	TT Model
Year from	1935	1934
Year to	1937	1935
Sprockets: rear (T)	46	46
O/A ratio: top	4·3	4·3
O/A ratio: 3rd	5	5
O/A ratio: 2nd	6·2	6·2
O/A ratio: 1st	7·8	7·8
Front tyre (in.)	3·00 × 20	3·00 × 20
Rear tyre (in.)	3·25 × 20	3·25 × 20
Rim front	WM1	WM1
Rim rear	WM2	WM2
Brake front dia. (in.)	7	7
Brake front width (in.)	$\frac{7}{8}$	$\frac{7}{8}$
Brake rear dia. (in.)	7	7
Brake rear width (in.)	$\frac{7}{8}$	$\frac{7}{8}$
Front suspension	Brampton	Brampton
Front movement (in.)	3 approx.	3 approx.

Model	TT Replica	TT Model
Year from	1935	1934
Year to	1937	1935
Rear type	pivoted fork	pivoted fork
Rear movement (in.)	4 approx.	4 approx.
Petrol tank (Imp. gal.)	5	5
Oil tank (Imp. pint)	8	8
Box capacity (Imp. pint)	$1\frac{1}{2}$	$1\frac{1}{2}$
Ignition system	BTH TT magneto	BTH TT magneto
Generator type	—	—
Output (watts)	—	—
Battery (volt)	—	—
Wheelbase (in.)	55	55
Dry weight (lb.)	335	335
Power: bhp	34	34
@ rpm	5800	5800

Series B

Model	Rapide	Black Shadow	Meteor
Year from	1946	1948	1948
Year to	1950	1950	1950
Bore (mm)	84	84	84
Stroke (mm)	90	90	90
Capacity (cc)	998	998	499
Above collar	0·310	0·310	0·310
Inlet valve stem dia (in.)			
Below collar	0·373	0·373	0·373
Above collar	0·310	0·310	0·310
Exhaust valve dia. (in.)			
Below collar	0·370	0·370	0·370
Comp. ratio (to 1)	6·45	7·3	6·45
Valve position	ohv	ohv	ohv
Valve timing:			
inlet opens BTDC	40–42	40–42	40–42
inlet closes ABDC	60–64	60–64	60–64
exhaust opens BBDC	72–70	72–70	72–70
exhaust closes ATDC	28–33	28–33	28–33
Valve cl/ce (cold) in. (in.)	nil	nil	nil
Valve cl/ce (cold) ex. (in.)	nil	nil	nil
Ignition timing degrees	39	38	38
Points gap (in.)	0·012	0·012	0·012

Model	**Rapide**	**Black Shadow**	**Meteor**
Year from	1946	1948	1948
Year to	1950	1950	1950
Primary drive chain (in.)	triplex 0·375	triplex 0·375	$\frac{1}{2} \times \frac{5}{16}$
Rear chain	$\frac{5}{8} \times \frac{3}{8}$	$\frac{5}{8} \times \frac{3}{8}$	$\frac{5}{8} \times \frac{3}{8}$
Sprockets: engine (T)	35	35	23
Sprockets: clutch (T)	56	56	40
Sprockets: gearbox (T)	21	21	18
Sprockets: rear (T)	46	46	48
O/A ratio: top	3·5	3·5	4·64
O/A ratio: 3rd	4·16	4·2	5·94
O/A ratio: 2nd	5·5	5·5	8·17
O/A ratio: 1st	9·1	7·25	12·4
Front tyre (in.)	3·00 × 20	3·00 × 20	3·00 × 20
Rear tyre (in.)	3·50 × 19	3·50 × 19	3·50 × 19
Rim front	WM1	WM1	WM1
Rim rear	WM2	WM2	WM2
Brake front dia. (in.)	7	7	7
Brake front width (in.)	$\frac{7}{8}$	$\frac{7}{8}$	$\frac{7}{8}$
Brake rear dia. (in.)	7	7	7
Brake rear width (in.)	$\frac{7}{8}$	$\frac{7}{8}$	$\frac{7}{8}$
Front suspension	Brampton	Brampton	Brampton
Front movement (in.)	3 approx.	3 approx.	3 approx.
Rear type	pivoted fork	pivoted fork	pivoted fork
Rear movement (in.)	4 approx.	4 approx.	4 approx.
Petrol tank (Imp. Gal.)	$3\frac{1}{2}$	$3\frac{1}{2}$	$3\frac{1}{2}$
Oil tank (Imp. pint)	6	6	6
Box capacity (Imp. pint)	2 approx.	2 approx.	$1\frac{1}{2}$
Chaincase (Imp. pint)	$\frac{1}{2}$ approx.	$\frac{1}{2}$ approx.	
Ignition system	Lucas magneto	Lucas magneto	Lucas magneto
Generator type	Miller dynamo	Miller dynamo	Miller dynamo
Output (watts)	50	50	50
Battery (volt)	6	6	6
Wheelbase (in.)	$56\frac{1}{2}$	$56\frac{1}{2}$	$55\frac{3}{4}$
Ground clear. (in.)	6	6	6
Seat height (in.)	$31\frac{1}{2}$	$31\frac{1}{2}$	$31\frac{1}{2}$
Width (bars) (in.)	$25\frac{3}{4}$	$25\frac{3}{4}$	$25\frac{3}{4}$
Length (in.)	$85\frac{1}{2}$	$85\frac{1}{2}$	$85\frac{1}{2}$
Dry weight (lb.)	455	458	386
Power: bhp	45	55	26
@ rpm	5300	5700	5300

The Norton/Vincent set up. Unfortunately, the power unit was 'chopped' and it was then impossible to mount the vee-twin in a Vincent frame.

This special features telescopic forks instead of Girdraulics and the Vincent back end converted to standard swinging arm (only the two trailing legs remain of the original rear frame member—not advisable)

Series C

Model	**Rapide**	**Black Shadow**	**Black Lightning**	**Comet**	**Grey Flash**	**Meteor**
Year from	1948	1948	1948	1948	1949	1950
Year to	1954	1954	1955	1954	1951	1950
Bore (mm)	84	84	84	84	84	84
Stroke (mm)	90	90	90	90	90	90
Capacity (cc)	998	998	998	499	499	499
above collar	0·310	0·310	0·310	0·310	0·310	0·310
Inlet valve dia. (in.)						
below collar	0·373	0·373	0·373	0·373	0·373	0·373
above collar	0·310	0·310	0·310	0·310	0·310	0·310
Exhaust valve dia. (in.)						
below collar	0·370	0·370	0·370	0·370	0·370	0·370
Comp. ratio (to 1)	6·45	7·3	to order	6·8	8	6·45
Valve position	ohv	ohv	ohv	ohv	ohv	ohv
inlet opens BTDC	40–42	40–42	55	40–42	55	40–42
inlet closes ABDC	60–64	60–64	68	60–64	68	60–64
exhaust opens BBDC	72–70	72–70	73	72–70	73	72–70
exhaust closes ATDC	28–33	28–33	50	28–33	50	28–33
Valve cl/ce (cold) in. (in.)	Nil	Nil	Nil	Nil	Nil	Nil
Valve cl/ce (cold) ex. (in.)	Nil	Nil	Nil	Nil	Nil	Nil
Ignition timing degrees	39	38	38	38	42	38
Points gap (in.)	0·012	0·012	0·010	0·012	0·010	0·012
Primary drive chain (in.)	triplex 0·375	triplex 0·375	triplex 0·375	$\frac{1}{2} \times \frac{5}{16}$	$\frac{1}{2} \times \frac{5}{16}$	$\frac{1}{2} \times \frac{5}{16}$
Rear chain	$\frac{5}{8} \times \frac{3}{8}$	$\frac{5}{8} \times \frac{3}{4}$	$\frac{5}{8} \times \frac{1}{4}$	$\frac{5}{8} \times \frac{3}{8}$	$\frac{5}{8} \times \frac{1}{4}$	$\frac{5}{8} \times \frac{3}{8}$
Sprockets: engine (T)	35	35	35	23	23	23
Sprockets: clutch (T)	56	56	56	40	42	40
Sprockets: gearbox (T)	21	21	22	18	18	18
Sprockets: rear (T)	46	46	45	48	48	48
O/A ratio: top (to 1)	3·5	3·5	3·27	4·64	4·87	4·64
O/A ratio: 3rd	4·16	4·16	3·89	5·94	5·75	5·94
O/A ratio: 2nd	5·64	5·64	5·26	8·17	6·82	8·17
O/A ratio: 1st	9·1	9·1*	6·77	12·4	10·36	12·4
*1:7·25 until (and inc.) eng. no. 7075.						
Front tyre (in.)	3·00 × 20	3·00 × 20	3·00 × 21	3·00 × 20	3·00 × 21	3·00 × 20
Rear tyre (in.)	3·50 × 19	3·50 × 19	3·50 × 20	3·50 × 19	3·50 × 20	3·50 × 19
Rim front	WM1	WM1	WM1	WM1	WM1	WM1
Rim rear	WM2	WM2	WM2	WM2	WM2	WM2
Brake front dia. (in.)	7	7	7	7	7	7
Brake front width (in.)	$\frac{7}{8}$	$\frac{7}{8}$	$\frac{7}{8}$	$\frac{7}{8}$	$\frac{7}{8}$	$\frac{7}{8}$
Brake rear dia. (in.)	7	7	7	7	7	7
Brake rear width (in.)	$\frac{7}{8}$	$\frac{7}{8}$	$\frac{7}{8}$	$\frac{7}{8}$	$\frac{7}{8}$	$\frac{7}{8}$
Front suspension	Girdraulic	Girdraulic	Girdraulic	Girdraulic	Girdraulic	Girdraulic
Front movement (in.)	3 approx.	3 approx.	3 approx.	3 approx.	3 approx.	3 approx.
Rear type	pivoted fork	pivoted fork	pivoted fork	pivoted fork	pivoted fork	pivoted fork

Model	Rapide	Black Shadow	Black Lightning	Comet	Grey Flash	Meteor
Year from	1948	1948	1948	1948	1949	1950
Year to	1954	1954	1955	1954	1951	1950
Rear movement (in.)	4 approx.	4 approx.	4 approx.	4 approx.	4 approx.	4 approx.
Petrol tank (Imp. gal.)	$3\frac{1}{2}$	$3\frac{1}{2}$	$3\frac{3}{8}$	$3\frac{1}{2}$	$3\frac{3}{8}$	$3\frac{1}{2}$
Oil tank (Imp. pint)	6	6	6	6	6	6
Box capacity (Imp. pint)	2 approx.	2 approx.	2	$1\frac{1}{2}$	$1\frac{1}{2}$	$1\frac{1}{2}$
Chaincase (Imp. pint)	$\frac{1}{2}$ approx.	$\frac{1}{2}$ approx.	$\frac{1}{2}$ approx.	$\frac{1}{2}$ approx.	$\frac{1}{2}$ approx.	$\frac{1}{2}$ approx.
Ignition system	Lucas magneto	Lucas magneto	Lucas TT magneto	Lucas magneto	BTH TT magneto	Lucas magneto
Generator type	Miller dynamo	Miller dynamo	—	Miller dynamo	—	Miller dynamo
Output (watts)	50	50	—	50	—	50
Battery (volt)	6	6	—	6	—	6
Wheelbase (in.)	$56\frac{1}{2}$	$56\frac{1}{2}$	$56\frac{1}{2}$	$55\frac{3}{4}$	$55\frac{3}{4}$	$55\frac{3}{4}$
Ground clear. (in.)	6	6	6	6	6	6
Seat height (in.)	32	32	32	32	32	32
Width (bars) (in.)	$25\frac{3}{4}$	$25\frac{3}{4}$	$25\frac{3}{4}$	$25\frac{3}{4}$	$25\frac{3}{4}$	$25\frac{3}{4}$
Length (in.)	$85\frac{1}{2}$	$85\frac{1}{2}$	$85\frac{1}{2}$	$85\frac{1}{2}$	$85\frac{1}{2}$	$85\frac{1}{2}$
Dry weight (lb)	455	458	380	390	330	386
Power: bhp	45	55	70	28	35	26
@ rpm	5300	5700	5600	5800	6200	5300

Series D

Model	Black Prince	Black Knight	Black Shadow	Rapide
Year from	1954	1954	1955	1955
Year to	1955	1955	1955	1955
Bore (mm)	84	84	84	84
Stroke (mm)	90	90	90	90
Capacity (cc)	998	998	998	998
Above collar	0·310	0·310	0·310	0·310
Inlet valve dia. (in.)				
Below collar	0·373	0·373	0·373	0·373
Above collar	0·310	0·310	0·310	0·310
Exhaust valve dia. (in.)				
Below collar	0·370	0·370	0·370	0·370
Comp. ratio (to 1)	7·3	6·45	7·3	6·45
Valve position	ohv	ohv	ohv	ohv
inlet opens BTDC	40–42	40–42	40–42	40–42
inlet closes ABDC	60–64	60–64	60–64	60–64
exhaust opens BBDC	72–70	72–70	72–70	72–70
exhaust closes ATDC	28–33	28–33	28–33	28–33
Valve cl/ce (cold) in. (in.)	nil	nil	nil	nil
Valve cl/ce (cold) (ex.) (in.)	nil	nil	nil	nil

A Series B frame with C vee-twin blown engine with four carbs! I've heard of tuning but this is ridiculous!

Model	Black Prince	Black Knight	Black Shadow	Rapide
Year from	1954	1954	1955	1955
Year to	1955	1955	1955	1955
Ignition timing degrees	38	39	38	39
Points gap (in.)	0·012	0·012	0·012	0·012
Primary drive chain (in.)	triplex 0·375	triplex 0·375	triplex 0·375	triplex 0·375
Rear chain	$\frac{5}{8} \times \frac{3}{8}$	$\frac{5}{8} \times \frac{3}{8}$	$\frac{5}{8} \times \frac{3}{8}$	$\frac{5}{8} \times \frac{3}{8}$
Sprockets: engine (T)	35	35	35	35
Sprockets: clutch (T)	56	56	56	56
Sprockets: gearbox (T)	21	21	21	21
Sprockets: rear (T)	46	46	46	46
O/A ratio: top	3·5	3·5	3·5	3·5
O/A ratio: 3rd	4·16	4·16	4·16	4·16
O/A ratio: 2nd	5·64	5·64	5·64	5·64
O/A ratio: 1st	9·1	9·1	9·1	9·1
Front tyre (in.)	3·50 × 19	3·50 × 19	3·50 × 19	3·50 × 19
Rear tyre (in.)	4·00 × 18	4·00 × 18	4·00 × 18	4·00 × 18
Rim front	WM2	WM2	WM2	WM2
Rim rear	WM3	WM3	WM3	WM3
Brake front dia. (in.)	7	7	7	7
Brake front width (in.)	$\frac{7}{8}$	$\frac{7}{8}$	$\frac{7}{8}$	$\frac{7}{8}$
Brake rear dia. (in.)	7	7	7	7
Brake rear width (in.)	$\frac{7}{8}$	$\frac{7}{8}$	$\frac{7}{8}$	$\frac{7}{8}$
Front suspension	Girdraulic	Girdraulic	Girdraulic	Girdraulic
Front movement (in.)	$3\frac{1}{2}$ approx	$3\frac{1}{2}$ approx.	$3\frac{1}{2}$ approx.	$3\frac{1}{2}$ approx.
Rear type	pivoted fork	pivoted fork	pivoted fork	pivoted fork
Rear movement (in.)	5 approx.	5 approx.	5 approx.	5 approx.
Petrol tank (Imp. gal.)	4	4	4	4
Oil tank (Imp. pint)	5	5	5	5
Box capacity (Imp. pint)	2 approx.	2 approx.	2 approx.	2 approx.
Chaincase (Imp. pint)	$\frac{1}{2}$ approx.	$\frac{1}{2}$ approx.	$\frac{1}{2}$ approx.	$\frac{1}{2}$ approx.
Ignition system	Lucas coil	Lucas coil	Lucas coil	Lucas coil
Generator type	Lucas dynamo	Lucas dynamo	Lucas dynamo	Lucas dynamo
Output (watts)	60	60	60	60
Battery (volt)	6	6	6	6
Wheelbase (in.)	$56\frac{1}{2}$	$56\frac{1}{2}$	$56\frac{1}{2}$	$56\frac{1}{2}$
Ground clear. (in.)	5	5	5	5
Seat height (in.)	$32\frac{1}{2}$	$32\frac{1}{2}$	$32\frac{1}{2}$	$32\frac{1}{2}$
Width (bars) (in.)	27	27	27	27
Length (in.)	89	89	86	86
Dry weight (lb)	460	460	447	447
Power: bhp	55	45	55	45
@ rpm	5700	5200	5700	5200

In the sixties, a Swizz, Fritz Egli, designed a special frame to take the Vincent vee-twin engine and the Egli-Vincent had several successes in competition against more modern machines

One of the 'production' Viscounts in the Isle of Man

Colours and insignia

Colours

In the introduction to the *Vincent H·R·D Gallery,* Philip Vincent is quoted as saying, 'There is no such thing as a standard Vincent!' Owners, therefore, should not be too dogmatic when refurbishing their motorcycles. When restoring a pre-war Vincent, it is advisable to check the specification of a particular machine with the factory records, held now by The Vincent HRD Owners' Club. Not all Vincents, by any means, were black.

The prototype Vincent vee-twin made in **1936** incorporated a hand-soldered, stainless steel petrol tank with black cellulose panels. Several of the **pre-war** production models, however, were finished in maroon instead of black. Wheel rims were usually chromium plated and unpainted.

Most of the **postwar** motorcycles incorporated steel petrol tanks stove-enamelled black using the Pinchin Johnson process. Some models were finished in Chinese red or blue for export to the USA.

Wheels were usually fitted with chromium plated rims with black enamel in the centre and thin red lines. Several **1952** rims were all-black due to the shortage of chromium.

The **Series C Grey Flash** petrol tank and cycle parts were finished in grey/green enamel.

Insignia

Before **1949** petrol tanks featured gold leaf lines curving from steeringhead lug to dualseat. The tank side transfers were *Vincent H·R·D* framed by gold leaf lines in the shape of a 'D'. As the gold leaf was done by hand, the thickness may vary slightly.

Small 'Mercury' transfers (*Vincent H·R·D*) were affixed to the top of the petrol tank (below the petrol cap) and the front of the steeringhead lug.

During and after **1949** with the introduction of the **Series C** range the *H·R·D* monogram was dropped and replaced by a new *Vincent* logo. The new *Vincent* scroll appeared either side of the petrol tank and a new 'Mercury' transfer was affixed to the steeringhead lug and top of the petrol tank bearing the caption: THE VINCENT "H·R·D" CO. LTD., STEVENAGE HERTS. From 1952, the caption read: VINCENT ENGINEERS (STEVENAGE) LTD., STEVENAGE, HERTS.

Three different transfers were used on the **Series D** enclosed models and they differed from those used on the **Series C** range. They were affixed to the front mudguard and the rear cowl with a smaller version sited on the dashboard in front of the steering damper knob.

The **Victor** had a large Vincent scroll with the word *Victor* below and a gloved fist above. The **Black Knight** featured the large Vincent scroll with *Black Knight* below and axe above. The **Black Prince** also portrayed the large Vincent scroll but with a helmet and the words *Black Prince* in capitals.

The **Series D Rapide** and **Black Shadow** featured the smaller Vincent scroll on each side of the petrol tank with a 'Mercury' transfer on the top and on the steeringhead lug. The petrol tank was also lined with gold leaf similar to the **Series C** style. The **Black Shadow** had another transfer on the top of the petrol tank—its name in longhand.

An Egli-Vincent being assembled in England. Many VOC members thought it sacrilegious to put the Vincent engine into a non-standard frame instead of the frame the works originally fitted

Heather Beynon won the Miss Federation title at the BMF Rally in 1969. A VOC member, she sits on her superbly restored Series C Comet

Engine and frame numbers

Pre-war singles

Model	Engine type symbol	Frame type symbol
Meteor	M	D
Comet	C	D
Comet Special	TTC	D
TT Replica	TTR	D . . . TTR

Pre-war vee twins

During production of the Series A Rapide vee twins (1936 to 1939) the engine numbers began at V1001 and finished at V1077—'V' denoted vee twin. Two engine numbers, to make matters complicated, were used twice—they were V1009 and V1041.

The engine number is located on the right-hand side of the machine—on the crankcase below the cambox.

The prototype engine (V1000) was fitted to a special frame originally made for Eric Fernihough.

Frame numbers of the production Rapides began at DV1229 (with engine number V1001) and finished at DV 1771—'DV' denoted 'diamond' frame for a vee twin engine. The frame number can be seen on the left-hand side of the motorcycle—on the steeringhead and again on the rear wheel spindle lug.

Postwar machines

Series B

Model	Engine type symbol	Frame type symbol
Meteor	F5AB/2	R/1
Rapide	F10AB/1	R
Black Shadow	F10AB/1B	R . . . . B

Series C

Model	Engine type symbol	Frame type symbol
Comet	F5AB/2A	RC/1
Rapide	F10AB/1	RC
Black Shadow	F10AB/1B	RC . . . . B
Grey Flash	F5AB/2B	RC/1A
Black Lightning	F10AB/1C	RC

Series D

Model	Engine type symbol	Frame type symbol
Black Prince	F10AB/2B	RD . . . . B/F
Black Knight	F10AB/2	RD . . . F
Victor	F5AB/3A	RD/1 . . F
Rapide	F10AB/2	RD
Black Shadow	F10AB/2B	RD . . B

(Speedway engine: F5AB/1 and F5MB/1)

Carburettor settings

Year	Model	Amal type	Size	Main	Throttle valve	Needle pos.	Needle jet
1935–39	**Series A**						
	Meteor[1]	76/022	$1\frac{1}{16}$	160–170	6/4	3	Std.
	Comet[2]	89/011	$1\frac{1}{8}$	170–180	29/4	3	Std.
	Rapide (front	6/301	$1\frac{1}{16}$	180	6/3	3	Std.
	(rear)[3]	76/022	$1\frac{1}{16}$	170	6/4	3	Std.
	Comet Special[4]	10TT	$1\frac{5}{32}$	330–350	6	4	.109
	TT Replica (petrol)[5]	10TT	$1\frac{5}{32}$	360	7	4	.109
1946–50	**Series B**						
	Meteor	276DQ/1DV	$1\frac{1}{16}$	170	6/4	3	Std.
	Rapide (front)	276CJ/1DO	$1\frac{1}{16}$	170	6/4	3	Std.
	(rear)	276CH/2DS	$1\frac{1}{16}$	170	6/4	3	Std.
	Black Shadow (front)	289M/1DO	$1\frac{1}{8}$	180	29/4	3	Std.
	(rear)	289N/2DS	$1\frac{1}{8}$	180	29/4	3	Std.
1949–54	**Series C**						
	Comet	229F/1DV	$1\frac{1}{8}$	200	29/3	3	Std.
	Rapide (front)	276DQ/1DV	$1\frac{1}{16}$	170	6/4	3	Std.
	(rear)	276CH/2DS	$1\frac{1}{16}$	170	6/4	3	Std.
	Black Shadow (front)	229E/1DV	$1\frac{1}{8}$	180	29/4	3	Std.
	(rear)[6]	289N/2DS	$1\frac{1}{8}$	180	29/4	3	Std.
	Grey Flash (alcohol)	10TT9	32 mm	1700	7	4	.120
	Black Lightning (petrol)[7]	10TT9	32 mm	360	7	4	.109
1954–55	**Series D**						
	Victor	389	$1\frac{1}{8}$	250	4	2	376/072
	Rapide	376	$1\frac{1}{16}$	220	4	2	376/072
	Black Knight	376	$1\frac{1}{16}$	220	4	2	376/072
	Black Shadow	389	$1\frac{1}{8}$	250	4	2	376/072
	Black Prince	389	$1\frac{1}{8}$	250	4	2	376/072

Footnotes

1 Clip fitting, 15° float chamber
2 Clip fitting, 15° float chamber
3 Clip fitting, 15° float chamber
4 With open exhaust pipe 460–500
5 Vary setting as required
6 29/3 throttle valve fitted from 1953 onwards
7 Same approximate settings apply for $1\frac{5}{32}$ in. and $1\frac{3}{16}$ in. carburettors. Vary settings as required

Top **Another Norvin. This hybrid has been very popular with amateur racers**

Below **This Series A TT Replica was raced at Cadwell Park in 1971. The rider has modified his front end the hard way**

John Surtees riding his modified Grey Flash at Brands Hatch in 1981

Prices

The ex-factory retail price sometimes changed during the year. The prices given here are those which usually appeared at 'motorcycle show' time. These can, therefore, be taken as the 'mean' for any one year and are given by way of comparison rather than for any other reason.

Date Pre-war singles	**Meteor**	**Comet**	**Comet Special**	**TT Replica**
1934	£79.10s.0d.	£86. 0s.0d.		
1935	£79.10s.0d.	£86. 0s.0d.	£95. 0s.0d.	£ 98. 0s.0d.
1936	£81. 0s.0d.	£87.10s.0d.	£96.10s.0d.	£105. 0s.0d.
1937	£89.10s.0d.	£96. 0s.0d.		£118. 0s.0d.*
1938	£79.10s.0d.	£85. 0s.0d.		
1939	£79.10s.0d.	£85. 0s.0d.		*Series B model

Series A	**Rapide**
1936	£138. 0s.0d.
1937	£138. 0s.0d.
1938	£142. 0s.0d.
1939	£128. 0s.0d.

Series B	**Rapide**	**Black Shadow**	**Meteor**
1946	£265. 5s.5d.		
1947	£293. 7s.5d.		
1948	£317. 0s.0d.	£381. 0s.0d.	£247.13s.0d.

Series C	**Rapide**	**Black Shadow**	**Black Lightning**	**Comet**	**Grey Flash**
1948	£336.11s.0d.	£400. 1s.0d.	£508. 0s.0d.	£273. 1s.0d.	
1949	£361.19s.0d.	£425. 9s.0d.	£533. 8s.0d.	£273. 1s.0d.	£381. 0s.0d.
1950	£323.17s.0d.	£375.18s.5d.	£501.13s.0d.	£241. 6s.0d.	£330. 4s.0d.
1951	£357.15s.7d.	£402.10s.0d.	£504.14s.5d.	£270.17s.9d.	
1952	£347.11s.1d.	£389.14s.5d.	£504.14s.5d.	£274.14s.5d.	
1953	£326. 8s.0d.	£366. 0s.0d.	£474. 0s.0d.	£258. 0s.0d.	

Series D	**Black Prince**	**Black Knight**	**Victor**	**Rapide**	**Black Shadow**
1954	£378. 0s.0d.	£348. 0s.0d.	£274.16s.0d.		
1955				£325. 0s.0d.	£355. 0s.0d.

The supercharged Vincent missile (alias *Super Nero*) ridden in the sixties by the leading sprinter George Brown, then in his fifties. The specially designed frame is rigid. Rather him than me

John Surtees astride his modified Black Lightning at Brands Hatch. The powerful boom of his exhaust note as he rode past was my favourite sound of 1981

Model recognition points

Pre-war
Series A, 1936–39

Easily recognizable because of mass of external pipes and stainless steel petrol tank with red or black panels. Essentially only two models produced, the 1000 vee twin and 500 single.

The prototype and other early Rapides were fitted with an 8-day clock which was later offered as an optional extra.

A Miller 'mag-dyno' was fitted to the prototype Rapide but most vee twins were fitted with a Lucas magneto.

Postwar
Series B, 1946–50

Black frame, forks, brake torque arms, brake drums and plates, brake arms, chainguard, mudguards stays, AVC unit, upper frame member, tool tray and holder, headlamp shell and stays, handlebars, speedometer case, speedometer gearbox, front and rear stands, number plates, rearlight case, spring boxes, dynamo end cap, engine plates, battery carrier, rear brake pedal and horn body.

Headlamp rim, exhaust pipes and silencer, kick start lever, speedometer rim, petrol and oil caps, front brake lever, clutch and valve lifter levers and horn rim were chromium plated.

H·R·D was embossed on the left-hand crankcase and timing cover. Inspection caps were embossed with a smaller version of the large tank transfer.

Wheel tommy bar (not steel axle), brake rods, small tommy bars holding the rear mudguard flap in position, chain adjuster sleeve, the long bolt and the metal strip holding the battery were made from stainless steel.

Water excluders (not fitted to very early machines), damper knobs, chain adjuster bodies and mudguards were aluminium.

Nuts, bolts, footrests, etc. were cadmium plated.

Tyres were 3·00 × 20 (front) and 3·50 × 19 (rear).

1946–47 Rapide
Clutch cover was circular with 6 screw holes drilled at 2, 4, 6, 8, 10 and 12 o'clock. Later clutch covers had a flat at the bottom (to give a better cornering angle) and the holes were repositioned accordingly. Tie-bolts and distance pieces at the rear of the petrol tank (to prevent splitting) were introduced from frame number 2416.

1948 Black Shadow
Black Shadow introduced with $1\frac{1}{8}$ in. carburettors, ribbed drum brakes, 150 mph 5 in. speedometer. Black engine. Engine number includes two capital 'B's (the Rapide has but one).

1948 Meteor
Meteor introduced. Front mudguard stay doubled as front stand. (The combined prop and front stand on the Rapide was not fitted to the Meteor.) Brampton forks. $1\frac{1}{16}$ in. carburettor. No hydraulic dampers.

Series C, 1948–54

Specifications much as the Series B except for the following points catagorized under the individual model headings and the following.

Basically, B Series machines were fitted with Brampton forks and a rear frame member incorporating straight lugs for the seat stays. Series C machines were fitted with Girdraulic front forks, hydraulic dampers fore and aft, and a rear frame member incorporating curved lugs for the seat stays. However, some Series C machines were fitted with rear frame members incorporating the straight (Series B) seat lug. The 1949 motorcycles were, therefore, a mixture of Series B and Series C components and much confusion has since arisen regarding the original specification of a particular machine.

The Vincent hydraulic damper introduced with the Girdraulic forks for the Series C range was fitted to the rear of the machine (between the spring boxes connecting the rear frame member to the rear of the oil tank). The outer casing was enamelled black, the lower aluminium body being left unpainted. The Girdraulics were enamelled black except for the inner spring boxes, which were cadium plated. (Series C rear spring inner cases were usually cadmium plated.)

Some 1949 crankcases, timing covers and inspection caps bear the *H·R·D* logo; some are simply plain, the logo having been ground off. Some carry the single word *Vincent* (*H·R·D* monogram dropped). First *Vincent* embossed crankcase from engine number 3090.

1948 Comet
Comet introduced with Girdraulic forks. Combined prop and front stand. Hydraulic dampers.

1948 Black Lightning
Black Lightning introduced. Amal TT carburettors fitted. Elektron brake plates (unpainted). Girdraulics lightened by cutting groove on inside of each fork blade. Rev counter. No speedometer or lighting equipment. Aluminium rims with 3·00 × 21 (front) and 3·50 × 20 (rear) tyres.

1948 Black Shadow and Rapide
Series C Black Shadow and Rapide introduced. Girdraulics replaced Bramptons. Vincent hydraulic dampers fitted.

1948 Touring Comet, Rapide and Black Shadow
New Touring specification model introduced with 3.50 × 19 (front) and 4.00 × 18 (rear) tyres. Steel mudguards enamelled black. 'Cowhorn' handlebars. Same prices as standard models.

1949 Grey Flash
Grey Flash introduced. Grooves in fork blades. TT carburettor. Early versions fitted with Burman gearbox; later ones used Albion box. Elektron brake plates. Lighting equipment and speedometer included on road-going version.

1951 All models
Gearchange lever changed to one-piece.

1953 All models
Furled knob replaced tommy bar holding rear mudguard flap in position.

Series D, 1954–55

Specifications as for the Series C but with the following changes. All models had one rear brake (with ribbed brake drum). Tyres were 3.50 × 19 (front) and 4.00 × 18 (rear). Steering damper knob higher than on the Series C. 3-inch speedometer. New water excluder on rear brake. Tommy bars omitted from front and rear wheels. Chain adjusters replaced by conventional nuts and bolts. New rear frame member fitted. The seat stays and friction dampers omitted. Single combined spring/damper unit fitted in place of Vincent damper and two spring boxes. Vincent damper replaced by Armstrong unit on Girdraulics. Coil ignition replaced magneto. Oil tank (5-pint) moved to position below dualseat. Combined front and side stands omitted; replaced by centre stand. Lucas stop/rear light unit replaced Miller. Monobloc carburettors.

1954 Victor, Black Knight and Black Prince
Only Black Prince had ribbed front brake drums. Fibreglass panels. No transfers on petrol tank. See 'Colours' section for details of lettering. Black Knight has 120 mph speedometer, the Black Prince 150 mph.

1955 Rapide and Black Shadow
Last two models introduced. Petrol tanks and lines similar to the previous year's Series C. No fibreglass fairings but they could be fitted if required.

Top **Ted Davis with his immaculate Grey Flash at Brands Hatch during the 'John Surtees Week-end' in 1981**

Right **A superb example of an early Vincent 1000 owned by Adolf Buhler and participating in the John Surtees Week-end at Brands**

Bibliography

Books

Vincent H.R.D Gallery by Roy Harper (Vincent Publishing)
Know thy Beast by E. M. G. Stevens (Vincent Publishing)
Vincent Motor Cycles by Paul Richardson
P.C.V.: An Autobiography by Philip Vincent (Vincent Publishing)
Black Smoke by Phil Irving (Research Publications)
Rich Mixture by Phil Irving (Vincent Publishing)
Restoring and Tuning Classic Motor Cycles by Phil Irving
Vincent H.R.D. Development 1931–1936 by Phil Irving (Vincent Publishing)
Vincent: 50 Years of the Marque by Philip Vincent (Vincent Publishing)
The Vincent H.R.D. Story by Roy Harper (Vincent Publishing)

Magazines

M.P.H. (Vincent HRD Owners' Club journal)
The Motor Cycle
Motor Cycling
Motorcycle Sport
Cycle World
Cycle
Classic Bike

Other Material

Factory catalogues
Factory produced rider's handbooks

Overleaf
Captain John Bradshaw is a Concorde pilot and a VOC member. He arranged a trip in Concorde for several club members who were delighted to travel at Mach 2 for seven minutes! With John are (L to R): Alan Lancaster, Bryan Phillips, Chris Lipscombe, Tony Wheatley and Ken Chamberlain

Pages 190 and 191
VOC members at the funeral of Philip Vincent (1979). Although he thought his motorcycles were old-fashioned, owners' enthusiasm for them was a continuous source of comfort to Philip in his last, difficult years. In a farewell message published posthumously in MPH, Philip wrote 'You have made me feel that my efforts on earth have been well worthwhile.'

British

VINCENT